Moment Methods in Antennas and Scattering

The Artech House Antenna Library

Helmut E. Schrank, *Series Editor*

Blake, L.V., Antennas

Corzine, Robert G., and Joseph A. Mosko, Four-Armed Spiral Antennas

Djordjevic, A.R., M.B. Bazdar, G.M. Bazdar, G.M. Vitosevic, T.K. Sarkar, and R.F. Harrington, Analysis of Wire Antennas and Scatterers: Software and User's Manual

Evans, Gary E., Antenna Measurement Techniques

Gupta, K.C., and A. Benalla, eds., Microstrip Antenna Design

Hansen, Robert C., ed., Momement Methods in Antennas and Scattering

Hirsch, Herbert L., and Douglas C. Grove, Practical Simulation of Radar Antennas and Radomes

Kitsuregawa, Takashi, Advanced Technology in Satellite Communication Antennas: Electrical and Mechanical Design

Kumar, A., and H.D. Hristov, Microwave Cavity Antennas

Law, Preston E., Jr., Shipboard Antennas

Li, S.T., J.C. Logan, J.W. Rockway, and D.W.S. Tam, The MININEC System: Microcomputer Analysis of Wire Antennas

Pozar, David M., Antenna Design Using Personal Computers

Scherer, James P., LAARAN: Linear Antenna Array Analysis Software and User's Manual

Scott, Craig, Modern Methods of Reflector Antenna Analysis and Design

Sletten, Carlye J., ed., Reflector and Lens Antennas: Analysis and Design Using Personal Computers

Sletten, Carlyle J., ed., Reflector and Lens Antennas: Software User's Manual and Example Book

Weiner, M.M., et al., Monopole Elements on Circular Ground Planes

Wolff, Edward A., Antenna Analysis, Second Edition

Yamashita, E., ed., Analysis Methods for Electromagnetic Wave Problems

Moment Methods in Antennas and Scattering

Robert C. Hansen
Editor

Artech
Boston • London

Library of Congress Cataloging-in-Publication Data

Moment methods in antennas and scattering / Robert C. Hansen, editor.
p. cm.
Includes bibliographical references and index.
ISBN 0-89006-466-0
1. Antennas (Electronics) 2. Electromagnetic waves--Scattering.
3. Moments methods (Statistics) I. Hansen, Robert C.
TK7871..6.M64 1990 90-36230
621.382'4--dc20 CIP

British Library Cataloguing in Publication Data

Moment methods in antennas and scattering
1. Electromagnetism. Statistical analysis. Moment methods
I. Hansen, Robert C.
537.072

ISBN 0-89006-466-0

685 Canton Street
Norwood, MA 02062

International Standard Book Number: 0-89006-466-0
Library of Congress Catalog Card Number: 90-36230

10 9 8 7 6 5 4 3 2 1

Contents

Introduction

Moment methods have transformed much of the study of antennas and scatterers from a mysterious art to an engineering science. This book presents a collection of the most important and useful papers, selected from the viewpoint of a user rather than that of a theoretician. Extensive use is made of non-IEEE resources, many of which are not readily available to practitioners. Although some consider any system involving solution of a set of simultaneous equations a *moment method* solution, the term herein is restricted to those solutions where an unknown field or current is represented by more than one mode or function, thus resulting in a matrix solution. The scope is limited to primarily spatial moment methods, simply because many excellent and important papers cannot be included, even with this narrowing of scope. Spectral methods are quite important; there is just not enough space to include them.

The papers are divided into six parts. Each part is briefly introduced, and a list of pertinent references is appended. References are through 1989. Early and general papers constitute the first part. Next is a group of wire papers including *bodies of revolution* (BOR). The third part includes conducting and dielectric bodies, the former with surface patch bases or wire gridding. Common to part 4 is a Sommerfeld integral representing the substrate of a printed circuit antenna, or the earth in an earth-based antenna. Part 5 covers apertures, slots, and arrays. Finally, hybrid methods, where moment method is combined with *geometric theory of diffraction* (GTD), harmonic functions, physical optics, *et cetera* are covered. Each part contains a bibliography of useful papers, grouped by subject.

To make the scope manageable, transmission line and waveguide component analyses have not been included. Papers concerned with statics or low frequencies have also been omitted. Many colleagues have made excellent suggestions on key papers, but the editor is solely responsible for the selection herein. Inevitably, someone's favorite paper has not been used. Books devoted to moment methods in electromagnetics are listed below.

REFERENCES

R. F. Harrington, *Field Computation by Moment Methods,* Macmillan, 1968; republished by Krieger, 1982.

J. Moore and R. Pizer, eds., *Moment Methods in Electromagnetics—Techniques and Applications,* Wiley Research Studies Press, 1984

B. J. Strait, *Applications of the Method of Moments to Electromagnetic Fields,* The SCEEE Press, 1980

B.D. Popovic, M.B. Dragovic and A.R. Djordjevic, *Analysis and Synthesis of Wire Antennas,* Wiley Research Studies Press, 1982

PART 1:
Early and General Papers

Although moment methods are older in some other disciplines, those in electromagnetics were used at least as early as 1960; Duncan and Hinchey use entire trigonometric bases for a dipole antenna. The next four papers are by the "founders" of modern electromagnetic moment methods. Mei and Van Bladel (1963) use pulse bases with point matching for rectangular conducting cylinders. Andreason (1964) also treats metallic cylinders, using collocation. Richmond (1965) attacks a dielectric cylinder with cubical cells. Harrington (1967) surveys moment methods. Note also, in Part 2, a 1966 paper on bodies of revolution by Andreason. The major start, then, occurred about 25 years ago.

Three papers of general interest complete this part. Wilton and Butler (1981) tutor on log-singular kernels, and the interaction with several basis sets. Sarkar, Siarkiewicz, and Stratton (1981) review iterative methods, including conjugate gradient. Sarkar (1983) establishes convergence rates for several formulations, including Galerkin.

BIBLIOGRAPHY

Additional Early Papers

V.H. Rumsey, "Reaction Concept in Electromagnetic Theory," *Phys. Rev.,* Vol. 94, No. 6, 1954, pp. 1483–1491.

K.K. Mei, "On the Integral Equations of Thin Wire Antennas," *Trans. IEEE,* Vol. AP-13, May 1965, pp. 374–378.

J.H. Richmond, "Digital Computer Solutions of the Rigorous Equations for Scattering Problems," *Proc. IEEE,* Vol. 53, August 1965, pp. 796–804.

J.H. Richmond, "Scattering by an Arbitrary Array of Parallel Wires," *Trans. IEEE,* Vol. MTT-13, July 1965, pp. 408–412.

J.H. Richmond, "TE-Wave Scattering by a Dielectric Cylinder of Arbitrary Cross Section Shape," *Trans. IEEE,* Vol. AP-14, July 1966, pp. 460–464.

J.H. Richmond, "A Wire-Grid Model for Scattering by Conducting Bodies," *Trans. IEEE,* Vol. AP-14, November 1966, pp. 782–786.

M.G. Andreasen, "Scattering from Cylinders with Arbitrary Surface Impedance," *Proc. IEEE,* Vol. 53, August 1965, pp. 812–817.

C.D. Taylor and E.A. Aronson, "An Iterative Solution of Fredholm Integral Equations of the First Kind," *Trans. IEEE,* Vol. AP-15, September 1967, pp. 695–696.

J.H. Richmond, "Scattering by Imperfectly Conducting Wires," *Trans. IEEE,* Vol. AP-15, November 1967, pp. 802–806.

E.A. Aronson and C.D. Taylor, "Matrix Methods for Solving Antenna Problems," *Trans. IEEE,* Vol. AP-15, September 1967, pp. 696–697.

Basis Function Choices and Comparisons

T.K. Sarkar, A.R. Djordjevic and E. Arvas, "On the Choice of Expansion and Weighting Functions in the Numerical Solution of Operator Equations," *Trans. IEEE,* Vol. AP-33, September 1985, pp. 988–996.

T.K. Sarkar, "A Note on the Choice of Weighting Functions in the Method of Moments," *Trans. IEEE,* Vol. AP-33, April 1985, pp. 436–441.

J.M. Bornholdt and L.N. Medgyesi-Mitschang, "Mixed-Domain Galerkin Expansions in Scattering Problems," *Trans. IEEE,* Vol. AP-36, February 1988, pp. 216–227.

D.R. Wilton and C.M. Butler, "Efficient Numerical Techniques for Solving Pocklington's Equation and their Relationships to Other Methods," *Trans. IEEE,* Vol. AP-24, January 1976, pp. 83–86.

D.K. Cohoon, "An Exact Formula for the Accuracy of a Class of Computer Solutions of Integral Equation Formulations of Electromagnetic Scattering Problems," *Electromagnetics,* Vol. 7, No. 2, 1987, pp. 153–165.

J.M. Arnold, "Modifications to the Method of Moments," *Electronics Letters,* Vol. 12, February 5, 1976, pp. 75–77.

R.D. Graglia, "The Use of Parametric Elements in the Moment Method Solution of Static and Dynamic Volume Integral Equations," *Trans. IEEE,* Vol. AP-36, May 1988, pp. 636–646.

C.A. Klein and R. Mittra, "Stability of Matrix Equations Arising in Electromagnetics," *Trans. IEEE,* Vol. AP-21, November 1973, pp. 902–905, Correction, July 1974, p. 636.

R. Mittra and C.A. Klein, "The Use of Pivot Ratios as a Guide to Stability of Matrix Equations Arising in the Method of Moments," *Trans. IEEE,* Vol. AP-23, May 1975, pp. 448–450.

Matrix Particulars

A.R. Djordjevic and T.K. Sarkar, "A Theorem on the Moment Method," *Trans. IEEE,* Vol. AP-35, March 1987, pp. 353–355.

N. Amitay and V. Galindo, "On Energy Conservation and the Method of Moments in Scattering Problems," *Trans. IEEE,* Vol. AP-17, November 1969, pp. 747–751.

T. De Mazancourt and D. Gerlic, "The Inverse of a Block-Circulant Matrix," *Trans. IEEE,* Vol. AP-31, September 1983, pp. 808–810.

T.R. Ferguson, T.H. Lehman and R.J. Balestri, "Efficient Solution of Large Moment Problems: Theory and Small Problem Results," *Trans. IEEE,* Vol. AP-24, March 1976, pp. 230–235.

C.Y. Wu and D.K. Cheng, "A Method for Symmetrizing Generalized Impedance Matrices," *Trans. IEEE,* Vol. EMC-19, May 1977, pp. 81–88.

D.K. Cheng and C. Liang, "Thinning Technique for Moment-Method Solutions," *Proc. IEEE,* Vol. 71, February 1983, pp. 265–266.

E.H. Newman, "Generation of Wide-Band Data from the Method of Moments by Interpolating the Impedance Matrix," *Trans. IEEE,* Vol. AP-36, December 1988, pp. 1820–1824.

Conjugate Gradient and Iterative Methods

T.K. Sarkar, X. Yang and E. Arvas, "A Limited Survey of Various Conjugate Gradient Methods for Solving Complex Matrix Equations Arising in Electromagnetic Wave Interactions," *Wave Motion,* Vol. 10, 1988, pp. 527–546.

C.H. Chan and R. Mittra, "Some Recent Developments in Iterative Techniques for Solving Electromagnetic Boundary Value Problems," *Radio Science,* Vol. 22, November 1987, pp. 929–934.

P.M. van den Berg, "Iterative Schemes Based on the Minimization of the Error in Field Problems," *Electromagnetics,* Vol. 5, Nos. 2–3, 1985, pp. 237–262.

M.M. Ney, "Method of Moments as Applied to Electromagnetic Problems," *Trans. IEEE,* Vol. MTT-30, October 1985, pp. 972–980.

L.W. Pearson, "A Technique for Organizing Large Moment Calculations for Use with Iterative Solution Methods," *Trans. IEEE,* September 1985, pp. 1031–1033.

D. Lesselier, D. Vuillet-Laurent, F. Jouvie and W. Tabbara, "Iterative Solution of Some Direct and Inverse Problems in Electromagnetics and Acoustics," *Electromagnetics,* Vol. 5, Nos. 2–3, 1985, pp. 147–189.

S.L. Ray and A.F. Peterson, "Error and Convergence in Numerical Implementations of the Conjugate Gradient Method," *Trans. IEEE,* Vol. AP-36, December 1988, pp. 1824–1827.

R. Mittra and C.H. Chan, "Iterative Approaches to the Solution of Electromagnetic Boundary Value Problems," *Electromagnetics,* Vol. 5, Nos. 2–3, 1985, pp. 123–146.

Review of Institutional Accomplishments

B.J. Strait and A.T. Adams, "On Contributions at Syracuse University to the Moment-Method," *Trans. IEEE,* Vol. EMC-22, November 1980, pp. 228–237.

E.H. Newman and R.J. Marhefka, "Overview of MM and UTD Methods at The Ohio State University," *Proc. IEEE,* Vol. 77, May 1989, pp. 700–708.

Cylindrical Antenna Theory [1]

R. H. Duncan and F. A. Hinchey

(February 15, 1960; revised May 2, 1960)

A partial survey of cylindrical antenna theory pertaining to a tubular model with a narrow gap is presented. The survey includes discussion of the theories of Hallén, King and Middleton, Storm, and Zuhrt. A conceptual relation between theory and experiment is described. The latter part of the article is concerned with a new Fourier series solution of the Hallén equation. This solution is developed in such a way that the expansion coefficients are the unknowns of a system of linear equations. The elements of the coefficient matrix are given by a highly convergent series. Numerical results are given for half and full wavelength antennas with half length to radius ratios of 60 and 500π. These results compare quite closely with those obtained from King-Middleton theory.

1. Introduction

Existing solutions to Hallén's integral equation for the current distribution on cylindrical antennas fall into two main categories [1, 2]: [2]

1. Iterative solutions which use an approximation to cylindrical antenna current as a starting point. Successive iterations generate improvements in the original assumption. Approximations are required at some stage in the process if tractable integrals are to be obtained. The approximations are not severe if h/a, the ratio of antenna half length to radius, is large. Impedances obtained from iterative solutions are in good agreement with experiments performed on thin antennas. Although successful, iterative solutions become laborious beyond second or third order, and the approximations become suspect in the case of thick structures.

2. Solutions in which the integral equation is converted into a set of linear simultaneous equations with Fourier coefficients of the current distribution as unknowns. Typical of these are the theories of Storm and Zuhrt. (Strictly speaking, Zuhrt did not solve Hallén's integral equation, but one derivable from a somewhat different point of view.) Storm approximated matrix elements in his set of equations so that his theory is limited to thin structures. In addition, Storm's theory contains two rather fundamental errors which, acting in concert, produce fortuitous results. Zuhrt obtained matrix elements by graphical integration, a sufficiently tedious process to limit his calculations to low order. Neither of these solutions fully exploits the Fourier series technique for obtaining the current distribution.

We have obtained a solution similar to Storm's in which computation of matrix elements can be easily done with high accuracy even for h/a ratios as low as 8 or 10. Matrix inversion is easily accomplished with modern digital computers so that solutions of high order are feasible. We have carried out calculations to 25th order for half and full wavelength antennas with h/a ratios of 60 and 500π. Results from the King-Middleton iterative solution compare favorably to ours so that their work for $h/a \geq 60$ has been checked by comparison with an exact theory.

Although this paper is principally concerned with developing a Fourier series solution, a reasonably complete treatment of iterative solutions is included in an attempt to provide a self-contained account of cylindrical antenna theory. Even with this aim in mind, the treatment of iterative solutions, as well as the theories of Storm and Zuhrt, is sufficiently involved that the reader is referred to the original work for many of the details.

2. Statement of the Problem

Vector potential as a function of position, $\overline{A}(\overline{r})$, is given in terms of a current distribution, $\overline{J}(\overline{r})$, by the expression

$$\overline{A}(\overline{r})=\frac{\mu}{4\pi}\iiint \overline{J}(\overline{r}_0)\,\frac{e^{-jk|\overline{r}-\overline{r}_0|}}{|\overline{r}-\overline{r}_0|}\,dV_0. \tag{2.1}$$

The vector integration in (2.1) must be taken over all sources of the vector potential field and it is this requirement that causes a direct solution of a realistic antenna problem to be exceedingly difficult. In the case of a cylindrical antenna the integration would have to include currents in the antenna, the feeding transmission line, and the oscillator which supplies power to the antenna-transmission line assembly. Every possible combination of antenna, transmission line and driving generator would have to be treated as a special case and a mathematical solution of any given case would by itself be formidable. An idealized problem can be extracted from this situation by consideration of an extremely thin walled tube of infinite conductivity with a narrow circumferential gap corresponding to the antenna terminal zone.

---

[1] Contribution from New Mexico State Univ., University Park, N. Mex. This work was carried out under P.O. 15-0079 from Sandia Corp.

[2] Figures in brackets indicate the literature references at the end of this paper.

Choosing the conventional cylindrical coordinates, (ρ, ϕ, z), the antenna is defined by $\rho=a$ and $|z|\leq h$.

A purely hypothetical generator is assumed such that the electric field in the gap is azimuthally symmetric. The voltage across the gap is defined by

$$V=-\int_{\text{gap}} E_s dz, \tag{2.2}$$

where E_s is the z-component of electric field at $\rho=a$. E_s is zero outside the gap since the simplifying assumption of perfectly conducting tube walls has been made. If the gap width is decreased as V is held constant, we must express E_s by

$$E_s=-V\delta(z) \text{ for } |z|\leq h, \tag{2.3}$$

where $\delta(z)$ is the usual delta function. Center-fed models will be considered here although this restriction can be removed.

The value of electric field anywhere in space is given by

$$E=-\text{grad } \Phi - j\omega \overline{A}, \tag{2.4}$$

where Φ is the scalar potential. (Time dependence proportional to $e^{j\omega t}$ is assumed in all of the relationships used here.) Scalar potential can be found from the Lorentz condition

$$\Phi=\frac{j}{\omega\mu\epsilon} \text{ div } \overline{A}. \tag{2.5}$$

The combination of a tubular model without end caps and the assumption of a symmetric field leads to the conclusion that only the z-component of $\overline{A}$ is different from zero since all of the current sources on the tubular surface will be in the z-direction. Equations (2.4) and (2.5) can be combined to give

$$E_z=-\frac{j}{\omega\mu\epsilon}(\partial^2 A/\partial z^2+k^2 A), \tag{2.6}$$

where the symbol A without the vector bar simply stands for A_z, and $k^2=\omega^2\mu\epsilon$. In general A is a function of both z and ρ. If the operator $(\partial^2/\partial z^2+k^2)$ is applied to $A(z, \rho)$ and then ρ is taken at the antenna surface, we obtain

$$\partial^2 A_s/\partial z^2+k^2 A_s=\frac{\omega\mu\epsilon}{j} V\delta(z), \tag{2.7}$$

an equation for the surface value of vector potential valid for $|z|\leq h$. Solutions of the homogeneous equation, $\partial^2 A_s/\partial z^2+k^2 A_s=0$, are simply cos kz, sin kz, e^{jkz} or e^{-jkz}. Linear combinations of these solutions may be used to build a solution to (2.7). Thus,

$$A_s=C_1 \cos kz+D_1 \sin k|z|, \tag{2.8}$$

or

$$A_s=C_2 \cos kz+D_2 e^{-jk|z|}. \tag{2.9}$$

These solutions are completely equivalent. However, we prefer (2.9) as a basis for studying certain properties of the infinite cylinder. D_1 and D_2 are evaluated by substituting these solutions into (2.7) and performing the indicated operations with due account being taken of the discontinuous derivatives of sin $k|z|$ and $e^{-jk|z|}$. The results are $D_1=(\omega\mu\epsilon V)/2jk$ and $D_2=(\omega\mu\epsilon V)/2k$. Then

$$A_s=C_1 \cos kz+\frac{\omega\mu\epsilon}{jk}\frac{V}{2} \sin k|z|, \tag{2.10}$$

$$A_s=C_2 \cos kz+\frac{\omega\mu\epsilon}{k}\frac{V}{2} e^{-jk|z|}. \tag{2.11}$$

Since $e^{-jk|z|}$ can be written as cos $kz-j$ sin $k|z|$, it is easily shown that $C_1=C_2+\omega\mu\epsilon V/2k$. Either expression may be used, the choice between them being only a matter of taste.

Under the assumptions being made, the current distribution of the general formula (2.1) degenerates to a surface distribution, $K(z_0)$. It must be borne in mind that the tube has both inner and outer surfaces, and $K(z_0)$ is the sum of current densities on both surfaces. The field point, $\overline{r}$, can be taken at the surface of the cylinder $\rho=a$ so that an alternate formula for A_s is

$$A_s=\frac{\mu}{4\pi}\int_{-h}^{+h}\int_{-\pi}^{\pi} K(z_0)\frac{e^{-jk|\overline{r}_s-\overline{r}_0|}}{|\overline{r}_s-\overline{r}_0|} a\,dz_0 d\phi_0, \tag{2.12}$$

where $\overline{r}_0$ ranges over the antenna surface during the course of the integration.

Several changes in notation are convenient at this point. The field point is (a, z, ϕ) and the source point has the coordinates (a, z_0, ϕ_0). Only the difference $\phi-\phi_0$ is of any significance because of the azimuthal symmetry. Therefore, we may set ϕ to zero, change ϕ_0 to ϕ and indicate the angular integration as already performed. The quantity z_0 is changed to ζ so that the subscripts may be avoided in future formulas. Total current is given by $I(\zeta)=2\pi aK(\zeta)$. The quantity $|\overline{r}_s-\overline{r}_0|$ is replaced simply by R. With these changes

$$A_s=\frac{\mu}{4\pi}\int_{-h}^{+h} I(\zeta)g(z-\zeta)d\zeta, \tag{2.13}$$

where

$$g(z-\zeta)=\frac{1}{2\pi}\int_{-\pi}^{\pi}\frac{e^{-jkR}}{R}d\phi \tag{2.14}$$

and

$$R=[4a^2 \sin^2\phi/2+(z-\zeta)^2]^{1/2}. \tag{2.15}$$

We now have two formulas for the surface value of vector potential. One has been derived from the generic formula for vector potential in terms of current sources by specializing to the geometry of the problem under consideration. As such it is a general formula for vector potential generated by a ϕ-symmetric current in the z-direction on a tubular conductor no matter what other conditions are to be imposed on the problem. On the other hand, (2.10) provides a vector potential which leads to the boundary values of electric field desired in the present problem. If these expressions are equated, an integral equation for the current distribution associated with the chosen boundary conditions results:

$$\frac{\mu}{4\pi}\int_{-h}^{+h} I(\zeta)g(z-\zeta)d\zeta = C_1 \cos kz + \frac{\omega\mu\epsilon}{jk}\frac{V}{2}\sin k|z|. \tag{2.16}$$

It is convenient to multiply the previous equation through by $jk/\omega\mu\epsilon$ and make the following definitions:

$$C=\frac{jkC_1}{\omega\mu\epsilon},\ \frac{jk}{\omega\mu\epsilon}\frac{\mu}{4\pi}=\frac{j}{4\pi}(\mu/\epsilon)^{1/2}=\frac{jZ_0}{4\pi}$$

and

$$\frac{jZ_0}{4\pi}I(\zeta)=f(\zeta).$$

No generality is lost by letting $V=1$. This completes the mathematical formulation of the problem. We are to consider the solution of

$$\int_{-h}^{+h} f(\zeta)g(z-\zeta)d\zeta = C\cos kz + \frac{1}{2}\sin k|z|,\ |z|\leq h. \tag{2.17}$$

The constant C must be determined by the boundary condition $f(\pm h)=0$.

There is no doubt that the integral equation corresponds exactly to the chosen model. It has already been pointed out that the model does not correspond to any physically realizable antenna. Physical antennas may be either solid or tubular conductors, and they may or may not be fed in such a way as to preserve ϕ-symmetry. Lack of symmetry (as exemplified by a linear antenna fed by a two wire line) can be rationalized to some extent if ka is small. However, the most serious point is the highly idealized nature of the generator region of the mathematical model. The infinitesimal gap is really a short circuit across which a hypothetical but finite voltage has been impressed. Thus, the input current and admittance of the model are certainly infinite.

It is not immediately clear that an infinite admittance model can be managed mathematically in such a way as to yield a physically significant finite result. Wu and King have discussed this point in a recent paper [3]. They have shown that the singularity in $I(z)$ near $z=0$ is logarithmic and of very short range. Thus, according to Wu and King, "since . . . the singularity actually gives a contribution to the current distribution only in an exceedingly small and physically meaningless distance of the order of magnitude $h \exp(-1/ka)$, it may, in principle simply be subtracted out." According to this line of reasoning iterative solutions of the integral equation are successful because they are started with a continuous approximating function and are carried to such low order that the singularity does not develop.

An important aspect of any theory is its relationship to experiment. It is possible to avoid the inherent singularity in the current distribution which is a solution to Hallén's integral equation and obtain a finite answer for the theoretical input admittance to a cylindrical antenna. The finiteness of the result does not, however, guarantee that it is physically significant. Experimental antennas must be provided with a realistic terminal zone which is connected to a transmission line of some sort. Measured impedances are then complex combinations of antenna and terminal zone effects. One way of extricating these effects is to make a theoretical correction for terminal zone effects on a sequence of experimental impedances and extrapolate the corrected data to the limit of a small terminal zone. The residual impedance is then supposed to be characteristic of the antenna itself and it is this idealized inference from experiment rather than raw data which is to be compared with theory.

An outline of a feasible experimental program will be helpful at this point. Consider the experimental arrangement shown in figure 1. Symmetry allows

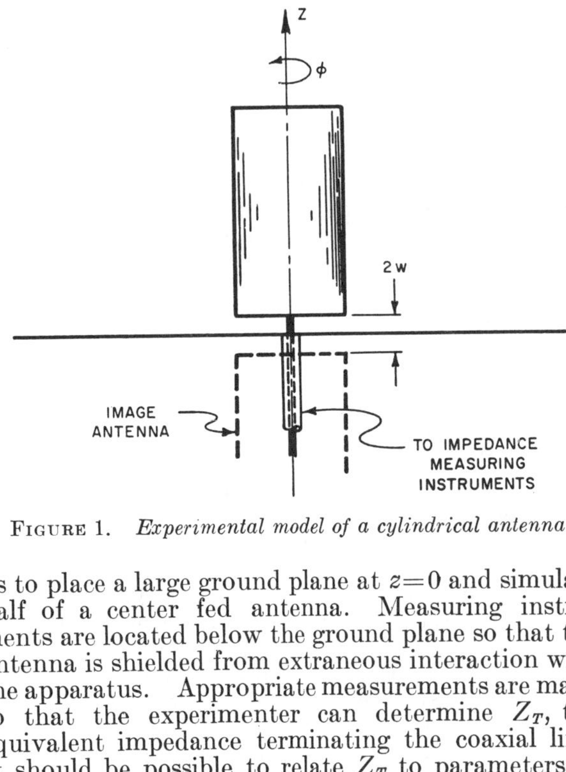

FIGURE 1. *Experimental model of a cylindrical antenna.*

us to place a large ground plane at $z=0$ and simulate half of a center fed antenna. Measuring instruments are located below the ground plane so that the antenna is shielded from extraneous interaction with the apparatus. Appropriate measurements are made so that the experimenter can determine Z_T, the equivalent impedance terminating the coaxial line. It should be possible to relate Z_T to parameters of the radial transmission line and to $Z_a(w)$, the input impedance of a cylindrical antenna with gap width 2 w. If so, one should be able to calculate $Z_a(w)$ given measurements of Z_T. $Z_a(w)$ can be extrapolated to obtain $Z_a(0)$. $Z_a(0)$ obtained in this way is independent of the terminal configuration actually used. We will not attempt to give a mathematical treatment of the experimental situation. It is sufficient for our purposes to establish a conceptual foundation for relating theory to experiment as a prelude to subsequent theoretical development.

Consider now the sense in which a Fourier series solution for $I(z)$ "subtracts the singularity." Since the unknown function in (2.17) is even, it is developed to finite order as

$$f_N(\zeta)=\frac{F_0}{2}+\sum_{n=1}^{n=N} F_n \cos n\pi\zeta/h. \tag{2.18}$$

A procedure for determining the F_n will be exhibited later. At this point it is sufficient to recall that the sum of a finite number of terms of a Fourier series fits the function being described in a least squares sense. Specifically,

$$q^2=\int_{-h}^{+h} |f_N(\zeta)-f(\zeta)|^2 d\zeta \tag{2.19}$$

is a minimum when the coefficients of $f_N(\zeta)$ are Fourier coefficients. Intense, short range variations cannot contribute appreciably to $f_N(\zeta)$ unless N is made so large that even the singular behavior of $f(\zeta)$ begins to develop.

A numerical estimate of the range of the singularity is worthwhile at this point. Consider an antenna for which $kh=\pi/2$ and $h=60a$. Then $ka\sim.025$ and $1/ka\sim40$. Substitution of these numbers into the range estimate of Wu and King leads to the conclusion that a continuous function can fit the current distribution except in a small distance equal to about he^{-40}. Even if the thickness is increased until $h=10a$ the singularity is important only over a range of about $he^{-6.5}\sim.0015h$. Thus it can be seen that $f(0)$ as given by

$$f(0)=\frac{F_0}{2}+\sum_{n=1}^{n=N} F_n \tag{2.20}$$

will be apparently well-behaved in calculations of practically feasible order even though theoretical considerations indicate that the infinite series must diverge.

An extraneous feature of the theoretical model becomes apparent when it is compared to a proposed experimental model. Current on the inner surface of the theoretical model near the feed point is replaced by current associated with a realizable terminal zone in the experimental model. If ka is less than $2.61a$, tube modes are below cutoff and are rapidly damped out [4]. Formulas for removing the inner current near $z=0$ from the total current, which is a solution to the integral equation, will be presented later. As might be expected, the correction makes a small difference in theoretical antenna admittance.

Once the singularity near $z=0$ and the tube current are understood and properly removed from the theory the remaining possibilities for refinement are somewhat limited and consist, for the most part, of removing the assumption of an infinitesimally thin, perfectly conducting tube. Such considerations would be supererogatory in view of the many successes of infinite conductivity models in electromagnetic theory.

3. Solution for the Infinite Cylinder

In subsequent work the current distribution on a finite antenna is expanded in a Fourier series. Such an expansion cannot be valid for a function containing a singularity unless the singularity is integrable. One expects that $I(z)$ is singular in the neighborhood of the delta generator and that the nature of the singularity is independent of antenna length. It can be imagined that the current on a finite antenna is composed of waves emanating from the feedpoint and waves reflected from the ends of the antenna. Only the outgoing waves are expected to be singular at $z=0$ since they are directly associated with the delta generator. It is instructive to consider the case of the infinite cylinder before proceeding with the solution for a finite antenna. This case can be solved exactly in integral form and the Fourier transform of the current identified from the solution. Asymptotic behavior of the transform gives a clue to the nature of $I(z)$ near $z=0$ even for the finite antenna.

Consider now the integral equation formed by equating (2.11) and (2.13),

$$\frac{\mu}{4\pi}\int_{-h}^{h} I(\zeta)g(z-\zeta)d\zeta=C_2\cos kz+\frac{\omega\mu\epsilon}{k}\frac{V}{2}e^{-jk|z|}. \tag{3.1}$$

If h becomes infinite there is no mechanism for the formation of standing waves on the cylinder. In that event C_2 may be set to zero. The integral equation becomes

$$\frac{Z_0}{2\pi}\int_{-\infty}^{+\infty} I(\zeta)g(z-\zeta)d\zeta=Ve^{-jk|z|};|z|<\infty. \tag{3.2}$$

Solution of this equation is easy if we are armed with the identities

$$g(z-\zeta)=-\frac{j}{2}\int_{-\infty}^{+\infty} J_0(\beta a)H_0^{(2)}(\beta a)e^{j\alpha(z-\zeta)}d\alpha, \tag{3.3}$$

and

$$e^{-jk|z|}=\frac{j}{2\pi}\int_{-\infty}^{+\infty}\frac{2ke^{j\alpha z}}{(\alpha+k)(\alpha-k)}d\alpha. \tag{3.4}$$

The parameter β is given by

$$\beta=(k^2-\alpha^2)^{1/2}. \tag{3.5}$$

Each integral is taken in the complex plane of α along the real axis from $-\infty$ to $+\infty$ with a downward identation at $\alpha=-k$ and an upward indentation at $\alpha=+k$.

The Fourier transform of $I(z)$ is

$$\hat{I}(\alpha)=\frac{1}{\sqrt{2\pi}}\int_{-\infty}^{+\infty} I(\zeta)e^{-j\alpha\zeta}d\zeta. \tag{3.6}$$

This definition and the identities for $g(z-\zeta)$ and

$e^{-jk|z|}$ lead to

$$\int_{-\infty}^{+\infty} J_0(\beta a)H_0^{(2)}(\beta a)\hat{I}(\alpha)e^{j\alpha z}d\alpha=\frac{4kV}{\sqrt{2\pi}Z_0}\int_{-\infty}^{+\infty}\frac{e^{-j\alpha z}d\alpha.}{\beta^2} \tag{3.7}$$

If (3.7) is to be true for all z,

$$\hat{I}(\alpha)=\frac{4kV}{\sqrt{2\pi}Z_0}\frac{1}{\beta^2 J_0(\beta a)H_0^{(2)}(\beta a)}\cdot \tag{3.8}$$

Then the solution of (3.2) is

$$I(z)=\frac{2kV}{\pi Z_0}\int_{-\infty}^{+\infty}\frac{e^{j\alpha z}}{\beta^2 J_0(\beta a)H_0^{(2)}(\beta a)}d\alpha\cdot \tag{3.9}$$

Since $e^{j\alpha z}$ is the only factor in the integrand containing an odd part, (3.9) becomes

$$I(z)=\frac{4kV}{\pi Z_0}\int_{0}^{+\infty}\frac{\cos \alpha z}{\beta^2 J_0(\beta a)H_0^{(2)}(\beta a)}d\alpha\cdot \tag{3.10}$$

We have defined β as follows:

$$\begin{aligned}&\beta^2=k^2-\alpha^2,\\ &\beta=|\beta| \text{ for } \alpha<k \text{ and}\\ &\beta=-j|\beta| \text{ for } \alpha>k.\end{aligned} \tag{3.11}$$

Therefore, for large α

$$\beta=-j\alpha. \tag{3.12}$$

Certain identities involving cylindrical functions are required:

$$\begin{aligned}&J_0(-j\alpha a)=I_0(-\alpha a)=I_0(\alpha a),\\ &H_0^{(2)}(-j\alpha a)=-H_0^{(1)}(j\alpha a)=-\frac{2}{j\pi}K_0(\alpha a),\\ &I_0(\alpha a)\to\frac{e^{\alpha a}}{(2\pi\alpha a)^{1/2}}, \text{ and}\\ &K_0(\alpha a)\to(\pi/2\alpha a)^{1/2}e^{-\alpha a}.\end{aligned} \tag{3.13}$$

For the definition of cosine transforms we take

$$\hat{I}_c(\alpha)=\sqrt{\frac{2}{\pi}}\int_0^{+\infty} I(z)\cos \alpha\cdot z dz. \tag{3.14}$$

Equations (3.10) through (3.14) imply that, for large α

$$\hat{I}_c(\alpha)\to 2j\frac{\sqrt{2\pi}kaV}{Z_0}\frac{1}{\alpha}\cdot \tag{3.15}$$

Inspection of tables of cosine transforms [5] reveals that when a function behaves as ln$z$ for small z, its cosine transform behaves as $-\sqrt{\pi/2}\ \alpha^{-1}$ for large α. It follows that for small enough z

$$I(z)\simeq -j\frac{4kaV}{Z_0}\ln kz\cdot \tag{3.16}$$

In the above expression k has been selected as a multiplier for $|z|$ to make the argument of the logarithmic function a pure number. The asymptotic behavior of $\hat{I}_c(\alpha)$ is of no help in deciding whether or not k is a proper choice for this parameter. However, k seems attractive since it is given by $k=2\pi/\lambda$, and the wavelength is a natural unit of length in radiation problems.

There is also a somewhat obscure reason for choosing k as a parameter to convert $|z|$ to nondimensional form. Consider another version of (3.15) in which the parameter k is retained as α is allowed to become large. As soon as α becomes greater than k, $\hat{I}_c(\alpha)$ becomes a pure imaginary, $J_0(\beta a)$ and $H_0^{(2)}(\beta a)$ go over to $I_0(|\beta|a)$ and $(-2/j\pi)\ K_0(|\beta|a)$. Then using asymptotic forms,

$$\hat{I}(\alpha>k)\to j\frac{2\sqrt{2\pi}kaV}{Z_0}\frac{1}{(\alpha^2-k^2)^{1/2}}\cdot \tag{3.17}$$

$\hat{I}_c(\alpha)$ can now be regarded as having been separated into two parts: $\hat{I}_c(\alpha<k)$ which is zero if $\alpha>k$ and $\hat{I}_c(\alpha>k)$ which is zero if $\alpha<k$. Only $\hat{I}_c(\alpha>k)$ is of interest at present since it is responsible for the singularity in $I(z)$. Now $\hat{I}_c(\alpha>k)$ is asymptotically proportional to the transform of the Neumann function, $Y_0(kz)$. The latter contains a singular part which is proportional to ln k z.

If the above arguments in favor of choosing (3.16) for the form of the singularity are acceptable, we can proceed with an estimate of the range over which this expression is a good approximation to antenna current. For this purpose we require an estimate of $I(z)$ outside the range in which the logarithmic function is dominant. It is difficult to obtain such an estimate from (3.10). However, an alternative argument can be constructed. It is known that an expression of the form $I_m \sin k(h-|z|)$ is a fair approximation to current on a finite antenna over most of its length. It is convenient, rather than necessary, to use knowledge of the finite length antenna in estimating the range of the singularity. We now arbitrarily establish the criterion that the logarithmic function is to be used in the range $|z|<\eta$ where η is given by

$$-\frac{4kaV}{Z_0}\ln k\eta\sim|I_m|. \tag{3.18}$$

$|I_m|$ is of the order of 0.01 when V=1. Z_0 is of the order of 400. An order of magnitude estimate for η is

$$\eta\sim\frac{\lambda}{2\pi}e^{-1/ka}. \tag{3.19}$$

This is a small fraction of a wavelength even for quite thick antennas. Moderately large changes in

$|I_m|$ do not affect this conclusion appreciably.

Further examination of this question would require analytic or numerical inversion of (3.10). Either approach is apparently formidable.

4. Iterative Solutions for Finite Antennas

A brief discussion of iterative solutions is provided here as background for the general reader. We are primarily interested in presenting a critique of the approach rather than compiling a report on the voluminous literature of the cylindrical antenna problem. Consequently, the presentation omits many points which are essential to a detailed understanding of the iterative method. It does present a few features which require comment by way of justifying additional consideration of finite antenna theory.

The first task is to cast eq (2.17) into a form suitable for iteration. To that end the quantity

$$f(z)\psi(z)=\int_{-h}^{h} f(z)w(z,\zeta)d\zeta \tag{4.1}$$

is added and subtracted to the left hand side of (2.17). It is convenient to abbreviate by letting

$$C\cos kz+\frac{1}{2}\sin k|z|=P(z). \tag{4.2}$$

With these changes and some elementary transpositions, (2.17) can be written

$$f(z)=\frac{1}{\psi(z)}\left\{P(z)-\int_{-h}^{h}[f\zeta)g(z-\zeta)-f(z)w(z,\zeta)]d\zeta\right\}. \tag{4.3}$$

Assuming that $w(z, \zeta)$ has been selected, a program for obtaining a solution is:

1. Substitute an approximation to $f(z)$ under the integral on the right hand side of (4.3). Denote this zeroth order approximation by $f_0(z)$. Carry out the indicated integrations to obtain $f_1(z)$ and adjust C so that $f_1(+h)=0$.
2. Repeat, using $f_1(z)$ to generate $f_2(z)$. Readjust C so that $f_2(+h)=0$.

In principle this process may be continued indefinitely with the formula for the Nth approximation being

$$f_N(z)=\frac{1}{\psi(z)}\left\{P(z)-\int_{-h}^{h}[f_{N-1}(\zeta)g(z-\zeta)-f_{N-1}(z)w(z,\zeta)]d\zeta\right\} \tag{4.4}$$

subject to $f_N(+h)=0$, which defines C_N, the Nth approximation to C.

Equation (4.4) is formally true for any $w(z, \zeta)$. However, it will clearly be to the advantage of the investigator to make a choice which results in rapid convergence of the iterative process. This matter is the raison d'etre of much of the literature of linear antenna theory. A choice which leads to manageable integrals is

$$w_H(z,\zeta)=\frac{1}{[a^2+(z-\zeta)^2]^{1/2}}. \tag{4.5}$$

Then

$$\psi_H(z)=\int_{-h}^{+h}\frac{d\zeta}{[a^2+(z-\zeta)^2]^{1/2}}. \tag{4.6}$$

The kernel $g(z\text{-}\zeta)$ as given by (2.14) and (2.15) is difficult to handle. A manageable but crude approximation is

$$g(z-\zeta)=\frac{e^{-jk|z-\zeta|}}{|z-\zeta|}. \tag{4.7}$$

If $g(z\text{-}\zeta)$ is approximated to this order it is appropriate to approximate $w_H(z\text{-}\zeta)$ to the same order when it is used in the integrals on the right hand side of (4.4), but not in the calculation of $\psi_H(z)$. Then

$$f_N(z)=\frac{1}{\psi_H(z)}\left\{P(z)-\int_{-h}^{h}\frac{f_{N-1}(\zeta)e^{-jk|z-\zeta|}-f_{N-1}(z)}{|z-\zeta|}d\zeta\right\} \tag{4.8}$$

becomes the fundamental equation of the iterative program. Equation (4.8) has been used by Hallén in the investigation of linear antenna theory [6].

Equation (4.8) can be criticized on two counts. First, the approximation involved in replacing R by simpy $|z\text{-}\zeta|$ is quite severe. Secondly, $w(z,\zeta)$ was chosen for its simplicity rather than according to the requirement that the iterative program produce good results in low order. King and Middleton improved the iterative procedure outlined above in that they used a kernel distance

$$R_1=\left[a^2+(z-\zeta)^2\right]^{1/2}. \tag{4.9}$$

This is a better approximation to R than is the quantity $|z\text{-}\zeta|$.

The second modification introduced by King and Middleton comes from considering the combination of integrals on the right hand side of (4.3). Denote these by

$$Q(z)=\int_{-h}^{h}[f(\zeta)g(z-\zeta)-f(z)w(z,\zeta)]d\zeta. \tag{4.10}$$

$Q(z)$ can be rewritten as

$$Q(z)=\int_{-h}^{h}[f(\zeta)-f(z)W(z,\zeta)]g(z-\zeta)d\zeta. \tag{4.11}$$

A $W(z, \zeta)$ which makes the integrand of (4.11) vanish is

$$W(z,\zeta)=f(\zeta)/f(z). \tag{4.12}$$

Of course, one cannot know the desired $W(z, \zeta)$ because $f(z)$ is not yet known. One can, however, approximate $W(z, \zeta)$ by making use of a fair low order approximation to $f(z)$. A suitable choice of $f(z)$ in this case is the sinusoidal approximation to antenna current.[3]

Instead of $W(z, \zeta)$ King and Middleton introduce

$$W_K(z, \zeta)=\frac{\sin k(h-|\zeta|)}{\sin k(h-|z|)}, \tag{4.13}$$

with the attendant

$$\psi_K(z)=\int_{-h}^{h} W_K(z, \zeta)\frac{e^{-jkR_1}}{R_1}d\zeta. \tag{4.14}$$

Using these definitions the interative program is based on

$$f_N(z)=\frac{1}{\psi_K(z)}\left\{P(z)-\int_{-h}^{h}[f_{N-1}(\zeta)-f_{N-1}(z)W_K(z, \zeta)]\frac{e^{-jkR_1}}{R_1}d\zeta\right\}. \tag{4.15}$$

Equations (4.8) and (4.15) are not quite the forms used by Hallén and King-Middleton in their computational programs. To appreciate the need for some improvement consider the boundary condition $f(+h)=0$ applied to (4.4) from which (4.8) and (4.15) were developed. The result of applying the boundary condition is

$$0=P(h)-\int_{-h}^{h} f_{N-1}(\zeta)g(h-\zeta)d\zeta. \tag{4.16}$$

(The term involving $f_{N-1}(z)$ drops out because the boundary condition is applied at each iteration.) Now, $P(h)$ contains $C\cos kh$ and if $kh=\pi/2$, the constant C disappears completely and (4.16) cannot be satisfied. A revision of the theory to overcome this defect can be made by subtracting (4.16) from (4.4) to obtain

$$f_N(z)=\frac{1}{\psi(z)}\left\{P(z)-P(h)+\int_{-h}^{h} f_{N-1}(\zeta)g(h-\zeta)d\zeta - \int_{-h}^{h}[f_{N-1}(\zeta)g(z-\zeta)-f_{N-1}(z)w(z, \zeta)]d\zeta\right\}. \tag{4.17}$$

The condition $f_N(+h)=0$ is always automatically satisfied by (4.17). Equation (4.16) is forced to hold for all N so that C is determined by

$$0=P(h)-\int_{-h}^{h} f_N(\zeta)g(h-\zeta)d\zeta. \tag{4.18}$$

A zeroth order approximation which satisfies boundary conditions can be obtained directly from (4.17) by omitting the integrals. Then

$$f_0(z)=\frac{P(z)-P(h)}{\psi(z)}. \tag{4.19}$$

Additional notation and definitions may be invented so that the result of the iterative process can be cast into series form. The series can be designed to lead off with $f_0(z)$ as given by (4.19). A sufficient basis for the remaining part of our discussion has been displayed at this point.

A noteworthy criticism of the use of R_1 has been present by Gans [7]. Gans correctly points out that

$$\int_{-h}^{h} f(\zeta)\frac{e^{-ikR_1}}{R_1}d\zeta=C\cos kz+\frac{1}{2}\sin k|z| \tag{4.20}$$

is not a true equation because the right hand side has discontinuous derivatives in z whereas the derivatives of the left hand side are continuous.

Hallén claims immunity from Gans' criticism on the grounds that he uses the distance $|z-\zeta|$. It seems to us that this practice raises another difficulty. A development similar to that used in examining the singularity in $f(\zeta)$ shows that $g(z-\zeta)$ is logarithmic near $z-\zeta=0$. The approximate kernel $|z-\zeta|^{-1}\exp(-jk|z-\zeta|)$ has an entirely different kind of singularity. It is extremely doubtful if the Hallén theory develops a solution to the original integral equation.

These considerations lead to a definite statement that

$$f(z)\neq\lim_{N\to\infty} f_N(z) \tag{4.21}$$

where $f_N(z)$ is taken from either the Hallén or King Middleton form of the theory and $f(z)$ is the correct solution of (4.3). Even though one is compelled to this conclusion, it is completely irrelevant because it is entirely possible for

$$f(z)\simeq f_N(z) \tag{4.22}$$

in low order. If any confusion exists about this matter it is because many writers (including some authors of senior and graduate level texts) begin their discussion directly with eq (4.20). The only correct procedure is to formulate a problem which is soluble in principle and to introduce judicious approximations as needed during the course of solution.

Iterative solutions have the disadvantage that they become tedious in second and third order even if approximations are made. High-order solutions with approximate kernels are not even desirable as they may have nothing to do with the original problem. When all is considered, one needs to know over what range of antenna parameters such solutions can be used with confidence. Obviously, the kernels used in Hallén and King-Middleton theory approximate $g(z-\zeta)$ over a range comparable to antenna length only if h/a is large. The point at which these theories

[3] A choice of $f(z)$ to serve in constructing $W(z, \zeta)$ and $\psi(z)$ need not influence the choice of $f_0(z)$ which is used to start the interative process.

break down is somewhat arbitrary since it must depend upon an arbitrarily selected amount of tolerable error.

5. Fourier Series Solutions of Storm and Zuhrt

One disadvantage of the iterative solutions discussed in section 1 is that they become extremely tedious beyond second or third order. Storm attempted to invent a theory which could be extended to higher order [8]. Unfortunately, he introduced the kernel distance $R_1=[a^2+(z-\zeta)^2]^{1/2}$ and studied

$$\int_{-h}^{h} f(\zeta)\frac{e^{-jkR_1}}{R_1}d\zeta=C\cos kz+\frac{1}{2}\sin k|z| \tag{5.1}$$

which has, in fact, no solution at all. "Solutions" to (5.1) are physically meaningful only if they apply to sufficiently thin antennas and if they are restricted to low order. In addition, Storm's theory contains fundamental errors which invalidate his solution no matter what kernel distance is used.

Storm expands the unknown function in the form

$$f(\zeta)=B\sin k(h-|\zeta|)+\sum_{n=0}^{n=N} F_n\cos (2n+1)\pi\zeta/2h. \tag{5.2}$$

Since $f(\zeta)$ is an even function the expansion also represents $f(\zeta)$ in the range $-h\leq\zeta\leq 0$.

Following Storm, we seek to determine the coefficients B and F_n of the above expansion. The result of substituting (5.2) into (5.1) is

$$M_0(z)B+\sum_{n=0}^{n=N} S_n(z)F_n=C\cos kz+\frac{1}{2}\sin k|z|, \tag{5.3}$$

where

$$M_0(z)=\int_{-h}^{h}\sin k(h-|\zeta|)\frac{e^{-jkR_1}}{R_1}d\zeta, \tag{5.4}$$

$$S_n(z)=\int_{-h}^{h}\cos\frac{(2n+1)\pi\zeta}{2h}\frac{e^{-jkR_1}}{R_1}d\zeta. \tag{5.5}$$

The integrals in (5.4) and (5.5) are somewhat difficult to evaluate unless approximations are made. Storm replaces the kernel with $|z-\zeta|^{-1}\exp(-jk|z-\zeta|)$ outside the range $z-5a<\zeta<z+5a$. Inside this range he replaces the kernel by $[a^2+(z-\zeta)^2]^{-1/2}$. These ranges are ambiguous if z is within $5a$ of the ends of the antenna since ζ must also be restricted to the range $-h\leq\zeta\leq h$. Presumably we are not to consider values of z too close to the ends of the antenna in what follows. Once the approximations are made, the integrations required in (5.4) and (5.5) can be performed in terms of elementary functions. We shall omit the details and return to consideration of (5.3).

Storm explicitly satisfies (5.3) at $N+2$ points and obtains $N+2$ equations in $N+2$ unknowns which are B, C, and N of the F_n. He has performed calculations with $N=0$, 1, 2, and 3 for both full and half wavelength antennas with $h/a=904$. The agreement with King-Middleton iterative theory is remarkable. A comparison between the impedances from the latter theory with those from Storm's five-point calculation is made in table 1. King-Middleton data used in this comparison are second order except for the resistance in the half wavelength case which is third order.

TABLE 1

kh	Storm: 5 point calculation	King-Middleton
$\pi/2$	$Z_s=81.5+j\,44.5$	$Z_{KM}=81.5+j\,43.4$
π	$Z_s=1162-j\,1354$	$Z_{KM}=1000-j\,1350$

In spite of the success of Storm's calculation, his theory breaks down in higher order or for smaller h/a ratios. The difficulties are made more evident if we specify the unknown function by the expansion

$$f(\zeta)=B\sum_{n=0}^{n=N}A_n\cos\frac{(2n+1)\pi\zeta}{2h}+\sum_{n=0}^{n=N}F_n\cos\frac{(2n+1)\pi\zeta}{2h}, \tag{5.6}$$

where the A_n are the first N terms of an expansion of $\sin k(h-|\zeta|)$. As such the A_n are known explicitly. It is clear that (5.6) approaches (5.2) as N becomes large. If (5.6) is used for $f(\zeta)$, (5.4) must be changed to

$$M_0'(z)=B\int_{-h}^{h}\frac{e^{-jkR_1}}{R_1}\left\{\sum_{n=0}^{n=N}A_n\cos\frac{(2n+1)\pi\zeta}{2h}\right\}d\zeta. \tag{5.7}$$

The definition of $S_n(z)$ does not change. The new equation to be satisfied is

$$M_0'(z)B+\sum_{n=0}^{n=N}S_n(z)F_n=C\cos kz+\frac{1}{2}\sin k|z|. \tag{5.8}$$

Set aside, for the moment, the question of how C is to be determined and consider the solution in terms of C. A set of linear simultaneous equations for the unknowns B, F_0, F_1, . . . F_N is obtained by satisfying (5.8) at $(N+1)$ values of z. The matrix of coefficients is

$$\begin{Vmatrix} M_0'(z_1) & S_0(z_1) & S_1(z_1) & \cdots & S_N(z_1) \\ M_0'(z_2) & S_0(z_2) & S_1(z_2) & \cdots & S_N(z_2) \\ \cdot & \cdot & \cdot & & \cdot \\ \cdot & \cdot & \cdot & & \cdot \\ \cdot & \cdot & \cdot & & \cdot \\ M_0'(z_N) & S_0(z_N) & S_1(z_N) & \cdots & S_N(z_N) \end{Vmatrix}$$

By comparing (5.5) and (5.7), one sees that the $M_0'(z_m)$ are linear combinations of the $S_n(z_m)$. The determinant of the above matrix is identically zero for any N. If the left hand column is replaced by $M_0(z_m)$ so that all of $B \sin k(h-|\zeta|)$ is used, the determinant of the matrix must approach zero as N becomes large. No high order solution can be obtained unless one of the unknowns is assigned an arbitrary value. A likely candidate for the assignment is B which can be selected from the elementary induced emf theory according to $B=(jZ_0/4\pi)I_m$.

Evidently, Storm did not become aware of these difficulties because he included C among the unknowns. If this is done and the matrix is augmented by appropriate bordering elements, a system of equations with an unique solution for every N results. However, the value of C which is obtained may not be the correct one. Storm's expansion for $f(z)$ is identically zero at the boundaries. If the coefficients, F_n, of that expansion are calculated in terms of a spurious C it may happen that the expansion (for large N) does not approach zero at the boundaries even though it is identically zero at $z=\pm h$. The only way out of this dilemma is to expand the current in functions which are not identically zero at $z=\pm h$ and then use the boundary condition $f(\pm h)=0$ to determine C.

An actual calculation to support these criticisms is worthwhile. In performing the calculation we used the theory of the next section which is equivalent to a corrected version of Storm's theory even though several of the technical details are quite different. Storm was followed to the extent that $f(\zeta)$ was expanded in the set $\{\cos (2n+1)\pi\zeta/2h\}$, and C was treated as an independent unknown. A structure with $h/a=60$ was selected for study. The results are shown in figure 2 which displays the real and

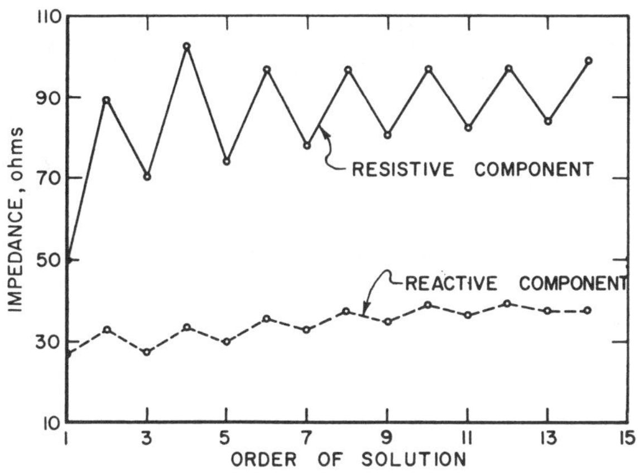

FIGURE 2. *Impedance versus order of solution (Storm's procedure).*

imaginary parts of antenna impedance as a function of the order of solution. Corresponding values of C are shown in figure 3. The result of mistreating the parameter C is easily observed from these curves. We have conjectured that the amplitude of oscillation in the results may decrease with increasing h/a.

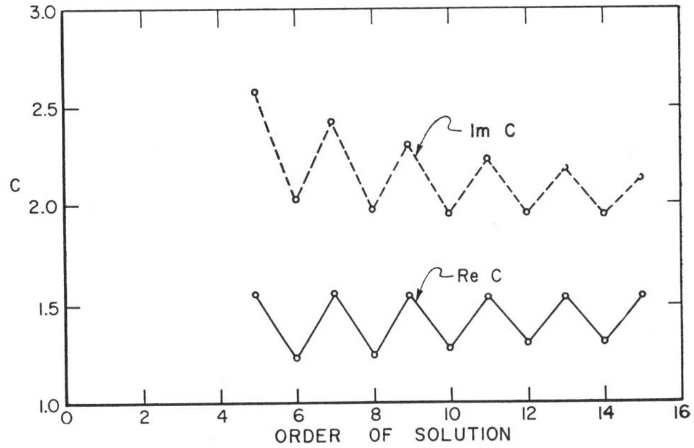

FIGURE 3. *The parameter C versus order of solution (Storm's procedure).*

If this is so it explains why Storm obtained good results with a defective theory.

Zuhrt considered the problem of developing $f(z)$ in a Fourier series from a somewhat different point of view [9]. Instead of considering a single isolated antenna, he formulated a boundary value problem for an infinite collinear array of such antennas spaced along the z-axis with centers at $z=\pm nd$, where $d>h$. Each unit is center-driven by a potential $V_n=(-1)^n V_0$ impressed across a gap of finite width. Ultimately he allows d to approach infinity and the gap width to become small. In this limit only the center unit remains and his theory represents the simple tubular model. This approach is unnecessarily intricate. Zuhrt's final equation can be derived directly from the Hallén integral equation. Equation (3.3) may be written

$$g(z-\zeta)=-j\int_0^\infty J_0(\beta a)H_0^{(2)}(\beta a)\cos\alpha(z-\zeta)d\alpha. \tag{5.9}$$

The term $\cos\alpha(z-\zeta)$ in the integrand can be expanded and the term $\sin\alpha z \sin\alpha\zeta$ omitted since $f(\zeta)$ is even for a center-fed antenna. Therefore, we can consider

$$\int_{-h}^{h} f(\zeta)K(z,\zeta)d\zeta=C\cos kz+\frac{1}{2}\sin k|z|, \tag{5.10}$$

where

$$K(z,\zeta)=-j\int_0^\infty J_0(\beta a)H_0^{(2)}(\beta a)\cos\alpha z\cos\alpha\zeta d\zeta. \tag{5.11}$$

Applying the operator $L_z=\partial^2/\partial z^2+k^2$ to both sides of (5.10), one obtains

$$\int_{-h}^{h} f(\zeta)L_zK(z,\zeta)d\zeta=k\delta(z). \tag{5.12}$$

Now expand $f(\zeta)$ as

$$f(\zeta)=\sum_{n=0}^{n=N} F_n \cos Hn\zeta, \tag{5.13}$$

where

$$\frac{(2n+1)\pi}{2h}=H_n. \tag{5.14}$$

Substituting the assumed current expansion into (5.12), one obtains

$$\sum_{n=0}^{n=N}\left\{\int_{-h}^{h} \cos H_n\zeta L_z K(z,\zeta)d\zeta\right\} F_n=k\delta(z). \tag{5.15}$$

Now multiply both sides of (5.15) by $\cos H_p z$, and integrate on z over the range $-h\leq z\leq h$ to obtain

$$\sum_{n=0}^{n=N}\left\{\frac{1}{k}\int_{-h}^{h} \cos H_p z dz \int_{-h}^{h} \cos H_n \zeta L_z K(z,\zeta)d\zeta\right\}F_n=1. \tag{5.16}$$

Equation (5.16) generates an infinite set of linear simultaneous equations as the indexing parameter, p, is allowed to range from zero to infinity. The coefficients are given by

$$Z_{pn}=\frac{1}{k}\int_{-h}^{h} \cos H_p z dz \int_{-h}^{h} \cos H_n \zeta L_z K(z,\zeta)d\zeta. \tag{5.17}$$

Further reduction is accomplished by substituting (5.11) for $K(z, \zeta)$ and carrying out the indicated operations. When this is done,

$$Z_{pn}=-4j(-1)^{p+n}H_pH_n \int_0^{\infty} \frac{J_0(\beta a)H_0^{(2)}(\beta a)(k^2-\alpha^2)\cos^2 \alpha k}{(H_p^2-\alpha^2)(H_n^2-\alpha^2)} d\alpha. \tag{5.18}$$

This is the same as Zuhrt's formula except for trivial differences in notation.

Zuhrt obtains an Nth order theory by truncating the infinite scheme at Nth order. The integral which defines the matrix elements is difficult to evaluate. Zuhrt resolves this difficulty by resorting to graphical integration. Each coefficient has a real and an imaginary part, so that an Nth order theory requires $2N^2$ graphical integrations, a formidable amount of labor even for small N.

6. Further Development of Fourier Series Solutions

The kernel of Hallén's integral equation,

$$g(z-\zeta)=\frac{1}{2\pi}\int_{-\pi}^{\pi} \frac{e^{-jkR}}{R} d\phi, \tag{6.1}$$

where

$$R=[4a^2 \sin^2 \phi/2+(z-\zeta)^2]^{1/2}, \tag{6.2}$$

represents the radiation at any value of z on a cylinder of radius a from a ring source of radius a located at ζ. As such it is meaningful in the entire domain $-\infty<z<+\infty$. However, the integral equation is valid only on the range $-h\leq z\leq +h$ and operations on the source coordinate, ζ, are restricted to the same range. Hence, a special expansion of $g(z-\zeta)$ for $-h\leq z,\zeta\leq +h$ is desirable. A frontal attack on the problem of obtaining such an expansion has been made by Bohn in an investigation of a theoretical model which is quite different from the one used here [10]. The point of interest at present is his method of handling $g(z-\zeta)$ which is the kernel of an integral equation occurring in his theory.

In our notation Bohn's expansion is

$$g(z-\zeta)=\sum_{n=-\infty}^{n=\infty}\sum_{m=-\infty}^{m=\infty} G_{nm}e^{j(n\pi z/h-m\pi\zeta/h)} \tag{6.3}$$

for $-h\leq z,\zeta\leq +h$, where the coefficients of the Fourier series are given by

$$G_{nm}=\frac{1}{4h^2}\int_{-h}^{h}\int_{-h}^{h} g(z-\zeta)e^{-j(n\pi z/h-m\pi\zeta/h)}dzd\zeta. \tag{6.4}$$

Substituting (5.9) for $g(z-\zeta)$, (6.4) becomes

$$G_{nm}=-2jh^2\int_{-\infty}^{+\infty} \frac{J_0(\beta a)H_0^{(2)}(\beta a)\sin(n\pi-\alpha h)\sin(m\pi-\alpha h)d\alpha}{(n\pi-\alpha h)(m\pi-\alpha h)}. \tag{6.5}$$

This expression exhibits the difficulties involved in a direct attack. The reader will appreciate that the integration is not trivial. Bohn evaluates (6.5) approximately by means of ingenious distortions of the contour of integration. Details will be omitted here, it being sufficient for our purpose to note that no investigator to date has been able to obtain the exact coefficients of a double Fourier series representation of $g(z-\zeta)$.

Fortunately, it is possible to avoid the double Fourier series representation entirely. It is only necessary to make use of the fact that z and ζ enter only as the square of their difference. Thus, $g(z-\zeta)$ is not a general function of (z, ζ). To proceed, let

$$\xi=(z-\zeta). \tag{6.6}$$

If z and ζ are separately in the range $-h\leq z,\zeta\leq h$, then ξ is in the range $-2h\leq\xi\leq 2h$. Since $g(\xi)$ contains only ξ^2, a cosine series in the range 0 to $2h$ will suffice. Thus, we seek an expansion of the form

$$g(\xi)=\frac{D_0}{2}+\sum_{m=1}^{m=\infty} D_m \cos m\pi\xi/2h, \text{ for } 0\leq\xi\leq 2h. \tag{6.7}$$

The fundamental formula for any coefficient is

$$D_m=\frac{1}{2h}\int_0^{2h} g(\xi) \cos m\pi\xi/2h d\xi. \tag{6.8}$$

Direct integration of (6.8) is difficult. An indirect method can be constructed by writing

$$g(\xi)=g_1(\xi)+g_2(\xi), \tag{6.9}$$

where $g_1(\xi)=g(\xi)$ in the range $0\leq\xi\leq 2h$, and is zero outside this range, $g_2(\xi)$ is equal to zero in the range $0\leq\xi\leq 2h$, and is identical to $g(\xi)$ in the range $2h\leq\xi\leq\infty$. Transposition of (6.9) gives

$$g_1(\xi)=g(\xi)-g_2(\xi). \tag{6.10}$$

The symbolic cosine transform of (6.10) is

$$G_1(\alpha)=G(\alpha)-G_2(\alpha). \tag{6.11}$$

The transform of $g_1(\xi)$ and its inverse are given by

$$G_1(\alpha)=\sqrt{\frac{2}{\pi}}\int_0^{2h} g_1(\xi)\cos\alpha\xi d\xi, \tag{6.12}$$

and

$$g_1(\xi)=\sqrt{\frac{2}{\pi}}\int_0^{\infty} G_1(\alpha)\cos\alpha\xi d\alpha. \tag{6.13}$$

Comparison of (6.12) and (6.8) shows that the coefficients of the cosine series expansion for $g_1(\xi)$ are simply proportional to sample values of its cosine transform, $G_1(\alpha)$. Thus,

$$D_m=\frac{1}{h}\sqrt{\frac{\pi}{2}}\left[G\left(\frac{m\pi}{2h}\right)-G_2\left(\frac{m\pi}{2h}\right)\right]. \tag{6.14}$$

$G(\alpha)$ can be immediately identified from (5.9) as

$$G(\alpha)=-j\sqrt{\frac{\pi}{2}}J_0(\beta a)H_0^{(2)}(\beta a). \tag{6.15}$$

Incidentally, an asymptotic expansion of $G(\alpha)$ shows that, for small ξ, $g(\xi)\sim\ln\xi$. Since the singularity in $g(\xi)$ is no worse than logarithmic, the integrals defining the D_m exist.

Determination of $G_2(\alpha)$ is tedious but not difficult. The definition of $G_2(\alpha)$ is

$$G_2(\alpha)=\sqrt{\frac{2}{\pi}}\int_{2h}^{\infty} g(\xi)\cos\alpha\xi d\xi, \tag{6.16}$$

which can be expanded in a highly convergent infinite series. If, for convenience, we set

$$y^2=4a^2\sin^2\phi/2 \tag{6.17}$$

and

$$u^2=y^2/\xi^2, \tag{6.18}$$

(6.1) becomes

$$g(\xi^2>y^2)=\frac{1}{2\pi}\int_{-\pi}^{\pi}\frac{e^{-jk\xi\sqrt{1+u^2}}}{\xi}(1+u^2)^{-1/2}d\phi. \tag{6.19}$$

When appropriate expansions of the integrand are made and the ϕ-integration is performed, there results

$$g(\xi^2>y^2)=\frac{e^{-jk\xi}}{\xi}\left[1-\frac{jka^2}{\xi}-\frac{a^2}{\xi^2}-\ldots\right]. \tag{6.20}$$

The above expression is to be used when $\xi\geq 2h$. The parameter k is of the same order as $1/h$. Therefore the terms retained in the square bracket are of order $(a/h)^2$. The first omitted term is of order $(a/h)^4$. We shall now omit all of (6.20) except the first term. It is not difficult to restore the small correction terms later.

With this omission, (6.16) becomes

$$G_2(\alpha)\simeq\sqrt{\frac{2}{\pi}}\int_{2h}^{\infty}\frac{e^{-jk\xi}\cos\alpha\xi}{\xi}d\xi. \tag{6.21}$$

Sine and cosine integral functions are defined by

$$Si(x)=\int_0^x\frac{\sin x}{x}dx=\frac{\pi}{2}-\int_x^{\infty}\frac{\sin x}{x}dx;$$
$$Ci(x)=-\int_x^{\infty}\frac{\cos x}{x}dx. \tag{6.22}$$

With the above, a few trigonometric identities, and a few elementary changes of variable, the integration of (6.21) follows almost by definition, it being only necessary to exercise a little care depending on whether $\alpha<k$ or $\alpha>k$. The case of $\alpha=k$ will require special attention. We have

$$G_2(\alpha<k)\simeq-\sqrt{\frac{1}{2\pi}}\{Ci[(k+)\alpha 2h]+Ci[(k-\alpha)2h]$$
$$+j\pi-jSi[(k+\alpha)2h]-jSi[(k-\alpha)2h]\}. \tag{6.23}$$

$$G_2(\alpha>k)\simeq-\sqrt{\frac{1}{2\pi}}\{Ci[(\alpha+k)2h]+Ci[(\alpha-k)2h]$$
$$-jSi[(\alpha+k)2h]+jSi[(\alpha-k)2h]\}. \tag{6.24}$$

It is convenient to rewrite $G(\alpha)$ from (6.15) separately for the cases $\alpha<k$ and $\alpha>k$. Let $|\beta|=b$. The phase of β has been defined so that $\beta=b$ if $\alpha<k$ and $\beta=-jb$ if $\alpha>k$. Then

$$G(\alpha<k)=-j\sqrt{\frac{\pi}{2}}J_0(ba)H_0^{(2)}(ba), \tag{6.25}$$

and

$$G(\alpha>k)=\sqrt{\frac{2}{\pi}}I_0(ba)K_0(ba). \tag{6.26}$$

Now $G_2(\alpha<k)$, $G_2(\alpha>k)$, $G(\alpha<k)$, $G(\alpha>k)$ are all singular at $\alpha=k$. In the first two functions the singularity comes from the cosine integral function, in the last two $H_0^{(2)}(ba)$ and $K_0(ba)$ become singular. Since these functions are to be sampled at the points

$m\pi/2h$, the singularities apparently give trouble if $h=m\lambda/4$. Actually we are concerned only with the difference $G(\alpha)-G_2(\alpha)$ and this turns out to be finite and independent of whether α approaches k from above or below. The special formula required for the $\alpha=k$ case is found from combining (6.23), (6.25) and making use of small argument formulas for the various functions involved. The latter are tabulated for the reader's convenience.

For small x,

$$J_0(x)\simeq 1, \quad Si(x)\simeq 0,$$

$$H_0^{(2)}(x)\simeq 1+j\frac{2}{\pi}\ln\frac{2}{\gamma x}, \quad Ci\,(x)\simeq \ln\gamma\, x. \tag{6.27}$$

The logarithmic singularities subtract off in $G(\alpha)-G_2(\alpha)$ and one obtains

$$G_1(\alpha=k)\simeq\frac{1}{\sqrt{2\pi}}\left\{\ln\frac{4h}{\gamma ka^2}+Ci\,(4kh)-jSi(4kh)\right\}. \tag{6.28}$$

The same result can be obtained by using (6.24), (6.26), and the small argument formulas for $I_0(ba)$ and $K_0(ba)$.

The degree of approximation in the above formulas may be improved by calculating the cosine transforms of the correction terms in (6.20). The next term to be included is

$$T_1(\alpha)=-jka^2\sqrt{\frac{2}{\pi}}\int_{2h}^{\infty}\frac{e^{-jk\xi}}{\xi^2}\cos\alpha\xi d\xi, \tag{6.29}$$

which reduces to trigonometric functions, sine integral functions and cosine integral functions. All higher order correction terms may be similarly treated.

Thus, an expansion of $g(z-\zeta)$ in the form (6.7) can be achieved and the coefficients can be calculated with any desired degree of accuracy. It will be convenient in what follows to re-define D_0 so that the leading term can be included under the summation sign. If this is done

$$g(z-\zeta)=\sum_{m=0}^{m=\infty}D_m[\cos m\pi z/2h\cos m\pi\zeta/2h + \sin m\pi z/2h\sin m\pi\zeta/2h]. \tag{6.30}$$

The sine terms are not needed in the treatment of a center-fed antenna. We are then led to consider

$$\int_{-h}^{h}f(\zeta)\,K(z,\zeta)\,d\zeta=C\cos kz+\frac{1}{2}\sin k|z|, \tag{6.31}$$

where

$$K(z,\zeta)=\sum_{m=0}^{m=\infty}D_m\cos m\,\pi\,z/2h\cos m\pi\,\zeta/2h, \tag{6.32}$$

subject to $f(\pm h)=0$.

We now expand $f(\zeta)$ as

$$f(\zeta)=\sum_{n=0}^{n=\infty}F_n\cos n\,\pi\,\zeta/h \tag{6.33}$$

and substitute into (6.31). After the ζ integration,

$$\sum_{n=0}^{n=\infty}\sum_{m=0}^{m=\infty}F_nD_m\gamma_{nm}\cos m\pi z/2h=C\cos kz+\frac{1}{2}\sin k|z|, \tag{6.34}$$

$$\gamma_{nm}=2\int_0^h\cos n\,\pi\,\zeta/h\cos m\,\pi\,\zeta/2\,h\,d\zeta. \tag{6.35}$$

An infinite set of linear simultaneous equations is obtained by multiplying both sides of (6.34) by $\{\cos\,(2p+1)\pi z/2h\}$ and integrating on z from $-h$ to $+h$. The result is

$$\sum_{n=0}^{n=\infty}\sum_{m=0}^{m=\infty}F_nD_m\gamma_{nm}\beta_{pm}=Cr_p+v_p, \tag{6.36}$$

where

$$\beta_{pm}=2\int_0^h\cos\,[(2p+1)\pi z/2h]\cos\,[m\pi z/2h]dz, \tag{6.37}$$

$$r_p=2\int_0^h\cos kz\cos\,[(2p+1)\pi z/2h]dz, \tag{6.38}$$

and

$$v_p=2\int_0^h\sin kz\cos\,[(2p+1)\pi z/2h]dz. \tag{6.39}$$

Equation (6.36) can be written

$$\sum_{n=0}^{n=\infty}\Gamma_{pn}F_n=Cr_p+v_p, \tag{6.40}$$

where

$$\Gamma_{pn}=\sum_{m=0}^{m=\infty}D_m\gamma_{nm}\beta_{pm}. \tag{6.41}$$

If (6.41) were actually an infinite sum, many terms would be required to satisfactorily approximate each Γ_{pn} and the theory would be laborious except in low order. However, (6.41) contains only two nonzero terms! To appreciate this fact consider the set of functions $\{\cos(m\pi z/2h)\}$ which appear in the expansion of $K(z,\,\zeta)$. These functions are complete on $0\leq z\leq 2h$. They appeared in the theory because we expanded a function of $(z-\zeta)$. Clearly since m is either even or odd, this basic set of functions can be divided into two subsets $\{\cos m\pi z/h\}$ and $\{\cos(2m+1)\pi z/2h\}$. Both of the subsets are complete and orthogonal on the range where they are actually used. In a sense we can refer to (6.32) as an overcomplete expansion of $K(z,\zeta)$. Therefore, the summation in (6.41) can be broken into two parts, one for m even, the other for m odd, and reduced to

$$\Gamma_{pn}=h[\beta_{p,2n}D_{2n}+\gamma_{n,2p+1}D_{2p+1}]. \tag{6.42}$$

The set of equations generated by (6.40) as p ranges from zero to infinity can be cast in matrix form as

$$\Gamma F = Cr + v, \qquad (6.43)$$

where Γ is a matrix of the Γ_{pn}; F, r, and v are column vectors.

The method of obtaining simultaneous equations from (6.34) used here is formally equivalent to any other method which might be used. Consider multiplying (6.34) by $X_p(z)$ and integrating on z from $-h$ to h, where $X_p(z)$ is arbitrary. Now $X_p(z)$ can be expanded in the set $\{\cos(2p+1)\pi z/2h\}$ so that the set of equations obtained by using $X_p(z)$ is a linear combination of the equations represented by (6.43). An Nth order colocation scheme is equivalent to choosing $X_p(z)$ from a set of N delta functions, $\delta(z-z_p)$ with $p=1, 2, 3, \ldots, N$.

The formal solution of (6.43) in terms of Γ^{-1} is simply

$$F = C\Gamma^{-1}r + \Gamma^{-1}v. \qquad (6.44)$$

We have not been able to discover a general form for Γ^{-1}. Consequently, it has been necessary to truncate the system of equations to finite order and invert finite matrices using digital computer methods.[4] This procedure is a cause of some concern in that there are apparently no mathematical theorems which justify the assumption that a sequence of such inverses converges to the inverse of the infinite matrix. However, there is evidence that the procedure being used here produces the correct solution of Hallén's integral equation. First of all, we have examined sequences of finite inverses up to 25th order for $h/a=60$ and $kh=\pi/2$. These are well-behaved and stable. For a rough definition of stability, we shall say that a stable Nth order inverse has been found if the elements of an Nth order matrix formed by truncating the inverse matrix of an $(N+M)$th order solution do not change appreciably as M is increased. M will be referred to as the stability margin. For the problem at hand our work indicates that $M\sim 3$.

The results obtained for finite antennas are thoroughly reasonable when compared to results obtained by other methods. From a pragmatic point of view there seems to be sufficient evidence that the matrix inversion procedure does indeed produce a finite number of terms of the correct solution to the original integral equation.

If the elements of the inverse of an Nth order truncation of Γ are designated as H^N_{np} the numerical solution is

$$F_n(C) = \sum_{p=0}^{p=N} H^N_{np}(Cr_p + v_p). \qquad (6.45)$$

[4] Numerical inversion of the required matrices was accomplished on the IBM 650, using a library program furnished by IBM Corporation.

The Nth approximation to $f(z)$ given by the computational program is

$$f_N(z) = \sum_{n=0}^{n=N} F_n(C)\cos(n\pi z/h). \qquad (6.46)$$

The leading terms of (6.46) are not a good approximation to antenna current. Consequently, this representation is slowly convergent. To improve convergence a good low order approximation to $f(z)$ was chosen, and expanded in a series

$$x(z) = \sum_{n=0}^{n=N} X_n \cos(n\pi z/h). \qquad (6.47)$$

Candidates for the role of $x(z)$ are the classical sinusoidal distribution, the zeroth order approximation from an iterative solution, the King-Middleton modified zeroth order approximation, or a new low order approximation by R. W. P. King referred to by its author as the "quasi-zeroth order approximation" [11]. We have used the latter. Once the choice has been made and X_n have been calculated, (6.46) is modified to read

$$f_N(z) = x(z) + \sum_{n=0}^{n=N} [F_n(C) - X_n]\cos(n\pi z/h). \qquad (6.48)$$

When the boundary condition $f(h)=0$ is imposed, one obtains an Nth order approximation to C from

$$\sum_{n=0}^{n=N} [F_n(C) - X_n](-1)^n = 0. \qquad (6.49)$$

By letting $N=0, 1, 2, \ldots$, N, (6.49) can be used to generate a sequence,

$$C_0, C_1, C_2, \ldots, C_N \qquad (6.50)$$

Now an Nth order solution should be regarded as the first N terms of a solution of infinite order, only the latter solution involves the Fourier series for $f(z)$. The Fourier series is unique and logically it must be in terms of C_∞. The most direct method of determining C_∞ is to plot the above sequence versus $1/n$ and extrapolate. Unfortunately, the sequence of C values obtained from the boundary condition oscillates and the extrapolation is subject to large error. This difficulty can be overcome by applying a Ceasaro transformation to the sequence of C values to form a new sequence which converges to the same limit. The transformed sequence plots a smooth curve against $1/n$, the extrapolated limit of which is taken to be C_∞. A graph of a typical treatment of C is shown in figure 4.

Our final expression for $f(z)$ is

$$f_N(z) = x(z) + \sum_{n=0}^{n=N} [F_n(C_\infty) - X_n]\cos(n\pi z/h). \qquad (6.51)$$

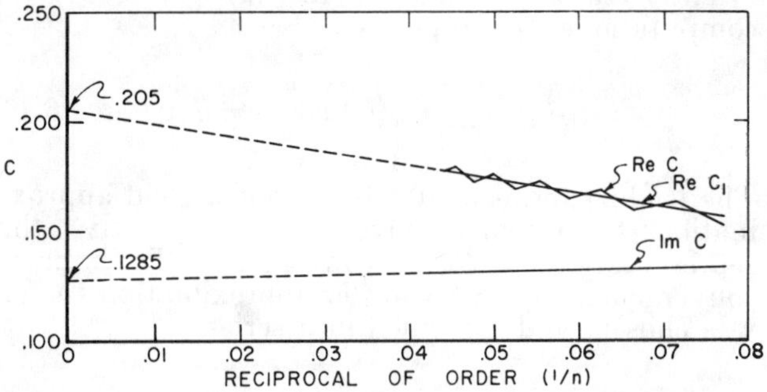

FIGURE 4. *Determination of parameter C.*

This expression does not quite satisfy the boundary condition $f(+h)=0$. However, except for truncation errors which affect only the last two or three of the F_n appreciably, it represents an estimate of the first N terms of a solution of infinite order. Readers who may prefer an expression which satisfies the boundary condition exactly in Nth order should be reminded once again that a finite number of terms of a Fourier series provides a least squares best fit over the entire range of the function being represented. This type of fit is to be preferred over one which is identically equal to the function at one point.

The admittance of the antenna is now simply

$$Y_N=\frac{4\pi}{jZ_0}\left\{x(0)+\sum_{n=0}^{n=N}[F_n(C_\infty)-X_n]\right\}. \tag{6.52}$$

The function $f(\zeta)$ is proportional to the sum of the currents on both the inner and outer surfaces of the tubular conductor. It was pointed out earlier that only the current on the outer surface and not the tube current is to be associated with the experimental admittance of the antenna.

The generic expression for the vector potential at a field point (ρ, ϕ, z) when $\rho_0=a$ is

$$A_z=\frac{\mu}{4\pi}\int_{-h}^{h}I(\zeta)g(\rho,z;a,\zeta)d\zeta, \tag{6.53}$$

where $g(\rho,\ z;\ a,\ \zeta)$ is given by either

$$g(\rho,z;a,\zeta)=\frac{1}{2\pi}\int_{-\pi}^{\pi}\frac{e^{-jkR}}{R}\,d\phi \tag{6.54}$$

with

$$R=[\rho^2+a^2-2\rho a\cos\phi+(z-\zeta)^2]^{1/2} \tag{6.55}$$

or

$$g(\rho,z;a,\zeta)=-j\int_0^\infty J_0(\beta a)H_0^{(2)}(\beta\rho)\cos\alpha(z-\zeta)d\alpha. \tag{6.56}$$

As a consequence of the ϕ-symmetry, only the ϕ-component of $\overline{H}$ is different from zero, and is given by

$$H_\phi=-\frac{1}{\mu}\,\partial A_z/\partial\rho. \tag{6.57}$$

The surface current density on the outer surface of the tube is obtained from the boundary condition $\overline{n}\times\overline{H}=\overline{K}$, and the current distribution along the outer surface is

$$I^0(z)\frac{1}{4\pi}\int_{-h}^{h}I(\zeta)g^0(\rho,z;a,\zeta)d\zeta \tag{6.58}$$

where

$$g^0(\rho,z;a,\zeta)=-2\pi a\left[\frac{\partial}{\partial\rho}\,g(\rho,z;a,\zeta)\right]_{\rho=a+}. \tag{6.59}$$

If $g(\rho,z;a,\zeta)$ is expanded in the Fourier series

$$g^0(\rho,z;a,\zeta)=\sum_{m=0}^{m=\infty}D_m^0\cos\frac{m\pi(z-\zeta)}{2h},\ 0\leq|z-\zeta|\leq h, \tag{6.60}$$

the coefficients D_m^0 may be determined by the same procedure used in expanding the kernel of Hallén's equation. Omitting details, the D_m^0 are given by sample values of

$$G_1^0(\alpha)=-j\beta aJ_0(\beta a)H_1^{(2)}(\beta a)+\left\{\frac{\partial}{\partial\rho}\int_{2h}^{\infty}\int_{-\pi}^{\pi}\frac{e^{-jkR}}{R}\cos\alpha(z-\zeta)d\phi d(z-\zeta)\right\}_{\rho=a+}, \tag{6.61}$$

where R is given by (6.55).

This leads to the following expression for Fourier coefficients of the exterior current:

$$I_m^0=\frac{4\pi}{jZ_0}D_m^0\left\{\int_{-h}^{h}x(\zeta)\cos(m\pi\zeta/2h)d\zeta+\sum_{n=0}^{n=N}[F_n-X_n]\int_{-h}^{h}\cos(n\pi\zeta/h)\cos(m\pi\zeta/2h)d\zeta\right\}. \tag{6.62}$$

7. Results

We have applied the theory to half and full wavelength antennas with $h/a=60$ and 500π, respectively. In each case the calculations were performed at 25th order. Correction terms from (6.30) were included in the calculation so that each matrix element is accurate to one part in 10^5. Graphs of total current and plots of total admittance versus order are shown (figs. 5 to 12). In the latter graphs, an nth order admittance is obtained by using the first n terms of a 25th order solution.

Comparison of these results with those of the King-Middleton theory in terms of impedance are shown in table 2.

King-Middleton impedances in table 2 were obtained by graphical interpolation of tables given in reference [2]. Resistances of the half wavelength structures are from third order solutions. Other quantities from King-Middleton theory are second order. Our own Z_T for the full wavelength antenna with $h/a=60$ is obtained by extrapolating the current graph to $z=0$. The current distribution on a full wavelength structure does not have zero slope at the origin as do the cosine terms used to describe it. Consequently, many cosine terms are required for high accuracy, especially in the treatment of a thick structure. The inner current correction is entirely negligible for structures with $h/a=500\ \pi$; it is still small for $h/a=60$.

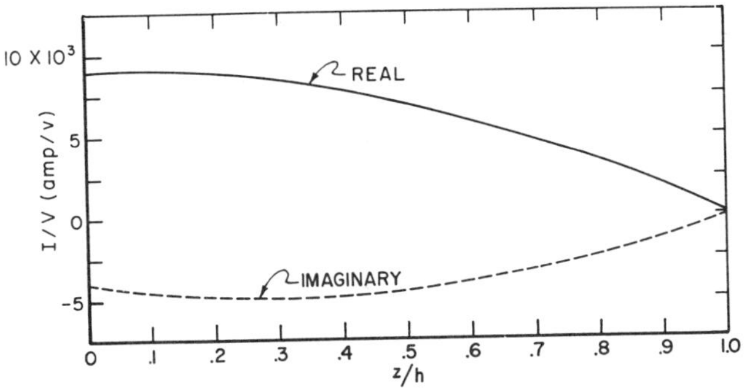

FIGURE 5. *Current distribution:* $kh=\pi/2$, $h/a=60$.

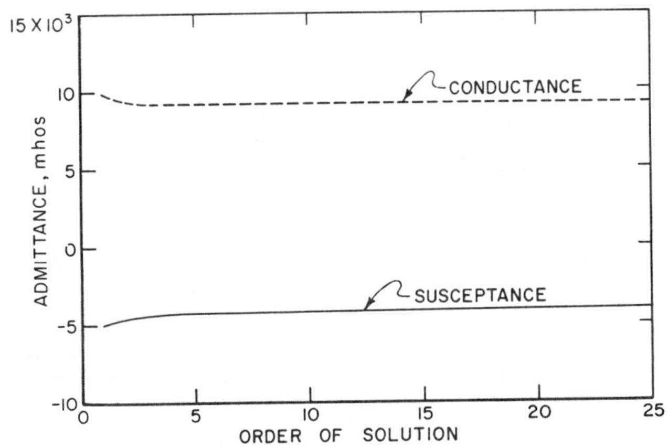

FIGURE 6. *Admittance versus order of solution:* $kh=\pi/2$, $h/a=60$.

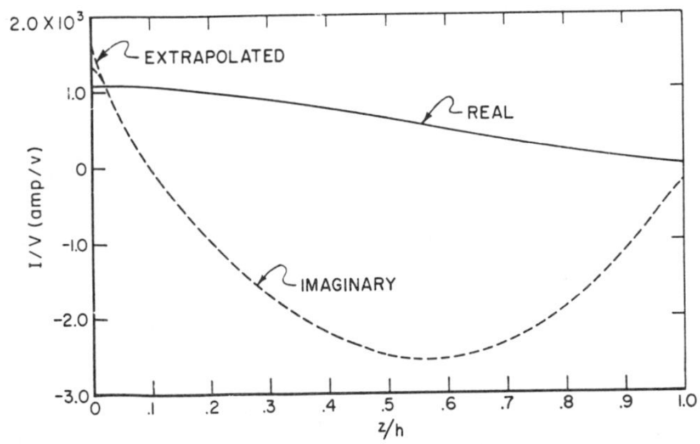

FIGURE 7. *Current distribution:* $kh=\pi$, $h/a=60$.

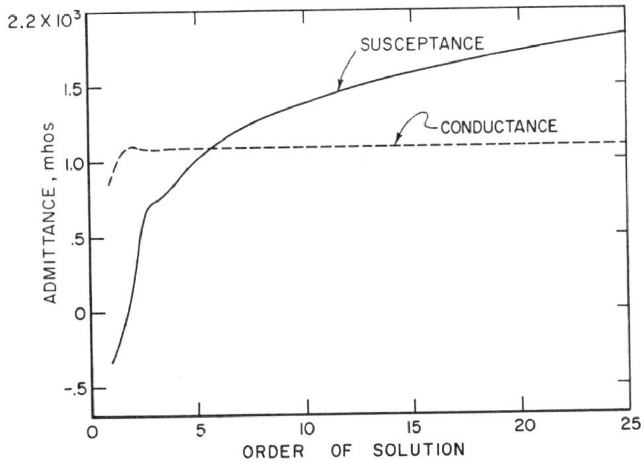

FIGURE 8. *Admittance versus order of solution:* $kh=\pi$, $h/a=60$.

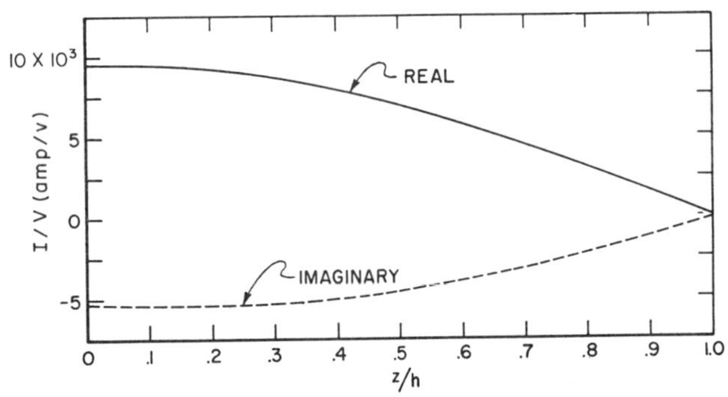

FIGURE 9. *Current distribution:* $kh=\pi/2$, $h/a=500\ \pi$.

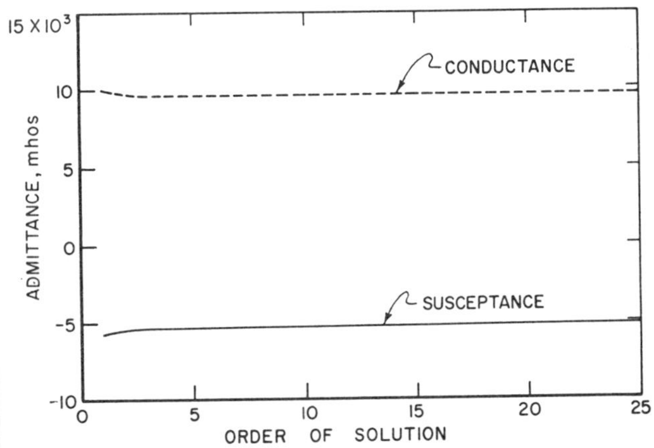

FIGURE 10. *Admittance versus order of solution:* $kh=\pi/2$, $h/a=500\ \pi$.

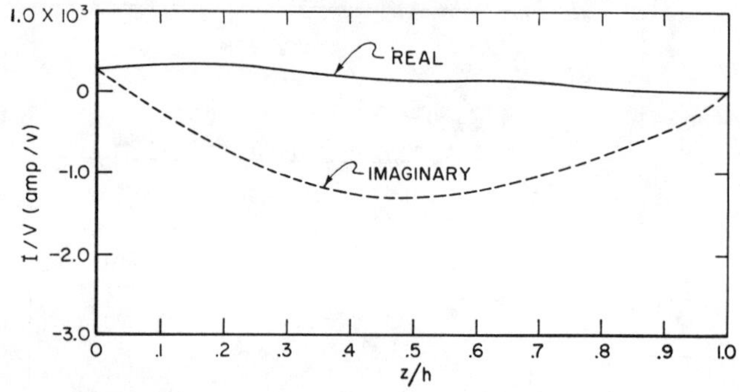

FIGURE 11. *Current distribution:* kh=π, h/a=500 π.

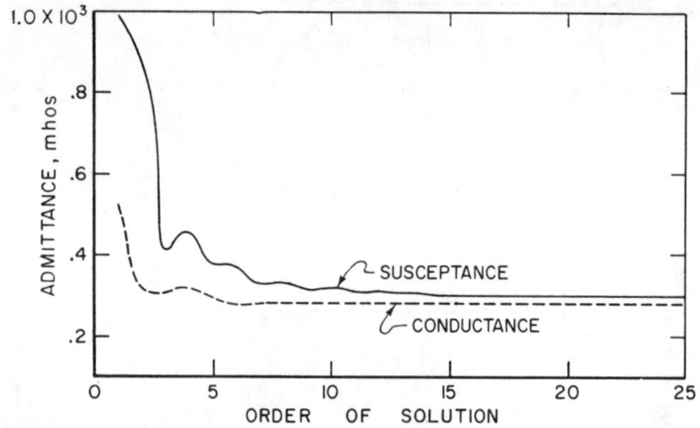

FIGURE 12. *Admittance versus order of solution:* kh=π, h/a=500 π.

It can be seen from table 2 that the amount of disagreement between this theory and the King-Middleton theory is only about two percent. One may expect disagreement of this order or less over the entire range of $h/a>60$.

TABLE 2

kh	h/a	Z_T	Z^0	Z_{KM}
$\pi/2$	60	$91.4+j\ 38.6$	$92.5+j\ 40.6$	$91.4+j\ 41.5$
π	60	$205\ -j\ 382$	$205\ -j\ 380$	$206\ -j\ 380$
$\pi/2$	500 π	$79.7-j\ 42.9$	----------	$80.3-j\ 43.4$
π	500 π	$1646\ -j\ 1768$	----------	$1625\ -j\ 1744$

It can be seen from the graphs of admittance versus order that, except for thick full wavelength structures, 25 terms are excessive for $h/a>60$. Thus, a great deal of margin for the study of thicker antennas is inherent in 25th order solutions. Studies of full wavelength antennas with $h/a<60$ will require more than 25 terms or further modification of some of the technical details of the theory.

The authors express their appreciation to T. R. Ferguson for his assistance in the calculations, to Walter Haas for the computer programs and to R. M. Conkling for fruitful discussions during the preparation of this paper.

8. References

[1] S. A. Schelkunoff and H. T. Friis, Antennas, theory and practice, chap. 8 (John Wiley and Sons, New York, N.Y., 1952).

[2] R. W. P. King, The theory of linear antennas (Harvard University Press, Cambridge, Mass., 1956).

[3] T. Wu and R. W. P. King, Driving point and input admittance of linear antennas, J. A. P., **30,** 1, 74, (1959).

[4] N. Marcuvitz, Waveguide Handbook, p. 66, (McGraw-Hill Book Co., New York, N.Y., 1951).

[5] A. Erdélyi, et al., Tables of integral transforms, Vol. I (McGraw-Hill Book Co., New York, N.Y., 1953).

[6] Erik Hallén, Exact Treatment of Antenna Current Wave Reflection at the End of a Tube Shaped Cylindrical Antenna, IRE Trans. on Ant. and Prop., **AP-4,** 3, 479, (1956).

[7] R. Gans, Zur Theorie der Geradlinigen Antenna, Arch. Electr. Ubertragung, **7,** 169, (1953).

[8] B. Storm, Investigations into modern aerial theory and a new solution of Hallén's integral equation for a cylindrical aerial, dissertation, Imperial College, London, 1953; Summary in wireless engineer (July 1953).

[9] H. Zuhrt, Eine Strenge Berechnung der Dipolantennen mit Röhrformigen Querschnitt, Frequenz **4,** 135 (1950).

[10] E. V. Bohn, The current distribution and input impedance of cylindrical antennas, IRE Trans. on Ant. and Prop. **AP-5,** 4, 343 (1957).

[11] R. W. P. King, Linear Arrays: Currents, Impedance, and Fields, U.R.S.I. Symposium on electromagnetic theory, University of Toronto (June 1959).

Scattering by Perfectly-Conducting Rectangular Cylinders*

K. K. MEI†, MEMBER, IRE, AND J. G. VAN BLADEL‡, SENIOR MEMBER, IRE

Summary—**The problem of determining the fields scattered by a perfectly-conducting rectangular cylinder is reduced to the solution of an integral equation. This equation is then solved by digital computer methods. Data are given for surface currents, radiation patterns and scattering cross sections for both *E*- and *H*-incident waves.**

Introduction

The determination of the field scattered by a perfectly-conducting cylinder reduces to the solution of the scalar wave equation

$$(\nabla^2 + k^2)\phi = 0. \tag{1}$$

This differential equation must be complemented by a suitable radiation condition and by Dirichlet or Neumann boundary conditions at the surface of the cylinder. The choice between these two conditions depends on the polarization of the incident field. Classical methods of solving partial differential equations allow solution of the scattering problem for a few simple geometries, such as the circular cylinder, the elliptical cylinder, the strip and the infinite wedge. For other geometries, approximate methods must be used. High-frequency and low-frequency approximations are well known. There is, however, a paucity of methods for the intermediate-frequency range where the wavelength is comparable to the characteristic dimension of scatterer. It is the purpose of the present paper to show that a method based on the solution of an integral equation is capable of giving excellent results in that range. This is particularly true when a digital computer is available for actual numerical computations. We have applied our method to the rectangular cross section because of the geometrical simplicity of that configuration.[1] Our objective is fourfold: 1) add another item to the catalog of solved configurations, 2) obtain data on the influence of the thickness of a strip on the scattering properties of that body, 3) provide those who are interested in high-frequency approximations with another set of data on which to test their pet theories 4) let the "scattering public" know that, given enough patience and computer time, it is perfectly feasible to solve the scattering problem for cylinders of arbitrary cross section.

* Received August 13, 1962; revised manuscript received, December 3, 1962.

† Department of Electrical Engineering, University of California, Berkeley. Calif. Formerly with Department of Electrical Engineering, University of Wisconsin, Madison, Wis.

‡ Department of Electrical Engineering, University of Wisconsin, Madison, Wis. Presently on leave at the Royal Institute of Technology, Stockholm, Sweden.

[1] The low-frequency limit of the same problem has already been discussed by K. Mei and J. Van Bladel in "Low Frequency Scattering by Rectangular Cylinders" IEEE Trans. on Antennas and Propagation, vol. AP-11, pp. 52–56; January, 1963. The high-frequency approximation will be discussed in a forthcoming paper by K. Mei.

Basic Integral Equations and Numerical Methods

The Dirichlet boundary condition, $\phi(\bar{r}) = 0$ on c, is appropriate for an incident wave whose *electric* field is polarized parallel to the axis of the cylinder (an E wave); the Neumann boundary condition, $\partial\phi/\partial n(\bar{r}) = 0$ on c, is appropriate for a wave whose magnetic field is polarized parallel to the axis of the cylinder (an H wave). Application of Green's second identity and the appropriate boundary conditions[2,3] yields the following expressions

$$E_z(\bar{r}) = E_z^{\text{inc}}(\bar{r}) + i\omega\mu \int_c G(\bar{r}, \bar{r}')K_z(\bar{r}')dc' \tag{2}$$

$$H_z(\bar{r}) = H_z^{\text{inc}}(\bar{r}) - \int_c \frac{\partial G(\bar{r}, \bar{r}')}{\partial n'} K_t(\bar{r}')dc' \tag{3}$$

for the fields outside the scatterer. The subscripts z and t denote the components in the axial and tangential directions, respectively; K_z and K_t are the surface current densities on the obstacle. The kernel

$$G(\bar{r}, \bar{r}') = \frac{i}{4} H_0^{(1)}(k \mid \bar{r} - \bar{r}' \mid)$$

is the two-dimensional free space Green's function, and $\bar{n}'$ is the unit vector perpendicular to the contour c (see Fig. 1). Integral equations for K_z and K_t can be obtained by allowing the point $\bar{r}$ in (2) and (3) to approach a point $\bar{r}_c$ on C. When $\bar{r}$ approaches $\bar{r}_c$, the left-hand side of (2) vanishes and that of (3) approaches $K_t(\bar{r}_c)$. The determination of the limit value of the integrals in the right-hand side of (3) is a fairly delicate

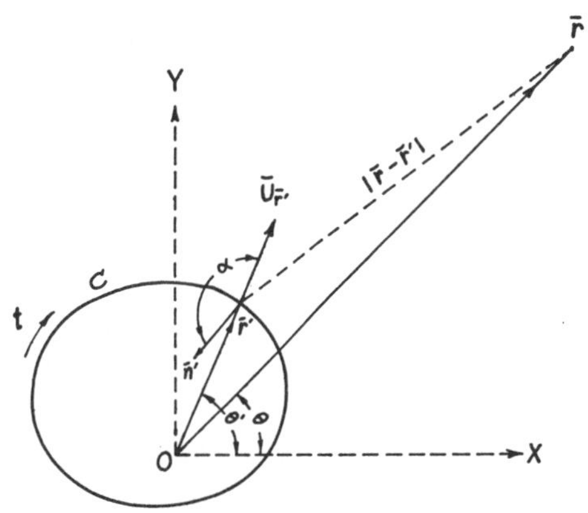

Fig. 1—Cylindrical scatterer and relevant coordinate system.

[2] K. Mei, "Scattering of Radio Waves by Rectangular Cylinders," Ph.D. dissertation, University of Wisconsin, Madison; 1962.

[3] C. H. Papas, "Diffraction by a cylindrical obstacle," *J. Appl. Phys.*, vol. 21, pp. 318–325; April, 1950.

task, which is performed in the Appendix. The final results are

$$0 = E_z^{\text{inc}}(\bar{r}_c) + \lim_{\sigma\to 0} i\omega\mu \int_{c-\sigma} G(\bar{r}_c, \bar{r}')K_z(\bar{r}')dc' \quad (4)$$

$$\tfrac{1}{2}K_t(\bar{r}_c) = H_z^{\text{inc}}(\bar{r}_c) - \lim_{\sigma\to 0} \int_{c-\sigma} \frac{\partial G(\bar{r}_c, \bar{r}')}{\partial n'} K_t(\bar{r}')dc' \quad (5)$$

where σ is a small segment of c containing the point $\bar{r}_c$. Eqs. (4) and (5) are recognizable as one dimensional Fredholm integral equations.

Several methods are available for the solution of Fredholm integral equations. When a high speed digital computer is available, the most suitable approach[4] is to expand the unknown in a set of algebraic functions $\phi_1, \phi_2, \cdots, \phi_n$, and to require to the integral equation be satisfied at n different points on the contour C. The solution of (5), for example, may be approximated by setting

$$K_t(\bar{r}_c) = \sum_{k=1}^{n} A_k\phi_k(\bar{r}_c).$$

Insertion of this expression in (5) gives

$$\sum_{k=1}^{n} A_k[\tfrac{1}{2}\phi_k(\bar{r}_{ci}) + \lim_{\sigma\to 0} \int_{c-\sigma} \frac{\partial G(\bar{r}_{ci}, \bar{r}')}{\partial n'} \phi_k(\bar{r}')dc'] = H_z^{\text{inc}}(\bar{r}_{ci}) \qquad (i = 1, 2, \cdots n) \quad (6)$$

from which A_k can be determined by solving a matrix equation of the type

$$[f][A] = [B]$$

where

$$f_{ik} = \tfrac{1}{2}\phi_k(r_{ci}) + \lim_{\sigma\to 0} \int_{c-\sigma} \frac{\partial G(\bar{r}_{ci}, \bar{r}')}{\partial n'} \phi_k(\bar{r}')dc'$$

$$B_i = H_z^{\text{inc}}(\bar{r}_{ci}).$$

For the solution of the problem at hand, we decided to divide the contour into n subintervals $\Delta_{ci}(i=1, 2, \cdots, n)$, and to choose a set of ϕ's consisting of discontinuous functions U_i (henceforth termed *steps*) defined by

$$U_i(\bar{r}_c) = \begin{cases} 1 \text{ in } \Delta C_i \\ 0 \text{ in } \Delta C_k \qquad i \neq K. \end{cases}$$

The theoretical justification for this choice has been discussed by Mei.[2] Once K_z and K_t are obtained, the field at any point $\bar{r}$ in space can be determined from (2) and (3). It is of interest to know that 10 subdivisions per wavelength are in general sufficient to give a convergent result for the current densities.

[4] F. B. Hildebrand, "Method of Applied Mathematics," Prentice Hall, Inc., Englewood Cliffs, N. J., pp. 444–461; 1954.

Numerical Results for E Waves

Surface Current Density

The behavior of the surface current near an edge has been analyzed by Meixner,[5] who shows that the current resulting from an E wave has a singularity at the edge. More specifically, at a small distance r from a 90° corner, K_z becomes infinite as $r^{-1/3}$. This type of discontinuity implies that the step function method does not give uniformly convergent results for K_z. However, since K_z is integrable at an edge, the approximation converges in the mean of order 1, that is, in such a manner that

$$\int_c | K_z(\bar{r}) - K_{zn}(\bar{r}) | \, dc < \epsilon \qquad \text{as } n\to\infty \text{ for any } \epsilon > 0 \quad (7)$$

where $K_{zn}(\bar{r})$ is the step function approximation for n steps. Inasmuch as the radiation pattern and scattering cross section are of more interest than the surface current density itself, and since both of them are found through integrations involving the surface current density, the type of convergence given in (7) is quite satisfactory for our problem.

Van Bladel[6] has shown that the surface current density in a low-frequency E wave is given by

$$| K_z | = \left| \frac{2\pi E^{\text{inc}}}{\omega\mu L \log\left(\dfrac{1}{kL}\right)} \left(\frac{f}{f_{\text{ave}}}\right) \right|$$

where f satisfies the integral equation

$$\frac{1}{2\pi}\int_c f(\bar{r}') \log \frac{L}{| \bar{r} - \bar{r}' |} \cdot \frac{dc'}{L} = 1$$

and where L is the length of the contour c'. The equation satisfied by $f(\bar{r})$ is recognizable as the integral equation satisfied by the charge density on a charged metallic cylinder. In Fig. 2 the actual value of K_z in a low-frequency E-wave is compared to the low-frequency limit f (curve 0). Each curve is normalized to make K_z equal to one at the center of the illuminated side. We notice that the convergence of K_z to f as $ka\to 0$ is excellent.

Typical curves for K_z at several intermediate frequencies are shown in Fig. 3, and data for the phase of K_z are given in Fig. 4. An interesting feature can be observed for both E and H waves, namely the appearance of a standing-wave pattern along the contour at certain frequencies. This pattern is very much like that observed on a short-circuited transmission line. The corresponding phase variation is similar to that ob-

[5] J. Meixner, "Die Kantenbedingung in der Theorie der Beugung electromagnetischer Wellen an volkommen leitenden ebenen Schirmen," *Ann. Phys.*, vol. 6, pp. 2–9; September, 1949.

[6] J. Van Bladel, "Low frequency scattering by cylindrical bodies," *Applied Scientific Res., Sect. B;* to be published.

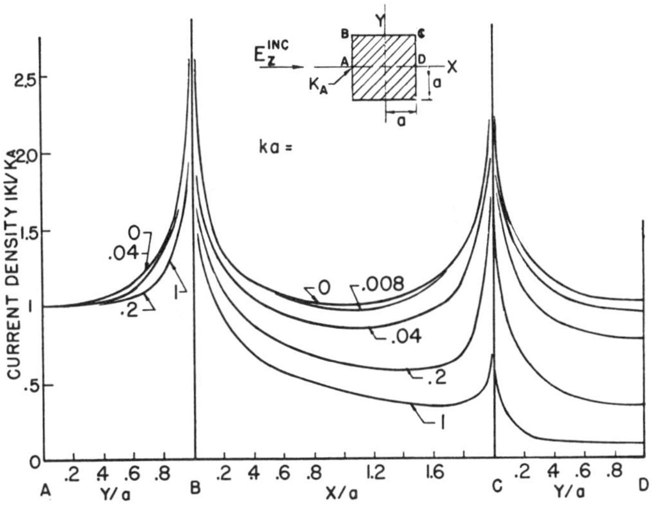

Fig. 2—Normalized surface current density on a square cylinder immersed in a very low-frequency E wave.

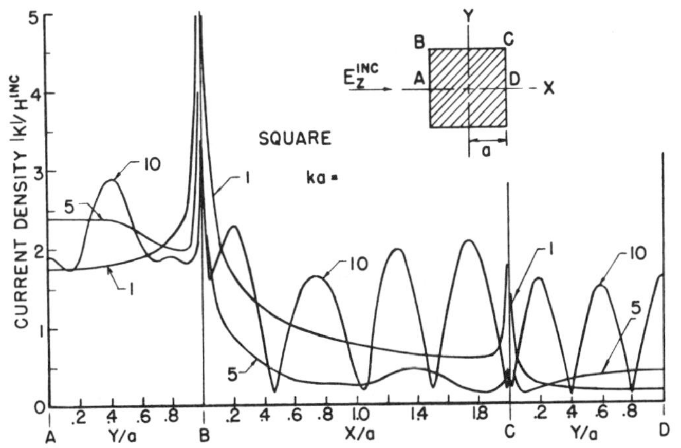

Fig. 3—Normalized surface current density on a square cylinder in an intermediate-frequency E wave.

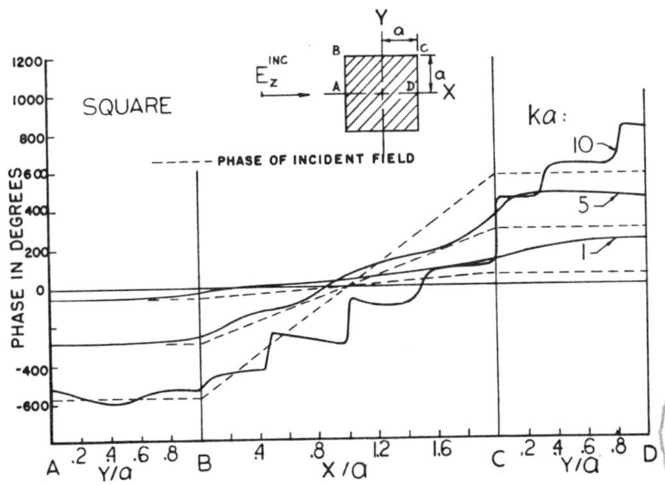

Fig. 4—Phase of surface current density on a square cylinder in an intermediate-frequency E wave.

tained from the interference of two waves propagating in opposite directions. We notice, in particular, that there is a 180° phase discontinuity at points where the amplitude is equal to zero. The existence of a standing-wave pattern is a characteristic of high-frequency scattering. Our data (only part of which are shown here) suggest that the pattern occurs only for values of ka larger than 5. The underlying mechanism of standing-wave formation will be discussed in a forthcoming paper by K. Mei.[2]

Radiation Pattern

The scattered field $E_z^{sc}(\bar{r})$ at any point in space can be obtained from the second term on the right-hand side of (3). At large distances, $E_z^{sc}(\bar{r})$ can be expressed as

$$E_z^{sc}(\bar{r}) = \frac{-\omega\mu}{4}\left(\frac{2}{\pi k r}\right)^{1/2} e^{ikr}\int_c e^{-ikr'\cos(\theta-\theta')}K_z(\bar{r}')dc'. \quad (8)$$

The power density is customarily expressed in terms of the average scattered-power density, which is

$$P_E = \frac{\int_c \left| E_z^{sc}(\bar{r}) \right|^2 dc}{2\pi r R_0}.$$

Here $R_0 = \sqrt{\mu_0/\epsilon_0}$ is the intrinsic impedance of free space, and the contour C is a large circle of radius r. The normalized power density, known as the "gain," is then

$$G_E = \frac{|E_z^{sc}(\bar{r})|^2}{P_E R_0} = \frac{|E_z^{sc}(\bar{r})|^2}{\frac{1}{2\pi}\int_c^{2\pi} |E_z^{sc}(\bar{r})|^2 d\theta}. \quad (9)$$

At low frequencies, the phase of $K_z(\bar{r})$ in an E wave is almost uniform over the entire contour. To a distant observer, the surface current looks like a line current oscillating at the frequency of the incident wave. The radiation pattern resulting from such a current is circular. In consequence, we should obtain almost circular gain patterns for low-frequency E waves. This is confirmed by Fig. 5. Typical gain patterns of rectangular cylindrical scatters are shown in Figs. 5 to 8. As the frequency increases, the patterns approach the geometrical optics limit, *viz.*, a strong lobe in the direction of reflection and a strong lobe in the direction of the shadow. At large distances, the scattered field in the forward direction ($\theta = 0$) is found to have an almost constant phase difference with respect to the incident field. This phenomenon allows near cancellation of the total field throughout that part of the shadow region. At $ka = 1$, for example, the phase difference between the scattered field and the incident field is about 169° for a square cylinder.

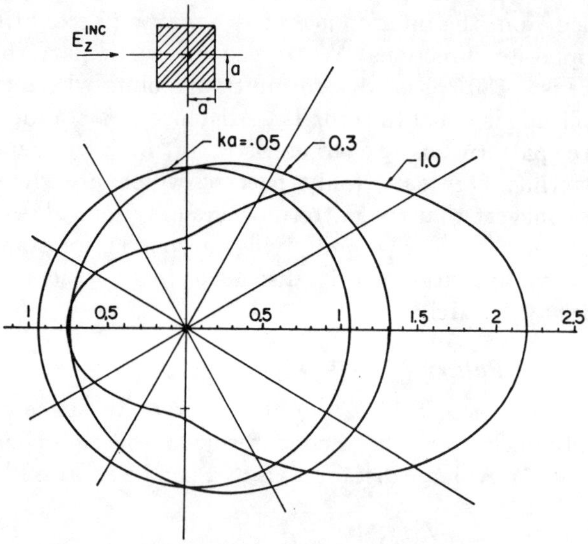

Fig. 5—Gain patterns for low-frequency E waves.

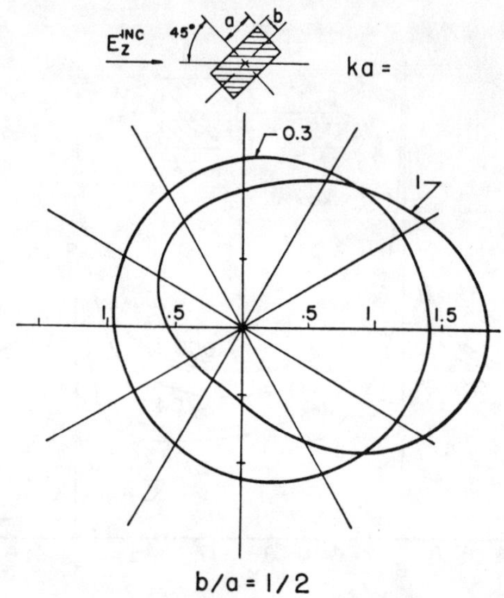

Fig. 7—Gain patterns for low-frequency E waves at 45° incidence.

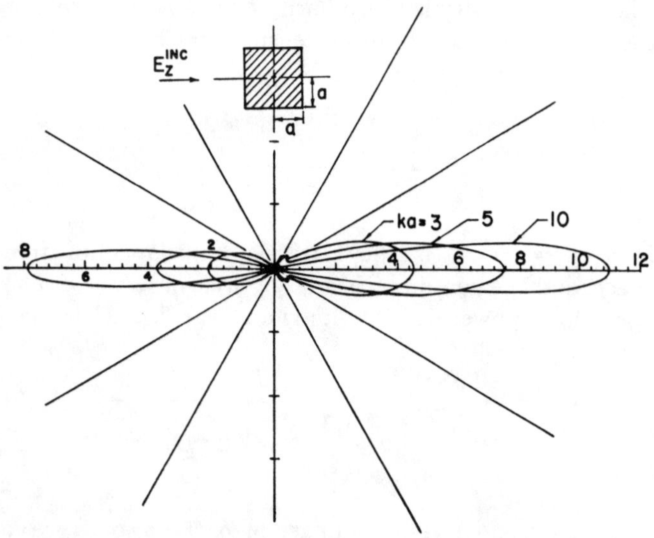

Fig. 6—Gain patterns for intermediate-frequency E waves.

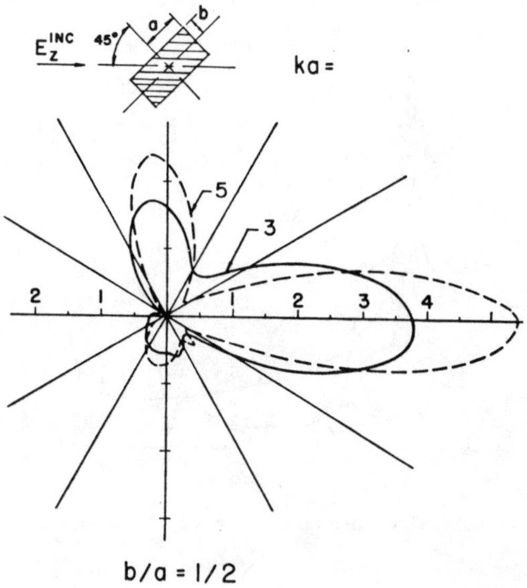

Fig. 8—Gain patterns for intermediate-frequency E waves at 45° incidence.

Scattering Cross Section

The scattering cross section of a cylindrical obstacle is defined as the ratio of the scattered power per unit axial length to the incident power density. The scattering cross section has the dimension of length. Thus,

$$\sigma_{sc} = \frac{\int_c \left| E_z^{sc}(\bar{r}) \right|^2 dc}{\left| E_z^{\text{inc}} \right|^2}. \tag{10}$$

In this expression, $E_z^{sc}(\bar{r})$ is the scattered field given in (8), and C is an arbitrary closed curve encircling the obstacle.

An alternative formula for the scattering cross section in a plane wave involves the value of the scattered field in the forward direction,[2,7] It is obtained as follows.

[7] R. W. P. King, and T. T. Wu, "The Scattering and Diffraction of Waves," Harvard University Press, Cambridge, Mass.; 1959.

Eq. (3) can be rewritten as

$$E_z^{sc} = \left[\frac{i}{4}\left(\frac{2}{k\pi r}\right)^{1/2} e^{ikr}e^{i(\pi/4)}\right] i\omega\mu \cdot \int_c e^{-ikr' \cos(\theta-\theta')} K_z(\bar{r}')dc'$$

$$= \left[\frac{i}{4}\left(\frac{2}{k\pi r}\right)^{1/2} e^{ikr}e^{i(\pi/4)}\right] f(\theta),$$

where

$$f(\theta) = i\omega\mu \int_c e^{-ikr' \cos(\theta-\theta')} K_z(\bar{r})dc'. \tag{11}$$

In the forward direction $(\theta = 0)$, we have

$$f(0) = i\omega\mu \int_c e^{-ikr' \cos\theta'} K_z(\bar{r}')dc'. \tag{12}$$

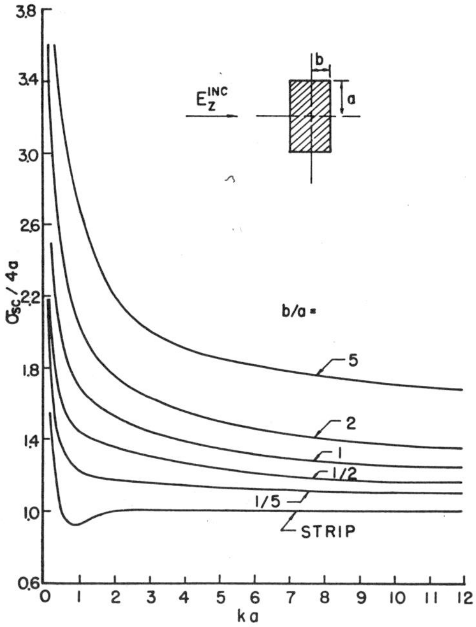

Fig. 9—Scattering cross sections for an E wave at 0° incidence.

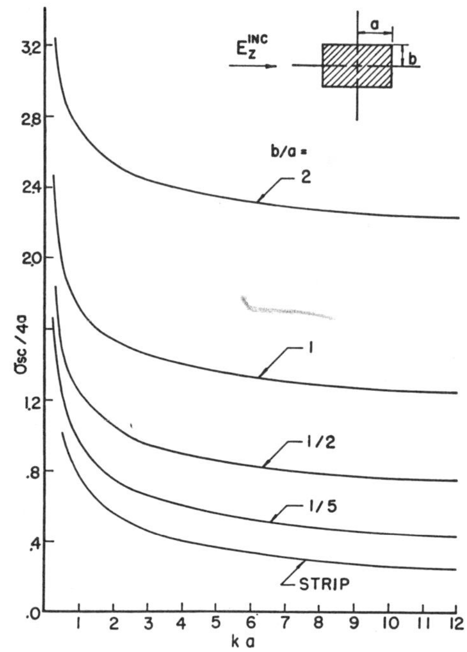

Fig. 10—Scattering cross sections for an E wave at 90° incidence.

The total scattering cross section is then found to be

$$\sigma_{sc} = \frac{\operatorname{Im} f(0)}{k} \tag{13}$$

where "Im" denotes the imaginary part. Typical curves of the scattering cross sections of rectangular cylinders are given in Figs. 9 to 13. The values of σ_{sc} calculated from (10) and (13) were found to agree up to seven significant figures for a square scatterer with thirty subdivisions per side in the step approximation. It is worthwhile to comment on the low-frequency value of the scattering cross section. In a low-frequency E wave the scattering cross section of a rectangular cylinder

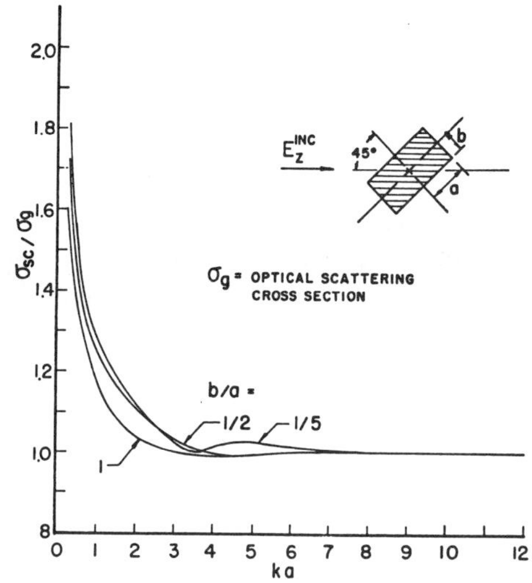

Fig. 11—Scattering cross sections for an E wave at 45° incidence.

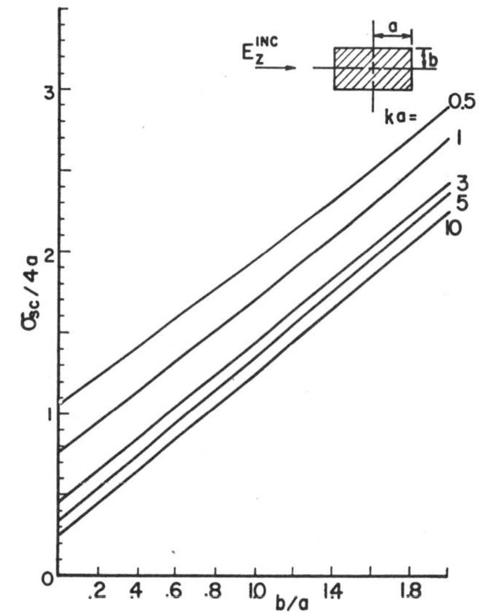

Fig. 12—Scattering cross sections for a 90° E wave as a function of the aspect ratio b/a.

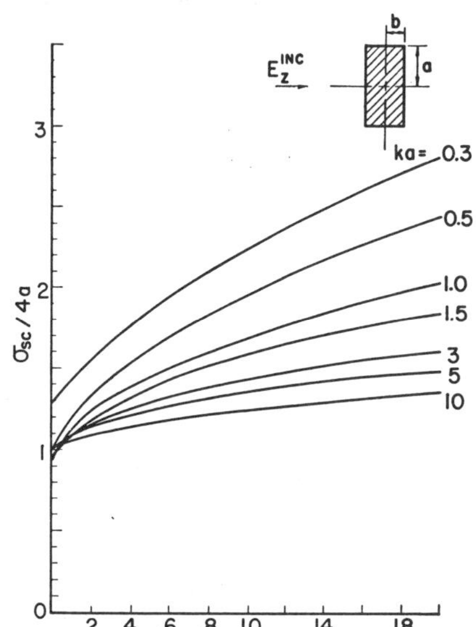

Fig. 13—Scattering cross sections for a 0° E wave as a function of the aspect ratio b/a.

is given by[6]

$$\sigma_{sc} = \frac{\pi^2}{k \log [4k(a+b)]}. \tag{14}$$

This formula, which is good at frequencies such that $4k(a+b) \ll 1$, implies that the low-frequency scattering cross section of a rectangular cylinder is independent of the latter's orientation. This is confirmed by Figs. 9 to 11, in which the scattering cross sections of rectangular cylinders are plotted for angles of incidence equal to 0°, 45°, and 90°. Our data shows that the scattering cross section remains practically independent of the orientation up to values of ka equal to 0.5.

Numerical Results for H Waves

Surface Current Density

The current density in a low-frequency H wave can be expressed by the two leading terms in its power expansion in (ik), *viz.*,

$$K_t = K_0 + ikK_1. \tag{15}$$

Here K_0 is the value of the static current necessary to cancel the low-frequency magnetic field H inside the cylinder. This value is equal to H_z^{inc}, and is a constant along the contour. The component ikK_1 is the capacitive current "ferrying" zeroth-order charges η_0 along the boundary. The equation of conservation of charge yields

$$\frac{\partial K_t}{\partial C} = i\omega\eta_0$$

which implies that

$$\frac{\partial K_1}{\partial C} = (\epsilon_0\mu_0)^{-1/2}\eta_0 = c\eta_0. \tag{16}$$

It should be noticed that η_0 is the charge density which appears on the surface of the obstacle when the latter is immersed in a (static) electric field $E_0 = R_0 H^{\text{inc}}$. The real and imaginary parts of K_t on a square cylinder are plotted in Fig. 14 for $ka = 0.3$. The curves confirm that the real part of K_t is almost constant, as expected for low values of ka, and that the derivative of the imaginary part in the C direction behaves like the static charge density on the cylinder.[1]

Typical curves for K_t at several intermediate frequencies are shown in Fig. 15, and the phase of K_t is plotted in Fig. 16.

Radiation Pattern

In an H wave the gain is

$$G_H = \frac{|H_z^{sc}(\bar{r})|^2}{\dfrac{1}{2\pi}\displaystyle\int_{c\to 0}^{2\pi} \left| H_z^{sc}(\bar{r}) \right|^2 d\theta}. \tag{17}$$

Figs. 17 to 20 show some typical gain patterns relative to an incident H wave. As the frequency increases, the

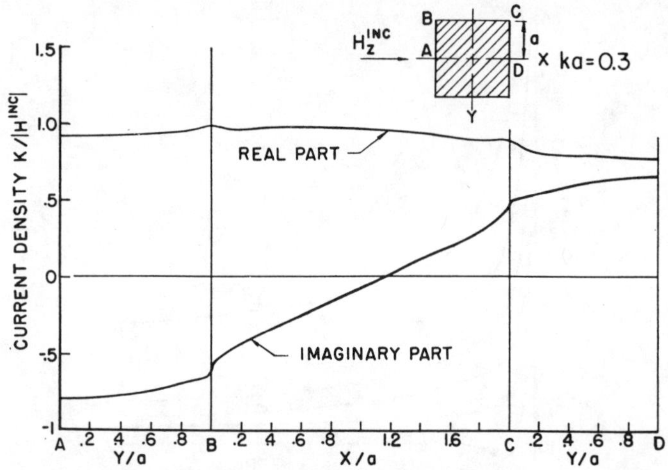

Fig. 14—Real and imaginary parts of the surface current on a square cylinder in a low-frequency H wave ($ka = 0.3$).

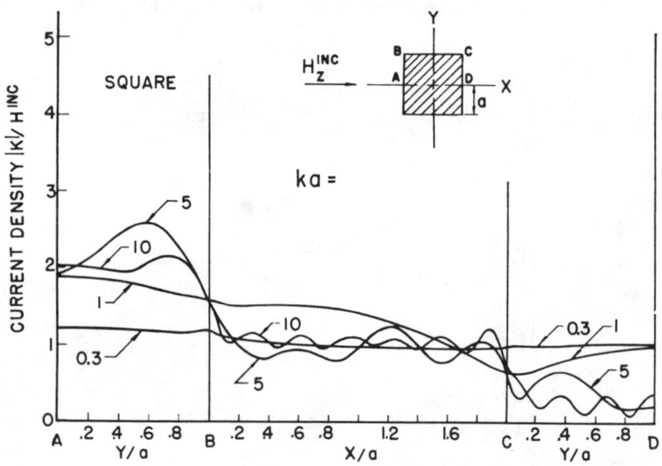

Fig. 15—Surface current density on a square cylinder in an H wave.

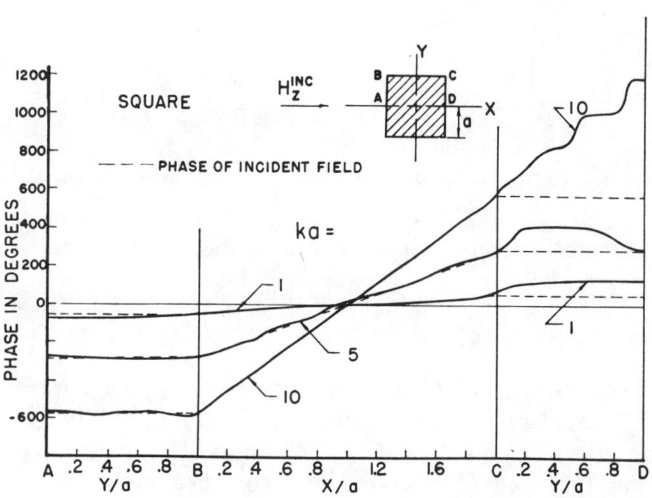

Fig. 16—Phase of the current density on a square cylinder in an H wave.

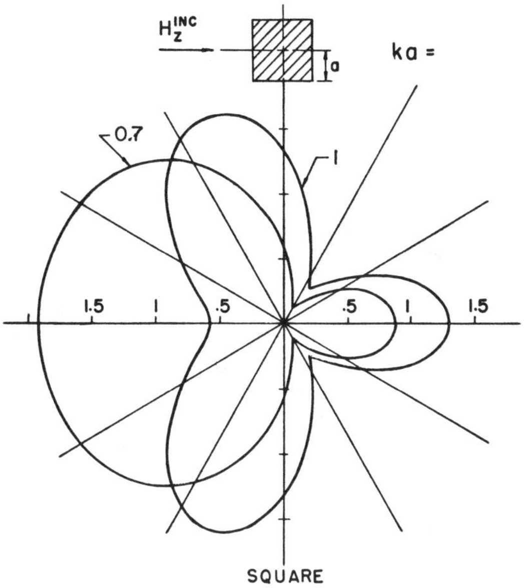

Fig. 17—Gain patterns for low-frequency H waves.

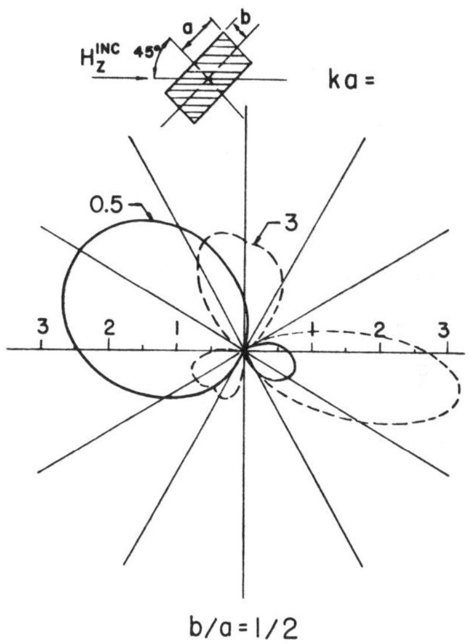

Fig. 19—Gain patterns for intermediate-frequency H waves at 45° incidence.

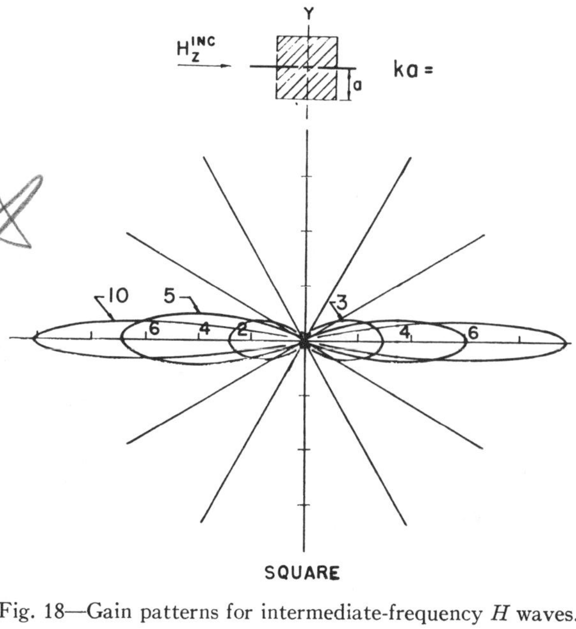

Fig. 18—Gain patterns for intermediate-frequency H waves.

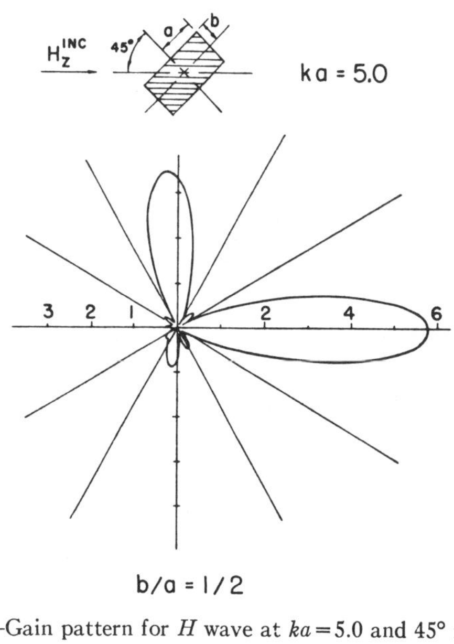

Fig. 20—Gain pattern for H wave at $ka = 5.0$ and 45° incidence.

patterns approach those predicted by geometrical optics.

Scattering Cross Section

The scattering cross section of a cylindrical obstacle immersed in an H wave is

$$\sigma_{sc} = \frac{\int_c \left| H_z^{sc}(\bar{r}) \right|^2 dc}{\left| H_z^{\text{inc}} \right|^2}. \tag{18}$$

Eq. (13) is still applicable in the case of an H wave, but $f(\theta)$ must now be defined by the expression

$$f(\theta) = ik \int_c e^{-ikr' \cos(\theta-\theta')} K_t(\bar{r}') \cos \alpha dc' \tag{19}$$

where α is the angle shown in Fig. 1.

Curves for σ_{sc} as a function of frequency are shown in Figs. 21 to 23. It is seen that the values of the low-frequency scattering cross-sections are in excellent agreement with those obtained from the low-frequency approximation.[1] The agreement is good up to values of ka of the order of 0.5.

Appendix

We wish to show that

$$\lim_{\bar{r}\to\bar{r}_c} \int_c \frac{\partial G(\bar{r}, \bar{r}')}{\partial n'} H_z(\bar{r}')dc'$$

$$= \lim_{\sigma\to 0} \int_{c-\sigma} \frac{\partial G(\bar{r}_c, \bar{r}')}{\partial n'} H_z(\bar{r}')dc' - \tfrac{1}{2}H(\bar{r}_c) \tag{20}$$

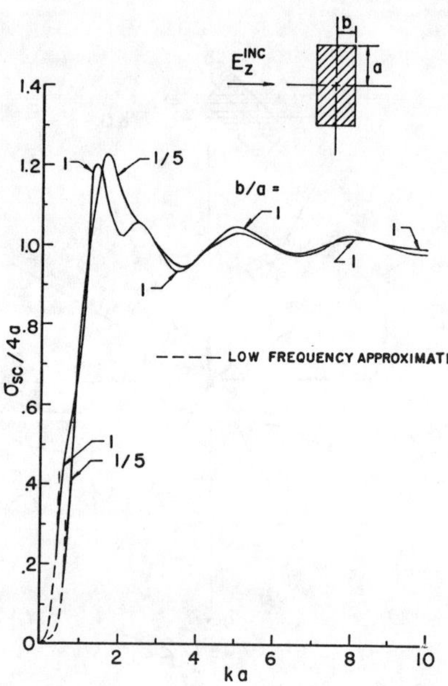

Fig. 21—Scattering cross sections for an H wave at 0° incidence.

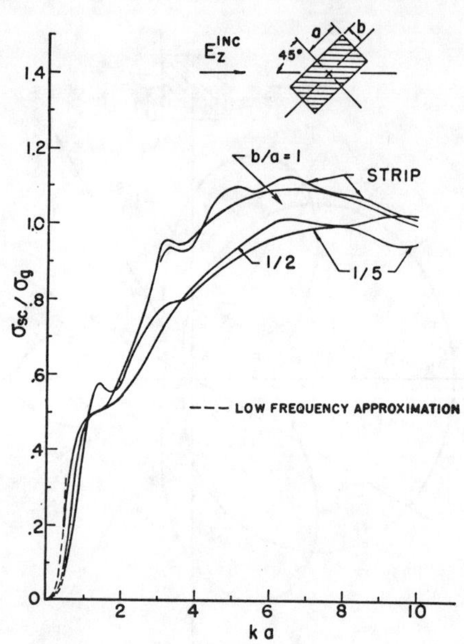

Fig. 23—Scattering cross sections for an H wave at 45° incidence.

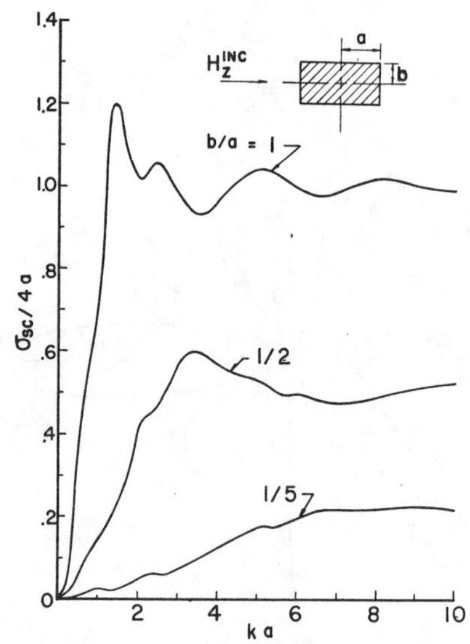

Fig. 22—Scattering cross sections for an H wave at 90° incidence.

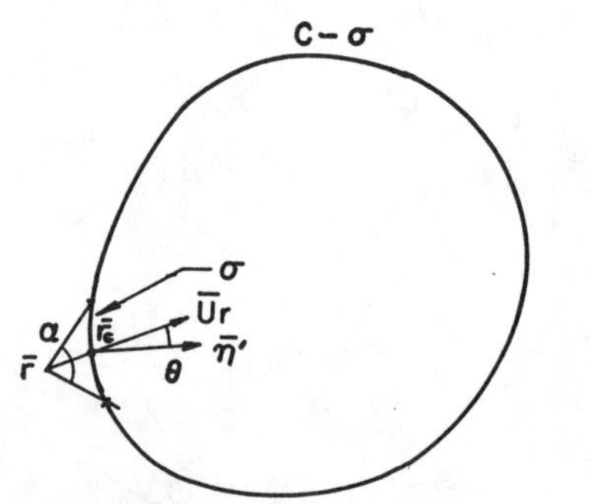

Fig. 24—Relevant to the derivation of the integral equation.

where $\bar{r}_c$ is a point on c,

$$G(\bar{r}, \bar{r}') = \frac{i}{4} H_0^{(1)}(k \,|\, \bar{r} - \bar{r}' \,|),$$

$\bar{n}'$ is the normal at $\bar{r}'$ as shown in Fig. 24, and σ is a small segment of c containing the point $\bar{r}_c$.

The left-hand member of (20) can be written as

$$\lim_{\bar{r}\to\bar{r}_c} \int_c \frac{\partial G}{\partial n'} (\bar{r}, \bar{r}') H_z(\bar{r}') dc'$$

$$= \lim_{\bar{r}\to\bar{r}_c} \left[\int_{c-\sigma} \frac{\partial G}{\partial n'} (\bar{r}, \bar{r}') H_z(\bar{r}') dc' + \int_\sigma \frac{\partial G}{\partial n'} (\bar{r}, \bar{r}') H_z(\bar{r}') dc' \right]. \quad (21)$$

As σ approaches zero, the right-hand member of (21) becomes

$$\lim_{\sigma\to 0} \int_{c-\sigma} \frac{\partial G}{\partial n'} (\bar{r}_c, \bar{r}') H_z(\bar{r}') dc' + \lim_{\bar{r}\to\bar{r}_c} \int_{\sigma\to 0} \frac{i}{4} \frac{\partial}{\partial n'} (i \log k \,|\, \bar{r} - \bar{r}' \,|) \frac{2}{\pi} H_z(\bar{r}') dc'. \quad (22)$$

But the second term in this expression can be rewritten as

$$\lim_{\bar{r}\to\bar{r}_c} -\frac{1}{2\pi} \int_\sigma \frac{\cos \theta H_z(\bar{r})}{|\, \bar{r} - \bar{r}' \,|} dc' = \lim_{\bar{r}\to\bar{r}_c} -\frac{1}{2\pi} \alpha H_z(\bar{r}') \quad (23)$$

where α is the angle subtended by σ at point $\bar{r}$, (Fig. 24). As $\bar{r}$ approaches a point $\bar{r}_c$ on σ, α approaches π. Therefore, the second member of (21) is $-\frac{1}{2}H_z(\bar{r}_c)$ and (20) is justified.

Acknowledgment

The present work was supported by the Research Committee of the Graduate School of the University of Wisconsin, Madison, from funds supplied by the Wisconsin Alumni Research Foundation and the National Science Foundation. The authors also want to acknowledge the help of the Numerical Analysis Laboratory of the University of Wisconsin.

Scattering from Parallel Metallic Cylinders With Arbitrary Cross Sections

MOGENS G. ANDREASEN, MEMBER, IEEE

Summary—**The integral equations for scattering by a set of parallel metallic cylinders, each cylinder of arbitrary cross section, are solved directly by means of a digital computer program giving the current distribution induced on the scatterer surfaces, the scattering cross section vs azimuthal angle, and the induced field ratio (IFR) for both parallel and perpendicularly polarized incident waves. The present program is suitable for cylinders up to 25λ to 40λ circumference. The validity of the computer program has been verified by comparing results 1) with exact calculations for circular cylinders, 2) with measured values for more than 40 rectangles and rectangles with feathered and rounded faces, and 3) by interchanging the direction of incidence and the direction of observation, whereby, according to the reciprocity theorem, the observed scattered field must remain the same.**

I. Introduction

A METHOD is described in this paper for solving the problem of scattering by metallic cylinders of arbitrary cross section [1], [2]. The techniques appear to have application to many other problems such as the accurate prediction of the radiation from reflector antennas.

When a characteristic dimension of an antenna or scatterer is within a large range, around the order of one wave length, the asymptotic methods are often of little value. Scatterers in this category include practically any object whose form will be arbitrarily specified and with a characteristic dimension in the above mentioned range. Antennas in this category include coupled dipole and slot antenna systems such as, for example, arrays and quasi-arrays of various types. Even for so commonly used an antenna as the Yagi antenna, a satisfactory theoretical solution does not exist except for very few elements.

While many aspects of the problems listed above can, of course, be solved experimentally, the need for solving these problems theoretically with high accuracy is nevertheless great. A theoretical solution is usually of much more general use and often provides more physical insight into the electromagnetic phenomena, thus leading eventually to the design of better and more economical systems. Moreover, in certain problems, such as the backscattering from bodies which have been deliberately shaped in attempts to minimize such scattering, experimental techniques are simply not accurate enough to yield the desired results. With the availability of fast electronic computers today, there is no reason why many of these problems cannot be solved relatively economically by exact theoretical methods.

Manuscript received October 8, 1963; revised May 22, 1964.
The author is with Technical Research Group-West, Menlo Park, Calif.

In the past, exact solutions of scattering problems have been obtained only for scatterers whose surfaces coincide with coordinate surfaces of orthogonal coordinate systems in which the vector wave equation is solvable by the method of separation of variables [3]. In these special cases, which include circular, elliptical and parabolic cylinders, wedges, spheres, and, under special circumstances, spheroids and paraboloids, the current distribution on the scatterer can be expressed as a sum of mode currents. While the mode theory is very useful for scatterers of the above mentioned form, and for these cases is certainly superior to any other known method when a characteristic dimension of the scatterer does not exceed about 25 wavelengths, it is useless for scatterers of any other form.

For a scatterer of arbitrary form, various asymptotic methods have been successfully used to predict the approximate current distribution and scattering patterns for scatterers whose dimensions are either very small or very large compared to one wavelength. When the maximum dimension of a scatterer is a small fraction of a wavelength, the classical Rayleigh scattering theory can be applied with good accuracy. When, on the other hand, characteristic dimensions of a scatterer are very large compared to one wavelength, the scattering problem can be solved with good accuracy by wave optics. Refinement methods have often been used to improve upon the straight wave-optical solution near the boundary of the optical shadow zone by the introduction of such concepts as creeping waves of various orders [4], [5].

When characteristic dimensions of the scatterer are between about one sixth and four wavelengths, none of the asymptotic methods will yield a very satisfactory solution of the scattering problem for a scatterer with arbitrary form. It is with problems in this category that the present paper is concerned. The problem of scattering of an arbitrarily polarized wave by an arbitrary number of parallel metallic cylinders, each of arbitrary cross section, is solved by an integral equation method. When the total circumferential length of the scatterer cross section does not exceed about 25 to 40 wavelengths, depending upon the complexity of the cross section, the integral equation can be solved numerically by means of a fast digital computer. Contrary to what might be expected, this method leads to considerable insight into the scattering mechanisms as well as to very accurate design information.

The integral equation for scattering by a metallic cylinder has been solved by a numerical method also by Mei and Van Bladel [6]. Their analysis was restricted to a single rectangular cross section, and the accuracy of their results seems greatly impaired by the simple integration scheme used [7]. In this paper, the number of cylinders is arbitrary, the cross section of each cylinder is arbitrary, and the integration scheme is sufficiently accurate to allow sampling the current distribution on smooth bodies at $\lambda/4$ intervals.

Similar techniques also have been applied by Banaugh and Goldsmith [8] for investigating an analogous acoustical problem.

II. Parallel Polarization

A. The Integral Equation

In this section, we shall derive and discuss the method used to solve the integral equation for the current density distribution induced on the surfaces of a set of parallel metallic cylinders, each with an arbitrary cross section, by an incident plane wave whose polarization is assumed to be parallel to the cylinder axis (see Fig. 1). Once the current distribution has been determined, other quantities of interest such as the scattering cross-section pattern and the total scattering cross section can be deduced by simple integrations.

For parallel polarization, all field components can be generated from the electric field component E_z along the cylinder axis. This field component must satisfy the homogeneous scalar wave equation with the boundary condition $E_z=0$ at the contour of the scatterers. Further, the radiation condition at infinity must be satisfied for the scattered field. Let $E_z(\bar{\rho})$ be the total z component of the E field at a point with radius vector $\bar{\rho}$. Then

$$E_z(\bar{\rho}) = E_z{}^i(\bar{\rho}) + E_z{}^s(\bar{\rho}), \tag{1}$$

where the right member terms are the incident and scattered fields, respectively. Using Green's formula, the radiation condition at infinity, the boundary condition $E_z=0$ at the surface of the scatterers, and the following relationship

$$\frac{\partial E_z}{\partial n} = -j\omega\mu K_z \tag{2}$$

between the inward normal derivative of the electric field, and the z-directed current density K_z, one arrives at the following integral expression for the scattered electric field [9]

$$E_z{}^s(\bar{\rho}) = -\frac{\zeta}{4}\int_C H_0{}^{(2)}[k(\bar{\rho}-\bar{\rho}')]K_z(\bar{\rho}')d(ks'), \tag{3}$$

where C is the total contour of the scatterers, and $H_0{}^{(2)}$ is the Hankel function of order zero and of the second kind. $k=2\pi/\lambda$ is the propagation constant and $\zeta=\sqrt{\mu/\epsilon}$ the intrinsic impedance of free space. The time factor $e^{j\omega t}$ has been suppressed.

If we let $\bar{\rho}$ approach a boundary point given by the vector $\bar{\rho}''$ (see Fig. 1), the left member of (3) approaches

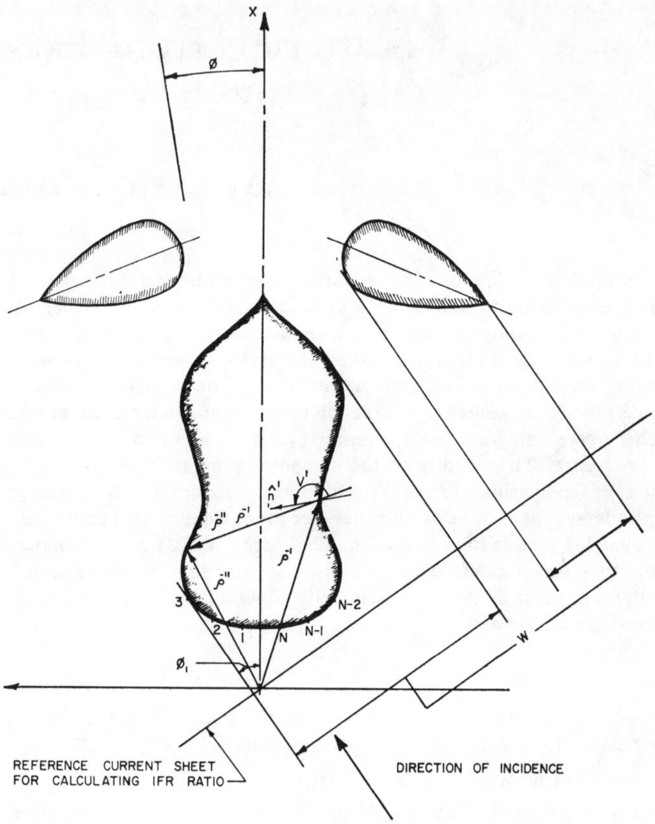

Fig. 1—Cross section of scatterers.

$-E_z{}^i(\bar{\rho}'')$ because of the boundary condition, and (3) is reduced to

$$\int_C H_0{}^{(2)}[k\,|\,\bar{\rho}''-\bar{\rho}'\,|\,]K_z(\bar{\rho}')d(ks') = 4\frac{E_z{}^i(\bar{\rho}'')}{\zeta}. \tag{4}$$

This is the desired integral equation for the current distribution. The numerical method used to solve this equation will be discussed in the following.

Eq. (4) is an inhomogeneous integral equation of the first kind. The integral in this equation can be approximated by a weighted sum of N sampled values of the current distribution. Assume the contour of the cylinders to be divided into a total of N intervals. Let the sampled current density in the center of the n'th interval be K_{zn}, and let this sampling point be described by the radius vector $\bar{\rho}_n$. When $\bar{\rho}''$ is chosen to assume the values $\bar{\rho}_p$, $p=1, 2, \cdots, N$, (4) can obviously be reduced to the following system of N simultaneous linear equations:

$$\sum_{n=1}^{N} B_{np}K_{zn} = A_p \qquad p = 1, 2, \cdots, N, \tag{5}$$

or, in matrix notation,

$$\{B\}\{K\} = \{A\}, \tag{6}$$

where

$$A_p = 4\frac{E_z{}^i(\bar{\rho}_p)}{\zeta}. \tag{7}$$

To find the N^2 coefficients B_{np}, which are weighting factors in the integration process, and the coefficient matrix of the linear equations, we consider the contour integral in (4) being a sum of N integrals, one over each interval. Over most of the intervals, the integral can be evaluated by a quadrature formula. In the pth interval, however, the Hankel function has a singularity. In this interval, it is, therefore, necessary to perform the integration analytically. The accuracy of the solution depends both on the number of sampling points chosen, how they are positioned along the contour, and on the order of the quadrature formula used to integrate over the intervals $n \neq p$. Within certain limits the same accuracy can be achieved with fewer sampling points when the order of the quadrature formula is increased. The computer time required to solve a set of N linear equations is proportional to N^3 and the computer time required to set up the coefficient matrix is only proportional to the product of N^2 and the order of quadrature formula used. A minimum of computer time is therefore achieved by using as few sampling points as possible and as high order of the quadrature formula as needed to give a satisfactory accuracy. There is, of course, a lower limit for the number of sampling points beyond which the accuracy does not improve by increasing the order of the quadrature formula. This lower limit can be estimated by simple physical considerations. In analogy to Shannon's communication theorem, the lower limit depends largely on the maximum space harmonic of the current distribution. For a smooth contour, the highest harmonic can be expected to have a period of approximately one-half wavelength along the cross-sectional contour of the scatterers. This harmonic will dominate the current distribution where standing-wave effects are pronounced and must be sampled at least two times per period in order to not greatly impair the accuracy of the solution. Thus, the integration interval must not exceed one-quarter wavelength. If the contour is not smooth, rapid changes of the current density can be expected in regions where the radius of curvature of the contour is very small. Additional sampling points must be introduced in such regions.

Sharp corners at which the current density has an integrable singularity must be given special consideration. In particular, one should not attempt to place a sampling point at a corner, but the current density should be sampled evenly to each side of the corner.

For a smooth contour, the order 2 of the quadrature formula (parabolic approximation of the integral) was found to be sufficient to give satisfactory accuracy of the solution even for an average interval length of one-quarter wave length. Using a parabolic approximation, the integral over each interval contributes to three coefficients in each row of the coefficient matrix of the equations. In the pth interval, where the singularity of the Hankel function is located, we assume a parabolic approximation for the current density and then integrate analytically. The Hankel function cannot be integrated directly analytically. However, since the argument is less than 1 in the pth interval, we can use the following approximation:

$$H_0^{(2)}(ks) \approx 1 - \frac{(ks)^2}{4} - j\frac{2}{\pi}\left[\left(1 - \frac{(ks)^2}{4}\right)\log\left(\frac{\gamma ks}{2} + \frac{(ks)^2}{4}\right)\right]$$
$$\gamma \approx 1.781072 \tag{8}$$

where s is the contour coordinate measured from the pth sampling point.

Introducing the incident field on the right-hand side of the equations through (5) and (6), these equations can be solved by the equation-solving subroutine used in connection with the scattering program. When the cross section of the scatterer configuration is symmetric with respect to an x axis (see Fig. 1), we can save a substantial amount of computer time by expanding the incident wave into a sum of an even function of y and an odd function of y and solve the equations for these waves independently. Since the computer time required to solve N linear equations is proportional to N^3, it takes approximately four times longer to solve N equations once than it does to solve $N/2$ equations twice. Let the expansion of the incident wave into a sum of the even mode and the odd mode be given by

$$E_z^i(y) = E_{ze}^i(y) + E_{zo}^i(y). \tag{9}$$

The even-mode current distribution K^e is an even function of y, *i.e.*,

$$K_{N+1-n}^e = K_n^e, \tag{10}$$

since the sampling point $\bar{\rho}_{N+1-n}$ is the image of $\bar{\rho}_n$ in the x axis. Similarly, the odd-mode current distribution has the symmetry property

$$K_{N+1-n}^0 = -K_n^0. \tag{11}$$

Making use of these symmetry relations in (5), we find the two systems of equations

$$\sum_{n=1}^{N/2} [B_{np} + B_{N+1-n,p}]K_n^e = A_p^e, \quad p = 1, 2, 3, \cdots \frac{N}{2} \tag{12}$$

and

$$\sum_{n=1}^{N/2} [B_{np} - B_{N+1-n,p}]K_n^0 = A_p^0, \quad p = 1, 2, 3, \cdots \frac{N}{2} \cdot \tag{13}$$

The total current distribution is then

$$K = \begin{cases} K_n^e + K_n^0 & \text{for } n = 1, 2, \cdots, \dfrac{N}{2} \\ K_{N+1-n}^e - K_{N+1-n}^0 & \text{for } n = \dfrac{N}{2} + 1, \\ & \dfrac{N}{2} + 2, \cdots, N. \end{cases} \tag{14}$$

The computer program was written for a normalized plane wave incident upon the scatterers from a direction ϕ_i (see Fig. 1),

$$\frac{E_z^i}{\zeta} = e^{-jk(X \cos \phi_i + y \sin \phi_i)}. \tag{15}$$

Expanding this in an even and an odd function of y, we find

$$A_p^e = 4e^{-jkx_p \cos \phi_i} \cos (ky_p \sin \phi_i) \tag{16}$$

and

$$A_p^0 = -j4e^{-jkx_p \cos \phi_i} \sin (ky_p \sin \phi_i), \tag{17}$$

where (x_p, y_p) are the coordinates of pth sampling point of the contour. The number of equations that can be solved is limited only by the capacity of the computer. The program was written for a CDC 1604 Computer and can solve a maximum of 90 equations. The total number of sampling points used on all scatterers is therefore limited to 180. For smooth contours, this limits the total contour length to about 40 wave lengths.

B. The Far-Zone Field

When the induced current distribution has been found, other quantities of interest such as the far-zone field, the IFR factor, and the scattering cross section, can be found by simple integrations over the section contour of the scattering cylinder.

Employing the asymptotic expansion of the Hankel function for large $|\bar{\rho}| = \rho$, we have

$$H_0^{(2)}(k|\bar{\rho} - \bar{\rho}'|) = \sqrt{\frac{2}{\pi k \rho}} \exp j\left[\frac{\pi}{4} - k|\bar{\rho} - \bar{\rho}'|\right]. \tag{18}$$

According to (3), the scattered magnetic far-zone field is then given by

$$H_\phi^s(\bar{\rho}) = \frac{1+j}{4} \frac{1}{\sqrt{\pi k \rho}} \int_C K_z(\bar{\rho}') e^{-jk|\bar{\rho}-\bar{\rho}'|} d(ks') \tag{19}$$

where ϕ is the direction of observation in Fig. 1. In Cartesian coordinates, this reduces to

$$H_\phi^s(\bar{\rho}) = \frac{1+j}{4} \frac{e^{-jk\rho}}{\sqrt{\pi k \rho}} \int_C K_z(\bar{\rho}') e^{jk(x' \cos \phi + y' \sin \phi)} d(ks'). \tag{20}$$

C. The Scattering Cross Section

The scattering cross section $\sigma_s(\phi)$ in direction ϕ is a quantity commonly of interest, especially the back-scattering cross section $\sigma_s(\pi)$. The scattering cross section is defined as the power per unit length of the scatterer that, radiated isotropically from the scatterer axis, would have to be supplied to produce the actually observed scattered field in the given direction when the power density of the incident wave is assumed to be unity. The scattering cross section so defined can be deduced from (20) and is found to be expressed by

$$\sigma_s(\phi) = \frac{1}{4k} \left| \frac{\int_{C'} K_z(\bar{\rho}') e^{jk(x' \cos \phi + y' \sin \phi)} dks'}{H^i} \right|^2. \tag{21}$$

The total scattering cross section is the average value of $\sigma_s(\phi)$ over all directions.

III. Perpendicular Polarization

A. The Integral Equation

Above we have considered scattering with parallel polarization. The problem of scattering with perpendicular polarization is solved in an analogous way.

For perpendicular polarization, all field components can be generated from the magnetic field component H_z along the scatterer axis. This field component must satisfy the homogeneous scalar wave equation with the boundary condition $\partial H_z / \partial n = 0$ at the scatterer contour. According to Green's theorem, and satisfying the radiation condition at infinity, the following integral representation is then found for H_z at a point with the radius vector $\bar{\rho}$

$$H_z(\bar{\rho}) = H_z^i(\bar{\rho}) + H_z^s(\bar{\rho}), \tag{22}$$

where $H_z^i(\bar{\rho})$ is the field of the incident wave, and [9]

$$H_z^s(\bar{\rho}) = -\frac{j}{4} \int_C \cos v' H_1^{(2)}(k|\bar{\rho} - \bar{\rho}'|) H_z(\bar{\rho}') dks' \tag{23}$$

is the scattered field represented by a line integral along the contour of the scatterers. v' is the angle between the vector $(\bar{\rho}' - \bar{\rho})$ and the normal unit vector $\bar{n}'$,

$$\cos v' = \frac{(\bar{\rho}' - \bar{\rho}) \cdot \bar{n}'}{|\bar{\rho}' - \bar{\rho}|}. \tag{24}$$

To arrive at an integral equation for the total magnetic field intensity at the surface of the scatterers, introduce (23) in (22) and let $\bar{\rho}$ approach a point $\bar{\rho}''$ on the scatterer contour. Unlike the case of parallel polarization, this operation requires evaluation of a limit since the Hankel function of order one develops a non-integrable singularity at the point $\bar{\rho} = \bar{\rho}'$. We assume that the boundary is smooth. Scatterers with corners (slope discontinuities) can be treated only if the corners are slightly rounded beforehand.

Let the general point $\bar{\rho}$ be a distance δ from the contour and let 0′ be the point on the contour at this distance, taken as an origin. For arc length s' on the contour (Fig. 2) consider a portion of the contour integral within a distance "a" of 0′ so small that the curvature of the contour is negligible over the arc length $2a$ about 0. Then assume δ is so small that $\delta \ll a$. We shall only seek, for a nonvanishing contribution from this portion of the contour integral as δ approaches zero. For the purpose, we may approximate the Hankel function by

$$H_1^{(2)}(k|\bar{\rho} - \bar{\rho}'|) = H_1^{(2)}(k\sqrt{\delta^2 + s'^2}) \simeq \frac{2j}{\pi k \sqrt{\delta^2 + s'^2}}, \quad k\sqrt{\delta^2 + s'^2} \ll 1, \tag{25}$$

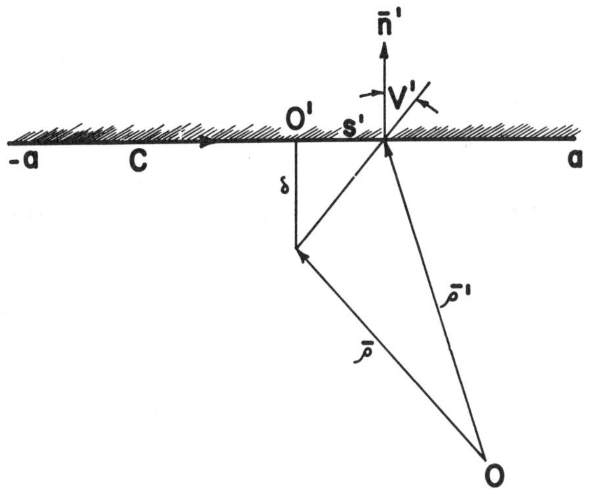

Fig. 2—Contour integration for perpendicular polarizations.

and cos v' by

$$\cos v' = \frac{\delta}{\sqrt{\delta^2 + s'^2}} \cdot \quad (26)$$

The right member of (23) becomes

$$\frac{\delta H_z(\bar{\rho}')}{2\pi} \int_{-a}^{a} \frac{ds'}{\delta^2 + s'^2} = \frac{H_z(\bar{\rho}')}{\pi} \tan^{-1} \frac{a}{\delta} \cdot \quad (27)$$

As δ approaches zero, $\bar{\rho}$ approaches $\bar{\rho}''$, and "a" may be allowed to approach zero at some slow rate so that a negligible part of the contour is involved in (27). In the limit then when $\bar{\rho} = \bar{\rho}''$, (23) becomes

$$\int_C \cos v' H_1^{(2)}(k \mid \bar{\rho}'' - \bar{\rho}' \mid) H_z(\bar{\rho}') dks' - 2jH_z(\bar{\rho}'') = -4jH_z^i(\bar{\rho}''). \quad (28)$$

This integral equation can be solved by the same numerical method as the one used to solve the integral equation (4). Also, for perpendicular polarization, the incident wave was expanded in an even mode and an odd mode, and the normal component of the electric field has an integrable singularity at sharp corners. The current density, however, remains continuous.

B. The Far-Zone Field

Introducing the asymptotic expression for the Hankel function for large values of the argument, and the asymptotic expression

$$\cos v' \simeq -\bar{\rho}_0 \cdot \bar{n}'; \qquad \bar{\rho}_0 = \frac{\bar{\rho}}{|\bar{\rho}|} \cdot \quad (29)$$

we find the far zone field

$$H_z^s(\bar{\rho}) = -\frac{1+j}{4} \frac{e^{-jk\rho}}{\sqrt{\pi k \rho}} \int_{C'} \bar{\rho}_0 \cdot \bar{n}' H_z(\bar{\rho}') \cdot e^{jk(x' \cos\phi + y' \sin\phi)} dks'. \quad (30)$$

C. The Scattering Cross Section

Using the analogy employed in the preceding section, we find the scattering cross section to be

$$\sigma_s(\phi) = \frac{1}{4k} \left| \frac{\int_C \bar{\rho}_0 \cdot \bar{n}' H_z(\bar{\rho}') e^{jk(x' \cos\phi + y' \sin\phi)} dks'}{H^i} \right|^2 . \quad (31)$$

III. Numerical Results

The validity of the scattering computer program has been verified by comparison against results obtained for circular cylinders using the harmonic series [1], [7]. For 0.1λ integration interval, the accuracy was found to be approximately 0.1 per cent. Even when the integration interval is 0.25λ, the accuracy is about 3 per cent. When the integration interval is increased beyond 0.25λ, the accuracy of the solution decreases rapidly.

The validity of the program was also verified by interchanging the direction of incidence and the direction of observation for several test cases [1], [7]. According to the reciprocity theorem, [9], [10] the observed scattered field must be the same in the two cases, and was found to be so to very high accuracy.

The results for forward scattering have been verified experimentally for more than 40 radome element shapes with agreement which appears to be limited only by the accuracy of the experimental measurements [1].

The scattering program was originally developed for application in connection with the design of space frame radomes. However, the program will solve virtually any scattering problem in two dimensions for which the total circumference does not exceed 25 to 40 wave lengths. Most of the cases for which numerical results are presented in this paper have, therefore, been chosen primarily to illustrate the generality of the program and to illustrate physical principles in scattering, rather than to offer design information for a specific application.

In the following figures the current density, normalized to unity for the magnetic field strength of the incident wave, is plotted vs perimeter distance "s" normalized to unity for the half perimeter. The distance variable s is taken equal to zero at the point on the center face nearest the source of the incident wave. Moving clockwise around the perimeter s increases in value reaching the value $s=1$ at the center face furthest from the source. The scattering patterns show the square root of the scattering width, measured in wavelengths, vs azimuth angle.

The elliptical cross section is the only finite-size cross section of a cylinder for which the scattering problem can be solved analytically by the method of separation of variables [11]. Fig. 3 shows the current distribution induced on an elliptic cylinder with a major axis of two wave lengths and an axial ratio of 4, computed by the scattering program. The incident plane wave is assumed to propagate in the direction of the minor axis of the ellipse. It is interesting to note that for both polarizations, the current density on the illuminated side of the cylinder is close to the asymptotic value in the limiting case of geometrical optics (2 for perpendicular polarization, and $2\bar{n}\times\bar{h}_i$ for parallel polarization where $\bar{h}_i$ is a unit vector in the direction of the magnetic field vector of the incident wave). The current distribution for perpendicular polarization is seen to exhibit standing-wave phenomena both on the illuminated side and on the shadow side of the ellipse. These standing waves are caused by creeping waves generated at the light-shadow boundaries at the ends of the ellipse. In the shadow region, the creeping waves will arrive in opposite phase at the midpoint of the rear face. This explains why the standing current wave exhibits a maximum for $s=1$. Since the creeping waves are attenuated due to radiation, the standing-wave amplitude will diminish as one moves away from $s=1$. The period of the standing wave is seen to be close to $\lambda/2$, as expected. On the illuminated side of the ellipse, the period of the standing wave appears to be greater than $\lambda/2$, due to interference between the creeping waves and the direct illumination on this side.

The current distribution for parallel polarization, as expected, decays much faster than for perpendicular polarization in the shadow region. Consequently, the current distribution for parallel polarization does not exhibit any standing wave effects either, but tends to have a small peak at the light-shadow boundary. For a strip, the current density here would approach infinity.

The scattering pattern for the elliptical cylinder is shown in Fig. 4. It exhibits a strong forward lobe representing the shadow behind the cylinder, and a strong back lobe representing the reflection from the front side of the ellipse. Optically, one would expect the forward scattering width to be

$$\sigma_s = WG \tag{32}$$

where W is the apparent width of the cylinder, and G is the power gain of a uniformly illuminated aperture of width W,

$$G = 2\pi \frac{W}{\lambda} \cdot \tag{33}$$

For the elliptical cylinder considered, we find $\sqrt{\sigma_s/\lambda} = \sqrt{8\pi} \approx 5$ which agrees closely with the results in Fig. 4.

Figs. 5 and 6 show the results for scattering by two elliptical cylinders. Some attempts have been made to treat this case analytically [12], [13]. Fig. 5 shows the current distribution on the upper cylinder (the distribution on the lower cylinder is the image of the distribution on the upper cylinder). It will be noticed that the current distribution is slightly asymmetric because of the coupling between the cylinders. The scattering patterns for parallel polarization (the pattern for perpendicular polarization is very similar) is quite interesting. It shows the array effect of the two cylinders. The centers of the cylinders being 4λ apart, one would expect the first grating lobes of the array pattern to appear at $\phi \sim \text{Arcsin}\ (0.25) = 14°.5$. From Fig. 6, they are seen to appear at 13°.5. A comparison between Fig. 4 and Fig. 6 shows that $\sqrt{\sigma_s/\lambda}$ is about twice as large for two cylinders as for one cylinder, and also that the element factor in Fig. 4 is essentially the envelope of the pattern in Fig. 6, both facts being expected.

To further illustrate the application of the scattering program to the solution of problems that can not be treated analytically, we have computed the scattering from a parabolic cylindrical reflector (see Fig. 7). The reflector is 0.2λ thick and the over-all dimension is 4.2λ. The focal length is chosen such that the reflector aperture is seen under an angle of 90° from the focal point. The induced current distribution is shown in Fig. 7. It shows similar characteristics as discussed in connection with the elliptical cylinder.

The scattering pattern for the parabolic reflector is shown in Fig. 8. The beam in the forward direction, representing the shadow behind the reflector, is seen to have a value close to 10.5, which is expected from (32). The first side lobe is seen to be nearly 13.2 db down from the main lobe which is what one would expect for the radiation pattern of a uniformly illuminated aperture. Optically, the rays reflected from the front side of the reflector will all pass through the focal point and be distributed nearly uniformly in a sector beam of width 2×45°. Consequently, the power gain in this sector beam is $G=360/90=4$, and from (32), it follows that $\sqrt{\sigma_s/\lambda}=2\sqrt{4.2}=4.1$. This value agrees fairly well with the actual pattern in Fig. 8. The sector beam in Fig. 8 is somewhat blurred due to the edge effect which is significant.

Figs. 9 and 10 illustrate scattering from a hemi-circular metallic tube. The current distributions on the illuminated convex side of the tube are fairly close to the optical distributions except very close to the edges of the tube. On the concave side of the tube, the current density is seen to drop to zero extremely rapidly for parallel polarization, while the expected standing wave effect is present for perpendicular polarization.

The scattering patterns are shown in Fig. 10. According to (32), the optical value of $\sqrt{\sigma_s/\lambda}$ is $1.2\sqrt{2\pi} \cong 3$ in the forward direction. This agrees fairly well with Fig. 10. The optical back scattering width equals the diameter of the cylinder, thus $\sqrt{\sigma_s/\lambda} \cong \sqrt{1 \cdot 2} \cong 1 \cdot 1$ which also agrees fairly well with Fig. 10.

In the above examples, the complex eigenfrequencies of the scatterer cross section are far removed from the real frequency of the exciting incident wave. A situation in which the complex eigenfrequency of the scatterer

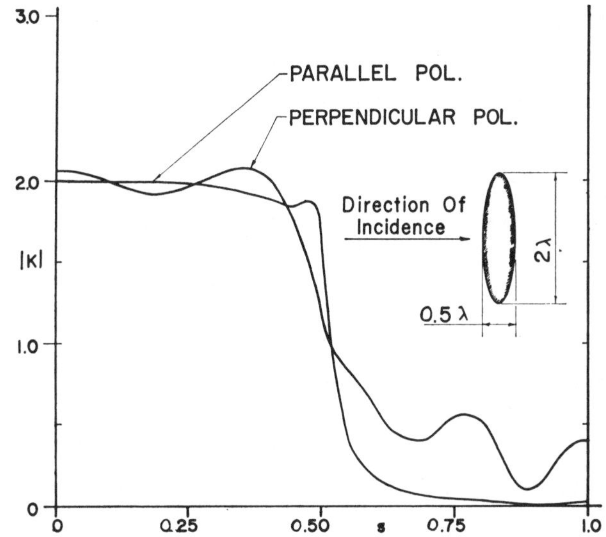

Fig. 3—Current distribution of elliptic cylinder.

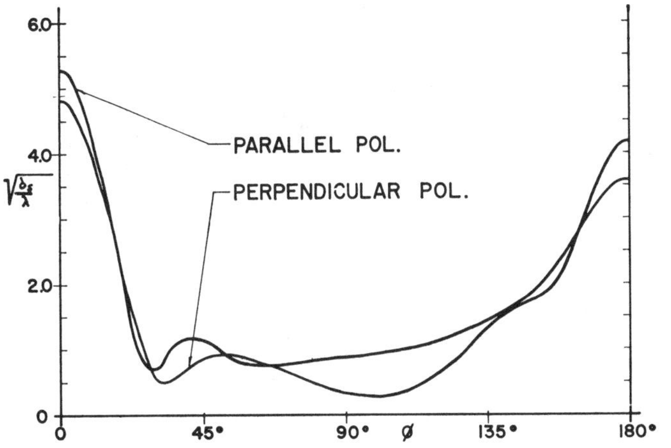

Fig. 4—Scattering pattern for elliptic cylinder.

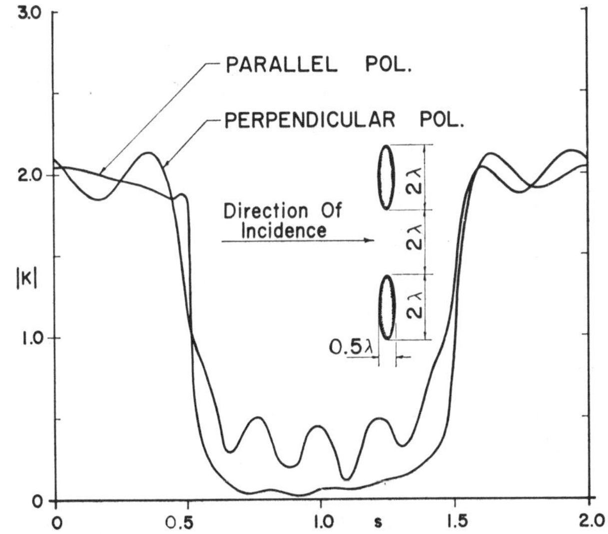

Fig. 5—Current distribution on two elliptic cylinders.

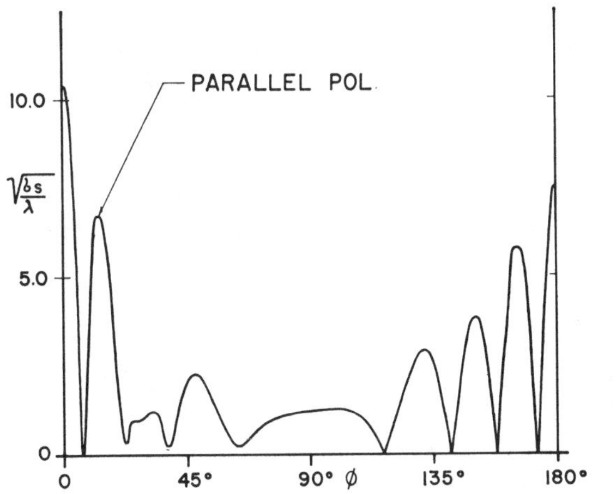

Fig. 6—Scattering pattern for two elliptic cylinders.

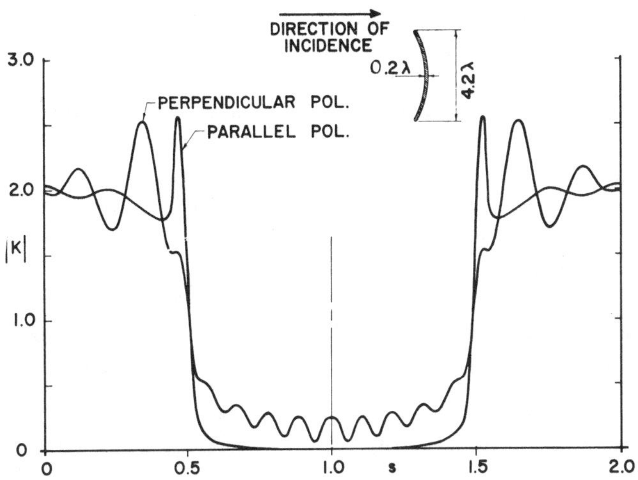

Fig. 7—Current distribution on parabolic cylinder.

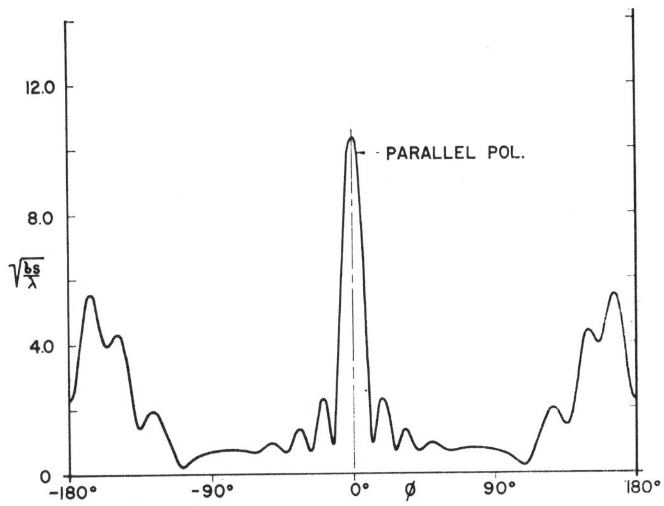

Fig. 8—Scattering pattern for parabolic cylinder.

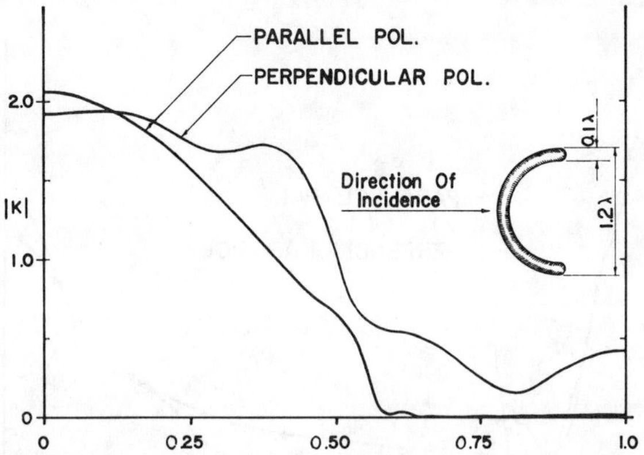

Fig. 9—Current distribution of semicircular cylindrical tube.

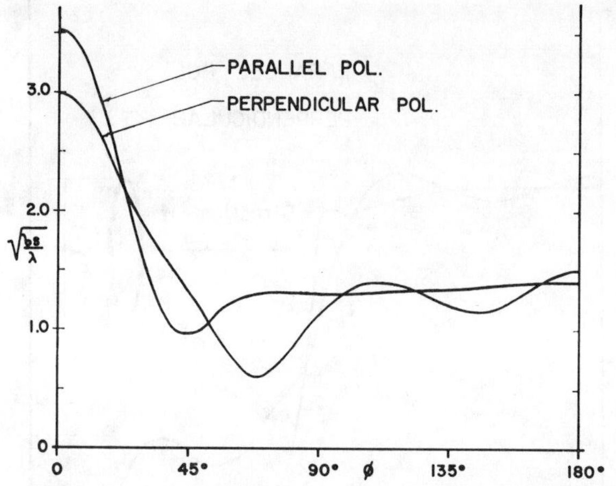

Fig. 10—Scattering pattern for semicircular cylindrical tube.

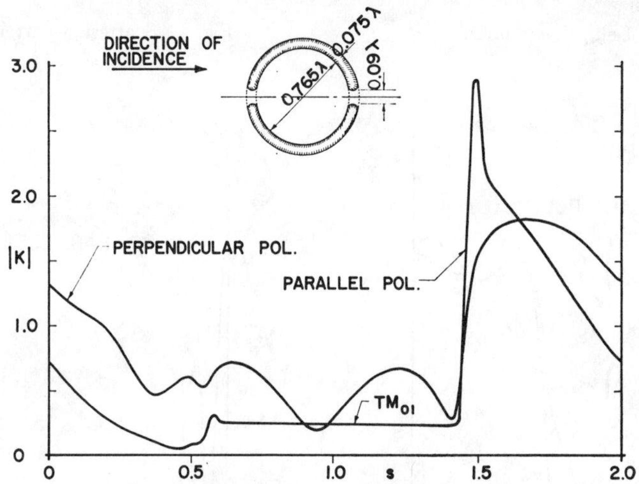

Fig. 11—Current distribution for circular cylindrical cavity with two slots excited at the eigenfrequency of the TM_{01}-mode.

is close to the frequency of the incident wave would constitute a severe test on the accuracy of the program. To conduct such a test, we consider a circular cylindrical cavity coupled to space by two slots (see Fig. 11). When the slots are closed, the cavity has real eigenfrequencies. The introduction of the slots provides for a coupling between the cavity and space. If this coupling is small, a complex eigenfrequency of the system is very close to the corresponding real eigenfrequency in the uncoupled case. The imaginary part of the eigenfrequency accounts for the energy radiated from the cavity through the slots. This energy is in balance with an equal amount of energy fed into the cavity from space to sustain the oscillations in the cavity. We wish to investigate the excitation of a cavity mode which is weakly coupled to space. The coupling between the cavity and space is fairly strong for perpendicular polarization because the slots interrupt the path of the transverse current. For parallel polarization, however, the coupling is fairly weak because the slots are then parallel to the current flow. Thus, we shall consider the excitation of TM modes which are excited only by a parallel polarized wave. The TM_{01} mode, which has no azimuthal variation, is resonant at a frequency that makes the internal diameter of the cavity 0.765λ. The current distribution at such a frequency is shown in Fig. 11 for a cavity with slots 0.09λ wide, and wall thickness 0.075λ. The distribution shown applies to the upper half of the cavity (the distribution on the lower half is the image of this distribution), the normalized perimeter distance "s" being measured clockwise from the uppermost point. The excitation of the TM_{01} mode inside the cavity by the parallel polarized wave is clearly demonstrated. The current density inside the cavity is seen to be constant with very good accuracy. For perpendicular polarization, several modes (TE) are seen to be excited in the cavity, all fairly strongly coupled to space.

Conclusion

The numerical method used to solve the general two-dimensional scattering problem for an arbitrary number of arbitrary metallic boundaries with a maximum perimeter length 25–40λ has demonstrated the importance of the use of computer techniques for solving complicated electromagnetic boundary value problems that cannot be solved by analytical methods. The method can be extended to three dimensions and appears to have application for the accurate prediction of current distributions and radiation patterns of antennas as well. Work in this general area is underway.

Acknowledgment

The author wishes to extend thanks to the Bell Telephone Laboratories, Whippany, N. J., who provided initial support for this report. The work was subsequently continued and extended by means of internal TRG support. The author also wishes to acknowledge the assistance and counsel provided by his colleagues at TRG. He is particularly indebted to Dr. A. F. Kay of the Boston division of TRG who suggested undertaking the research and, in addition, offered valuable suggestions regarding specific technical details. The resolution of the incident wave in the manner described was one such suggestion. Dr. F. B. Harris, Jr. of TRG-West, Menlo Park, provided invaluable help in the form of specific computer subroutines essential to the successful operation of the program.

References

[1] M. G. Andreasen and A. F. Kay, "Scattering from a Metallic Cylinder of Arbitrary Cross Section," TRG, Incorporated, East Boston, Mass., TRG Rept. FR-220-2, Contract DA 30-069 ORD-1955; November, 1962.

[2] M. G. Andreasen, "Scattering from a Wedge Capped by a Circular Cylinder," TRG-West Menlo Park, Calif., Final Rept., TRG-West Project No. W-208; December, 1962.

[3] J. A. Stratton, "Electromagnetic Theory," McGraw-Hill Book Co., Inc., New York, N. Y.; 1941.

[4] W. Franz and K. Deppermann, "Theory of diffraction by a cylinder as affected by the surface wave," *Ann. der Phy.*, vol. 6, p. 361; 1952.

[5] J. B. Keller, "Diffraction by a Convex Cylinder," New York University, New York, N. Y., Research Rept. No. EM-94; 1956.

[6] K. Mei and J. VanBladel, "Low-frequency scattering by rectangular cylinders," IEEE Trans. on Antennas and Propagation, vol. PGAP-11, pp. 52–56; January, 1963.

[7] M. G. Andreasen and K. Mei, "Comments on scattering by conducting rectangular cylinders," IEEE Trans. on Antennas and Propagation, vol. PGAP-12, pp. 235–236; March, 1964.

[8] R. P. Banaugh and W. Goldsmith, "Diffraction of steady acoustic waves by surfaces of arbitrary shape," *J. Acous. Soc. Am.*, vol. 35, pp. 1626–1632; October, 1963.

[9] F. E. Borgnis and Papas, "Randwertprobleme der Mikrowellenphysik," Springer-Verlag, Berlin, Germany; 1955.

[10] A. F. Stevenson, "Relations between the transmitting and receiving properties of antennas," *Quart. Appl. Math.*, vol. 5, pp. 369–384; 1948.

[11] G. Sinclair, "The patterns of antennas located near cylinders of elliptical cross section," Proc. IRE, vol. 39, pp. 660–668; June, 1951.

[12] K. Saermark, "On Some Two-Dimensional Diffraction Problems," Thesis, University of Copenhagen, Denmark; 1960.

[13] V. Twersky, "Scattering of waves by two objects" in "Electromagnetic Waves," ed. R. E. Langer, Ed., University of Wisconsin Press, Madison, Wis.; 1962.

Scattering by a Dielectric Cylinder of Arbitrary Cross Section Shape

JACK H. RICHMOND, SENIOR MEMBER, IEEE

***Abstract*—The theory and equations are developed for the scattering pattern of a dielectric cylinder of arbitrary cross section shape. The harmonic incident wave is assumed to have its electric vector parallel with the axis of the cylinder, and the field intensities are assumed to be independent of distance along the axis. Solutions are readily obtained for inhomogeneous cylinders when the permittivity is independent of distance along the cylinder axis.**

Although other investigators have approximated the field within the dielectric body by the incident field, we treat the total field as an unknown function which is determined by solving a system of linear equations.

In the case of the dielectric cylindrical shell of circular cross section, this technique yields results which agree accurately with the exact classical solution. Scattering patterns are also presented in graphical form for a dielectric shell of semicircular cross section, a thin homogeneous plane dielectric sheet of finite width, and an inhomogeneous plane sheet. The effects of surface-wave excitation and mutual interaction among the various portions of the dielectric shell are included automatically in this solution.

Manuscript received December 9, 1963; revised November 9, 1964. This work was supported in part by the Research and Technology Division, Air Force Systems Command, Wright-Patterson Air Force Base, Ohio, under Contract AF 33(616)-7614 with The Ohio State University Research Foundation.

The author is with the Dept. of Electrical Engineering, Antenna Lab., The Ohio State University, Columbus, Ohio.

I. Introduction

Although rigorous solutions are available for scattering by homogeneous dielectric cylinders of circular and elliptical cross section, only approximate solutions exist for homogeneous or inhomogeneous cylinders of other shapes with finite cross section dimensions. A ray-optics technique is commonly employed by radome designers. Each ray passing through the dielectric body is assumed to undergo the same reflection and phase delay that is experienced by a plane wave passing through an infinitely wide plane sheet of the same thickness and with the same angle of incidence. This ray-optics method often provides reasonably accurate results for slightly curved dielectric shells, but it is inadequate for rapidly curving shells and the edge region of a truncated shell.

The ray-optics solution has been refined and extended considerably by Kouyoumjian, Peters and Thomas. [1] Their technique has proven quite successful with circular dielectric cylinders, spheres, and a few other shapes. However, the method becomes somewhat complicated in the general case and it does not always provide accurate results for small or irregular dielectric bodies.

Rhodes [2] has developed an iteration technique for dielectric scattering problems. The first-order solution is obtained by approximating the total field in the dielectric body by the incident field, and then calculating the scattered field by considering the equivalent volume currents in the dielectric region to radiate in unbounded free space. Useful results are obtained for thin dielectric shells having a dielectric constant near unity. By a similar technique, Andreasen [3] has obtained data for thin spherical shells; Stickler [4] has calculated the scattering patterns of dielectric cylinders of rectangular cross section and Philipson [5] has made an analysis of the scattering properties of thin dielectric rings.

Using a variational formulation, Cohen [6] has obtained accurate solutions for the circular dielectric cylinder. This approach becomes rather complicated, however, and the calculations become lengthy when the dielectric body has an arbitrary shape and dielectric constant.

The technique developed in this paper is based on the integral equation for the field of a harmonic source in the presence of a dielectric cylinder of arbitrary cross section shape. The dielectric cylinder is divided into square cells which are small enough so that the electric field intensity is nearly uniform in each cell. The total electric field intensity within each cell is initially considered to be an unknown quantity. A system of linear equations is obtained by enforcing at the center of each cell the condition that the total field must equal the sum of the incident and scattered fields. This system of equations is solved with the aid of a digital computer to evaluate the electric field intensity in each cell. It is then a rather simple procedure to calculate the scattered field at any other point in space. The technique has the following advantages.

1) The solution approaches the exact solution if a sufficiently large number of cells is employed.

2) Solutions are obtained for a dielectric shell of arbitrary shape as quickly and systematically as for a circular shell.

3) Simply by inserting the appropriate equations for the incident field, one obtains the solution for any two-dimensional source (such as a line source, any array of line sources, or a plane-wave source) in the presence of a dielectric cylinder.

4) Solutions are readily obtained for dielectric shells of tapered thickness and inhomogeneous dielectric shells.

5) The effects of surface-wave excitation and interaction among the various parts of the dielectric cylinder are included automatically in the solution.

6) Accurate solutions are obtained for dielectric cylinders having cross section dimensions up to several wavelengths. The ray-tracing or geometrical-optics methods often fail when the cross section dimensions are on the order of one wavelength or less.

7) Once a solution has been obtained for any particular source location, a relatively simple calculation will provide the new solution corresponding to a rotated or translated source.

The following sections develop the theory and the equations involved in this technique and present numerical results for homogeneous and inhomogeneous dielectric cylinders of various cross section shapes.

II. The Basic Theory

Consider a harmonic wave incident in free space on a dielectric cylinder of arbitrary cross section as suggested in Fig. 1. The time-factor $e^{j\omega t}$ is understood. It is assumed that the incident electric field intensity $\boldsymbol{E}^i$ has only a z component and it is not a function of z, where the z axis is taken to be parallel with the axis of the cylinder. That is,

$$\boldsymbol{E}^i = \hat{z}E^i(x, y) \tag{1}$$

The dielectric cylinder is assumed to have the same permeability as free-space ($\mu = \mu_0$). The dielectric material is assumed to be linear and isotropic, but it may be inhomogeneous with respect to the transverse coordinates as follows:

$$\epsilon = \epsilon(x, y) \tag{2}$$

where ϵ represents the complex permittivity.

Let $\boldsymbol{E}$ represent the total field; that is, the field set up by the source in the presence of the dielectric cylinder. The "scattered field" is defined to be the difference between the total and the incident fields. Thus,

$$\boldsymbol{E} = \boldsymbol{E}^i + \boldsymbol{E}^s \tag{3}$$

Under the assumed conditions, the total and the scattered electric field intensities will have only z components.

The scattered field $\boldsymbol{E}^s$ may be generated by an equivalent electric current J radiating in unbounded free space, where

$$J = j\omega(\epsilon - \epsilon_0)\boldsymbol{E} \tag{4}$$

with ω representing the angular frequency $2\pi f$. This equivalent current density is often called the "polarizarion current."

The field of an electric current filament dI parallel with the z axis in free space is given by

$$d\boldsymbol{E}^s = -\hat{z}(\omega\mu/4)H_0^{(2)}(k\rho)dI \tag{5}$$

where $H_0^{(2)}(k\rho)$ is the Hankel function of order zero, ρ is the distance from the current filament to the observation point and $k = \omega\sqrt{\mu_0\epsilon_0} = 2\pi/\lambda$. The free-space wavelength is denoted by λ. The increment of electric current which generates the scattered field is given by

$$dI = J\,dS = j\omega(\epsilon - \epsilon_0)E\,dS \tag{6}$$

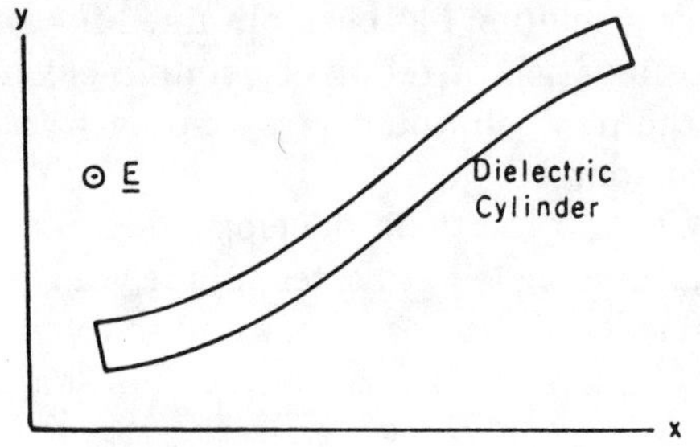

Fig. 1. Cross section of a dielectric cylinder showing the coordinate system.

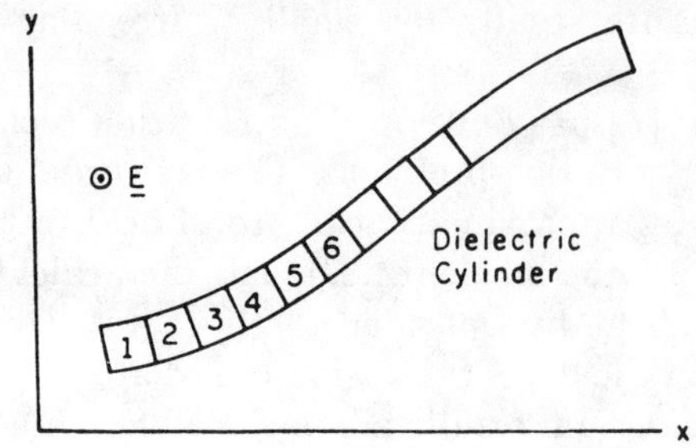

Fig. 2. The cross section of the dielectric cylinder is divided into small cells.

where dS is the increment of surface area on the cross section of the dielectric cylinder. From (5) and (6), the scattered field is given by

$$E^s(x, y) = -(jk^2/4)\iint(\epsilon_r - 1)E(x', y')H_0^{(2)}(k\rho)dx'dy' \quad (7)$$

where (x, y) and (x', y') are the coordinates of the observation point and the source point, respectively, ϵ_r is the complex relative dielectric constant ($\epsilon_r = \epsilon/\epsilon_0$) and

$$\rho = \sqrt{(x - x')^2 + (y - y')^2} \quad (8)$$

The integration in (7) is to be performed over the cross section of the dielectric cylinder. In the inhomogeneous case the relative dielectric constant is considered to be a function of the source coordinates

$$\epsilon_r = \epsilon_r(x', y')$$

Equation (7) is valid for the scattered field at any point inside or outside the dielectric region. The integral equation for the total field E is obtained from (3) and (7),

$$E(x, y) + (jk^2/4)\iint(\epsilon_r - 1)E(x', y')H_0^{(2)}(k\rho)dx'dy' = E^i(x, y) \quad (9)$$

Let us divide the cross section of the dielectric cylinder into cells sufficiently small so that the dielectric constant and the electric field intensity are essentially constant over each cell. The division into cells is indicated in Fig. 2. If (9) is enforced at the center of cell m, the following expression is obtained:

$$E_m + (jk^2/4)\sum_{n=1}^{N}(\epsilon_n - 1)E_n \cdot \iint_{\text{cell } n} H_0^{(2)}(k\rho)dx'dy' = E_m^i \quad (10)$$

where ϵ_n and E_n represent the complex relative dielectric constant and the electric field intensity at the center of cell n and

$$\rho = \sqrt{(x' - x_m)^2 + (y' - y_m)^2} \quad (11)$$

By taking $m = 1, 2, 3, \cdots, N$, (10) yields a system of N linear equations, where N represents the total number of cells. These can be solved to determine the total electric field intensity at the center of each cell ($E_1, E_2, E_3, \cdots, E_N$). Having thus determined the total field $E(x, y)$ in the dielectric region, it is then possible to calculate the scattered field of the dielectric cylinder at any point in space by means of (7). The details of the solution are described in the following section.

III. Surface Integrals of the Hankel Function

The surface integrals in (10) can be evaluated by numerical integration formulas such as the trapezoidal rule and Simpson's rule. These methods have been employed successfully, but it was found that the calculations are quite lengthy, and care must be exercised in integrating through the singularity that exists when the observation point is at the center of cell n. The region of integration (over cell n) is square or rectangular in the simplest case, and a closed-form solution for this integral is not known. A simple solution is, however, available for the integral of the zero-order Hankel function over a circular region. It is given by

$$(jk^2/4)\underset{\text{cell } n}{\int_0^{2\pi}\int_0^a} H_0^{(2)}(k\rho)\rho'd\rho'd\phi'$$
$$= (j/2)[\pi kaH_1^{(2)}(ka) - 2j] \qquad \text{if } m = n$$
$$= (j\pi ka/2)J_1(ka)H_0^{(2)}(k\rho_{mn}) \qquad \text{if } m \neq n \quad (12)$$

where ρ is given by (11) and ρ' and ϕ' are polar coordinates based on a coordinate origin at the center of cell n. The first solution given in (12) applies if the observation point is at the center of the circular region (i.e., if $m = n$). The second solution applies if the observation point is at a distance ρ_{mn} from the center of the circular region, where ρ_{mn} is greater than the radius a of the circular region. Numerical calculations have shown that little error is incurred in approximating square cells with circular cells of the same cross section area to take advantage of the simple expressions given in (12). The distance between the centers of cells m and n is given by

$$\rho_{mn} = \sqrt{(x_m = x_n)^2 + (y_m - y_n)^2} \tag{13}$$

Now the system of linear equations represented by (10) can be written in the following form:

$$\sum_{n=1}^{N} C_{mn}E_n = E_m^i \quad \text{with } m = 1, 2, \cdots, N \tag{14}$$

If a_n represents the radius of the equivalent circular cell which has the same cross section area as cell n, the coefficients C_{mn} are given by

$$C_{mn} = 1 + (\epsilon_m - 1)(j/2)[\pi k a_m H_1^{(2)}(ka_m) - 2j] \quad \text{if } n = m \tag{15}$$

$$C_{mn} = (j\pi k a_n/2)(\epsilon_n - 1)J_1(ka_n)H_0^{(2)}(k\rho_{mn}) \quad \text{if } n \neq m \tag{16}$$

IV. Formulating the Scattered Field

Once the system of linear equations (14) has been solved, the scattered field of the dielectric cylinder can be calculated at any point in space by means of (7). To simplify the surface integral in (7), it is convenient again to divide the cylinder into small cells whose cross section shape is approximately square. From (7) and (12), the scattered field at any point outside the dielectric body is given by

$$E^s(x, y) = -j(\pi k/2) \sum_{n=1}^{N} (\epsilon_n - 1)E_n a_n J_1(ka_n)H_0^{(2)}(k\rho_n) \tag{17}$$

where ϵ_n represents the average dielectric constant over cell n, a_n is the radius of the circle having the same area as cell n and ρ_n is the distance from the observation point to the center of cell n,

$$\rho_n = \sqrt{(x - x_n)^2 + (y - y_n)^2} \tag{18}$$

The distant scattering pattern of the dielectric cylinder is obtained by employing the asymptotic form for the Hankel function of large argument and taking

$$\rho_n = \rho_0 - x_n \cos\phi - y_n \sin\phi \tag{19}$$

where ρ_0 and ϕ are the polar coordinates of the distant observation point. The distant scattered field is given by

$$E^s(\rho_0, \phi) = -j(\pi k/2)\sqrt{2j/\pi k\rho_0} \cdot e^{-jk\rho_0} \sum_{n=1}^{N} (\epsilon_n - 1)E_n a_n J_1(ka_n) \cdot e^{jk(x_n \cos\phi + y_n \sin\phi)} \tag{20}$$

The plane-wave scattering properties of any cylindrical body of infinite length are conveniently described in terms of the echo width [7] which is denoted by $W(\phi)$ and is defined as follows:

$$W(\phi) = \lim_{\rho_0 \to \infty} 2\pi\rho_0 \left| \frac{E^s(\rho_0, \phi)}{E^i} \right|^2 \tag{21}$$

From (20) and (21), the echo width of a dielectric cylinder of arbitrary cross section shape is given as follows:

$$W(\phi) = \frac{\pi^2 k}{|E^i|^2} \left| \sum_{n=1}^{N} (\epsilon_n - 1)E_n a_n J_1(ka_n) e^{jk(x_n \cos\phi + y_n \sin\phi)} \right|^2 \tag{22}$$

V. Numerical Results for the Circular Cylindrical Shell

Using (14), (15), (16), (20) and (22), numerical solutions were obtained for a circular dielectric cylindrical shell. Figure 3 shows the electric field distribution in the dielectric shell set up by an incident plane wave, and Fig. 4 shows the distant scattering pattern. Figure 4 also shows the exact classical solution which involves an infinite series of cylindrical mode functions. It may be noted that the integral-equation technique yields results which show excellent agreement with the exact solution. In view of the nature of the integral-equation formulation, it is believed that highly accurate results can also be obtained for dielectric cylinders of other cross section shapes where an exact solution is not available.

In Fig. 3 it may be noted that the electric field intensity in the dielectric region varies from 0.8 to 1.58 in magnitude. The incident plane wave has an electric field intensity of 1.0. Thus, poor accuracy is to be expected if the electric field intensity in the dielectric region is approximated by the incident electric field intensity.

In addition to the plane-wave solution previously described, the integral-equation technique readily yields solutions when the incident wave is the field of a line source or any array of line sources of infinite length. The line sources are assumed to be parallel with the axis of the dielectric cylinder. Figure 5 shows the scattering pattern of a circular dielectric cylindrical shell when the incident field is generated by a nearby line source having an omnidirectional pattern. In Fig. 5 the echo width is referred to the incident power density at the center of the dielectric cylindrical shell. Similar calculations show that the scattering pattern approaches the plane-wave solution shown in Fig. 4 when the line source is at a great distance from the dielectric shell. This provides one check on the accuracy and validity of the integral-equation technique as applied to the dielectric cylinder in the presence of a line source.

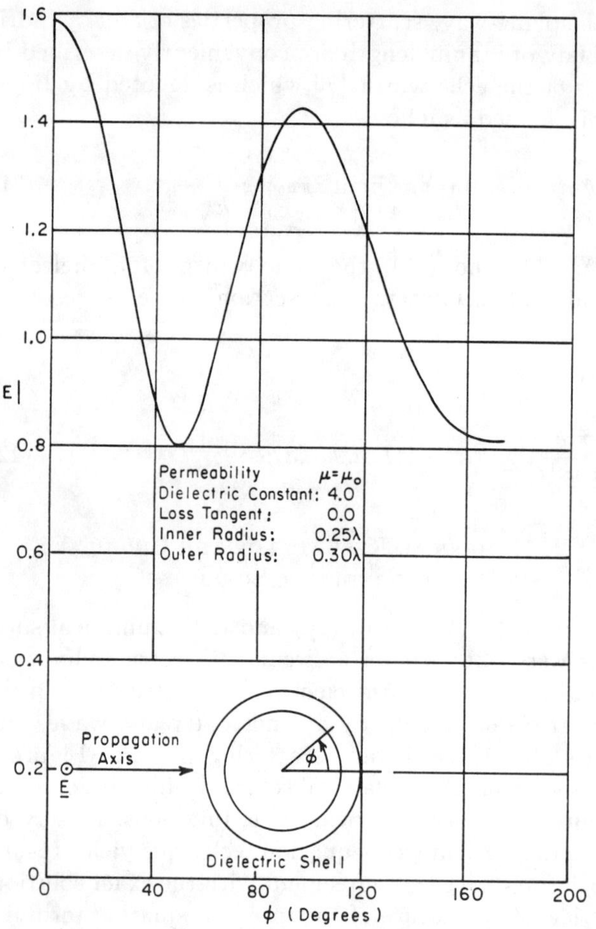

Fig. 3. Electric field distribution in circular dielectric cylindrical shell with plane-wave incident, calculated with the integral-equation technique.

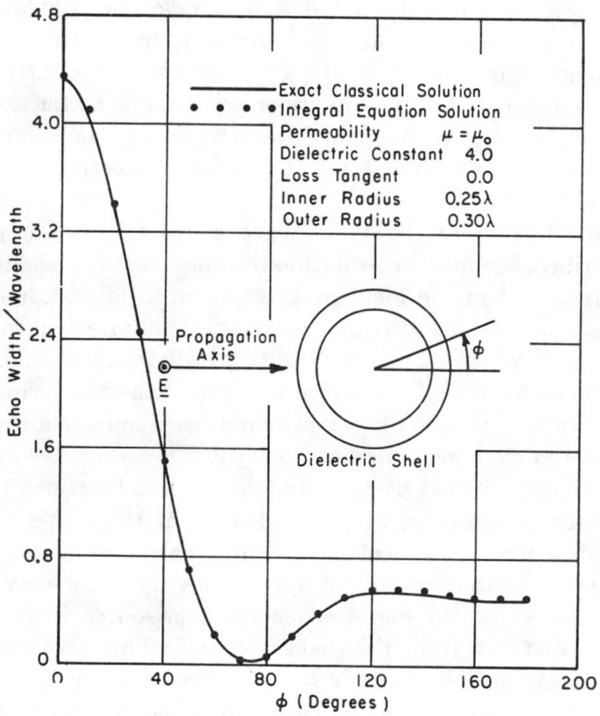

Fig. 4. Distant scattering pattern of circular dielectric cylindrical shell with plane-wave incident.

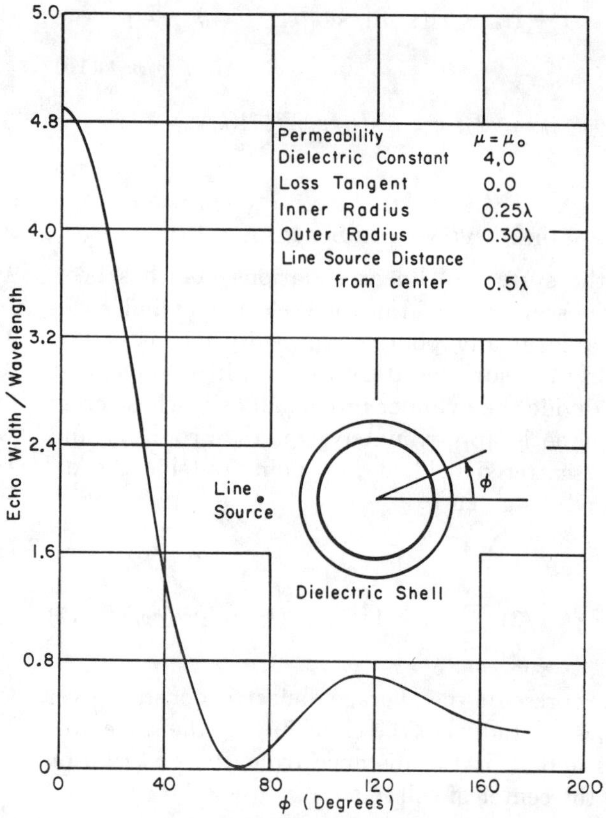

Fig. 5. Scattering pattern of a circular dielectric cylindrical shell in the presence of a nearby parallel line source, calculated with the integral-equation technique.

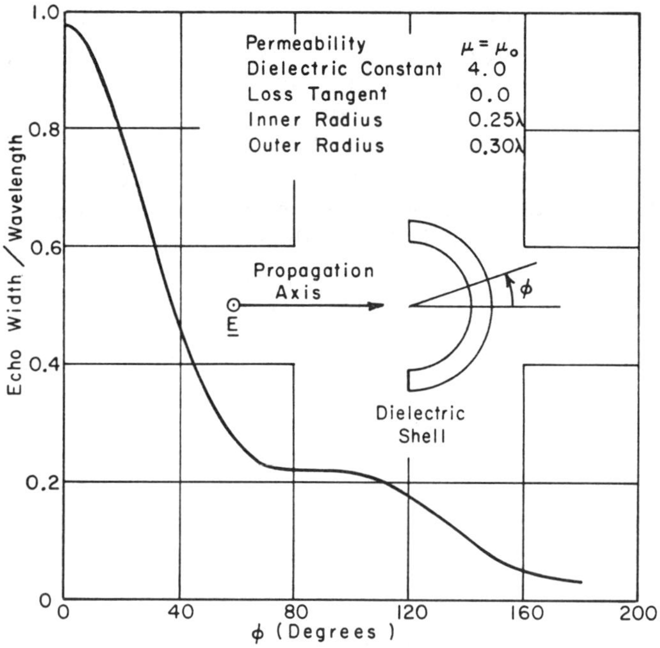

Fig. 6. Scattering pattern of a semicircular dielectric cylindrical shell with plane-wave incident, calculated with the integral-equation technique.

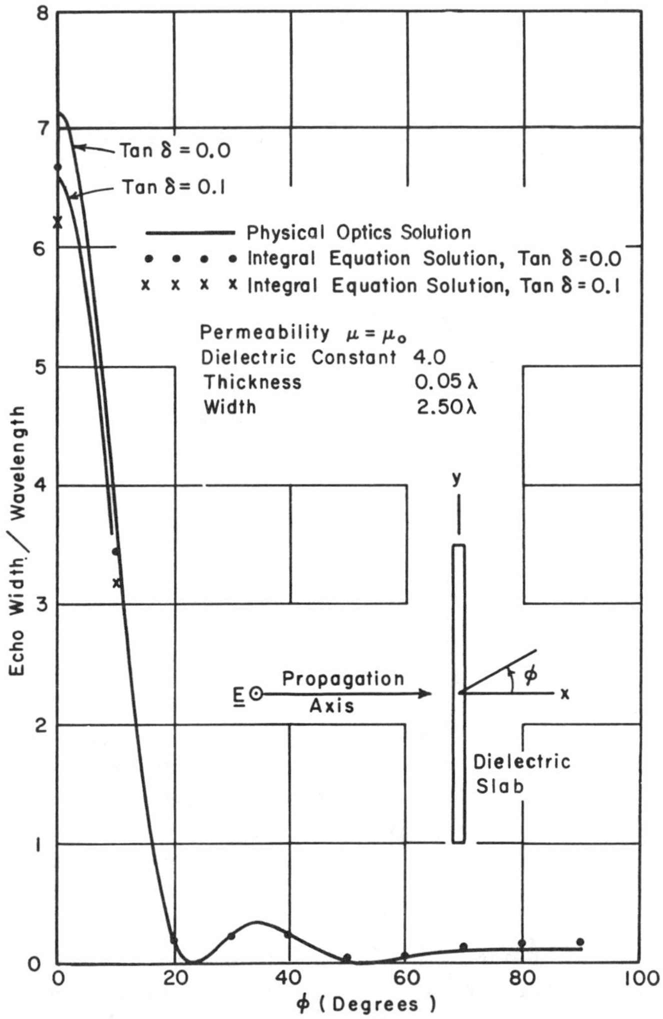

Fig. 7. Calculated scattering patterns of a homogeneous plane dielectric slab of finite thickness and width with a plane wave having normal incidence.

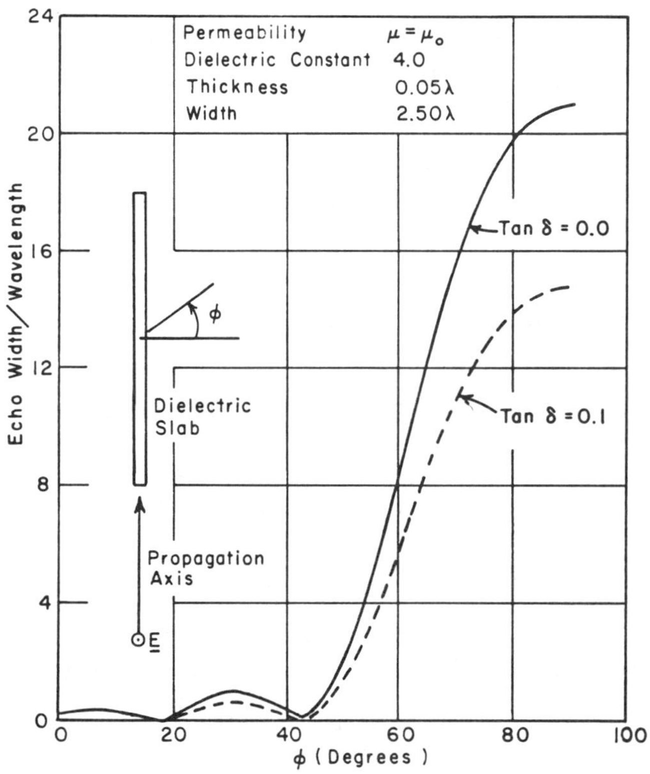

Fig. 8. Calculated scattering patterns of a homogeneous plane dielectric slab with a plane wave at grazing incidence.

VI. Numerical Results for Semicircular Cylindrical Shell

Figure 6 shows the plane-wave scattering pattern of a semicircular cylindrical shell, computed with the integral-equation technique developed in Sections II and III. This particular semicircular dielectric shell has much smaller echo width than the corresponding circular shell illustrated in Fig. 4. However, it is possible to obtain a considerable increase in the backscatter echo width of a semicircular shell (compared with the circular shell of the same radii) by proper choice of the inner and outer radii and the wavelength. No exact solution is available for comparison with the results shown in Fig. 6, and the geometric-optics solution is not likely to be accurate for a shell of such small cross section dimensions.

It should be pointed out that the solution for the semicircular shell (or any other cross section shape) is as simple, straightforward and systematic as that for the circular cylindrical shell when the integral-equation technique is employed. This represents a distinct advantage over the boundary-value solution, the geometric-optics solutions and the variational solutions.

VII. Numerical Results for Plane Homogeneous Dielectric Slabs

Figure 7 shows the calculated scattering patterns of a plane dielectric slab of finite thickness and width for the case where a plane wave is incident normally on the slab. Results are shown both for lossless and dissipative homogeneous slabs.

For comparison, the physical optics solutions are also shown. The physical optics solution is based on the approximation that the electric field intensity within the dielectric body is the same as that in a dielectric slab of infinite width. Thus, the edge effects and surface-wave phenomena are usually neglected in the physical-optics approximation in contrast with the integral-equation solution. Although accurate agreement is not to be expected between the two methods, reasonably good agreement is observed in Fig. 7.

It should be mentioned that slightly dissipative dielectric cylinders are handled as easily as lossless dielectric bodies with the integral-equation technique. In problems involving perfectly conducting or highly conducting scatterers, a somewhat different formulation has been employed with good success. Some modifications must be made in the techniques developed here to study the scattering properties of highly dissipative dielectric bodies, since in this case it is not reasonable to assume the electric field intensity is nearly uniform over each cell of the type employed here.

Figure 8 shows the scattering patterns of the same homogeneous dielectric slab for grazing incidence. Results are shown both for the lossless and dissipative cases. The physical-optics solution is not shown for this case simply because the optical solutions do not yield useful results for grazing incidence unless considerable effort is made to include the effects of surface-wave excitation in the dielectric slab. It may be noted in Fig. 8 that the forward scattering intensity is much greater for grazing incidence than for normal incidence, and that the results are much more sensitive to the loss tangent (tan δ) for grazing incidence.

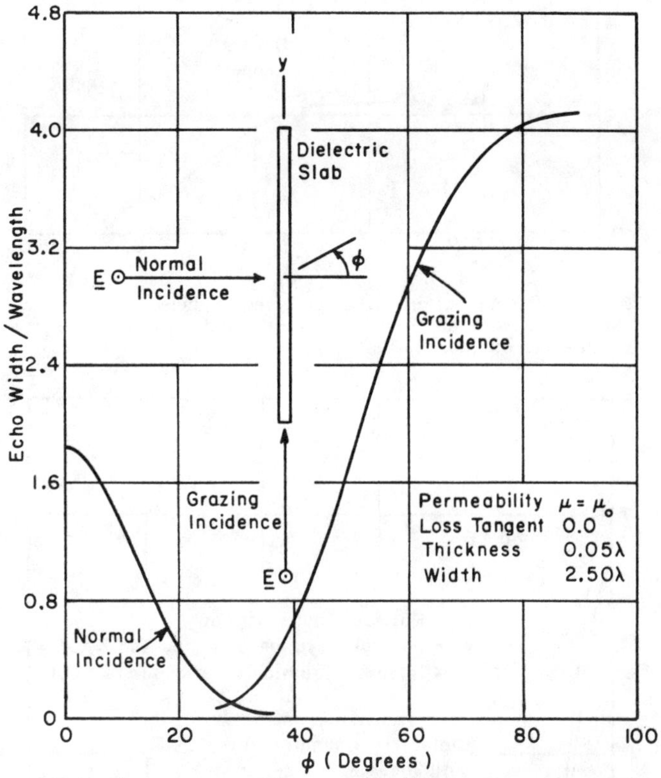

Fig. 9. Calculated scattering patterns of an inhomogeneous plane dielectric slab with a plane wave at normal and grazing incidence. Dielectric constant varies linearly from 4.0 at center to 1.0 at edges of slab.

VIII. Numerical Results for Plane Inhomogeneous Dielectric Slabs

Figure 9 shows the scattering patterns for a lossless inhomogeneous dielectric slab for normal incidence and grazing incidence. The dielectric constant is assumed to vary linearly from 4.0 at the center of the slab to 1.0 at the edges, being a function of the y coordinate only. A physical-optics solution for this problem would be somewhat involved. With the integral-equation technique the solution is straightforward and systematic.

IX. Conclusions

A technique is developed for calculating the scattered fields of a dielectric cylinder or cylindrical shell of arbitrary cross section shape. The solution is based on the integral equation for the problem, and it involves the solution of a system of linear equations. The first step in the solution is to determine the electric field distribution within the dielectric body. This field distribution is often of considerable interest in itself, since it promotes an understanding of the phenomena which are important in the scattering process. For example, any surface waves which may be excited in the dielectric body are revealed in the standing waves in the field distribution, and the edge effects are evident in the field perturbations near the edges of a dielectric shell.

The solution is best accomplished with the aid of a digital computer. This integral-equation solution is more systematic, more general and more accurate than the optical methods or the variational solutions. For an inhomogeneous dissipative dielectric shell of any cross section shape, one merely inserts in the equations the coordinates which define the shape of the shell, and the appropriate dielectric constant and loss tangent associated with each point in the dielectric body. The incident wave may be a plane wave or any combination of plane waves, or it may be the field generated by a parallel line source or any array of parallel line sources. One merely inserts the value of the incident field associated with each point in the dielectric body. The effects of surface waves and edge diffraction are included automatically in the solution.

Examples are included to show the scattering patterns of circular and semicircular dielectric cylindrical shells and plane dielectric slabs of finite thickness and width. Results are illustrated for inhomogeneous and dissipative dielectric slabs as well as their simpler counterparts.

The technique can be extended to apply to incident waves having arbitrary polarization, to dielectric bodies having a permeability differing from that of free space, and even to three-dimensional scattering problems. There will, however, be a corresponding increase in the complexity and the computational costs.

This integral-equation technique is applicable primarily to problems in which the cross section area of the dielectric cylinder is not too large. The dielectric cross section is divided into cells which are square or nearly so. For accurate results, the edge dimensions of each cell should not exceed $0.2/\sqrt{\epsilon_r}$ wavelengths. If, for example, the dielectric constant is 4.0, a solution can be obtained at reasonable cost for a dielectric shell of thickness up to 0.1 wavelength and width up to 10 wavelengths. This will involve the solution of 100 linear equations, which is not an unreasonable number. By employing interpolation formulas, it is possible to handle a shell of thickness 0.1 wavelength and width up to 40 wavelengths by solving a set of 100 linear equations. The computation time is determined by the cross section area of the dielectric cylinder. It is believed that a cross section area up to four square wavelengths can be handled at reasonable cost and without difficulty by this technique with the aid of a computer such as the IBM 7094.

References

[1] Kouyoumjian, R. G., L. Peters, Jr., and D. T. Thomas, A modified geometrical optics method for scattering by dielectric bodies, *IEEE Trans. on Antennas and Propagation*, vol AP-11, Nov 1963, pp 690–703

[2] Rhodes, D. R., On the theory of scattering by dielectric bodies, Rept. 475-1, Antenna Lab., The Ohio State University, Columbus, Jul 1953.

[3] Andreasen, M. G., Back-scattering cross section of a thin, dielectric, spherical shell, *IRE Trans. on Antennas and Propagation*, vol AP-5, Jul 1957, pp 267–270.

[4] Stickler, D. C., Electromagnetic diffraction by dielectric strips, *IRE Trans. on Antennas and Propagation*, vol AP-6, Jan 1958, pp 148–151.

[5] Philipson, L. L., An analytical study of scattering by thin dielectric rings, *ibid.*, pp 3–8.

[6] Cohen, M. H., Application of the reaction concept to scattering problems, *IRE Trans. on Antennas and Propagation*, vol AP-3, Oct 1955, pp 193–199.

[7] Harrington, R. F., Time-Harmonic Electromagnetic Fields. New York: McGraw-Hill, 1961, pp 358–359.

Matrix Methods for Field Problems

ROGER F. HARRINGTON, SENIOR MEMBER, IEEE

Abstract—**A unified treatment of matrix methods useful for field problems is given. The basic mathematical concept is the method of moments, by which the functional equations of field theory are reduced to matrix equations. Several examples of engineering interest are included to illustrate the procedure. The problem of radiation and scattering by wire objects of arbitrary shape is treated in detail, and illustrative computations are given for linear wires. The wire object is represented by an admittance matrix, and excitation of the object by a voltage matrix. The current on the wire object is given by the product of the admittance matrix with the voltage matrix. Computation of a field quantity corresponds to multiplication of the current matrix by a measurement matrix. These concepts can be generalized to apply to objects of arbitrary geometry and arbitrary material.**

I. Introduction

THE USE of high-speed digital computers not only allows one to make more computations than ever before, it makes practicable methods too repetitious for hand computation. In the past much effort was expended to analytically manipulate solutions into a form which minimized the computational effort. It is now often more convenient to use computer time to reduce the analytical effort. Almost any linear problem of analysis can be solved to some degree of approximation, depending upon the ingenuity and effort expended. In other words, the methods are known, but much work remains to be done on the details.

It is the purpose of this paper to give a brief discussion of a general procedure for solving linear field problems, and to apply it to some examples of engineering interest. The procedure is called a *matrix method* because it reduces the original functional equation to a matrix equation. The name *method of moments* has been given to the mathematical procedure for obtaining the matrix equations. Sometimes the procedure is called an approximation technique, but this is a misnomer when the solution converges in the limit. It is only the computational time for a given accuracy which differs from other solutions, as, for example, an infinite power series. Of course, the method can also be used for truly approximate solutions, that is, ones which do not converge in the limit.

The mathematical concepts are conveniently discussed in the language of linear spaces and operators. However, an attempt has been made to minimize the use of this language, so that readers unfamiliar with it may better follow the discussion. Those concepts which are used are defined as they are introduced. Detailed expositions of linear spaces and operators may be found in many textbooks [1]–[3].

In this paper, only equations of the inhomogeneous type

$$L(f) = g \tag{1}$$

will be considered. Here L is a *linear operator*, g is the *excitation* or *source* (known function), and f is the *field* or *response* (unknown to be determined). The problem is said to be *deterministic* if the solution is unique, that is, if only one f is associated with each g. The problem of *analysis* involves determining f when L and g are given. The problem of *synthesis* involves determining L when f and g are specified. This paper deals only with analysis.

The method of moments gives a general procedure for treating field problems, but the details of solution vary widely with the particular problem. The examples of this paper have been chosen not only because they illustrate these details, but also because they are problems of engineering interest. It is hoped that these examples will allow the reader to solve similar problems, and also will suggest extensions and modifications suitable for other types of problems. While the examples are all taken from electromagnetic theory, the procedures apply to field problems of all kinds.

II. Formulation of Problems

Given a deterministic problem of the form (1), it is desired to identify the operator L, its domain (the functions f on which it operates), and its range (the functions g resulting from the operation). Furthermore, one usually needs an *inner product* $\langle f, g\rangle$, which is a scalar defined to satisfy[1]

$$\langle f, g\rangle = \langle g, f\rangle \tag{2}$$

$$\langle \alpha f + \beta g, h\rangle = \alpha\langle f, h\rangle + \beta\langle g, h\rangle \tag{3}$$

$$\begin{aligned} \langle f^*, f\rangle &> 0, \quad \text{if } f \neq 0 \\ &= 0, \quad \text{if } f = 0 \end{aligned} \tag{4}$$

where α and β are scalars, and * denotes complex conjugate. The *norm* of a function is denoted $\|f\|$ and defined by

$$\|f\| = \sqrt{\langle f, f^*\rangle}. \tag{5}$$

It corresponds to the Euclidean vector concept of length. The *metric* d of two functions is

$$d(f, g) = \|f - g\| \tag{6}$$

Manuscript received September 15, 1966. This invited paper is one of a series planned on topics of general interest.—*The Editor*. This work was supported partly by Contract AF 30(602)-3724 from the Rome Air Development Center, Griffiss Air Force Base, N. Y., and partly by the National Science Foundation under Grant GK-704.

The author is with the Department of Electrical Engineering, Syracuse University, Syracuse, N. Y.

[1] The usual definition of inner product in Hilbert space corresponds to $\langle f^*, g\rangle$ in our notation. For this paper it is more convenient to show the conjugate operation explicitly wherever it occurs, and to define the adjoint operator without conjugation.

and corresponds to the Euclidean vector concept of distance between two points. It is important for discussing the convergence of solutions.

Properties of the solution of (1) depend on properties of the operator L. The *adjoint operator* L^a and its domain are defined by

$$\langle Lf, g\rangle = \langle f, L^a g\rangle \tag{7}$$

for all f in the domain of L. An operator is *self adjoint* if $L^a = L$ and the domain of L^a is that of L. An operator is *real* if Lf is real whenever f is real. An operator is *positive definite* if

$$\langle f^*, Lf\rangle > 0 \tag{8}$$

for all $f \neq 0$ in its domain. It is *positive semidefinite* if $>$ is replaced by $\geq$ in (8), *negative definite* if $>$ is replaced by $<$ in (8), etc. Other properties of operators will be identified as they are needed.

If the solution to $L(f) = g$ exists and is unique for all g, then the *inverse operator* L^{-1} exists such that

$$f = L^{-1}(g). \tag{9}$$

If g is known, then (9) represents the solution to the original problem. However, (9) is itself an inhomogeneous equation for g if f is known, and its solution is $L(f) = g$. Hence, L and L^{-1} form a pair of operators, each of which is the inverse of the other.

Facility in formulating problems using the concepts of linear spaces comes only with practice, which will be provided by the examples in later sections. For the present, a simple abstract example will be considered, so that mathematical concepts may be illustrated without bringing physical concepts into the picture.

Example: Given $g(x)$, find $f(x)$ in the interval $0 \leq x \leq 1$ satisfying

$$-\frac{d^2 f}{dx^2} = g(x) \tag{10}$$

and

$$f(0) = f(1) = 0. \tag{11}$$

This is a boundary value problem for which

$$L = -\frac{d^2}{dx^2}. \tag{12}$$

The range of L is the space of all functions g in the interval $0 \leq x \leq 1$ which are being considered. The domain of L is the space of those functions f in the interval $0 \leq x \leq 1$, satisfying the boundary conditions (11), and having second derivatives in the range of L. The solution to (10) is not unique unless appropriate boundary conditions are included. In other words, both the differential operator and its domain are required to define the operator.

A suitable inner product for this problem is

$$\langle f, g\rangle = \int_0^1 f(x)g(x)\,dx. \tag{13}$$

It is easily shown that (13) satisfies the postulates (2) to (4), as required. Note that the definition (13) is not unique. For example,

$$\int_0^1 w(x)f(x)g(x)\,dx \tag{14}$$

where $w(x) > 0$ is an arbitrary weighting function, is also an acceptable inner product. However, the adjoint operator depends on the inner product, and it can often be chosen to make the operator self adjoint.

To find the adjoint of a differential operator, form the left-hand side of (7), and integrate by parts to obtain the right-hand side. For the present problem

$$\begin{aligned}\langle Lf, g\rangle &= \int_0^1 \left(-\frac{d^2 f}{dx^2}\right) g\,dx \\ &= \int_0^1 \frac{df}{dx}\frac{dg}{dx}\,dx - \left[\frac{df}{dx}g\right]_0^1 \\ &= \int_0^1 f\left(-\frac{d^2 g}{dx^2}\right)dx + \left[f\frac{dg}{dx} - g\frac{df}{dx}\right]_0^1.\end{aligned} \tag{15}$$

The last terms are boundary terms, and the domain of L^a may be chosen so that these vanish. The first boundary terms vanish by (11), and the second vanish if

$$g(0) = g(1) = 0. \tag{16}$$

It is then evident that the adjoint operator to (12) for the inner product (13) is

$$L^a = L = -\frac{d^2}{dx^2}. \tag{17}$$

Since $L^a = L$ and the domain of L^a is the same as that of L, the operator is self adjoint.

It is also evident that L is a real operator, since Lf is real when f is real. That L is a positive definite operator is shown from (8) as follows:

$$\begin{aligned}\langle f^*, Lf\rangle &= \int_0^1 f^*\left(-\frac{d^2 f}{dx^2}\right)dx \\ &= \int_0^1 \frac{df^*}{dx}\frac{df}{dx}\,dx - \left[f^*\frac{df}{dx}\right]_0^1 \\ &= \int_0^1 \left|\frac{df}{dx}\right|^2 dx.\end{aligned} \tag{18}$$

Note that L is a positive definite operator even if f is complex.

The inverse operator to L can be obtained by standard Green's function techniques.[2] It is

$$L^{-1}(g) = \int_0^1 G(x, x')g(x')\,dx' \tag{19}$$

where G is the Green's function

$$G(x, x') = \begin{cases} x(1 - x'), & x < x' \\ (1 - x)x', & x > x' \end{cases}. \tag{20}$$

[2] See, for example, Friedman [2], ch. 3.

One can verify that (19) is the inverse operator by forming $f=L^{-1}(g)$, differentiating twice, and obtaining (10). Note that no boundary conditions are needed on the domain of L^{-1}, which is characteristic of most integral operators. That L^{-1} is self adjoint follows from the proof that L is self adjoint, since

$$\langle Lf_1, f_2\rangle = \langle g_1, L^{-1}g_2\rangle. \tag{21}$$

Of course, the self-adjointness of L^{-1} can also be proved directly. It similarly follows that L^{-1} is positive definite whenever L is positive definite, and vice versa.

III. Method of Moments

A general procedure for solving linear equations is the *method of moments* [4]. Consider the deterministic equation

$$L(f) = g \tag{22}$$

where L is a linear operator, g is known, and f is to be determined. Let f be expanded in a series of functions $f_1, f_2, f_3, \cdots$ in the domain of L, as

$$f = \sum_n \alpha_n f_n \tag{23}$$

where the α_n are constants. The f_n are called *expansion functions* or *basis functions*. For exact solutions, (23) is usually an infinite summation and the f_n form a complete set of basis functions. For approximate solutions, (23) is usually a finite summation. Substituting (23) into (22), and using the linearity of L, one has

$$\sum_n \alpha_n L(f_n) = g. \tag{24}$$

It is assumed that a suitable inner product $\langle f, g\rangle$ has been determined for the problem. Now define a set of *weighting functions*, or *testing functions*, $w_1, w_2, w_3, \cdots$ in the range of L, and take the inner product of (24) with each w_m. The result is

$$\sum_n \alpha_n \langle w_m, Lf_n\rangle = \langle w_m, g\rangle \tag{25}$$

$m=1, 2, 3, \cdots$. This set of equations can be written in matrix form as

$$[l_{mn}][\alpha_n] = [g_m] \tag{26}$$

where

$$[l_{mn}] = \begin{bmatrix} \langle w_1, Lf_1\rangle & \langle w_1, Lf_2\rangle & \cdots \\ \langle w_2, Lf_1\rangle & \langle w_2, Lf_2\rangle & \cdots \\ \cdots & \cdots & \cdots \end{bmatrix} \tag{27}$$

$$[\alpha_n] = \begin{bmatrix} \alpha_1 \\ \alpha_2 \\ \vdots \end{bmatrix} \qquad [g_m] = \begin{bmatrix} \langle w_1, g\rangle \\ \langle w_2, g\rangle \\ \vdots \end{bmatrix}. \tag{28}$$

If the matrix $[l]$ is nonsingular its inverse $[l^{-1}]$ exists. The α_n are then given by

$$[\alpha_n] = [l_{nm}^{-1}][g_m] \tag{29}$$

and the solution for f is given by (23). For concise expression of this result, define the matrix of functions

$$[\tilde{f}] = [f_1 \;\; f_2 \;\; f_3 \;\; \cdots] \tag{30}$$

and write

$$f = [\tilde{f}_n][\alpha_n] = [\tilde{f}_n][l_{nm}^{-1}][g_m]. \tag{31}$$

This solution may be exact or approximate, depending upon the choice of the f_n and w_n. The particular choice $w_n=f_n$ is known as *Galerkin's method* [5], [6].

If the matrix $[l]$ is of infinite order, it can be inverted only in special cases, for example, if it is diagonal. The classical eigenfunction method leads to a diagonal matrix, and can be thought of as a special case of the method of moments. If the sets f_n and w_n are finite, the matrix is of finite order, and can be inverted by known computational algorithms.

One of the main tasks in any particular problem is the choice of the f_n and w_n. The f_n should be linearly independent and chosen so that some superposition (23) can approximate f reasonably well. The w_n should also be linearly independent and chosen so that the products $\langle w_n, g\rangle$ depend on relatively independent properties of g. Some additional factors which affect the choice of f_n and w_n are a) the accuracy of solution desired, b) the ease of evaluation of the matrix elements, c) the size of the matrix that can be inverted, and d) the realization of a well-conditioned matrix $[l]$.

Example: Consider again the problem stated by (10) and (11). For a power-series solution, choose

$$f_n = x^{n+1} - x \tag{32}$$

$n=1, 2, 3, \cdots, N$, so that the series (23) is

$$f = \sum_{n=1}^{N} \alpha_n(x^{n+1} - x). \tag{33}$$

Note that the term $-x$ is needed in (34), else the f_n will not be in the domain of L, that is, the boundary conditions will not be satisfied. For testing functions, choose

$$w_n = f_n = x^{n+1} - x \tag{34}$$

in which case the method is that of Galerkin. In Section V it is shown that the w_n should be in the domain of the adjoint operator. Since L is self adjoint for this problem, the w_n should be in the domain of L, as are those of (34). Evaluation of the matrix (27) for the inner product (13) and L given by (12) is straightforward. The resultant elements are

$$l_{mn} = \langle w_m, Lf_n\rangle = \frac{mn}{m+n+1}. \tag{35}$$

A knowledge of the matrix elements (35) is fully equivalent to the original differential equation. Hence, a matrix formulation for the problem has been obtained. For any particular excitation g, the matrix excitation $[g_m]$ has elements given by

$$g_m = \langle w_m, g\rangle = \int_0^1 (x^{m+1} - x)g(x)\,dx \tag{36}$$

and a solution to the boundary value problem is given by (31). This solution is a power series, exact if f can be expressed as a power series. In general, it is an infinite power-series solution, in which case a finite number of terms gives an approximate solution. The nature of the approximation is discussed in Section V.

IV. Special Techniques

As long as the operator equation is simple, application of the method of moments gives solutions in a straightforward manner. However, most field problems of engineering interest are not so simple. The physical problem may be represented by many different operator equations, and a suitable one must be chosen. Even then the form of L may be very complicated. There are an infinite number of sets of expansion functions f_n and testing functions w_n that may be chosen. Finally, there are mathematical approximations that can be made in the evaluation of the matrix elements of l_{mn} and g_m. In this section a number of special techniques, helpful for overcoming some of these difficulties, will be discussed in general terms. Some of these concepts will be used in the electromagnetic field problems considered later.

Point-Matching: The integration involved in evaluating the $l_{mn}=\langle w_m, Lf_n\rangle$ of (27) is often difficult to perform in problems of practical interest. A simple way to obtain approximate solutions is to require that (24) be satisfied at discrete points in the region of interest. This procedure is called a *point-matching method.* In terms of the method of moments, it is equivalent to using Dirac delta functions as testing functions.

Subsectional Bases: Another approximation useful for practical problems is the *method of subsections.* This involves the use of basis functions f_n each of which exists only over subsections of the domain of f. Then each α_n of the expansion (23) affects the approximation of f only over a subsection of the region of interest. This procedure often simplifies the evaluation and/or the form of the matrix $[l]$. Sometimes it is convenient to use the point-matching method of the preceding section in conjunction with the subsection method.

Extended Operators: As noted earlier, an operator is defined by an operation (for example, $L=-d^2/dx^2$) plus a domain (space of functions to which the operation may be applied). We can *extend the domain* of an operator by redefining the operation to apply to new functions (not in the original domain) so long as this extended operation does not change the original operation in its domain. If the original operator is self adjoint, it is desirable to make the extended operator also self adjoint. By this procedure we can use a wider class of functions for solution by the method of moments. This becomes particularly important in multivariable problems (fields in multidimensional space) where it is not always easy to find simple functions in the domain of the original operator.

Approximate Operators: In complex problems it is sometimes convenient to approximate the operator to obtain solutions. For differential operators, the finite difference approximation has been widely used [7]. For integral operators, an approximate operator can be obtained by approximating the kernel of the integral operator [5]. Any method whereby a functional equation is reduced to a matrix equation can be interpreted in terms of the method of moments. Hence, for any matrix solution using approximation of the operator there will be a corresponding moment solution using approximation of the function.

Perturbation Solutions: Sometimes the problem under consideration is only slightly different (perturbed) from a problem which can be solved exactly (the unperturbed problem). A first-order solution to the perturbed problem can then be obtained by using the solution to the unperturbed problem as a basis for the method of moments. This procedure is called a *perturbation method.* Higher-order perturbation solutions can be obtained by using the unperturbed solution plus correction terms in the method of moments. Sometimes this is done as successive approximations by including one correction term at a time, but for machine computations it is usually easier to include all correction terms at once.

V. Variational Interpretation

It is known that Galerkin's method ($w_n=f_n$) is equivalent to the Rayleigh-Ritz variational method [5], [6]. The method of moments is also equivalent to the variational method, the proof being essentially the same as that for Galerkin's method. The application of these techniques to electromagnetic field problems is known as the reaction concept [8], [9].

An interpretation of the method of moments in terms of linear spaces will first be given. Let $\mathcal{S}(Lf)$ denote the range of L, $\mathcal{S}(Lf_n)$ denote the space spanned by the Lf_n, and $\mathcal{S}(w_n)$ denote the space spanned by the w_n. The method of moments (25) then equates the projection of Lf onto $\mathcal{S}(w_n)$ to the projection of the approximate Lf onto $\mathcal{S}(w_n)$. In other words, both the approximate Lf and the exact Lf have equal components in $\mathcal{S}(w_n)$. The difference between the approximate Lf and the exact Lf is the error, which is orthogonal to $\mathcal{S}(w_n)$. Because of this orthogonality, a first-order change in the projection produces only a second-order change in the error. In Galerkin's method, $\mathcal{S}(w_n)=\mathcal{S}(f_n)$, and the distance from the approximate Lf to the exact Lf is minimized. In general, the method of moments does not minimize the distance from the approximate f to the exact f, although it may in some special cases.

The variational approach to the same problem is as follows. Given an operator equation $Lf=g$, it is desired to determine a functional of f (number depending on f)

$$\rho(f)=\langle f, h\rangle \tag{37}$$

where h is a given function. If h is a continuous function, then $\rho(f)$ is a *continuous linear functional.* Now let L^a be the adjoint operator to L and define an adjoint function f^a (adjoint field) by

$$L^a f^a = h. \tag{38}$$

By the calculus of variations, it can then be shown that [6]

$$\rho = \frac{\langle f, h\rangle\langle f^a, g\rangle}{\langle Lf, f^a\rangle} \tag{39}$$

is a variational formula for ρ with stationary point (37) when f is the solution of $Lf=g$ and f^a the solution to (38). For an approximate evaluation of ρ, let

$$f = \sum_n \alpha_n f_n \qquad f^a = \sum_m \beta_m w_m. \tag{40}$$

Substitute these into (39), and apply the Rayleigh-Ritz conditions $\partial\rho/\partial\alpha_i = \partial\rho/\partial\beta_i = 0$ for all i. The result is that the necessary and sufficient conditions for ρ to be a stationary point are (25), [6]. Hence, the method of moments is identical to the Rayleigh-Ritz variational method. Sometimes the method of moments is called a *direct method*, in contrast to variational approaches which are often rather circuitous.

The above variational interpretation can be used to give additional insight in how to choose the testing functions. It is evident from (38) and (40) that the w_n should be chosen so that some linear combination of them can closely represent the adjoint field f^a. When we calculate f itself by the method of moments, h of (37) is a Dirac delta function, ρ of (37) is no longer a continuous linear functional, and f^a of (38) is a Green's function. This implies that some combination of the w_n must be able to approximate the Green's function. Since a Green's function is usually poorly behaved, one should expect computation of a field by the method of moments to converge less slowly than computation of a continuous linear functional. This is found to be the case.

VI. Electrostatics

This section is a general discussion of electrostatic problems according to the operational formulation. The static electric intensity $\boldsymbol{E}$ is conveniently found from an electrostatic potential ϕ according to

$$\boldsymbol{E} = -\nabla\phi \tag{41}$$

where ∇ is the gradient operator. In a region of constant permittivity ε and volume change density ρ, the electrostatic potential satisfies the *Poisson equation*

$$-\varepsilon\nabla^2\phi = \rho \tag{42}$$

where ∇^2 is the Laplacian operator. For unique solutions, boundary conditions on ϕ are needed. In other words, the domain of the operator must be specified.

For now, consider fields from charges in unbounded space, in which case

$$r\phi \to \text{constant as } r \to \infty \tag{43}$$

for every ρ of finite extent, where r is the distance from the coordinate origin. The differential operator formulation is therefore

$$L\phi = \rho \tag{44}$$

where

$$L = -\varepsilon\nabla^2 \tag{45}$$

and the domain of L is those functions ϕ whose Laplacian exists and which have $r\phi$ bounded at infinity according to (43). The well-known solution to this problem is

$$\phi(x, y, z) = \iiint \frac{\rho(x', y', z')}{4\pi\varepsilon R}\, dx'dy'dz' \tag{46}$$

where $R=\sqrt{(x-x')^2+(y-y')^2+(z-z')^2}$ is the distance from a source point (x', y', z') to a field point (x, y, z). Hence, the inverse operator to L is

$$L^{-1} = \iiint dx'dy'dz' \frac{1}{4\pi\varepsilon R}. \tag{47}$$

It is important to keep in mind that (47) is inverse to (45) only for the boundary conditions (43). If the boundary conditions are changed, L^{-1} changes. Also, the designation of (45) as L and (47) as L^{-1} is arbitrary, and the notation could be reversed if desired.

A suitable inner product for electrostatic problems is

$$\langle\phi, \psi\rangle = \iiint \phi(x, y, z)\psi(x, y, z)\, dxdydz \tag{48}$$

where the integration is over all space. That (48) satisfies the required postulates (2), (3), and (4) is easily verified. It will now be shown that L is self adjoint for this inner product. From the left-hand side of (7)

$$\langle L\phi, \psi\rangle = \iiint (-\varepsilon\nabla^2\phi)\psi\, d\tau \tag{49}$$

where $d\tau = dxdydz$. Green's identity is

$$\iiint_V (\psi\nabla^2\phi - \phi\nabla^2\psi)\, d\tau = \oint\!\!\oint_S \left(\psi\frac{\partial\phi}{\partial n} - \phi\frac{\partial\psi}{\partial n}\right) ds \tag{50}$$

where S is the surface bounding the volume V and n is the outward direction normal to S. Let S be a sphere of radius r, so that in the limit $r\to\infty$ the volume V includes all space. For ϕ and ψ satisfying boundary conditions (43), $\psi\to C_1/r$, and $\partial\phi/\partial n \to C_2/r^2$ as $r\to\infty$. Hence, $\psi\partial\phi/\partial n \to C/r^3$ as $r\to\infty$, and similarly for $\phi\partial\psi/\partial n$. Since $ds = r^2 \sin\theta\, d\theta d\phi$ increases only as r^2, the right-hand side of (50) vanishes as $r\to\infty$. Equation (50) then reduces to

$$\iiint \psi\nabla^2\phi\, d\tau = \iiint \phi\nabla^2\psi\, d\tau \tag{51}$$

from which it is evident that the adjoint operator L^a is

$$L^a = L = -\varepsilon\nabla^2. \tag{52}$$

Since the domain of L^a is that of L, the operator L is self adjoint. The concept of self adjointness in this case is related to the physical concept of reciprocity.

It is evident from (45) and (47) that L and L^{-1} are real operators. It will now be shown that they are also positive definite, that is, they satisfy (8). As discussed in Section II, this need be shown only for L or L^{-1}. For L, form

$$\langle \phi^*, L\phi \rangle = \iiint \phi^*(-\varepsilon \nabla^2 \phi)\, d\tau \tag{53}$$

and use the vector identity $\phi\nabla^2\phi = \nabla \cdot (\phi \nabla \phi) - \nabla\phi \cdot \nabla\phi$ plus the divergence theorem. The result is

$$\langle \phi^*, L\phi \rangle = \iiint_V \varepsilon \nabla\phi^* \cdot \nabla\phi\, d\tau - \oiint_S \varepsilon \phi^* \nabla\phi \cdot ds \tag{54}$$

where S bounds V. Again take S a sphere of radius r. For ϕ satisfying (43), the last term of (54) vanishes as $r \to \infty$ for the same reasons as in (50). Then

$$\langle \phi^*, L\phi \rangle = \iiint \varepsilon |\nabla\phi|^2\, d\tau \tag{55}$$

and, for ε real and $\varepsilon > 0$, L is positive definite. In this case positive definiteness of L is related to the concept of electrostatic energy.

VII. Charged Conducting Plate

Consider a square conducting plate 2a meters on a side and lying on the $z = 0$ plane with center at the origin, as shown in Fig. 1. Let $\sigma(x, y)$ represent the surface charge density on the plate, assumed to have zero thickness. The electrostatic potential at any point in space is

$$\phi(x, y, z) = \int_{-a}^{a} dx' \int_{-a}^{a} dy' \frac{\sigma(x', y')}{4\pi\varepsilon R} \tag{56}$$

where $R = \sqrt{(x - x')^2 + (y - y')^2 + z^2}$. The boundary condition is $\phi = V$ (constant) on the plate. The integral equation for the problem is therefore

$$V = \int_{-a}^{a} dx' \int_{-a}^{a} dy' \frac{\sigma(x', y')}{4\pi\varepsilon\sqrt{(x - x')^2 + (y - y')^2}} \tag{57}$$

$|x| < a$, $|y| < a$. The unknown to be determined is the charge density $\sigma(x, y)$. A parameter of interest is the capacitance of the plate

$$C = \frac{q}{V} = \frac{1}{V} \int_{-a}^{a} dx \int_{-a}^{a} dy\, \sigma(x, y) \tag{58}$$

which is continuous linear functional of σ.

A straightforward development of a subsection and point-matching solution [10] will first be given, and later it will be interpreted in terms of more general concepts. Consider the plate divided into N square subsections, as shown in Fig. 1. Define functions

$$f_n = \begin{cases} 1 & \text{on } \Delta s_n \\ 0 & \text{on all other } \Delta s_m \end{cases} \tag{59}$$

and let the charge density be represented by

$$\sigma(x, y) \approx \sum_{n=1}^{N} \alpha_n f_n. \tag{60}$$

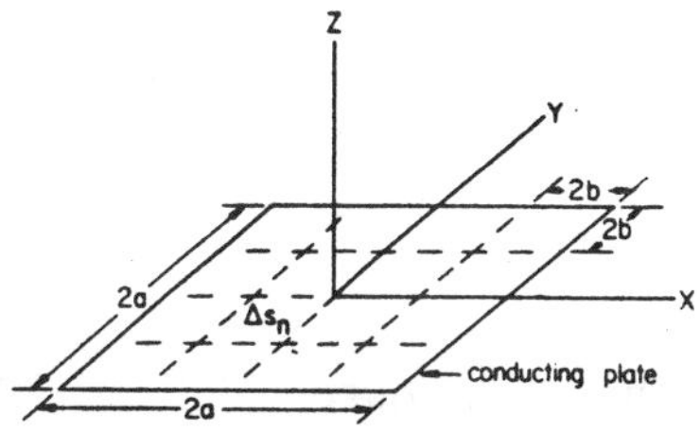

Fig. 1. A square conducting plate.

Substituting (60) into (57) and satisfying the resultant equation at the midpoint (x_m, y_m) of each Δs_m, one obtains the set of equations

$$V = \sum_{n=1}^{N} l_{mn} \alpha_n, \qquad m = 1, 2, \cdots, N \tag{61}$$

where

$$l_{mn} = \int_{\Delta x_n} dx' \int_{\Delta y_n} dy' \frac{1}{4\pi\varepsilon\sqrt{(x_m - x')^2 + (y_m - y')^2}}. \tag{62}$$

Note that l_{mn} is the potential at the center of Δs_m due to a uniform charge density of unit amplitude over Δs_n. A solution to the set (61) gives the α_m, in terms of which the charge density is approximated by (60). The corresponding capacitance of the plate, approximating (58), is

$$C \approx \frac{1}{V} \sum_{n=1}^{N} \alpha_n \Delta s_n = \sum_{mn} l_{nm}^{-1} \Delta s_n. \tag{63}$$

This result can be interpreted as stating that the capacitance of an object is the sum of the capacitances of all its subsections plus the mutual capacitances between every pair of subsections.

To translate the above results into the language of linear spaces and the method of moments, let

$$f(x, y) = \sigma(x, y) \tag{64}$$

$$g(x, y) = V, \quad |x| < a, \quad |y| < a \tag{65}$$

$$L(f) = \int_{-a}^{a} dx' \int_{-a}^{a} dy' \frac{f(x', y')}{4\pi\varepsilon\sqrt{(x - x')^2 + (y - y')^2}}. \tag{66}$$

Then $L(f) = g$ is equivalent to (57). A suitable inner product, satisfying (2) to (4), for which L is self adjoint, is

$$\langle f, g \rangle = \int_{-a}^{a} dx \int_{-a}^{a} dy\, f(x, y)\, g(x, y). \tag{67}$$

To apply the method of moments, use the function (59) as a subsectional basis, and define testing functions as

$$w_m = \delta(x - x_m)\delta(y - y_m) \tag{68}$$

which is the two-dimensional Dirac delta function. Now the elements of the $[l]$ matrix (27) are those of (62) and the $[g]$ matrix of (28) is

$$[g_m] = \begin{bmatrix} V \\ V \\ \vdots \\ V \end{bmatrix}. \tag{69}$$

The matrix equation (26) is, of course, identical to the set of equations (61). In terms of the inner product (67), the capacitance (58) can be written as

$$C = \frac{\langle \sigma, \phi \rangle}{V^2} \tag{70}$$

since $\phi = V$ on the plate. Equation (70) is the conventional stationary formula for the capacitance of a conducting body [11].

For numerical results, the l_{mn} of (62) must be evaluated. Let $2b = 2a/\sqrt{N}$ denote the side length of each Δs_n. The potential at the center of Δs_n due to unit charge density over its own surface is

$$\begin{aligned} l_{nn} &= \int_{-b}^{b} dx \int_{-b}^{b} dy \frac{1}{4\pi\varepsilon\sqrt{x^2+y^2}} \\ &= \frac{2b}{\pi\varepsilon} \ln(1+\sqrt{2}) = \frac{2b}{\pi\varepsilon}(0.8814). \end{aligned} \tag{71}$$

This derivation uses Dwight 200.01 and 731.2 [12]. The potential at the center of Δs_m due to unit charge over Δs_n can be similarly evaluated, but the formula is complicated. For most purposes it is sufficiently accurate to treat the charge on Δs_n as if it were a point charge, and use

$$l_{mn} \approx \frac{\Delta s_n}{4\pi\varepsilon R_{mn}} = \frac{b^2}{\pi\varepsilon\sqrt{(x_m-x_n)^2+(y_m-y_n)^2}} \quad m \neq n. \tag{72}$$

This approximation is 3.8 percent in error for adjacent subsections, and has less error for nonadjacent ones. Table I shows capacitance, calculated by (63) using the α's obtained from the solution of (61), for various numbers of subareas. The second column of Table I uses the approximation (72), the third column uses an exact evaluation of the l_{mn}. A good estimate of the true capacitance is 40 picofarads. Figure 2 shows a plot of the approximate charge density along the subareas nearest the center line of the plate, for the case $N = 100$ subareas. Note that σ exhibits the well-known square root singularity at the edges of the plate.

TABLE I

CAPACITANCE OF A SQUARE PLATE (PICOFARADS PER METER)

No. of subareas	$C/2a$ approx. l_{mn}	$C/2a$ exact l_{mn}
1	31.5	31.5
9	37.3	36.8
16	38.2	37.7
36	39.2	38.7
100		39.5

VIII. ELECTROMAGNETIC FIELDS

The operator formulation of electromagnetic fields is analogous to that of electrostatic fields, but considerably more complicated. For the time-harmonic case, $e^{j\omega t}$ variation, the Maxwell equations are[3]

[3] Only the case of electric sources is considered in this paper. The more general case of electric and magnetic sources is treated by the reaction concept [8], [9].

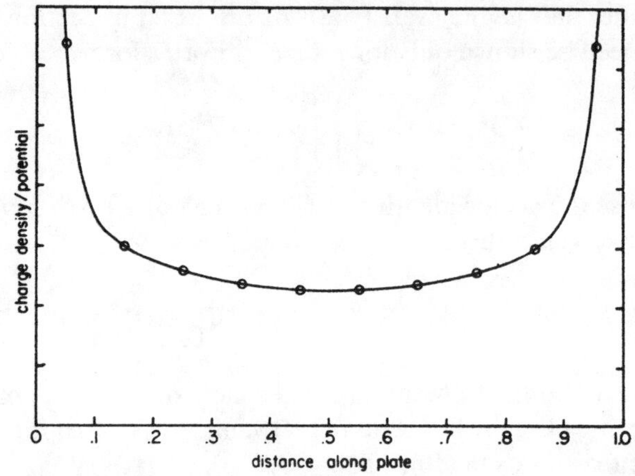

Fig. 2. Approximate charge density on subareas closest to the centerline of a square plate.

$$\begin{aligned} \nabla \times \boldsymbol{E} &= -j\omega\mu\boldsymbol{H} \\ \nabla \times \boldsymbol{H} &= j\omega\varepsilon\boldsymbol{E} + \boldsymbol{J} \end{aligned} \tag{73}$$

where $\boldsymbol{E}$ is the electric field, $\boldsymbol{H}$ the magnetic field, and $\boldsymbol{J}$ the electric current density. Equations (73) can be combined into a single equation for $\boldsymbol{E}$ as

$$\frac{-1}{j\omega} \nabla \times (\mu^{-1} \nabla \times \boldsymbol{E}) - j\omega\varepsilon\boldsymbol{E} = \boldsymbol{J}. \tag{74}$$

This is of the form

$$L(\boldsymbol{E}) = \boldsymbol{J} \tag{75}$$

where the operator L is evident from (74). For a specific case, let the permittivity and permeability be that of free space, that is, $\varepsilon = \varepsilon_0$ and $\mu = \mu_0$. The domain of L must be restricted by suitable differentiability conditions on $\boldsymbol{E}$, and boundary conditions on $\boldsymbol{E}$ must be given. To be specific, let these boundary conditions be the radiation condition, that is, the field must represent outward traveling waves at infinity.

The inverse operator is the well-known potential integral solution to (74), which is

$$\boldsymbol{E} = L^{-1}(\boldsymbol{J}) = -j\omega\boldsymbol{A} - \nabla\Phi \tag{76}$$

where

$$\boldsymbol{A}(\boldsymbol{r}) = \mu \iiint \boldsymbol{J}(\boldsymbol{r}') \frac{e^{-jk|\boldsymbol{r}-\boldsymbol{r}'|}}{4\pi|\boldsymbol{r}-\boldsymbol{r}'|} d\tau' \tag{77}$$

$$\Phi(\boldsymbol{r}) = \frac{1}{\varepsilon} \iiint \rho(\boldsymbol{r}') \frac{e^{-jk|\boldsymbol{r}-\boldsymbol{r}'|}}{4\pi|\boldsymbol{r}-\boldsymbol{r}'|} d\tau' \tag{78}$$

$$\rho = \frac{-1}{j\omega} \nabla \cdot \boldsymbol{J}. \tag{79}$$

These equations can be combined into a single equation

$$\boldsymbol{E} = L^{-1}(\boldsymbol{J}) = \iiint \Gamma(\boldsymbol{r}, \boldsymbol{r}') \cdot \boldsymbol{J}(\boldsymbol{r}') d\tau' \tag{80}$$

where Γ is the dyadic Green's function. However, the derivation of (80) involves an interchange of integration and differentiation which restricts the domain of L^{-1} more than necessary [13]. It is often better to consider (76) to (78) as the basic equations, with (80) as symbolic of them.

A suitable inner product for electromagnetic field problems is

$$\langle \mathrm{E}, \mathrm{J} \rangle = \iiint \boldsymbol{E} \cdot \boldsymbol{J}\, d\tau \tag{81}$$

which is the quantity called *reaction*. Note that (81) satisfies postulates (2), (3), and (4). The concept of reciprocity is a statement of the self-adjointness of L^{-1}, that is,

$$\langle L^{-1}J_1, J_2 \rangle = \langle J_1, L^{-1}J_2 \rangle. \tag{82}$$

The operator L is also self adjoint, since (82) can be written as

$$\langle E_1, LE_2 \rangle = \langle LE_1, E_2 \rangle. \tag{83}$$

Other properties of L can be determined as the need arises.

IX. Wires of Arbitrary Shape

An important engineering problem is the electromagnetic behavior of thin wire objects. A general analysis of such objects according to the method of moments is presented in this section. The impressed field is considered arbitrary, and hence both the antenna and scatterer problems are included in the solution. The distinction between antennas and scatterers is primarily that of the location of the source. If the source is at the object it is viewed as an antenna; if the source is distant from the object it is viewed as a scatterer.

So that the development of the solution may be easily followed, it is given with few references to the general theory. Basically, it involves a) an approximation of the exact equation for conducting bodies by an approximate equation valid for thin wires, b) replacement of the derivatives by finite difference approximations, yielding an approximate operator, c) use of pulse functions for expansion functions, to give a step approximation to the current and charge, and d) the use of point-matching for testing.

A particularly descriptive exposition of the solution can be made in terms of network parameters. To effect a solution, the wire is considered as N short segments connected together. The end points of each segment define a pair of terminals in space. These N pairs of terminals can be thought of as forming an N port network, and the wire object is obtained by short-circuiting all ports of the network. One can determine the impedance matrix for the N port network by applying a current source to each port in turn, and calculating the open circuit voltages at all ports. This procedure involves only current elements in empty space. The admittance matrix is the inverse of the impedance matrix. Once the admittance matrix is known, the port currents (current distribution on the wire) are found for any particular voltage excitation (applied field) by matrix multiplication.

An integral equation for the charge density σ_s and current $\boldsymbol{J}_s$ on a conducting body S in a known impressed field $\boldsymbol{E}^i$ is obtained as follows. The scattered field $\boldsymbol{E}^s$, produced by σ_s and $\boldsymbol{J}_s$, is expressed in terms of retarded potential in-

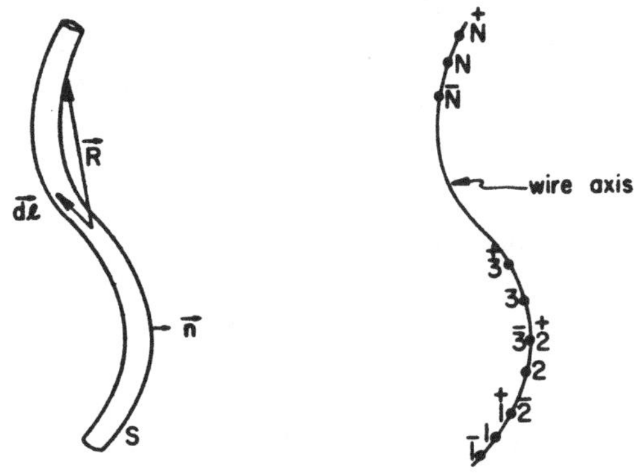

(a) (b)

Fig. 3. (a) A wire scatterer. (b) The wire axis divided into N segments.

tegrals, and the boundary condition $\boldsymbol{n} \times (\boldsymbol{E}^i + \boldsymbol{E}^s) = 0$ on S is applied. This is summarized by

$$\boldsymbol{E}^s = -j\omega \boldsymbol{A} - \nabla \phi \tag{84}$$

$$\boldsymbol{A} = \mu \oiint_S \boldsymbol{J}_s \frac{e^{-jkR}}{4\pi R}\, dS \tag{85}$$

$$\phi = \frac{1}{\varepsilon} \oiint_S \sigma_s \frac{e^{-jkR}}{4\pi R}\, dS \tag{86}$$

$$\sigma_s = \frac{-1}{j\omega} \nabla_s \cdot \boldsymbol{J}_s \tag{87}$$

$$\boldsymbol{n} \times \boldsymbol{E}^s = -\boldsymbol{n} \times \boldsymbol{E}^i \quad \text{on } S. \tag{88}$$

Figure 3(a) represents an arbitrary thin-wire scatterer, for which the following approximations are made. a) The current is assumed to flow only in the direction of the wire axis. b) The current and charge densities are approximated by filaments of current $\boldsymbol{I}$ and charge σ on the wire axis. c) The boundary condition (88) is applied only to the axial component of $\boldsymbol{E}$ at the wire surface. To this approximation, (84) to (88) become

$$-E_l^i = -j\omega A_l - \frac{\partial \phi}{\partial l} \quad \text{on } S \tag{89}$$

$$\boldsymbol{A} = \mu \int_{\text{axis}} \boldsymbol{I}(l) \frac{e^{-jkR}}{4\pi R}\, dl \tag{90}$$

$$\phi = \frac{1}{\varepsilon} \int_{\text{axis}} \sigma(l) \frac{e^{-jkR}}{4\pi R}\, dl \tag{91}$$

$$\sigma = \frac{-1}{j\omega} \frac{dI}{dl} \tag{92}$$

where l is the length variable along the wire axis, and R is measured from a source point on the axis to a field point on the wire surface.

A solution to the above equations is obtained as follows. Integrals are approximated by the sum of intergrals over N small segments, obtained by treating I and q as constant over each segment. Derivatives are approximated by finite differences over the same intervals used for integration. Figure 3(b) illustrates the division of the wire axis into N segments, and defines the notation. If a wire terminates, the boundary condition $I = 0$ is taken into account by starting the first segment 1/2 interval in from the end of the wire. This is suggested in Fig. 3(b) by the extra 1/2 interval shown at each end. The nth segment is identified by its starting point $\overset{-}{n}$, its midpoint n, and its termination $\overset{+}{n}$. An increment Δl_n denotes that between $\overset{-}{n}$ and $\overset{+}{n}$, $\Delta l_{\overset{-}{n}}$ and $\Delta l_{\overset{+}{n}}$ denote increments shifted 1/2 segment minus or plus along l. The desired approximations for (89) to (92) are then

$$-E_l^i(m) \approx -j\omega A_l(m) - \frac{\phi(\overset{+}{m}) - \phi(\overset{-}{m})}{\Delta l_m} \quad (93)$$

$$A(m) = \mu \sum_n I(n) \int_{\Delta l_n} \frac{e^{-jkR}}{4\pi R}\, dl \quad (94)$$

$$\phi(\overset{+}{m}) \approx \frac{1}{\varepsilon} \sum_n \sigma(\overset{+}{n}) \int_{\Delta l_{\overset{+}{n}}} \frac{e^{-jkR}}{4\pi R}\, dl \quad (95)$$

$$\sigma(\overset{+}{n}) \approx \frac{-1}{j\omega}\left[\frac{I(n+1) - I(n)}{\Delta l_{\overset{+}{n}}}\right] \quad (96)$$

with equations similar to (95) and (96) for $\phi(\overset{-}{m})$ and $\sigma(\overset{-}{n})$.

The σ's are given in terms of the I's by (96), and hence (93) can be written in terms of the $I(n)$ only. One can view the N equations represented by (93) as the equations for an N port network with terminal pairs $(\overset{+}{n}, \overset{-}{n})$. The voltages applied to each port are approximately $\boldsymbol{E}^i \cdot \Delta \boldsymbol{l}_n$. Hence, by defining

$$[I] = \begin{bmatrix} I(1) \\ I(2) \\ \vdots \\ I(N) \end{bmatrix} \qquad [V] = \begin{bmatrix} \boldsymbol{E}^i(1)\cdot \boldsymbol{l}_1 \\ \boldsymbol{E}^i(2)\cdot \boldsymbol{l}_2 \\ \vdots \\ \boldsymbol{E}^i(N)\cdot \boldsymbol{l}_N \end{bmatrix} \quad (97)$$

one can rewrite (93) in matrix form as

$$[V] = [Z][I]. \quad (98)$$

This corresponds to the method of moment representation (26), with $[Z]$ corresponding to $[l]$, $[V]$ to $[g]$, and $[I]$ to $[\alpha]$. The elements of the matrix $[Z]$ can be obtained by substituting (94) through (96) into (93) and rearranging into the form of (98). Alternatively, one can apply (93) through (96) to two isolated elements and obtain the impedance elements directly. This latter procedure will be used because it is somewhat easier to follow.

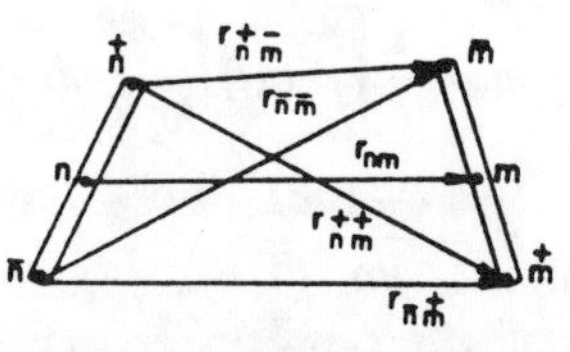

Fig. 4. Two segments of a wire scatterer.

Consider two representative elements of the wire scatterer, as shown in Fig. 4. The integrals in (94) and (95) are of the same form, and are denoted by

$$\psi(n, m) = \frac{1}{\Delta l_n} \int_{\Delta l_n} \frac{e^{-jkR_{mn}}}{4\pi R_{mn}}\, dl_n. \quad (99)$$

Symbols + and − are used over m and n when appropriate. Evaluation of the ψ in general is considered in the Appendix. Let element n of Fig. 4 consist of a current filament $I(n)$, and two charge filaments of net charge

$$q(\overset{+}{n}) = \frac{1}{j\omega} I(n) \qquad q(\overset{-}{n}) = \frac{-1}{j\omega} I(n) \quad (100)$$

where $q = \sigma \Delta l$. The vector potential at m due to $I(n)$ is, by (94),

$$\boldsymbol{A} = \mu I(n) \Delta \boldsymbol{l}_n \psi(n, m). \quad (101)$$

The scalar potentials at $\overset{+}{m}$ and $\overset{-}{m}$ due to the charges (100) are, by (95)

$$\begin{aligned} \phi(\overset{+}{m}) &= \frac{1}{j\omega\varepsilon}\left[I(n)\psi(\overset{+}{n}, \overset{+}{m}) - I(n)\psi(\overset{-}{n}, \overset{+}{m})\right] \\ \phi(\overset{-}{m}) &= \frac{1}{j\omega\varepsilon}\left[I(n)\psi(\overset{+}{n}, \overset{-}{m}) - I(n)\psi(\overset{-}{n}, \overset{-}{m})\right]. \end{aligned} \quad (102)$$

Substituting from (101) and (102) into (93), and forming $Z_{mn} = \boldsymbol{E}^i(m) \cdot \Delta \boldsymbol{l}_m / I(n)$, one obtains

$$\begin{aligned} Z_{mn} &= j\omega\mu \Delta \boldsymbol{l}_n \cdot \Delta \boldsymbol{l}_m \psi(n, m) \\ &+ \frac{1}{j\omega\varepsilon}\left[\psi(\overset{+}{n}, \overset{+}{m}) - \psi(\overset{-}{n}, \overset{+}{m}) - \psi(\overset{+}{n}, \overset{-}{m}) + \psi(\overset{-}{n}, \overset{-}{m})\right]. \end{aligned} \quad (103)$$

This result applies for self impedances $(m = n)$ as well as for mutual impedances. When the two current elements are widely separated, a simpler formula based on the radiation field from a current element can be used.

The wire object is completely characterized by its impedance matrix, subject, of course, to the approximations involved. The object is defined by $2N$ points on the wire axis, plus the wire radius. The impedance elements are calculated by (103), and the voltage matrix is determined by the impressed field, according to (97). The current at N points on the scatterer is then given by the current matrix, obtained from the inversion of (98) as

$$[I] = [Y][V] \qquad [Y] = [Z]^{-1}. \quad (104)$$

Once the current distribution is known, parameters of interest such as field patterns, input impedances, echo areas, etc., can be calculated by numerically evaluating the appropriate formulas.

X. Wire Antennas

A wire antenna is obtained when the wire is excited by a voltage source at one or more points along its length. Hence, for an antenna excited in the nth interval, the applied voltage matrix (97) is

$$[V^s] = \begin{bmatrix} 0 \\ \vdots \\ V_n \\ \vdots \\ 0 \end{bmatrix} \quad (105)$$

i.e., all elements zero except the nth, which is equal to the source voltage. The current distribution is given by (104), which for the $[V]$ of (105) becomes

$$[I] = V_n \begin{bmatrix} Y_{1n} \\ Y_{2n} \\ \vdots \\ Y_{Nn} \end{bmatrix}. \tag{106}$$

Hence, the nth column of the admittance matrix is the current distribution for a unit voltage source applied to the nth interval. Inversion of the impedance matrix therefore gives simultaneously the current distributions when the antenna is excited in any arbitrary interval along its length. The diagonal elements Y_{nn} of the admittance matrix are the input admittances of the wire object fed in the nth interval, and the Y_{mn} are the transfer admittances between a port in the mth interval and one in the nth interval.

The radiation pattern of a wire antenna is obtained by treating the antenna as an array of N current elements $I(n)\Delta l_n$. By standard formulas, the far-zone vector potential is given by

$$A = \frac{\mu e^{-jkr_0}}{4\pi r_0} \sum_n I(n)\Delta l_n e^{jkr_n \cos \xi_n} \tag{107}$$

where r_0 and r_n are the radius vectors to the distant field point and to the source points, respectively, and ξ_n are the angles between r_0 and r_n. The far-zone field components are

$$E_\theta = -j\omega A_\theta \qquad E_\phi = -j\omega A_\phi \tag{108}$$

where θ and ϕ are the conventional spherical coordinate angles.

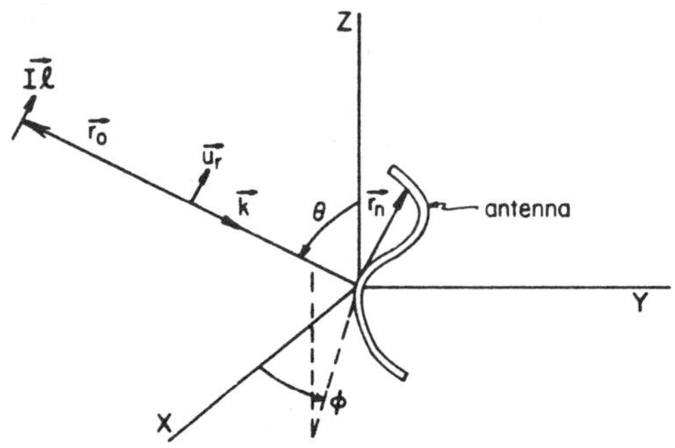

Fig. 5. A wire antenna and distant dipole.

An alternative derivation of the radiation pattern can be obtained by reciprocity. Figure 5 represents a distant current element Il_r (subscripts r denote "receiver"), adjusted to produce the unit plane wave

$$E^r = u_r e^{-jk_r \cdot r_n} \tag{109}$$

in the vicinity of the antenna. Here u_r is a unit vector specifying the polarization of the wave, k_r is a wave number vector pointing in the direction of travel of the wave, and r_n is the radius vector to a point n on the antenna. By reciprocity,

$$E_r = \frac{1}{Il} \int_{\text{antenna}} E^r \cdot I \, dl \tag{110}$$

where E_r is the u_r component of E from the antenna, and I is the current on the antenna. The constant $1/Il$ is that needed to produce a plane wave of unit amplitude at the origin, which is

$$\frac{1}{Il} = \frac{\omega\mu e^{-jkr_0}}{j4\pi r_0}. \tag{111}$$

A numerical approximation to (110) is obtained by defining a voltage matrix

$$[V^r] = \begin{bmatrix} E^r(1) \cdot \Delta l_1 \\ E^r(2) \cdot \Delta l_2 \\ \vdots \\ E^r(N) \cdot \Delta l_N \end{bmatrix} \tag{112}$$

where E^r is given by (109), and expressing (110) as the matrix product

$$E_r = \frac{\omega\mu e^{-jkr_0}}{j4\pi r_0} [\tilde{V}^r][I] = \frac{\omega\mu e^{-jkr_0}}{j4\pi r_0} [\tilde{V}^r][Y][V^s] \tag{113}$$

where $[\tilde{V}]$ denotes the transpose of $[V]$. Note that $[V^r]$ is the same matrix for plane-wave excitation of the wire. Equation (113) remains valid for an arbitrary excitation $[V^s]$; it is not restricted to the single source excitation (105).

The power gain pattern for the u_r component of the radiation field is given by

$$g(\theta, \phi) = \frac{4\pi r_0^2}{\eta} \frac{|E_r(\theta, \phi)|^2}{P_{\text{in}}} \tag{114}$$

where $\eta = \sqrt{\mu/\varepsilon}$ is the intrinsic impedance of space, and P_{in} is the power input to the antenna (* denotes conjugate)

$$P_{\text{in}} = \text{Re}\{[\tilde{V}^s][I^*]\} = \text{Re}\{[\tilde{V}^s][Y^*][V^{s*}]\}. \tag{115}$$

For the special case of a single source, (105), P_{in} becomes simply Re $(|V_n|^2 Y_{nn})$. Using (113) and (115) in (114), one has

$$g(\theta, \phi) = \frac{\eta k^2}{4\pi} \frac{|[\tilde{V}^r(\theta, \phi)][Y][V^s]|^2}{\text{Re}\{[\tilde{V}^s][Y^*][V^{s*}]\}} \tag{116}$$

where $[V^r(\theta, \phi)]$ is given by (112) for various angles of incidence θ, ϕ. Equation (116) gives the gain pattern for only a single polarization of the radiation field. If the total power gain pattern is desired, the g's for two orthogonal polarizations may be added together.

Computations for linear wire antennas have been made using the formulas of this section, and good results obtained. For far-field quantities, such as radiation patterns, as few as 10 segments per wavelength give accurate results. (Radiation patterns are continuous linear functionals, that is, they depend on the weighted integral of the antenna current.) For the current itself, convergence was slower. A typical result for a half-wave antenna was about four percent change in going from 20 to 40 segments, less for other lengths. Faster convergence can be obtained by going from a step approximation to a piecewise-linear approximation to the current. This modification was used for most of the computations, of which Fig. 6 is typical. It shows the input admittance to a center-fed linear antenna with length-to-diameter ratio 74.2 ($\Omega = 2 \log L/a = 10$) using 32 segments.[4]

[4] Because of the extra 1/2 interval at each wire end, this corresponds to an $N = 31$ solution.

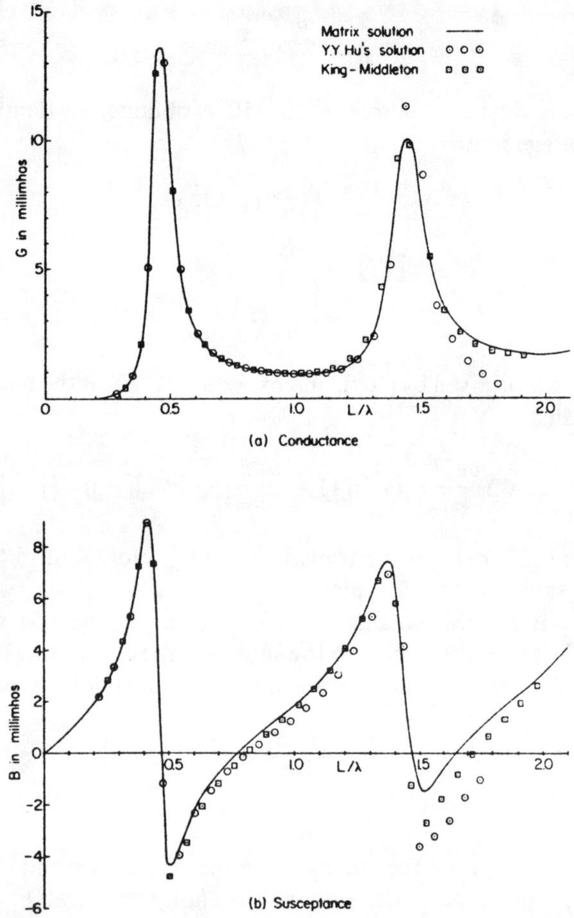

Fig. 6. Input admittance for a center-fed linear antenna of length L and diameter $L/72.4$.

For the points tested, it was almost identical to the 64 segment solution using (103). It is compared to the second-order variational solution of Y. Y. Hu [14], and to the second iteration of Hallén's equation by King and Middleton [15]. The conductances are in close agreement except for Hu's solution $L > 1.3\lambda$, in which case her trial functions are inadequate. The input susceptances are in poorer agreement, which is to be expected because each solution treats the gap differently. The matrix solution of this paper treats it as if it were one segment in length. Hu's solution contains no trial function which can support a singularity in current at the gap, hence gives a low gap capacitance. The King-Middleton method is an iterative procedure, and hence B depends on the number of iterations. Many more computations, as well as a description of the piecewise-linear modification for the current, can be found in the original report [16].

XI. Wire Scatterers

Consider now the field scattered by a wire object in a plane wave incident field. Figure 7 represents a scatterer and two distant current elements, $\boldsymbol{Il}_t$ at the transmitting point, $\boldsymbol{r}_t$, and $\boldsymbol{Il}_r$ at the receiving point $\boldsymbol{r}_r$. The $\boldsymbol{Il}_t$ is adjusted to produce a unit plane wave at the scatterer

$$\boldsymbol{E}^i = \boldsymbol{u}_t e^{-j\boldsymbol{k}_t \cdot \boldsymbol{r}_n} \tag{117}$$

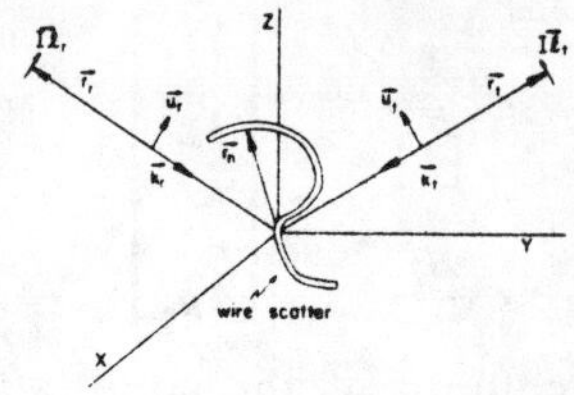

Fig. 7. Definitions for plane-wave scattering.

where the notation is analogous to that of (109). The voltage excitation matrix (97) is then

$$[V^i] = \begin{bmatrix} \boldsymbol{E}^i(1) \cdot \Delta \boldsymbol{l}_1 \\ \boldsymbol{E}^i(2) \cdot \Delta \boldsymbol{l}_2 \\ \vdots \\ \boldsymbol{E}^i(N) \cdot \Delta \boldsymbol{l}_N \end{bmatrix} \tag{118}$$

and the current $[I]$ is given by (104) with $[V] = [V^i]$. The field produced by $[I]$ can then be found by conventional techniques.

The distant scattered field can also be evaluated by reciprocity, the same in the antenna case. A dipole $\boldsymbol{Il}_r$ at the receiving point is adjusted to produce the unit plane wave (109) at the scatterer. The scattered field is then given by (113) with $[V^s]$ replaced by $[V^i]$, that is,

$$E_r = \frac{\omega \mu e^{-jkr_r}}{j4\pi r_r} [\tilde{V}^r][Y][V^i]. \tag{119}$$

A parameter of interest is the bistatic scattering cross section σ, defined as that area for which the incident wave contains sufficient power to produce the field E_r by omnidirectional radiation. In equation form, this is

$$\begin{aligned} \sigma &= 4\pi r_r^2 |E_r|^2 \\ &= \frac{\eta^2 k^2}{4\pi} |[\tilde{V}^r][Y][V^i]|^2. \end{aligned} \tag{120}$$

For the monostatic cross section, set $[V^r] = [V^i]$ in (120). The cross section depends on the polarization of the incident wave and of the receiver. A better description of the scatterer can be made in terms of a scattering matrix.

Another parameter of interest is the total scattering cross section σ_t, defined as the ratio of the total scattered power to the power density of the incident wave. The total power radiated by $[I]$ is given by (115) for any excitation; therefore the scattered power is given by (115) with $[V^s]$ replaced by $[V^i]$. The incident power density is $1/\eta$, hence

$$\sigma_t = \eta \operatorname{Re} [\tilde{V}^i][Y^*][V^{i*}]. \tag{121}$$

Note that σ_t is dependent on the polarization of the incident wave.

Computations for linear wire scatterers have been made using the same $[Y]$ matrix as for antennas. Again far-field quantities, such as echo areas, converged rapidly, with good results obtained with as few as 10 segments per wavelength. Computation of the current converged less rapidly than far-field quantities, but more rapidly than did computation of the current on antennas. This is because the impressed field

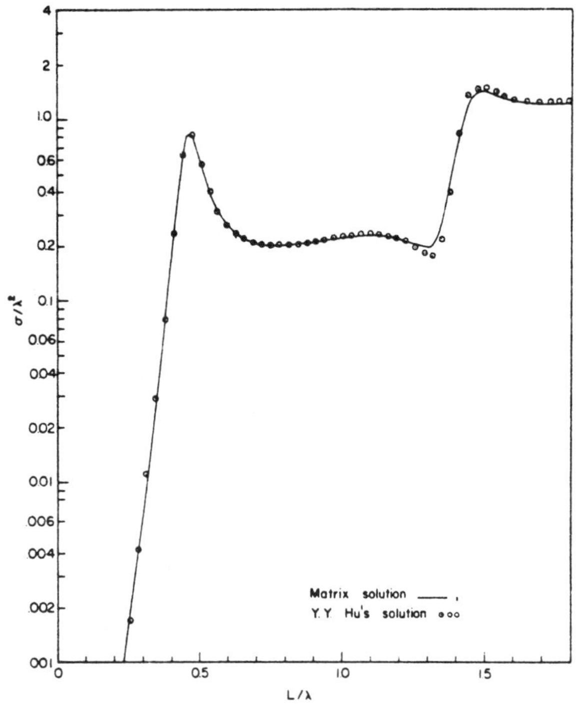

Fig. 8. Echo area of a wire scatterer of length L and diameter $L/72.4$, broadside incidence.

E^i for plane-wave scattering is a well-behaved function, compared with the impulsive impressed field of the antenna problem. Figure 8 shows the echo area for the same wire as was used for an antenna in Fig. 6. Hence, length-to-diameter ratio is 74.2, and a 32 segment piecewise-linear current approximation was used. It is compared to Hu's second-order variational solution [14]. Again good agreement is obtained in the range $L<1.3\lambda$, for which Hu's trial functions are adequate, and a slight discrepancy shows up for $L>1.3\lambda$. Additional computations for linear wire scatterers are given in the original report [16].

XII. Discussion

The method of reducing a functional equation to a matrix equation, and then inverting the matrix for a solution, is particularly well suited to machine computation. Furthermore, the inverted matrix is a representation of the system for arbitrary excitation, hence all responses are solved for at once. As demonstrated by the treatment of wire objects of arbitrary shape, one can also obtain solutions for classes of systems.

In electromagnetic theory, the interpretation of the solution in terms of generalized network parameters is quite general, and applies to bodies of arbitrary shape and arbitrary material. This generalization has been discussed in another paper [17]. The network representation is also useful for the treatment of loaded bodies, both with lumped loads [18] and with continuous loading. Examples of continuously loaded bodies are dielectric coated conductors, magnetic coated conductors, and imperfect conductors.

The solution for wires of arbitrary shape, Section IX, is a first-order solution to the appropriate integrodifferential equation. Higher-order solutions can be obtained by using better-behaved expansion and/or testing functions, and by taking into account the curvature of the wire within the elementary segments. For a general solution, it appears to be more convenient to use a numerical procedure than an analytical procedure. This numerical procedure can be implemented by further subdividing each wire segment, and summing the contributions from the finer subdivisions to obtain the elements of $[Z]$.

As the order of solution is increased, much of the complication comes from the treatment of singularities. The derivative of the current (i.e., charge) is discontinuous at wire ends and at any voltage source along the wire. In the first-order solution this problem has not been accurately treated, and computations appear to justify that this procedure is permissible. For example, at the end of a wire the solution (103) treats the charge as an equivalent line segment extending 1/2 interval beyond the current. The actual charge is singular (or almost so), and could be treated by a special subroutine. While this modification is simple, a similar modification for voltage sources along the wire is not practicable for a general program. This is because the impedance matrix would then depend on the location of the source instead of being a characteristic of the wire object alone. On the basis of experience, it appears that a first-order solution with no special treatment of singularities is adequate for most engineering purposes. This is particularly true for far-zone quantities, such as radiation patterns and echo areas, which are relatively insensitive to small errors in the current distribution.

A number of other electromagnetic field problems have been treated in the literature by procedures basically the same as the method of moments with point matching. Some of these problems are scattering by conducting cylinders [19], [20], scattering by dielectric cylinders [21], [22], and scattering by bodies of revolution [23]. Also available in the literature is an alternative treatment of linear wire scatterers, using sinusoidal expansion functions [24], and an alternative treatment of wire antennas of arbitrary shape, using an equation of the Hallén type [25].

Appendix—Evaluation of ψ

An accurate evaluation of the scalar ψ function of (99) is desired. Let the coordinate origin be located at the point n, and the path of integration lie along the z axis. Then

$$\psi(m, n) = \frac{1}{8\pi\alpha}\int_{-\alpha}^{\alpha} \frac{e^{-jkR_{mn}}}{R_{mn}}\, dz' \tag{122}$$

where

$$2\alpha = \Delta l_n \tag{123}$$

$$R_{mn} = \begin{cases} \sqrt{\rho^2 + (z - z')^2} & m \neq n \\ \sqrt{a^2 + (z')^2} & m = n \end{cases} \tag{124}$$

and a = wire radius. The geometry for these formulas is given in Fig. 9.

One approximation to the ψ's can be obtained by expanding the exponential in a Maclaurin series, giving

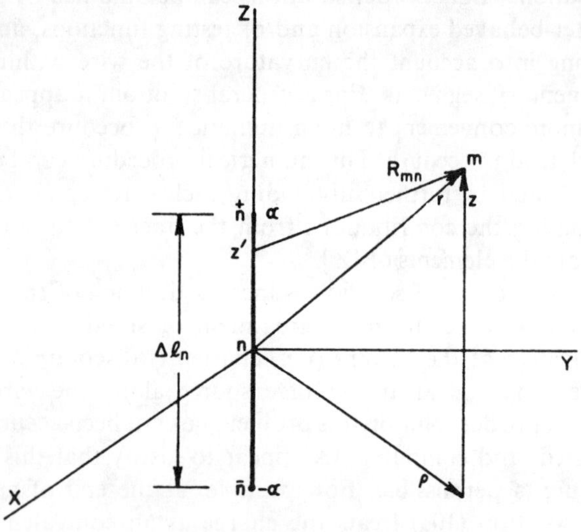

Fig. 9. Geometry for evaluating $\psi(m, n)$.

$$\psi = \frac{1}{8\pi\alpha}\int_{-\alpha}^{\alpha}\left(\frac{1}{R_{mn}} - jk - \frac{k^2}{2}R_{mn} + \cdots\right)dz'. \quad (125)$$

The first term is identical with the static potential of a filament of charge. The second term is independent of R_{mn}. Hence, a two-term approximation of (122) is

$$\psi(m, n) \approx \frac{1}{8\pi\alpha}\log\left[\frac{z + \alpha + \sqrt{\rho^2 + (z + \alpha)^2}}{z - \alpha + \sqrt{\rho^2 + (z - \alpha)^2}}\right] - \frac{jk}{4\pi}. \quad (126)$$

If $r = \sqrt{\rho^2 + z^2}$ is large and $\alpha \ll \lambda$, then

$$\psi(m, n) \approx \frac{e^{-jkr}}{4\pi r}. \quad (127)$$

For a first-order solution, one can take (126) as applying for small r, say $r \leq 2\alpha$, and (127) for large r, say $r > 2\alpha$.

For higher-order approximations, more rapid convergence can be obtained by taking a phase term e^{-jkr} out of the integrand. Then

$$\begin{aligned}\psi &= \frac{e^{-jkr}}{8\pi\alpha}\int_{-\alpha}^{\alpha}\frac{e^{-jk(R_{mn} - r)}}{R_{mn}}dz' \\ &= \frac{e^{-jkr}}{8\pi\alpha}\int_{-\alpha}^{\alpha}\left(\frac{1}{R_{mn}} - \frac{jk(R_{mn} - r)}{R_{mn}} - \frac{k^2(R_{mn} - r)^2}{2R_{mn}} + \cdots\right)dz'. \quad (128)\end{aligned}$$

Term by term integration gives

$$\psi(m, n) = \frac{e^{-jkr}}{8\pi\alpha}\left[I_1 - jk(I_2 - rI_1) - \frac{k^2}{2}(I_3 - 2rI_2 + r^2 I_1) + j\frac{k^3}{6}(I_4 - 3rI_3 + 3r^2 I_2 - r^3 I_1) + \cdots\right] \quad (129)$$

where

$$I_1 = \log\left[\frac{z + \alpha + \sqrt{\rho^2 + (z + \alpha)^2}}{z - \alpha + \sqrt{\rho^2 + (z - \alpha)^2}}\right] \quad (130)$$

$$I_2 = 2\alpha \quad (131)$$

$$I_3 = \frac{\alpha + z}{2}\sqrt{\rho^2 + (\alpha + z)^2} + \frac{\alpha - z}{2}\sqrt{\rho^2 + (z - \alpha)^2} + \frac{\rho^2}{2}I_1 \quad (132)$$

$$I_4 = 2\alpha\rho^2 + \frac{2\alpha^3 + 6\alpha z^2}{3}. \quad (133)$$

An expansion of the type (129) is theoretically valid for all r, but it fails numerically for large r because it involves subtractions of almost equal numbers. For $\rho < a$, one should set $\rho = a$ in the expansion.

An expression suitable for large r is obtained by expanding (122) in a Maclaurin series in z' as

$$\psi = \frac{1}{8\pi\alpha}\int_{-\alpha}^{\alpha}\left[f(0) + f'(0)z' + \frac{1}{2!}f''(0)(z')^2 + \cdots\right]dz' \quad (134)$$

where

$$f(z') = \frac{e^{-jk\sqrt{\rho^2 + (z - z')^2}}}{\sqrt{\rho^2 + (z - z')^2}}.$$

When a five-term expansion of (134) is integrated term by term, there results

$$\psi = \frac{e^{-jkr}}{4\pi r}\left[A_0 + jk\alpha A_1 + (k\alpha)^2 A_2 + j(k\alpha)^3 A_3 + (k\alpha)^4 A_4\right] \quad (135)$$

where

$$\begin{aligned}A_0 &= 1 + \frac{1}{6}\left(\frac{\alpha}{r}\right)^2\left[-1 + 3\left(\frac{z}{r}\right)^2\right] + \frac{1}{40}\left(\frac{\alpha}{r}\right)^4\left[3 - 30\left(\frac{z}{r}\right)^2 + 35\left(\frac{z}{r}\right)^4\right] \\ A_1 &= \frac{1}{6}\left(\frac{\alpha}{r}\right)\left[-1 + 3\left(\frac{z}{r}\right)^2\right] + \frac{1}{40}\left(\frac{\alpha}{r}\right)^3\left[3 - 30\left(\frac{z}{r}\right)^2 + 35\left(\frac{z}{r}\right)^4\right] \\ A_2 &= -\frac{1}{6}\left(\frac{z}{r}\right)^2 - \frac{1}{40}\left(\frac{\alpha}{r}\right)^2\left[1 - 12\left(\frac{z}{r}\right)^2 + 15\left(\frac{z}{r}\right)^4\right] \\ A_3 &= \frac{1}{60}\left(\frac{\alpha}{r}\right)\left[3\left(\frac{z}{r}\right)^2 - 5\left(\frac{z}{r}\right)^4\right] \\ A_4 &= \frac{1}{120}\left(\frac{z}{r}\right)^4. \end{aligned} \quad (136)$$

For accuracy of better than one percent, one can use (129) for $r < 10\alpha$ and (135) for $r \geq 10\alpha$.

An alternative derivation of the type of (135) can be obtained as follows. For $r > z'$, one has the expansion

$$\frac{e^{-jkR_{mn}}}{-jkR_{mn}} = \sum_{n=0}^{\infty}(2n + 1)j_n(kz')h_n^{(2)}(kr)P_n\left(\frac{z}{r}\right) \quad (137)$$

where j_n are the spherical Bessel functions of the first kind, $h_n^{(2)}(kr)$ are the spherical Hankel functions of the second kind, and $P_n(z/r)$ are the Legendre polynomials. If (137) is substituted into (122) and integrated term by term, there results.

$$\psi(m, n) = \frac{1}{4\pi j} \sum_{n=0}^{\infty} b_n h_n^{(2)}(kr) P_n\left(\frac{z}{r}\right) \tag{138}$$

where

$$b_n = \frac{2n+1}{2\alpha} \int_{-k\alpha}^{k\alpha} j_n(x)\, dx. \tag{139}$$

Equation (138) can be rearranged into the form of (135), although the recurrence formulas for $h_n^{(2)}$ and P_n make computation directly from (138) almost as easy.

Acknowledgment

The computer programming for the numerical results was done by J. Mautz.

References

[1] B. Z. Vulikh, *Introduction to Functional Analysis for Scientists and Technologists*, I. N. Sneddon, trans. Oxford: Pergamon, 1963.

[2] B. Friedman, *Principles and Techniques of Applied Mathematics*. New York: Wiley, 1956.

[3] J. W. Dettman, *Mathematical Methods in Physics and Engineering*. New York: McGraw-Hill, 1962.

[4] L. Kantorovich and G. Akilov, *Functional Analysis in Normed Spaces*, D. E. Brown, trans. Oxford: Pergamon, 1964, pp. 586–587.

[5] L. Kantorovich and V. Krylov, *Approximate Methods of Higher Analysis*, C. D. Benster, trans. New York: Wiley, 1964, ch. 4.

[6] D. S. Jones, "A critique of the variational method in scattering antennas," *IRE Trans. on Antennas and Propagation*, vol. AP-4, pp. 297–301, July 1956.

[7] Forsythe and Wasov, *Finite Difference Methods for Partial Differential Equations*. New York: Wiley, 1960.

[8] V. H. Rumsey, "The reaction concept in electromagnetic theory," *Phys. Rev.*, ser. 2, vol. 94, pp. 1483–1491, June 15, 1954.

[9] R. F. Harrington, *Time-Harmonic Electromagnetic Fields*. New York: McGraw-Hill, 1961, pp. 340–345.

[10] D. K. Reitan and T. J. Higgins, "Accurate determination of the capacitance of a thin rectangular plate," *Trans. AIEE (Communication and Electronics)*, pt. I, vol. 75, pp. 761–766, 1956 (Jan. 1957 section).

[11] J. Van Bladel, *Electromagnetic Fields*. New York: McGraw-Hill, 1964, p. 96.

[12] H. B. Dwight, *Tables of Integrals and Other Mathematical Datc*. New York: Macmillan, 1947.

[13] J. Van Bladel, "Some remarks on Green's dyadic for infinite space," *IRE Trans. on Antennas and Propagation*, vol. AP-9, pp. 563–566, November 1961.

[14] Yueh-Ying Hu, "Back-scattering cross section of a center-loaded cylindrical antenna," *IRE Trans. on Antennas and Propagation*, vol. AP-6, pp. 140–148, January 1958.

[15] R. W. P. King, *The Theory of Linear Antennas*. Cambridge, Mass.: Harvard University Press, 1956, p. 172.

[16] R. F. Harrington et al., "Matrix methods for solving field problems," Rome Air Development Center, Griffiss AFB, N. Y., final rept. under Contract AF 30(602)-3724, March 1966.

[17] R. F. Harrington, "Generalized network parameters in field theory," *Proc. Symposium on Generalized Networks*, MRIS series, vol. 16. Brooklyn, N. Y.: Polytechnic Press, 1966.

[18] ——, "Theory of loaded scatterers," *Proc. IEE (London)*, vol. 111, pp. 617–623, April 1964.

[19] K. K. Mei and J. G. Van Bladel, "Scattering by perfectly-conducting rectangular cylinders," *IEEE Trans. on Antennas and Propagation*, vol. AP-11, pp. 185–192, March 1963.

[20] M. G. Andreasen, "Scattering from parallel metallic cylinders with arbitrary cross sections," *IEEE Trans. on Antennas and Propagation*, vol. AP-12, pp. 746–754, November 1964.

[21] J. H. Richmond, "Scattering by a dielectric cylinder of arbitrary cross section shape," *IEEE Trans. on Antennas and Propagation*, vol. AP-13, pp. 334–341, May 1965.

[22] ——, "TE-wave scattering by a dielectric cylinder of arbitrary cross-section shape," *IEEE Trans. on Antennas and Propagation*, vol. AP-14, pp. 460–464, July 1966.

[23] M. G. Andreasen, "Scattering from bodies of revolution," *IEEE Trans. on Antennas and Propagation*, vol. AP-13, pp. 303–310, March 1965.

[24] J. H. Richmond, "Digital computer solutions of the rigorous equations for scattering problems," *Proc. IEEE*, vol. 53, pp. 796–804, August 1965.

[25] K. K. Mei, "On the integral equations of thin wire antennas," *IEEE Trans. on Antennas and Propagation*, vol. AP-13, pp. 374–378, May 1965.

EFFECTIVE METHODS FOR SOLVING INTEGRAL AND INTEGRO-DIFFERENTIAL EQUATIONS

Donald R. Wilton and Chalmers M. Butler, *Department of Electrical Engineering, University of Mississippi, University, MS 38677*

ABSTRACT

Effective numerical methods for solving simple integral and integro-differential equations with logarithmically singular kernels are presented tutorially in this paper. Fundamental features and interrelationships of the methods are discussed. Procedures for applying the techniques to practical equations of electromagnetics are suggested.

1. INTRODUCTION

The development of numerical methods for solving integral equations in electromagnetics has been the subject of intensive research for more than fifteen years. During these years, careful analysis has paved the way for the development of efficient and effective numerical methods and, of equal importance, has provided a solid foundation for a thorough understanding of the techniques. Progress in both understanding and procedures has been so great that equations which could be solved only by researchers ten years ago now can be solved handily by undergraduate engineering students.

In this paper are described simple but highly effective methods for solving integral and integro-differential equations of a class having principal properties usually found in equations encountered in electromagnetics. Features of the methods are discussed and their interrelationships are delineated. The presentation is tutorial and should be readily accessible to an applications-oriented audience, but it is hoped that the reader more deeply steeped in numerical techniques would benefit from it also.

To illustrate numerical techniques, the authors outline sample methods for solving simple one-dimensional equations whose kernels exhibit logarithmic singularities. These equations are referred to as *generic* equations and are selected for consideration here for two reasons. First, practically important one-dimensional electric field integral equations and integro-differential equations of electromagnetics possess logarithmically singular kernels and, since the nature of the kernel dictates major properties of an integral equation, the salient features of the simpler generic equations replicate those of the electric field equations. Second, all but one of the generic equations have known exact solutions which provide a basis for ascertaining the accuracy of numerical solutions. Exactly how techniques for solving the generic equations can be applied to practical equations is discussed.

2. SIMPLE SOLUTION OF INTEGRAL EQUATION

As an introductory example, a simple but effective numerical method is developed for the elementary integral equation

$$-\frac{1}{2\pi}\int_{-w}^{w} u(x')\ln|x-x'|dx' = f(x), \qquad x\varepsilon(-w,w) \tag{1}$$

in which u(x) is the unknown, f(x) is the forcing function, and $-\frac{1}{2\pi}\ln|x-x'|$ is the kernel. In the solution procedure u is approximated by a linear combination of known, linearly independent basis functions. In this example, pulses [1] are employed which means that u(x) is represented in (-w,w) by a piecewise-constant approximation as

$$u(x) \approx \sum_{n=1}^{N} U_n \Pi_n(x) \tag{2}$$

in which the U_n's are unknown constants and the pulse functions Π_n's are defined by

$$\Pi_n(x) = \begin{cases} 1, & x\varepsilon(x_n - \Delta/2,\ x_n + \Delta/2) \\ 0, & \text{otherwise} \end{cases} \tag{3}$$

The approximation is illustrated in Fig. 1. The interval (-w,w) is divided into N segments of equal length $\Delta = 2w/N$ with their centers at

$$x_n = -w + \Delta(n - \tfrac{1}{2}),\ n = 1,2,\ldots,\ N. \tag{4}$$

Subject to the above approximation of u(x), (1) becomes

$$\sum_{n=1}^{N} U_n s_n(x) \approx f(x) \quad , \quad x\varepsilon(-w,w) \tag{5}$$

where

$$s_n(x) = -\frac{1}{2\pi}\int_{x_n-\Delta/2}^{x_n+\Delta/2} \ln|x-x'|dx' \tag{6}$$

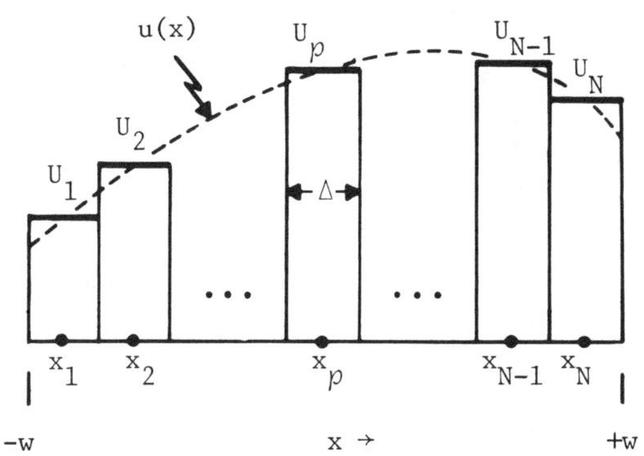

Fig. 1. Weighted pulse-function representation of u(x): $u(x_n) \approx U_n\Pi_n(x)$.

which can be integrated readily. Eq. (5) is enforced exactly at N points x_m, $m = 1,2,\ldots,N$, in the interval (-w,w), as illustrated in Fig. 1, to obtain the following set of N equations in N unknowns $\{U_n\}$:

$$\sum_{n=1}^{N} U_n S_{mn} = F_m,\ m = 1,2,\ldots,N. \tag{7}$$

In (7), $F_m = f(x_m)$ and $S_{mn} = s_n(x_m)$ or

$$S_{mn} = \frac{\Delta}{2\pi}\left\{1 - \ell n\Delta - \tfrac{1}{2}\,\ell n\left|(m-n)^2 - \tfrac{1}{4}\right| - (m-n)\,\ell n\left(\frac{|m-n+\frac{1}{2}|}{|m-n-\frac{1}{2}|}\right)\right\} . \qquad (8)$$

In matrix notation (7) is

$$\begin{bmatrix} S_{mn} \end{bmatrix}\begin{bmatrix} U_n \end{bmatrix} = \begin{bmatrix} F_m \end{bmatrix} . \qquad (9)$$

The solution of (7) or (9) yields $\{U_n\}$, knowledge of which enables one to approximate u(x) by means of (2).

Approximate solutions of (1) with $f = 1$ and $f = x$, determined by the numerical method outlined above, are presented graphically in Fig. 2 for various values of N together with the exact solutions (Table 1) for comparison. For $N = 5$ the actual pulse approximation is plotted but only values of U_n are indicated at the pulse centers for $N = 9$ and $N = 20$. A few points are omitted to avoid obscuration. One should note the improvement in the solution as N is increased and should observe that the numerical results exhibit approximately the edge condition near $x = \pm w$.

From (8) one sees that S_{mn} depends on the difference (m-n) rather than on m and n individually which means that $[S_{mn}]$ is a *Toeplitz* matrix and that its elements depend on a single index (m-n): $S_{mn} = S_{m-n}$. Since $[S_{mn}]$ also is observed to be *symmetric*,

$$S_{mn} = S_{m-n} = S_{n-m} = S_{|m-n|} , \qquad (10)$$

from which we conclude that there are only N distinct elements in

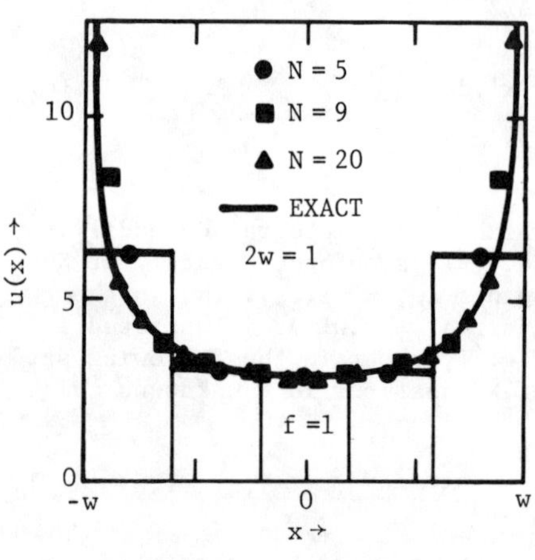

Fig. 2(a). Solution of Eq. (1) with f(x) = 1 for different N (pulse expansion/point matching).

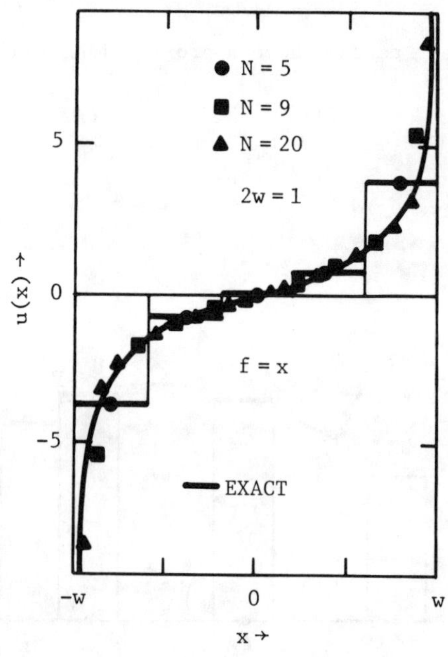

Fig. 2(b). Solution of Eq. (1) with f(x) = x for different N (pulse expansion/point matching).

TABLE 1

INTEGRAL EQUATION SOLUTION

Case i	Unknown $u_i(x)$	Forcing Function $f_i(x)$
0	$\frac{2}{w\ell n(2/w)} \frac{1}{\sqrt{1-(x/w)^2}}$	1
1	$\frac{2}{w}(x/w)\frac{1}{\sqrt{1-(x/w)^2}}$	(x/w)
2	$\frac{2}{w}\left[2(x/w)^2 + \frac{1}{2\ell n\left(\frac{2}{w}\right)} - 1\right]\frac{1}{\sqrt{1-(x/w)^2}}$	$(x/w)^2$
3	$\frac{3}{w}\left[2(x/w)^3 - (x/w)\right]\frac{1}{\sqrt{1-(x/w)^2}}$	$(x/w)^3$
Integral Equation: $-\frac{1}{2\pi}\int_{-w}^{w} u_i(x')\ell n\lvert x-x'\rvert dx' = f_i(x),\ x\varepsilon(-w,w);\quad w \neq 2$		

$[S_{mn}]$. All elements can be found from knowledge of those of the first (or last) row or of the first (or last) column, an observation that can be used to reduce the time and labor needed to compute the matrix elements. If we choose to compute the elements of the first row, then all others can be determined from

$$S_{mn} = S_{1,|m-n|+1} \ . \tag{11}$$

The properties above are direct results of (*i*) dependence of the kernel upon $|x-x'|$, (*ii*) even-function nature of $\Pi_n(x)$ about its center, (*iii*) equal-width pulses, and (*iv*) location of match points at pulse centers. A third interesting property of $[S_{mn}]$ is that it is *diagonally strong*. That is, the magnitude of the main diagonal elements is greater than that of any of the off-diagonal elements $\left(|S_{nn}| > |S_{mn}|,\ m \neq n\right)$. Usually computing the inverse of such a matrix or solving the associated set of linear equations is a very stable process.

3. SOLUTION OF INTEGRAL EQUATION—NUMERICALLY COMPUTED MATRIX ELEMENTS

In the numerical solution of (1) the matrix elements $[S_{mn}]$ are determined from (8), but, in solving practical integral equations, integrals which are encountered can rarely be evaluated analytically. In such instances numerical integration methods are necessary but they must be employed with great care, especially when the integral is improper. An integral equation with such a kernel is

$$\frac{1}{4\pi}\int_{-h}^{h} u(z')k(z-z')dz' = f(z), \quad z\varepsilon(-h,h) \tag{12}$$

in which the kernel is

$$k(z-z') = \frac{1}{2\pi}\int_{-\pi}^{\pi} \frac{1}{R}\, d\phi' \tag{13}$$

with

$$R = \left[(z-z')^2 + 4a^2 \sin^2 \frac{\phi'}{2}\right]^{\frac{1}{2}} . \tag{14}$$

With f(z) a constant, (12) is essentially the equation for the charge density on a conducting tube of length 2h and radius 2a which is charged to a constant potential.

Below, we outline a procedure for solving (12) by the pulse expansion/ point-matching technique. Eqs. (1) and (12) are of the same form but have different kernels so the sole difference in the solution procedures lies in the computation of the elements of the coefficient matrices. In the present case the matrix is denoted $[S_{mn}]$ and its elements are

$$S_{mn} = \frac{1}{4\pi}\int_{z_n-\Delta/2}^{z_n+\Delta/2} k(z_m - z')dz' = \frac{1}{4\pi}\int_{-\Delta/2}^{\Delta/2} k(z_m - z_n - z')dz' \tag{15}$$

where $\Delta = 2h/N$ is the length of the N segments into which the interval (-h,h) is partitioned and where z_n and z_m designate pulse centers and match points, respectively:

$$z_p = -h + \Delta(p - \tfrac{1}{2}), \quad p = 1,2,\ldots, N \quad . \tag{16}$$

A procedure to solve (12) must incorporate a method to perform numerically the double integration of (15), which, unless done with care, can be quite laborious. The difficulty in evaluating (15) numerically is due to the singular behavior of $k(z-z')$ at $z' = z$. To circumvent these problems we observe that

$$k(z-z') \xrightarrow[\left(\frac{z-z'}{2a}\right)\to 0]{} -\frac{1}{\pi a}\,\ell n\left(\frac{|z-z'|}{8a}\right) \tag{17}$$

and we replace $k(z-z')$ by

$$k(z-z') = \left\{k(z-z') + \frac{1}{\pi a}\,\ell n\left(\frac{|z-z'|}{8a}\right)\right\} - \frac{1}{\pi a}\,\ell n\left(\frac{|z-z'|}{8a}\right) \tag{18}$$

which is formed by subtracting (17) from and adding it to $k(z-z')$. Clearly the right side of (18) is equal to $k(z-z')$ and it possesses two significant properties which we use to advantage. First, the term in the braces taken compositely is very slowly varying and therefore can be integrated numerically with ease. Second, the remaining term, which contains the troublesome singularity, is analytically integrable. Hence, with (18) in (15) we arrive at

$$S_{mn} = \frac{\Delta}{4\pi^2 a}\,\ell n(8a) + \frac{1}{2\pi a}\,\mathrm{S}_{mn} + \frac{1}{4\pi}\int_{-\Delta/2}^{\Delta/2}\left\{k(\Delta[m-n]-z') + \frac{1}{\pi a}\,\ell n\left(|\Delta[m-n]-z'|/8a\right)\right\} dz' \tag{19}$$

where S_{mn} is given in (8). Even though the integrand in (19) is very smooth, $k(\Delta[m-n]-z')$ still is an integral and hence evaluation of (19) requires in part double numerical integration. Fortunately, this difficulty can be overcome also by converting (19) to

$$S_{mn} = \frac{1}{4\pi^2 a}\left[\Delta\ell n(8a) + 2\pi\,\mathrm{S}_{mn}\right] + \frac{1}{4\pi^2 a}\int_{-\Delta/2}^{\Delta/2}\left\{\beta_{mn}K(\beta_{mn}) + \ell n\left(|\Delta[m-n]-z'|/8a\right)\right\} dz' \tag{20}$$

in which K is the *complete elliptic integral of the first kind* [2] with modulus

$$\beta_{mn} = \left\{1 + \left(\frac{\Delta[m-n]-z'}{2a}\right)^2\right\}^{-\frac{1}{2}} . \tag{21}$$

The advantage of (20) over (19) is that there exist accurate approximations for $K(\beta)$ [3] of simple form which are easy to integrate. The elements S_{mn} can be determined from (20) for all combinations of indices and for any radius a. The numerical integration method employed in the computation must be one which does not require evaluation of the integrand at $z' = 0$ (identically), since for m=n both the elliptic integral and the logarithm function in the braces are unbounded as $z' \to 0$. Even though (20) may be employed for all values of m and n, some increase in efficiency can be realized by employing the approximation

$$S_{mn} \approx \frac{1}{4\pi}\,\ell n\left(\frac{|m-n| + \frac{1}{2}}{|m-n| - \frac{1}{2}}\right) , \quad \Delta|m-n| >> 2a . \tag{22}$$

whenever $\Delta|m-n| >> 2a$.

Observe that $[S_{mn}]$ is symmetric and Toeplitz,

$$S_{mn} = S_{|m-n|} , \tag{23}$$

and hence all its elements can be determined from those of the first (or last) row or column as in (11). Also $[S_{mn}]$ is diagonally strong.

With a technique for computing $[S_{mn}]$ now in hand, the procedure for solving (12) is identical to that for solving (1)—in fact, the same computer program may be employed but with a subroutine for computing S_{mn} replaced by one for computing S_{mn}. Sample solutions of (12) are displayed in Fig. 3. Notice that u(z) exhibits the known edge behavior proportional to $[1 - |z/h|]^{-\frac{1}{2}}$ at interior points near +h and -h.

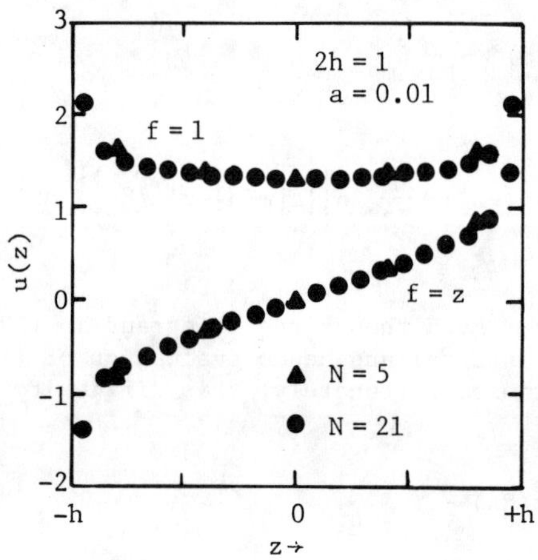

Fig. 3. Solution of Eq. (12) for different forcing functions.

4. SOLUTION OF FIRST ORDER INTEGRO-DIFFERENTIAL EQUATION

In this section techniques for solving an integro-differential equation are presented. The simple equation

$$-\frac{1}{2\pi}\frac{d}{dx}\int_{-w}^{w} u(x')\ell n|x-x'|dx' = f(x), \quad x\varepsilon(-w,w) \tag{24}$$

is to be solved subject to the constraint

$$\int_{-w}^{w} u(x)\,dx = 0 \quad . \tag{25}$$

By integration, (24) can be converted to

$$-\frac{1}{2\pi}\int_{-w}^{w} u(x')\ell n|x-x'|dx' = \mathfrak{f}(x) + C, \quad x\varepsilon(-w,w) \tag{26}$$

where C is an unknown constant of integration and where

$$\mathfrak{f}(x) = \int_{0}^{x} f(x')dx'. \tag{27}$$

Eq. (24) reminds one of a differential equation with the needed constraining condition (25) while the pair (26) and (25) can be viewed as two simultaneous equations with unknowns u and C. Apart from C, (26) is the familiar equation (1) but the presence of the differential operator in (24) is a feature which warrants careful investigation.

The role of the constant C in (26) can be clarified by a simple example. Let $f = 1$, then $\mathfrak{f} = x$ and the right side of (26) becomes $C + x$. From Table 1, it is clear that $u(x) = C\, u_0(x) + w\, u_1(x)$. To satisfy (25) C must be zero, so the solution is $u(x) = w\, u_1(x)$.

Because (26) is so similiar to (1), a method for solving it is considered before (24) is addressed. By expanding u in pulses (2) and enforcing (26) at match points x_m at pulse centers, one obtains

$$\sum_{n=1}^{N} U_n S_{mn} - C = \delta(x_m) \quad , \quad m = 1,2,\ldots,N \tag{28}$$

where S_{mn} are the elements (8). Also, with u approximated by (2), the constraining equation (25) becomes

$$\Delta \sum_{n=1}^{N} U_n = 0 \quad . \tag{29}$$

Eqs. (28) and (29) can be written as a matrix equation

$$\begin{bmatrix} S_{11} & S_{12} & S_{13} & \cdots & S_{1N} & -1 \\ S_{21} & S_{22} & S_{23} & \cdots & S_{2N} & -1 \\ \vdots & \vdots & \vdots & & \vdots & \vdots \\ S_{N1} & S_{N2} & S_{N3} & \cdots & S_{NN} & -1 \\ \Delta & \Delta & \Delta & & \Delta & 0 \end{bmatrix} \begin{bmatrix} U_1 \\ U_2 \\ \vdots \\ U_N \\ C \end{bmatrix} = \begin{bmatrix} \delta(x_1) \\ \delta(x_2) \\ \vdots \\ \delta(x_N) \\ 0 \end{bmatrix} \tag{30}$$

in which the role of C as an unknown is clear.

We now outline a method of solving (24) and (25) directly. With u expanded in pulses (24) can be expressed as

$$\sum_{n=1}^{N} U_n \frac{d}{dx} \mathscr{s}_n(x) \approx f(x), \quad x \varepsilon (-w,w) \quad . \tag{31}$$

Since there are N unknowns $\{U_n\}$, N-1 equations must be obtained from the numerical processing of (31) to be solved simultaneously with (29). To obtain N-1 equations from (31), one might be tempted to place match points at $x_n \pm \Delta/2$ midway between pulse centers, but $\left\{\frac{d}{dx} \mathscr{s}_n(x)\right\}_{x = x_n \pm \Delta/2}$ is unbounded. Hence, the simple point-matching scheme applied to (1) cannot be applied to (24), if we insist that u be represented by pulses. Perhaps it has occurred to the reader that there are other deficiencies in the point-matching method. A simple example is the case of an equation whose forcing function happens to pass through zero at several of the selected match points. One remedy for this difficulty is to replace enforcement of an equation at a match point by enforcement of its average at several nearby points. That is, the mth equation of (7) would be replaced by one which equates the average of the left member of (5) to that of the right member, with both averages determined from values computed at points near x_m. Contributions from more and more points in a fixed interval about x_m can be included in the averaging process until, in the limit, one is in effect equating the integrals over the interval of the two sides of (5). Below we apply this procedure of equating averages to Eq. (31) and we discover that it effectively replaces the differential operator by a difference operator.

N-1 equations are to be obtained from the above-mentioned enforcements of equality of averages of the two sides of (31). To this end, we multiply both sides of (31) by $\Pi_{m+\frac{1}{2}}(x)$, which is a pulse of width Δ centered at $x_{m+\frac{1}{2}}$ midway between centers of expansion pulses, and integrate the resulting product over (-w,w) to obtain

$$\Delta \sum_{n=1}^{N} U_n S'_{mn} = F_m , \quad m = 1,2,\ldots,N-1 \tag{32}$$

where

$$F_m = \int_{x_m}^{x_{m+1}} f(x)\,dx \tag{33}$$

and

$$S'_{mn} = \int_{x_m}^{x_{m+1}} \frac{d}{dx} \delta_n(x)\,dx = \frac{1}{\Delta}\left[\delta_n(x_{m+1}) - \delta_n(x_m)\right] = \frac{1}{\Delta}\left[S_{m+1,n} - S_{mn}\right]. \tag{34}$$

The so-called testing pulses $\Pi_{m+\frac{1}{2}}$ and expansion pulses Π_n are illustrated in Fig 4. Often f(x) is sufficiently smooth over (x_m, x_{m+1}) to permit the approximation $F_m \approx \Delta\, f(x_{m+\frac{1}{2}})$ in which case (32) becomes

$$\sum_{n=1}^{N} U_n S'_{mn} \approx f(x_{m+\frac{1}{2}}), \quad m = 1,2,\ldots,N-1. \tag{35}$$

Eq. (29) and the N-1 equations (32) can be solved for the N coefficients U_n or, when f is sufficiently smooth, one solves (29) and (35). In matrix form (35) is

$$\begin{bmatrix} S'_{11} & S'_{12} & \cdots & S'_{1N} \\ S'_{21} & S'_{22} & \cdots & S'_{2N} \\ \vdots & \vdots & & \vdots \\ S'_{N-1,1} & S'_{N-1,2} & \cdots & S'_{N-1,N} \\ \Delta & \Delta & \cdots & \Delta \end{bmatrix} \begin{bmatrix} U_1 \\ U_2 \\ \vdots \\ \\ U_N \end{bmatrix} \approx \begin{bmatrix} f(x_{3/2}) \\ f(x_{5/2}) \\ \vdots \\ f(x_{N-\frac{1}{2}}) \\ 0 \end{bmatrix}. \tag{36}$$

For larger and larger N, taken to achieve a converged solution, the approximation $F_m \approx \Delta\, f(x_{m+\frac{1}{2}})$ becomes better and better in general whenever f is well behaved, thereby justifying ultimate utilization of (35) or (36).

Even though $\Delta S'_{mn}$ results directly from integrating $\Pi_{m+\frac{1}{2}}(x)\frac{d}{dx}\delta_n(x)$ over (-w,w) without approximation, S'_{mn} can be interpreted as $\frac{d}{dx}\delta_n(x)$ at $x_{m+\frac{1}{2}}$ with $\frac{d}{dx}$ replaced by the finite-difference operator. This interpretation is clear from (34). Hence, (35) can be viewed as (31) with the derivative replaced by the difference and enforced at the points $x_{m+\frac{1}{2}}$. Or, as suggested by (34), (32) can be obtained directly by subtracting the mth equation from the $(m+1)^{th}$ equation of (28).

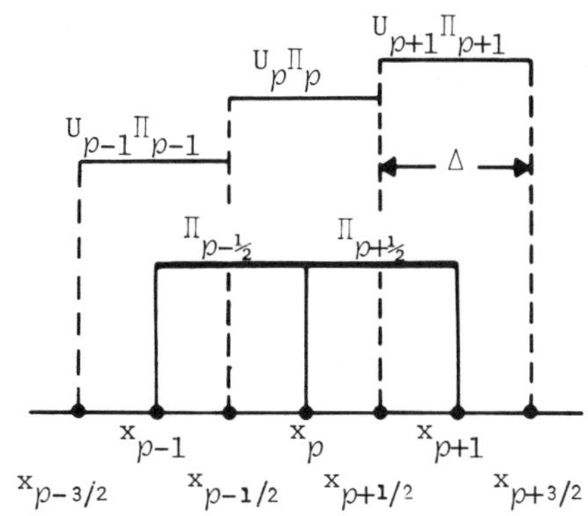

Fig. 4. Illustration of the weighted expansion pulses $U_p\Pi_p$ and the pulses $\Pi_{p\pm\frac{1}{2}}$ used in the enforcement procedure.

5. SOLUTION OF A SECOND ORDER INTEGRO-DIFFERENTIAL EQUATION

In this section we consider techniques for solving the integro-differential equation

$$-\frac{1}{2\pi}\frac{d^2}{dx^2}\int_{-w}^{w} u(x')\ln|x-x'|dx' = f(x), \quad x\varepsilon(-w,w) \tag{37}$$

where the unknown u(x) must satisfy the boundary conditions

$$u(w) = u(-w) = 0 \quad . \tag{38}$$

Our interest in (37) stems from the fact that it has features analogous to those frequently found in integral equations of electromagnetics. For the sake of definiteness, however, (37) may be viewed as a low-frequency approximation of the integro-differential equation for the current induced on a conducting strip of width 2w excited by TE illumination [4]. The boundary condition (38) corresponds to the requirement that the current u(x) must vanish at the strip edges.

Paralleling the integration of (24) to obtain (26), we eliminate the derivatives in (37) by regarding the equation as a differential equation with the integral as the unknown and solve the equation to obtain

$$-\frac{1}{2\pi}\int_{-w}^{w} u(x')\ln|x-x'|dx' = A + Bx + \int_{-w}^{x} f(x')(x-x')dx', \quad x\varepsilon(-w,w) \ . \tag{39}$$

If A and B were known, (39) would have the same form as (1) and hence u(x) could be found by any procedure used to solve the latter. Since A and B are unknown, however, u(x) can only be determined in terms of A and B, and the additional information necessary to fix these constants must come from the boundary conditions (38). This may be illustrated by solving (39) explicitly for f(x) = a + bx. In this case, (39) becomes

$$-\frac{1}{2\pi}\int_{-w}^{w} u(x')\ln|x-x'|dx' = A' + B'x + a\frac{x^2}{2} + b\frac{x^3}{6} \qquad (40)$$

where A' and B' are new constants which absorb contributions of the particular integral to the homogeneous solution of (37). Using Table 1, one may obtain the solution of (40) by superposition of the responses due to individual driving terms on the right hand side:

$$u(x) = \frac{1}{\sqrt{w^2 - x^2}} (C + Dx + 2ax^2 + bx^3) \quad . \qquad (41)$$

Since (40) has the same kernel as (1) and (26) considered earlier, it is not surprising that the solution *appears* to be singular at $x = \pm w$. However, the boundary conditions (38) require that u(x) vanish at $x = \pm w$ and this is possible only if the quantity in parentheses in (41) vanishes at these points. Enforcing this condition yields

$$u(x) = \sqrt{w^2 - x^2}\,(-2a - xb) \ , \qquad (42)$$

in which it is seen that u(x) vanishes as $\sqrt{w - |x|}$ as x approaches the endpoints $\pm w$. The *derivative* of u(x), however, is singular at and varies as $(w - |x|)^{-\frac{1}{2}}$ near the endpoints.

5.1. Solution of Integral Equation

In this section we develop a numerical approach for solving (39) subject to (38). The similarity of (39) to (1) and (26) suggests that we again employ a pulse function representation of u(x) and use the point-matching procedure to enforce (39). In contrast to the solution procedures for (1) and (26), however, u(x) is now required to satisfy (38) as well. This is accomplished by dividing the interval (-w,w) into N+1 segments of equal length $\Delta = 2w/(N+1)$ with segment endpoints located at (Fig. 5)

$$x_n = -w + n\Delta \ , \ n = 0,\ldots,N+1 \ , \qquad (43)$$

and by representing u(x) in the piecewise-constant approximation

$$u(x) \approx \sum_{n=1}^{N} U_n \Pi_n(x) \qquad (44)$$

with $\Pi_n(x)$ defined as in (3). Note that the points x_n in (43) and Fig. 5 are located such that no pulse exists in the intervals of length $\Delta/2$ about the endpoints $x = \pm w$. By this construction the constraint (38) is automatically satisfied by the piecewise constant representation (44).

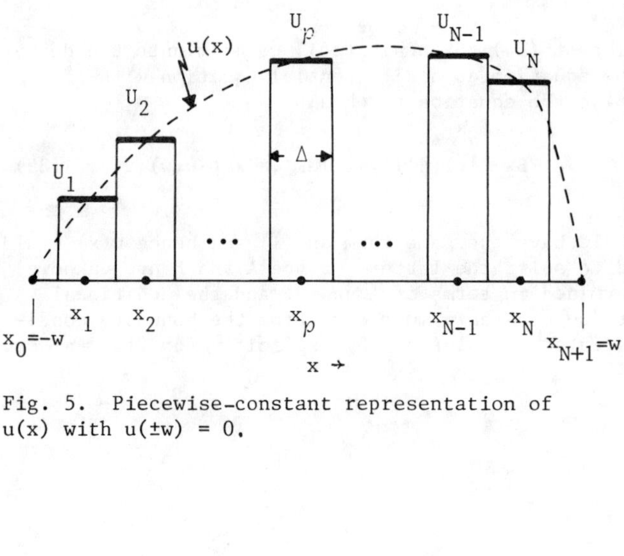

Fig. 5. Piecewise-constant representation of u(x) with $u(\pm w) = 0$.

If (44) is substituted into (39) and the resulting equation is point-matched at $x = x_0,\ldots,x_{N+1}$, one obtains (N+2) linear equations in the (N+2) unknowns $U_1,\ldots,U_N$, A, and B:

$$\sum_{n=1}^{N} U_n S_{mn} - A - Bx_m = F_m , \quad m = 0,1,\dots,N+1 \quad , \tag{45a}$$

or in matrix form

$$\begin{bmatrix} S_{01} & S_{02} & \cdots & S_{0N} & -1 & -x_0 \\ S_{11} & S_{12} & \cdots & S_{1N} & -1 & -x_1 \\ \vdots & \vdots & & \vdots & \vdots & \vdots \\ S_{N1} & S_{N2} & \cdots & S_{NN} & -1 & -x_N \\ S_{N+1,1} & S_{N+1,2} & \cdots & S_{N+1,N} & -1 & -x_{N+1} \end{bmatrix} \begin{bmatrix} U_1 \\ U_2 \\ \vdots \\ U_N \\ A \\ B \end{bmatrix} = \begin{bmatrix} F_0 \\ F_1 \\ \vdots \\ F_N \\ F_{N+1} \end{bmatrix} \tag{45b}$$

in which

$$F_m = \int_{-w}^{x_m} f(x')(x_m - x')dx' \approx \Delta \sum_{p=0}^{m-1} f(x_p)(x_m - x_p) \quad . \tag{46}$$

The elements S_{mn} are defined as in (8), and the approximation in (46) may be used if f(x) is slowly varying over each subinterval. Fig. 6 shows numerically computed solutions of (45) for various values of N with f(x) = 1 and x. For comparison the exact solution obtained from (42) is also shown.

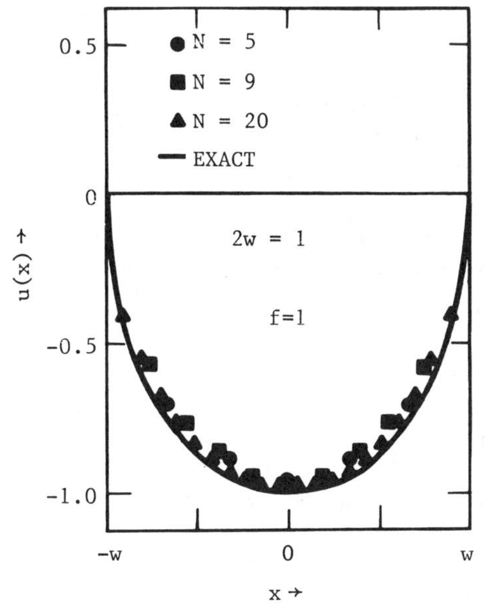

Fig. 6(a). Solution of Eq. (45) with f(x) = 1 for different N (pulse expansion/point matching).

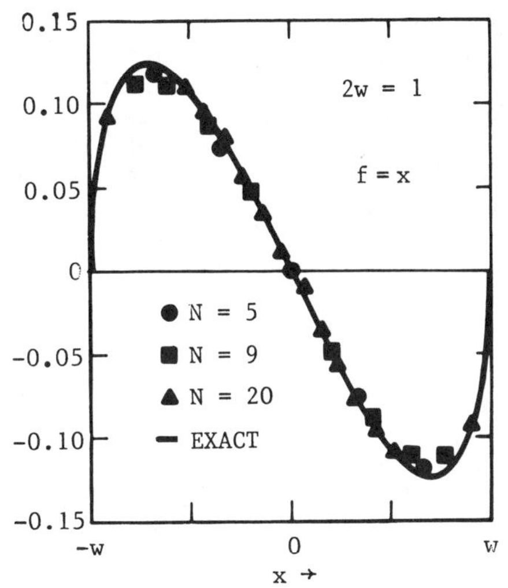

Fig. 6(b). Solution of Eq. (45) with f(x) = x for different N (pulse expansion/point matching).

5.2. Solution of Integro-Differential Equation

In this section we examine several procedures for directly solving the integro-differential equation (37) subject to (38). An advantage of solving (37) directly is that the superfluous constants A and B in (39) are not needed. In practical cases involving coupled equations, a large number of such constants may appear, and avoiding them may substantially save computer storage and computation time. Furthermore, in problems involving junctions of several structures methods for the direct solution of the integro-differential equations are much simpler than those for the corresponding integral equations. That the direct approach imposes little or no penalty in numerical efficiency is demonstrated by revealing approximate equivalences among the numerically processed integral equation and the several integro-differential equations at various stages of the numerical developments.

Triangle expansion and pulse testing functions. The direct solution procedure developed to solve (24) suggests a method for solving (37) if we put the latter equation into a form similar to that of the former. To this end, one of the derivatives in (37) is brought under the integral sign and onto the kernel. Since

$$\frac{d}{dx}\ln|x-x'| = \frac{1}{x-x'} = -\frac{d}{dx'}\ln|x-x'| , \tag{47}$$

the resulting integral is convergent only if interpreted in a Cauchy principal value sense. (Both derivatives in (37) cannot be brought onto the kernel since the resulting integral would not converge.) The last form in (47) permits the resulting integral to be integrated by parts which, with (38), yields

$$-\frac{1}{2\pi}\frac{d}{dx}\int_{-w}^{w}\frac{d}{dx'}u(x')\ln|x-x'|\,dx' = f(x), \quad x\varepsilon(-w,w) . \tag{48}$$

Note that the unbounded derivative of $u(x)$ at $x = \pm w$ does not allow us to repeat this procedure with the remaining derivative outside the integral. Eq. (48) now has the same form as (24) but with the derivative $u'(x)$ rather than $u(x)$ itself playing the role of the unknown. Indeed, as (41) illustrates, $u'(x)$ in the present problem has the same singular edge behavior as did $u(x)$ in (24). Thus, paralleling the solution of (24), we represent $u(x)$ in (48) in such a way that its *derivative* is piecewise constant and test the equation with pulse functions. But if $u'(x)$ is approximated as piecewise constant, then $u(x)$ itself must be approximated as piecewise linear. This may be conveniently done through use of the *triangle functions*

$$\Lambda_n(x) = \begin{cases} 1 - \dfrac{|x - x_n|}{\Delta}, & x\varepsilon(x_{n-1}, x_{n+1}) \\ 0, \text{ otherwise} \end{cases} \tag{49}$$

which are of unit height and defined on intervals of width 2Δ. With these basis functions $u(x)$ may be approximated as

$$u(x) \approx \sum_{n=1}^{N} U_n\Lambda_n(x) \quad . \tag{50}$$

The piecewise linear approximation of $u(x)$ by triangles and the resulting piecewise constant approximation of $u'(x)$ are illustrated in Fig. 7. Note that the boundary conditions $u(\pm w) = 0$ are automatically satisfied in this representation.

It is easily demonstrated that the derivative of $\Lambda_n(x)$ is

$$\Lambda_n'(x) = -\frac{1}{\Delta}\left[\Pi_{n+\frac{1}{2}}(x) - \Pi_{n-\frac{1}{2}}(x)\right] \tag{51}$$

and hence from (50)

$$u'(x) \approx -\frac{1}{\Delta}\sum_{n=1}^{N} U_n\left[\Pi_{n+\frac{1}{2}}(x) - \Pi_{n-\frac{1}{2}}(x)\right] = \sum_{n=0}^{N} \frac{U_{n+1} - U_n}{\Delta}\,\Pi_{n+\frac{1}{2}}(x) \tag{52}$$

where $\Pi_{n+\frac{1}{2}}(x)$ is a unit pulse of width Δ centered at $x_{n+\frac{1}{2}}$, midway between the points x_n and x_{n+1}, and $U_0 = U_{N+1} \equiv 0$. If (52) is substituted into (48) and the resulting equation is integrated against $\Pi_m(x)$, i.e., tested with $\Pi_m(x)$, for $m = 1,\ldots,N$, property (10) permits the resultant equation to be manipulated into the form

$$\Delta \sum_{n=1}^{N} U_n S''_{mn} = \int_{-w}^{w} f(x)\,\Pi_m(x)\,dx \approx \Delta f(x_m) \quad , \quad m = 1,2,\ldots,N \tag{53}$$

where the elements S''_{mn} are defined as

$$S''_{mn} = \frac{1}{\Delta^2}\left[S_{m+1,n} - 2S_{mn} + S_{m-1,n}\right] \equiv \frac{\Delta^2 S_{mn}}{\Delta x^2} \tag{54}$$

The notation $\Delta^2/\Delta x^2$ is introduced to denote the operation of forming a second central finite difference, which approximates a second derivative in terms of finite differences. The approximation of the right hand side in (53) improves with increasing N and may be made if f(x) is sufficiently smooth. Eq. (53) constitutes a linear system of N equations in N unknowns which can be solved for the unknown coefficients $\{U_n\}$.

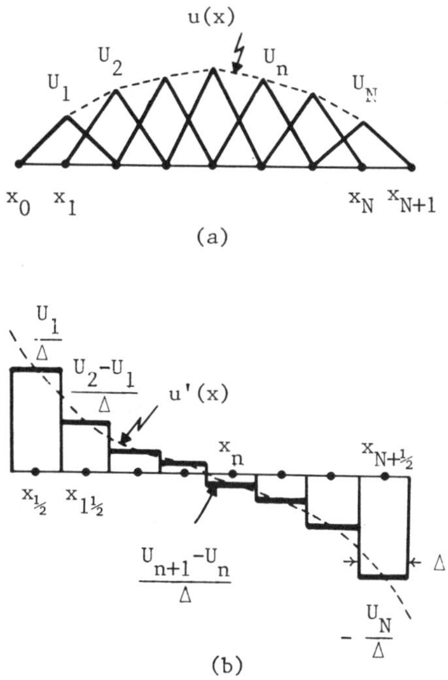

Fig. 7. The use of triangle basis functions yields (a) a piecewise linear approximation of u(x) and (b) a piecewise constant approximation of u'(x).

Eqs. (45) and (53) are two different linear systems of equations from which approximate solutions of (37) and (38) may be determined. Remarkably, however the system (53) with the approximated right hand side is derivable from (45) if its right hand side is approximated as in (46). Hence the two systems with approximated right hand sides are equivalent and must have identical solutions. To establish this equivalence we apply the operator $\Delta(\Delta^2/\Delta x^2)$ to (45) with approximated right hand side (46). That is, we multiply the mth equation of (45) by $-2/\Delta$ and add to it $1/\Delta$ times the $(m+1)^{th}$ and $(m-1)^{th}$ equations for $m = 1,2,\ldots,N$. The left hand side of the resultant equation is identical to the left hand side of (53) whereas the right hand side is identical to the approximate form of the right hand side of (53) since

$$\sum_{p=0}^{m} f(x_p)(x_{m+1}-x_p) - 2\sum_{p=0}^{m-1} f(x_p)(x_m-x_p) + \sum_{p=0}^{m-2} f(x_p)(x_{m-1}-x_p)$$

$$= f(x_{m-1})(x_{m+1}-x_{m-1}) + f(x_m)(x_{m+1}-x_m) - 2f(x_{m-1})(x_m-x_{m-1})$$

$$= \Delta f(x_m) \quad . \tag{55}$$

Since the system (53) can be obtained from a linear combination of equations (45), the two systems must have identical solutions. Hence Fig. 6 also serves to illustrate the solutions of (53) for $f(x) = 1$ and x.

Before proceeding to develop other approaches for solving (37), we pause to reflect on the procedures developed to solve both (24) and (37). We chose pulse functions to represent the unknown in (24) but found that the resultant integral was not differentiable at the desired match points. Hence we smoothed the derivative by averaging the equation, i.e., testing it with pulse functions, in each subinterval containing the derivative singularity. In (37), however, the presence of the additional derivative dictated the choice of the smoother triangle functions to represent the unknown, thus sufficiently smoothing the integral that the pulse testing approach could again be used. It seems plausible, however, that pulse functions could have been used to represent the unknown if smoother testing functions—for example, triangle functions—had been chosen. After all, testing involves multiplication of the equation by a testing function followed by an integration of the product, and integration by parts should permit us to transfer some of the burden of differentiability to the testing function. This idea is examined further in the next section.

Pulse expansion and triangle testing. In this section we outline a numerical procedure for solving (37) and (38) using pulse functions to represent $u(x)$ and triangle functions to test the equation. The effect of testing a second derivative with a triangle function may be seen by evaluating the following integral involving the second derivative of an arbitrary function $h(x)$ defined on $(-w,w)$ and tested with $\Lambda_m(x)$:

$$\int_{-w}^{w} \frac{d^2h}{dx^2}\Lambda_m\,dx = \Delta\left[\frac{h(x_{m+1}) - 2h(x_m) + h(x_{m-1})}{\Delta^2}\right] = \Delta\,\frac{\Delta^2 h_m}{\Delta x^2} \quad . \tag{56}$$

This result is easily obtained by one integration by parts over the various subintervals defining $\Lambda_m(x)$ in (49). Now let $u(x)$ be represented by pulse functions as in (44), substitute the representation into (37), and test the resulting equation with $\Lambda_m(x)$ for $m = 1,2,\ldots,N$. Use of (56) allows the left hand side to be written as

$$\Delta\sum_{n=1}^{N} U_n \frac{\Delta^2}{\Delta x^2} s_n(x_m) = \Delta\sum_{n=1}^{N} U_n \frac{\Delta^2 S_{mn}}{\Delta x^2} = \Delta\sum_{n=1}^{N} U_n S''_{mn} \tag{57}$$

whereas the right hand side is

$$\int_{-w}^{w} f(x)\Lambda_m(x)\,dx \approx \Delta\, f(x_m) \quad . \tag{58}$$

We have already seen that application of the finite difference operator $\Delta(\Delta^2/\Delta x^2)$ to the left hand side of (45a) yields (57). Application of the operator to the right hand side (with exact F_m) yields

$$\frac{1}{\Delta}\int_{-w}^{x_{m-1}} f(x')(x_{m-1} - x')dx' - \frac{2}{\Delta}\int_{-w}^{x_m} f(x')(x_m - x')dx' + \frac{1}{\Delta}\int_{-w}^{x_{m+1}} f(x')(x_{m+1} - x')dx'$$

$$= \frac{1}{\Delta}\int_{x_{m-1}}^{x_m} f(x')(x - x_{m-1})dx' + \frac{1}{\Delta}\int_{x_m}^{x_{m+1}} f(x')(x_{m+1} - x')dx'$$

$$= \int_{-w}^{w} f(x')\Lambda_m(x')dx' \approx \Delta\, f(x_m) \qquad (59)$$

which is identical to (58). Since (57) and (58) can be obtained (without approximation) from linear combinations of equations (45a), the solutions to the two systems must be identical. Further, with the approximations on the right hand side, the solution of these two systems is equivalent to that of (53).

We have seen in this section that smooth testing functions can be chosen to compensate for the choice of less well-behaved expansion functions. On the other hand, if we had chosen a smoother expansion function to begin with, perhaps we could have eliminated testing with a pulse function and used the simpler point-matching procedure instead. This possibility is explored next.

Spline function expansion and point-matching. In this section we investigate the use of a set of expansion functions which are smoother than triangles. These functions, called *spline functions*, are continuous and have a continuous first derivative. They are defined as

$$\Omega_n(x) = \begin{cases} \frac{1}{2}\left(\frac{x - x_{n-3/2}}{\Delta}\right)^2 , & x\varepsilon(x_{n-3/2}, x_{n-1/2}) \\ \frac{3}{4} - \left(\frac{x - x_n}{\Delta}\right)^2 , & x\varepsilon(x_{n-1/2}, x_{n+1/2}) \\ \frac{1}{2}\left(\frac{x - x_{n+3/2}}{\Delta}\right)^2 , & x\varepsilon(x_{n+1/2}, x_{n+3/2}) \\ 0, \text{ otherwise} , & \end{cases} \qquad (60)$$

and are shown in Fig. 8. From the figure it is seen that $\Omega_1(x)$ and $\Omega_N(x)$ extend past the ends of the interval $(-w,w)$. This deficiency can be remedied by defining special basis functions for the end intervals, but we are not presently concerned with such details. We shall merely expand $u(x)$ as

$$u(x) = \sum_{n=1}^{N} U_n\, \Omega_n(x) \qquad (61)$$

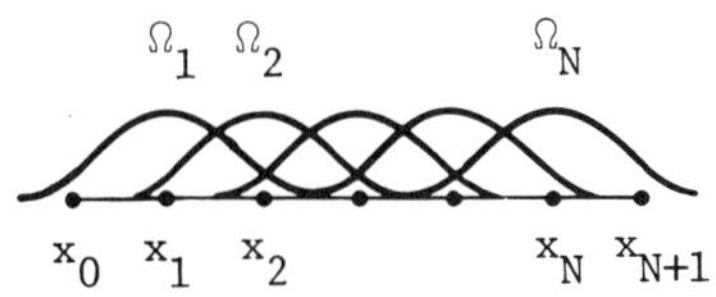

Fig. 8. The spline functions $\Omega_n(x)$.

and extend the domain of integration in (37) to the interval $(-w-\Delta/2,\ w+\Delta/2)$. The smoothness of the representation (61) renders the integral in (37) sufficiently smooth that it can be differentiated twice throughout the solution interval. Hence we substitute (61) into (37), point-match the resulting equation at x_m, $m=1,\ldots,N$, and multiply both sides by Δ to obtain a linear system of equations for U_n. The matrix elements are easily computed by bringing the derivatives inside the integral and integrating by parts, yielding

$$-\frac{\Delta}{2\pi}\frac{d^2}{dx^2}\int_{w-\Delta/2}^{w+\Delta/2}\Omega_n(x')\ln|x-x'|dx'\Bigg|_{x=x_m}$$

$$=-\frac{\Delta}{2\pi}\int_{w-\Delta/2}^{w+\Delta/2}\frac{d^2}{dx'^2}\Omega_n(x')\ln|x_m-x'|dx'$$

$$=\Delta\, S''_{mn}\quad . \tag{62}$$

The right hand side of the mth equation is just $\Delta f(x_m)$. Again we obtain a system of equations identical to the approximate form of (53). However, though the values $\{U_n\}$ are the same as those found by the other procedures with the approximated right hand sides, the values of $u(x_m)$ as computed from (61) are now $u(x_m) \approx (U_{m-1} + 6U_m + U_{m+1})/8$ whereas (44) and (50) yielded $u(x_m) \approx U_m$. This implies that the present scheme yields a somewhat smoother solution than the previous solutions because each point $u(x_m)$ in the present scheme consists of a weighted average of three adjacent values of the previous solutions.

We observe that the point-matching procedure applied to an integro-differential equation may be interpreted as testing the equation with delta functions $\Delta\delta(x-x_m)$, $m=1,2,\ldots,N$, which, by their sifting property, merely select the points x_m at which equality is enforced. In the next section we carry to extreme the idea that smoothness in the expansion functions can be traded for smoothness in the testing functions by choosing a delta-function representation of the unknown combined with spline testing functions.

Delta-function expansion and spline testing. If the spline functions are chosen as testing functions, it seems reasonable to expect that the resulting smoothing of the left hand side of (37) may allow us to use expansion functions less well-behaved than the pulse functions. Accordingly, let us attempt to approximate u(x) by delta-functions as

$$u(x) \approx \sum_{n=1}^{N} U_n\Delta\,\delta(x-x_n)\quad , \tag{63}$$

where $U_n\Delta$ is the total moment of u(x) in the interval $(x_{n-\frac{1}{2}}, x_{n+\frac{1}{2}})$. Next (63) is substituted into (37) and the resulting equation is tested with $\Omega_m(x)$ (again the domain of the testing interval is extended to $(w-\Delta/2,\ w+\Delta/2)$). The resulting left hand side is

$$\sum_{n=1}^{N} U_n\int_{-w-\Delta/2}^{w+\Delta/2}\frac{d^2}{dx^2}\left[-\frac{1}{2\pi}\int_{-w}^{w}\Delta\delta(x'-x_n)\ln|x-x'|dx'\right]\Omega_m(x)dx = \Delta\sum_{n=1}^{N}U_n\,S''_{mn} \tag{64}$$

and the right hand side can be approximated as

TABLE 2

EXPANSION/TESTING COMBINATIONS YIELDING IDENTICAL SOLUTIONS OF EQ. (37)

Expansion functions	Testing functions
$\Delta\delta(x - x_n)$	$\Omega_m(x)$
$\Pi_n(x)$	$\Lambda_m(x)$
$\Lambda_n(x)$	$\Pi_m(x)$
$\Omega_n(x)$	$\Delta\delta(x - x_m)$

$$\int_{-w}^{w} f(x)\Omega_m(x)dx \approx f(x_m) \int_{-w-\Delta/2}^{w+\Delta/2} \Omega_m(x)dx = \Delta f(x_m) . \quad (65)$$

Hence the system of equations so obtained is again equivalent to (53).

In summary, we have seen in these last four subsections that with the approximations

$$\begin{aligned} \int_{-w}^{w} f(x)\Delta\delta(x - x_m)dx &\approx \int_{-w}^{w} f(x)\Pi_m(x)dx \\ \approx \int_{-w}^{w} f(x)\Lambda_m(x)dx &\approx \int_{-w-\Delta/2}^{w+\Delta/2} f(x)\Omega_m(x)dx \approx \Delta f(x_m) \end{aligned} \quad (66)$$

involving the right hand side of (37), each combination of expansion and testing functions selected yielded an identical system of equations with identical solution coefficients $\{U_m\}$. The various combinations selected are summarized in Table 2, where one notes that as one decreases the smoothness of the expansion functions, one must correspondingly increase the smoothness of the testing functions in order to obtain a comparable solution. While other considerations may dictate the actual choice of basis and testing functions in more advanced problems, with minor modifications the general observations of this section remain valid in these cases also.

6. OBSERVATIONS

In this final section we briefly relate some of the properties of the generic equations of previous sections to corresponding properties of electromagnetic equations. To this end, consider the electric field integral equation for scattering by a perfectly conducting surface S illuminated by a time-harmonic incident field $\bar{E}^{inc}$:

$$\frac{1}{j\omega\mu\varepsilon}\left[(k^2 + \nabla\nabla\cdot)\bar{A}\right]_{tan} = -\bar{E}^{inc}_{tan}(\bar{r}) \;,\; \bar{r} \in S \quad (67)$$

where

$$\bar{A}(\bar{r}) = \frac{\mu}{4\pi}\int_S \bar{J}(\bar{r}')\,\frac{e^{-jkR}}{R}\,dS' \quad (68)$$

and

$$R = |\bar{r} - \bar{r}'| \;. \quad (69)$$

Eqs. (67) and (68), together with the boundary condition

$$\bar{J}\cdot\hat{n} = 0 \quad (70)$$

at each point on the boundary of S with $\hat{n}$ a unit vector tangent to S and normal to the boundary, constitute an integro-differential equation for the unknown induced current $\bar{J}$ on S. Although (67) is a more complicated equation in that it is a vector equation which applies on a two dimensional surface, it has a number of features common to (37). The most important of these features are the following:

1) Both (37) and (67) have singular kernels, the former of logarithmic type, the latter of the type 1/R. When the problem has symmetry such that $\bar{J}$ is independent of one dimension, 1/R can be integrated in that dimension to obtain a one-dimensional integral equation with a logarithmically singular kernel closely corresponding to (37). With or without symmetry, the singular nature of the kernel determines the behavior, i.e., the edge conditions, of the current in the neighborhood of any boundaries and/or discontinuities of S.

2) In both (37) and (67), second derivative operators act on the integral. Thus, as in (37), (67) requires a certain degree of smoothness in representing $\bar{J}$ before the derivative operator may be applied to $\bar{A}$, or a testing procedure (or finite differences) must be used to effect the smoothing. As was found in solving (37), smoothness in the expansion functions may be traded for smoothness in the testing functions.

If (67) is specialized to a thin tubular cylinder parallel to the z-axis and of length 2h and radius a, the thin wire assumptions lead to [5]

$$\frac{1}{j\omega\varepsilon}\left(\frac{d^2}{dz^2}+k^2\right)\int_{-h}^{h} I(z')K(z-z')dx' = -E_z^{inc}(z) \ , \quad z\varepsilon(-h,h), \tag{71}$$

where the total wire current I(z) satisfies the boundary conditions

$$I(\pm h) = 0 \tag{72}$$

and where

$$K(z-z') = \frac{1}{2\pi}\int_{-\pi}^{\pi}\frac{e^{-jkR}}{R}\,d\phi = k(z-z') + K_1(z-z') \tag{73}$$

with

$$R = \sqrt{(z-z')^2 + 4a^2\sin^2\frac{\phi}{2}} \ . \tag{74}$$

Note in (73) that the term $k(z-z')$, defined in (13), contains a logarithmic singularity. The term $K_1(z-z')$ is non-singular (in fact, vanishes at $z=z'$) and may be viewed as a perturbation on $k(z-z')$. Matrix elements involving integration over the singular point in K may be computed as in (18)-(20) with an additional contribution from the integral of

$$K_1(z-z') = \frac{1}{2\pi}\int_{-\pi}^{\pi}\left(\frac{e^{-jkR}-1}{R}\right)d\phi \ . \tag{75}$$

The resulting double integral is easily integrated numerically.

Our analysis of the previous section suggests that any of the expansion and testing function combinations of that section as summarized in Table 2 may also be used to solve (71). Note that the additional integral arising from testing the integral associated with the k^2 term in (71) could be eliminated by using one of the approximations (66) since the integral in (71) (but not its derivative) is relatively smooth.

With minor modifications, observations similar to the above may be made for all electric field integral equation formulations in scattering and radiation, including, for example, scattering and radiation by strips [6], wires [7], and bodies of revolution [6]. Integral and integro-differential equations which are mathematically equivalent to (67) also arise in analyses of apertures [8,9]. Equations in investigations of scattering by dielectric bodies also involve the integro-differential operator of (67) and, hence, similar techniques are applicable [6].

7. ACKNOWLEDGEMENT

The authors gratefully acknowledge the expert assistance cheerfully provided by Ms. Tina Begley in the preparation of the manuscript.

REFERENCES

1. Harrington, R. F. (1968), *Field Computations by Moment Methods*, 229 pp, Macmillan, New York.

2. Dwight, H. B. (1961), *Tables of Integrals and other Mathematical Data*, p. 180, Macmillan, New York.

3. Abramowitz, M., and I. A. Stegun (1964), *Handbook of Mathematical Functions*, p. 591, U. S. Government Printing Office, Washington, D. C.

4. Butler, C. M., and D. R. Wilton (1980), General analysis of narrow strips and slots, *IEEE Trans. Antennas Propagat.*, AP-28(1), 42-48.

5. Butler, C. M., and D. R. Wilton (1975), Analysis of various numerical techniques applied to thin-wire scatterers, *IEEE Trans. Antennas Propagat.*, AP-23(4), 534-540.

6. Glisson, A. W., and D. R. Wilton (1980), Simple and efficient numerical methods for problems of electromagnetic radiation and scattering from surfaces, *IEEE Trans. Antennas Propagat.*, AP-28(5), 593-603.

7. Wilton, D. R., and C. M. Butler (1976), Efficient numerical techniques for solving Pocklington's equation and their relationships to other methods, *IEEE Trans. Antennas Propagat.*, AP-24(1), 83-86.

8. Butler, C. M., Y. Rahmat-Samii, and R. Mittra (1978), Electromagnetic penetration through apertures in conducting surfaces, *IEEE Trans. Antennas Propagat.*, AP-26(1), 82-93.

9. Butler, C. M., and K. R. Umashankar (1976), Electromagnetic penetration through an aperture in an infinite, planar screen separating two half spaces of different electromagnetic properties, *Radio Sci.*, 11(7), 611-619.

Survey of Numerical Methods for Solution of Large Systems of Linear Equations for Electromagnetic Field Problems

TAPAN K. SARKAR, SENIOR MEMBER, IEEE, KENNETH R. SIARKIEWICZ, SENIOR MEMBER, IEEE, AND ROY F. STRATTON

Abstract—**Many of the popular methods for the solution of large matrix equations are surveyed with the hope of finding an efficient method suitable for both electromagnetic scattering and radiation problems and system identification problems.**

I. INTRODUCTION

THE PROBLEM of radiation and scattering from electromagnetic structures may be formulated in terms of the E-field, the H-field, or the combined field integral equations. The integral equations are then reduced to matrix equations by the method of moments. Hence the maximum size of an electromagnetic field problem that can be solved by this technique depends on how efficiently solutions of a set of simultaneous equations are obtained.

In system identification, on the other hand, the problem is formulated in terms of a convolution integral. When any of the standard techniques is utilized to identify the system, one again encounters a set of simultaneous equations. The only difference between the two cases is that in the former, one often encounters a matrix which has large elements on the diagonal whereas in the latter case, the matrix may be nearly singular.

The objective of this paper is to survey many of the popular methods for the solution of large matrix equations with the hope of finding an efficient method suitable for both electromagnetic scattering and radiation problems and system identification problems.

In our discussions of the various methods, only references which are directly relevant are noted. No attempt has been made to cite the earliest sources. In many cases additional references may be found in the papers mentioned.

II. DIRECT METHODS FOR SOLVING MATRIX EQUATIONS [1]-[3]

In this section we present all the direct methods. These include Cramer's rule and two versions of Gaussian elimination [LU decomposition and the compact method]. The Cramer's rule is too well-known to elaborate and too cumbersome for practical use. The other two methods of Gaussian elimination consist in solving $\mathbf{AX} = \mathbf{Y}$ in two parts. This is achieved by decomposing $\mathbf{A}$ into $\mathbf{LU}$ where $\mathbf{L}$ is a lower triangle matrix and $\mathbf{U}$ is an upper triangle matrix. Thus $\mathbf{LUX} = \mathbf{Y}$ represents two triangular systems

$$\mathbf{LG} = \mathbf{Y} \tag{1}$$

$$\mathbf{UX} = \mathbf{G} \tag{2}$$

which can very easily be solved. The calculation of $\mathbf{L}$ and $\mathbf{U}$ together with the solution of $\mathbf{LG} = \mathbf{Y}$ is usually called the forward elimination and the solution of $\mathbf{UX} = \mathbf{G}$ is the backward substitution. It is important to note that this method is really optimum in obtaining an exact solution (assuming there is no truncation or roundoff error) if and only if one is interested in handling the elements of the matrices by rows or by columns. Under this condition Klyuyev and Kokovkin-Scherbak [4] have proved that no general system of linear equations can be solved with fewer arithmetic operations than are required by Gaussian elimination. Recently Volker Strassen [5] has shown that it is possible to solve a general system of linear equation with $\theta(N^p)$ arithmetic operations, where in this case $p = \log_2 7 = 2.807$. Here θ is defined as terms of the order of N^p. However it is not known whether this value of p is the minimum exponent.

The advantage of obtaining a solution in a finite number of steps is offset by the building of truncation and round-off errors in direct methods. It can be shown [6] that if $\Delta\mathbf{A}$ are the uncertainties associated with $\mathbf{A}$ and $\Delta\mathbf{Y}$ are the inaccuracies associated with the representation of $\mathbf{Y}$, then the uncertainty $\Delta\mathbf{Y}$ in the solution $\mathbf{X}$ is given by

$$\frac{\|\Delta\mathbf{X}\|_2}{\|\mathbf{X}\|_2} \leqslant \frac{\text{cond}\,[\mathbf{A}]}{1-\sqrt{N}\cdot\text{cond}\,[\mathbf{A}]\cdot 2^{-t}} \cdot \left[\frac{\|\Delta\mathbf{Y}\|_2}{\|\mathbf{Y}\|_2} + \frac{\|\Delta\mathbf{A}\|_2}{\|\mathbf{A}\|_2}\right] \leqslant \frac{2^{-t}\,[\sqrt{N}+1]\,\text{cond}\,[\mathbf{A}]}{1-\sqrt{N}\cdot\text{cond}\,[\mathbf{A}]\cdot 2^{-t}} \tag{3}$$

where t is the number of binary digits with which computation is actually carried out in the computer, N is the dimension of $\mathbf{A}$ and cond $[\mathbf{A}]$ represents the condition number of the matrix A, which is the ratio between the maximum and the minimum eigenvalues of $\mathbf{A}$. *Thus there is absolutely no way to recognize an accurate* solution $\mathbf{X}$ given by Gaussian elimination unless [6]

$$\sqrt{N}\cdot 2^{-t}\cdot\text{cond}\,[\mathbf{A}] \ll 1. \tag{4}$$

As an example, consider the solution of the following problem $\mathbf{AX} = \mathbf{Y}$ by Gaussian elimination. Let $\mathbf{A}$ be the ill-conditioned Hilbert matrix and $\mathbf{Y}$ be that vector for which $\mathbf{X} = \{1, 2, 3, \cdots, N\}$. The problem then is to find $\mathbf{X}$ given $\mathbf{A}$ and $\mathbf{Y}$. We solved this problem on the Xerox Sigma-9 computer where computation is carried out using 24 binary digits. For a fourth order Hilbert matrix the condition number is obtained as cond $[\mathbf{A}_4] = 1.55 \times 10^4$ [2]. Thus for a fourth order Hilbert matrix, (3) reduces to

$$\frac{\|\Delta\mathbf{X}\|_2}{\|\mathbf{X}\|_2} \leqslant 0.00278$$

Manuscript received March 10, 1980; revised January 5, 1981. This work was supported by the RADC post-doctoral program.

T. K. Sarkar is with the Department of Electrical Engineering, Rochester Institute of Technology, Rochester, NY 14623.

K. Siarkiewicz and R. Stratton are with RADC/RBCT, Griffiss Air Force, Rome, NY 13441.

and hence a very good accuracy in the results is expected. However for a fifth order Hilbert matrix

$$\frac{\|\Delta \mathbf{X}\|_2}{\|\mathbf{X}\|_2} \leqslant 0.09825$$

since cond $[\mathbf{A}_5] = 4.77 \times 10^2$ [2]. This is reflected in the following results:

$$\mathbf{X} \to 0.996;\ 2.067;\ 2.708;\ 4.442;\ 4.783$$

$$\frac{\|\Delta \mathbf{X}\|_2}{\|\mathbf{X}\|_2} \to 0.07772.$$

Note that the theortetical error bound is large. For a sixth order Hilbert matrix we have

$$\mathbf{X} \to 1.007;\ 1.786;\ 4.509;\ -0.0341;\ 9.543;\ 4.182$$

$$\mathbf{X}_{\text{exact}} \to 1.0;\ 2.0;\ 3.0;\ 4.0;\ 5.0;\ 6.0.$$

In this case

$$\frac{\|\Delta \mathbf{X}\|_2}{\|\mathbf{X}\|_2} \leqslant -2.59!! \qquad \text{(from (3))}$$

since cond $[\mathbf{A}_6] = 1.5 \times 10^7$ from [2].

These results prompt us to look for alternate methods in which we could reduce the effects of round-off error in solving a system of equations. The effect of round-off error may become pronounced not only for very ill-conditioned matrices but also for large systems of equations in which a large number of arithmetic operations must be carried out. Iterative methods are good alternatives to rectify this problem of roundoff error. For example, for a seventh order Hilbert matrix for which we know the direct method would not work, we obtained this result by the conjugate gradient method at the end of seven iterations:

$$\mathbf{X} \to 0.993;\ 2.090;\ 2.768;\ 4.034;\ 5.208;\ 6.121;\ 6.780$$

$$\mathbf{X}_{\text{exact}} \to 1.0;\ 2.0;\ 3.0;\ 4.0;\ 5.0;\ 6.0;\ 7.0.$$

This is because in iterative methods, the roundoff error is limited only to the most recent stage of iteration.

III. ITERATIVE METHODS FOR SOLVING MATRIX EQUATIONS

The basic philosophy of the iterative methods is discussed in this section. It can be shown [7], [8] that the solution of the set of equations $\mathbf{AX} = \mathbf{Y}$ is equivalent to the maximization/minimization of the functional $F(\mathbf{X}) = 1/2 \langle \mathbf{AX}, \bar{\mathbf{X}}^* \rangle - \langle \mathbf{Y}, \mathbf{X}^* \rangle$ if $\mathbf{A}$ is negative/positive definite, the asterisk denotes the conjugate, and T denotes the transpose conjugate. The contours of constant $F(\mathbf{X})$ are generally N-dimensional ellipsoids. Also, the residuals $\mathbf{R}_n (= \mathbf{AX}_n - \mathbf{Y})$ at the end of each step are normals to the ellipsoid at $\mathbf{X}_n$. The paths $\mathbf{P}$ by which one reaches the center point of the ellipsoid (which is the solution $\mathbf{X}_{\text{exact}}$) are different for the different iterative methods. An iterative process is called linear if the present estimate $\mathbf{X}_n$ is a linear combination of the past estimates $\mathbf{X}_0, \mathbf{X}_1, \mathbf{X}_2, \cdots, \mathbf{X}_{n-1}$. Otherwise the iterative process is nonlinear. A process is called a stationary iterative process if the rule by which $\mathbf{X}_n$ is determined does not change from iteration to iteration. Otherwise the iterative process is called nonstationary. Nonstationary methods are not pursued in this presentation because some ideas are needed about the magnitude of the maximum and the minimum eigenvalues of $\mathbf{A}$ for these methods to be effective. This information is seldomly available.

A. Linear Iterative Methods

In this section we present Gauss's hand relaxation method, Jacobi's cyclical iteration method (simultaneous displacement method), Seidel's method (successive displacement method), back and forth Seidel's method, and the successive overrelaxation method (SOR).

1) *Gauss's Hand Relaxation Method [3]:* This method was developed by Gauss for hand calculation and is mostly of historical significance. In this method the elements of the search direction vector $\mathbf{P}_n$ are chosen corresponding to greatest residual in absolute value. Mathematically, the iterations are obtained as

$$X_{n+1}{}^k = X_n{}^k - \frac{R_n{}^k}{A^{kk}} \cdot P_n{}^k. \tag{5}$$

This method is not very suitable for automatic computation as it is a very laborious process to search for the largest absolute element in the residual.

2) *Jacobi's Cyclical Iteration Method (Simultaneous Displacement Method [9]:* Here the search directions $\mathbf{P}$ now run cyclically through the coordinate directions regardless of the residuals. Thus the ith element of $\mathbf{X}$ at the $n+1$ iteration is refined by

$$X_{n+1}{}^i = \frac{1}{A^{ii}} \left[Y^i - \sum_{\substack{j=1 \\ j \neq i}}^{N} A^{ij} \cdot X_n{}^j \right] \qquad \text{for } i = 1, 2, \cdots, N. \tag{6}$$

3) *Seidel's Method (Successive Displacement Method) [9] (often incorrectly called the Gauss–Seidel method):* The rate of convergence of Jacobi's method was improved by Seidel in modifying (6) in the following way

$$X_{n+1}{}^k = \frac{1}{A^{kk}} \left\{ Y^k - \sum_{q=1}^{k-1} A^{kq} X_{n+1}{}^q - \sum_{q=k+1}^{N} A^{kq} X_n{}^q \right\} \tag{7}$$

or

$$(\mathbf{D} + \mathbf{L})\mathbf{X}_{n+1} + \mathbf{UX}_n = \mathbf{Y}$$

where $\mathbf{A} = \mathbf{D} + \mathbf{L} + \mathbf{U}$, where $\mathbf{D}, \mathbf{L}, \mathbf{U}$ are the diagonal, lower, and upper triangular matrices of $\mathbf{A}$, respectively. This is achieved by updating the column vector $\mathbf{X}_n$ as soon as one of its components has been calculated and using that updated value in the same iteration rather than waiting for the next iteration as was done by Jacobi. Thus

$$\begin{aligned} \mathbf{X}_{n+1} &= -[\mathbf{D} + \mathbf{L}]^{-1}\mathbf{UX}_n + [\mathbf{D} + \mathbf{L}]^{-1}\mathbf{Y} \\ &\triangleq \mathbf{QX}_n + \mathbf{S}. \end{aligned} \tag{8}$$

TABLE I
VARIOUS COMPONENTS OF X AT THE END OF EACH ITERATION[1]

i	X_0	X_1	X_2	X_3	X_4	X_5	X_6	X_7
1	1.00	−1.60	−0.83	−0.41	−0.18	−0.062	−0.0080	0.01220
2	1.00	0.32	0.00	−0.12	−0.13	−0.107	−0.0761	0.04927
3	1.00	0.80	0.56	0.36	0.21	0.115	0.0577	0.02561
4	1.00	0.21	0.11	0.06	0.03	0.012	0.0034	−0.0007

i	X_8	X_9	X_{10}
1	0.01656	0.014576	0.010786
2	−0.02953	−0.016425	−0.008403
3	0.00907	0.001421	0.001527
4	−0.00127	−0.001514	−0.001164

[1] Note that the ratios do not approach any limit. This is shown in Table II.

The necessary and sufficient condition for the convergence of (8) is that the largest eigenvalue of **Q** be less than unity in magnitude. Other convergence criteria will be discussed later on.

The major drawback of the linear iterative schemes in general and particularly that of Seidel's method is that the convergence is quite irregular if the dominant eigenvalue of **Q** in (8) is complex. As an example, consider the following problem.

Example 1: Let **A** be the matrix

$$\begin{bmatrix} 1.0 & 0.7 & 0.7 & 0.2 \\ 0.7 & 1.0 & 0.7 & 0.1 \\ 0.7 & 0.7 & 1.0 & 0.1 \\ 0.2 & 0.1 & 0.1 & 1.0 \end{bmatrix},$$

and we wish to solve the matrix equation

$$\mathbf{AX} = 0.$$

Since det $[\mathbf{A}] \neq 0$, the only possible solution is $\mathbf{X} = 0$. The various iterates are given in Table I. Such erratic behavior is to be expected since the eigenvalues of the matrix **Q** in (8) are complex. The eigenvalues of **Q**

$$-\mathbf{Q} = [\mathbf{D}+\mathbf{L}]^{-1}[\mathbf{U}] = \begin{bmatrix} 0 & 0 & 0 & 0 \\ 0 & 0.49 & 0.147 & 0.0763 \\ 0 & 0.147 & 0.5341 & 0.0439 \\ 0 & 0.0763 & 0.0439 & 0.0228 \end{bmatrix}$$

are $\lambda = 0,\ 0.028333,\ 0.566733 \pm j0.157158$. That is, the dominant eigenvalues of **Q** are complex conjugates. Thus although the Seidel scheme is convergent in this case, the convergence is erratic. This erratic behavior of the Seidel process is remedied by the "back and forth" Seidel process (see Table II).

4) Back and Forth Seidel Process [9]: The back and forth Seidel process was designed by Aitken and Rosser to overcome the irregular convergence of the Seidel process. This is achieved by making all the eigenvalues of the iterative matrix **Q** real. It proceeds as follows. Start with a first approximation vector $\mathbf{X}_0$ and then obtain $\overline{\mathbf{X}}_1$ by the regular Seidel process as

$$\overline{\mathbf{X}}_1 = -[\mathbf{D}+\mathbf{L}]^{-1}\mathbf{U}\mathbf{X}_0 + [\mathbf{D}+\mathbf{L}]^{-1}\mathbf{Y}. \tag{9}$$

Then find the next iterate by applying the Siedel process to the equations in reverse order, i.e.,

$$\begin{aligned}\mathbf{X}_1 &= -[\mathbf{D}+\mathbf{U}]^{-1}\mathbf{L}\overline{\mathbf{X}}_1 + [\mathbf{D}+\mathbf{U}]^{-1}\mathbf{Y} \\ &= [\mathbf{D}+\mathbf{U}]^{-1}\mathbf{L}[\mathbf{D}+\mathbf{L}]^{-1}\mathbf{U}\mathbf{X}_0 \\ &\quad - [\mathbf{D}+\mathbf{U}]^{-1}\mathbf{L}[\mathbf{D}+\mathbf{L}]^{-1}\mathbf{Y} + [\mathbf{D}+\mathbf{U}]^{-1}\mathbf{Y}.\end{aligned} \tag{10}$$

Thus we see that for this process, the iteration matrix is

$$\mathbf{S} = [\mathbf{D}+\mathbf{U}]^{-1}\mathbf{L}[\mathbf{D}+\mathbf{L}]^{-1}\mathbf{U}. \tag{11}$$

If **A** is assumed to be symmetric $\mathbf{L} = \mathbf{U}^T$ and $\mathbf{U} = \mathbf{L}^T$ (here T denotes transpose) and so the iteration matrix can be rewritten as

$$\begin{aligned}\mathbf{S} &= [\mathbf{D}+\mathbf{U}]^{-1}\mathbf{U}^T[\mathbf{D}+\mathbf{U}^T]^{-1} \\ &= [\mathbf{D}+\mathbf{U}]^{-1}\mathbf{U}^T[\{\mathbf{D}+\mathbf{U}\}^{-1}]^T\mathbf{U} = MM^T\end{aligned} \tag{12}$$

where $\mathbf{M} = [\mathbf{D}+\mathbf{U}]^{-1}\mathbf{U}^T$. Thus **S** is similar to a nonnegative matrix. Since det $[\mathbf{M}]$ = det $[\mathbf{U}] = 0$, $\mathbf{MM}^T$ is semi-definite. So we can conclude that all the eigenvalues of **S** are on the real half line $x \geqslant 0$ and hence the dominant eigenvalue of **S** is unique, though possibly it may be a multiple root.

Example 2: We now apply the back and forth Seidel process to Example 1. We again start with the same initial guess. The results are summarized in Table III.

The results from Table IV indicate that these ratios are tending to 0.67, i.e., that the dominant eignevalues of **S** in this case are 0.65611, 0.36368, 0.02720, and 0. The dominant eigenvalue is real and is equal to

$$\underset{n\to\infty}{\text{limit}} \left\{ \frac{X_{n+1}{}^i}{X_n{}^i} \right\}.$$

In this example, the convergence of the back and forth Seidel process, while slower than that of the ordinary Seidel process, is much more regular than the ordinary Seidel process. However the back and forth Seidel process can easily be accelerated. It is not certain, however, which process would give the most accuracy per unit of labor.

Also in most method of moments problems, we encounter a matrix **A** whose eigenvalues are often complex. The use of the back and forth Seidel process in these problems can be justified; but, as we shall show later, there are faster schemes to treat these problems.

TABLE II
RATIOS OF X_{n+1}^i/X_n^i FOR SEIDEL'S PROCESS.

i	$\frac{X_1^i}{X_0^i}$	$\frac{X_2^i}{X_1^i}$	$\frac{X_3^i}{X_2^i}$	$\frac{X_4^i}{X_3^i}$	$\frac{X_5^i}{X_4^i}$	$\frac{X_6^i}{X_5^i}$	$\frac{X_7^i}{X_6^i}$	$\frac{X_8^i}{X_7^i}$	$\frac{X_9^i}{X_8^i}$	$\frac{X_{10}^i}{X_9^i}$
1	−1.60	0.52	0.49	0.44	0.344	0.1290	−1.525	1.3574	0.8802	0.739983
2	0.32	0.00	−∞	1.08	0.823	0.7112	0.6474	0.5994	0.5562	0.511597
3	0.80	0.70	0.64	0.58	0.548	0.5017	0.4438	0.3542	0.1567	−1.074603
4	0.21	0.52	0.55	0.05	0.400	0.2833	−0.0205	18.1429	1.1114	0.822614

TABLE III
VARIOUS ITERATES OF BACK AND FORTH SEIDEL PROCESS[1]

i	X_0	$\bar{X}_1$	X_1	$\bar{X}_2$	X_2	$\bar{X}_3$	X_3	$\bar{X}_4$	X_4	$\bar{X}_5$	X_5	$\bar{X}_6$	X_6
1	1.0	−1.60	−0.99	−0.99	−0.64	−0.64	−0.42	−0.417	−0.277	−0.27650	−0.18437	−0.18437	−0.12332
2	1.0	0.32	0.48	0.06	0.23	−0.01	0.12	−0.032	0.070	−0.02835	0.04305	−0.02144	0.02746
3	1.0	0.80	0.88	0.63	0.64	0.43	0.45	0.305	0.309	0.20780	0.20966	0.14033	0.14157
4	1.0	0.21	0.21	0.13	0.13	0.09	0.09	0.056	0.056	0.03736	0.03736	0.02499	0.2499

1 The ratios X_{k+1}/X_k are given in Table IV.

TABLE IV
RATIO OF X_{n+1}/X_n FOR BACK AND FORTH SEIDEL PROCESS

i	X_1/X_0	X_2/X_1	X_3/X_2	X_4/X_3	X_5/X_4	X_6/X_5
1	−0.99	0.65	0.66	0.660	0.66560	0.66887
2	0.48	0.48	0.52	0.583	0.61500	0.63787
3	0.88	0.73	0.70	0.687	0.67851	0.67524
4	0.21	0.62	0.70	0.622	0.66714	0.66890

5) The Method of Successive Over/Under Relaxation [1], [9], [12]-[16]: In the case of large systems of equations, the Jacobi or Seidel process converges poorly when the maximum absolute eigenvalue (often referred to as the spectral radius) of the iteration matrix (for example **Q** in (8) lies close to unity. Convergence of the Seidel process could be improved if instead of reaching the minimum point of $F(\mathbf{X})$ we go beyond this point by a certain amount. It seems paradoxical at first to refrain from minimizing the quadratic functional at each iteration step with the goal of achieving better convergence. Let us now modify (8) in the following way:

$$(\mathbf{D}+\mathbf{L})\mathbf{X}_{n+1} + \mathbf{U}\mathbf{X}_n = \mathbf{Y}$$

or

$$\mathbf{D}(\mathbf{X}_{n+1} - \mathbf{X}_n) = \mathbf{Y} - \mathbf{U}\mathbf{X}_n - \mathbf{D}\mathbf{X}_n - \mathbf{L}\mathbf{X}_{n+1}.$$

We now introduce the parameter ω and define the new iteration

$$\omega^{-1}\mathbf{D}(\mathbf{X}_{n+1} - \mathbf{X}_n) = \mathbf{Y} - \mathbf{U}\mathbf{X}_n - \mathbf{D}\mathbf{X}_n - \mathbf{L}\mathbf{X}_{n+1}$$

or

$$\mathbf{X}_{n+1} = [\mathbf{I} - (\omega^{-1}\mathbf{D}+\mathbf{L})^{-1}\mathbf{A}]\mathbf{X}_n + (\omega^{-1}\mathbf{D}+\mathbf{L})^{-1}\mathbf{Y}.$$

In this case the iteration matrix is $[\mathbf{I} - (\omega^{-1}\mathbf{D}+\mathbf{L})^{-1}\mathbf{A}]$ where **I** is the identity matrix. We are now interested in determining ω so as to give this matrix a small maximum eigenvalue. It is interesting to note that in symmetric definite systems of equations, the relaxation methods converge to the solution for any fixed value of ω in the range $0 < \omega < 2$. For $0 < \omega < 1$, the method is referred to as underrelaxation and for $1 < \omega < 2$, the method is called overrelaxation. It has been observed by Kahan and Young [11]-[13] that values of $\omega < 1$ tends to reduce the rate of convergence whereas $\omega > 1$ accelerates the rate of convergence. For $\omega > 1$ we overcorrect the solution vectors and hence we speak of overrelaxation methods. Unfortunately for a given problem it is difficult to find the optimum choice of the relaxation parameter ω. For this, additional information about the structure of matrix **A** is necessary. Nonetheless we can say that the worse the condition of the matrix, the closer the optimum value of ω lies to 2. In such a case, at each iterative step we jump far beyond the minimum point to a new approximation which leaves the quadratic functional $F(\mathbf{X})$ almost as large as it was before. Hence the strategy of making the best improvement in each individual iterative step by going to the minimum point is not the best way of achieving the optimum long-term result.

The successive overrelaxation method has found wide application in the solution of boundary value problems by the finite difference method. In this particular type of problem one often encounters a very sparse matrix. For such matrix equations optimum values of ω have been given by Kahan and Young [11]-[13]. In the case of a full matrix it is difficult to find the optimum ω theoretically unless there is a certain structure to the matrix. Otherwise for each individual problem the optimum value of ω has to be obtained experimentally.

B. Nonlinear Iterative Methods

In nonlinear iterative methods, the refined estimate is no longer a linear function of the past estimates. Newton's method, because of its quadratic convergence $\{\|\mathbf{X}_n - \mathbf{X}_0\| \leqslant c\|\mathbf{X}_{n-1} - \mathbf{X}_0\|^2\}$, is mathematically the most preferred of the several known nonlinear methods for the solution of systems of equations. Practically however a very important limitation on Newton's method is that it does not generally converge to some solution for an arbitrary starting point. Thus Newton's method may fail to converge if the initial estimate is not sufficiently close to the solution.

The size of the domain of convergence depends upon the system of equations. For real algebraic equations, the size of the domain of convergence is generally inversely related to the degree and the number of equations. Therefore one finds that for two simultaneous second degree equations, almost any initial estimate will lead to one of the solutions; while for eight simultaneous tenth degree equations, the domain becomes much smaller, and it may be very difficult to obtain an initial estimate for which the iteration converges. Kantorovich thus modified Newton's method for optimization problems to become a rapidly converging descent method. This is done by selecting the direction vectors according to Newton's method but moving along them to a point that minimizes $F(\mathbf{X})$ in that direction.

1) Method of Steepest Descent [14]: From a geometric point of view, the steepest descent method involves describing a piecewise linear path with right angled corners in N-dimensional Euclidian space, with the path terminating at the minimum of the quadratic functional $F(\mathbf{X})$. This is achieved by generating

$$\mathbf{X}_{n+1} = \mathbf{X}_n + t_n\mathbf{A}^T \cdot \mathbf{R}_n \tag{13}$$

where

$$\mathbf{R}_n = \mathbf{A}X_n - \mathbf{Y} \quad \text{and} \quad t_n = \frac{\langle \mathbf{A}^T\mathbf{R}_n, \ (\mathbf{A}^T\mathbf{R}_n)^*\rangle}{\langle \mathbf{A}\mathbf{A}^T\mathbf{R}_n, \ (\mathbf{A}^T\mathbf{R}_n)^*\rangle}. \tag{14}$$

Unfortunately it turns out that the choice of the best local direction along the largest residual reduction of $F(\mathbf{X})$ in each iteration does not always yield good convergence. This is illustrated by solving the same problem as presented in example 1. The various iterates are in Table V.

2) Conjugate Direction Method [1]: Conjugate direction methods are based on the generation of a set of A-orthogonal vectors and then minimizing successively in the direction of each of them. A set of vectors $\{\mathbf{P}_n\}$, $n = 1, 2, \cdots, N$ is said to be A-conjugate or A-orthogonal if they satisfy

$$\langle \mathbf{A}\mathbf{P}_i, \mathbf{P}_j^*\rangle = 0, \qquad \text{for } i \neq j.$$

Geometrically the method of conjugate directions is equivalent to that of finding the center of an N-dimensional ellipsoid when the starting point is on the surface of the ellipsoid. Thus the minimum point lies on a line parallel to a fixed nonnull vector $\mathbf{P}_k$ which lies on the $(N-1)$ dimensional hyperplane

$$\langle \mathbf{P}_k^*, \mathbf{A}\mathbf{X} - \mathbf{Y}\rangle = 0 \tag{15}$$

whose normal is $\mathbf{AP}$. This $(N-1)$ dimensional plane contains the center point $\mathbf{X}_{\text{exact}} = A^{-1}\mathbf{Y}$ of the ellipsoid in the given space and is said to be conjugate to the vector $\mathbf{P}_k$. Thus the conjugate direction methods are finite step methods. That is, theoretically they all yield the exact solution at the end of a finite number of steps ($\leqslant N$), assuming no truncation and roundoff error.

The finite number of steps is equivalent to the number of independent eigenvalues of $\mathbf{A}$ provided the dependent eigenvalues do not constitute a Jordan canonical form. Thus if the eigenvalues are equal, $\mathbf{A}$ is proportional to an identity matrix and hence convergence would be obtained in one step.

But the conjugate direction method does not specify how to compute the vector $\mathbf{P}_k$. When the vectors $\mathbf{P}_k$ are obtained by A-orthogonalization of the unit coordinate vectors this particular conjugate direction method yields the popular Gaussian elimination. When the vectors $\mathbf{P}_k$ are obtained by A-orthogonalization of the residual vectors $\mathbf{R}_k$, a conjugate gradient method results. Thus the conjugate gradient method applies more constraints on the iteration process than those imposed by Gaussian elimination. Hence the conjugate gradient method may yield acceptable results under conditions where Gaussian elimination fails.

3) Conjugate Gradient Method [15]-[17]: For the solution of $\mathbf{AX} = \mathbf{Y}$, the conjugate gradient method starts with an initial guess $\mathbf{X}_0$ and obtains

$$\mathbf{P}_0 = -\mathbf{A}^T\mathbf{R}_0 = -\mathbf{A}^T[\mathbf{A}\mathbf{X}_0 - \mathbf{Y}] \tag{16}$$

and then develops each successive approximation by

$$\mathbf{X}_{n+1} = \mathbf{X}_n + t_n\mathbf{P}_n \tag{17}$$

where

$$t_n = -\frac{\langle \mathbf{A}\mathbf{P}_n, \mathbf{P}_n^*\rangle}{\langle \mathbf{A}\mathbf{P}_n, (\mathbf{A}\mathbf{P}_n)^*\rangle} = \frac{\|\mathbf{A}^T\mathbf{R}_n\|^2}{\|\mathbf{A}\mathbf{P}_n\|^2}. \tag{18}$$

The residuals are generated as

$$\mathbf{R}_{n+1} = \mathbf{R}_n + t_n \cdot \mathbf{A}\mathbf{P}_n. \tag{19}$$

The direction vectors are obtained iteratively as

$$\mathbf{P}_{n+1} = -A^T\mathbf{R}_{n+1} + q_n\mathbf{P}_n \tag{20}$$

where

$$q_n = \frac{\langle \mathbf{A}\mathbf{P}_n, \ (\mathbf{A}\mathbf{A}^T\mathbf{R}_{n+1})^*}{\langle \mathbf{A}\mathbf{P}_n, \ (\mathbf{A}\mathbf{P}_n)^*\rangle} = \frac{\|\mathbf{A}^T\mathbf{R}_{n+1}\|^2}{\|\mathbf{A}^T\mathbf{R}_n\|^2}. \tag{21}$$

TABLE V
RESULTS OF VARIOUS ITERATIONS BY METHOD OF STEEPEST DESCENT[1]

i	X_0	X_1	X_2	X_3	X_4	X_5	X_6
1	1.0	−0.08125	−0.02817	−0.04386	−0.01563	−0.01713	−0.00997
2	1.0	−0.03967	0.04264	0.01892	0.01358	0.00739	0.00690
3	1.0	−0.03967	0.04264	0.01892	0.01358	0.00739	0.00690
4	1.0	0.41779	0.02524	0.01373	−0.00251	−0.00099	0.00202
		X_7	X_8	X_9	X_{10}		
		−0.01002	−0.00584	−0.00587	−0.00342		
		0.00408	0.00404	0.00238	0.00237		
		0.00408	0.00404	0.00238	0.00237		
		0.00122	0.00119	0.00072	0.00069		

[1] Note that the method of steepest descent converges much faster than either of the Seidel methods. Once again the tactic of seeking the most efficient goal by choosing the best local option does not lead to the best overall strategy. The rate of convergence of the method of steepest descent is discussed in Section-III.

TABLE VI
RESULTS OF VARIOUS ITERATIONS GIVEN BY THE CONJUGATE GRADIENT METHOD

i	X_0	X_1	X_2	X_3	X_4
1	1.0	−0.8125	−0.04848	0.6012×10^{-5}	0.7966×10^{-7}
2	1.0	−0.03966	0.02373	0.6217×10^{-5}	0.2905×10^{-6}
3	1.0	−0.03966	0.02373	0.6217×10^{-5}	0.2905×10^{-6}
4	1.0	0.41779	0.00599	0.1281×10^{-5}	-0.9905×10^{-6}

The conjugate gradient method is applied to solve the same problem presented in Example 1. The various iterates for the solution are shown in Table VI.

Observe that there is a sharp increase in the accuracy of the solutions at $\mathbf{X}_3$. One has obtained essentially an exact solution after three iterations. This is because the four eigenvalues of the matrix $\mathbf{A}$ are 2.4372, 0.9725, 0.300, and 0.2903. Note that there are approximately three independent eigenvalues of $\mathbf{A}$. Thus one would expect excellent results at the end of three steps. Hence the conjugate gradient method might converge quite rapidly for a large system of equations if the matrix has quite a few eigenvalues bunched together. This generally happens in matrices which have dominant diagonals (as in the magnetic field integral equation).

Next we derive the various theoretical rates of convergence of the various iterative schemes and show how the conjugate gradient method converges much faster than the others.

IV. ANALYSIS OF CONVERGENCE OF VARIOUS ITERATIVE METHODS [6]

The rates of convergence of the various iterative schemes, both linear and nonlinear, are discussed in this section. We show that for the linear iterative schemes, the rate at which the $\mathbf{X}_n$ approach the exact solution is linear and the $\mathbf{X}_n$ converge geometrically. The nonlinear iterative schemes on the other hand have a geometrical rate of convergence to begin with and possess "superlinear" convergence. The rates of convergence have been given when $\mathbf{A}$ is positive/negative definite.

A. Rate of Convergence for the Linear Iterative Schemes

Let a sequence $\{s_n\}$ converge to α. Then the sequence $\{s_n\}$ is said to have linear convergence if

$$e_{i+1} = (\beta + \sigma_i)e_i \tag{22}$$

where $e_i = \alpha - s_i$ and for a constant β, $|\beta| < 1$ and $\sigma_i \to 0$ as $i \to \infty$. The sequence $\{s_n\}$ is said to have geometric convergence if

$$e_{i+1} = \beta e_i \tag{23}$$

for $|\beta| < 1$. Thus geometric convergence is a special case of convergence in which all $\sigma = 0$. For a large number of iterations the linear iterative schemes converge geometrically as for sufficiently large m and $k > 0$

$$||\mathbf{X}_{m+k} - \mathbf{X}_0|| < |\lambda_1|^k \, ||\mathbf{X}_m - \mathbf{X}_0|| \tag{24}$$

where λ_1 is the dominant eigenvalue of the iteration matrix $\mathbf{T}$ in the linear iterative schemes

$$\mathbf{X}_{n+1} = \mathbf{T}\mathbf{X}_n + \mathbf{W}. \tag{25}$$

Thus the smaller the eigenvalue of $\mathbf{T}$, the faster the convergence of the linear iterative schemes. Conversely, when the magnitude of the dominant eigenvalue of $\mathbf{T}$ is close to unity, many iterations are necessary. Hence the number of iterations necessary to reduce the error $||\mathbf{X}_m - \mathbf{X}_0||$ by a factor of ten is approximately inversely proportional to $-1/\{\log_{10}\lambda_1\}$. Thus to gain an additional significant decimal place in $\mathbf{X}_m$ demands k iterations from (24).

So the fastest rate of convergence that can be achieved by the linear iterative schemes can at best be geometric and the successive approximations always converge for a definite system of equations. Equivalently the latter condition may also be stated by saying $||\mathbf{T}|| < 1$. This is also true if the diagonals of the matrix $\mathbf{T}$ are diagonally dominant.

B. Rate of Convergence for Nonlinear Iterative Schemes

1) Method of Steepest Descent: If the matrix $\mathbf{A}$ has N eigenvalues which are ordered as

$$B = \lambda_1 \geqslant \lambda_2 \geqslant \lambda_3 \cdots \geqslant \lambda_N = b > 0, \tag{26}$$

then it has been shown [14]-[18]

$$F(\mathbf{X}_k) - F(\mathbf{X}_{\text{exact}}) \leqslant \left(\frac{B-b}{B+b}\right)^{2k} \cdot \{F(X_0) - F(\mathbf{X}_{\text{exact}})\} \tag{27}$$

where B and b are the largest and the smallest eigenvalues of **A**, respectively. Also it can be shown that [14]

$$\|\mathbf{X}_k - \mathbf{X}_{\text{exact}}\|^2 \leqslant \frac{2}{b}\{F(\mathbf{X}_k) - F(\mathbf{X}_{\text{exact}})\}$$
$$\leqslant \frac{F(\mathbf{X}_0) - F(\mathbf{X}_{\text{exact}})}{0.5 \times b} \cdot \left(\frac{B-b}{B+b}\right)^{2k}. \tag{28}$$

Thus the method of steepest descent converges geometrically to the exact solution. For the case when **A** has N distinct eigenvalues Akaike [19] has shown that this is the best possible estimate.

It has been shown by Daniel [17] and Hayes [9] that whenever **A** is a Legendre operator (i.e., A is a sum of positive definite bounded self-adjoint operator plus a completely continuous operator) the method of steepest descent converges faster than that of a geometric series with ratio greater than $(B - b)/(B + b)^2$. This type of convergence is referred to as "superlinear" convergence. Thus the method of steepest descent converges at worst like a geometric series, and in most cases the convergence is superlinear. For a finite dimensional case **A** is always a Legendre operator.

2) Method of Conjugate Gradient: The method of conjugate gradient generally requires a little more computation than that of steepest descent. However this slight increase in computation required leads to a significant improvement in the rate of convergence over that of steepest descent. It can be shown that [6]

$$\frac{\|\mathbf{X}_k - \mathbf{X}_{\text{exact}}\|}{\|\mathbf{X}_0 - \mathbf{X}_{\text{exact}}\|} \leqslant \frac{2}{\left(\frac{\sqrt{B}+\sqrt{b}}{\sqrt{B}-\sqrt{b}}\right)^k + \left(\frac{\sqrt{B}-\sqrt{b}}{\sqrt{B}+\sqrt{b}}\right)^k} \tag{29}$$

$$\leqslant 2\left(\frac{\sqrt{B}-\sqrt{b}}{\sqrt{B}+\sqrt{b}}\right)^k. \tag{30}$$

Observe that (29) is a better estimate than (30). As is well known [19], (29) cannot be improved upon.

Also as before, if **A** is a Legendre operator then the method of conjugate gradient converges faster than a geometric series with ratio greather than $(\sqrt{B} - \sqrt{b})/(\sqrt{B} + \sqrt{b})^2$ [19]. In an N-dimensional quadratic problem the error tends to zero in one step with Newton's method and within M steps with the conjugate gradient method, where M is the number of independent eigenvalues of **A**. Hence it is generally stated that the conjugate gradient method yields "quadratic convergence" $[\|\mathbf{X}_{m+1} - \mathbf{X}_0\| \leqslant c\|\mathbf{X}_m - \mathbf{X}_0\|^2]$ in the sense that it converges in M steps for an N-dimensional problem. So it seems appropriate to term this "$1/M$ quadratic convergence" since it requires M steps to achieve the effect of one step of a method with a true quadratic convergence rate.

V. ROUNDOFF ERRORS ASSOCIATED WITH ITERATIVE SCHEMES

In iterative methods the condition number of **A** has very little influence on roundoff error. The roundoff error in iterative methods is confined to the last stage of iteration only.

In a system of linear equations that arises from a practical problem, the elements of matrix **A** and **Y** may not be sharply defined. It is now assumed that all that is known about the typical element A^{ij} of matrix **A** or Y^i for the matrix **Y** is that they are within certain intervals. Here the subscript E denotes exact quantities

$$A_E{}^{ij} - \Delta A^{ij} \leqslant A^{ij} \leqslant A_E{}^{ij} + \Delta A^{ij} \tag{31}$$

$$Y_E{}^i - \Delta Y^i \leqslant Y^i \leqslant Y_E{}^i + \Delta Y^i. \tag{32}$$

It is also assumed that ΔA^{ij} and ΔY^i are independent distinct quantities due to round-off error and that one does not depend on the other. Thus

$$\mathbf{A}_E\mathbf{X}_E = \mathbf{Y}_E \tag{33}$$

and denote

$$\mathbf{R} = \mathbf{A}_E\mathbf{X} - \mathbf{Y}_E. \tag{34}$$

We check whether

$$\sum_j \Delta A^{ij} \cdot |X^j| + \Delta Y^i \geqslant |R^i| \tag{35}$$

for $i = 1, 2, \cdots, n$. As was shown by Oettli and Prager [21] the inequality (35) is a necessary and sufficient condition for **X** to be a solution of **AX** = **Y** under (31) and (33). So in any iteration method where the residuals are computed routinely, if for a certain residual the inequality (35) is satisfied, we have obtained an excellent solution under the conditions (31) and (32). Note that the introduction of the conditions (31) and (32) does not make the solution of **AX** = **Y** unique. In fact, there are many solutions of **AX** = **Y**. But only those solutions are acceptable which satisfy the inequality (35). Thus the upper and lower bounds for a certain component of the solution X are obtained by solving the linear programming problem [22]

$$\left.\begin{array}{l} \min \rightarrow \\ X^j \\ \max \rightarrow \end{array}\right| \begin{array}{l} R^i - \sum_j \Delta A^{ij} \cdot |X^j| - \Delta Y^i \leqslant 0 \\ \\ -R^i - \sum_j \Delta A^{ij} \cdot X^j - \Delta Y^i \leqslant 0 \end{array}$$

$$\text{for } i = 1, 2, \cdots, N. \tag{36}$$

Iterative methods may be quite advantageous for large systems of matrices or for ill-conditioned matrices as compared to a direct method like Gaussian elimination. This is because *cond (A) does not arise in round-off error analysis of iterative methods*; and this is the reason we have been able to solve a 7 × 7 system of equations when **A** is a Hilbert matrix by the conjugate gradient method where Gaussian elimination has failed.

VI. CORE STORAGE REQUIRED FOR VARIOUS METHODS [1]

The core storage required for various methods is listed in order of the amount of core storage required, starting with the method requiring the least core storage.

Method	*Core Storage*
Gaussian elimination	$N^2 + 2N$
Seidel's iterative method	$N^2 + 2N$

Gaussian elimination with complete pivoting	$N^2 + 3N$
Jacobi's iterative method	$N^2 + 3N$
Method of Steepest Descent	$N^2 + 4N + 2$
Conjugate gradient method	$N^2 + 6N + 3$.

Note that Gaussian elimination with no pivoting and Seidel's method require the least amount of storage and that the conjugate gradient method requires the largest amount of storage.

VII. OPERATIONS REQUIRED FOR VARIOUS METHODS

The number of divisions, multiplications, and additions/subtractions provide a rough estimate of the efficiency of the algorithm. For each method it is possible to estimate the number of arithmetic operations as a function of N, the order of the matrix. Such functions could be discontinuous if N is large enough that auxiliary storage is required. In the total number of arithmetic operations we have *not* included the timings taken for recording of intermediate results and the time taken in searching for the pivotal element in Gaussian elimination.

Method	Number of Arithmetic Operations		
	Division $\div$	Multiplications $\times$	Additions $+$
Gaussian elimination	N	$\frac{N^3}{3} + N^2 - \frac{N}{3}$	$\frac{N^3}{3} + \frac{N^2}{2} - \frac{5N}{6}$ (total).

(Gaussian elimination with complete pivoting requires $N^3/3 + N^2/2 - 5N/6$ comparisons in addition to the above arithmetic operations)

Jacobi and Seidel	N	N^2	$N^2 - \mathrm{N}$ per iteration

(We could do with N divisions only once rather than per iteration at the expense of N more storage spaces.)

Steepest descent for (unsymmetric) matrix **A**	1	$2N^2 + 3N$	$2N^2 + 4N$ per iteration
Conjugate gradient for (unsymmetric) matrix **A**	2	$2N^2 + 6N$	$2N^2 + 6N$ per iteration

For symmetric matrices the work is reduced by half.

For vary large values of N, an iterative method applied to a full symmetric matrix would need to converge in fewer than $N/3$ steps to bring its operations count down to that of a direct method.

VIII. A SPECIAL NOTE ON THE CONJUGATE GRADIENT METHOD

An iterative method called the banded matrix iterative scheme has recently been applied by Ferguson [23] to solve large electromagnetic field problems by the method of moments. The characteristic features of the method applied by Ferguson are as follows.

1) The convergence of the iterative scheme is sensitive to the choice of the numbering scheme used.
2) Because of (1) it requires a person with certain technical background to run the program.
3) The rate of convergence is irregular and sometimes the solution diverges.
4) The banded matrix iterative scheme applied by Ferguson is basically a Jacobi type of iterative scheme and hence it converges slowly [23, p. 16]
5) Theoretically, the method requires an infinite number of steps to converge to the exact result if there is no round-off error.

As we have seen from the previous sections the conjugate gradient method is a nonlinear iterative scheme, in contrast to the linear Jacobi method. The conjugate gradient method also converges at a faster rate than that of a geometric series. Moreover it is highly insensitive to the choice of the initial guess for the solution. Since the conjugate gradient method yields an exact result (assuming no round-off errors) in at most M steps (where M is the number of independent eigenvalues of the $N \times N$ matrix), it has the good points of both an iterative method and a direct method of solution. It has the advantage of an iterative scheme in that roundoff error is limited only to the final step of the solution. It has the advantage of a direct method in that it converges in a finite number of steps.

As a first example consider a wire 3-m in length and 0.01-m in radius. The wire is charged to a potential of $4\pi\epsilon$ volts. The objective is to find the charge distribution on the wire. A method of moments formulation has been employed and the wire is divided into 30 segments. The moment matrix formed by this problem is a typical one which often occurs in the method of moments. The results are presented in Table VII. The first three columns indicate the charge distribution on the wire obtained by the conjugate gradient method. The third column represents the charge distribution corresponding to the segment numbers appearing on column two. The first column states that this result has been obtained at the end of three iterations. The next three columns indicate the charge distribution obtained after eight iterations by the conjugate gradient method. Finally the seventh column gives the result due to Gaussian elimination. As is clear from the data presented in Table VII the conjugate gradient method yields a result better than one percent after three iterations ($M = N/10$).

If for this problem the banded matrix technique is used to yield an accuracy of one percent in one interation, a bandwidth of approximately 15 may be necessary [23, table 10, p. 34]. Hence for the same accuracy, the conjugate gradient method is faster by a factor of 2.5. Also, if the same problem is to be solved by the symmetric Cholesky decomposition, it would have required approximately $N^3/6$ multiplications. The conjugate gradient method required approximately $3N^2$ multiplications as compared to $5N^2$ for Gaussian elimination. Also note that an essentially exact result has been obtained (accuracy better than 10^{-5} in the residuals) after only eight iterations.

As a second example, consider the same problem as above but now the wire is 25-m long. So this time **A** is a 100×100 matrix. Again we obtained an essentially exact result (better than 10^{-5} in the residuals) *after only nine iterations*. This implies that in this type of problem the number of independent eigenvalues is approximately eight or nine. Note that the number of independent distinguishable eigenvalues does not increase as the order of the system is increased considerably. This is an interesting property of diagonally dominant matrices which could easily be exploited by the conjugate gradient method.

As a third example, consider **A** as a 20×20 Hilbert matrix

TABLE VII
COMPARISON OF THE DIFFERENT SOLUTIONS FOR THE CHARGED WIRE

①	②	③	④	⑤	⑥	⑦
3	1	0.1310801E+00	8	1	0.1319899E+00	0.1319883E 00
3	2	0.1079367E+00	8	2	0.1062454E+00	0.1062447E 00
3	3	0.9741676E−01	8	3	0.9698462E−01	0.9698367E−01
3	4	0.9380448E−01	8	4	0.9409392E−01	0.9409255E−01
3	5	0.9250486E−01	8	5	0.9345561E−01	0.9345549E−01
3	6	0.8838725E−01	8	6	0.8868694E−01	0.8868611E−01
3	7	0.8869338E−01	8	7	0.8924234E−01	0.8824180E−01
3	8	0.8770591E−01	8	8	0.8796191E−01	0.8796018E−01
3	9	0.8790541E−01	8	9	0.8807307E−01	0.8807319E−01
3	10	0.8776426E−01	8	10	0.8779728E−01	0.8779657E−01
3	11	0.8714312E−01	8	11	0.8701491E−01	0.8701420E−01
3	12	0.8617383E−01	8	12	0.8591259E−01	0.8592147E−01
3	13	0.8600593E−01	8	13	0.8575714E−01	0.8575720E−01
3	14	0.8614635E−01	8	14	0.8587742E−01	0.8587623E−01
3	15	0.8584630E−01	8	15	0.8543098E−01	0.8542997E−01
3	16	0.8631968E−01	8	16	0.8599222E−01	0.8599192E−01
3	17	0.8616120E−01	8	17	0.8590782E−01	0.8590662E−01
3	18	0.8592218E−01	8	18	0.8567208E−01	0.8567178E−01
3	19	0.8541530E−01	8	19	0.8520442E−01	0.8520377E−01
3	20	0.8547300E−01	8	20	0.8517402E−01	0.8517450E−01
3	21	0.8801979E−01	8	21	0.8832079E−01	0.8832002E−01
3	22	0.8718151E−01	8	22	0.8732998E−01	0.8732986E−01
3	23	0.8758122E−01	8	23	0.8772111E−01	0.8772045E−01
3	24	0.8841532E−01	8	24	0.8893603E−01	0.8893472E−01
3	25	0.8897913E−01	8	25	0.8934522E−01	0.8934456E−01
3	26	0.9207904E−01	8	26	0.9296322E−01	0.9296477E−01
3	27	0.9305668E−01	8	27	0.9314167E−01	0.9314030E−01
3	28	0.9872329E−01	8	28	0.9863120E−01	0.9863114E−01
3	29	0.1081076E+00	8	29	0.1062285E+00	0.1062284E 00
3	30	0.1315722E+00	8	30	0.1324828E+00	0.1324818E 00
	Conjugate gradient method after three iterations			Conjugate gradient method after eight iterations		Gaussian elimination

and let **Y** be chosen in such a way that the solution vector has components 1–20. The problem then is to find **X** given **A** and **Y**. The philosophy behind choosing **A** to be a Hilbert matrix is that nearly singular matrices are often encountered in a system identification problem. So if the conjugate gradient method can efficiently solve such an ill-conditioned problem, then this method may easily be applicable to system-identification problems. The results obtained by two different methods are shown in Table VIII. Calculations were carried out using double precision.

It is clear from Table VIII that the conjugate gradient method yields good results at the end of eight steps. The largest error is only 2.25 percent. The Gaussian elimination method for the same problem completely breaks down. (Note: the Hilbert matrix is extremely ill-conditioned. The condition number of a 20 × 20 Hilbert matrix is of the order of $e^{3.5N} = 2.5 \times 10^{30}$ from [2].)

As a final example, consider the solution of the two components of the current density of a 1λ square metal plate irradiated by a plane wave. When the total number of unknowns for the complex current is 71, we have to solve a 71 × 71 matrix equation. The total time taken for the solution of the complete problem utilizing various techniques is as follows. Gaussian elimination: 27 s (CPU time); conjugate gradient method: 30 s (CPU time) (with 1 percent accuracy in the residual).

Observe that the conjugate gradient method is quite inefficient in this case. However as the dimension of the problem is increased from 71 to 180, the time required by various methods to solve the complete problem is as follows. Gaussian elimination: 500 s (CPU time); conjugate gradient method: for 10^{-2} accuracy in the residual–220 s (CPU time); for 10^{-3} accuracy in the residual–290 s (CPU time); for 10^{-4} accuracy in the residual–390 s (CPU time); for 10^{-5} accuracy in the residual–520 s (CPU time).

TABLE VIII
COMPARISON OF GAUSSION ELIMINATION AND CONJUGATE GRADIENT METHOD FOR THE SOLUTION VECTOR $\mathbf{X} = [\mathbf{A}]^{-1}\mathbf{Y}$.

Gaussian Elimination	Exact Solution	Conjugate Gradient at the End of Eight Steps
0.9999954	1	1.000289
2.000349	2	1.990388
2.978199	3	3.056398
4.103616	4	3.909776*
4.355928	5	4.981514
55.54727	6	6.056422
−20.17007	7	7.066274
−391.4351	8	8.030005
1050.932	9	8.982147
−397.4495	10	9.947065
212.3952	11	10.93533
1407.415	12	11.94698
−1800.392	13	12.97573
−682.3352	14	14.01228
2770.054	15	15.04653
−1692.660	16	16.06884
637.3160	17	17.07074
559.0285	18	18.04519
78.80465	19	18.98660
97.65299	20	19.8907

* The largest error is about 2.25 percent.

So for large systems of equations the conjugate gradient method may prove to be quite useful, especially if one is interested in obtaining an accuracy of 10^{-3} to 10^{-4} in the solutions.

In summary, it is argued that the application of the conjugate gradient method to the analysis of large bodies by method of moments would yield stable, reliable, consistent, and accurate results faster than any methods currently used to obtain a solution. The same is true for problems in system identification. However there may be some buildup of the roundoff error if the residuals are computed iteratively by (19) rather than directly from $\mathbf{AX}_k = \mathbf{Y}$, which would be more time consuming.

NOMENCLATURE

A	$N \times N$ square matrix.
XY	$N \times 1$ matrices.
A^{ij}	element belonging to the ith row and jth column.
x^i	ith element of **X**.
$\mathbf{X}_n$	matrix obtained after n iterations in interative methods and in direct methods it is just another matrix obtained after processing it n times.
$\|\|X\|\|$	norm of X.
cond [**A**]	condition number of **A** $\triangleq$ (largest eigenvalue of (**A**))/(minimum eigenvalue of (**A**)).

REFERENCES

[1] J. Westlake, *A Handbook of Numerical Matrix Inversion and Solution of Linear Equations*. New York: Wiley, 1968.

[2] G. Forsythe and C. B. Moler, *Computer Solution of Linear Algebraic Systems*. Englewood Cliffs, NJ: Prentice–Hall, 1967.

[3] D. K. Fadeev and V. N. Fadeeva, *Computational Methods of Linear Algebra*. San Francisco: W. H. Freeman, 1963, (translated from Russian by R. C. Williams).

[4] V. V. Klyuyev and N. I. Kokovkin—Scherbak, "On the minimization of the number of arithmetic operations for the solution of linear algebraic systems of equations," Comp. Sci. Dept., Stanford University, CA, Tech. Rep.-CS24, 1965, (translated by G. J. Tee).

[5] Volker Strassen, "Gaussian elimination is not optimal," *Numer. Math.*, vol. 13, pp. 354–365, 1969.

[6] T. K. Sarkar *et al.*, "Solution of large systems of linear equations," RADC-TR-81-103, Rochester, NY, Sept. 1979.

[7] M. R. Hestenes, "Applications of the theory of quadratic forms in Hilbert space to the calculus of variations," *Pacific J. Math.*, vol. 1, pp. 525–581, 1951.

[8] R. M. Hayes, "Iterative methods of solving linear problems in Hilbert space," in *Contributions to the solution of systems of linear equations and the determination of eigenvalues*, O. Taussky, ed. Nat. Bur. Standards Appl. Math. Ser., vol. 39, pp. 71–104, 1954.

[9] G. E. Forsythe, "Solving linear equations can be interesting," *Bull. of Am. Math. Society*, pp. 299–329, 1953.

[10] ——, "Theory of selected methods of finite matrix inversion and decomposition," U.S. Dept. of Commerce, INA 52-5.

[11] W. Kahan, "Gauss–Seidel methods for solving large systems of linear equations," Ph.D. dissertation, University of Toronto, 1958.

[12] D. Young, "On the solution of linear systems by iterations," in *Am. Math. Soc. Numerical Analysis Proceedings of Symposia in Applied Mathematics*, J. H. Curtiss, Ed. New York: McGraw—Hill, 1956.

[13] ——, "On the solution of large systems of linear algebraic equations with sparse positive definite matrices," in *Numerical solution of Nonlinear Algebraic equations*, G. D. Byrne, Ed. New York: Academic, 1974.

[14] L. V. Kantorovich, "Functional analysis and applied mathematics," *Uspekhi Matematicheskikh Nauk*, vol. III, no. 6, pp. 89–185, 1948.

[15] J. W. Daniel, "The conjugate gradient method for linear and nonlinear operator equations," *SIAM J. Numer. Anal.*, vol. 4, no. 1, pp. 10–26, 1967.

[16] M. Hestenes and E. Stiefel, "Method of conjugate gradients for solving linear systems," *J. Res. Nat. Bur. Standards*, vol. 49, pp. 409–436, 1952.

[17] J. W. Daniel, "The conjugate gradient method for linear and nonlinear operator equations," Ph.D. dissertation, Stanford University, CA, 1965.

[18] M. Z. Nashed, "Steepest descent for singular linear operator equations," *SIAM J. Numer. Anal.*, vol. 7, no. 3, pp. 358–362, 1970.

[19] H. Akaike, "On a successive transformation of the probability distribution and its application to the analysis of the optimum gradient method," *Ann. Inst. Statistics Math.*, vol. 11, pp. 1–16, Tokyo, 1959.

[20] W. J. Kammerer and M. Z. Nashed, "On the convergence of the conjugate gradient method for singular linear operator equations," *SIAM J. Numer. Anal.*, vol. 9, no. 1, pp. 165–181, 1972.

[21] W. Oettli and W. Prager, "Compatibility of approximate solution of linear equations with given error bounds for coefficients and right hand sides," *Num. Math.*, vol. 6, pp. 405–409, 1964.

[22] W. Oettli, "On the solution set of a linear system with inaccurate coefficients," *J. SIAM Numer. Anal.*, Ser. B., vol. 2, no. 1, pp. 115–119, 1965.

[23] T. R. Ferguson, "The EMCAP iterative techniques in the method of moments," RADC-TR-75-121, May 1975.

[24] M. Marcus, "Basic theorems in matrix theory," U.S. Dept. of Commerce, NBS, *Appl. Math.* ser. 57, 1959.

Tapan K. Sarkar (S'69–M'76–SM'81), for a photograph and biography please see page 379 of the March 1981 issue of this TRANSACTIONS.

Kenneth R. Siarkiewicz (SM'77) received the B.E.E. degree from the University of Detroit, Detroit, MI, in 1965, and the M.S.E. degree from the University of Michigan, Ann Arbor, in 1966.

He has been employed by the Air Force at the Rome Air Development Center for 16 years. During that time, he has conducted many programs whose results have provided an insight into the role played by electromagnetic coupling in the generation of electromagnetic interference. For the past nine years he has been engaged in the development of computerized analysis tools to characterize the electromagnetic phenomena associated with radiating and scattering systems. These computer codes, based on the method of moments formulation, provide a capability to model a large area with a fine resolution. The general electromagnetic model for the analysis of complex systems (GEMACS) is now a part of the Air Force Electromagnetic Compatibility/Intrasystem Analysis Program (EMC/IAP). He is also the manager of the interface mechanism responsible for the technology transfer of the EMC/IAP. In this role, he directs the dissemination of the EMC/IAP technical base to the user community and translates user needs into specifications and objectives for the Air Force R&D program in EMC.

Mr. Siarkiewicz is a member of Tau Beta Pi and Eta Kappa Nu.

Roy F. Stratton received the B.S. degree in physics from Southwestern at Memphis, in 1951, the M.S. and Ph.D. degrees in physics, in 1953 and 1957, respectively, and the M.S.E.E. degree from the Georgia Institute of Technology, in 1975.

He worked at Oak Ridge National Laboratory (1958–1970), and Pikeville College (1970–1973) before coming to the Rome Air Development Center (RADC) in 1975. At RADC, his area of expertise is in Electromagnetic Compatibility. Much of his work at RADC has had to do with the electrical properties of composite materials.

A note on the variational method (Rayleigh-Ritz), Galerkin's Method, and the method of least squares

Tapan K. Sarkar

Department of Electrical Engineering, Rochester Institute of Technology

(Received May 6, 1982; revised May 16, 1983; accepted June 8, 1983.)

The variational method (Rayleigh-Ritz), Galerkin's method and the method of least squares have been utilized extensively in electromagnetic theory to solve various field problems. In most cases, one makes a formal use of the techniques. In this paper, a tutorial approach has been taken to bring out the mathematical distinctions between the variational method (Rayleigh-Ritz), Galerkin's method and the method of least squares. Examples have been presented to illustrate the strong and weak points of each of the techniques. Finally, the rates of convergence of the three techniques have been derived. It is hoped that with the appearance of this paper, application of a particular technique to solve an electromagnetic field problem would be more rigorous rather than formal.

1. INTRODUCTION

Most problems of electromagnetic theory (as referenced in the appendix) can be formulated in terms of an operator equation

$$Ax = f \tag{1}$$

where A is often a linear integro-differential operator. The problem is to find x for a given f. From (1) we see that the solution can be written symbolically as

$$x = A^{-1}f \tag{2}$$

We shall assume that f is in the range of A. This would ensure that $A^{-1}f$ can be defined for a given f, so solution to (1) exists.

We shall also assume for our present discussions that the homogeneous equation

$$Ax = 0 \tag{3}$$

associated with (1) has only a trivial solution, i.e., $x \equiv 0$. This will guarantee uniqueness of the solution of (1).

Finally, we assume that the inverse operator is bound. This implies

$$\| A^{-1}f\| \leq C \cdot \|f\| \qquad \|f\| = 1 \tag{4}$$

where C is a constant for all $f \varepsilon R(A)$ (R(A) = range of A). It is understood that

$$\|f\| \triangleq \langle f,f\rangle^{1/2} \triangleq \{\int_a^b f(z)\bar{f}(z)dz\}^{1/2} = 1 \tag{5}$$

where the overbar over f denotes complex conjugate.

The three assumptions made above guarantee that the problem in (1) is well posed. If any of the above conditions are violated, then the problem is ill posed. For example, if the inverse operator is unbounded (i.e., (4) is not satisfied), then small perturbations in f would produce large perturbations in the solution x. The problem of obtaining an approximate solution for a class of ill posed problems has been discussed in detail by Sarkar et al. [1981].

So in the present discussions we assume

that the operator equation in (1) is well posed. We next investigate the nature of the solution yielded by the three techniques - variational method (Rayleigh-Ritz), Galerkin's method and the method of least squares. We also present the conditions under which the three techniques yield meaningful results and their rates of convergence. The definitions of the various mathematical terms are available in the appendix.

In our discussions of the various approaches, only references which are directly relevant are noted. No attempt has been made to cite the earliest sources. In many cases, additional references may be found in papers mentioned.

In our discussions, we present only Galerkin's method rather than the method of moments. For the general case of the method of moments, it is extremely difficult to develop any general conclusions regarding convergence.

All of the three above techniques proceed by assuming the nature of the solution a priori in terms of known expansion functions. They start with the idea that the approximate solution y_N of x can be expressed in the following form:

$$y_N = \sum_{n=1}^{N} a_n u_n \tag{6}$$

where a_n are certain unknown constants to be solved for. The expansion functions u_n are a complete set of functions in an N dimensional space. In other words, the sequence $\{u_n\}_{n=1}^{N}$ is a linearly independent set and thereby forms a basis for the N dimensional space in which the approximate solution y_N is sought. It is not necessary that $\{u_n\}_{n=1}^{N}$ form an orthogonal set of functions. The exact solution x $\{=\Sigma_{n=1}^{\infty} a_n u_n\}$ lies in an infinite dimensional space. By the term convergence then we imply convergence in the mean

$$\underset{N\to\infty}{\text{Lt}} \|y_N - x\| \to 0 \tag{7}$$

i.e., in the limit N approaches infinity, the mean squared difference between y_N and x approaches zero. In the remaining sections, we discuss the restrictions that each of the three methods impose on the expansion functions u_n and how an approximate solution y_N is obtained. We also address the question of convergence of $y_N \to$ x as $N\to\infty$ for the three methods.

2. VARIATIONAL METHOD (RAYLEIGH-RITZ)

In a variational method, the solution to (1) is attempted by minimizing the functional F(y) defined by [Mikhlin, 1964,1965a,b; Finlayson and Scriven, 1966,1967; Rektorys, 1972]

$$F(y) = \langle Ay,y\rangle - \langle f,y\rangle - \langle y,f\rangle \tag{8}$$

In (8) it is assumed that the operator A is a self-adjoint operator (i.e., A=A* (adjoint operator)) and D(A) = domain of A = D(A*)). So by definition, a self-adjoint operator satisfies the following equality:

$$\langle Ax,z\rangle = \langle x,Az\rangle \tag{9}$$

For non-self-adjoint operators, A and f in (8) may be replaced by A*A and A*f, respectively, and the following discussion still holds. The technique then becomes the method of least squares.

Observe that (8) can be rewritten as

$$F(y) = \langle A(y-x),(y-x)\rangle - \langle f,x\rangle \tag{10}$$

Since $\langle f,x\rangle$ is a constant, minimization of F(y) directly implies minimization of $\|y-x\|$ if and only if the operator A is positive definite. This is because

$$\|y-x\|^2 \le \langle A(y-x),(y-x)\rangle \tag{11}$$

This is the philosophy behind solving (1) in a roundabout fashion by minimizing the functional F(y) of (8).

We now replace y in (8) by y_N (as defined in (6)) to obtain

$$F(y_N) = \langle \sum_{n=1}^{N} a_n Au_n, \sum_{i=1}^{N} a_i u_i\rangle - \langle f, \sum_{n=1}^{N} a_n u_n\rangle - \langle \sum_{i=1}^{N} a_i u_i, f\rangle \tag{12}$$

We solve for the unknowns a_n by minimizing $F(y_N)$ of (12). This is achieved by taking the partial derivatives $\partial F(y_N)/\partial a_i$ for i=1,2,...,N and equating each of the partial derivatives to zero. This results in a set of equations

$$\sum_{n=1}^{N} a_n \langle Au_n, u_k \rangle = \langle f, u_k \rangle \qquad k=1,2,...,N \tag{13}$$

The unknowns a_n are solved from (13).

In the following subsections we develop the finer structures and limitations of this technique.

2.1. Does a functional F(y) always exist?

According to Stakgold [1981], the solution x of a boundary value problem can also be characterized as the element for which a related functional F(y) is stationary. Such a stationary characterization is a variational principle whose Euler-Lagrange equation is the differential equation of the original boundary value problem. The variational problem may also provide guidance to the existence questions for the boundary value problem.

If Ax=f has a solution x_0, this solution minimizes the functional F(y) (as defined in equation (10)). Conversely, if there exists an element $x_0 \varepsilon D(A)$ that minimizes the functional F(y), then x_0 satisfies Ax=f. This method of solving a boundary value problem that consists in replacing Ax=f by the problem of minimizing a functional F(y) is also called the energy method (as it minimizes the energy of the system). This implies that a variational method finds a solution with finite energy, or mathematically $\langle f, x_0 \rangle$ or $\langle Ax_0, x_0 \rangle$ is bounded.

As a first example, condiser the self-adjoint operator A as

$$Ax = -\frac{d^2x}{dz^2} \qquad x(z=0) = 0 \qquad x(z=1) = 0 \tag{14}$$

It has been shown by Mikhlin [1964] that this operator is positive definite and has a bounded inverse. Moreover, the homogeneous equation has the trivial solution. Now if we select

$$f = -\frac{1}{4} z^{-3/2}$$

then the exact solution is given by

$$x = z - \sqrt{z}$$

We observe

$$F(x) = \infty$$

as

$$\langle f, x \rangle = \int_0^1 \left(-\frac{z^{-1.5}}{4}\right)(z - \sqrt{z})dz = \infty$$

Hence, the variational method is defined only for a class of problems in which $\langle f, x \rangle$ and $\langle f, y_N \rangle$ are bounded.

The discussion so far does not address the question of existence of the functional F(y) nor does it show how a solution x_0 can be obtained. However, such information may be obtained if the operator A is positive/negative definite or for a semidefinite operator if $\langle x_0, f \rangle$ is bounded. In both cases it is assumed that f is square integrable. Thus, if $\langle x_0, f \rangle$ is unbounded, a variational method cannot be set up for the solution of Ax=f, as the functional F(y) become infinite. This is true for any self-adjoint operator. As an analogy, consider the vector field $\vec{E}$. Can $\vec{E}$ be derived from a potential V? If the curl of the vector is zero, i.e., $\vec{\nabla} x \vec{E} = 0$, then only the vector can be represented as a potential $\vec{E} = \vec{\nabla} V$. When the curl is zero, the integral of $\vec{E}$ along a path in space Z is independent of the actual path taken and depends only on the end points z_1 and z_2, i.e.,

$$\int_{z_1}^{z_2} \vec{E} \cdot d\vec{z} = V_2 - V_1$$

This provides an equivalent method of determining whether a vector field can be derived from a potential. So from these elementary concepts of vector calculus it follows that if we regard the Euler equation in a variational method as the gradient of a functional F(y) analogous to a potential, then we should not expect every operator equation to be derivable from a potential independent of the path integral.

Consider again the second order differential operator given by (14). We know that the domain of the operator D(A) consists of the space of twice differentiable and square integrable functions which satisfy the boundary conditions. Now if we consider the solution to Ax=f, where A is the second order differential operator and f is a square integrable function, our starting point for solution is F(y) as given by (8). In this case, F(y) is bounded below on D(A) so that F(y) has a minimum, say d, on D(A). If there is an element $x_0 \varepsilon D(A)$, for which $F(x_0)=d$, the infimum is attained on D(A), so that the problem for F(y) on D(A) has the solution x_0. There are two reasons for considering the minimum of F(y) on a larger domain M (as yet unspecified). First, F(y) may not have a minimum on D(A). If f is only piecewise continuous, and A is defined by (14), then the solution x_0 to Ax=f is not in the space of square integrable, twice differentiable functions, so that Ax=f has no solution. Hence, F(y) does not have a minimum on D(A). The trouble here is simply that D(A) is not a suitable domain on which to analyze the boundary value problem when f is not continuous. The second reason is more subtle. Even if F(y) has a minimum on D(A), it may be hard to prove the fact directly. It turns out to be easier to prove the existence of a minimum on the larger domain M and then to show that the minimizing element lies in D(A). Thus, we shall try to enlarge D(A) so that the minimum for F(y) can be guaranteed on the new domain. It is reasonable to hope for such an existence since the expression for F(y) involves only the first derivative of x, whereas an element of D(A) has a continuous derivative of order 2.

We now complete D(A) to form the larger domain M. This is achieved by completing D(A) with respect to a new inner product

$$\langle u,u\rangle_A \triangleq \langle Au,u\rangle \qquad u \varepsilon M$$

The completion of D(A) with respect to this inner product is a new domain M. Thus, all elements of M are with finite energy as <Au,u> is bounded. If the operator A is positive definite, than all elements of M satisfy

$$\langle Au,u\rangle \geq c^2 \langle u,u\rangle = c^2 \cdot \|u\|^2$$

where c is a predetermined positive constant. Therefore, for a positive definite operator,

$$\|u\|^2 = \langle u,u\rangle \leq \frac{\langle Au,u\rangle}{c^2} = \frac{\langle u,u\rangle_A}{c^2} \tag{15}$$

and all elements of M are in D(A). But in general, M is larger than D(A).

Now we investigate the conditions for the existence of the functional F(y) on M. For a positive definite operator, A is defined by (14) and

$$\langle u,f\rangle^2 \leq \|f\|^2 \cdot \|u\|^2$$

$$\leq \frac{\|f\|^2}{c^2} \langle Au,u\rangle \qquad u\varepsilon M \tag{16}$$

which implies that for any square integrable function f, <u,f> is always bounded on M. Thus, the variational method exists for a positive definite operator and the operator equation has a solution for any square integrable function f.

However, when A is semidefinite, then inequalities like (15) and (16) do not exist. Hence, for the variational method to exist, it is required that <u,f> be bounded for a functional F(y) to exist.

Since <u,f> is now bounded, then according to Stakgold [1981] there exists a unique element $u_0 \varepsilon M$ such that $\langle u,f\rangle = \langle u,Au_0\rangle$ for $u\varepsilon M$. This element u_0 also minimizes the functional F(y) and the minimum value is given by

$$F(u_0) = -\langle u_0,u_0\rangle_A = -\langle Au_0,u_0\rangle$$

For a positive definite operator it is also possible to obtain bounds for F(y) on both sides. Let u_0 be the exact solution and y_N be the Nth approximation; then it can be shown that

$$F(y_N) - \frac{\|Ay_N - f\|^2}{L_A} \leq F(u_0)$$

$$\leq F(y_N) - \frac{\|Ay_N - f\|^2}{U_A} \tag{17}$$

where L_A and U_A are the lower and upper bounds for A, i.e.,

$$U_A \cdot ||x||^2 \geq <Ax,x> \geq L_A ||x||^2$$

When the solution u_0 constructed by the variational method lies in the subset D(A) of M, and satisfies Ax=f, then we call u_0 the classical solution. It may so happen that u_0 which minimizes the functional F(y) on the domain M may not be in D(A). In that case u_0 is referred to as the generalized solution.

These mathematical concepts can be explained by the following example. Let us view the minimization of a functional from a point of generating a soap film on a deformed wire for producing soap bubbles. The wire is bent in any odd shape and is dipped into a soap solution and slowly removed. The equation to the surface of the soap film generated usually gives the solution to the problem of minimizing a functional. The functional in general represents the energy associated with the soap film. There are three possible alternatives:

1. The case where the surface of the soap film minimizes the energy and also the functional is representable in terms of known functions. The solution to this problem is straightforward and yields the classical solution.

2. In this case, the surface minimizes the energy of the film, but the solution to the surface is not representable by elementary known functions in the domain of the operator. That is, we know the functional $F(y_N)$ exists, but we do not know how to obtain y_N. In this case we obtain the generalized solution of the problem.

3. For this case, we consider a surface of the soap film, where it is difficult to see that the film indeed minimizes the energy; this occurs, for example, when three surfaces meet at a point. For this class of problems, the variational method is not clearly defined.

In this section we have shown that case 3 occurs when <f,x> is unbounded. When the element y_N which minimizes $F(y_N)$ does not lie in the domain of the operator, then case 2 occurs. When y_N is in the domain of A, then we get case 1.

For a non self-adjoint operator, if we do not intend to replace A by A*A and f by A*f in (8) and proceed with the classical approach, then serious problems may arise.

2.2. <u>Does minimization of $F(y_N)$ yield a set of approximate solutions y_N which converges monotonically in the mean to x(i.e., $\mathrm{Lt}_{N\to\infty} ||y_N - x|| \to 0$)?</u>

The answer is yes, only if A is a definite operator. Otherwise there is no guarantee that the error in the solution $||y_N - x||$ becomes smaller as the order of the approximation is increased. We illustrate this by an example.

As a second example, consider the solution of the following 3x3 matrix equation.

$$\begin{bmatrix} 3 & -4 & 2 \\ -4 & 3 & 6 \\ 2 & 6 & 2 \end{bmatrix} \begin{bmatrix} x_1 \\ x_2 \\ x_3 \end{bmatrix} = \begin{bmatrix} 1 \\ 5 \\ 10 \end{bmatrix} \tag{18}$$

In this problem, A is the 3x3 square matrix and f is the 3x1 column matrix. We would like to obtain approximate solutions y_N by the variational method. Of course when N=3 we should get the exact solution of (18). In this problem we study the nature of the various orders of approximations. By the terms of the problem, we form the functional

$$F(x) = [x_1\ x_2\ x_3][A] \begin{bmatrix} x_1 \\ x_2 \\ x_3 \end{bmatrix} - 2[x_1\ x_2\ x_2] \begin{bmatrix} 1 \\ 5 \\ 10 \end{bmatrix} \tag{19}$$

$$F(x) = 3x_1^2 - 8x_1x_2 + 3x_2^2 + 4x_1x_3 + 12x_2x_3 + 2x_3^2 - 2x_1 - 10x_2 - 20x_3 \tag{20}$$

We now choose the expansion functions u_n (orthonormal) as

$$u_1 = \begin{bmatrix}1\\0\\0\end{bmatrix} \quad u_2 = \begin{bmatrix}0\\1\\0\end{bmatrix} \quad u_3 = \begin{bmatrix}0\\0\\1\end{bmatrix}$$

We now seek the first approximate solution to x as

$$y_1 = au_1 = \begin{bmatrix}a\\0\\0\end{bmatrix} \tag{21}$$

and solve for y_1 by the variational method (by utilizing (13)) to obtain

$$y_1 = \begin{bmatrix}1/3\\0\\0\end{bmatrix} \tag{22}$$

Since the exact solution to the problem is $x_i = 1$ for i=1,2,3, the error is

$$||x - y_1|| = \{(1 - 1/3)^2 + (1 - 0)^2 + (1-0)^2\}^{1/2}$$
$$= 1.56 \tag{23}$$

We now solve for y_2 which is the second order approximation to x. So

$$y_2 = a_1u_1 + a_2u_2 = \begin{bmatrix}a_1\\0\\0\end{bmatrix} + \begin{bmatrix}0\\a_2\\0\end{bmatrix} = \begin{bmatrix}a_1\\a_2\\0\end{bmatrix} \tag{24}$$

and we obtain the solution for y_2 by utilizing (13) as

$$y_2 = \begin{bmatrix}-23/7\\-19/7\\0\end{bmatrix} \tag{25}$$

Observe that the error in this case is

$$||x-y_2|| = \{(1+23/7)^2 + (1+19/7)^2 + 1^2\}^{1/2}$$
$$= 5.76 \tag{26}$$

From (23) and (26) it is clear that there is a 269% increase in error as we have increased the order of the approximation from N=1 to N=2. Also note that the residuals

$$||Ay_1 - f|| = 11.28$$

$$||Ay_2 - f|| = 32.86$$

do not decrease either as N increases. However, when N=3 we would obtain the exact solution.

The error in the solution increases, as we increased the order of the approximation, because A is not a positive definite matrix even though $||A^{-1}||$ is bounded. So, if in a problem the operator A is self-adjoint but not positive definite, then the variational method will not yield more accurate answers as the order of the approximation is increased. Only in the scalar problem of charge distribution on a wire or a plate is the operator self-adjoint and semidefinite. In almost all other cases, A is a complex symmetric operator. Hence, unless we have an operator which is positive definite, increasing the order of approximation in a variational method may not yield a more accurate solution. It is important to note that neither the operator in Pocklington's nor that in Hallen's integral equation for the solution of the currents on a thin wire is positive/negative definite. Hence there is no guarantee that as we increase the order of the approximation, we are in fact obtaining a more accurate solution.

It is important to point out that the symmetric inner product defined by Harrington [1981] is a mere mathematical artifact and does not give rise to a natural definition of the norm. Also, one cannot comment on the eigenvalues of a complex symmetric operator. The symmetric product is more akin to the definition of reciprocity and the normal inner product is related to the definition of power, which is a physical quantity. Thus by utilizing the symmetric product it is very difficult to comment on the properties of the operator, as a symmetric product does not naturally give rise to a norm and hence one cannot talk about convergence.

2.3. What are the restrictions on the expansion functions u_n?

The expansion functions need not be in the domain of the operator. However, it is essential that the expansion functions form a complete set with respect to the following inner product $<Au_i, u_j>$. It is totally immaterial whether the expansion functions form a complete set with respect to the inner product $<u_i, u_j>$. For a positive definite second order differential operator A, physically the above requirements imply that the expansion functions must be capable of representing the solution x, as well as its first derivative. Thus the expansion functions need not be required to approximate the second derivative.

As a third example, consider the following problem of Rektorys [1972]

$$Ax = -\frac{d^2x}{dz^2} = \cos z = f$$

$$\left.\frac{dx}{dz}\right|_{z=0} = 0 = \left.\frac{dx}{dz}\right|_{z=\pi} \tag{27}$$

The operator A in (27) is positive definite [Rektorys, 1972] on the set of functions which are twice differentiable, satisfy the boundary conditions and must have

$$\int_0^\pi y_N(z)dz = 0 \tag{28}$$

where $y_N(z)\varepsilon D(A)$.

Equation (28) is necessary to make the operator A positive definite. The solution to (27) is unique and exists for all f that satisfy $\int_0^\pi f(z)dz=0$. Under (28), (27) has a unique solution given by

$$x = \cos z \tag{29}$$

Let us choose the expansion functions as

$$u_j(z) = \sin 2jz \tag{30}$$

Note that this particular choice of expansion functions does not satisfy the boundary conditions, but satisfies (28). However, we still proceed with the above expansion functions and use the following form of the inner product:

$$<Au,v> = \int_0^\pi -\frac{d^2u}{dz^2} \cdot v(z)dz = \int_0^\pi \frac{du}{dz}\frac{dv}{dz}dz \tag{31}$$

The reason we use (31) is because in (31) neither u nor v need satisfy the boundary conditions for the operator A. So if

$$y_N = \sum_{j=1}^{N} a_j \sin 2jz$$

we solve for the a_j by utilizing (13) and obtain

$$y_N(z) = \frac{2}{\pi}\sum_{j=1}^{N}\frac{\sin 2jz}{j(4j^2-1)} \tag{32}$$

Observe that (32) is nowhere near the exact solution (29). We did not expect it to be - but why? Where is the error? We now show that the error is not in the fact that sin 2jz does not satisfy the boundary conditions. The problem is that sin 2jz is not complete with respect to the inner product $<Au_i, u_j>$. However, sin 2jz is complete with respect to the inner product $<u_i, u_j>$. To show that sin 2jz is not complete with respect to the inner product $<Au_i, u_j>$ we utilize (31). So if sin 2jz is not complete with respect to the inner products in (31), there must be a function $\phi(z)$ which satisfies the orthogonality relationship

$$\int_0^\pi \frac{d\phi}{dz} \cdot \frac{d}{dz}(\sin 2jz) = \int_0^\pi \frac{d\phi}{dz} \cdot 2j \cdot \cos 2jz = 0 \tag{33}$$

Thus (33) implies

$$\frac{d\phi}{dz} = C_1$$

where C_1 is a constant, and hence

$$\phi = C_1 z + C_2 \tag{34}$$

Equation (34) reveals that indeed sin 2jz is not complete with respect to the inner product of (31). Let us now choose the expansion functions as

$$u_o = z - \frac{\pi}{2} \qquad u_j = \sin 2jz \qquad j=1,2,\ldots,N \tag{35}$$

The addition of $u_o(z)$ is done to make the expansion functions a complete set with respect to the inner product defined by (31). The function $u_o(z)$ is chosen in this particular form so that $\int_0^\pi u_o(z)dz=0$. We now express y_N as

$$y_N = \sum_{i=0}^{N} a_i u_i$$

We solve for the a_i by utilizing (13) to obtain

$$y_N(z) = 1 - \frac{2z}{\pi} + \frac{2}{\pi}\sum_{j=1}^{N} \frac{\sin 2jz}{j(4j^2-1)} \tag{36}$$

In the limit $N\to\infty$, (36) is indeed the expansion for cos z for all $0 \le z \le \pi$.

2.4. Does minimization of the functional $F(y_N)$ guarantee that the corresponding residual $[Ay_N-f]$ is also minimized?

The question can be restated mathematically as follows: We know that when A is positive definite and $F(y_{N+1}) < F(y_N)$, then we have

$$||x-y_{N+1}|| < ||x-y_N||$$

Does this imply

$$||Ay_{N+1}-f|| < ||Ay_N-f|| \tag{37}$$

The answer is that there is no guarantee whatsoever that (37) would be satisfied. In fact, it is not even certain that the residuals will decrease to zero as N increases. This is because, in defining the expansion functions u_n, we have not specified that they belong to the domain of the operator. If the expansion functions do not belong to the domain of the operator, they do not satisfy all the boundary conditions and the requisite number of differentiability conditions for the problem. Hence if $u_n \notin D(A)$ {domain of A}, then Au_n and hence Ay_N lacks any meaning. So in a variational method it is not at all necessary that

$$\underset{N\to\infty}{\text{Lt}} ||Ay_N - f|| \to 0 \tag{38}$$

If A is a second order differential operator, physically this section implies that the variational method yields a solution such that the solution y_N and its first derivative y'_N converge in the mean to the exact solution x and its derivative x', respectively, as $N\to\infty$. However, the second derivative y''_N may not converge to x" in any sense. However, if the sequences Au_n are bounded, then Mikhlin [1965a,b] has shown that the residual Ay_N converges weakly (see appendix for definition) to f.

2.5. Epilogue

Observe that in the above sections we have talked of convergence with respect to the norm defined by (5) (i.e., convergence in the mean). Thus the class of problems for which the convergence of a variational method is mathematically justified is the class of problems which has a square integrable solution x. The problems then for which the convergence of a variational method cannot be physically or mathematically justified are the charge distribution on a rectangular plate, the current distribution on an open-ended cylinder irradiated by a plane wave and the like. These are problems which do not have a square integrable solution.

However, a variational method may be extended to problems which do not have a square integrable solution x, by redefining the inner product in (5) as

$$\langle u_i,u_j\rangle_1 \triangleq \int_a^b w(z)u_i(z)u_j(z)dz \tag{39}$$

where w(z) is generally a nonnegative weight function. If we find a solution $\tilde{y}_N(z)$ with the redefined inner products of (39), then we know from the above discussions that we cannot guarantee either of these two conditions, i.e.,

$$\underset{N\to\infty}{\text{Lt}} \int_a^b [\tilde{y}_N(z) - x(z)]^2 dz \to 0 \tag{40a}$$

$$\underset{N\to\infty}{\mathrm{Lt}} \int_a^b [A\tilde{y}_N(z) - f]^2 dz \to 0 \qquad (40b)$$

So the application of a variational method to a problem with non-square integrable solutions with the redefined norm of (39) may yield a mathematical solution which may not have any correlation to the actual physical problem.

The other important point to remember is that in a variational method, we minimize the functional F(y) in (8). Moreover, we have not chosen the expansion functions to be in the domain of the operator (i.e., satisfy both the differentiability conditions and the boundary conditions). So there may be a possibility that the element u_o, which minimizes the functional F(y), may not even be in the domain of the operator. Often this happens [Rektorys, 1972]. Under those circumstances, the solution obtained by a variational method is called the generalized solution to the problem. [A generalized solution is one which obeys $<\phi, Ay_N - f> = 0$ for all possible ϕ in a given class.]

In summary, the crux of a variational method lies in setting up the functional F(y), as it may not always exist. The proper selection of a complete set of expansion functions u_n is very important. Since the expansion functions need not be in the domain of the operator, there is no guarantee that

$$\underset{N\to\infty}{\mathrm{Lt}} \|Ay_N - f\| \to 0$$

If $F(y_N)$ exists, then

$$\underset{N\to\infty}{\mathrm{Lt}} \|y_N - x\| \to 0$$

if and only if $<f,x>$ and $<f,y_N>$ are bounded for all N.

3. GALERKIN'S METHOD

In Galerkin's method we select a complete set of expansion functions u_n which are in the domain of the operator, to approximate y_N in (6). This implies that the set $\{u_n\}$ satisfies not only all the boundary conditions but also the differentiability conditions of the problem. We now substitute (6) in (1) to form

$$\sum_{i=1}^{N} a_i A u_i = f$$

Now Au_i has a meaning as $u_i \varepsilon D(A)$. Next the residual $\sum_{i=1}^{N} a_i A u_i - f$ is formed and it is equated to zero with respect to the functions u_k, k=1,2,...,N. This yields a set of equations [Rektorys, 1972; Finlayson and Scriven, 1966; Duncan, 1937,1938]

$$<A \sum_{i=1}^{N} a_i u_i, u_k> = <f,u_k> \qquad k=1,2,\ldots,N \quad (41)$$

Observe that in (41) A can even be nonlinear and non-self-adjoint. A variational formulation for a nonlinear operator is quite complex. When A is a linear operator, then (41) simplifies to

$$\sum_{i=1}^{N} a_i <Au_i, u_k> = <f,u_k> \qquad k=1,2,\ldots,N \quad (42)$$

from which the unknowns a_i are solved for. The only requirement in (42) is that $<f,u_k>$ be finite. Since in (42) we weigh the residuals to zero, Galerkin's method is also known as the method of weighted residuals. Even though (13) of the variational method is identical in form to (42), there are significant differences in obtaining them. For example, in (13) A needs to be a self-adjoint operator, whereas in (42) A can be non-self-adjoint. Equation (13) is not applicable to non-self-adjoint operators. For a non-self-adjoint operator a variational principle has to be formulated not only for the original operator, but also for its adjoint [Mikhlin, 1965a,b; Rektorys, 1972]. A physical interpretation for the mathematical differences between the variational method and Galerkin's method is given by Singer [1960].

If the expansion functions u_n for Galerkin's method are chosen such that the boundary conditions are not satisfied, then additional constraints have to be imposed on the solution procedure. For example, if for the operator equation, $Ax = f$, the boundary conditions are

$$x(z=g) = 0 \qquad x(z=h) = 0$$

now if we choose the approximation for x as

$$x \simeq y_N = \sum_{i=1}^{N} a_i u_i$$

and the u_i do not satisfy the boundary conditions, then we must have the following equations [Brebbia and Walker, 1980]:

$$\sum_{i=1}^{N} a_i \langle Au_i, u_k\rangle = \langle f, u_k\rangle$$

$$\sum_{i=1}^{N} a_i \langle u_i\big|_{z=g}, u_k\rangle = 0$$

$$\sum_{i=1}^{N} a_i \langle u_i\big|_{z=h}, u_k\rangle = 0$$

for $k = 1,2,\ldots,N$.

3.1. <u>Is it always possible to find a set of approximate solutions y_N such that $\lim_{N\to\infty} \langle Ay_N - f, u_n\rangle \to 0$ for $n=1,2,\ldots,N$?</u>

In general, (42) of Galerkin's method guarantees weak convergence of the residuals in a finite dimensional space, i.e., $\langle Ay_N - f, u_n\rangle = 0$ for all u_n; $n=1,2,\ldots,N$. The question then is raised if $N\to\infty$, whether this weak convergence of the residuals be still guaranteed. It has been shown by Mikhlin [1960,1964,1965a,b] that, if

$$\|Ay_N\| \leq C \tag{43}$$

where C is a constant independent of N for a particular choice of expansion functions u_n, then as $N\to\infty$, weak convergence of the residuals is guaranteed. This is illustrated in the fourth example in the next subsection.

3.2. <u>Does $\lim_{N\to\infty} \langle Ay_N - f, u_n\rangle \to 0$, for $n=1,2,\ldots,N$ guarantee that $\lim_{N\to\infty} \|Ay_N - f\| \to 0$?</u>

In other words, does weak convergence of the residuals imply strong convergence (i.e., convergence in the norm) of the residuals? Again Mikhlin [1960, 1964] has shown that if A is a bounded operator, then strong convergence of the residuals is always guaranteed. However, if A is an unbounded operator, then for the strong convergence of the residuals, either of the following two conditions must be satisfied:

1. The projection operator P_N has to be bounded. The projection operator is defined as

$$P_N[Ay-f] = Ay_N - P_N f = \sum_{i=1}^{N} b_i u_i \tag{44}$$

It is the operator that projects the problem from an infinite dimensional space to a finite dimensional space. Thus the properties of the projection operator depend on the choice of the expansion function u_i. So the boundedness of the projection operator implies

$$\|P_n\| \leq C \tag{45}$$

where C is a constant for all N. It is easy to show that a projection operator is always bounded if it is an orthogonal projection.

In that case, it is easy to show that [Mikhlin, 1965 a,b; Bogarayan, 1961]

$$\|P_N\| = 1 \quad \text{for all } N \tag{46}$$

However, if P_N is not an orthogonal projection as in the case of a polynomial basis function, i.e.,

$$P_N[Ay-f] = \sum_{i=0}^{N} b_i x^i \tag{47}$$

then it can be shown that the norm of the projection operator in (47) is given by [Mikhlin, 1964]

$$\|P_N\| \leq C \ln N \tag{48}$$

In this case, the projection operator is unbounded.

2. If it is not possible to obtain a bounded projection operator, then select an operator B which is similar to the

original operator A with D(B) = D(A). For example, if

$$A = a\frac{d^2x}{dz^2} + b\frac{dx}{dz} + cx \qquad x(z=0)=0=x(z=1)$$

and if

$$B = \frac{d^2x}{dz^2} \qquad x(z=0)=0=x(z=1)$$

then operators A and B are similar.

We select the operator B in such a way that B has discrete eigenvalues. Then if the eigenfunctions of B are chosen as the expansion functions u_n, the residuals will converge to zero even though the operator A is unbounded. This implies that the expansion functions in this case are complete with respect to $<Bu_i,u_j>$; i.e., the expansion functions are capable of representing not only the exact solution, but also its first derivative.

The following example of Stephens [1976] illustrates the above principles.

As a fourth example, consider the operator equation

$$Ax = -\frac{d^2x}{dz^2} = f(z) \qquad x(z=-1)=0=x(z=1) \tag{49}$$

Here D(A) is the space of functions with the requisite boundary conditions having an absolutely continuous first derivative and satisfies

$$\int_{-1}^{+1} \frac{d^2x}{dz^2}\, dz < \infty$$

Let us choose the expansion functions as Legendre polynomials. The Legendre polynomials form a complete set in [-1,1] and an nth order Legendre polynomial $P_n(z)$ satisfies

$$|P_n(z)| < 1$$

So for all f, belonging to the space of square integrable functions, we have

$$f(z) = \sum_{i=0}^{\infty} a_i P_i(z) \qquad -1 \leq z \leq 1 \tag{50}$$

where

$$a_i = \frac{2i+1}{2} <f,P_i>$$

$$<f,P_i> = \int_{-1}^{+1} f(z)P_i(z)dz$$

It is easy to show that

$$||f||^2 = \sum_{i=0}^{\infty} \frac{2}{2i+1} a_i^2$$

by utilizing the orthogonal properties of the Legendre polynomials

$$\int_{-1}^{+1} P_i(z)P_j(z)dz = \frac{2\delta_{ij}}{2i+1}$$

where δ_{ij} is the Kronecker delta function.

Next we choose the expansion functions ϕ_k as

$$\phi_k(z) = \int_{-1}^{z} P_k(t)dt$$

$$= \frac{1}{2k+1}\{P_{k+1}(z) - P_{k-1}(z)\} \tag{51}$$

Observe that these expansion functions satisfy the boundary conditions

$$\phi_k(-1) = \phi_k(1) = 0$$

Since

$$<A\phi_j,\phi_k> = \frac{2\delta_{jk}}{2k+1}$$

it is clear that the Galerkin approximation of the operator equation in (49) is given by

$$y_N(z) = \sum_{k=1}^{N} \frac{2k+1}{2} <f,\phi_k> \phi_k(z) \tag{52}$$

We now obtain Ay_N as

$$Ay_N = -\frac{d^2y_N}{dz^2}$$

$$= \sum_{k=0}^{N-1} a_k P_k(z) - \sum_{\substack{k=0 \\ k \text{ even}}}^{N-1} \frac{a_{N+1}}{2N+3}(2k+1)P_k(z)$$

$$- \sum_{\substack{k=1 \\ k \text{ odd}}}^{N-2} \frac{a_N}{2N+1}(2k+1)P_k(z) \qquad (53)$$

This is the result for N odd. In obtaining the above result we have utilized the identity

$$P_k'(z) = (2k-1)P_{k-1}(z) + (2k-5)P_{k-3}(z)$$

$$+ \ldots + \begin{cases} 3P_1(z) \rightarrow k \text{ even} \\ P_0(z) \rightarrow k \text{ odd} \end{cases}$$

A similar expression for Ay_N may be obtained for N even. In this problem, however, we shall focus our attention on N odd. Next we observe

$$\|Ay_N\|^2 = \sum_{k=0}^{N-1} \frac{2a_k^2}{2k+1} - \frac{4a_{N+1}}{2N+3}\sum_{\substack{k=0 \\ k \text{ even}}}^{N-1} a_k$$

$$- \frac{4a_N}{2N+1}\sum_{\substack{k=1 \\ k \text{ odd}}}^{N-2} a_k + \frac{2a_{N+1}^2}{(2N+3)^2}\sum_{\substack{k=0 \\ k \text{ even}}}^{N-1} (2k+1)$$

$$+ \frac{2a_N^2}{(2N+1)^2}\sum_{\substack{k=1 \\ k \text{ odd}}}^{N-2} (2k+1) \qquad (54)$$

Now we shall work on two special cases. In the first special case, let

$$a_k = m \qquad k = m^4 \qquad m \text{ an integer}$$
$$a_k = 0 \qquad \text{otherwise} \qquad (55)$$

For this particular choice of a_i, we find that both f and the approximate solution y_N are square integrable. However, for a sequence of N, the first three terms of $\|Ay_N\|^2$ converge and the last two terms diverge. Hence $\|Ay_N\| \rightarrow \infty$ and the residual (Ay_N-f) does not converge to zero. So this is an example where the sequence of approximate solutions may converge to the exact solution as $N \rightarrow \infty$; however, the residual may not converge to zero. This is a typical characteristic of both the variational (Rayleigh-Ritz) and the Galerkin methods.

In the second special case, let

$$a_k = 1 \qquad k = m^2 \qquad m \text{ an integer}$$
$$a_k = 0 \qquad \text{otherwise} \qquad (56)$$

For this choice of a_k we find that $\|Ay_N\| \leq M$ (a constant) independent of N. One can also show that the residuals weakly converge to zero, i.e.,

$$\lim_{N\rightarrow\infty} \langle Ay_N - f, \phi_k \rangle = 0$$

However, we can show that

$$\|Ay_N\| \neq \|f\|$$

and hence the residuals do not converge strongly to zero. This is the illustration for this subsection.

It can be shown that the residuals Ay_N-f do not converge at all for all f that are square integrable, if the expansion functions are chosen of the form

$$\phi_k(z) = (1-z^2)\sum_{j=1}^{k-1} a_j z^j$$

So what we have demonstrated by the two special cases is that for the weak convergence of the residuals for the Galerkin method, it is necessary that $\|Ay_N\| \leq M$ for all N with a particular choice of expansion functions.

In the fourth example, we could have chosen the expansion function as

$$\cos\frac{(2n+1)\pi z}{2} \qquad \sin n\pi z$$

These functions satisfy the boundary conditions as they become zero at $z = \pm 1$. Observe that they are also the eigenfunctions for the second order differential operator. Hence for this particular choice of expansion functions, one would have strong convergence of the residuals, even though the operator is unbounded. It is shown in the next section that for this particular choice of expansion functions, Galerkin's method becomes equivalent to the method of least squares, which always guarantees strong convergence of the residuals.

3.3. Does the sequence of solutions $||y_N - x||$ converge to zero as $N \to \infty$?

We observe that

$$||y_N - x|| \le ||A^{-1}|| \cdot ||Ay_N - Ax||$$
$$\le ||A^{-1}|| \cdot ||Ay_N - f|| \qquad (57)$$

So if for a particular problem $||Ay_N - f||$ go to zero and the inverse operator is bounded, then the sequence of approximate solutions converges to the exact solution. As is clear from the above equation, if we have no guarantee about the convergence of the residuals and we do not know the exact solution, then it is extremely difficult, if not impossible, to comment on the accuracy of the approximate solutions obtained by Galerkin's method.

3.4. Is Galerkin's method always applicable?

If the same expansion functions do not form the basis of both the domain and the range of the operator, then meaningful results may not be obtained by Galerkin's method. However, for some self-adjoint operators, Galerkin's method reduces to the method of least squares

As a fifth example, consider the solution of the following problem:

$$Ax = -\frac{d^2x}{dz^2} = \cos z = f \qquad x(z=0)=0=x(z=\pi) \qquad (58)$$

We observe that Galerkin's method for this self-adjoint operator yields the least squares solution as we choose the expansion functions to be eigenfunctions.

A complete set of expansion functions for this problem would be sin nz. These expansion functions are the eigenfunctions for the operator and satisfy the boundary conditions for this problem. If

$$y_N = \sum_{n=1}^{N} a_n \sin nz \qquad (59)$$

then

$$Ay_N = -\sum_{n=1}^{N} n^2 a_n \sin nz$$

Observe that sin nz in the expression for Ay_N cannot span cos z. Indeed it does not, if we talk about pointwise (uniform) convergence. Hence one may jump to the conclusion that Galerkin's method may not be applicable for this problem. However, they do converge in the mean. The convergence in the mean is guaranteed by this subsection, and we shall show that it does. In general, it is difficult, if not impossible, to know whether a certain set of expansion functions will span the range of the operator.

By solving for the unknowns a_n by Galerkin's method we get

$$y_N = \frac{2}{\pi} \sum_{j=1}^{N} \frac{\sin 2jz}{j(4j^2 - 1)} \qquad (60)$$

The exact solution to the problem is given by

$$x = -1 + \frac{2z}{\pi} + \cos z \qquad (61)$$

It is quite straightforward to show that in the limit as $N \to \infty$, both y_N and dy_N/dz converge to x and dx/dz uniformly in the region of $0 \le z \le \pi$. However, $||Ay_N - f||^2$ does not converge uniformly but converges in mean to zero if $N \to \infty$. This is clear from the following equation:

$$\lim_{N\to\infty} ||Ay_N - f||^2$$
$$= \lim_{N\to\infty} \left\{\frac{\pi}{2} - \frac{32}{\pi} \sum_{j=1}^{N} \frac{j^2}{(4j^2 - 1)^2}\right\}^2 \to 0$$

3.5. Epilogue

In summary, the convergence of the residuals is guaranteed only if the expansion functions are chosen in a particular fashion. And since the inverse operator is bounded, the convergence of the residuals also implies that the sequence of solutions converges to the exact solution. In Galerkin's method the expansion functions must satisfy all the boundary conditions and the differentiability conditions of the problem. This may not be an easy thing to do for differential operators. So variational methods are suitable for differential operators (as the expansion functions do not have to satisfy any boundary conditions) and Galerkin's method is suitable for integral operators as the latter are often bounded and hence strong convergence of the residuals is always guaranteed.

4. METHOD OF LEAST SQUARES

In the variational method (Rayleigh-Ritz) and Galerkin's method the method of convergence of the residuals

$$\lim_{N\to\infty} [Ay_N - f] \to 0$$

is guaranteed only under special circumstances. The method of least squares on the other hand always yields a strong convergence of the residuals even when $||A||$ is unbounded. However, the expansion functions u_n are required to be in the domain of A, and Au_n must span the range of A. So for a second order operator, the expansion functions must be able to approximate second derivatives in the mean, whereas in the variational method it was only necessary to approximate first derivatives. However, if A is bounded, then any complete set of expansion functions will be adequate.

In the method of least squares the solution of $Ax = f$ is attempted by trying to minimize the functional $F^1(y_N)$ which is given by

$$F^1(y_N) = ||Ay_N - f||^2 = \langle Ay_N - f, Ay_N - f\rangle \qquad (62)$$

We now substitute (6) in (62) and solve for the unknowns a_n by equating the partial derivatives of $F^1(y_N)$ with respect to a_n equal to zero. This results in a set of equations

$$\sum_{n=1}^{N} a_n \langle Au_n, Au_k\rangle = \langle f, Au_k\rangle \qquad k=1,2,\ldots,N \qquad (63)$$

By comparing (63) to (13) and (42), it is clear that the method of least squares requires more work than either Galerkin's or the Rayleigh-Ritz method. However, the reward of doing more work lies in obtaining strong convergence of the residuals (i.e., as $N\to\infty$ the residuals converge to zero in a mean square sense).

It is interesting to note that when the expansion functions u_n are chosen to be the eigenfunctions of the operator A, then (63) reduces to (42) of Galerkin's method. Under the above restrictions the two methods yield identical sets of equations.

In the following subsections the finer points of the method of least squares are explained.

4.1. What are the characteristics of the expansion functions?

The expansion functions u_n must be such that Au_n must form a linearly independent set and therefore should span f. In other words u_n must be complete with respect to the $\langle Au_n, Au_k\rangle$ norm. It does not make any difference whether u_n forms a complete set with respect to the $\langle u_n, u_k\rangle$ norm. The reason behind this is that if the residuals must converge to zero, then a linear combination of Au_n should be able to reproduce f. This is the subtlety behind the method of least squares.

As a sixth example, consider the operator equation

$$\frac{dx}{dz} = 1 \qquad x(z=0) = 0 \qquad 0 \le z \le 1$$

If we choose the expansion functions as $\sin n\pi z/2$, then we see

$$\frac{dy_N}{dz} = \frac{d}{dz}\sum_{n=1}^{N} a_n \sin\frac{n\pi z}{2}$$

$$= \sum_{n=1}^{N} \frac{a_n n\pi}{2} \cos\frac{n\pi z}{2} \tag{64}$$

In this case, the method of least squares yields a solution which is the Fourier series solution. The solution of the above operator equation is obtained from (63) as

$$x = \sum_{n=1}^{\infty} \frac{8}{n^2\pi^2} \sin\frac{n\pi z}{2} \tag{65}$$

Observe that the above solution converges uniformly (pointwise) to the exact solution

$$x = z$$

However,

$$\lim_{N\to\infty} \frac{dy_N}{dz}$$

does not converge uniformly to 1, but converges in the mean, i.e.,

$$\lim_{N\to\infty} \int_0^1 [1 - \frac{dy_N}{dz}]^2 \, dz = 0$$

$$\lim_{N\to\infty} \frac{dy_N}{dz} \neq 1 \tag{66}$$

So even though $||A||$ is unbounded, we have obtained strong convergence of the residuals.

4.2. Do the residuals converge monotonically as $N \to \infty$?

In the method of least squares it is always true that the residuals $Ay_N - f$ converge in the mean to zero monotonically even when A is unbounded. This is seen from (62). As the order of the approximation y_N is increased, $F^1(y_N)$ in (62) will always decrease monotonically.

4.3 Does the sequence of approximate solutions monotonically converge to the exact solution?

Since

$$||x-y_N|| \leq ||A^{-1}|| \cdot ||Ay_N - f|| \tag{67}$$

and because $||A^{-1}||$ is bounded, monotonic convergence of the residuals will always guarantee monotonic convergence of the sequence of approximate solutions y_N to the exact solution. Therefore, it is always true that

$$\lim_{N\to\infty} ||x-y_N|| \to 0 \tag{68}$$

4.4 Epilogue

In summary, the method of least squares always generates monotonic convergence of the residuals (in the mean) if the expansion functions are in the domain of A and if Au_n span f. The disadvantage of this method is that it requires more work than the other two methods. Finally, the sequence of approximate solutions y_N monotonically approaches x as $N \to \infty$.

5. COMPARISON OF RATES OF CONVERGENCE OF THE VARIOUS METHODS

To compare the rate of convergence between the variational method (Rayleigh-Ritz) and the method of least squares, we assume that we have utilized the same expansion functions u_n in solving $Ax - y$. Let y_N^V and y_N^L be the approximate solutions obtained by the variational method and the method of least squares, respectively, so that

$$y_N^V = \sum_{n=1}^{N} a_n^V u_n \tag{69}$$

$$y_N^L = \sum_{n=1}^{N} a_n^L u_n \tag{70}$$

Observe that for the same number of expansion functions, the approximate solutions are different. This is because the variational method minimizes the functional

$$F(y_N^V) = \langle Ay_N^V, y_N^V\rangle - \langle f, y_N^V\rangle - \langle y_N^V, f\rangle \tag{71}$$

whereas the method of least squares minimizes

$$F^1(y_N^L) = ||Ay_N^L - f||^2 = \langle Ay_N^L - f, Ay_N^L - f\rangle \tag{72}$$

So by the terms of the problem we have

$$F(y_N^V) \leq F(y_N^L) \tag{73}$$

By utilizing (10) we get

$$\langle A(y_N^V - x), (y_N^V - x)\rangle \leq \langle A(y_N^L - x), (y_N^L - x)\rangle \tag{74}$$

Hence the rate of convergence of the approximate solutions for the variational method is at least as fast as the method of least squares, if not better. The equality in the rates of convergence is achieved when the expansion functions are the eigenfunctions of the operator A.

So even though the method of least squares requires more work, it has a slower rate of convergence than the variational method. On the other hand, the method of least squares guarantees monotonic convergence of the residuals. Hence when the exact solution is not known, it is much easier with the method of least squares to establish relative error estimates of the solution.

Observe, when the expansion functions are the eigenfunctions of the operator, the method of least squares and Galerkin's method become identical.

6. CONCLUSION

An attempt has been made to bring out the mathematical differences between the variational method (Rayleigh-Ritz), Galerkin's method and the method of least squares.

It is seen that for a variational method the expansion functions need not be in the domain of the operator whereas for Galerkin's and the method of least squares they are required to be in the domain of the operator. Moreover, for Galerkin's method the expansion functions have to span both the domain and range of the operator, whereas for the method of least squares the operator operating on the expansion functions must span the range of the operator. For a variational method, however, the expansion functions have to be complete with respect to the inner product $\langle Au_i, u_j\rangle$.

Both Galerkin's method and the method of least squares are applicable to non-self-adjoint operators, whereas for the Rayleigh-Ritz method one has to formulate the variational principle not only for the original operator but also for its adjoint. For a non-self-adjoint operator either Galerkin's method or the variational method may not give monotonic convergence for the solution.

Only the method of least squares guarantees the convergence of the residuals as the order of the approximation gets higher. This advantage is, however, offset by the fact that the method of least squares requires considerably more computations than the other two methods. For Galerkin's method, the convergence of the residual is guaranteed only under special circumstances.

For a variational method, the problem of setting up the functional F(y) is a very tedious process and often the functional may not be defined.

Finally, it is shown that the variational method converges as fast as, if not better than, the method of least squares.

APPENDIX

In the appendix we present the definitions of the commonly used mathematical terms in the manuscript.

In this paper we wish to solve the operator equation

$$Ax - f = 0 \tag{75}$$

The solution x is called a classical solution if (75) is obeyed everywhere in the space S = domain of A. A generalized solution x is one which satisfies

$$\langle Ax - f, \phi\rangle = 0 \tag{76}$$

for all possible ϕ in a given class.

There are four types of convergence:

1. Uniform convergence (pointwise convergence implies that given a fixed $\varepsilon > 0$, we can find a number N such that the absolute value

$$|x(z) - y_N(z)| < \varepsilon \quad (77)$$

2. Strong convergence (convergence in the norm or convergence in the mean) implies

$$\|x(z) - y_N(z)\| < \varepsilon \quad (78)$$

i.e.,

$$\left[\int_a^b \{x(z) - y_N(z)\}^2 dz\right]^{1/2} < \varepsilon \quad (79)$$

3. Convergence in energy requires that

$$\langle A(x-y_N),(x-y_N)\rangle < \varepsilon \quad (80)$$

($\langle Ay,y\rangle$ in some physical problems represent energy; hence the name energy norm), while convergence in the mean implies that the converging sequence y_N may not approach the limit function x at every point in the domain of A, but the region in which the converging sequence y_N differs from x becomes arbitrarily small as $N \to \infty$. Convergence in energy implies similar things to happen to the derivative of the sequence as well.

4. Weak convergence (convergence in a distributional sense) implies that y_N converges weakly to x if

$$\lim_{N\to\infty} \langle y_N,\phi\rangle = \langle x,\phi\rangle \quad (81)$$

for all ϕ in the space. The Galerkin method often yields solutions which converge weakly to a generalized solution.

A set of functions u_i is said to be complete in a space if any function x in that space can be expanded in terms of the set $\{u_i\}$,

$$\left\|x - \sum_{i=1}^{\infty} a_i u_i\right\| < \varepsilon \quad (82)$$

A set of functions complete in one space need not be complete for another space.

A set of functions is said to be complete in energy if

$$\langle A(x - \sum_{i=1}^{\infty} a_i u_i),(x - \sum_{i=1}^{\infty} a_i u_i)\rangle < \varepsilon \quad (83)$$

for a positive definite operator A.

An operator is said to be self-adjoint if

$$\langle Ax,y\rangle = \langle x,Ay\rangle \quad (84)$$

for all functions in the domain of A. A self-adjoint operator is said to be positive definite if

$$\langle Ay,y\rangle > 0 \qquad y \not\equiv 0 \quad (85)$$

Acknowledgments. The author is greatly indebted to R. W. P. King, who encouraged him to look into the mathematical aspects of various techniques extensively used in electromagnetic theory while the author was a visiting fellow at Harvard University. The author is also very grateful to R. F. Harrington for initiating him into the method of moments and for having numerous stimulating discussions over the last decade. Thanks are also due to the reviewers whose criticisms helped to improve this manuscript. This work has been supported in part by the Office of Naval Research under contract N00014-79-C-0598 and in part by the Air Force Office of Scientific Research under AFOSR-76-3073.

REFERENCES

Bogarayan, O. K., Convergence of the residual of the Bubnov-Galerkin and Ritz methods, Sov. Math. 2, Engl. Transl., 1413-1415, 1961.

Brebbia, C. A., and S. Walker, Boundary element techniques in engineering, Newnes-Butterworth, Boston, Mass., 1980.

Duncan, W. J., Galerkin's method in mechanics and differential equations, Tech. Rep. R&M 1978, Aeronaut. Res. Comm., Great Britain, 1937.

Duncan, W. J., The principles of the Galerkin method, Tech. Rep. R&M 1848 vol. 11, Aeronaut. Res., Great Britain, 1938.

Finlayson, B. A., and L. E. Scriven, The method of weighted residuals - A review,

Appl. Mech. Rev., 19,(9), 735-748, 1966.

Finlayson, B. A., and L. E. Scriven, On the search for variational principles, Int. J. Heat Mass Transfer, 10, 799-821, 1967.

Harrington, R. F., Field Computation by Moment Methods, McMillan, New York, 1968

Mikhlin, S. G., The stability of the Ritz method, Sov. Math., Engl. Transl., 1, 1230-1233, 1960.

Mikhlin, S. G., Variational Methods in Mathematical Physics, Pergamon, New York, 1964.

Mikhlin, S. G., The Problem of the Minimum of a Quadratic Functional, Holden-Day, San Francisco, Calif., 1965a.

Mikhlin, S. G., Stability of some computing processes, Sov. Math., Engl. Transl., 6, 931-933, 1965b.

Rektorys, K., Variational Methods in Mathematical Sciences and Engineering, D. Reidel, Hingham, Mass., 1972.

Sarkar, T. K., D. D. Weiner, and V. K. Jain, Some mathematical considerations in dealing with the inverse problem, IEEE Trans. Antennas Propag., AP-29, 373-379, 1981.

Singer, J., On the equivalence of the Galerkin and the Rayleigh-Ritz methods, J. R. Aeronaut. Soc., 66, 592, 1960.

Stakgold, Ivar, Green's Functions and Boundary Value Problems, John Wiley and Sons, New York, 1979.

Stephens, A. B., The convergence of residuals for projective approximations, SIAM J. Numer. Anal., 13, (4), 607-614, 1976.

Tapan K. Sarkar, Department of Electrical Engineering, Rochester Institute of Technology, Rochester, NY 14623.

Eigenvalues of the Moment-Method Matrix and Their Effect on the Convergence of the Conjugate Gradient Algorithm

A. F. PETERSON, MEMBER, IEEE, C. F. SMITH, AND R. MITTRA, FELLOW, IEEE

Abstract—**A theory that related eigenvalues of a continuous operator to those of the moment-method matrix operator is discussed and confirmed by examples. This theory suggests reasons for ill conditioning when certain types of basis and testing functions are employed. In addition, the effect of eigenvalue location on the convergence of the conjugate gradient (CG) method is studied. The convergence rate of the CG method is dependent on the eigenvalues of the iteration matrix as well as on the number of eigenvectors of the iteration matrix needed to represent the right side of the equation. These findings explain the previously reported convergence behavior of the CG method when applied to electromagnetic scattering problems.**

I. Eigenvalues and the Discretization Process

The approximate solution of integral equations using a moment-method type of discretization is well documented [1]. The numerical stability of the matrix solution algorithm is often of interest as the direct methods of solution fail for very ill-conditioned equations. For this reason, if would be desirable to select the specific discretization procedure in such a manner to produce a well-conditioned matrix (a formal definition of the condition number of a general matrix is presented in [2]). It is sometimes stated that subsectional basis and testing functions usually yield a well-conditioned matrix equation and thus are preferred over the entire-domain type of expansion (for example, see [3, p. 410]). The relationship between the condition number of the matrix and the choice of basis and testing functions is investigated. In addition, an understanding of the convergence rates of iterative solution algorithms requires knowledge of the relationship between the eigenvalues of the matrix operator and those of the original continuous operator. This relationship is delineated.

If the continuous operator equation is of the form

$$Lf = g \tag{1}$$

the moment-method procedure produces the matrix equation

$$\bar{\mathbf{L}}\mathbf{f} = \mathbf{g} \tag{2}$$

where the elements of the Nth order matrix are

$$l_{mn} = \langle T_m, LB_n \rangle \tag{3}$$

Manuscript received October 29, 1986; revised August 5, 1987.

The authors are with the Electromagnetic Communication Laboratory, University of Illinois, 1406 W. Green St., Urbana, IL 61801.

IEEE Log Number 8821176.

for some suitable inner product, a set of basis functions $\{B_i\}$, and a set of testing functions $\{T_i\}$. If we consider the eigenvalue equation

$$Le = \lambda e \tag{4}$$

and discretize using the same basis and testing functions employed earlier, the result is the matrix equation [1]

$$\bar{\mathbf{S}}^{-1}\bar{\mathbf{L}}\mathbf{e} = \lambda \mathbf{e} \tag{5}$$

where the elements of $\bar{\mathbf{L}}$ are defined in (3) and the elements of $\bar{\mathbf{S}}$ are

$$s_{mn} = \langle T_m, B_n \rangle. \tag{6}$$

Equation (5) involves the same eigenvalues $\{\lambda\}$ appearing in (4) and suggests that the eigenvalues of the product matrix $\bar{\mathbf{S}}^{-1}\bar{\mathbf{L}}$ should approximate the eigenvalue spectrum of the original continuous operator (at least to the degree that the basis functions can represent the operator's eigenfunctions).

For instance, consider a transverse magnetic (TM) wave incident on an infinite perfectly conducting cylinder. The moment-method formulation of this problem in terms of the electric-field integral equation (EFIE), pulse basis functions, and Dirac delta testing functions has been discussed by Harrington [1] and will not be repeated here. However, if the cylinder is circular, the eigenvalues of the EFIE are easily found analytically and given by

$$\lambda_n^{\text{TM}} = \frac{\eta \pi k a}{2} J_n(ka) H_n^{(2)}(ka) \tag{7}$$

where k is the wavenumber, η is the intrinsic impedance, a is the radius of the cylinder, and J and H represent the nth order Bessel and Hankel functions. Since these are exact, they can be compared with the matrix eigenvalues to test the foregoing theory. In the case where each cell in the cylinder model is of identical size, the scaling matrix $\bar{\mathbf{S}}$ is an identity matrix and the eigenvalues of $\bar{\mathbf{L}}$ should directly approximate those of (7). Table I compares the first 11 distinct eigenvalues of the EFIE with the eigenvalues of the moment-method matrix constructed using 30 expansion functions for a circular cylinder having one-wavelength circumference. Excellent agreement is obtained between the exact and numerical results.

In the case of TE wave excitation the moment-method formulation for the EFIE discretized with subsectional triangle basis functions and pulse testing functions has been discussed by Glisson and Wilton [4]. For perfectly conducting circular cylinders, the eigenvalues of the continuous EFIE are discrete and given as

$$\lambda_n^{\text{TE}} = \frac{\eta \pi k a}{2} J_n'(ka) H_n^{(2)\prime}(ka) \tag{8}$$

where the primes denote differentiation with respect to the arguments. Table II compares the first 11 distinct eigenvalues of the EFIE with those of the moment-method matrix $\bar{\mathbf{L}}$ and the product matrix $\bar{\mathbf{S}}^{-1}\bar{\mathbf{L}}$, for the order-30 system representing a cylinder having one-wavelength circumference. Clearly, the eigenvalues of $\bar{\mathbf{S}}^{-1}\bar{\mathbf{L}}$ are a better approximation to the exact values than those of $\bar{\mathbf{L}}$.

The relationship between the $\bar{\mathbf{S}}$ matrix, $\bar{\mathbf{L}}$ matrix, and the continuous operator suggests that the condition number of $\bar{\mathbf{S}}^{-1}\bar{\mathbf{L}}$ will be relatively independent of the basis and testing functions. This implies that if the basis and testing functions are such that the $\bar{\mathbf{S}}$-matrix is poorly conditioned, the $\bar{\mathbf{L}}$-matrix will also be poorly conditioned to compensate. For certain entire-domain basis functions, the condition numbers of $\bar{\mathbf{S}}$ and $\bar{\mathbf{L}}$ often degrade rapidly as additional basis functions are employed. This in turn suggests that to avoid producing a poorly conditioned $\bar{\mathbf{L}}$ matrix, it is advisable to

TABLE I

FIRST 11 DISTINCT EIGENVALUES OF THE EFIE COMPARED TO THOSE OF THE MOMENT-METHOD MATRIX EQUATION FOR A ONE-WAVE-LENGTH CIRCUMFERENCE CYLINDER, TM POLARIZATION; THE $\bar{S}$-MATRIX IN THIS CASE IS AN IDENTITY AND THE EIGENVALUES SHOWN ARE THOSE OF THE $\bar{L}$-MATRIX

EFIE (7)	30 × 30 Matrix
346.495 − j 39.964	346.4 − j 41.6
114.592 + j 203.433	114.5 + j 201.7
7.813 + j 112.240	7.82 + j 110.4
0.226 + j 67.395	0.23 + j 65.6
0.004 + j 48.773	0.00 + j 47.1
0.0 + j 38.488	0.01 + j 37.0
0.0 + j 31.854	0.02 + j 30.5
0.0 + j 27.194	0.02 + j 26.1
0.0 + j 23.735	0.00 + j 23.0
0.0 + j 21.062	−0.01 + j 20.8
0.0 + j 18.932	0.00 + j 19.0

TABLE II

FIRST 11 EIGENVALUES OF THE EFIE COMPARED TO THOSE OF THE MOMENT-METHOD $\bar{L}$-MATRIX AND THE PRODUCT MATRIX $\bar{S}^{-1}\bar{L}$ FOR A ONE-WAVELENGTH CIRCUMFERENCE CYLINDER, TE POLARIZATION; THE ORDER OF $\bar{S}$ AND $\bar{L}$ IS 30

EFIE (8)	$\bar{L}$-matrix	$\bar{S}^{-1}\bar{L}$ Matrix
114.592 + j 203.43	111.6 + j 200.1	112.2 + j 201.1
62.562 − j 167.30	61.7 − j 166.6	62.4 − j 168.4
26.157 − j 313.54	25.6 − j 306.4	26.3 − j 314.9
1.870 − j 526.05	1.83 − j 500.5	1.93 − j 528.5
0.055 − j 727.42	0.05 − j 668.9	0.06 − j 733.2
0.001 − j 921.87	0.0 − j 813.0	0.0 − j 934.3
0.0 − j 1113.9	0.0 − j 935.6	0.0 − j 1137
0.0 − j 1304.7	0.0 − j 1037	0.0 − j 1344
0.0 − j 1494.9	0.0 − j 1120	0.0 − j 1555
0.0 − j 1684.7	0.0 − j 1185	0.0 − j 1770
0.0 − j 1874.1	0.0 − j 1234	0.0 − j 1984

select basis and testing functions that produce a well-conditioned $\bar{S}$-matrix. Subsectional functions usually accomplish this goal.

II. EIGENVALUES AND THE SOLUTION PROCESS

The convergence rate of iterative solution algorithms is linked to the eigenvalue spectrum of the iteration matrix. Consider the approximate solution of an integral equation, such as the EFIE, using a moment-method type of discretization coupled with the iterative conjugate gradient (CG) method [5]. Because the CG method requires the iteration matrix to be Hermitian, it is usually necessary to convert the original moment-method system

$$\bar{\mathbf{A}}\mathbf{x}=\mathbf{b} \tag{9}$$

into the so-called "normal equations"

$$\bar{\mathbf{A}}^*\bar{\mathbf{A}}\mathbf{x}=\bar{\mathbf{A}}^*\mathbf{b} \tag{10}$$

where the star denotes the transpose-conjugate matrix. The CG method can be considered in the context of minimizing an error functional at each iteration step. At the nth iteration step, the CG method produces an estimate $\mathbf{x}_n$ that attempts to minimize the error norm

$$E_n=\langle \mathbf{x}-\mathbf{x}_n,\ \bar{\mathbf{A}}^*\bar{\mathbf{A}}(\mathbf{x}-\mathbf{x}_n)\rangle \tag{11}$$

where the conventional Euclidean inner product is implied. The matrix $\bar{\mathbf{A}}^*$ is the adjoint of $\bar{\mathbf{A}}$ under this inner product. In the following, we show the relationship of the error norm of (11) to the eigenvalues of the matrix $\bar{\mathbf{A}}^*\bar{\mathbf{A}}$.

In the CG method, the user is required to supply an initial guess $\mathbf{x}_0$, from which a residual vector

$$\mathbf{r}_0=\mathbf{b}-\bar{\mathbf{A}}\mathbf{x}_0 \tag{12}$$

is computed. The conjugate gradient method involves the process of generating new estimates of the solution according to the recipe

$$\mathbf{p}_0=\bar{\mathbf{A}}^*\mathbf{r}_0 \tag{13}$$

$$\alpha_n=\frac{\|\bar{\mathbf{A}}^*\mathbf{r}_n\|^2}{\|\bar{\mathbf{A}}\mathbf{p}_n\|^2} \tag{14}$$

$$\mathbf{x}_{n+1}=\mathbf{x}_n+\alpha_n\mathbf{p}_n \tag{15}$$

$$\mathbf{r}_{n+1}=\mathbf{r}_n-\alpha_n\bar{\mathbf{A}}\mathbf{p}_n \tag{16}$$

$$\beta_n=\frac{\|\bar{\mathbf{A}}^*\mathbf{r}_{n+1}\|^2}{\|\bar{\mathbf{A}}^*\mathbf{r}_n\|^2} \tag{17}$$

$$\mathbf{p}_{n+1}=\bar{\mathbf{A}}^*\mathbf{r}_n+\beta_n\mathbf{p}_n. \tag{18}$$

It is not difficult to show that the residual at the nth iteration step can be expressed as

$$\mathbf{r}_n=R_n(\bar{\mathbf{A}}\bar{\mathbf{A}}^*)\mathbf{r}_0 \tag{19}$$

where $R_n(\bar{\mathbf{A}}\bar{\mathbf{A}}^*)$ is a polynomial of order n in the matrix $\bar{\mathbf{A}}\bar{\mathbf{A}}^*$, i.e.,

$$R_n(\bar{\mathbf{A}}\bar{\mathbf{A}}^*)=\sum_{k=0}^{n}\xi_{nk}(\bar{\mathbf{A}}\bar{\mathbf{A}}^*)^k. \tag{20}$$

The coefficients in (20) are functions of the previous scalars α and β from (14) and (17) and can be shown to be real valued [6].

To show the desired relationship, let $\{\lambda_i\}$ represent the eigenvalues of the matrix $\bar{\mathbf{A}}\bar{\mathbf{A}}^*$ (these are also the eigenvalues of the matrix $\bar{\mathbf{A}}^*\bar{\mathbf{A}}$), and let $\{\mathbf{u}_i\}$ denote the corresponding orthonormal eigenvectors of $\bar{\mathbf{A}}\bar{\mathbf{A}}^*$. The initial residual may be expanded in terms of the eigenvectors to produce

$$\mathbf{r}_0=\sum_{i=1}^{N}\langle \mathbf{u}_i,\ \mathbf{r}_0\rangle\mathbf{u}_i. \tag{21}$$

Using this expansion in (19), replacing $(\bar{\mathbf{A}}\bar{\mathbf{A}}^*)^k u_i$ by $(\lambda_i)^k u_i$, and using the orthonormal property of the eigenvectors yields

$$\langle \mathbf{r}_n,\ \mathbf{r}_n\rangle=\sum_{i=1}^{N}|\langle u_i,\ r_0\rangle|^2[R_n(\lambda_i)]^2 \tag{22}$$

which is recognized as (11).

Thus the CG process can be interpreted as generating the "residual polynomial" R_n in such a manner that upon termination of the CG process with $E_n=0$, $R_n(\lambda)$ has zeros at each of the eigenvalues λ_i. The number of iterations required to generate this polynomial is equal to the number of distinct eigenvalues of the iteration matrix $\bar{\mathbf{A}}\bar{\mathbf{A}}^*$.

Equation (22) also illustrates the effect of the right side and the initial estimate of the solution on the error norm being minimized by the conjugate gradient method. The ith term in the summation for the error norm depends on the ith coefficient from the expansion of (21). This shows that if the initial residual could be represented in terms of a single eigenvector, the CG method would converge in one step regardless of the number of distinct eigenvalues appearing in the matrix.

The foregoing analysis of the CG method is not new, having been presented long ago in the literature [5]–[7]. However, when coupled with the discussion concerning the eigenvalues of the type of matrix equation under consideration, it lends new insight into the behavior of the CG algorithm when applied to electromagnetic scattering problems.

It was previously reported for a variety of electromagnetic scattering problems that the residual norm produced by the CG method after each iteration step appeared to be converging as the order of the moment-method matrix equation was increased for the same scatterer [8]. If the eigenvalues of the moment-method matrix "converge" in some sense as the order of the discretization is improved, as implied by the previous discussion, and if the initial residual (which in many cases is equal to the right side) can be adequately represented by the eigenvectors corresponding to those eigenvalues at a given order discretization, then the CG method should require a fixed number of iterations to converge as the order of the moment-method system is increased. This occurs in spite of the fact that additional eigenvalues are projected from the original operator as the order of the matrix is increased because the CG method minimizes a norm involving only those eigenvalues corresponding to eigenvectors needed to represent the initial residual. In addition, it appears from (8) that the eigenvalues of the original operator are sometimes distributed throughout the complex plane. This suggests that the eigenvalues of the moment-method matrix may spread in the complex plane as more and more are included in the matrix description of the problem, which degrades the condition number of the matrix without degrading the convergence behavior of the CG method.

These findings support one conclusion of [8], namely, that the CG method can provide useful feedback concerning the discretization employed to construct the moment-method matrix equation. In other words, the convergence behavior for several different discretizations can indicate how well the important part of the eigenvalue spectrum has been represented by the matrix operator. While this is by no means a foolproof criterion, an analyst can use the additional information to help determine the validity of a given numerical solution. The results also indicate that, as the order of a given discretization is increased, the CG approach should remain a numerically stable solution algorithm, despite the degrading condition of the matrix. This is in contrast to direct methods, which may fail because of round-off errors if the condition number of the matrix becomes too large in the process of refining the discretization.

REFERENCES

[1] R. F. Harrington, *Field Computation by Moment Methods.* Malabar, FL: Krieger, 1982 (reprint).

[2] R. Mittra and C. A. Klein, "Stability and convergence of moment method solutions," in *Numerical and Asymptotic Techniques in Electromagnetics,* R. Mittra, Ed. New York: Springer-Verlag, 1975.

[3] J. R. Mautz and R. F. Harrington, "Radiation and scattering from bodies of revolution," *Appl. Sci. Res.,* vol. 20, pp. 405–435, June 1969.

[4] A. W. Glisson and D. R. Wilton, "Simple and effective numerical methods for problems of electromagnetic radiation and scattering from surfaces," *IEEE Trans. Antennas Propagat.,* vol. AP-28, pp. 593–603, Sept. 1980.

[5] M. R. Hestenes and E. Stiefel, "Methods of conjugate gradients for solving linear systems," *J. Res. Nat. Bur. Stand.,* vol. 49, pp. 409–435, 1952.

[6] E. L. Stiefel, "Kernel polynomials in linear algebra and their numerical applications," *Nat. Bur. Stand. Appl. Math. Ser.,* vol. 49, pp. 1–22, 1958.

[7] A. Jennings, "Influence of the eigenvalue spectrum on the convergence rate of the conjugate gradient method," *J. Inst. Math. Appl.,* vol. 20, pp. 61–72, 1977.

[8] A. F. Peterson and R. Mittra, "Convergence of the conjugate gradient method when applied to matrix equations representing electromagnetic scattering problems," *IEEE Trans. Antennas Propagat.,* vol. AP-34, pp. 1447–1454, Dec. 1986.

Part 2:
Wires and Bodies Of Revolution

Wire antennas were among the first applications of the moment method, and they are still important, as the techniques have been extended to more complex structures. Butler and Wilton (1975) compare convergence rates of various basis functions applied to thin wires. An important early application by Thiele (1969) is to Yagi-Uda antennas; moment methods allow a complete analysis and design of log-periodic and Yagi-Uda arrays. Piecewise sinusoidal mutual-impedance algorithms are developed by Richmond and Geary (1970) and Hansen and Brunner (1979). Polynomial basis functions are introduced by Popovic (1970). Recent work by Canning (1989) uses long and partially overlapped basis functions that are selected to reduce significantly many matrix elements, producing a sparse matrix. The bases are like traveling wave sources. Otherwise, short bases can be used with a matrix preprocessor and postprocessor that incorporates the traveling wave directionality.

Because a body of revolution (BOR) only requires basis functions along a generator, a great reduction in the complexity of the moment method solution occurs. The BOR work also started early (e.g., a paper by Andreason (1965)). Mautz and Harrington (1969, 1978) apply piecewise triangular, pulse functions to BOR problems. Extension to include impedance boundaries is by Medgyesi-Mitschang and Eftimiu (1979).

BIBLIOGRAPHY

Thin Wire Integral Equations

F.M. Tesche, "The Effect of the Thin-Wire Approximation and the Source Gap Model on the High-Frequency Integral Equation Solution of Radiating Antennas," *Trans. IEEE,* Vol. AP-20, March 1972, pp. 210–211.

C.M. Butler, "Evaluation of Potential Integral at Singularity of Exact Kernel in Thin-Wire Calculations," *Trans. IEEE,* AP-23, March 1975, pp. 293–295.

D.S. Jones, "Note on the Integral Equation for a Straight Wire Antenna," *Proc. IEE-MOA,* Vol. 128, April 1981, pp. 114–116.

T.K. Sarkar, "A Study of the Various Methods for Computing Electromagnetic Field Utilizing Thin Wire Integral Equations," *Radio Science,* Vol. 18, January-February 1983, pp. 29–38.

T.K. Sarkar, "The Application of the Conjugate Gradient Method for the Solution of Operator Equations Arising in Electromagnetic Scattering from Wire Antennas," *Radio Science,* Vol. 19, September-October 1984, pp. 1156–1172.

L.W. Pearson and C.M. Butler, "Inadequacies of Collocation Solutions to Pocklington-Type Models of Thin-Wire Structures," *Trans. IEEE,* Vol. AP-23, 1975, pp. 295–298.

F.M. Tesche, "Evaluation of the Surface Integral Occurring in the E-field Integral Equations for Wire Antennas," *Trans. IEEE,* Vol. EMC-16, November 1974, pp. 209–210.

E.K. Miller G.J. Burke and E.S. Selden, "Accuracy-Modeling Guidelines for Integral-Equation Evaluation of Thin-Wire Scattering Structures," *Trans. IEEE,* Vol. AP-19, July 1971, pp. 534–536.

C.A. Klein and R. Mittra, "The Effect of Different Testing Functions in the Moment Method Solution of Thin Wire Antenna Problems," *Trans. IEEE,* Vol. AP-23, 1975, pp. 258–261.

D.A. McNamara and D.J.J. van Rensburg, "A Rapidly Convergent Galerkin-Collocation (GC) Method for the Analysis of Electromagnetic Scattering from Thin Wire Structures," *Microw. & Opt. Tech. Ltrs.,* Vol. 2, October 1989, pp. 355–357.

Dipoles

J.H. Richmond, "Coupled Linear Antennas with Skew Orientation," *Trans. IEEE,* Vol. AP–18, September 1970, pp. 694–696.

R.C. Hansen, "Formulation of Echelon Dipole Mutual Impedance for Computer," *Trans. IEEE,* Vol. AP-20, November 1972, pp. 780–781.

R.E. Collin, "Equivalent Line Current for Cylindrical Dipole Antennas and its Asymptotic Behavior," *IEEE Trans.,* Vol. AP-32, February 1984, pp. 200–204.

E.P. Sayre, "Junction Discontinuities in Wire Antenna and Scattering Problems," *Trans. IEEE,* Vol. AP-21, 1973, pp. 216–217.

W.A. Imbriale and P.G. Ingerson, "On Numerical Convergence of Moment Solutions of Moderately Thick Wire Antennas Using Sinusoidal Basis Functions," *Trans. IEEE,* Vol. AP-21, May 1973, pp. 363–366.

J.H. Richmond, "On the Edge Mode in the Theory of Thick Cylindrical Monopole Antennas," *Trans. IEEE,* Vol. AP-28, November 1980, pp. 916–921.

J.H. Richmond, "Monopole Antenna on a Circular Disk," *Trans. IEEE,* Vol. AP-32, December 1984, pp. 1282–1287.

J.H. Richmond and E.H. Newman, "Dielectric Coated Wire Antennas," *Radio Science,* Vol. 11, January 1976, pp. 13–20.

B.D. Popovic, "On Polynomial Approximation of Current Along Thin Asymmetrical Cylindrical Dipoles," *Trans. IEEE,* Vol. AP-19, January 1971, pp. 117–120.

B.D. Popovic, "Theory of Imperfectly Conducting Thin Cylindrical Dipoles," *Proc. IEE,* Vol. 117, December 1970, pp. 2205–2208.

B.D. Popovic and Z.D. Popovic, "Imperfectly Conducting Cylindrical Antenna: Variational Approach," *Trans. IEEE,* Vol. AP-19, May 1971, pp. 435–436.

B.D. Popovic and J.V. Surutka, "A Variational Solution to the Problem of an Asymmetrical Cylindrical Dipole," *Trans. IEEE,* Vol. AP-19, January 1971, pp. 17–22.

B.D. Popovic, "Analysis of Two Identical Parallel Arbitrarily Located Thin Asymmetrical Antennas," *Proc. IEEE,* Vol. 117, September 1970, pp. 1735–1740.

B.D. Popovic, "Theory of Cylindrical Antennas with Lumped Impedance Loadings," *Radio and Electronic Engineer,* Vol. 43, April 1973, pp. 243–248.

B.D. Popovic and T.S. Gavrilov, "Simple Method for Analysis of Cylindrical Antennas at the Interface Between Two Media," *Radio and Electronic Engineer,* Vol. 46, November 1976, pp. 553–554.

B.D. Popovic, A.R. Djordjevic and N.M. Kircanski, "Simple Method for Analysis of Dielectric-Coated Wire Antennas," *Radio and Electronic Engineer,* Vol. 51, March 1981, pp. 141–145.

E. Mendelovicz, A.T. Adams, K. Siarkiewicz and A. Mendelovicz, "Feedline Interference with Dipole Performance Theory and Experiment," *Trans. IEEE,* Vol. EMC-16, May 1974, pp. 90–97.

C.D. Taylor, "Thin Wire Receiving Antennas in a Parallel Plate Waveguide," *Trans. IEEE,* Vol. AP-15, July 1967, pp. 572–574.

R.B. Mack, "Phase Properties of Backscattered Fields from Thin Rods," *Trans. IEEE,* Vol. AP-19, May 1971, pp. 450–451.

R. Mittra and W.L. Ko, "A Finite Difference Approach to the Wire Junction Problem," *Trans. IEEE,* Vol. AP-25, May 1975, pp. 435–438.

T.L. Simpson, J.C. Logan and J.W. Rockway, "Equivalent Circuits for Electrically Small Antennas Using LS-Decomposition with the Method of Moments," *Trans. IEEE,* Vol. AP-37, December 1989, pp. 1632–1635.

J.H. Richmond, L.M. Schwab and R.G. Wickliff, "Tumble-Average Radar Backscatter of Some Thin-Wire Chaff Elements," *Trans. IEEE,* Vol. AP-22, January 1974, pp. 124–126.

Complex Wire Antennas

R.F. Harrington and J.R. Mautz, "Straight Wires with Arbitrary Excitation and Loading," *Trans. IEEE,* Vol. AP-15, July 1967, pp. 502–515.

B.J. Strait and K. Hirasawa, "On Long Wire Antennas with Multiple Excitations and Loadings," *Trans. IEEE,* Vol. AP-18, September 1970, pp. 699–700.

S.A.S. Saoudy and M. Hamid, "Optimal Design of Multiply Fed Dipole Antennas," *Trans. IEEE,* Vol. AP-35, September 1987, pp. 1001–1009.

M.C. Elias and C.T. Carson, "Direct Analysis of Folded Dipole by Method of Moments," *Electronics Letters,* Vol. 9, November 1, 1973, pp. 520–521.

J.E. Jones, "Analysis of the Symmetric Center-Fed V-dipole Antenna," *Trans. IEEE,* Vol. AP-24, May 1976, pp. 316–322.

R. Neri, "Moment-Method Solution of Directed Dipole Antennas," *Electronics Letters,* Vol. 16, June 5, 1980, pp. 453–455.

D.K. Cheng and C.H. Liang, "Shaped Wire Antenna with Maximum Directivity," *Electronics Letters,* Vol. 18, September 16, 1982, pp. 816–817.

T.L. Simpson, "The Theory of Top-Loaded Antennas: Integral Equations for the Currents," *Trans. IEEE,* Vol. AP-19, March 1971, pp. 186–190.

E.K. Miller, F.J. Deadrick and W.O. Henry, "Computer Evaluation of Large Low-Frequency Antennas," *Trans. IEEE,* Vol. AP-21, May 1973, pp. 386–389.

D.L. Smith, "The Trap-Loaded Cylindrical Antenna," *Trans. IEEE,* Vol. AP-23, January 1975, pp. 20–27.

H. Nakano, J. Yamauchi, K. Kawashima and K. Hirose, "Effects of Arm Bend and Asymmetric Feeding on Dipole Antennae," *Int. J. Electronics,* Vol. 55, 1983, pp. 353–364.

H. Nakano, K. Hirose and J. Yamauchi, "Effects of Arm Bend on One-Wavelength Dipole Antenna with Asymmetric Feeding," *Int. J. Electronics,* Vol. 56, 1984, pp. 121–125.

H. Kawakami, G. Sata and R.W. Masters, "Characteristics of TV Transmitting Batwing Antennas," *Trans. IEEE,* Vol. AP-32, December 1984, pp. 1318–1326.

G. Sato, H. Kawakami and F. Funatsu, "Theoretical Analysis and Some Experiments on Vierergruppe Antennae," *Trans. IEEE,* Vol. AP-31, May 1983, pp. 413–418.

P.K. Agrawal, G.A. Richards, G.A. Thiele and J.H. Richmond, "Analysis and Design of TEM-Line Antennas," *Trans. IEEE,* Vol. AP-20, September 1972, pp. 561–568.

B.J. Strait and A.T. Adams, "Analysis and Design of Wire Antennas with Applications to EMC," *Trans. IEEE,* Vol. EMC-12, May 1970, pp. 45–54.

R.M. Bevensee, J.N. Brittingham, F.J. Deadrick, T.H. Lehman, E. K. Miller and A.J. Poggio, "Computer Codes for EMP Interaction and Coupling," *Trans. IEEE,* Vol. AP-26, January 1978, pp. 156–165.

S. Lin and K.K. Mei, "Numerical Solution of Dipole Radiation in a Compressible Plasma," *Trans. IEEE,* Vol. AP-16, March 1968, pp. 235–241.

S.H. Lin and K.K. Mei, "On the Effect of Sheath, Collision and Absorptive Surface on the Performance of a Linear Antenna in a Compressible Plasma," *Trans. IEEE,* Vol. AP-18, September 1970, pp. 672–679.

H.S. Lu and K.K. Mei, "Cylindrical Antennas in Gyrotropic Media," *Trans. IEEE,* Vol. AP-19, September 1971, pp. 669–674.

Loops

A. Baghdasarian and D.J. Angelakos, "Scattering from Conducting Loops and Solution of Circular Loop Antennas by Numerical Methods," *Proc. IEEE,* Vol. 53, August 1965, pp. 818–822.

R.F. Harrington and J. Mautz, "Electromagnetic Behaviour of Circular Wire Loops with Arbitrary Excitation and Loading," *Proc. IEE,* Vol. 115, January 1968, pp. 68–77.

C.D. Taylor and C.W. Harrison, "On Thin-Wire Multiturn Loop Antennas," *Trans. IEEE,* Vol. AP-22, May 1974, pp. 407–413.

D.E. Barrick, "Miniloop Antenna Operation and Equivalent Circuit," *Trans. IEEE,* Vol. AP-34, January 1986, pp. 111–114.

Spirals and Log-Periodics

Y. Yeh and K.K. Mei, "Theory of Conical Equiangular-Spiral Antennas, Part I—Numerical Technique," *Trans. IEEE,* Vol. AP-15, September 1967, pp. 634–639.

A.E. Atia and K.K. Mei, "Analysis of Multiple-Arm Conical Log-Spiral Antennas," *Trans. IEEE,* Vol. AP-19, May 1971, pp. 320–331.

H. Nakano and J. Yamauchi, "Characteristics of Modified Spiral and Helical Antennas," *Proc. IEE-MOA,* Vol. 129, October 1982, pp. 232–237.

H. Nakano, J. Yamauchi and S. Hashimoto, "Numerical Analysis of 4-Arm Archimedean Spiral Antenna," *Electronics Letters,* Vol. 19, February 3, 1983, pp. 78–79.

H. Nakano, T. Yamane and J. Yamauchi, "Directive Properties of Parasitic Helix and its Application to Circularly Polarized Antenna," *Proc. IEE-MOA,* Vol. 130, October 1983, pp. 391–396.

M. Hilbert, M.A. Tilston and K.G. Balmain, "Resonance Phenomena of Log-Periodic Antennas," *Trans. IEEE,* Vol. AP-37, October 1989, pp. 1224–1234.

A.T. Adams, R.K. Greenough, R.F. Wallenberg, A. Mendelovicz and C. Lumjiak, "The Quadrifilar Helix Antenna," *Trans. IEEE,* Vol. AP-22, March 1974, pp. 173–178.

Slots and Strips

R.F. Harrington and D.T. Auckland, "Electromagnetic Transmission Through Narrow Slots in Thick Conducting Screens," *Trans. IEEE,* Vol. AP-28, September 1980, pp. 616–622.

C.M. Butler and D.R. Wilton, "General Analysis of Narrow Strips and Slots," *Trans. IEEE,* Vol. AP-28, January 1980, pp. 42–48.

E.H. Newman, "Strip Antennas in a Dielectric Slab," *Trans. IEEE,* Vol. AP-26, September 1979, pp. 647–653.

J.H. Richmond, "On the Edge Mode in the Theory of TM Scattering by Strip or Strip Grating," *Trans. IEEE,* Vol. AP-28, November 1980, pp. 883–887.

Bodies of Revolution—Conducting

P.L.E. Uslenghi, "Computations of Surface Currents on Bodies of Revolution," *Alta Frequenza,* Vol. 39, August 1970, pp. 213–224.

R.F. Harrington and J.R. Mautz, "Radiation and Scattering from Loaded Bodies of Revolution," *Appl. Sci. Res.,* Vol. 26, June 1971, pp. 209–217.

R.F. Harrington and J.R. Mautz, "Green's Functions for Surfaces of Revolution," *Radio Science,* Vol. 7, May 1972, pp. 603–611.

R.M. Kalafus, "Broadband Dipole Antenna Design Using the Method of Moments," *Trans. IEEE,* Vol. AP-19, November 1971, pp. 771–773.

L.N. Medgyesi-Mitschang and J.H. Mullen, "Radiation and Scattering from Asymmetrically Excited Bodies of Revolution," *Trans. IEEE,* Vol. AP-24, January 1976, pp. 90–93.

A.W. Glisson and C.M. Butler, "Analysis of a Wire Antenna in the Presence of a Body of Revolution," *Trans. IEEE,* Vol. AP-28, September 1980, pp. 604–609.

J.F. Shaeffer and L.N. Medgyesi-Mitschang, "Radiation from Wire Antennas Attached to Bodies of Revolution: The Junction Problem," *Trans. IEEE,* Vol. AP-29, May 1981, pp. 479–487.

J.F. Shaeffer, "EM Scattering from Bodies of Revolution with Attached Wires," *Trans. IEEE,* Vol. AP-30, May 1982, pp. 426–431.

D.B. Davidson and D.A. McNamara, "Moment Method Formulation for Asymmetrically Excited Conducting Bodies of Revolution," *Electronics Letters,* Vol. 24, September 29, 1988, pp. 1280–1282.

R.A. Perala, "Integral Equation Solution for Induced Surface Currents on Bodies of Revolution," *Trans. IEEE,* Vol. EMC-16, August 1974, pp. 172–177.

D.B. Davidson and D.A. McNamara, "Comparisons of the Application of Various Conjugate-Gradient Algorithms to Electromagnetic Radiation from Conducting Bodies of Revolution," *Microw. & Opt. Tech. Letters,* Vol. 1, September 1988, pp. 243–246.

Bodies of Revolution—Dielectric and Lossy

T.K. Wu and L.L. Tsai, "Scattering from Arbitrarily-Shaped Lossy Dielectric Bodies of Revolution," *Radio Science,* Vol. 12, September 1977, pp. 709–718.

J.R. Mautz and R.F. Harrington, "Electromagnetic Scattering from a Homogeneous Material Body of Revolution," *AEU,* Vol. 33, February 1979, pp. 71–80.

T.K. Wu, "Electromagnetic Field and Power Deposition in Body-of-Revolution Models of Man," *Trans. IEEE,* Vol. MTT-27, March 1979, pp. 279–293.

F. Stenger, M. Hagmann and J. Schwing, "An Algorithm for the Electromagnetic Scattering Due to an Impedance Boundary Condition," *Jour. of Math. Analysis and Applications,* Vol. 78, December 1980, pp. 531–573.

J.R. Mautz and R.F. Harrington, "Electromagnetic Coupling to a Conducting Body of Revolution with a Homogeneous Material Region," *Electromagnetics,* Vol. 2, October-December 1982, pp. 257–308.

L.N. Medgyesi-Mitschang and J.M. Putnam, "Electromagnetic Scattering from Axially Inhomogeneous Bodies of Revolution," *Trans. IEEE,* Vol. AP-32, August 1984, pp. 797–806.

A.A. Kishk, Y.M.M. Antar, L. Shafai and L.E. Allan, "Electromagnetic Scattering from Dielectric Bodies of Revolution: Theoretical and Experimental Results," *Electromagnetics,* Vol. 7, No. 1, 1987, pp. 51–60.

R.D. Graglia and P.L.E. Uslenghi, "Surface Currents on Impedance Bodies of Revolution," *Trans. IEEE,* Vol. AP-36, September 1988, pp. 1313–1317.

Bodies of Translation

L.N. Medgyesi-Mitschang and J.M. Putnam, "Scattering from Finite Bodies of Translation: Plates, Curved Surfaces, and Noncircular Cylinders," *Trans. IEEE,* Vol. AP-31, November 1983, pp. 847–852.

L.N. Medgyesi-Mitschang and J.M. Putnam, "Formulation for Wire Radiators on Bodies of Translation with and without End Caps," *Trans. IEEE,* Vol. AP-31, November 1983, pp. 854–862.

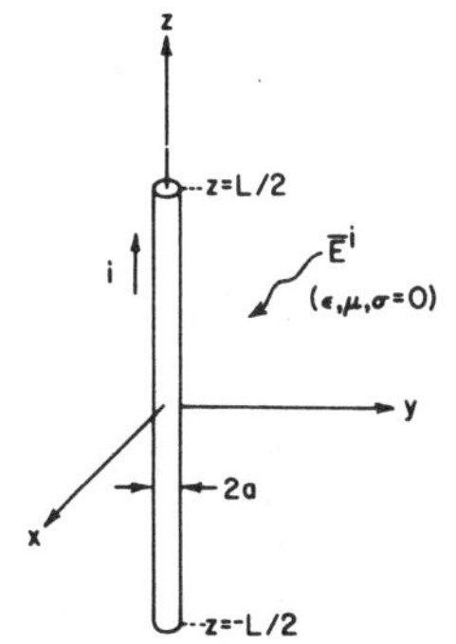

Fig. 1. Straight wire subject to incident illumination.

Analysis of Various Numerical Techniques Applied to Thin-Wire Scatterers

CHALMERS M. BUTLER AND D. R. WILTON

***Abstract*—Several numerical schemes for solving Pocklington's and Hallén's equations for thin-wire scatterers are investigated. Convergence rates of solutions obtained from seven methods are given and reasons for different rates are delineated.**

INTRODUCTION

In moment method solutions of antenna and scattering problems associated with thin-wire structures, both Pocklington (electric field) and Hallén (magnetic vector potential) type integral equations are commonly used. From either, one may obtain solutions for the current on a wire antenna or a scatterer and subsequently calculate all other quantities of interest. Even though the two equations are intimately related [1], each exhibits distinct advantages and disadvantages. The authors present the findings of their investigations of several numerical schemes for solving wire problems. They discuss the relative merits of various techniques, point out pitfalls to be avoided, and summarize their findings.

Manuscript received August 1, 1974; revised December 23, 1974. This work was supported in part by the National Science Foundation under Grant GU3833.

The authors are with the School of Engineering, University of Mississippi, University, Miss. 38677.

In the numerical methods, different basis sets for representing the unknown wire current are employed and relative convergences are investigated. So that one may focus attention on the generic feature of the capacity of a given basis set to represent adequately, and converge to, the correct current on the wire, only the case of a scatterer subject to normally incident illumination is considered here. With illumination constant over the length of the scatterer, the problem of inadequate sampling [2] of the integral equation's driving term has no bearing on convergence. Also, in the interest of simplicity and of addressing only the fundamental question of convergence in the sense of how well the actual wire current is represented by the solution, the wire junction problem is not treated nor is any record of computer times given for the various methods.

THIN-WIRE EQUATIONS

From basic electromagnetic theory, one may readily obtain the following fundamental integro-differential equation:

$$\left(\frac{d^2}{dz^2} + k^2\right) \int_{\zeta=-L/2}^{L/2} i(\zeta) K(z-\zeta)\, d\zeta = -j4\pi\omega\varepsilon E_z^i(z) \quad (1)$$

which relates the unknown total axial current i on a cylinder to the known incident electric field having an axial component E_z^i on the surface of the scatterer. The scatterer is perfectly conducting and resides in a homogeneous space characterized by (μ, ε, $\sigma = 0$), and, as suggested in Fig. 1, it is a tube of length L and radius a. The kernel in (1) is

$$K(\xi) = \frac{1}{2\pi} \int_{\phi'=-\pi}^{\pi} \frac{e^{-jk[\xi^2 + 4a^2 \sin^2(\phi'/2)]^{1/2}}}{[\xi^2 + 4a^2 \sin^2(\phi'/2)]^{1/2}}\, d\phi' \quad (2)$$

where k is 2π/wavelength at the angular frequency ω of the suppressed harmonic time variation $e^{j\omega t}$. For present purposes, the wire radius is looked upon as being very small relative to the wavelength λ as well as to the cylinder length. Such restrictions, common in thin-wire analyses, assure one that the current on the cylinder is circumferentially independent and that it can be accounted for by the total axial current i. The thin-wire assumptions also lead to the so-called reduced kernel approximation to $K(\xi)$

$$K(\xi) \doteq \frac{e^{-jk[\xi^2 + a^2]^{1/2}}}{[\xi^2 + a^2]^{1/2}}. \quad (3)$$

From (1), one may readily derive Hallén's equation

$$\int_{\zeta=-L/2}^{L/2} i(\zeta)K(z-\zeta)\,d\zeta = C\cos kz + B\sin kz - j\frac{4\pi}{\eta}\int_{\zeta=0}^{z} E_z^i(\zeta)\sin k(z-\zeta)\,d\zeta \quad (4)$$

where η ($=\sqrt{\mu/\varepsilon}$) is the intrinsic impedance of the medium and where C and B are constants of integration which must be consistent with the boundary condition that the current be zero at the wire ends ($\pm L/2$). In the special case of normally incident plane wave illumination, E_z^i is constant over $z \in (-L/2, L/2)$ allowing one to perform analytically the integration on the right side of (4), and $i(z)$ is an even function of z allowing one to eliminate the constant B. The exact kernel (2) may be used in (4) or, where warranted by thin-wire conditions, the approximation (3) often suffices.

Having accepted the approximation (3), one may interchange at will the integral and differential operators in (1) and, thereby, convert it to any of several equivalent forms usually referred to as Pocklington-type equations. For the present discussion, the form which is most useful is

$$\int_{\zeta=-L/2}^{L/2}\left[\left(\frac{d^2}{d\zeta^2}+k^2\right)i(\zeta)\right]K(z-\zeta)\,d\zeta + \left[i(\zeta)\frac{\partial}{\partial\zeta}K(z-\zeta) - \frac{d}{d\zeta}i(\zeta)K(z-\zeta)\right]_{\zeta=-L/2}^{L/2} = -j4\pi\omega\varepsilon E_z^i(z). \quad (5)$$

Bases

The unknown current $i(z)$ is represented in a numerical solution technique by a linear combination of known elements $i_n(z)$ of a basis set $\{i_n(z)\}$ selected to approximate the current as

$$i(z) \doteq \sum_n I_n i_n(z). \quad (6)$$

With i in the integral equations replaced by its approximation of (6), Hallén's and Pocklington's equations become, respectively,

$$\sum_n I_n \int_{\zeta=l_n}^{u_n} i_n(\zeta)K(z-\zeta)\,d\zeta = C\cos kz + B\sin kz - j\frac{4\pi}{\eta}\int_{\zeta=0}^{z} E_z^i(\zeta)\sin k(z-\zeta)\,d\zeta \quad (7a)$$

and

$$\sum_n I_n \left\{\int_{\zeta=l_n}^{u_n}\left[\left(\frac{d^2}{d\zeta^2}+k^2\right)i_n(\zeta)\right]K(z-\zeta)\,d\zeta + \left[i_n(\zeta)\frac{\partial}{\partial\xi}K(z-\zeta) - \frac{d}{d\zeta}i_n(\zeta)K(z-\zeta)\right]_{\zeta=l_n}^{u_n}\right\} = -j4\pi\omega\varepsilon E_z^i(z). \quad (7b)$$

One chooses the set $\{i_n\}$ for its capacity to represent the current well and from the viewpoint of its utility in the numerical procedure. The basis sets considered here are illustrated in Fig. 2 and are defined subsequently where one sees them to be of the subdomain type, since each element i_n differs from zero over only a single subdomain of the total domain of interest $z \in (-L/2, L/2)$.

Piecewise Sinusoidal:

$$i_n(z) = \begin{cases} \dfrac{\sin k(\Delta - |z-z_n|)}{\sin k\Delta}, & z \in (z_{n-1}, z_{n+1}) \\ 0, & z \notin (z_{n-1}, z_{n+1}) \end{cases} \quad (8a)$$

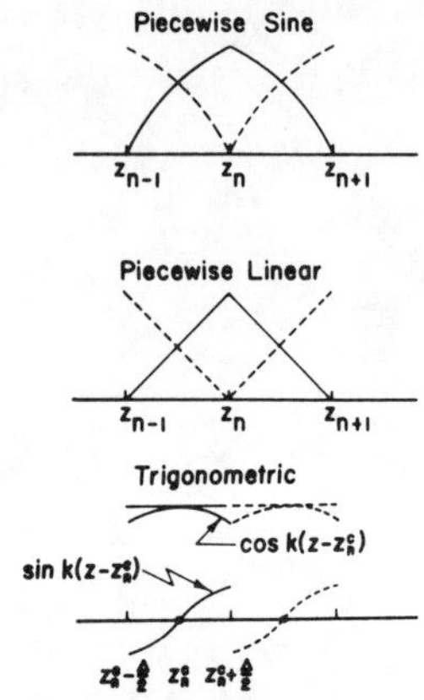

Fig. 2. Bases.

Piecewise Linear:

$$i_n(z) = \begin{cases} \dfrac{\Delta - |z-z_n|}{\Delta}, & z \in (z_{n-1}, z_{n+1}) \\ 0, & z \notin (z_{n-1}, z_{n+1}) \end{cases} \quad (8b)$$

Trigonometric:

$$i_n(z) = \begin{cases} 1 + C_n\cos k(z-z_n^c) + B_n\sin k(z-z_n^c), & z \in (z_n^c - \Delta/2,\ z_n^c + \Delta/2) \\ 0, & z \notin (z_n^c - \Delta/2,\ z_n^c + \Delta/2). \end{cases} \quad (8c)$$

Entire domain basis sets where each element exists over the full domain $(-L/2, L/2)$ are useful in special cases, but, since their application is limited, they are not treated in this paper.

Implicit in the use of (8) in (7) is the division of the scatterer length L into N subintervals of length $\Delta = L/N$. In all cases the boundary conditions $i(\pm L/2) = 0$ are satisfied by (6) *a priori* which, in using (8a) and (8b), means that one sets the I_n associated with the i_n over the intervals $(L/2 - \Delta, L/2 + \Delta)$ and $(-L/2 - \Delta, -L/2 + \Delta)$ equal to zero whereas with (8c) the C_n and B_n of the end-most intervals are adjusted so that $i(\pm L/2) = 0$. For piecewise linear and piecewise sinusoidal basis sets, the interval (l_n, u_n) must be partitioned into (z_{n-1}, z_n) plus (z_n, z_{n+1}) and the terms on the left side of (7b) must be evaluated in each of these two open intervals and summed to obtain the contribution from (l_n, u_n). This partitioning is necessary due to the derivative discontinuity at z_n exhibited by both (8a) and (8b). For the trigonometric set, $l_n = z_n^c - \Delta/2$ and $u_n = z_n^c + \Delta/2$, where z_n^c is the center of the nth subdomain.

As discussed in detail subsequently, the convergence rate of a solution method can be enhanced by modification of (6) through certain constraints applied to $\{i_n\}$. Two such constraints are employed by the authors in methods discussed forthwith. In one case they adjust the C_n and B_n of (8c) to force i of (6) plus its derivative to be continuous at each common boundary point of adjacent subdomains. Subject to these adjustments of the coefficients in (8c), a basis set is obtained which renders the current and its first derivative continuous everywhere in $(-L/2, L/2)$. In another case employing (8c), $i_n(z)$, the current in the nth subdomain, is required to satisfy[1] $i_n(z_{n-1}^c) = i_{n-1}(z_{n-1}^c)$ and $i_n(z_{n+1}^c) = i_{n+1}(z_{n+1}^c)$. One might refer to this as *extrapolated continuity*, and it is evident that the requirements do not force the current or its derivative to be continuous at the boundary of subdomains.

[1] See [3].

Testing

Substitution of (6) into (4) and (5) yields (7a) and (7b), respectively, in which all integrations and differentiations operate on known functions but into which the additional unknown constants I_n are introduced.

According to the method of moments [4], one can transform (7a) or (7b) into a system of linear algebraic equations with unknowns I_n by testing both sides of either equation with members of a suitable testing set $\{W_m\}$. For example, testing of (7b) yields

$$\sum_n I_n \left\langle \int_{\zeta=l_n}^{u_n} \left[\left(\frac{d^2}{d\zeta^2} + k^2\right) i_n(\zeta)\right] K(z-\zeta)\, d\zeta \right.$$
$$\left. + \left[i_n(\zeta)\frac{\partial}{\partial\zeta} K(z-\zeta) - \frac{d}{d\zeta} i_n(\zeta) K(z-\zeta)\right]_{\zeta=l_n}^{u_n}, W_m(z)\right\rangle$$
$$= -j4\pi\omega\varepsilon \langle E_z^i(z), W_m(z)\rangle, \qquad m = 1,2,\cdots,M \tag{9}$$

where

$$\langle f,W\rangle = \int_{z=-L/2}^{L/2} f(z)W(z)\, dz.$$

In the preceding (9) is a system of linear equations which can be solved for the I_n for proper M.

Discussion of Methods

In this section the seven solution methods in Table I are discussed in the order of their listing, and relative convergence data are given in Figs. 3–12. The value of current at the center of the scatterer is the quantity upon which the study of convergence is based, since it was ascertained that the convergence rate at points other than the scatterer center was comparable to that of the center current. In all cases, data are normalized with respect to that obtained from Galerkin's method with 31 piecewise sinusoids. Normalized results of each method are plotted against $1/N$ where one recalls N to be the total number of subdomains into which $(-L/2, L/2)$ is divided. In Methods I, II, III, and VI the number of unknown I_n to be determined is $(N-1)$, while in Method V the number of unknown current coefficients is N. After constraining the C_n and B_n of (8c) in Method IV, one has N unknown I_n to calculate but these constraining equations either augment the system of equations (9) or necessitate that one solve an auxiliary difference equation. Finally, the number of unknown I_n in Method VII is $(N-1)$ but the two constants C and B are unknown *a priori* so that, effectively, one must solve for $N+1$ quantities.

Data are given for two radii and for selected wire lengths so that they include special cases of practical interest, e.g., $L = \lambda/2$ and $L = \lambda$. Trends inferred from data given should be representative of typical cases encountered in practice except, of course, for very long wires.

The real and imaginary parts of the center current are investigated separately and each part is normalized with respect to the corresponding part obtained from Galerkin's method as mentioned. The real and imaginary normalization factors, C_R and C_I, are given in each figure. Normalization of each part separately has the advantage that small differences in a part can be independently observed in the data. However, on the other hand, when C_R and C_I differ appreciably, deviations from the norm of data normalized with respect to the smaller may be in large measure due to round-off errors; such a deviation appears extreme to an observer when actually it is of no consequence upon comparison with $\sqrt{C_R^2 + C_I^2}$.

TABLE I
Numerical Solution Methods

Method	Equation	Basis Set	Testing Method	Key
I	Pocklington	Piecewise sinusoidal	Collocation	P-PS-C
II	Pocklington	Piecewise sinusoidal	Collocation/ Galerkin	P-PS-C/G
III	Pocklington	Piecewise sinusoidal	Galerkin	P-PS-G
IV	Pocklington	Trigonometric with continuous current and continuous derivative	Collocation	P-T/C-C
V	Pocklington	Trigonometric with extrapolated continuity	Collocation	P-T/E-C
VI	Pocklington (Difference Equation)	Piecewise linear	Collocation	P-PL-DE
VII	Hallén	Piecewise linear	Collocation	H-PL-C

Method I (Pocklington-Piecewise Sinusoid-Collocation)

Since no numerical integration is needed to calculate terms in (9), Pocklington's equation subject to collocation ($W_m(z) = \delta(z - z_m)$) with the piecewise sinusoids (8a) in the current approximation (6) appears, superficially, to be a very attractive method. No numerical integration is needed in (9), since the harmonic differential operator in the integrand applied to (8a) yields zero for $\zeta \in (z_n, z_{n+1})$, leaving one with simply

$$\frac{k}{\sin k\Delta}\sum_n I_n[K(z_m - z_{n-1}) - 2\cos k\Delta\, K(z_m - z_n)$$
$$+ K(z_m - z_{n+1})] = -j4\pi\omega\varepsilon E_z^i(z_m)$$

where $z_m = m\Delta$, $m = 0, \pm 1, \pm 2, \cdots, \pm(N-2)/2$, are the match points in $(-L/2, L/2)$ at which (9) is enforced. Unfortunately, the obviation of numerical integration not withstanding, this procedure is unsatisfactory except for large radius and/or resonant length wires. Under close scrutiny the method is not generally satisfactory even for a large radius wire, since the solution converges rapidly for relatively small N, but with further increase in N the system of linear equations (9) which one must solve to determine the coefficients I_n becomes ill-conditioned [5]. These comments are supported by the data of Figs. 3–12 where one sees that for $a = 0.01\lambda$ the method enjoys somewhat better convergence than for $a = 0.001\lambda$ except when $L = 0.5\lambda$; also, P-PS-C exhibits better convergence for $L = 0.5\lambda$ and $a = 0.001\lambda$ than for other lengths and this radius.

Method II (Pocklington-Piecewise Sinusoid-Collocation/Galerkin)

Initially the authors attributed the poor convergence of P-PS-C to the fact that the integral equation is enforced at the interior points (match points z_m) but not at the end points ($\pm L/2$) with the consequence that the resulting system (9) may characterize a scatterer of length $L - 2\Delta$ rather than of L, the correct value. As a remedy, they elected to form (9) by testing over $(-L/2, -L/2 + 2\Delta)$ and $(L/2 - 2\Delta, L/2)$ with the piecewise sinusoids of (8a) and point-matching at the interior points (match points z_m). This hybrid testing procedure (Method II) is designated P-PS-C/G. Deceptively, it appears to converge for small N but is little better than P-PS-C for larger N as is seen from the data of Figs. 3–12.

Further investigation of (8) and (9) reveals that the poor convergence rates of P-PS-C and P-PS-C/G are due to the rapidly

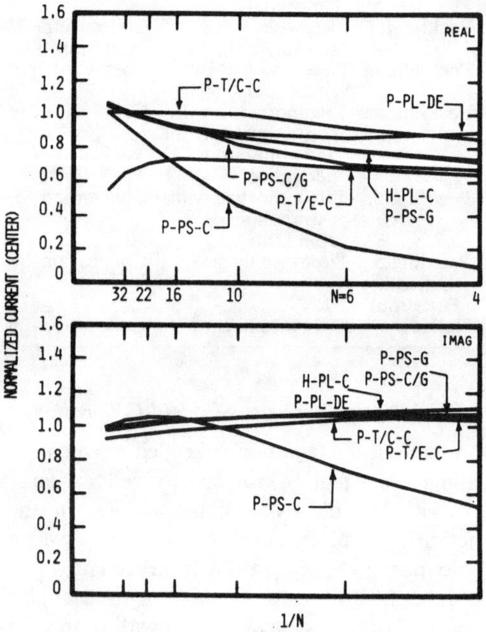

Fig. 3. Convergence data: normalized current at center of scatterer versus 1/number of subdomains ($l = 0.4\lambda$, $a = 0.01\lambda$, $C_R = 3.007$, $C_I = 2.310$).

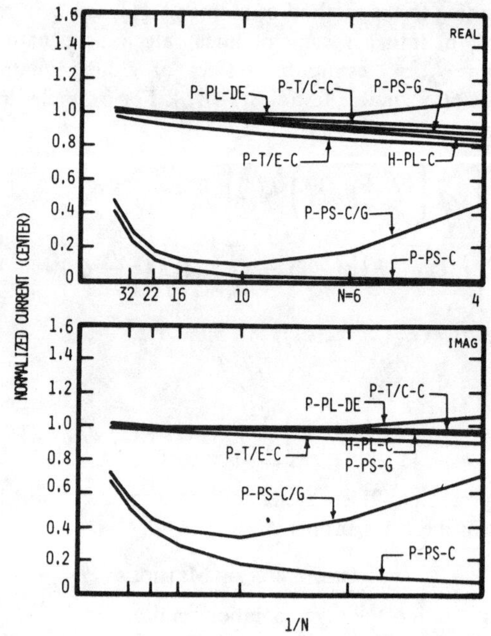

Fig. 4. Convergence data: normalized current at center of scatterer versus 1/number of subdomains ($l = 0.4\lambda$, $a = 0.001\lambda$, $C_R = 0.5498$, $C_I = 1.566$).

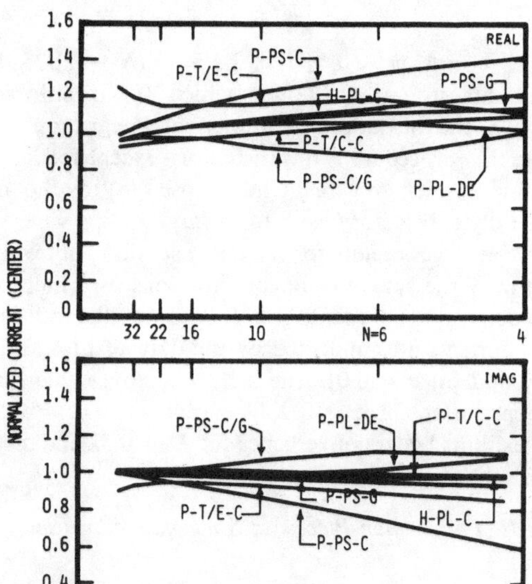

Fig. 5. Convergence data: normalized current at center of scatterer versus 1/number of subdomains ($l = 0.5\lambda$, $a = 0.01\lambda$, $C_R = 2.808$, $C_I = -0.908$).

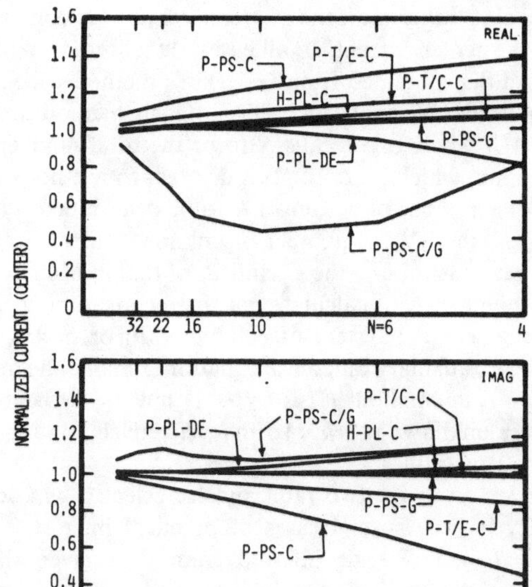

Fig. 6. Convergence data: normalized current at center of scatterer versus 1/number of subdomains ($l = 0.5\lambda$, $a = 0.001\lambda$, $C_R = 2.999$, $C_I = -1.929$).

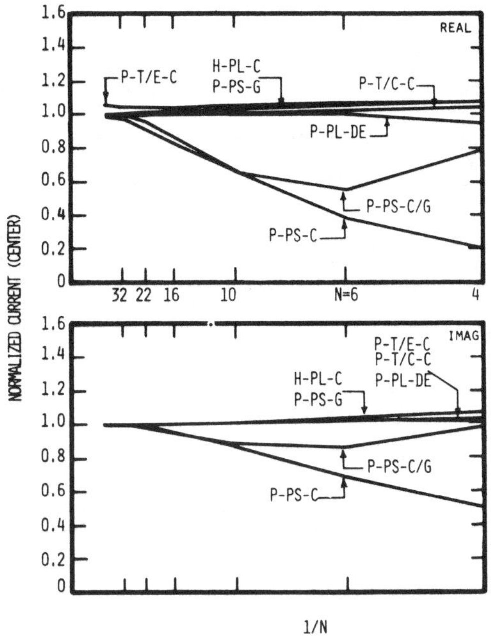

Fig. 7. Convergence data: normalized current at center of scatterer versus 1/number of subdomains ($l = 0.667\lambda$, $a = 0.01\lambda$, $C_R = 1.038$, $C_I = -1.619$).

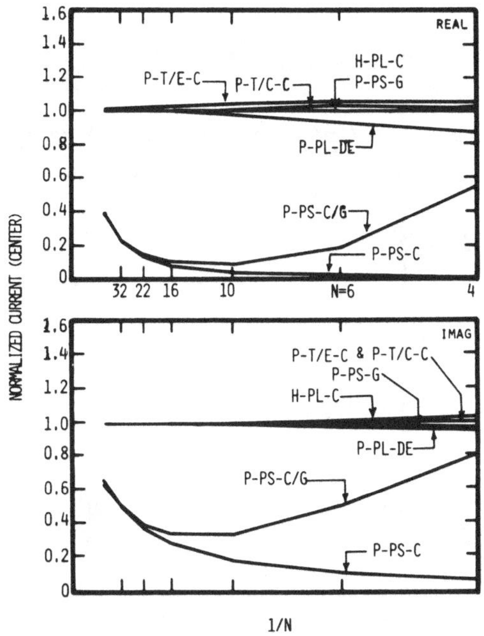

Fig. 8. Convergence data: normalized current at center of scatterer versus 1/number of subdomains ($l = 0.667\lambda$, $a = 0.001\lambda$, $C_R = 0.4658$, $C_I = -1.210$).

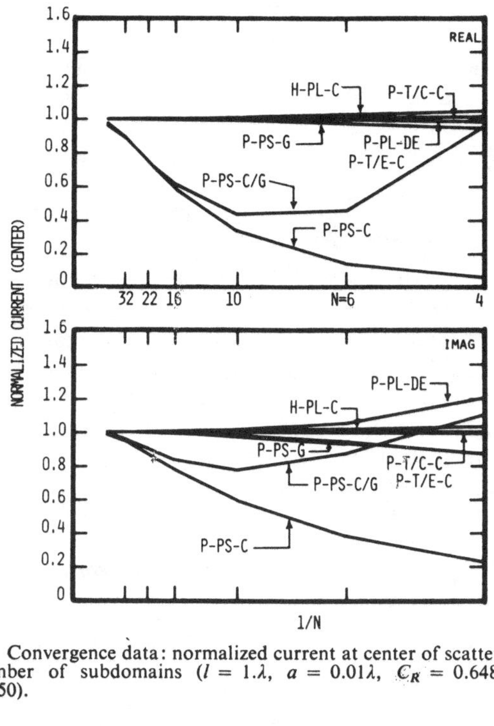

Fig. 9. Convergence data: normalized current at center of scatterer versus 1/number of subdomains ($l = 1.\lambda$, $a = 0.01\lambda$, $C_R = 0.6482$, $C_I = -1.450$).

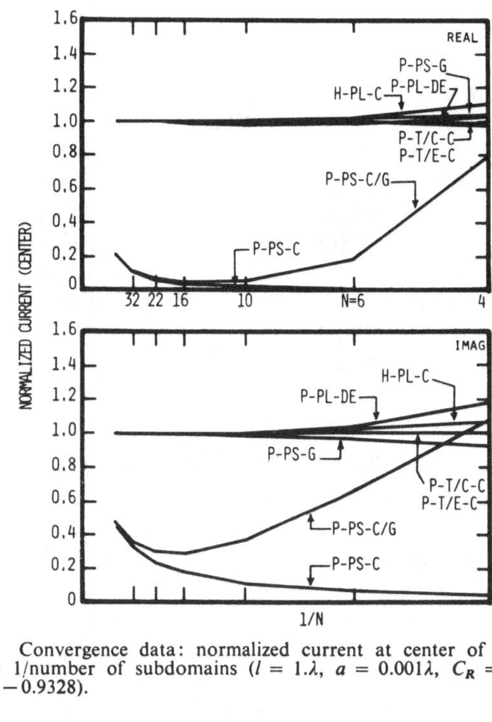

Fig. 10. Convergence data: normalized current at center of scatterer versus 1/number of subdomains ($l = 1.\lambda$, $a = 0.001\lambda$, $C_R = 0.2479$, $C_I = -0.9328$).

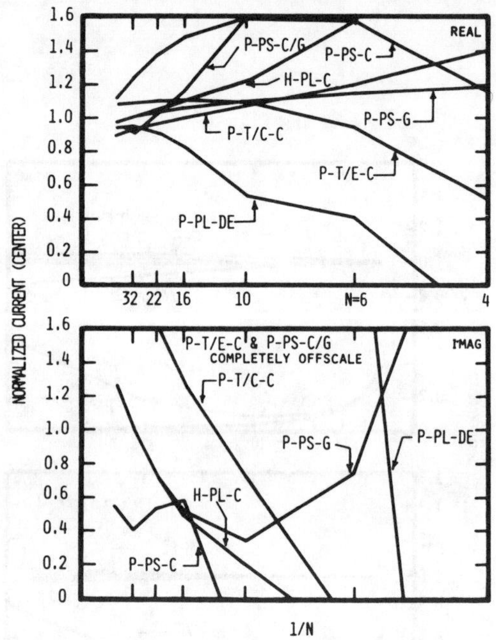

Fig. 11. Convergence data: normalized current at center of scatterer versus 1/number of subdomains ($l = 1.5\lambda$, $a = 0.01\lambda$, $C_R = -1.711$, $C_I = 0.0788$).

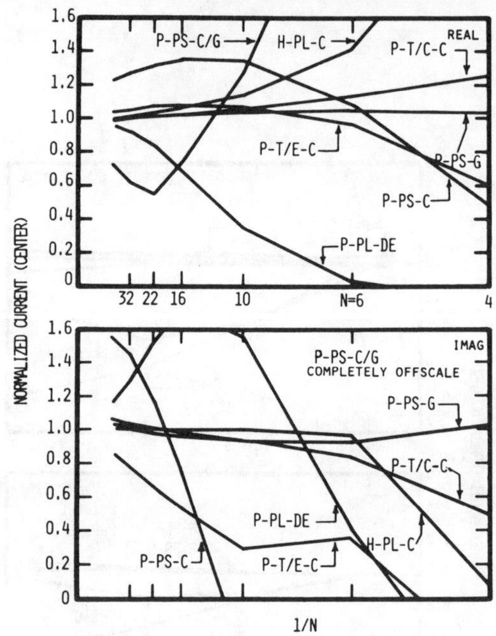

Fig. 12. Convergence data: normalized current at center of scatterer versus 1/number of subdomains ($l = 1.5\lambda$, $a = 0.001\lambda$, $C_R = -2.175$, $C_I = 0.4404$).

varying contributions to the scattered electric field arising from the unnatural discontinuities in derivative of the current approximation formed from the piecewise sinusoidal basis set [2]. Notice, however, in Figs. 5 and 6, the acceptable convergence rates of P-PS-C for resonant-length elements, which are due to the fact that resonant current is almost sinusoidal and the discontinuities in derivative almost vanish [1], [2].

A discontinuity in current or its derivative is unphysical and any approximation (6) exhibiting either produces a rapidly varying scattered electric field [2]. In particular, at a discontinuity in current, there is a contribution to scattered field proportional to $(\partial/\partial\zeta)K(z-\zeta)_{z=\zeta}$ as is seen from (5), whereas at a discontinuity of derivative of current there is a contribution proportional to $K(0)$. With either discontinuity, a highly peaked, rapidly varying scattered electric field is produced along the scatterer, which according to (5) must be equal to the constant $-E_z{}^i$ (normal incidence). Of course, equating of a constant to a rapidly varying function at discrete points z_m is unlikely to give rise to equality at points other than z_m over the interval $(-L/2, L/2)$.

Method III (*Pocklington-Piecewise Sinusoid-Galerkin*)

Failure to satisfy (7b) in some sense at all points on the scatterer implies, of course, a poor solution and, in P-PS-C and P-PS-C/G, can be attributed to the unnatural variation of the scattered electric field caused by unphysical discontinuities in the derivative of (6), the representation of current. In an attempt to improve solution methods employing basis sets which introduce discontinuities in the approximation to the current or its derivative, one may equate weighted averages of both sides of (7b) in hopes of achieving a better representation of the scattered field over the entire range $(-L/2, L/2)$ compared with its values at discrete z_m. Galerkin's method, which is mentioned under TESTING, effects such averaging. P-PS-G is a technique for solving Pocklington's equation by Galerkin's method with piecewise sinusoids both for representing the current and for testing. Solutions obtained by means of P-PS-G are seen from Figs. 3–12 to converge rapidly. Note that the Galerkin procedure requires two integrations, one over the basis element and another over the testing function. Richmond, who calls this procedure "piecewise sinusoidal reaction matching," has shown, however, that these integrations can be performed analytically for piecewise sinusoidal basis and testing sets [6], [1].

Notice that Methods I, II, and III incorporate (8a) in (6), each with different testing schemes for solving Pocklington's equation. Since, for $\Delta \ll \lambda$, (8b) closely approximates (8a), comments for the piecewise sinusoid also hold for the piecewise linear basis set.

Method IV (*Pocklington-Trigonometric/Continuous Current and Derivative-Collocation*)

In keeping with the desire to maintain continuity of current and its derivative over the scatterer, the authors investigated the use of the trigonometric set (8c) in Pocklington's equation, but, in addition to point-matching at the center of each subdomain, i and $(di/d\zeta)$ were forced to be equal at the common boundary points of adjacent subdomains.[2] Note that such a current representation may be called a trigonometric spline function. Solutions obtained by this method P-T/C-C are seen to converge rapidly (Figs. 3–12) as one would expect.

Method V (*Pocklington-Trigonometric/Extrapolated Continuity-Collocation*)

Because the introduction in P-T/C-C of the equations to maintain continuity of (6) and its derivative adds significant complexity to the method, an alternate scheme, P-T/E-C, for suppressing the discontinuities was investigated. Subject to *extrapolated continuity*[3] of P-T/E-C, the discontinuities are not zero but are reduced to levels at which their contributions do not dominate the scattered field. In contrast to the unacceptable

[2] See last paragraph under BASES.
[3] See last paragraph under BASES and [3].

complexities encountered in P-T/C-C in constraining the C_n and B_n, P-T/E-C is not difficult to implement, yet one observes from Figs. 3–12 that the convergence rate of solutions obtained by the latter method is near that of those obtained by the former. However, for larger and larger N, P-T/E-C solutions are seen to exhibit trends of diverging. This divergence is understandable, since in P-T/E-C the discontinuities are not zero, and, as $\Delta \to 0$ for larger and larger N, the unnatural contributions due to $K(\Delta/2)$ and $(\partial/\partial\zeta)K(z - \zeta)_{z-\zeta=\Delta/2}$ become more and more significant.

Modifications of P-T/E-C and P-T/C-C where the trigonometric functions of (8c) are replaced by other analytic functions, e.g., three-term power series, also lead to satisfactory solution techniques.

Method VI (Pocklington-Piecewise Linear-Difference Equation)

Methods I–V exhibit convergence rates which strongly depend upon how one handles discontinuities of (6) and its derivative. As an alternative to placing continuity requirements upon (6) to suppress the peaked field variations, one may relax or smooth the derivatives in (1) that operate on

$$\int_{\zeta=-L/2}^{L/2} i(\zeta) K(z - \zeta)\, d\zeta \qquad (10)$$

to produce these unnatural contributions to scattered electric field at the discontinuities. In particular in P-PL-DE the harmonic operator of (1) may be replaced by its corresponding difference operator; the difference operator is insensitive to the local variation of (10) due to discontinuities of (6) but does correctly account for desired global derivatives. Specifically, in P-PL-DE, the derivative discontinuities of the piecewise linear representation, (8b) in (6), do not manifest themselves in sharp peaks of the scattered field. Moreover, the difference operator equation approximation to (1) is amenable to use with basis sets which cause discontinuities in (6) as well as in its derivative. The convergence of P-PL-DE is given in Figs. 3–12 [1].

Method VII (Hallén-Piecewise Linear-Collocation)

Pocklington's equation relates the electric field to the sources on the scatterer in such a manner that the value of field is very sensitive to both the current and its derivative and, consequently, any efficient method for solving this equation must include special treatment of either the derivatives or of the discontinuities. On the other hand, Hallén's equation is based on the vector potential and the electric field enters the relationship only through integration. In addition, the vector potential is less sensitive to local variation of current than is its second derivative so Hallén's equation yields to solution methods in which the current is approximated (6) by unphysical currents possessing discontinuities. Good convergence is seen in solutions of Hallén's equation when the basis set is piecewise linear (H-PL-C) and in which no attempt is made to lessen the effects of the discontinuous derivatives (Figs. 3–12).

Conclusions

Seven methods for determining the current on a scatterer are presented and the relative convergence rates of the solutions obtained by the methods are investigated. Reasons for differences in rates are delineated. It is shown that, if acceptable convergence rates are to be attained, solution methods applied to Pocklington's equation must incorporate means of suppressing discontinuities in the current approximation and its derivative, or, on the other hand, the deleterious effects of these discontinuities must be circumvented by rendering the equation insensitive to them. The latter can be accomplished by averaging techniques applied to both sides of the equation, e.g., Galerkin's method, or by smoothing the derivatives, e.g., use of difference operator. Hallén's equation governs quantities which are less sensitive to discontinuities and it may be successfully solved numerically with almost any reasonable basis set, even one which causes the approximate current to be discontinuous. For a given basis set, the convergence rate of solutions to Hallén's equation obtained by point-matching is as high as that of solutions to Pocklington's equation with any testing scheme. Solutions by the difference equation method attain a high rate of convergence essentially identical to that of the point-matched Hallén equation. Of further importance, the difference equation procedure is simple, and it is highly amenable to numerical implementation, even when applied to multiple-wire structures.

References

[1] D. R. Wilton and C. M. Butler, "Use of difference equations in conjunction with moment methods," in *USNC/URSI 1974 Spring Meeting*, Atlanta, Ga., June 1974.

[2] L. W. Pearson and C. M. Butler, "Inadequacies of collocation solutions to Pocklington-type models of thin-wire structures," *IEEE Trans. Antennas Propagat.* (Commun.), vol. AP-23, pp. 295–298, Mar. 1975.

[3] Y. S. Yeh and K. K. Mei, "Theory of conical equiangular-spiral antennas, Part I—numerical technique," *IEEE Trans. Antennas Propagat.*, vol. AP-15, pp. 634–639, Sept. 1967.

[4] R. F. Harrington, *Field Computation by Moment Methods.* New York: Macmillan, 1968.

[5] C. D. Taylor and D. R. Wilton, "The extended boundary condition solution of the dipole antenna of revolution," *IEEE Trans. Antennas Propagat.* (Commun.), vol. AP-20, pp. 772–776, Nov. 1972.

[6] J. H. Richmond and N. H. Geary, "Mutual impedance between coplanar-skew dipoles," *IEEE Trans. Antennas Propagat.* (Commun.), vol. AP-18, pp. 414–416, May 1970.

Analysis of Yagi–Uda-Type Antennas

GARY A. THIELE, MEMBER, IEEE

Abstract—A method of analyzing Yagi–Uda-type antennas is presented. Since the Yagi–Uda array is a fairly well-known antenna, it is used as an example to demonstrate the application and accuracy of the method. However, the method of solution is not limited to a planar array, such as the Yagi, but can be applied to arrays of nonplanar linear elements. The approach taken in analyzing Yagi–Uda antennas is based on rigorous equations for the electric field radiated by the elements in the array. All interactions are taken into account. Calculated results are presented for the Yagi–Uda array that show excellent agreement with experimental results reported in the literature. In addition, the dependence of the far-field patterns on the phase velocity is shown. It is also demonstrated that the phase velocity is generally nonuniform along the array.

Introduction

In 1927 and 1928, respectively, Uda and Yagi published their papers on what today is known as the Yagi–Uda antenna [1], [2]. Much is known about the Yagi–Uda array, but this has been due mostly to experimental data rather than a method of theoretical investigation. There are, however, some exceptions. One is a paper published by Wilkinshaw in 1946 which applied only to short (4 directors) Yagi–Uda antennas [3]. A theoretical and experimental paper by Green gives design data for several practical Yagi–Uda arrays [4]. His method was the classical one in which the Yagi–Uda array is viewed as a special kind of center-driven dipole array in which all but the actual exciter element are short-circuited at the terminals. Currents are induced in the parasitic elements via the mutual impedances with the exciter, and radiation occurs as from a set of discrete sources. Green has analyzed arrays with up to ten elements using this method with good results.

Ehrenspeck and Poehler have published an interesting experimental paper in which they describe a method for obtaining maximum gain from a Yagi–Uda array [5]. Their method is based on achieving the proper phase velocity along the array, but the resulting effect of the phase velocity on the far-field patterns is not shown. They imply equal phase delay from director to director. However, calculations made with the technique described in this paper indicate that the phase progression along the array is nonuniform.

Serracchioli and Levis theoretically investigated the phase velocity of long endfire uniform dipole arrays [6]. Their curves provide information relating element radius, length, and spacing to the phase velocity. This information is useful in the initial selection of these parameters when using the technique presented here. The longer the Yagi under consideration, the more applicable are their curves.

Mailloux has developed a method for approximately analyzing the behavior of finite Yagi arrays [7]. His method essentially consists of truncating a semi-infinite array and then calculating the reflection coefficient at the end of the array. He presents results for a 20-element reflectorless array. More recently King has applied his three-term theory to a uniform Yagi–Uda array [8].

The following discussion describes a method of theoretically studying the complex current distributions on all the elements of a Yagi–Uda array, the phase velocity and the corresponding radiation patterns. The technique is based on rigorous equations for the electric field radiated by the elements in the array. All interactions are taken into account. There appears to be no restriction imposed by the method itself on how few or how many elements may be in the array. However, the size of the digital computer employed to do the computations will place an upper limit on the number of elements that may be considered. Typically, it has been found that the storage capacity of the IBM 7094 is sufficient to enable 27 elements to be simultaneously included in the calculations.

Excellent agreement between calculated results obtained by this method and experimental results reported in the literature were obtained. In addition, these results enable one to show clearly the dependence of the far-field patterns on the phase velocity along the array.

Formulation of the Problem

The approach taken in formulating this method of solving the Yagi–Uda-type antenna problem is based on an integral equation for the electric field of the array. The point-matching technique is then used to satisfy the integral equation at discrete points on the axis of each element rather than attempting to satisfy this equation everywhere on the surface of every element [9]–[13]. Thus a system of linear algebraic equations is generated in terms of the complex coefficients in the Fourier series expansion of the currents on the elements. Inversion of the matrix yields the value of these coefficients from which the current distributions, phase velocity, and far-field patterns may readily be obtained. Experience has shown that if one chooses a sufficient number of points at which to match boundary conditions, then one can obtain solutions to problems, such as this one, heretofore not easily solvable.

In the case of linear elements, such as those in Fig. 1, it has been found that an efficient representation for the current on element p is

$$I_p(z) = \sum_{n=1}^{N} I_{np} \cos (2n-1) \frac{\pi z}{L}. \qquad (1)$$

Manuscript received May 8, 1968; revised September 11, 1968.
The author is with the ElectroScience Laboratory, Department of Electrical Engineering, Ohio State University, Columbus, Ohio 43212.

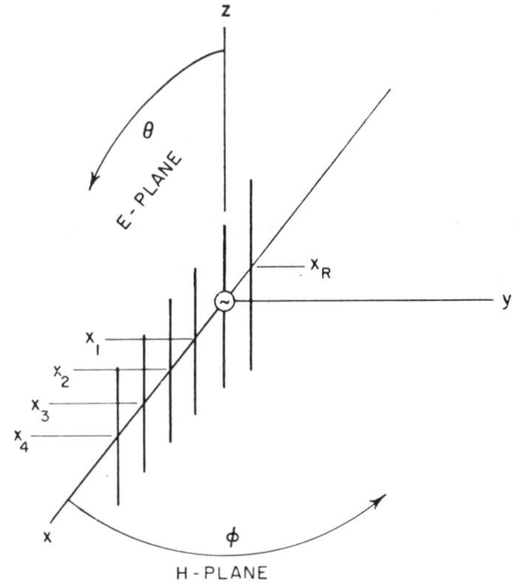

Fig. 1. Coordinate system used to analyze Yagi–Uda array.

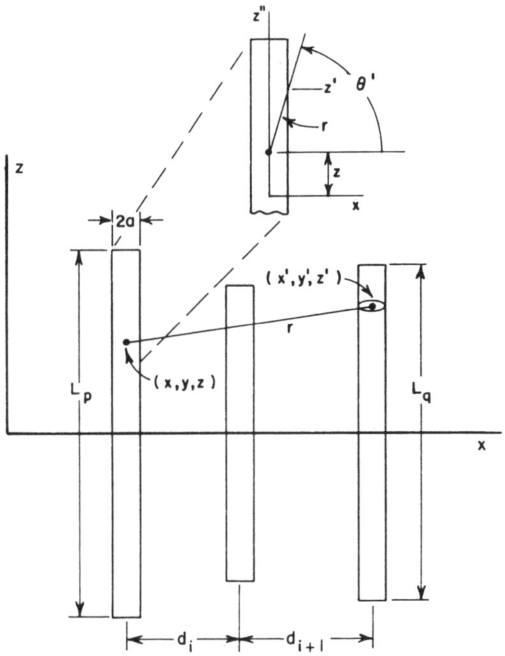

Fig. 2. Diagram showing distance from a matching point (observation point) on pth element to source region on qth element. Insert shows relationship between z' and θ' when observation point and source region are on the same element.

This series of odd-ordered even modes is chosen such that the current goes to zero at the ends of element p. This is a suitable approximation for elements whose diameter is small in terms of the wavelength. Also, this series has been used by the author in treating the thin linear antenna or dipole [13].

From the vector potential may be derived an integral expression for the electric field radiated by the array. This integral expression is

$$E_z(x,y,z) = \frac{\lambda\sqrt{\mu/\epsilon}}{8\pi^2 j} \sum_{p=1}^{D+2} \sum_{n=1}^{N} I_{np} \int_{-L_p/2}^{L_p/2} G(x,y,z|x',y',z') \cdot \cos\,(2n-1)\frac{\pi z'}{L_p}\,dz' \quad (2)$$

where

$$G(x,y,z|x',y',z') = \frac{\exp\,[-jkr]}{r^5}\,[(1+jkr)(2r^2-3a^2) + k^2a^2r^2] \quad (3)$$

and

$$r = \sqrt{(x'-x)^2 + (y'-y)^2 + a^2 + (z'-z)^2} \quad (4)$$

as shown in Fig. 2.

It has been found that z' is an efficient variable of integration when the current on wire p is observed at a matching point on element q, $p \neq q$. However, when $p = q$, it is more efficient to use θ' as the variable of integration since the integrand varies rapidly when $z' \approx z$. Details of this change of variables may be found in [9]. For either variable of integration, the integral in (2) is carried out numerically.

System of Linear Equations

Considerable insight into the point-matching technique, when applied to a problem as complex as the Yagi–Uda array, can be obtained by looking at the system of linear equations and the resulting matrix representation. For the moment, let us consider an array composed of D directors, a reflector, and a driven element. Let us have N modes on each element, but let each element be of different length. Using (2), the first part of the system of equations is then of the form

$$\sum_{p=1}^{D+2} \sum_{n=1}^{N} C_{m,np} I_{np} = 0, \quad m = 1, 2, \cdots, N \times D. \quad (5)$$

These equations are generated by requiring that tangential E be zero at N points on each director. That is to say, tangential E is zero at a total of $N \times D$ points on the directors. The matching points on any one director are illustrated in Fig. 3.

The next N equations are similar to the previous $N \times D$ equations since tangential E vanishes at N points on the reflector element, as shown in Fig. 3. Thus

$$\sum_{p=1}^{D+2} \sum_{n=1}^{N} C_{m,np} I_{np} = 0, \quad m = (N \times D) + 1, \cdots, N \times (D+1). \quad (6)$$

The last N equations are generated by using the boundary condition on the driven element, as shown in Fig. 4. That is,

$$\sum_{p=1}^{D+2} \sum_{n=1}^{N} C_{m,np} I_{np} = 0, \quad m = N \times (D+1) + 1, \cdots, N \times (D+2) - 1 \quad (7)$$

and

$$\sum_{n=1}^{N} I_{ne} = 1, \quad e = D+2, \quad m = N \times (D+2). \quad (8)$$

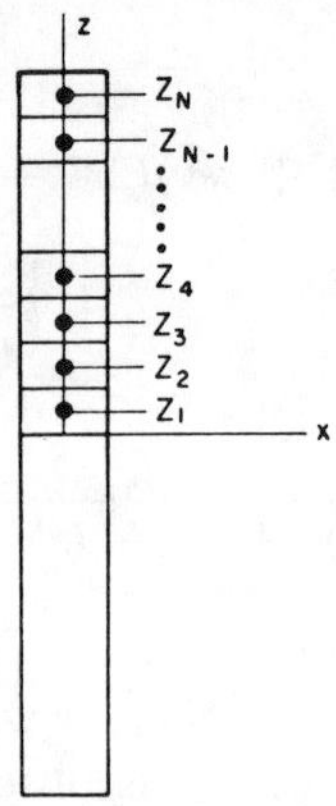

Fig. 3. Parasitic element with N matching points along its axis.

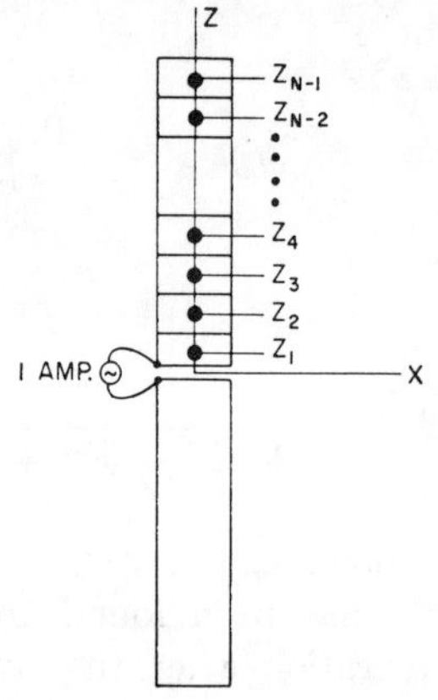

Fig. 4. Driven element with $N - 1$ matching points along its axis.

On the driven element the tangential E boundary condition is only enforced at $N - 1$ points even though there are N modes. The Nth equation on the exciter arises from the constraint on the terminal current value [13].

Computer Program[1]

A discussion of the computer program is, perhaps, facilitated by considering a simple example. Consider an array composed of a director, a reflector, and a driven element. Further, assume only two modes on each element. Thus the complete system for this example would be similar to

$$\begin{bmatrix} C_{11} & C_{12} & C_{13} & C_{14} & C_{15} & C_{16} \\ C_{21} & C_{22} & C_{23} & C_{24} & C_{25} & C_{26} \\ & & & & & \\ C_{31} & C_{32} & C_{33} & C_{34} & C_{35} & C_{36} \\ C_{41} & C_{42} & C_{43} & C_{44} & C_{45} & C_{46} \\ & & & & & \\ C_{51} & C_{52} & C_{53} & C_{54} & C_{55} & C_{56} \\ 0 & 0 & 0 & 0 & 1 & 1 \end{bmatrix} \cdot \begin{bmatrix} I_1 \\ I_2 \\ \\ I_3 \\ I_4 \\ \\ I_5 \\ I_6 \end{bmatrix} = \begin{bmatrix} 0 \\ 0 \\ \\ 0 \\ 0 \\ \\ 0 \\ 1 \end{bmatrix}.$$

The physical interpretation of the matrix elements is as follows: at point z_1 on element 1 (the director), C_{11} is the "field" generated by mode 1, and C_{12} is the field generated at the same point by mode 2. C_{21} is the field generated by mode 1 at point z_2 on the director, and C_{22} is that of mode 2. In a similar manner we can interpret C_{33}, C_{34}, C_{43}, and C_{44}.

If we write the above matrix in a submatrix representation,

$$\begin{bmatrix} S_{11} & S_{12} & S_{13} \\ & & \\ S_{21} & S_{22} & S_{23} \\ & & \\ S_{31} & S_{32} & S_{33} \end{bmatrix} \cdot \begin{bmatrix} I_1 \\ I_2 \\ I_3 \\ I_4 \\ I_5 \\ I_6 \end{bmatrix} = \begin{bmatrix} 0 \\ 0 \\ 0 \\ 0 \\ 0 \\ 1 \end{bmatrix}$$

then it is apparent that, regardless of the number of submatrices, those on the main diagonal of submatrices will represent the field generated by the current on the element at which the tangential E boundary condition is being enforced. Hence if there are D identical directors, the first D submatrices on the main diagonal will all be identical. This fact is utilized to shorten the running time of the program but is no restriction on the method itself which could be used on tapered or modulated structures.

For elements off the main diagonal of submatrices the following reasoning applies. The quantity C_{35} represents the field at point 1 on element 2 (the reflector) due to the first mode (mode 5) on the driven element (element 3). A similar interpretation applies to C_{36}, C_{45}, and C_{46}. Thus submatrices S_{qr}, $q \neq r$, represent the interaction between elements q and r. If all the directors are of the same length, then $S_{qr} = S_{rq}$ for director submatrices. Further, for uniform director spacing there will be several inter-director spacings or distances that will be the same. Consider the 4-director Yagi in Fig. 1. The distance between directors 1 and 3 is the same as that between 2 and 4, for instance. Hence $S_{13} = S_{24}$, assuming the directors are of the same length.

Thus we see that it may not be necessary to calculate all the submatrices individually, and, consequently, a significant savings in computer time is realized when working with long and complex problems that require large amounts of numerical integration. The program developed for this problem utilizes the above mentioned techniques. However, the program is not restricted to uniform director spacing. It simply takes advantage of geometrical symmetries that may exist.

Once the entire matrix has been calculated, it is easily solved by standard methods for the coefficients I_{np} in the Fourier series representation of the currents on the various elements. Using these currents, the far-field patterns may be obtained in a straightforward manner.

Far-Field Patterns

The far-field pattern of a single element of the Yagi antenna in Fig. 1 is given by

$$E_\theta(\theta) = \frac{j\omega\mu}{4\pi r_0} \exp\left[-jkr_0\right] \sin\theta \int_{-L/2}^{L/2} I(z') \cdot \exp\left[jkz\cos\theta\right] dz'. \quad (9)$$

Since the current is expressed as a Fourier series, we obtain

$$E_\theta(\theta) = \frac{-jL\sqrt{\mu/\epsilon}}{\pi r_0} \exp\left[-jkr_0\right] \sin\theta \cdot \sum_{n=1}^{N} (-1)^n \frac{(2n-1)I_n \cos(\pi L' \cos\theta)}{(2n-1)^2 - (2L'\cos\theta)^2} \quad (10)$$

where $L' = L/\lambda$. Since

$$r_0 \approx r - (x\sin\theta\cos\phi + y\sin\theta\sin\phi + z\cos\theta) \quad (11)$$

[1] Requests for additional information on the computer program may be addressed as follows: Director, ElectroScience Laboratory, Ohio State University, 1320 Kinnear Road, Columbus, Ohio 43212.

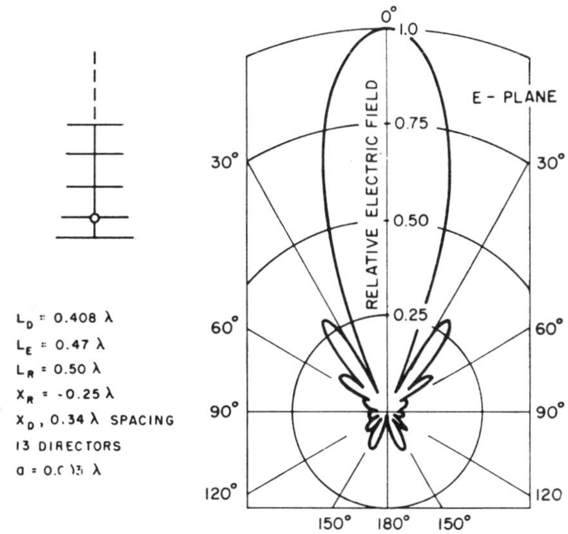

Fig. 5. *E*-plane pattern of 15-element Yagi–Uda array.

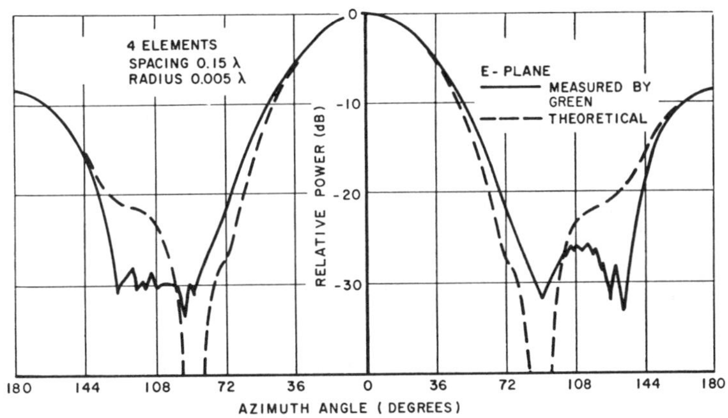

Fig. 6. *E*-plane pattern of 4-element Yagi–Uda array.

we can define a pattern factor

$$F(\theta,\phi) = L_p \sin\theta \exp\,[+jk(x_p \sin\theta\cos\phi + y_p \sin\theta\sin\phi + z_p\cos\theta)] \sum_{n=1}^{N} (-1)^n \cdot \frac{(2n-1)I_{np}\cos(\pi L'\cos\theta)}{(2n-1)^2 - (2L'\cos\theta)^2}. \tag{12}$$

Thus if we have W elements each with N modes, we may write for the total pattern factor $F_T(\theta,\phi)$,

$$F_T(\theta,\phi) = \sin\theta \sum_{p=1}^{W} L_p \exp\,[+jk(x_p\sin\theta\cos\phi + y_p\sin\theta\sin\phi + z_p\cos\theta)] \sum_{n=1}^{N} (-1)^n \cdot \frac{(2n-1)I_{np}\cos(\pi L'\cos\theta)}{(2n-1)^2 - (2L'\cos\theta)^2}, \quad W = D + 2. \tag{13}$$

If the number of modes on each element is not the same, then the above expression is necessarily more complicated. In practice, it has been found desirable to retain more modes on the driven element than are required for the parasitic elements. Hence the computer program actually allows for K modes on the former and N modes on the latter.

Results

Excellent results were obtained with the point-matching method applied to the Yagi–Uda antenna problem. The patterns calculated by this method agree quite well with experimental patterns published in the literature [4], [14]. For example, Fig. 5 shows a calculated *E*-plane pattern for a 15-element Yagi–Uda array. The pattern agrees so well with an experimental one published by Fishenden and Wiblin [14] that there is no reason to attempt to distinguish between them in the figure. In this and subsequent figures L_D, L_E, and L_R are, respectively, the director length, exciter length, and reflector length. As implied by Fig. 1, X_R is the x coordinate of the reflector, and X_D indicates the uniform director spacing. Fig. 6 shows an experimental power plot of a 4-element array published by Green [4]. Good agreement is seen to exist between the experimental pattern and the theoretical

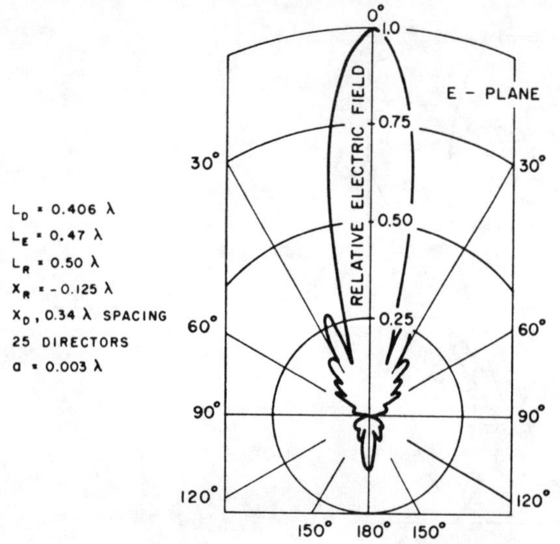

Fig. 7. *E*-plane pattern of 27-element Yagi–Uda array.

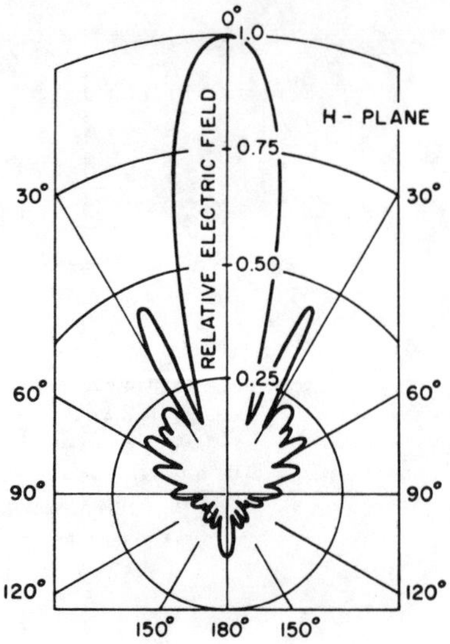

Fig. 8. *H*-plane pattern for array o Fig. 7.

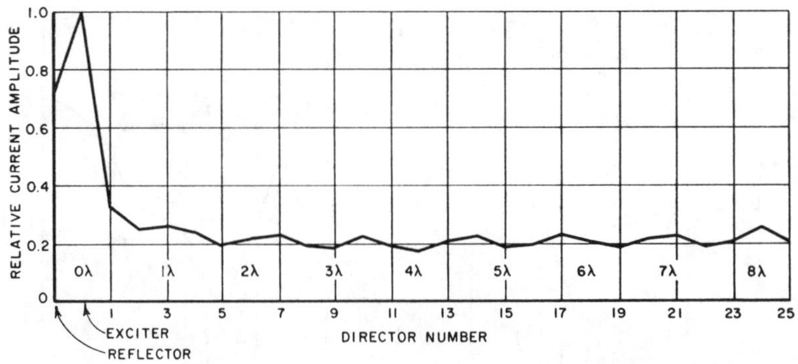

Fig. 9. Relative current amplitudes for array of Fig. 7.

one obtained with the point-matching method. In fact, the front-to-back ratio is precisely the same.

Figs. 7 and 8 show calculated *E*-plane and *H*-plane patterns, respectively, for a 27-element array. This was the largest number of elements that could be handled conveniently by the computer program on the IBM 7094. With a computer of larger memory more elements could be accommodated. It is interesting to note that the *E*-plane pattern of this 27-element array is similar to a measured pattern also published by Fishenden and Wiblin for a 32-element array with a slightly longer director length of 0.408λ. All other parameters were the same. Fig. 9 shows a plot of the current amplitudes on all of the 27 elements. This plot clearly shows that the director currents are not equal nor are they smoothly tapered in the forward direction. It will be shown later that the progressive phasing of the currents is also not uniform, even for uniform director spacing.

There are two principle ways in which the far-field pattern of a Yagi–Uda array may be "tuned" or adjusted for a particular frequency. One way is to vary the director spacing while holding the element length and reflector spacing constant. The second method is to vary the director length while holding all other parameters fixed. Either one of these methods will have a considerable effect on the far-field patterns. On the other hand, the usual effects of altering the reflector spacing or length are mostly to change the level of the back lobe and to control the impedance of the array. Further, changing the length of the driven element will have a negligible effect on the pattern. Of course, as one would expect, altering the exciter length will change the input impedance of the array. A method of calculating the array input impedance is under consideration.

All of the aforementioned changes that affect the far-field pattern do so because they alter the phase velocity along the array. Hence it is reasonable to use this as a design criteria rather than simply observing the change in far-field patterns due to a change in director length and/or spacing, etc.

The Hansen–Woodyard condition for an endfire array states that for increased directivity in the forward direction it is necessary to have a total phase delay of almost 180 degrees along the array. That is to say, the phase velocity needs to be slower than the corresponding free-space phase velocity. The Hansen–Woodyard condition assumes equal amplitudes in the array elements. However, as seen in Fig. 9, this is not the situation for a Yagi–Uda array. Nevertheless, it is possible to draw a vector diagram which will show when the array has been prop-

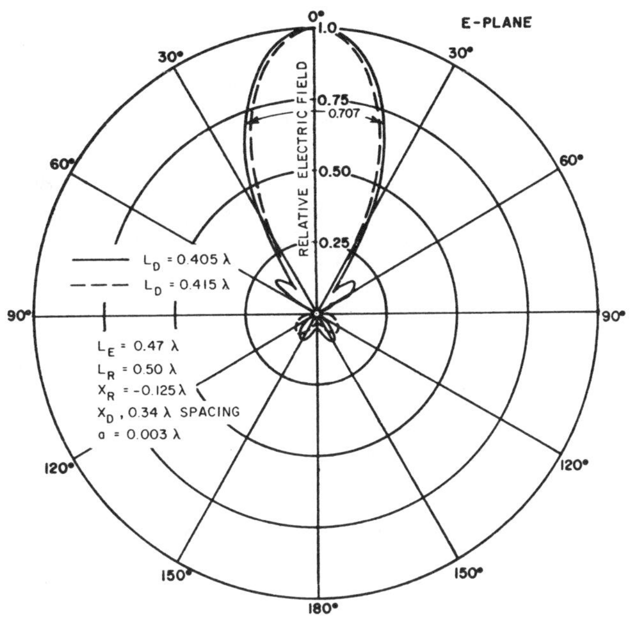

Fig. 10. *E*-plane patterns of two 8-element arrays.

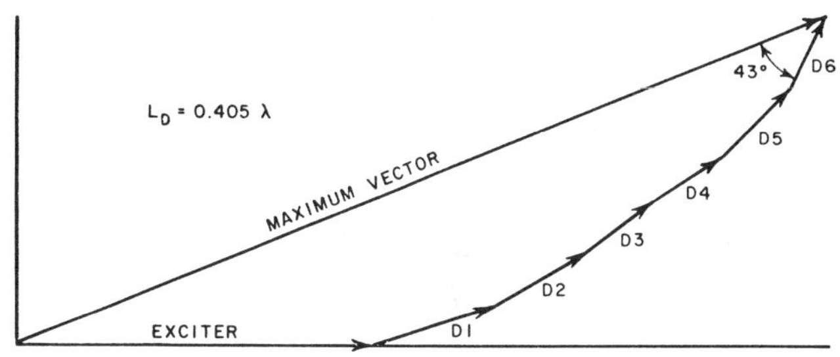

Fig. 11. Phase velocity diagram for 8-element array with $L_D = 0.405\lambda$.

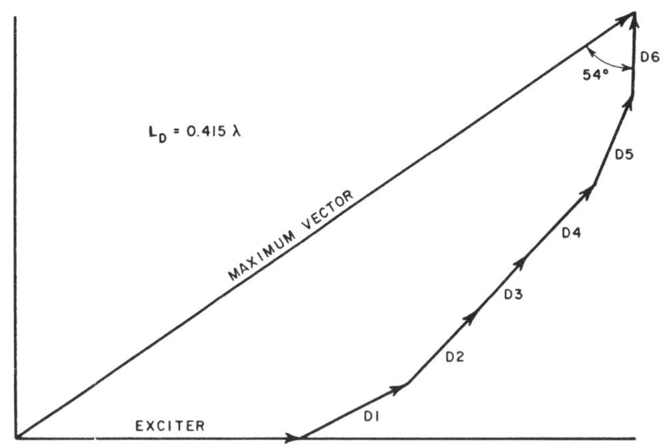

Fig. 12. Phase velocity diagram for 8-element array with $L_D = 0.415\lambda$.

erly adjusted for increased directivity in the forward direction. The vector diagram will, of course, take into account the unequal amplitudes in the various elements. Furthermore, the diagram will afford much greater insight into the behavior of the array. Such diagrams have been published by Ehrenspeck and Poehler [5]. However, they indicate equal phase delay from director to director. Calculations with the point-matching method show that this is not the case; a result which is not illogical when one considers the complexity of the finite array and the fact that the current amplitudes are unequal.

Fig. 10 shows two patterns of an 8-element array. We observe that an increase in the director length has caused a narrowing of the main beam without an increase in the side lobes. Thus we might expect that the phase velocity on the $L_D = 0.415\lambda$ array is somewhat slower than that for $L_D = 0.405\lambda$. Inspection of Figs. 11 and 12 shows this to be true. Fig. 11 shows a plot of the current amplitudes on the exciter and directors. The change in direction from one vector to the next is determined by taking the difference between the calculated phase progression and that which would exist if the phase velocity were that of free space. Hence if the phase velocity were that of free space, all the vectors would lie along the horizontal axis. Further, if the phase velocity were uniform, the change in direction from one vector to the next would be the same for any two adjacent vectors (assuming uniform director spacing). Examination of Fig. 12, for instance, shows that such is not the case. Thus we may conclude that the phase velocity is generally nonuniform. In principle, this finding is in agreement with Damon's results for endfire dipole arrays [15].

As additional phase delay is achieved, the tip of the last vector (*D*6 in this situation) moves away from the horizontal axis and toward the vertical axis. The maximum vector is defined to be that vector from the origin to the tip of the component vector farthest from the origin. Thus we see in Figs. 10–12 that by increasing the director lengths we have slightly increased the directivity in the forward direction because the phase velocity has been decreased.

It should be mentioned that although the point-matching method permits one to obtain more accurate results than are generally obtainable by other methods, one should not infer that the director lengths given here are meaningful in practice to more than two decimal places. What is meaningful, however, is the change in phase velocity due to a given change in director lengths, as discussed in the following paragraphs.

Figs. 13 and 14 show the vector diagrams for successively longer director lengths. We note that in Fig. 14 the maximum vector is not to the tip of the last vector *D*6. When this situation occurs, it indicates that the phase delay is excessive, and, consequently, that the far-field pattern will start to deteriorate. To see that this is indeed the case, we refer to Fig. 15. Here we see the pattern for the case $L_D = 0.4275\lambda$. Also indicated are the back and side lobe levels for several other cases. Comparing Fig. 15 with Fig. 10, we immediately see that the main beam in the former is significantly less than in the latter. This is consistent with what the vector diagrams indicate.

In Figs. 13–15 we see that as the maximum vector approaches the point where it no longer extends to the tip of *D*6, but to *D*5 instead, the lobes in the pattern start to come up quickly, especially the back lobe. At the same time the beamwidth is decreased only a very

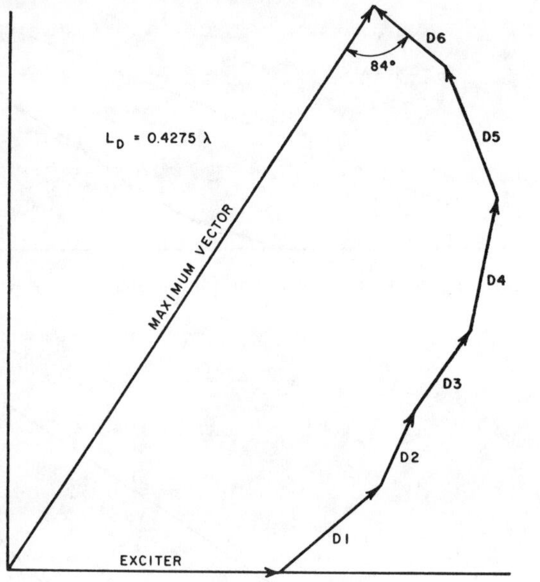

Fig. 13. Phase velocity diagram for 8-element array with L_D = 0.4275λ.

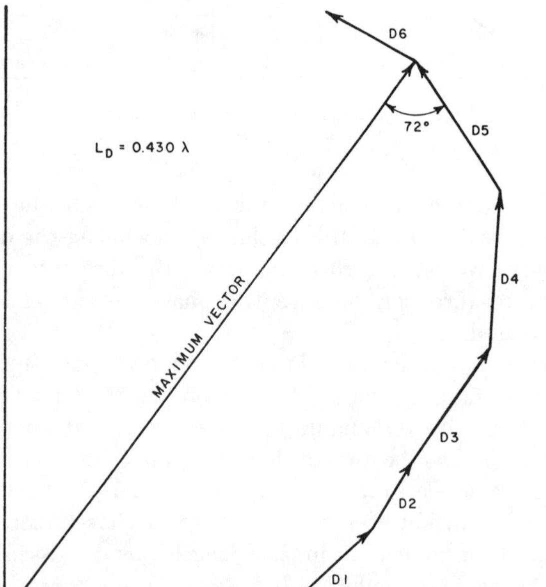

Fig. 14. Phase velocity diagram for 8-element array with L_D = 0.430λ.

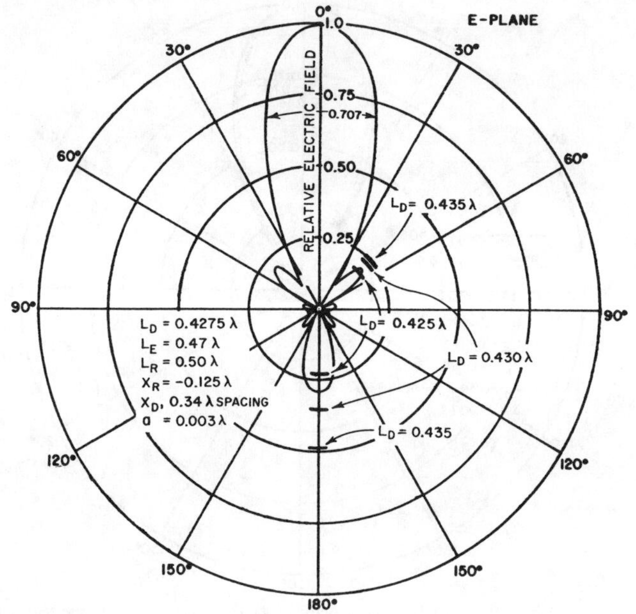

Fig. 15. *E*-plane pattern of optimum 8-element array with side lobe levels indicated for several other versions of array.

slight amount. Specifically, for $L_D = 0.4250\lambda$, it is about 2 degrees wider than the pattern for $L_D = 0.4275\lambda$, about 2 degrees narrower for $L_D = 0.4300\lambda$, and about 4 degrees narrower than indicated in Fig. 15 for $L_D = 0.4350\lambda$.

Consequently, it would appear that in the design of a Yagi–Uda array one should strive to keep the phase velocity at or above that which would produce a vector diagram similar to Fig. 13. From another viewpoint we can interpret the slower phase velocity as giving rise to a backward surface wave which is evidenced most prominently in an increasing back lobe. In fact, the computer printout shows an increasing current amplitude on the reflector element in such cases.

Conclusion

A method of analyzing the Yagi–Uda array by using the point-matching or linear equation technique has been presented. The method is general in that it may be applied to nonplanar arrays with an arbitrary number of elements having arbitrary length and with arbitrary spacing between them. While there is essentially no minimum number of elements that may be considered, the maximum number appears to be limited only by the size of the digital computer available. For instance, 27 elements were accommodated by the program on the IBM 7094 by using three modes on each of the parasitic elements and seven on the driven element. In addition, this method of analysis permits one to obtain accurate information not only about the far-field patterns, but also on the current distributions and phase velocity as well. One could also investigate the scattering behavior of such arrays by removing the generator and exciting the antenna with a plane wave.

This method need not be limited to the one reflector, *D*-director-type of Yagi–Uda array depicted in Fig. 1, but it may easily be extended to other antenna configurations using linear elements or elements that may be subdivided into linear elements such as in a vee configuration. Consequently, the various elements would not need to be parallel with one another as they are in the Yagi–Uda array. In such cases it would be necessary to consider the contribution to the tangential electric field at a matching point by the radial component of the electric field in addition to the z component given here. Further, current distributions that are not symmetrical about the center of the elements may be handled by the inclusion of the modes $\sin 2n\pi z/L_p$ in the current expansion for $I_p(z)$.

Finally, the use of a computerized approach in the investigation and design of Yagi–Uda-type antennas presents several distinct advantages over an experimental approach. Among these are the obvious savings in the expenditure of time and money and also the possibility of obtaining antennas optimized with respect to directivity, side lobe levels, bandwidth, or any weighted combination of these.

Acknowledgment

The author wishes to acknowledge the support of the Ohio State University Numerical Computation Laboratory, which provided the computer time for the Yagi–Uda antenna calculations. The author also wishes to thank Profs. C. H. Walter and J. H. Richmond for their helpful comments.

References

[1] S. Uda, "Wireless beam of short electric waves," *J. IEE* (Japan), pp. 273–282, March 1926, and pp. 1209–1219, November 1927.

[2] H. Yagi, "Beam transmission of ultra short waves," *Proc. IRE*, vol. 16, pp. 715–741, June 1928.

[3] W. Wilkinshaw, "Theoretical treatment of short Yagi aerials," *Proc. IEE* (London), pt. 3, vol. 93, p. 598, 1946.

[4] H. E. Green, "Design data for short and medium length Yagi–Uda arrays," *Elec. Engrg. Trans. Inst. Engrs.* (Australia), pp. 1–8, March 1966.

[5] H. W. Ehrenspeck and H. Poehler, "A new method for obtaining maximum gain from Yagi antennas," *IRE Trans. Antennas and Propagation*, vol. AP-7, pp. 379–386, October 1959.

[6] F. Serracchioli and C. A. Levis, "The calculated phase velocity of long end-fire uniform dipole arrays," *IRE Trans. Antennas and Propagation*, vol. AP-7, pp. S424–S434, December 1959.

[7] R. J. Mailloux, "The long Yagi–Uda array," *IEEE Trans. Antennas and Propagation*, vol. AP-14, pp. 128–137, March 1966.

[8] R. W. P. King, "The linear antenna—eighty years of progress," *Proc. IEEE*, vol. 55, pp. 2–16, January 1967.

[9] J. H. Richmond, "Digital computer solutions of the rigorous equations for scattering problems," *Proc. IEEE*, vol. 53, pp. 796–804, August 1965.

[10] R. F. Harrington, "Matrix methods for field problems," *Proc. IEEE*, vol. 55, pp. 136–149, February 1967.

[11] Y. S. Yeh and K. K. Mei, "Theory of conical equiangular-spiral antennas. Part I—Numerical technique," *IEEE Trans. Antennas and Propagation*, vol. AP-15, pp. 634–639, September 1967.

[12] V. R. Arens, C. T. Elfving, U. R. Emlory, and D. L. Johnstone, "Integral equation formulation of a log periodic antenna," *1967 IEEE G-AP Internat'l Symp. Dig.* (Ann Arbor, Mich., October 17–19), pp. 55–63.

[13] G. A. Thiele, "Calculation of the current distribution on a thin linear antenna," *IEEE Trans. Antennas and Propagation (Communications)*, vol. AP-14, pp. 648–649, September 1966.

[14] R. M. Fishenden and E. R. Wiblin, "Design of Yagi aerials," *Proc. IEE* (London), pt. 3, vol. 96, pp. 5–13, January 1949.

[15] E. K. Damon, "The near fields of long end-fire dipole arrays," *IRE Trans. Antennas and Propagation*, vol. AP-10, pp. 511–523, September 1962.

Mutual Impedance Between Coplanar-Skew Dipoles

***Abstract*—The induced EMF formulation is employed to develop a closed-form expression for the mutual impedance between coplanar-skew dipoles. Numerical results are presented in graphical form.**

I Introduction

King [1] has determined the mutual impedance between parallel dipoles, and Lewin [2] has analyzed the coplanar-skew dipoles. Lewin's analysis, however, is restricted to half-wave center-fed dipoles. In this paper we derive an expression for the mutual impedance between coplanar-skew dipoles with arbitrary lengths and terminal positions. This expression is relatively simple and convenient for computer programming.

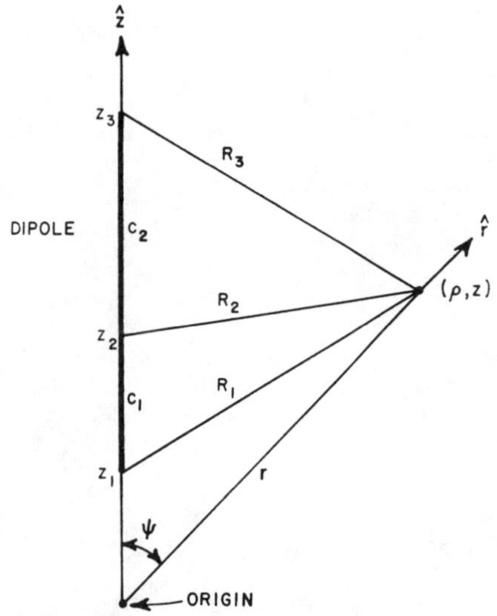

Fig. 1. Linear dipole and coordinate system.

Manuscript received October 1, 1969. This work was supported in part by the Department of the Army, Ballistic Research Laboratory, Aberdeen Proving Ground, Md., under Contract DAAD05-69-C-0031 with the Ohio State University Research Foundation.

In Section II we present a new expression for the near-zone field of a linear dipole, in a form most convenient for the mutual-impedance analysis.

II. The Near-Zone Field of the Linear Dipole

As shown in Fig. 1, we consider a dipole located on the z axis and let z_1 and z_3 denote the endpoints and z_2 the terminals. In the induced EMF method, the dipole current is given by

$$I(z) = \hat{z} I_1 \frac{\sin k(z - z_1)}{\sin kc_1}, \quad z_1 < z < z_2 \tag{1}$$

$$I(z) = \hat{z} I_1 \frac{\sin k(z_3 - z)}{\sin kc_2}, \quad z_2 < z < z_3. \tag{2}$$

The time dependence $\exp(j\omega t)$ is suppressed, c_1 and c_2 denote the dipole arm lengths, $\hat{z}$ is a unit vector, I_1 is the terminal current, and $k = 2\pi/\lambda$. The field generated in free space by this dipole (or sinusoidal line source) is determined most readily from the expressions of Schelkunoff and Friis [3]. The cylindrical components of the electric field intensity are

$$E_\rho = \frac{j30I_1}{\rho}\left[\frac{(z - z_1)\exp(-jkR_1)}{R_1 \sin kc_1} - \frac{(z - z_2)\exp(-jkR_2)\sin kc}{R_2 \sin kc_1 \sin kc_2} + \frac{(z - z_3)\exp(-jkR_3)}{R_3 \sin kc_2}\right] \tag{3}$$

$$E_z = j30I_1\left[-\frac{\exp(-jkR_1)}{R_1 \sin kc_1} + \frac{\exp(-jkR_2)\sin kc}{R_2 \sin kc_1 \sin kc_2} - \frac{\exp(-jkR_3)}{R_3 \sin kc_2}\right] \tag{4}$$

where c is the dipole length and R_1, R_2, and R_3 are the distances defined in Fig. 1.

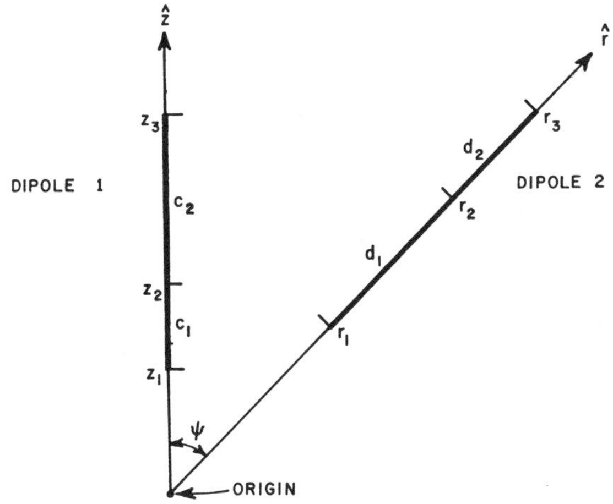

Fig. 2. Coplanar-skew dipoles.

The radial component of the field is obtained from a linear combination of E_ρ and E_z:

$$E_r = \frac{-j30I_1}{r} \sum_{m=1}^{3} C_m z_m \frac{\exp(-jkR_m)}{R_m} \tag{5}$$

where

$$C_1 = 1/(\sin kc_1) \tag{6}$$

$$C_2 = -(\sin kc)/(\sin kc_1 \sin kc_2) \tag{7}$$

$$C_3 = 1/(\sin kc_2). \tag{8}$$

Equation (5) is believed to be a new and useful form for the rigorous near-zone field of the linear dipole.

III. The Mutual Impedance

For the coplanar-skew dipoles shown in Fig. 2, the mutual impedance is

$$Z_{12} = -\frac{1}{I_1 I_2} \int_{r_1}^{r_3} \boldsymbol{I}_2(r) \cdot \boldsymbol{E}_1(r)\, dr. \tag{9}$$

The current on dipole 2 is

$$\boldsymbol{I}_2(r) = \hat{r} \frac{I_2 \sin k(r - r_1)}{\sin kd_1}, \quad r_1 < r < r_2 \tag{10}$$

$$\boldsymbol{I}_2(r) = \hat{r} \frac{I_2 \sin k(r_3 - r)}{\sin kd_2}, \quad r_2 < r < r_3. \tag{11}$$

From (5) and (9),

$$Z_{12} = \frac{j30}{I_2} \sum_{m=1}^{3} z_m C_m \int_{r_1}^{r_3} \frac{I_2(r) \exp(-jkR_m)\, dr}{rR_m}. \tag{12}$$

These integrals are given in terms of the sine and cosine integrals as follows:

$$z_m \int_{r_1}^{r_2} \frac{\exp(\pm jr) \exp(-jR_m)\, dr}{rR_m} = \Big[\exp(-jz_m) E(R_m - z_m \mp r) - \exp(jz_m) E(R_m + z_m \mp r)\Big]_{r=r_1}^{r=r_2} \tag{13}$$

where

$$E(x) = Ci(|x|) - jSi(x). \tag{14}$$

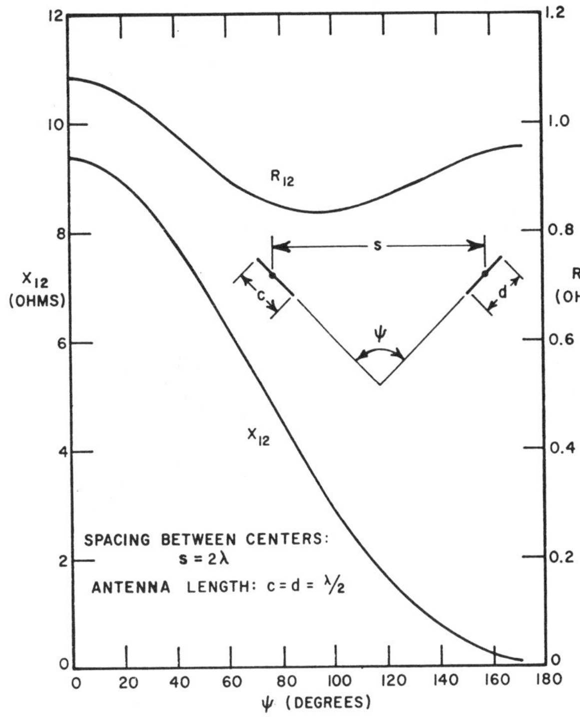

Fig. 3. Mutual impedance between coplanar-skew dipoles.

From (12) and (13), the mutual impedance is

$$Z_{12} = -15 \sum_{m=1}^{3} \sum_{n=1}^{3} C_m D_n \sum_{p=-1}^{1} \sum_{q=-1}^{1} pq \exp[jk(pz_m + qr_n)] \cdot E(kR_{mn} + kpz_m + kqr_n) \tag{15}$$

where p and q assume only the values ± 1, and R_{mn} is the distance from point z_m on dipole 1 to point r_n on dipole 2

$$R_{mn} = (z_m^2 + r_n^2 - 2z_m r_n \cos \psi)^{1/2}. \tag{16}$$

The coefficients D_n have the same form as the C_m

$$D_1 = 1/(\sin kd_1) \tag{17}$$

$$D_2 = -(\sin kd)/(\sin kd_1 \sin kd_2) \tag{18}$$

$$D_3 = 1/(\sin kd_2). \tag{19}$$

The coordinates z_m and r_n are measured from the coordinate origin at the intersection of the axes of the coplanar-skew dipoles. For center-fed half-wave dipoles, (15) reduces to (15) of Lewin's paper [2].

For parallel dipoles, z_m and r_n go to infinity and it is not difficult to show that (15) reduces to the expression published by H. E. King [1].

IV. Numerical Results

Fig. 3 illustrates the mutual impedance between half-wave center-fed dipoles as a function of the angle ψ between their axes. Using (15), the calculations require 0.15 seconds on an IBM 7094 computer for each value of ψ.

Although the numerical approach of Baker and LaGrone [4] is efficient when the dipoles are adequately separated, (15) is preferable for closely spaced dipoles.

V. Conclusion

A convenient new form is presented for the rigorous near-zone field of the linear dipole with sinusoidal current distribution. This is employed to derive a closed-form expression for the mutual impedance between coplanar-skew dipoles.

J. H. Richmond
N. H. Geary
ElectroScience Laboratory
Dept. of Elec. Engrg.
Ohio State University
Columbus, Ohio 43212.

References

[1] H. E. King, "Mutual impedance of unequal length antennas in echelon," *IRE Trans. Antennas and Propagation*, vol. AP-5, pp. 306–313, July 1957.
[2] L. Lewin, "Mutual impedance of wire aerials," *Wireless Engr.*, vol. 28, pp. 352–355, December 1951.
[3] S. A. Schelkunoff and H. T. Friis, *Antennas, Theory and Practice*. New York: Wiley, 1952, pp. 370, 401.
[4] H. C. Baker and A. H. LaGrone, "Digital computation of the mutual impedance between thin dipoles," *IEEE Trans. Antennas and Propagation*, vol. AP-10, pp. 172–178, March 1962.

Dipole Mutual Impedance for Design of Slot Arrays

R. C. HANSEN
Consulting Engineer, Tarzana, CA
GARY BRUNNER
Goodyear Aerospace Corporation, Litchfield Park, AZ

In the design of waveguide slot arrays it is necessary to include the effects of external mutual coupling between the slots. The mutual admittance between a slot and each other slot is appropriately added to the self admittance of the slot. When the slots are in the broad wall, slot mutual admittance is related to mutual impedance between corresponding strip dipoles through Babinet's Principle, if the slot array is sufficiently large that the slots may be assumed to be in a flat infinite ground plane. Kurtz and Elliott[1] have developed a design methodology for resonant broad wall slot arrays. The flat strip dipoles have the same dimensions of the slots. Since slots are usually thin (width much less than length), the strip dipoles may be replaced by cylindrical dipoles of the same length, with radius equal to one fourth the strip (slot) width. For thin dipoles the current distribution is approximately sinusoidal, and for this case the formulas of Howard King,[2] based on the original work of Carter[3] can be used. However these formulas contain 24 Sine and Cosine Integral terms, of various complicated and diverse arguments. This paper presents a computer subroutine in Fortran IV for mutual impedance between echelon dipoles of unequal lengths, in which the Sine and Cosine Integrals are compactly and efficiently arranged. Actually the formula is written in terms of the Exponential Integral[5] which is just the complex combination of S_1, C_1.

Although the results of King could be re-arranged and grouped into a computer algorithm, it is simpler to directly derive the formula, using the exact near electric field formulations of Schelkunoff and Friis.[4] They show that the exact field from a sinusoidal current is written as three spherical waves, although the field components are in cylindrical coordinates. Mutual impedance is then the integral of this electric field (from each half dipole) times the current distributions at the other dipole. With the geometry of **Figure 1** the mutual impedance expression is:

$$Z = \frac{-j30}{S_1 S_2} \int_0^{d_1} \left[\psi_1 - 2 \cos kd_2\, \psi_2 + \psi_3 + \psi_4 - 2\, \psi_5 \cos kd_2 + \psi_6 \right] \cdot \sin k(d_1 - x)\, dx \qquad (1)$$

where $k = 2\pi/\lambda$, $S_1 = \sin kd_1$,

$$S_2 = \sin kd_2, \quad \psi_i = \frac{\exp(-jkR_i)}{R_i}$$

The R values are:

$$R_1^2 = y_o^2 + (x_o + d_2 - x)^2$$

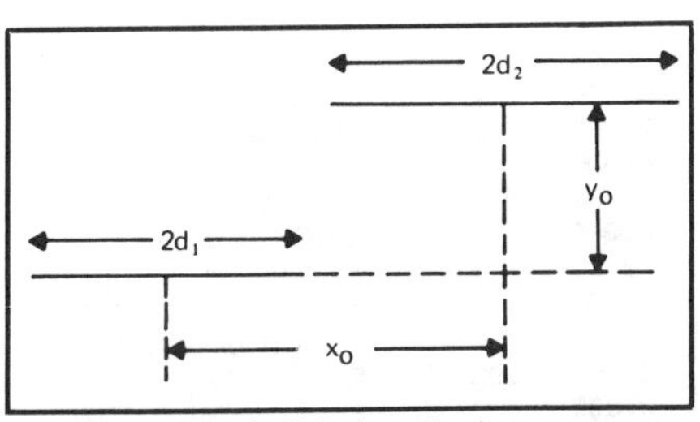

Fig. 1 Echelon dipoles.

$$R_2^2 = y_o^2 + (x_o - x)^2$$

$$R_3^2 = y_o^2 + (x_o - d_2 - x)^2$$

$$R_4^2 = y_o^2 + (x_o - d_2 + x)^2$$

$$R_5^2 = y_o^2 + (x_o + x)^2$$

$$R_6^2 = y_o^2 + (x_o + d_2 + x)^2$$

The integral could be evaluated numerically, e.g. by Romberg integration, and for large slot separations this method is preferable. However, the adjacent and nearby slots are most important, and for these an exact evaluation is necessary.

The exact solution can be written as two sums:

$$Z = \frac{-15}{S_1 S_2} \sum_{m=-1}^{1} \sum_{n=-1}^{1,2} C_m \left\{ e^{jkU} \left[E(kA + kU) - E(kB + kV) \right] + e^{-jkU} \left[E(kA - kU) - E(kB - kV) \right] \right\} \qquad (2)$$

The coefficients are $C_{-1} = C_1 = 1$, $C_o = -2 \cos kd_2$; $E(x)$ is the exponential integral [5], with arguments:

$$U = d_1 + n(x_o + md_2)$$
$$V = n(x_o + md_2)$$
$$A^2 = y_o^2 + (x_o + nd_1 + md_2)^2$$
$$B^2 = y_o^2 + (x_o + md_2)^2$$

Equation (2) is computed by subroutine ECHELU, given in

Figure 2. The routine requires 24 calls to a subroutine to calculate sine and cosine integrals.[5] This subroutine, called SICINT, is generally available in library subroutine packages, uses polynominal approximations[5,7] and generally is accurate to five or six significant figures; results should be checked against tables in.[5]

Computer storage requirements and execution times will be machine dependent. A comparison between a 16-bit, byte-oriented minicomputer and a 36-bit main frame machine is shown in Table I; the execution time is for a test case with dipole lengths of 0.2 and 0.3 wavelengths, parallel and perpendicular spacings of 0.1 and 0.2 wavelengths, and $Z = 8.788 - j3.379\Omega$.

A companion subroutine, ECHEL, developed previously for equal length dipoles,[6] is given in Figure 3. Comparative storage requirements and execution times are given in Table I for a test case with dipole lengths of 0.25 wavelengths, parallel and perpendicular spacings of 0.25 and 0.001 wavelengths, and $Z = 10.473 + j37.598$ ohms.

Both subroutines, as written, require double precision arithmetic if the argument of the exponential integral given in Equation (2) is close to zero. This requirement has been shown to exist on machines as large as a 36-bit mainframe computer; a user with access to a computer having a longer length word may wish to verify the need for double precision. A test case that can establish the need for double precision arithmetic and that is valid for either subroutine uses dipole lengths of 0.5 wavelength, parallel and perpendicular spacings of 0.6 and 0.0001 wavelengths, and $Z = 14.6743 - j4.0143$.
to be invaluable in the analysis and design of dipole and slotted waveguide antenna arrays. They provide accurate and efficient means of calculating the mutual impedance terms that must be included in any practical array design.

REFERENCES

1. Elliott, R. S., and L. A. Kurtz, "The Design of Small Slot Arrays," *Trans.*

TABLE I

Computer	**PDP 11/34**	**DECsystem-2040**
Word size	**16 bits (2 bytes)**	**36 bits**
Storage requirements (memory and code)		
ECHEL	**1500 bytes**	**236 words**
ECHELU	**2066 bytes**	**289 words**
SICINT	**1370 bytes**	**185 words**
Execution time (including SICINT)		
ECHEL	**820 ms**	**17 ms**
ECHELU*	**750 ms**	**27 ms**

*** Approximately 700 multiplies and 100 trig function evaluations.**

```
      SUBROUTINE ECHELU (SL1,SL2,GSX,GSY,ZP)
C
C     MUTUAL IMPEDANCE OF ECHELON DIPOLES OF UNEQUAL LENGTH
C     USES SINE AND COSINE INTEGRAL SUBROUTINE
C     SL1,SL2 = DIPOLE HALF-LENGTHS IN WAVELENGTHS
C     GSX = PARALLEL CNTR-TO-CNTR SPACING IN WAVELENGTHS,
C     (E-PLANE)
C     GSY = PERPENDICULAR SPACING OR DIPOLE RADIUS
C     IN WAVELENGTHS (H-PLANE)
C     ZP = RP+(J)XP (DIPOLE IMPEDANCE, OHMS)
C     ALL ARGUMENTS IN CALL STATEMENT ARE SINGLE PRECISION
C     TEST CASE:  SL1=0.1, SL2=0.15, GSX=0.1, GSY=0.2,
C                 ZP=(8.788, -3.379)
C     TEST CASE:  SL1=0.25, SL2=0.25, GSX=0.6, GSY=0.0001,
C                 ZP=(14.6743, -4.0143)
C
      DIMENSION C(3)
      COMPLEX E(4),SUM,ZP
      DOUBLE PRECISION A,B,G2,G3,Q,T(4),U,V,P,F,SL12,SL22
      DOUBLE PRECISION COMPLEX F1,F2
      P=6.283185307179586476925287
      F=1,570796326794896619231322
      G=.57721566
      SL12=DBLE(SL1)
      SL22=DBLE(SL2)
      G2=DBLE(GSY*GSY)
      G3=DBLE(GSX)
      D1=DSIN(P*SL12)
      D2=DSIN(P*SL22)
      C(1)=1.
      C(3)=1.
      C(2)=-2.*DCOS(P*SL22)
      SUM=(0.,0.)
      DO 505 M=-1,1
      DO 505 N=-1,1,2
      V=N*(G3+4*SL22)
      U=V+SL12
      A=DSQRT(G2+(G3+4*SL12+M*SL22)**2)
      B=DSQRT(G2+(G3+M*SL22)**2)
      T(1)=P*(A+U)
      T(2)=P*(A-U)
      T(3)=P*(B+V)
      T(4)=P*(B-V)
      DO 511 I=1,4
      Q=DABS(T(I))
      IF(Q.GT..0001)GO TO 510
      S1=T(I)
      C1=G+DLOG(Q)-.25*Q**2
      GO TO 520
  510 RA=T(I)
      CALL SICINT(C,S1,RA)
      S1=S1+F
  520 E(I)=CMPLX(C1,-S1)
  511 CONTINUE
      F1=CDEXP(DCMPLX(O.DO,P*U) )
      F2=CDEXP(DCMPLX(O.DO,-P*U) )
      SUM=SUM+C(M+2)*(F1*(E(1)-E(3) )+F2*(E(2)-E(4) ) )
  505 CONTINUE
      ZP=15.*SUM/(D1*D2)
      RETURN
      END
```

Fig. 2 Subroutine ECHELU.

```
      SUBROUTINE ECHEL (SL,GSX,GSY,ZP)
C
C     MUTUAL IMPEDANCE OF ECHELON DIPOLES OF EQUAL LENGTH
C     OR SELF IMPEDANCE OF SINGLE DIPOLE
C     USES SINE AND COSINE INTEGRAL SUBROUTINE
C     SL = HALF DIPOLE LENGTH IN WAVELENGTHS
C     GSX = PARALLEL CENTER-TO-CENTER SPACING IN WAVELENGTHS
C     (E-PLANE)
C     GSY = PERPENDICULAR SPACING OR DIPOLE RADIUS IN
C     WAVELENGTHS (H-PLANE)
C     ZP = RP+(J)XP  (DIPOLE IMPEDANCE, OHMS)
C     ALL ARGUMENTS IN CALL STATEMENT ARE SINGLE PRECISION
C     TEST CASE:  SL=0.125, GSX=0.25, GSY=0.001
C                 ZP=(10.473, 37.598)
C     SEE HANSEN, AP-20,NOV. 1972,P780
C
      COMPLEX ZP
      DIMENSION C(5)
      DOUBLE PRECISION ARG, CO, SO, R, Q,P
      P=6.2831853071795864769252 87
      DN = SIN(P*SL)**2
      B = COS(P*SL)
      C(1)  = 1.
      C(5)  = 1.
      C(2)  = -4.*B
      C(4)  = -4.*B
      C(3)  = 2.*(1.+2.*B**2)
      SUMR = 0.
      SUMX = 0.
      DO 505 M = -2, 2
      ARG = GSX+M*SL
      DO 505 N = -1, 1, 2
      CO = DCOS(P*N*ARG)
      SO = DSIN(P*N*ARG)
      R = P*(DSQRT(GSY**2+ARG**2)-N*ARG)
      Q = DABS(R)
      IF(Q.GT. . 0001) GO TO 506
      S1 = R
      C1 = .57721566+DLOG(Q)-.25*Q**2
      GO TO 504
  506 R1 = R
      CALL SICINT(C1,S1,RA)
      S1 = S1+.25*P
  504 SUMR = SUMR+C(M+3)*(SO*S1-CO*C1)
      SUMX = SUMX+C(M+3)*(SO*C1+CO*S1)
  505 CONTINUE
      ZP = -15.*CMPLX(SUMR,SUMX)/DN
      RETURN
      END
```

Fig. 3 Subroutine ECHEL.

IEEE, Vol. AP-26, March 1978, pp. 214-219.

2. King, H. E., "Mutual Impedance of Unequal Length Antennas in Echelon," *Trans. IRE*, Vol. AP-5, July 1957, pp. 300-313.
3. Carter, P. S., "Circuit Relations in Radiating Systems and Applications to Antenna Problems," *Proc. IRE*, Vol. 20, June 1932, pp. 1004-1041.
4. Schelkunoff, S. A., and H. T. Friis, *Antenna Theory and Practice*, John Wiley, NY, 1942, pp. 368 and 401.
5. M. Abramowitz, and I. A. Stegun, *Handbook of Mathematical Functions,* NBS, AMS-55, 1970, EQS. 5.25, 5.2.2 and 5.2.36-39.
6. Hansen, R. C., "Formulation of Echelon Dipole Mutual Impedance for Computer," *Trans. IEEE*, Vol. AP-20, November 1972, pp. 780-781.
7. Wimp, Jet, "Polynominal Approximations to Integral Transforms," *Mathematical Tables and Other Aids to Computation*, Vol. 15, 1961, Issue 74, pp. 174-178.

Dr. R. C. Hansen received his BSEE degree from the Missouri School of Mines and his PhD degree from the U. of Illinois in 1955. From 1949 to 1955, he worked in the Antenna Laboratory of the U. of Illinois on ferrite loops, streamlined airborne antennas, and DF and homing systems. In 1960, he became Senior Staff Member in the Telecommunications Laboratory of Hughes Aircraft Co., working on surface wave antennas, slot arrays, near-dynamic antennas. From 1964 to 1966, Dr. Hansen formed and was director of the Test Mission Analysis Office, responsible for computer programs for planning and control of classified AF satellites. From 1966-67, he was Operations Group Director of the Manned Orbiting Laboratory Systems Engineering Office of the Aerospace Corp., with responsibilities for: flight crew training, simulator, software system for mission profiles, and mission control center equipment and displays. Since 1971, he is a consulting engineer for antennas and systems-related problems. Projects for the DoD and prime contractors have covered frequencies from ELF to 10.7 microns.

Gary D. Brunner received the BSEE degree in 1965 and the MSEE degree in 1967, both from the University of Missouri, Rolla. He is currently Section Head, Antennas, in the Research and Development Department of Goodyear Aerospace Corporation, Litchfield Park, AZ. While at Goodyear Aerospace, Mr. Brunner has been responsible for a number of high-performance slotted waveguide arrays and reflector antennas for application to synthetic aperture radar programs. He is currently involved in design and implementation of airborne electronically scanned arrays.

Polynomial approximation of current along thin symmetrical cylindrical dipoles

B. D. Popović, D.Sc.

Indexing terms: Dipole antennas, Current distribution, Admittance

Abstract

Current distributions and admittances of an isolated symmetrical cylindrical dipole, and of two non-staggered parallel dipoles, are analysed using a polynomial approximation for current along the dipoles. For antennas of lengths less than or not greatly exceeding the free-space wavelength, a polynomial of only second order approximates the current distribution quite closely, and results in self and mutual admittances which are, on average, as accurate as those according to the Chang-King 5-term theory. For longer antennas, of lengths up to approximately $1\cdot5\lambda$, a polynomial of only third order is an excellent approximation.

1 Introduction

Historically, the sinusoidal current distribution along thin cylindrical antennas was the first approximate current distribution utilised for analysing the radiation properties of such structures. This simple distribution was used for designing probably the majority of the existing antennas and antenna arrays. However, the deficiencies of this approximation are well known. The radiation patterns predicted by sinusoidal theory agree more or less satisfactorily with the real patterns, but the properties of an antenna as a circuit element (i.e. its input admittance and impedance), are, in general, very much in error.

To remedy these deficiencies, in the existing improved theories various correction terms are added to the 'leading' sinusoidal term.[1–7] These correction terms are usually in the form of some other trigonometric functions. The current distributions calculated according to these theories agree much better with experimentally determined distributions. However, this agreement is in no way perfect,[8] which shows that the types of the approximate current distributions assumed may not be the best choice possible. In fact, there are no particularly strong arguments, except the historic ones, in favour of trigonometric functions. Nevertheless, it seems that there were no attempts to use an approximate current distribution of a radically different form from a sum of trigonometric functions.

In the present paper, the current distribution and admittance of linear antennas are analysed using a basically different approximation of current from the sinusoidal one. The analysis is restricted to antennas shorter than approximately $1\cdot5\,\lambda$. These cases are the most important, in practice, and extensive experimental data about current distributions and admittances of such dipoles, isolated or in arrays, are already available,[8] so that the accuracy of the theoretical results can be estimated.

Mathematically, the simplest function with which to approximate the current-distribution function along thin cylindrical dipoles is a polynomial with complex coefficients. It is well known that, for a reasonable range of values of the independent variable, any well behaved function can be approximated very accurately by a polynomial of a relatively low order. There is no reason why the real current distribution along a cylindrical dipole should not be able to be approximated with a high accuracy by a polynomial. Indeed, inspecting the measured current distributions along antennas of reasonable lengths (shorter than $1\cdot5\,\lambda$),[8] it is not difficult to see that a current distribution in the form of a polynomial of a quite low order may be at least as good an approximation to the true distribution as a sum of trigonometric functions. The present paper discusses this possibility in the cases of an isolated symmetrical cylindrical dipole and of two such non-staggered coupled dipoles.

Paper 6155 E, first received 30th December 1969 and in revised form 23rd February 1970

Dr. Popović is with the Department of Electrical Engineering, University of Belgrade, Yugoslavia

To determine the unknown complex coefficients, Hallén's integral equation for current is satisfied at as many points as necessary to obtain a sufficient number of linear equations in these coefficients and in the unknown constant which appears in Hallén's equation. For antenna lengths less than approximately λ, excellent results, on average as accurate as those according to the Chang-King 5-term theory,[7] are obtained with polynomial currents of only second order. For antennas of lengths between λ and $1\cdot5\,\lambda$, polynomial currents of only third order appear to give excellent approximation.

2 Current distribution and admittance of an isolated dipole

2.1 Integral equation for current and polynomial current approximation

Consider an isolated dipole of radius a and length $2h$, centre-fed by an idealised slice or delta-function generator of voltage V and frequency ω. Let the z axis of the co-ordinate system to be used for analysis coincide with the dipole-axis, and let the origin $z = 0$ be at the dipole centre. We assume that $h \gg a$ and $a \ll \lambda$, λ being the free-space wavelength of the radiated wave, and that the dipole is made of a perfect conductor.

As the current along the dipole is in the z direction, the complex magnetic vector potential A at all points, and hence also along the antenna itself, has only the A_z component. Along the antenna, A_z is a function of the co-ordinate z alone. A differential equation for A_z along the antenna is obtained from the boundary conditions, as follows. The z component E_z of the electric-field-intensity vector, which can be expressed in terms of A_z, must be zero at all points of the antenna surface outside the slice generator. Within the belt region of the generator, E_z must be equal in magnitude and opposite in direction to the field E_{iz} equivalent to nonelectric forces acting in the belt. As, by definition,

$$V = \lim_{\delta \to 0} \int_{-\delta}^{\delta} E_{iz} dz \quad \ldots \quad (1)$$

it follows that $E_{iz} = V\delta(z)$, $\delta(z)$ representing the Dirac delta function. Recalling that

$$E_z = -j\omega\left(1 + \frac{1}{k^2}\frac{\partial^2}{\partial z^2}\right)A_z \quad \ldots \quad (2)$$

where $k = \omega(\epsilon_0\mu_0)^{1/2}$ is the free-space propagation coefficient the differential equation for the vector potential A_z along the antenna becomes

$$j\omega\left(1 + \frac{1}{k^2}\frac{\partial^2}{\partial z^2}\right)A_z = V\delta(z) \quad \ldots \quad (3)$$

The solution of this equation for $V = 0$ (homogeneous equation) is

$$A_z' = C_1 \cos kz + C_2 \sin kz \quad \ldots \quad (4)$$

As the current, and thus also A_z', are symmetric with respect to $z = 0$, $C_2 = 0$.

To obtain the particular solution of eqn. 3, we first note that, for $z > 0$ and $z < 0$, it must be of the form shown in eqn. 4. Denote by $\phi(z)$ the electric scalar potential along the antenna. Then it must be

$$V = \lim_{z \to 0} \{\phi(z) - \phi(-z)\} \quad \ldots \ldots \quad (5)$$

Making use of the Lorentz condition for potentials

$$\phi(z) = \frac{j\omega}{k^2} \frac{\partial A_z}{\partial z} \quad \ldots \ldots \quad (6)$$

eqn. 5 becomes

$$V = \frac{j\omega}{k^2} \lim_{z \to 0} \left(\frac{\partial A_z}{\partial z} - \frac{\partial A_{-z}}{\partial z} \right) \quad \ldots \ldots \quad (7)$$

It follows that the particular solution of eqn. 3 is

$$A_z'' = \frac{k}{j\omega} \frac{V}{2} \sin k|z| \quad \ldots \ldots \quad (8)$$

The solution of eqn. 3 therefore reads

$$A_z = C_1 \cos kz + \frac{kV}{2j\omega} \sin k|z| \quad \ldots \ldots \quad (9)$$

The vector potential A_z at z can also be expressed approximately in terms of the (unknown) current $I(z)$ along the antenna, assumed along the antenna axis:

$$A_z = \frac{\mu_0}{4\pi} \int_{-h}^{h} I(z') \frac{e^{-jkr}}{r} dz' \quad \ldots \ldots \quad (10)$$

where $r = \{(z - z')^2 + a^2\}^{1/2} \quad \ldots \ldots \quad (11)$

Equating the right-hand sides of eqns. 9 and 10, we obtain the approximate form of the Hallén integral equation[9] for current along the dipole:

$$\int_{-h}^{h} I(z') \frac{e^{-jkr}}{r} dz' = C \cos kz + \frac{V}{j60} \sin k|z| \quad \ldots \quad (12)$$

where $C = 4\pi C_1/\mu_0$ and $1/j60 = 4\pi k/(2j\omega\mu_0)$

$$= \{2\pi(\epsilon_0/\mu_0)^{1/2}\}/j.$$

A method for solving eqn. 12 for $I(z')$ exactly is not known. Several methods[1-7, 9, 10] exist for solving it approximately. One of the methods is the method of undetermined coefficients.[3, 4, 7] According to that method, $I(z')$ is represented as a series of known functions with unknown complex coefficients. These coefficients are then determined by stipulating that eqn. 12 be satisfied at as many points along the antenna as are needed to determine these coefficients and the unknown constant C of integration in eqn. 12.

So far, only series of trigonometric functions[3, 4] or of modified trigonometric functions[7] have been used. Here we shall assume that $I(z')$ can be represented in the form of a polynomial of the nth order:

$$I(z') = \sum_{m=1}^{n} I_m (1 - |z'|/h)^m \quad \ldots \ldots \quad (13)$$

where I_m are complex coëfficients to be determined. With this current-distribution function, eqn. 12 becomes

$$\sum_{m=1}^{n} I_m \int_{-h}^{h} (1 - |z'|/h)^m \frac{e^{-jkr}}{r} dz'$$

$$= C \cos kz + \frac{V}{j60} \sin k|z| \quad (14)$$

2.2 Determination of coefficients I_m and antenna admittance

To determine the unknown complex coefficients I_m of the approximate polynomial current distribution given by eqn. 13, we chose any $n + 1$ convenient points Z_p, $p = 1, 2, \ldots, (n + 1)$ along the antenna, and stipulate that eqn. 14 be satisfied at these points. This results in $n + 1$ complex linear equations in $n + 1$ complex unknowns, $I_1, I_2, \ldots, I_n$ and C:

$$\sum_{m=1}^{n} I_m F_m(z_p) - C \cos kz_p = \frac{V}{j60} \sin k|z_p|$$

$$p = 1, 2, \ldots, (n + 1) \quad (15)$$

where $F_m(z_p) = \int_{-h}^{h} (1 - |z'|/h)^m \frac{e^{-jkr_p}}{r_p} dz' \quad \ldots \quad (16)$

and $r_p = \{(z_p - z')^2 + a^2\}^{1/2} \quad \ldots \ldots \quad (17)$

Using high-speed electronic computers, it is easy to evaluate integrals $F_m(z_p)$ numerically, and then solve the system of linear equations (eqns. 15) for $I_1, I_2, \ldots, I_n$ and, if desired, for the constant C (which is not usually required).

The choice of points z_p is, in a sense, arbitrary. However, a simple analysis shows that we must adhere to certain principles to obtain, presumably, the best results with the polynomial of a given order. Note that by stipulating eqn. 15 to be satisfied at points $z_{p'}$ at the same time we approximately determine $I(z')$ and $dI(z')/dz'$ in the vicinity of these points. This obviously also imposes restrictions on the $I(z')$ curve at points different from z_p. To obtain the optimal average $I(z')$ along the antenna, it is most natural to choose equidistant z_p points. The point $z = 0$ must be one of these points if an accurate value of the input admittance is to be obtained. The point $z = h$, or a point close to it, must also be included, as it plays a predominant role in determining the slope of $I(z')$ at the antenna end. Therefore, the simplest, and at the same time the best, choice of the z_p points appears to be

$$z_p = (p - 1)h/n \quad p = 1, 2, \ldots, (n + 1) \quad \ldots \quad (18)$$

Note that z_p points symmetric with respect to the origin result in the identical equation, because of the system symmetry. This is the reason why only positive values of z_p are indicated in eqn. 18. The quantitative influence of shifting the points between $z = 0$ and $z = h$ toward $z = 0$ or $z = h$ will be discussed briefly in Section 2.3.

Once the coefficients $I_1, I_2, \ldots, I_n$ have been determined, the complex current along the antenna corresponding to the given voltage V is obtained from eqn. 13. In particular, the complex current intensity at $z = 0$ is

$$I(0) = \sum_{m=1}^{n} I_m \quad \ldots \ldots \quad (19)$$

so that the antenna input admittance is given by

$$Y = \frac{I(0)}{V} = \frac{1}{V} \sum_{m=1}^{n} I_m \quad \ldots \ldots \quad (20)$$

2.3 Numerical results

All numerical results in this Section refer to symmetrical dipoles of radius $a = 0{\cdot}007022\,\lambda$. This is because the best available measured current distributions and admittances of isolated dipoles are obtained for this antenna thickness.[8] Table 1 shows input admittances of dipoles, with $h = 0{\cdot}250\,\lambda$, $h = 0{\cdot}375\,\lambda$ and $h = 0{\cdot}500\,\lambda$, for different orders n of the current polynomial. Shown also are experimental values of admittances. It can be seen that, for $h \leqslant 0{\cdot}500\,\lambda$, a polynomial of only second order yields excellent admittance values.

Note that the susceptance B increases (i.e. decreases in magnitude if negative) with increasing n. This effect is to be expected, for, with a polynomial of a very high order, B must tend to large positive values, because the capacitance between the two cylindrical antenna parts divided by the infinitesimal gap is theoretically infinite. Fortunately, this effect is not pronounced for any reasonable order of polynomial, or any other well behaved approximate current distribution. This precisely explains why the theoretical results are in excellent agreement with experiment, in spite of the idealised driving conditions assumed.

A polynomial current distribution was also found to be an excellent current approximation along the length of the antenna. For $h = 0{\cdot}250\,\lambda$ and $h = 0{\cdot}500\,\lambda$, the agreement of the second-order polynomial currents with measured distributions was found to be very close. Somewhat poorer agreement was observed for $h = 0{\cdot}375\,\lambda$ (approximate antiresonant length for the antenna thickness considered), $n = 2$,

Table 1

ADMITTANCES OF CYLINDRICAL DIPOLES OF RADIUS $a = 0{\cdot}007022\,\lambda$ FOR INCREASING ORDER OF POLYNOMIAL CURRENT APPROXIMATION

h/λ	Admittance			
	Order of polynomial approximation			Measured (Mack)
	2	3	4	
	mA/V	mA/V	mA/V	mA/V
0·250	$9{\cdot}16 - j3{\cdot}57$	$9{\cdot}16 - j3{\cdot}55$	$8{\cdot}81 - j3{\cdot}31$	$8{\cdot}90 - j3{\cdot}46$
0·375	$1{\cdot}52 - j0{\cdot}36$	$1{\cdot}54 - j0{\cdot}27$	$1{\cdot}53 - j0{\cdot}10$	$1{\cdot}58 - j0{\cdot}17$
0·500	$0{\cdot}98 + j1{\cdot}54$	$0{\cdot}96 + j1{\cdot}58$	$0{\cdot}97 + j1{\cdot}67$	$1{\cdot}02 + j1{\cdot}68$

and for $h = 0{\cdot}625\,\lambda$, $n = 3$. Figs. 1 and 2 show the theoretical polynomial current distributions compared with the measured distributions in these two cases. For comparison, King–Middleton approximate second-order currents[8] are also shown. It may be concluded that overall agreement of the polynomial distribution with the experimentally determined distribution is very good, in fact better than the King–Middleton approximate second-order distribution. Polynomials of only one order higher approximate the current distribution still better. These results are not shown, because the accuracy of the simple second- and third-order polynomial currents is seen to be very satisfactory.

Table 2 shows computed admittance values for a number of antenna lengths for which experimental data are available.[8] The second-order polynomial is used for $h \leqslant 0{\cdot}5528\,\lambda$, and the third-order approximation for $h \geqslant 0{\cdot}500\,\lambda$ (for these two values of h, both the second- and the third-order admittances are computed). Fig. 3 shows real and imaginary parts G and B of these admittances, together with experimental points. It is seen that agreement of the theory with experimental data is again excellent.

Finally, the influence of shifting the middle point, $z_2 = h/2$, in the second-order approximation was investigated. It was found that, by shifting it to $z_2' = h/4$, more accurate input admittances were generally obtained but, simultaneously, the current-distribution function was, on average, a somewhat poorer approximation to the true distribution. By shifting the middle point to $z_2'' = 3h/4$, the input admittance and the current distribution were found to be, on average, somewhat worse than for $z_2 = h/2$. Both quantitative conclusions thus confirm the general reasoning of Section 2.2.

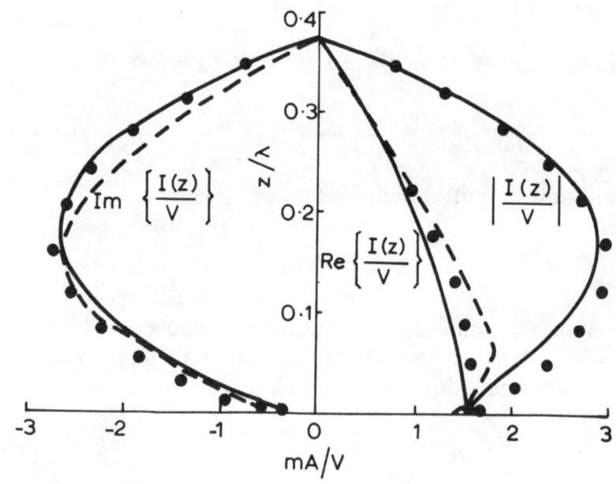

Fig. 1
Current distribution along an isolated dipole of length $2h = 0{\cdot}750\,\lambda$ and of radius $a = 0{\cdot}007022\,\lambda$
.... experimental (Mack)
--- King-Middleton, approximate second-order
—— polynomial distribution, second-order

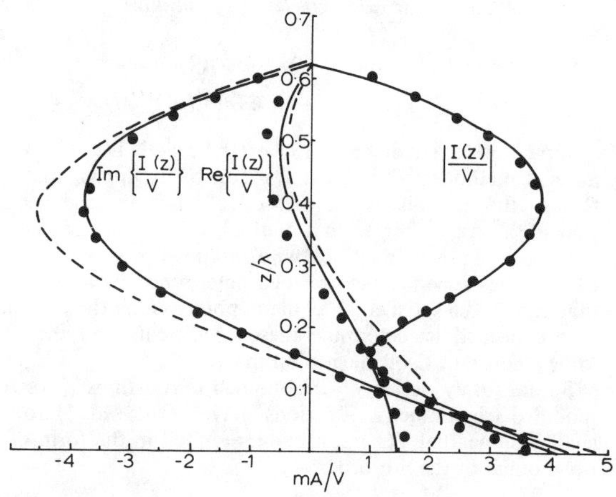

Fig. 2
Current distribution along an isolated dipole of length $2h = 1{\cdot}250\,\lambda$ and radius $a = 0{\cdot}007022\,\lambda$
.... experimental (Mack)
--- King-Middleton, approximate second-order
—— polynomial distribution, third-order

3 Self and mutual admittances of two identical coupled dipoles

3.1 Polynomial approximation of current in symmetric and antisymmetric driving conditions

The theory presented in Section 2.1 can easily be adapted for determining self and mutual admittances of two identical centre-fed parallel dipoles. Consider two such dipoles of radius a and length $2h$ centred in the plane $z = 0$ and parallel to the z axis. Let the distance between the dipole axes be b, and let the driving voltages of the two slice generators be V_1 and V_2. Following the process analogous to that outlined in Section 2.1, the following simultaneous integral equations, corresponding to eqn. 12, for currents $I_1(z')$ and $I_2(z')$ along the two dipoles are obtained:

$$\int_{-h}^{h}\{I_1(z')K_{11}(z-z') + I_2(z')K_{12}(z-z')\}dz' = C_1 \cos kz + \frac{V_1}{j60}\sin k|z| \quad (21)$$

$$\int_{-h}^{h}\{I_1(z')K_{21}(z-z') + I_2(z')K_{22}(z-z')\}dz' = C_2 \cos kz + \frac{V_2}{j60}\sin k|z| \quad (22)$$

where C_1 and C_2 are complex constants to be determined, and

$$K_{11}(z-z') = K_{22}(z-z') = \frac{\exp(-jkr_s)}{r_s} \quad (23)$$

$$K_{12}(z-z') = K_{21}(z-z') = \frac{\exp(-jkr_m)}{r_m} \quad (24)$$

$$r_s = \{(z-z')^2 + a^2\}^{1/2} \quad (25)$$

$$r_m = \{(z-z')^2 + b^2\}^{1/2} \quad (26)$$

Approximate solution of simultaneous integral eqns. 21 and 22, corresponding to arbitrary driving voltages V_1 and V_2, assuming currents $I_1(z')$ and $I_2(z')$ in the form of polynomials with complex coefficients to be determined, can be done in exactly the same way as in Section 2.2. The only difference is that now we have to solve a system of $2(n+1)$ complex linear equations in $2(n+1)$ complex unknowns, instead of

Table 2

ADMITTANCES OF CYLINDRICAL DIPOLES OF RADIUS $a = 0{\cdot}007022\lambda$

h/λ	$Y = G + jB$	h/λ	$Y = G + jB$	h/λ	$Y = G + jB$
Second-order polynomial			mA/V		mA/V
	mA/V	0·3096	2·768 − *j*1·984	0·7097	9·358 + *j*4·730
0·1574	0·898 + *j* 6·349	0·3750	1·524 − *j*0·356	0·7252	9·937 + *j*1·040
0·1822	2·734 + *j* 8·661	0·4423	1·115 + *j*0·714	0·7296	9·561 + *j*0·101
0·2021	7·241 + *j*10·332	0·5000	0·979 + *j*1·542	0·7318	9·313 − *j*0·308
0·2123	11·165 + *j* 9·213	0·5528	0·981 + *j*2·426	0·7363	8·724 − *j*1·008
0·2189	13·426 + *j* 6·877	Third-order polynomial		0·7500	6·784 − *j*2·115
0·2233	14·271 + *j* 4·745	0·5000	0·963 + *j*1·580	0·7631	5·276 − *j*2·264
0·2322	13·820 + *j* 0·384	0·5528	0·945 + *j*2·401	0·7852	3·668 − *j*1·835
0·2410	11·602 − *j* 2·397	0·6249	1·416 + *j*4·189	0·8073	2·783 − *j*1·268
0·2499	9·184 − *j* 3·565	0·6635	2·846 + *j*5·902		
0·2764	4·800 − *j* 3·281	0·6967	7·112 + *j*6·511		

$(n + 1)$ equations in $(n + 1)$ unknowns, which does not represent any special problem.

However, if we are interested only in determining self and mutual admittances of the coupled dipoles, the problem can be solved essentially by the single-antenna technique. For that purpose, we consider the two dipoles in two special driving conditions:

(*a*) when they are fed with equal voltages $V_1 = V_2 = V^s$ (symmetric driving conditions)

(*b*) when they are fed with equal, but opposite, driving voltages $V_1 = -V_2 = V^a$ (antisymmetric driving conditions).

By symmetry, in the symmetric case currents along the two dipoles are equal, $I_1(z') = I^s(z') = I_2(z')$, and, in the antisymmetric case, they are equal and opposite, $I_2(z') = -I_2(z') = I^a(z')$. As a result, in both cases, eqns. 21 and 22 reduce to the same equation, of the form

$$\int_{-h}^{h} I(z')\{K_{11}(z - z') \pm K_{12}(z - z')\}dz' = C \cos kz + \frac{V}{j60} \sin k|z| \quad (27)$$

which is the same as eqn. 12, except for a different kernel. The upper sign in braces corresponds to the symmetric case and the lower sign to the antisymmetric case. The currents $I^s(z)$ and $I^a(z)$ can thus be determined by the same procedure as for an isolated dipole. Once the coefficients I_m^s and I_m^a of the polynomial currents are determined, symmetric and antisymmetric admittances are obtained by analogy to eqn. 20:

$$Y^s = \frac{1}{V^s} \sum_{m=1}^{n} I_m^s \quad (28)$$

$$Y^a = \frac{1}{V^a} \sum_{m=1}^{n} I_m^a \quad (29)$$

The self and mutual admittances $Y_{11} = Y_{22}$ and $Y_{12} = Y_{21}$ are now obtained as follows. They are defined by the equations

$$I_1(0) = V_1 Y_{11} + V_2 Y_{12} \quad (30)$$

$$I_2(0) = V_1 Y_{21} + V_2 Y_{22} \quad (31)$$

In the symmetric case, both these equations become simply

$$I^s(0) = V^s(Y_{11} + Y_{12}) \quad (32)$$

and in the antisymmetric case both read

$$I^a(0) = V^a(Y_{11} - Y_{12}) \quad (33)$$

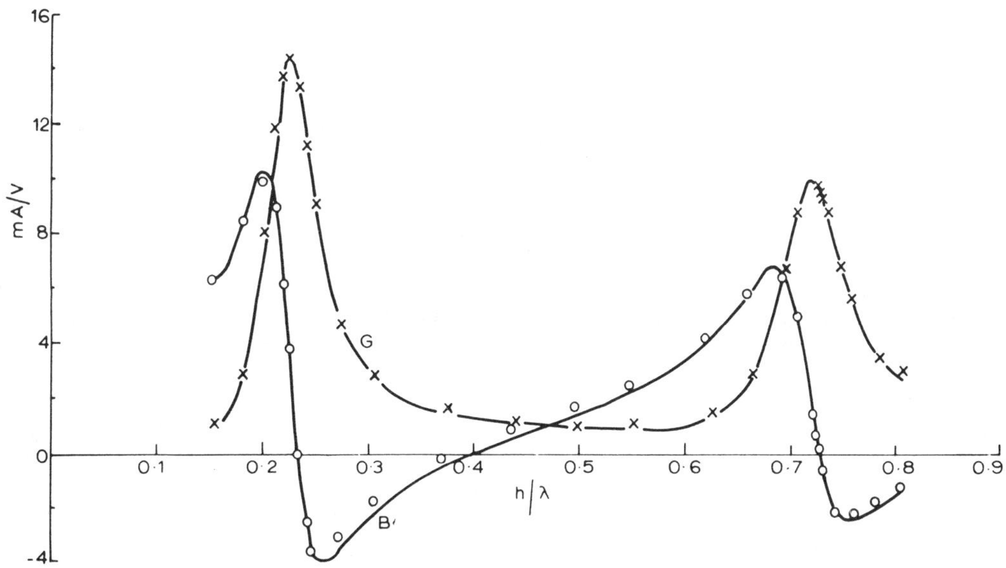

Fig. 3

Real and imaginary parts of admittances of isolated dipoles of radius $a = 0{\cdot}007022\lambda$, for different lengths $2h$ of dipoles

× ○ experimental (Mack)
—— polynomial distribution, second-order for $b/\lambda < 0{\cdot}5$, third-order for $h/\lambda > 0{\cdot}5$

Table 3

SELF AND MUTUAL ADMITTANCES OF TWO PARALLEL CYLINDRICAL DIPOLES OF RADIUS $a = 0{\cdot}007022\,\lambda$: SECOND-ORDER POLYNOMIAL APPROXIMATION OF CURRENT

b/λ	$h = 0{\cdot}250\,\lambda$		$h = 0{\cdot}375\,\lambda$		$h = 0{\cdot}500\,\lambda$	
	Y_{11}	Y_{12}	Y_{11}	Y_{12}	Y_{11}	Y_{12}
	mA/V	mA/V	mA/A	mA/V	mA/V	mA/V
0·125	5·111 − j9·501	0·734 + j9·412	0·983 − j0·399	0·686 + j0·919	0·670 + j1·580	0·478 + j0·230
0·250	6·278 − j3·944	1·487 + j4·753	1·430 − j0·097	0·812 + j0·338	0·953 + j1·692	0·534 + j0·027
0·375	7·999 − j2·049	2·608 + j3·188	1·694 − j0·264	0·753 − j0·219	1·080 + j1·564	0·424 − j0·275
0·500	10·174 − j2·342	4·056 + j1·010	1·597 − j0·479	0·327 − j0·609	0·994 + j1·465	0·100 − j0·435
0·625	10·065 − j4·386	3·028 − j2·281	1·439 − j0·413	−0·162 − j0·552	0·922 + j1·529	−0·195 − j0·322
0·750	8·645 − j4·188	0·375 − j2·881	1·474 − j0·297	−0·417 − j0·262	0·965 + j1·586	−0·315 − j0·095
0·875	8·659 − j3·266	−1·257 − j2·074	1·567 − j0·312	−0·441 + j0·074	1·015 + j1·556	−0·273 + j0·134
1·000	9·369 − j3·132	−2·160 − j0·701	1·562 − j0·390	−0·245 + j0·332	0·992 + j1·511	−0·094 + j0·268
1·125	9·529 − j3·747	−1·875 + j1·002	1·497 − j0·388	0·053 + j0·368	0·953 + j1·529	0·112 + j0·235
1·250	9·021 − j3·866	−0·513 + j1·792	1·496 − j0·335	0·263 + j0·209	0·967 + j1·562	0·224 + j0·086
1·375	8·914 − j3·473	0·742 + j1·481	1·540 − j0·332	0·310 − j0·026	0·996 + j1·553	0·209 − j0·085
1·500	9·233 − j3·355	1·458 + j0·547	1·545 − j0·369	0·192 − j0·220	0·989 + j1·526	0·086 − j0·197
1·625	9·349 − j3·636	1·340 − j0·618	1·512 − j0·375	−0·017 − j0·271	0·966 + j1·532	−0·072 − j0·188
1·750	9·104 − j3·734	0·456 − j1·275	1·507 − j0·346	−0·187 − j0·169	0·970 + j1·553	−0·171 − j0·080
1·875	9·018 − j3·529	−0·509 − j1·137	1·532 − j0·341	−0·237 + j0·007	0·989 + j1·550	−0·171 + j0·058
2·000	9·194 − j3·445	−1·095 − j0·446	1·537 − j0·363	−0·156 + j0·162	0·986 + j1·533	−0·078 + j0·153

But, we also have that

$$I^s(0) = Y^s V^s \quad \quad (34)$$

$$I^a(0) = Y^a V^a \quad \quad (35)$$

From eqns. 32 to 35, we easily obtain that

$$Y_{11} = \tfrac{1}{2}(Y^s + Y^a) \quad \quad (36)$$

$$Y_{12} = \tfrac{1}{2}(Y^s - Y^a) \quad \quad (37)$$

It is possible to show that current distributions and admittances corresponding to any set of driving conditions (i.e. voltages V_1 and V_2) can be found if they are known in symmetric and antisymmetric driving conditions. It should be noted, however, that the solution for symmetric and antisymmetric cases are only approximate, and that, in combining the two cases, results which are less accurate than in the separate cases may be obtained. Therefore, for a precise analysis it is advisable to analyse the system in arbitrary driving conditions starting from fundamental eqns. 21 and 22.

3.2 Numerical results

Using the method explained in Section 3.1, the self and mutual admittances were computed for $h = 0{\cdot}250\,\lambda$, $h = 0{\cdot}375\,\lambda$ and $h = 0{\cdot}500\,\lambda$ for a number of distances b between the dipoles. The radius was again assumed to be $a = 0{\cdot}007022\,\lambda$ because of the excellent set of experimental results available for this radius.[8] Table 3 shows the results, computed in all cases using the simplest, second-order polynomial approximation. By comparing the results with those obtained by Mack,[8] it can be seen that agreement is very satisfactory, in spite of the only second-order polynomial distribution used. As an illustration, self and mutual admittances $Y_{11} = G_{11} + jB_{11}$ and $Y_{12} = G_{12} + jB_{12}$ are shown in Fig. 4 for $h = 0{\cdot}500\lambda$. Experimental points and admittances according to the Chang–King 5-term theory[7] are also shown in Fig. 4. We see that the results from the present theory with polynomial distribution of the second order are, on average, as accurate as those according to the Chang–King 5-term theory.

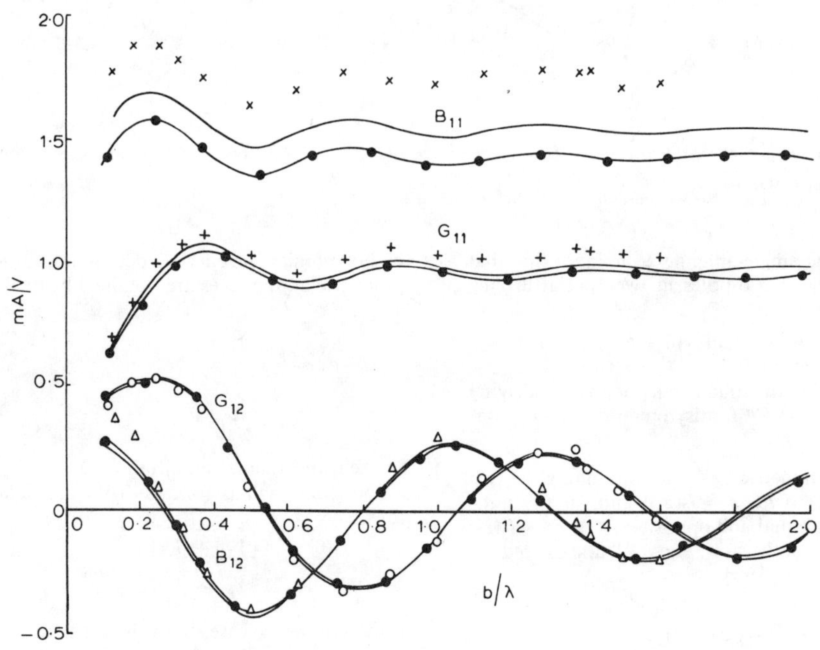

Fig. 4

Real and imaginary parts of self and mutual admittances of two parallel dipoles of length $2h = \lambda$ and radius $a = 0{\cdot}007022\lambda$, for different distances b between the dipoles

× + ○ △ experimental (Mack)
—●—●— Chang–King 5-term theory
——— second-order polynomial distribution

4 Conclusions

The existing theories of thin cylindrical dipoles and of arrays of such dipoles are aimed at improving the leading sinusoidal current term by some convenient additional terms, mostly in the form of some other trigonometric functions. However, by inspecting the experimental current distribution curves, it may be seen that a low-order polynomial with complex coefficients might be at least as good an approximation as the conventional sum of the leading sinusoidal term and several correction terms. In the present paper, a polynomial current distribution is used for analysing thin cylindrical symmetrical dipoles. The unknown complex coefficients of the polynomial are determined by stipulating that Hallén's integral equation for current be satisfied at a sufficient number of points along the dipole. It was found that it was best to choose equidistant points, one of which must be at the antenna centre, and the second near the antenna end.

The cases of an isolated symmetrical dipole and of two such parallel identical dipoles are analysed using the polynomial current distribution. In the case of an isolated dipole, a polynomial of only second order appears to be an excellent approximation for the current, provided that $2h \leqslant \lambda$, and a polynomial of third order is excellent if $2h \leqslant 1 \cdot 5\,\lambda$. Admittances of dipoles resulting from these simple distributions are in almost perfect agreement with experimental results. In the case of two coupled dipoles with $2h \leqslant \lambda$, agreement between experimentally determined self and mutual admittances and theoretical admittances with the second-order polynomial distribution is again excellent. A comparison of the results according to the present theory with experimental results, and with those obtained by the Chang–King 5-term theory, shows that self and mutual admittances obtained with the second-order polynomial distribution are, on average, as accurate as those calculated according to the Chang–King theory. Coupled dipoles of lengths greater than λ were not considered.

The numerical results indicate that, if $2h \leqslant 1 \cdot 5\,\lambda$, the polynomial current distribution of second or higher order is superior to any other distribution of the same order so far used. For longer antennas, it is known that the current distribution exhibits a certain sinusoidal-type quasiperiodicity. Therefore the polynomial distribution, though possible (in fact, there is no difference in the computer programs for shorter and longer antennas), would require more than three terms, and thus loses its main advantage: a short computer time needed for obtaining very accurate results. In practice, however, such long antennas appear to be of minor importance.

Another important advantage of the polynomial current distribution over the other distributions used is that, owing to its simple programming logic, it is possible to obtain higher-order (i.e. more accurate) solutions easily. The present method thus has an important property of permitting a certain insight into the accuracy of the results, by simply changing a datum in the computer program.

The polynomial distribution can also be used for analysing other types of linear antenna and antenna array. The accuracy of the results may be anticipated as of the same order as those for an isolated dipole and of two parallel dipoles. Several such cases will be considered in future papers.

5 Acknowledgment

The author is grateful for the computer time made available to him for this project in the Computing Centre of the Department of Electrical Engineering, University of Belgrade, Yugoslavia.

6 References

1 KING, R. W. P.: 'The theory of linear antennas' (Harvard University Press, 1956)

2 STORER, J. E.: 'Solution to thin wire antenna problems by variational methods', doctoral dissertation, Harvard University, 1951

3 STORM, B.: 'Investigation into modern aerial theory and a new solution of Hallén's integral equation for a cylindrical aerial', dissertation, Imperial College, London, 1953

4 DUNCAN, R. H., and HINCHEY, F.: 'Cylindrical antenna theory', *J. Res. Nat. Bur. Stand.*, 1960, **64D**, pp. 569–584

5 KING, R. W. P., and WU, T. T.: 'The cylindrical antenna with arbitrary driving point', *IEEE Trans.*, 1965, **AP–13**, pp. 710–718

6 KING, R. W. P.: 'Linear arrays: currents, impedances and fields—I', *IRE Trans.*, 1959, **AP–7**, pp. S440–S457

7 CHANG, V. W. H., and KING, R. W. P.: 'On two arbitrarily located identical parallel antennas', *IEEE Trans.*, 1968, **AP–16**, pp. 309–317

8 MACK, R. B.: 'A study of circular arrays', Cruft Laboratory, Harvard University, technical reports 382 and 383, 1963

9 HALLÉN, E.: 'Theoretical investigation into the transmitting and receiving qualities of antennae', *Nova Acta Soc. Sci. Upsal.*, 1938, pp. 1–44

10 MEI, K. K.: 'On the integral equations of thin wire antennas' *IEEE Trans.*, 1965, **AP–13**, pp. 374–378

REDUCING MOMENT METHOD STORAGE FROM ORDER N^2 TO ORDER N

Indexing terms: EM waves, Scattering, EM field theory

Previously, integral equation-based calculations of electromagnetic scattering and radiation (e.g. moment method calculations) have required a full $N \times N$ matrix when there are N unknowns. However, in the letter novel basis and testing functions are presented, for which the resulting matrix is approximately sparse. That is, nearly all of its elements are nearly zero. This kind of technique is expected to reduce typical storage and execution time requirements by orders of magnitude.

Introduction: Most methods for numerically calculating electromagnetic fields are limited to small objects (in terms of wavelength) due to both the storage and execution time these methods require. The method of moments (MOM), which is based on integral equations, has the advantage of requiring unknowns only on the scattering body. However, to date N unknowns have required an $N \times N$ MOM matrix, where all N^2 elements of that matrix are significant. In the case of surface integral equations (as for conductors), there generally has been no way around explicitly calculating and using all (N^2) of these matrix elements. An exception occurs for special geometries, such as strips and circular cylinders (where a convolutional structure may be utilised). In this letter frequency-domain calculations for general conducting bodies in two dimensions are considered. A method for reducing the required storage from N^2 to a constant multiple of N (say $100N$) is presented. This reduction requires that the majority of the body, while of general shape, is smooth on the scale of several wavelengths. The methods presented here immediately generalise to three dimensions, although the $100N$ is then increased to a larger multiple of N.

Consider a MOM problem, which is stated in the generic matrix form:

$$ZJ = V \tag{1}$$

In this equation Z is an $N \times N$ matrix (the 'MOM matrix'), while J and V are N-dimensional vectors. To arrive at this finite-dimensional problem, the 'current' $J_c(s)$, which is a function of arc length s along the body, is approximated by

$$J_c(s) = \sum_{\beta=1}^{N} J_\beta b_\beta(s) \tag{2}$$

for some appropriate choice of N 'basis functions' $b_\beta(s)$.[1] N testing functions $t_\alpha(s')$ are used to enforce the integral equation. Elements of the resulting matrix Z are given by integrals of the basis and testing functions multiplied by the kernel of the integral operator.

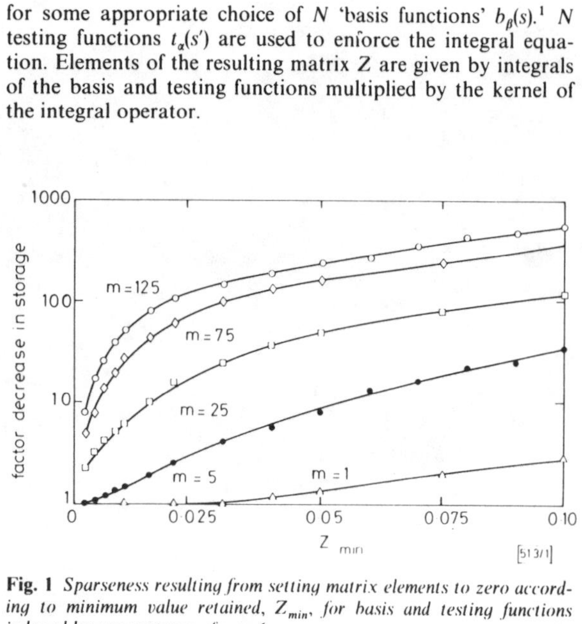

Fig. 1 *Sparseness resulting from setting matrix elements to zero according to minimum value retained, Z_{min}, for basis and testing functions indexed by parameter m of eqn. 6*

Novel basis and testing functions: As an example of what is possible with creative choices of these functions, consider the electric field integral equation (EFIE) for a two-dimensional transverse magnetic (TM) scattering problem. For incident electric field of wave number k,

$$E_{inc}(s') = k\eta/4 \int J_c(s) H_0^{(2)}(kR) ds \tag{3}$$

where $H_0^{(2)}$ is the second-kind Hankel function, R is the distance between source and field points, and η is the impedance of free space. The elements of the matrix Z are then given by

$$(Z)_{\alpha,\beta} = k\eta/4 \int H_0^{(2)}(kR) b_\beta(s) t_\alpha(s') ds\, ds' \tag{4}$$

Most commonly, the basis and testing functions are chosen to be local functions (e.g. pulses), while global functions (e.g. Fourier modes) are sometimes also used. We consider an intermediate type, which is nonzero over a distance of several wavelengths. As discussed below, if the scatterer is smooth over this region, these functions may be chosen so that the matrix Z is effectively sparse.

N is chosen so that it may be factored into the product of integers n and m, with m odd. β may then be decomposed as

$$\beta = (q - 1)m + p \tag{5}$$

where q and p are integers with $1 \le p \le m$. That is, β is the pth point in the qth group of points. Now, an orthonormal set of basis functions may be defined by

$$b_\beta(s) = \begin{cases} (n/s_0)^{1/2} \exp\,[j2\pi sn(p - (m+1)/2)/s_0] & \text{if } (q-1)/n < s/s_0 < q/n \\ 0 & \text{otherwise} \end{cases} \tag{6}$$

where s_0 is the total arc length of the body. The testing functions are now chosen to be the same as the basis functions.

Assuming that the surfaces of the body within each block are smooth, one can devise simple estimates of the matrix elements' size. For the elements of Z where the basis and testing functions are from different blocks (i.e. the q are different), these formulas provide some interesting information. When m is large, the elements of Z fall into two groups. A few elements are quite large, and most are much smaller. In fact, the ratio of sizes is roughly

$$\text{abs}\,[(Z)_{\alpha,\beta}|_{large}]/\text{abs}\,[(Z)_{\alpha,\beta}|_{typical}] \simeq m^2 \tag{7}$$

In contrast, for pulse basis and testing functions the matrix elements would be all nearly the same size. The basis and testing functions considered here may be interpreted as ones for which the preponderance of the interactions represented within the matrix become grouped into a relatively small number of matrix elements. This raises the possibility of discarding (i.e. setting equal to zero) most of the matrix elements, and still obtaining result of acceptable accuracy.

Numerical results: Accurate numerical results using the basis and testing functions of eqn. 6 have been obtained for a pefectly conducting 'modified' NACA0012 airfoil, which has a shape defined by

$$y = \pm 0{\cdot}6[0{\cdot}2969x^{1/2} - 0{\cdot}1281x - 0{\cdot}3516x^2 + 0{\cdot}2843x^3 - 0{\cdot}1015x^4] \tag{8}$$

and a cord length of 37 wavelengths. The elements of Z are calculated using eqns. 5 and 6 with N fixed at 750, which gives a sampling density of about 10 points per wavelength along the body. Several values of the width of the blocks, m, are considered in the range from 1 to 125.

Fig. 1 illustrates the effective sparseness that occurs for

large widths m. For each value of m considered, the magnitude of each matrix element is divided by the magnitude of the largest matrix element. When this ratio is less than a value called Z_{min}, that matrix element is set to zero. Then, the original number of matrix elements (N^2) is divided by the number of elements that are now nonzero, to give a 'factor decrease in storage'. This number is the factor by which the storage necessary for this matrix is reduced. Most iterative solution techniques require Z to be multiplied by a trial solution at each iteration. Since this requires only as many 'operations' as the number of nonzero elements of Z, the 'factor decrease in storage' is also a 'factor increase in speed' for the resulting computer program. This is of great practical value, especially with the increased interest in iterative techniques due to their suitability on modern vector processing computers.

When $m = 1$, eqn. 6 defines pulse basis functions. The $m = 1$ curve on Fig. 1 thus corresponds to pulse basis and testing functions, and shows that none of the resulting matrix elements has a relative size of less than 0·02. Setting any of the elements of this matrix equal to zero would in most cases cause a large error. However, when m is increased, the situation changes significantly. When $m = 125$, nearly 99 out of 100 elements had a relative size of less than 0·02. Calculations (to be reported elsewhere) show that using $Z_{min} = 0{\cdot}002$ gives scattered field in reasonable agreement with exact results, while Fig. 1 shows it also gives nearly an order of magnitude savings in both storage and execution time.

The basis functions of eqn. 6 are just one example of what may be achieved. A general method for generating alternative basis functions according to a variety of criteria is presently being developed. A somewhat more complicated set of basis and testing functions changes the estimate in eqn. 7 from m^2 to m^6. A computer program using these recently constructed functions is presently under development.

F. X. CANNING

Rockwell International Science Center
Thousand Oaks, CA 91360, USA

19th July 1989

Reference

1 HARRINGTON, R. F.: 'Field computation by moment methods' (R. E. Krieger Publishing Co., Malabar, FL, 1983)

Scattering from Bodies of Revolution

MOGENS G. ANDREASEN, MEMBER, IEEE

Abstract—The problem of scattering of a plane electromagnetic wave from an arbitrary metallic body of revolution is solved by a theoretical method for arbitrary incidence and polarization. The method permits numerical computations by high-speed digital computers, and examples are given. The incident wave is expanded in cylindrical modes, and an integral equation is solved for the induced current distribution of each mode. The scattering cross section, including the back-scattering or radar cross section, is found by summation of the mode scattered fields. The method is limited to a maximum perimeter length of twenty wavelengths.

While the cases discussed in the paper pertain to perfectly conducting bodies, other surface boundary conditions, an arbitrary surface impedance or coatings by lossy dielectrics, can also be treated with equal precision.

I. Introduction

THIS WORK is concerned with a theoretical evaluation of the electromagnetic scattering from an arbitrary metallic body of revolution. This is a problem which has been given considerable attention over the past twenty years because of its significance for numerous applications in radar techniques in general. More specifically, the problem has become important in recent years for tracking and discriminating between space vehicles and objects.

The general scattering problem is a very complicated mathematical boundary-value problem which so far has resisted exact analytical treatment except in such special cases as the sphere and the spheroid. In these cases the surface of the scatterer coincides with a coordinate surface of an orthogonal coordinate system in which the vector wave equation can be solved by the method of separation of variables. The general scattering problem for an arbitrary metallic body of revolution is solved here by an integral equation method which is, in principle, an exact method. The method takes its starting point in an expansion of the incident wave in a set of orthogonal cylindrical TE and TM modes propagating along the symmetry axis of the body considered. Each of these incident modes will induce a current distribution on the surface of the body. Due to the mode orthogonality, the total induced current distribution is represented by the sum of the induced mode current distributions. Each mode current distribution satisfies two coupled integral equations. These integral equations are extremely complicated for an analytical treatment. However, they can be reduced to a set of simultaneous linear complex equations which can be solved by a digital computer. The accuracy thus obtained depends upon the number of equations used, but can, in principle, be increased as much as desired by increasing the number of equations sufficiently.

The only limitation of the numerical integral equation method for solving the general scattering problem is the speed of presently available computers. At the present state-of-the-art, the general problem cannot be solved economically if it is necessary to solve more than about one hundred linear equations. Since the mode current density does always have two orthogonal components to be determined, this means that each mode current distribution can be sampled a maximum of about fifty times. Most often, the sampling points along the perimeter of the cross section of the body can be spaced as much as one sixth of a wavelength apart without greatly impairing the accuracy of the solution. As a consequence, the method permits treating bodies with a maximum cross-sectional perimeter of about twenty wavelengths.

As indicated above, the general scattering problem for a metallic body of revolution is solved by expanding the incident wave in cylindrical modes and solving the scattering problem for each mode independently. This approach may seem to be an unnecessary complication of the problem. The scattering problem could have been formulated without expanding the incident wave, assuming the current distribution to be sampled in a sufficiently large number of points on the surface of the body. However, with a maximum average distance between neighboring sampling points of about one sixth of a wavelength and a maximum of fifty sampling points, the maximum linear dimension of a body that can be treated by this method is less than one wavelength. By expanding the incident wave in cylindrical modes, each of which induces a current distribution with known azimuthal symmetry properties, the scattering problem can be solved for a much larger body as described above.

Moreover, even if the digital computer would permit the solution of the scattering problem without expanding the incident wave, it would still be of a great advantage economically to perform the expansion. The reason for this is that the computer time needed to solve a set of simultaneous linear equations is proportional to the cube of the number of equations. Accordingly, it is more economical to solve several integral equations with a moderate number of sampling points, as is done when the expansion is carried out, than it is to solve a single integral equation with a large number of sampling points, as is necessary when no expansion is used.

Manuscript received May 21, 1964; revised August 11, 1964, and August 31, 1964. The work reported here was carried out at TRG-West, Menlo Park, Calif., and was supported in part by the U. S. Air Force under Contract No. AF 19(628)-500 on Subcontract No. 294 from Lincoln Laboratory, Massachusetts Institute of Technology, Lexington, Mass.

The author is with TRG-West, a division of TRG, Inc., Menlo Park, Calif.

II. The Integral Equations

The integral equations for scattering by a rotationally symmetric body will be derived from the vector potential of the current distribution. Assuming the surface current distribution on the scatterer surface to be $\overline{K}$, the vector potential of the scattered field is[1]

$$\overline{A}^s(\bar{r}') = \frac{\mu}{4\pi}\int_A \overline{K}(\bar{r})\,\frac{e^{-jkR}}{R}\,da \tag{1}$$

where the integration is extended over the surface of the body. R is the distance between the source point with radius vector $\bar{r}$ and the observation point with radius vector $\bar{r}'$, both points being on the surface of the body (see Fig. 1). $k = 2\pi/\lambda$ is the free-space propagation constant. The time factor exp $(j\omega t)$ is suppressed.

The scalar potential is defined by

$$\psi^s(\bar{r}') = \frac{1}{4\pi\epsilon}\int_A \eta(\bar{r})\,\frac{e^{-jkR}}{R}\,da \tag{2}$$

where $\eta(\bar{r})$ is the surface charge density on the scatterer. The continuity equation requires the following relation between $\overline{K}$ and η to be satisfied

$$\eta(\bar{r}) = -\frac{\nabla_t\cdot\overline{K}(\bar{r})}{j\omega} \tag{3}$$

where ∇_t is the differential operator tangential to the scatterer surface.

The scattered electric field can now be expressed by

$$\overline{E}^s(\bar{r}') = -j\omega\overline{A}^s(\bar{r}') - \nabla'\psi^s(\bar{r}') \tag{4}$$

Let the electric field of the incident wave be $\overline{E}^i$. The boundary condition at the surface of the scatterer is then

$$\hat{t}_n'\cdot[j\omega\overline{A}^s(\bar{r}') + \nabla_t'\psi^s(\bar{r}')] = \hat{t}_n'\cdot\overline{E}^i(\bar{r}'), \qquad n = 1, 2 \tag{5}$$

where $(\hat{t}_1, \hat{t}_2)$ are unit vectors of a local orthogonal coordinate system on the surface of the scatterer. The integral equations (5) can be reduced as follows by introducing the expressions (1) and (2) for the potentials

$$\hat{t}_n'\cdot\overline{E}^i(\bar{r}') = \frac{j\omega\mu}{4\pi}\,\hat{t}_n'\cdot\left[\int_A \overline{K}(\bar{r})\,\frac{e^{-jkR}}{R}\,da + \frac{1}{k^2}\nabla_t'\int \frac{e^{-jkR}}{R}\,\nabla_t\cdot\overline{K}(\bar{r})da\right] \qquad n = 1, 2 \tag{6}$$

To obtain a further reduction of this expression for a body of revolution, let us introduce the local coordinate system (t, ϕ) (see Fig. 1). We assume that the incident plane wave has been expanded in a set of azimuthal modes and consider the scattering of the mth mode. The electric field components along $\hat{t}$ and $\hat{\phi}$ of the mth mode of the incident wave can be expressed as a sum of an even mode and an odd mode as follows:

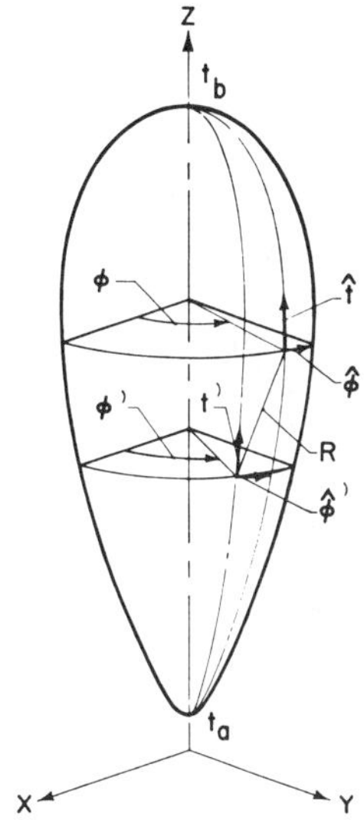

Fig. 1. Coordinates used on scatterer surface.

$$E_t = E_1(t)\,\begin{matrix}\cos\\ \sin\end{matrix}\,(m\phi) \tag{7}$$

$$E_\phi = E_2(t)\,\begin{matrix}\sin\\ -\cos\end{matrix}\,(m\phi) \tag{8}$$

The current distribution on the scatterer generated by this incident field is of the form

$$\overline{K} = K_1(t)\,\begin{matrix}\cos\\ \sin\end{matrix}\,(m\phi)\hat{t} + K_2(t)\,\begin{matrix}\sin\\ -\cos\end{matrix}\,(m\phi)\hat{\phi} \tag{9}$$

To obtain the reduction of (6), introduce

$$\nabla_t = \hat{t}\,\frac{\delta}{\delta t} + \frac{\hat{\phi}}{\rho}\,\frac{\delta}{\delta\phi} \tag{10}$$

and (9) and (10) in (6). Also, introducing the angle v between $\hat{t}$ and $\hat{z}$, and the following expressions for the unit vectors

$$\left.\begin{aligned}\hat{t} &= \sin v\cos\phi\hat{x} + \sin v\sin\phi\hat{y} + \cos v\hat{z}\\ \hat{t}' &= \sin v'\cos\phi'\hat{x} + \sin v'\sin\phi'\hat{y} + \cos v'\hat{z}\\ \hat{\phi} &= -\sin\phi\hat{x} + \cos\phi\hat{y}\\ \hat{\phi}' &= -\sin\phi'\hat{x} + \cos\phi'\hat{y}\end{aligned}\right\} \tag{11}$$

we find after tedious calculations the following pair of integral equations to be satisfied by the current components $K_1(t)$ and $K_2(t)$:

$$\begin{aligned}-j\frac{4\pi}{\zeta}E_1(t') &= \int_{kt_a}^{kt_b}\{[\sin v\sin v'(G_{m-1}+G_{m+1}) + 2\cos v\cos v' G_m]\\ &\quad\cdot K_1(t) - \sin v'(G_{m-1}-G_{m+1})K_2(t)\}\,k\rho d(kt)\\ &\quad + 2\frac{d}{d(kt')}\int_{kt_a}^{kt_b}\left[\frac{dK_1(t)}{d(kt)} + \frac{\sin v}{k\rho}K_1(t) + \frac{m}{k\rho}K_2(t)\right]G_m k\rho d(kt)\end{aligned} \tag{12}$$

[1] Stratton, J. A., *Electromagnetic Theory*, New York: McGraw-Hill, 1941.

and

$$-j\frac{4\pi}{\zeta}E_2(t')$$

$$=\int_{kt_a}^{kt_b}[-\sin v(G_{m-1}-G_{m+1})K_1(t)+(G_{m-1}+G_{m+1})K_2(t)]$$

$$\cdot k\rho d(kt)-2\frac{m}{k\rho'}\int_{kt_a}^{kt_b}\left[\frac{dK_1(t)}{d(kt)}+\frac{\sin v}{k\rho}K_1(t)\right.$$

$$\left.+\frac{m}{k\rho}K_2(t)\right]G_m k\rho d(kt) \tag{13}$$

where $\zeta=\sqrt{\mu/\epsilon}$ is the intrinsic impedance of free space. In these equations we have introduced a Green's function defined by

$$G_m=\int_0^{\pi}\frac{e^{-jkR}}{kR}\cos(m\phi)d\phi \tag{14}$$

This Green's function can be interpreted physically as the scalar field radiated by a scalar ring source with harmonically varying source strength along the ring. The integral in (14) cannot be reduced analytically except in special cases. However, for $t\neq t'$, the integrand is a regular function and the integral can be evaluated by numerical integration. For $t=t'$, the Green's function has an integrable singularity. The integration along the t axis [see (12) and (13)] therefore must be done analytically in the immediate vicinity of $t=t'$. The details of this evaluation[2] will not be dealt with here.

The reduction of the integral equations [(12) and (13)] to a set of simultaneous linear equations is trivial. The integrals in (12) and (13) can be approximated by a weighted sum of sampled values of the integrand. That is, by a weighted sum of N-sampled values K_{1n} and K_{2n} of the current components $K_1(t)$ and $K_2(t)$, as follows:

$$\sum_{n=1}^{N}[A_{pn}K_{1n}+B_{pn}K_{2n}]=-j\frac{4\pi}{\zeta}E_1(t_p') \tag{15}$$

$$\sum_{n=1}^{N}[C_{pn}K_{1n}+D_{pn}K_{2n}]=-j\frac{4\pi}{\zeta}E_2(t_p') \tag{16}$$

Requiring these relations to hold in the N-sampling points along the t axis, $p=1, 2, \cdots, N$, equations (15) and (16) represent a system of $2N$-simultaneous linear equations from which the N values of $K_1(t)$ and the N values of $K_2(t)$ can be determined. The choice of the weighting factors A_{pn}, B_{pn}, C_{pn} and D_{pn} is, of course, very important for the accuracy of the solution. For $p\neq n$, the weighting factors can be found using the trapezoidal or the Simpson rule of integration. For $p=n$, the weighting factors have to be evaluated analytically because of the singularity at $t=t'$.

[2] Andreasen, M. G., Scattering from rotationally symmetric metallic bodies, Final Rept, Subcontract 294, AF 19(628)-500, TRG-West, Menlo Park, Calif. Apr 1964.

III. Expansion of the Incident Wave

In the derivation of the integral equations [(12) and (13)] we have tacitly assumed that the plane wave incident upon the metallic body of revolution can be expanded in a set of azimuthal modes. According to (7) and (8), the tangential electrical field component of the mth incident mode at the surface of the metallic body was assumed to be of the form

$$\bar{E}_{is}{}^{m}=\hat{t}E_1(t)\begin{matrix}\cos\\ \sin\end{matrix}(m\phi)+\hat{\phi}E_2(t)\begin{matrix}\sin\\ -\cos\end{matrix}(m\phi) \tag{17}$$

where the upper function symbol applies for parallel polarization and the lower symbol applies for perpendicular polarization. We shall now show that the plane incident wave can, in fact, be expanded in a discrete set of cylindrical modes, each mode having a different azimuthal mode number, while all modes propagating with the same velocity along the symmetry axis of the body.

The incident plane wave is assumed to propagate in the direction $\theta=\pi-\theta_i$, $\phi=0$, and to be polarized in a direction making an angle θ_p with the unit vector $-\hat{\theta}$ (see Fig. 2). Let the amplitude of the magnetic field of the incident wave be unity. The amplitude of the electric field is then $E_0=\zeta=120\pi$. To simplify the expansion of the plane wave, let us separate the wave into a wave polarized parallel to the x-z plane of incidence, and a wave polarized perpendicular to this plane, as follows:

$$\bar{E}_i=-E_i^{\parallel}\hat{\theta}+E_i^{\perp}\hat{y} \tag{18}$$

where

$$E_i^{\parallel}=\zeta\cos\theta_p(\cos\theta_i\hat{x}+\sin\theta_i\hat{z})e^{-jk(z\cos\theta_i-x\sin\theta_i)} \tag{19}$$

and

$$E_i^{\perp}=\zeta\sin\theta_p\hat{y}e^{-jk(z\cos\theta_i-x\sin\theta_i)} \tag{20}$$

The parallel-polarized wave is a TM wave (no magnetic field component along the z axis) and is expandable in cylindrical TM modes. The perpendicular-polarized wave is expandable in cylindrical TE modes (no electric field component along the z axis). The last exponential in (19) and (20) can be expanded in a Fourier series[1]

$$e^{jkx\sin\theta_i}=e^{jk\rho\cos\phi\sin\theta_i}$$

$$=e^{-jkz\cos\theta_i}\sum_{m=0}^{\infty}\epsilon_m j^m J_m(k\rho\sin\theta_i)\cos(m\phi) \tag{21}$$

where J_m is the Bessel function or order m, and ϵ_m is Neumann's number

$$\epsilon_m=\begin{cases}1 & \text{for } m=0\\ 2 & \text{for } m>0\end{cases} \tag{22}$$

Defining the tangential electric field at the surface of the body by a sum of mode fields of the form (17), we find for the parallel-polarized wave

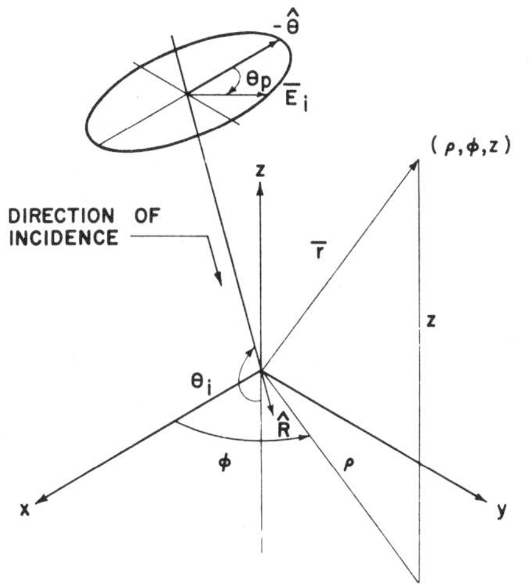

Fig. 2. Coordinates for the incident wave.

$$E_1(t) = -\cos\theta_p e^{-jkz\cos\theta_i}\epsilon_m j^{m+1}$$
$$\cdot[j\sin\theta_i\cos vJ_m(k\rho\sin\theta_i) \tag{23}$$
$$+\cos\theta_i\sin vJ_m'(k\rho\sin\theta_i)] \tag{24}$$

$$E_2(t) = \cos\theta_p\cos\theta_i e^{-jkz\cos\theta_i}\epsilon_m j^{m+1}m\frac{J_m(k\rho\sin\theta_i)}{k\rho\sin\theta_i} \tag{25}$$

where (ρ, ϕ, z) is a point on the surface of the body of revolution.

For the perpendicular-polarized wave we find similarly

$$E_1(t) = -\sin\theta_p\sin ve^{-jkz\cos\theta_i}\epsilon_m j^{m+1}m\frac{J_m(k\rho\sin\theta_i)}{k\rho\sin\theta_i} \tag{26}$$

$$E_2(t) = \sin\theta_p e^{-jkz\cos\theta_i}\epsilon_m j^{m+1}J_m'(k\rho\sin\theta_i) \tag{27}$$

This completes the derivation of the expansion formula for the incident plane wave. Because of the rotational symmetry of the body considered, the cylindrical modes are uncoupled. That is, a particular incident mode induces no mode current with a mode number different from the mode number of the incident mode. This is the reason why the scattering problem can be solved mode for mode, with subsequent application of the superposition principle.

While the mode expansion is an infinite series, the series can be truncated in practice. Due to the asymptotic behavior of the Bessel function for small arguments, the series can be truncated approximately for

$$m = k\rho_{max}\sin\theta_i + 6 \tag{28}$$

where ρ_{max} is the maximum radial dimension of the body. This formula holds only for $k\rho_{max}\sin\theta_i \gtrsim 3$. For small values at $k\rho_{max}\sin\theta_i$ m approaches 1 for $\theta_i = 0$. In the computer program developed, m is actually determined by searching a table of the Bessel function.

IV. The Scattering Cross Section

So far, we have been concerned only with the calculation of the current distribution induced on the surface of the metallic body. The determination of the current distribution is only a convenient intermediate step for finding the scattering cross section of the body, but it also provides valuable physical insight in the scattering phenomena. Once the current distribution has been found, it is a relatively simple matter to determine the scattering cross section of any direction in space by integration of each mode current distribution and summation of the scattered mode fields. The details of this calculation will be dealt with only briefly.

The scattering cross section in a particular direction (θ_0, ϕ_0) can be defined by

$$\sigma_s(\theta_0, \phi_0) = 4\pi R_0^2\left|\frac{E_s(R_0, \theta_0, \phi_0}{E_0}\right|^2 \tag{29}$$

where E_s is the scattered electric field in the direction (θ_0, ϕ_0) at a large distance R_0 from the metallic body, and E_0 is the electric field intensity of the incident wave. The scattered field is a sum of mode fields, each mode field being radiated by a mode current distribution. The relation between the scattered mode field and the mode current is the well-known integral expression[3]

$$\frac{1}{\zeta}\bar{E}_s(\bar{r}') = -\frac{j}{4\pi}\int_A \bar{K}(\bar{r})\cdot[\epsilon - \hat{R}\hat{R}]\frac{e^{-jkR}}{kR}k^2da \tag{30}$$

where $\bar{K}$ is the current density, ϵ the unit dyad, $\hat{R}$ the unit vector pointing in the direction (θ_0, ϕ_0), and R the distance from the point $\bar{r}$ on the metallic body to the observation point $\bar{r}'$. This expression can be reduced considerably by introducing the current distribution (9) and separating the ϕ dependence. After tedious calculations, the scattering cross section in the direction (θ_0, ϕ_0) is then found to be expressed by

$$\frac{\sigma_s(\theta_0, \phi_0)}{\lambda^2} = \frac{1}{16\pi}\left[\left|\sum_{m=0}^{\infty} j^m \begin{matrix}\cos\\ \sin\end{matrix}(m\phi_0)T_{1m}(\theta_0)\right|^2 + \left|\sum_{m=0}^{\infty} j^m \begin{matrix}\sin\\ -\cos\end{matrix}(m\phi_0)T_{2m}(\theta_0)\right|^2\right] \tag{31}$$

where

$$T_{1m}(\theta_0) = \int_{kt_a}^{kt_b}\{K_{1m}(t)[\cos\theta_0\sin v(J_{m+1}(w) - J_{m-1}(w))$$
$$+j2\sin\theta_0\cos vJ_m(w)]$$
$$+K_{2m}(t)\cos\theta_0(J_{m+1}(w) + J_{m-1}(w))\}k\rho$$
$$\cdot e^{jkz\cos\theta_0}d(kt) \tag{32}$$

[3] Silver, S., *Microwave Antenna Theory and Design*, MIT Radiation Lab Ser No. 12, New York: McGraw-Hill, 1949.

and

$$T_{2m}(\theta_0) = \int_{kt_a}^{kt_b} \{K_{1m}(t) \sin v[J_{m+1}(w) + J_{m-1}(w)] + K_{2m}(t)[J_{m+1}(w) - J_{m-1}(w)]\} k\rho \cdot e^{jkz \cos \theta_0} d(kt) \qquad (33)$$

where we have introduced the abbreviation

$$w = k\rho \sin \theta_0 \qquad (34)$$

The integrals in (32) and (33) are most easily evaluated numerically by a digital computer.

V. Numerical Results

The integral equations [(12) and (13)] have been programmed for a digital computer. The resulting computer program will solve the general scattering problem for virtually any cross section of a metallic body of revolution. The validity of the program has been checked by comparing the results for the induced current distribution on a sphere with exact analytical results which can be derived for this special case by the method of separation of variables. These results are shown in Fig. 3 for a 1.0-wavelength diameter sphere. The two components K_1 and K_2 of the current density [see (19)] are plotted as functions of a normalized surface coordinate s which has the value 0 on the front of the sphere and the value 1 on the back of the sphere. The agreement between the results obtained by the two methods is seen to be quite good.

Figure 4 shows the cross section of a cylindrical rod with hemispherical caps at the ends. The rod is 0.4-wavelength diameter and is 1.5 wavelengths long. The current distribution induced on this rod by a plane wave incident along the axis of symmetry of the rod was computed by the scattering program. The numerical results are shown in Fig. 5. The circumferential current component K_2 is seen to be very small on the cylindrical part of the rod, as expected, while the current component K_1 along the contour of the rod is seen to oscillate about the value 2. The oscillation of the current component K_1 represents a standing wave that is due to reflection from the end of the rod and has a period of one-half wavelength, as expected.

The cross section of another body used to test the scattering program is shown in Fig. 6. The body consists of a cone capped by a sphere in each end. The body is 2 wavelengths long and has a maximum diameter of 1.2 wavelengths. The current distribution induced on this body was computed by the scattering program both when the plane wave is incident upon the fat end of the conesphere (direction of incidence 1 in Fig. 6) and when it is incident from the opposite side (direction of incidence 2 in Fig. 6). The numerical results are shown in Figs. 7 and 8, respectively. It is interesting to note that the shadow effect is much less pronounced when the plane wave is incident upon the fat end of the conesphere than when it is incident from the opposite end. This is expected and can be explained qualitatively by the dependence of the decay exponent of the creeping waves in the shadow zone upon the princiapl radii of curvature of the surface of the body. The circumferential current component K_2 is seen to be relatively small at the small end of the conesphere as expected. The current distributions also exhibit standing-wave effects due to reflection from the far end of the conesphere.

Figure 9 shows the cross section of a circular disk with rounded edge. The disk is 1.1 wavelengths in diameter and is 0.1 wavelength thick. The current distribution induced on this body was computed by the scattering program and is shown in Fig. 10. The scattering results for the disk are particularly interesting because they exhibit physical phenomena not present on the sphere, the rod or the conesphere treated above. One notices that the circumferential current component K_2 has a sharp peak at the edge of the disk. This peak was, of course, expected. In the limit of zero thickness of the disk, the circumferential current component will have an integrable singularity at the edge. The circumferential current component K_2 also is seen, as expected, to be attenuated much faster than the radial component K_1 in the shadow zone.

Figures 3, 5, 7, 8 and 10 show only the magnitude of the current components. The phase distributions were, of course, also computed but have not been plotted.

The determination of the induced current distribution is the principal and most difficult task in any scattering problem. Once this current distribution has been found, it is a relatively simple matter to determine the scattering cross section for any direction in space by integration of each mode current distribution and summation of the scattered mode fields according to the formulas given in Section IV.

The scattering patterns for the three bodies considered previously, the rod, the conesphere and the disk, are plotted in Figs. 11–14. These figures show the scattering cross section in the plane of incidence ($\phi = 0°$) and in the plane perpendicular to the plane of incidence ($\phi = 90°$) vs. the angle of observation θ (0° for the forward direction, 180° for the backward direction).

The scattering pattern for the cylindrical rod is shown in Fig. 11. The pattern exhibits several lobes due to interference between the forward and backward traveling waves on the rod. It may be interesting to compare the value of the scattering cross section in the forward direction with the value that can be predicted optically. Optically, for the forward scattered field, the rod is equivalent to an aperture with the same area as the rod cross section. The scattering cross section in the forward direction can then be expressed by

$$\sigma = GA \qquad (35)$$

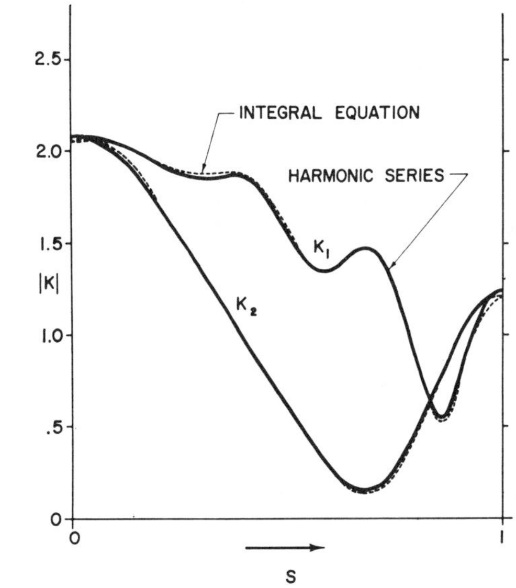

Fig. 3. Current distribution on sphere 1.0λ diameter

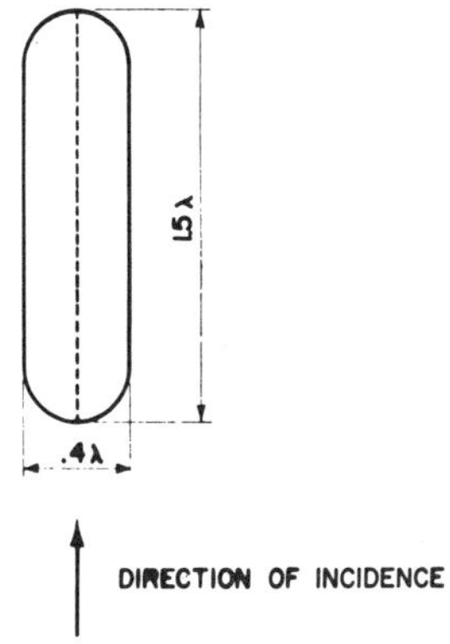

Fig. 4. Cylindrical rod with rounded ends.

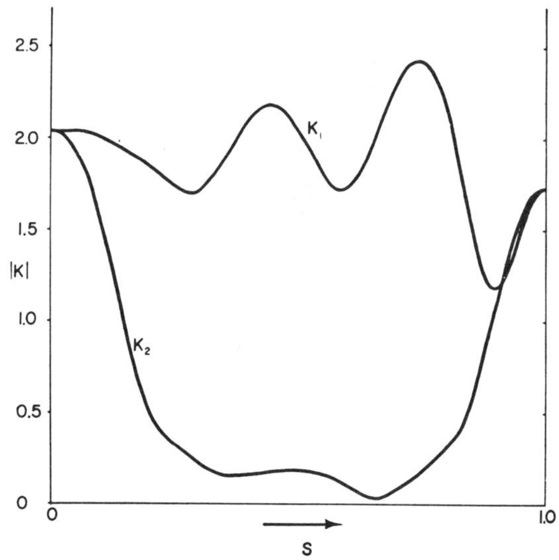

Fig. 5. Current distribution induced on cylindrical rod shown in Fig. 4.

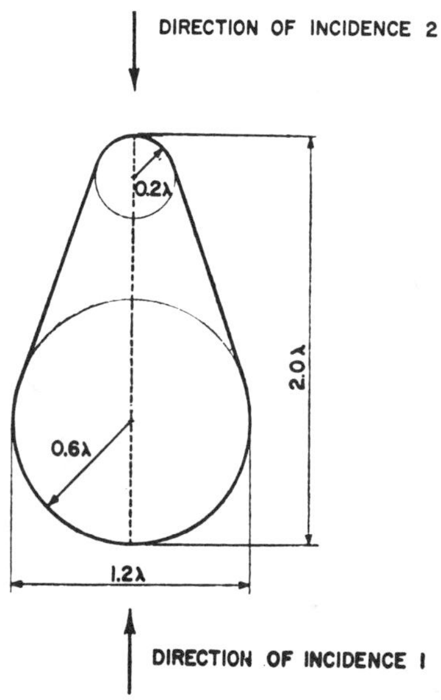

Fig. 6. Cone with spherical caps.

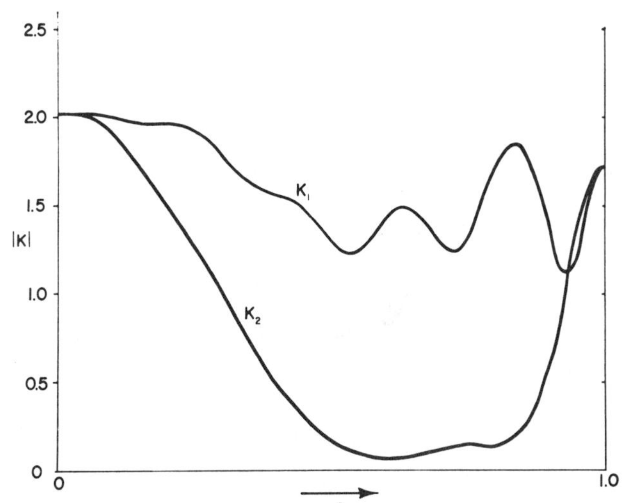

Fig. 7. Current distribution induced on cone-sphere for direction of incidence 1 in Fig. 6.

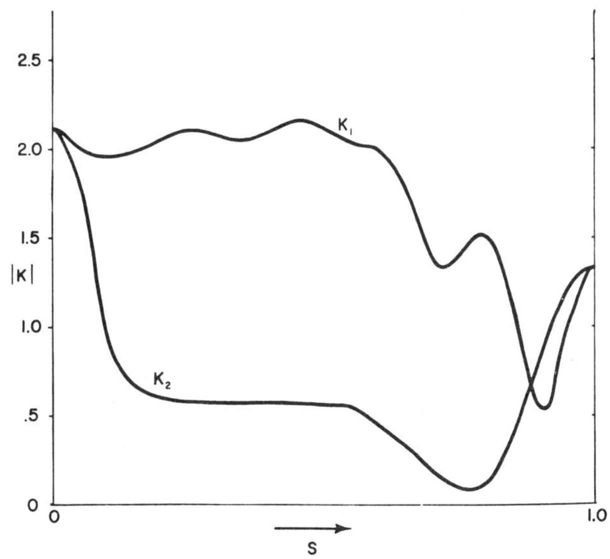

Fig. 8. Current distribution induced on cone-sphere for direction of incidence 2 in Fig. 6.

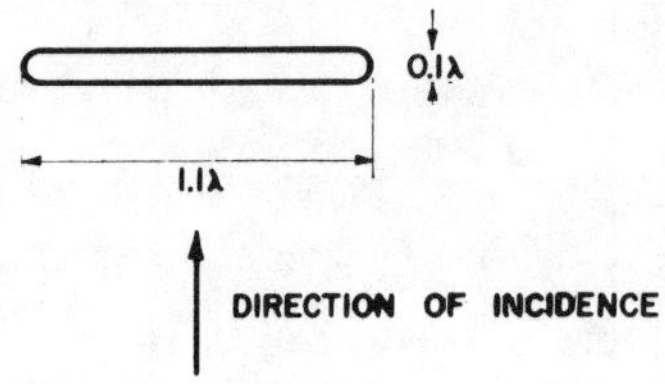

Fig. 9. Circular disk with rounded edge.

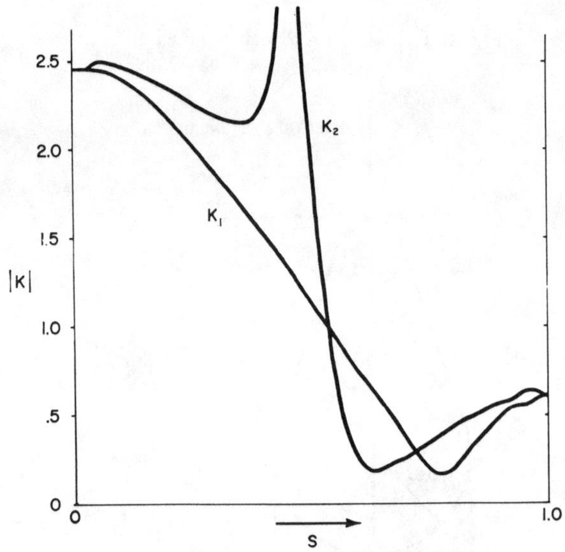

Fig. 10. Current distribution induced on circular disk shown in Fig. 9.

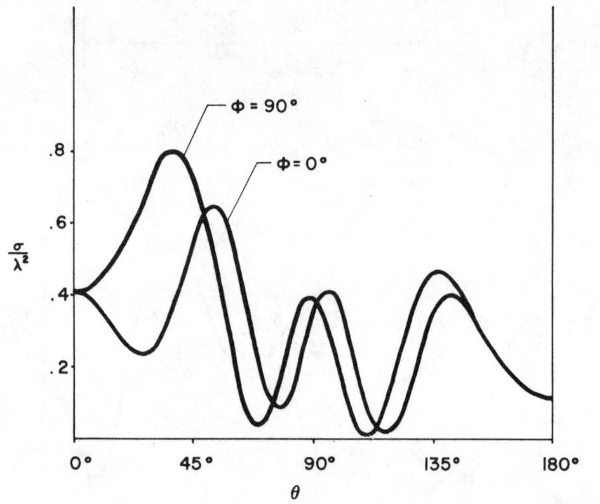

Fig. 11. Scattering pattern for cylindrical rod shown in Fig. 4.

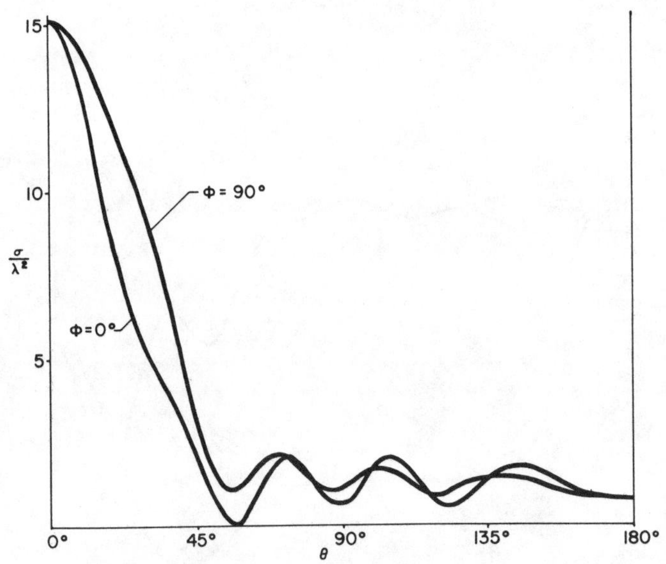

Fig. 12. Scattering pattern for cone-sphere for direction of incidence 1 in Fig. 6.

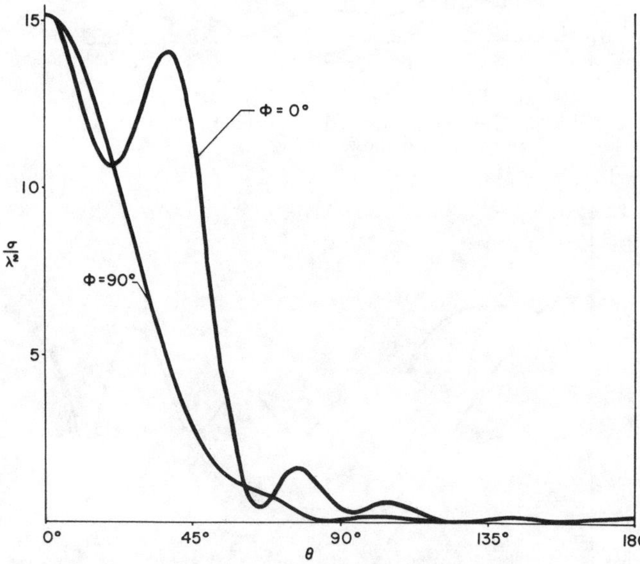

Fig. 13. Scattering pattern for cone-sphere for direction of incidence 2 in Fig. 6.

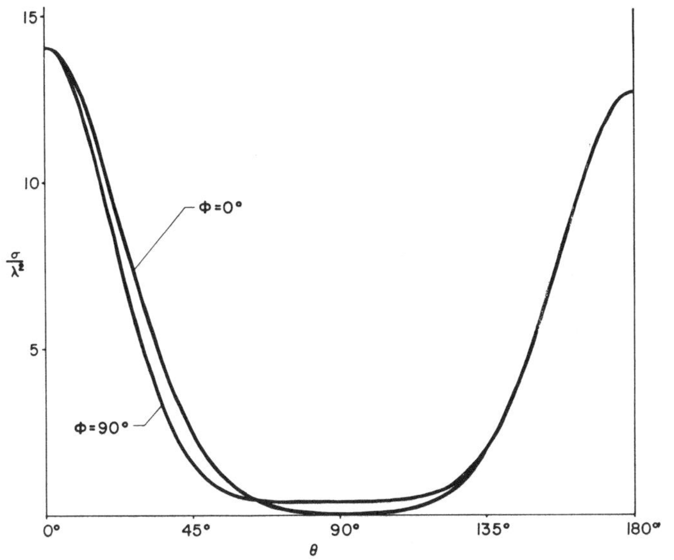

Fig. 14. Scattering pattern for circular disk shown in Fig. 9.

where A is the aperture area and G is the power gain of the aperture

$$G = \frac{4\pi}{\lambda^2} A \tag{36}$$

Accordingly, we find $\sigma/\lambda^2 = 0.12$ which agrees very poorly with the value 0.41 read from Fig. 11. The reason for this discrepancy is that the rod is only 0.4λ diameter and that scattering from the far end of the rod is, therefore, not at all negligible.

The scattering patterns for the conesphere are shown in Figs. 12 and 13. As expected, they exhibit a strong main lobe due to the shadowing. It is interesting to note that for direction of incidence 2, the pattern has a strong sidelobe for $\theta = 37.5°$. This lobe is the result of reflection from the conical part of the body. Optically, the sidelobe should appear for $\theta = 37°$. According to the reciprocity theorem, the forward scattering cross section must be the same for directions of incidence 1 and 2. Although the induced current distributions for the two cases are entirely different (see Figs. 7 and 8), the com puted values of the forward scattering cross section were found to agree to 0.4 per cent, or 0.02 dB. The optical scattering cross section in the forward direction according to (35) and (36) is $\sigma/\lambda^2 = 16.1$ which agrees reasonably well with the value 15.2 read from Figs. 12 and 13.

The scattering patterns for the disk are shown in Fig. 14. As expected, the patterns exhibit a strong lobe both in the forward direction and in the backward direction. The optical scattering cross section in the forward direction is $\sigma/\lambda^2 = 11.4$ according to (35) and (36). The actual value read from Fig. 14 is 14.1. The discrepancy is principally due to the strong edge effect of the disk.

The numerical results shown so far are for axial incidence only. To illustrate the use of the computer program for nonaxial incidence, we have computed the back-scattering cross section vs. the angle of incidence for a prolate spheroid 2.833λ long and 1.417λ overall diameter. The results are shown in Fig. 15. It is interesting to note that for perpendicular polarization the back-scattering cross section exhibits oscillations. These oscillations can be ascribed to interference between the specularly reflected wave and creeping waves and show up much more strongly for perpendicular polarization than for parallel polarization because the decay exponent for the creeping waves is much smaller for perpendicular polarization than it is for parallel polarization.

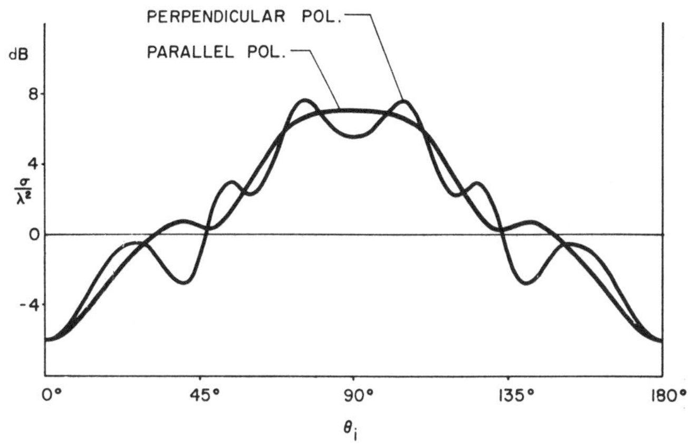

Fig. 15. Back-scattering cross section of prolate spheroid vs. angle of incidence.

The bistatic scattering cross section for the spheroid was computed for several directions. The cross sections for any two directions were found to agree by better than 0.15 dB.

VI. Conclusion

The integral equation method presented here permits the exact evaluation of the scattering of a plane electromagnetic wave by a metallic body of revolution, for any direction of incidence, any polarization and any longitudinal cross section of the body. An analytical solution of this general problem does not exist and there is little hope that such solution can be found. The solution obtained here has been made possible by extensive use of a fast digital computer.

Although the results discussed specifically in this paper pertain to scattering from perfectly conducting bodies, the methods described can be applied with relative ease, and equal effectiveness, to bodies having other boundary conditions, such as coating of lossy dielectric materials.

Acknowledgment

The author wishes to acknowledge many helpful discussions with his colleagues at TRG.

RADIATION AND SCATTERING FROM BODIES OF REVOLUTION

J. R. MAUTZ and R. F. HARRINGTON

Syracuse University,
Syracuse (N.Y.), U.S.A.

Abstract

The problem of electromagnetic radiation and scattering from perfectly conducting bodies of revolution of arbitrary shape is considered. The mathematical formulation is an integro-differential equation, obtained from the potential integrals plus boundary conditions at the body. A solution is effected by the method of moments, and the results are expressed in terms of generalized network parameters. The expansion functions chosen for the solution are harmonic in ϕ (azimuth angle) and subsectional in t (contour length variable). Because of rotational symmetry, the solution becomes a Fourier series in ϕ, each term of which is uncoupled to every other term.

Illustrative computations are given for radiation from apertures and plane wave scattering from bodies of revolution. The impedance elements, currents, radiation patterns, and scattering patterns for a conducting sphere are computed both from the general solution and from the classical eigenfunction solution. The agreement obtained serves to check the general solution. Similar computations for a cone-sphere illustrate the application of the general solution to problems not solvable by classical methods.

§ 1. Introduction

Determination of the behavior of a conducting body in a known impressed field is a fundamental problem of applied electromagnetic theory. Specific examples are scattering by conducting objects and radiation from conducting antennas. A general procedure for treating such problems is given by the method of moments [1, 2]. This is a procedure for reducing the functional problem to a finite dimensional matrix problem. In electromagnetic theory, the resultant matrices can be identified as generalized network parameters [3]. Once the generalized impedance matrix of a body is known, the behavior of that body for arbitrary excitation is easily calculated by matrix manipulations. Furthermore, the effect of impedance loading can be accounted for by constraints on the matrix equations [2, 4, 5].

The problem is formulated in the conventional way as follows. Let $\boldsymbol{E}^{\mathrm{i}}$ denote the known impressed field and $\boldsymbol{E}^{\mathrm{s}}$ the scattered field due to currents on the body. Then the total field $\boldsymbol{E}$ is the sum of the impressed and scattered fields, that is,

$$\boldsymbol{E} = \boldsymbol{E}^{\mathrm{i}} + \boldsymbol{E}^{\mathrm{s}} \tag{1}$$

The scattered field can be expressed in terms of a vector potential $\boldsymbol{A}$ and a scalar potential Φ as

$$\boldsymbol{E}^{\mathrm{s}} = -\mathrm{j}\omega\boldsymbol{A} - \nabla\Phi \tag{2}$$

where

$$\boldsymbol{A} = \mu \iint_S \boldsymbol{J} \frac{\mathrm{e}^{-\mathrm{j}kR}}{4\pi R} \, \mathrm{d}s \tag{3}$$

$$\Phi = \frac{1}{\varepsilon} \iint_S \sigma \frac{\mathrm{e}^{-\mathrm{j}kR}}{4\pi R} \, \mathrm{d}s. \tag{4}$$

Here S is the surface of the conductor, R is the distance from a source point to the field point, $\boldsymbol{J}$ is the surface current on S, and σ is the surface charge on S. The current and the charge are related by the equation of continuity

$$\nabla \cdot \boldsymbol{J} = -\mathrm{j}\omega\sigma. \tag{5}$$

The boundary condition requires that the tangential component of total $\boldsymbol{E}$ vanish on S. Hence,

$$\boldsymbol{E}^{\mathrm{i}}_{\mathrm{tan}} = -\boldsymbol{E}^{\mathrm{s}}_{\mathrm{tan}} \tag{6}$$

where the subscript 'tan' denotes tangential component on S. The problem can now be stated succinctly as

$$\mathrm{L}(\boldsymbol{J}) = \boldsymbol{E}^{\mathrm{i}}_{\mathrm{tan}} \tag{7}$$

where L is the integro-differential operator

$$\mathrm{L}(\boldsymbol{J}) = [\mathrm{j}\omega\boldsymbol{A} + \nabla\Phi]_{\mathrm{tan}}. \tag{8}$$

A solution of (7) gives the current $\boldsymbol{J}$ on S. Usually we are interested in some functional of $\boldsymbol{J}$ which can be computed once $\boldsymbol{J}$ is known.

This report considers the special case of bodies which are rotationally symmetric about an axis. Because of this symmetry a Fourier series expansion in the angle of rotational symmetry reduces the problem to a system of independent modes. This is important from the standpoint of computation, because it is faster to invert several small matrices instead of one large one. The particular case of scattering by conducting bodies of revolution has been treated previously by Andreasen [6]. His solution is conceptually similar to ours, but differs in detail.

§ 2. Method of solution

To effect a solution of (7) we use the method of moments, which is closely related to Galerkin's method [1, 2]. This procedure approximates (7) by a matrix equation, which can then be inverted by known algorithms. The matrix so obtained is a generalized impedance matrix for the body [3]. The excitation of the body is represented by a voltage matrix, and the resultant current on the body is represented by a current matrix.

For the method of moments, let the inner product be defined as

$$\langle \boldsymbol{W}, \boldsymbol{J} \rangle = \iint_S \boldsymbol{W} \cdot \boldsymbol{J} \, \mathrm{d}s \tag{9}$$

where $\boldsymbol{W}$ and $\boldsymbol{J}$ are tangential vectors on S. A set of expansion functions $\{\boldsymbol{J}_j\}$ is next defined, and the current on S approximated by

$$\boldsymbol{J} = \sum_j I_j \boldsymbol{J}_j \tag{10}$$

where I_j are constants to be determined. Equation (10) is substituted into (7), which, because of the linearity of L, reduces to

$$\sum_j I_j \, \mathrm{L}\boldsymbol{J}_j = \boldsymbol{E}^{\mathrm{i}}_{\mathrm{tan}}. \tag{11}$$

A set of testing functions $\{\boldsymbol{W}_i\}$ is defined, and the inner product of (11) with each $\boldsymbol{W}_i$ is taken. The result is

$$\sum_j I_j \langle \boldsymbol{W}_i, \mathrm{L}\boldsymbol{J}_j \rangle = \langle \boldsymbol{W}_i, \boldsymbol{E}^{\mathrm{i}} \rangle \quad i = 1, 2, 3, \ldots \tag{12}$$

The subscript 'tan' has been dropped from $\boldsymbol{E}^{\mathrm{i}}$ because the inner product involves only tangential components. We now define the generalized network matrices

$$[Z] = [\langle \boldsymbol{W}_i, \mathrm{L}\boldsymbol{J}_j \rangle] \tag{13}$$

$$[V] = [\langle \boldsymbol{W}_i, \boldsymbol{E}^{\mathrm{i}} \rangle] \tag{14}$$

$$[I] = [I_i] \tag{15}$$

and rewrite the set (12) as

$$[Z][I] = [V], \tag{16}$$

$[Z]$ is the generalized impedance matrix and $[Y] = [Z]^{-1}$ is the generalized admittance matrix. The inverse of (16)

$$[I] = [Y][V] \tag{17}$$

gives the coefficients I_j of the current expansion (10), and hence is an approximate solution to the problem.

The impedance elements of (13) are explicitly

$$Z_{ij} = \iint_S \boldsymbol{W}_i \cdot (\mathrm{j}\omega \boldsymbol{A}_j + \nabla \Phi_j)\, \mathrm{d}s \tag{18}$$

where we have used (8) and (9). The subscript j denotes that $\boldsymbol{A}_j$ and Φ_j are the potentials due to $\boldsymbol{J}_j$ and σ_j. If the two-dimensional divergence theorem is applied to the vector $\Phi \boldsymbol{W}_i$ on a closed surface, the following identity results:

$$\iint_S \nabla \Phi \cdot \boldsymbol{W}_i \, \mathrm{d}s = - \iint_S \Phi \nabla \cdot \boldsymbol{W}_i \, \mathrm{d}s. \tag{19}$$

If $\boldsymbol{W}$ is thought of as a current, the charge associated with it is

$$\sigma_i = \frac{-1}{\mathrm{j}\omega} \nabla \cdot \boldsymbol{W}_i. \tag{20}$$

Now (18) can be written as

$$Z_{ij} = \mathrm{j}\omega \iint_S (\boldsymbol{W}_i \cdot \boldsymbol{A}_j + \sigma_i \Phi_j)\, \mathrm{d}s. \tag{21}$$

Equation (21) is more convenient for computation than (18) because the gradient operation on Φ has been eliminated.

So far the discussion has been for an arbitrary conducting body. Henceforth we restrict the consideration to surfaces S generated by revolving a plane curve about the z-axis. The surface and coordinate system are shown in Fig. 1. Here ρ, ϕ, z are the usual cylindrical coordinate variables, and t is a length variable along the curve generating S. We desire the expansion (10) to be general enough to approximate an arbitrary $\boldsymbol{J}$ on S. Hence, independent sets of functions are defined as

$$\begin{aligned} \boldsymbol{J}^t_{mj} &= \boldsymbol{u}_t f_j(t)\, \mathrm{e}^{\mathrm{j}m\phi} \\ \boldsymbol{J}^\phi_{mj} &= \boldsymbol{u}_\phi f_j(t)\, \mathrm{e}^{\mathrm{j}m\phi} \end{aligned} \tag{22}$$

where $\boldsymbol{u}_t$ and $\boldsymbol{u}_\phi$ are t-directed and ϕ-directed unit vectors. We have chosen the f_j in both sets to be the same, but it is not necessary to do so. The current expansion (10) now becomes

$$\boldsymbol{J} = \sum_{m,j} (I^t_{mj}\boldsymbol{J}^t_{mj} + I^\phi_{mj}\boldsymbol{J}^\phi_{mj}). \tag{23}$$

For testing functions, we choose

$$\begin{aligned} \boldsymbol{W}^t_{ni} &= \boldsymbol{u}_t f_i(t)\, \mathrm{e}^{-jn\phi} \\ \boldsymbol{W}^\phi_{ni} &= \boldsymbol{u}_\phi f_i(t)\, \mathrm{e}^{-jn\phi} \end{aligned} \tag{24}$$

which differ from (22) only in the sign of the exponent. The $\boldsymbol{W}_n$ are orthogonal to $\boldsymbol{J}_m$, $m \neq n$, over 0 to 2π on ϕ, and also to $\mathrm{L}\boldsymbol{J}_m$ (the field from $\boldsymbol{J}_m$). Hence, all impedance elements are zero except those for which $m = n$, and each mode can be treated separately. This is the major simplification introduced by the rotational symmetry of the body.

The use of (22) and (24) to evaluate the elements of (21) results in the partitioned matrix equation

$$\begin{bmatrix} [Z^{tt}_n] & [Z^{t\phi}_n] \\ [Z^{\phi t}_n] & [Z^{\phi\phi}_n] \end{bmatrix} \begin{bmatrix} [I^t_n] \\ [I^\phi_n] \end{bmatrix} = \begin{bmatrix} [V^t_n] \\ [V^\phi_n] \end{bmatrix}. \tag{25}$$

Here the elements of the submatrices Z are

$$\begin{aligned} (Z^{tt}_n)_{ij} &= \langle \boldsymbol{W}^t_{ni}, \mathrm{L}\boldsymbol{J}^t_{nj} \rangle \\ (Z^{t\phi}_n)_{ij} &= \langle \boldsymbol{W}^t_{ni}, \mathrm{L}\boldsymbol{J}^\phi_{nj} \rangle \\ (Z^{\phi t}_n)_{ij} &= \langle \boldsymbol{W}^\phi_{ni}, \mathrm{L}\boldsymbol{J}^t_{nj} \rangle \\ (Z^{\phi\phi}_n)_{ij} &= \langle \boldsymbol{W}^\phi_{ni}, \mathrm{L}\boldsymbol{J}^\phi_{nj} \rangle, \end{aligned} \tag{26}$$

the elements of the submatrices V are

$$\begin{aligned} (V^t_n)_i &= \langle \boldsymbol{W}^t_{ni}, \boldsymbol{E}^{\mathrm{i}} \rangle \\ (V^\phi_n)_i &= \langle \boldsymbol{W}^\phi_{ni}, \boldsymbol{E}^{\mathrm{i}} \rangle \end{aligned} \tag{27}$$

and the elements of the submatrices I are the coefficients in (23). We note that, for N terms in the Fourier series on ϕ, there are N sets of matrix equations (25).

The solution to (25) can also be written in partitioned form as

$$\begin{bmatrix} [I^t_n] \\ [I^\phi_n] \end{bmatrix} = \begin{bmatrix} [Y^{tt}_n] & [Y^{t\phi}_n] \\ [Y^{\phi t}_n] & [Y^{\phi\phi}_n] \end{bmatrix} \begin{bmatrix} [V^t_n] \\ [V^\phi_n] \end{bmatrix}. \tag{28}$$

The submatrices Y must in general be obtained after inversion of the entire Z matrix, and are not the inverses of corresponding Z submatrices. However, the $-n$ mode matrices are related to the $+n$ mode matrices by

$$\begin{bmatrix} [Y^{tt}_{-n}] & [Y^{t\phi}_{-n}] \\ [Y^{\phi t}_{-n}] & [Y^{\phi\phi}_{-n}] \end{bmatrix} = \begin{bmatrix} [Y^{tt}_{n}] & [-Y^{t\phi}_{n}] \\ [-Y^{\phi t}_{n}] & [Y^{\phi\phi}_{n}] \end{bmatrix}. \tag{29}$$

Hence, only the $n \geq 0$ mode matrices need be inverted. The proof of (29) follows from the fact that the Z matrices satisfy the same equation (see next section), and this symmetry survives matrix inversion.

Finally, for an explicit solution we must choose the $f_i(t)$, the t expansion functions. It is known that subsectional expansions, using pulse functions or triangle functions, give well conditioned matrices [2]. The current must be differentiated to obtain the charge; hence it is preferable to use triangle functions. Furthermore, a triangle expansion gives a piecewise-linear approximation which converges about twice as fast as a step approximation [1, 2]. Finally, if $\rho \boldsymbol{J}$ is expanded in triangles instead of $\boldsymbol{J}$, the divergence formula and the treatment of the end points of t become simpler. Hence, we choose

$$f_i(t) = \frac{1}{\rho} T(t - t_i) \tag{30}$$

where T is the triangle function

$$T(t) = \begin{cases} 1 - |t|, & |t| < 1. \\ 0 & |t| > 1 \end{cases} \tag{31}$$

When using these functions, distance and frequency are scaled so that the t_i are one unit apart. As in Fig. 1, t is zero at the lower pole and N at the upper pole. There are $N - 1$ triangle functions with peaks at $1, 2, 3, \ldots, N - 1$.

Numerical evaluation of the impedance elements is quite difficult because of the complexity of the formulas. An approximate evaluation was obtained by approximating each triangle function by four steps. The potential integrals due to each step are evaluated at the center of all steps by a numerical integration. The second integration, represented by (18), is then approximated by the value of the integrand at the center of each step, summed over the four

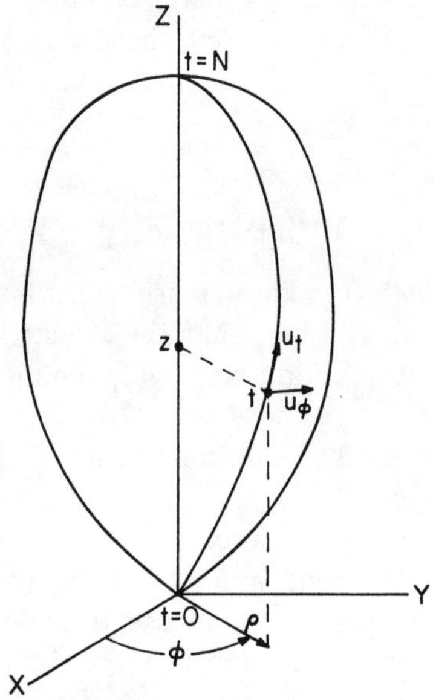

Fig. 1. Body of revolution and coordinate system.

step approximation to the testing triangle. The details of the solution are given in the next section, and the computer program and instructions for using it are given in a research report [7].

§ 3. Evaluation of the impedances

The generalized impedances for a body are given by (21), which can be written in greater detail as

$$Z_{ij} = \iint_S \mathrm{d}s' \iint_S \mathrm{d}s \left[\mathrm{j}\omega\mu \boldsymbol{W}_i \cdot \boldsymbol{J}_j + \frac{1}{\mathrm{j}\omega\varepsilon} (\boldsymbol{\nabla}' \cdot \boldsymbol{W}_i)(\boldsymbol{\nabla} \cdot \boldsymbol{J}_j) \right] \frac{\mathrm{e}^{-\mathrm{j}kR}}{4\pi R} . \tag{32}$$

This is valid for bodies of arbitrary shape. For bodies of revolution,

$$\iint_S \mathrm{d}s = \int_0^N \mathrm{d}t \int_0^{2\pi} \mathrm{d}\phi \, \rho(t) \tag{33}$$

$$\boldsymbol{\nabla} \cdot \boldsymbol{J} = \frac{1}{\rho} \frac{\partial}{\partial t} (\rho J_t) + \frac{1}{\rho} \frac{\partial}{\partial \phi} (J_\phi) \tag{34}$$

$$R = \{\rho^2 + \rho'^2 - 2\rho\rho' \cos(\phi - \phi') + (z - z')^2\}^{\frac{1}{2}}. \tag{35}$$

Four types of impedances are defined by (26). To evaluate them, we use (22) and (24) to obtain the $\boldsymbol{W} \cdot \boldsymbol{J}$ terms in (32) as

$$W^{p}_{ni} \cdot J^{q}_{nj} = e^{jn(\phi - \phi')} f_i(t')\, f_j(t)\, u'_p \cdot u_q, \tag{36}$$

where p and q represent permutations of t and ϕ. The unit vector dot products, in terms of the body coordinates defined by Fig. 1, are

$$\begin{aligned} u'_t \cdot u_t &= \sin v \sin v' \cos(\phi - \phi') + \cos v \cos v' \\ u'_t \cdot u_\phi &= -\sin v' \sin(\phi - \phi') \\ u'_\phi \cdot u_t &= \sin v \sin(\phi - \phi') \\ u'_\phi \cdot u_\phi &= \cos(\phi - \phi'). \end{aligned} \tag{37}$$

Here v is the angle between the t direction and the z-axis, being positive if u_t points away from the z-axis and negative if u_t points toward it. Changing $\phi - \phi'$ to a new variable, and expressing the sine and cosine terms of (37) as exponentials, one ϕ integration in (32) can be performed. The remaining ϕ integration defines the Green function [6]

$$g_n = \int_0^\pi \frac{e^{-jkR_0}}{R_0} \cos n\phi \, d\phi, \tag{38}$$

where R_0 is given by (35) with $\phi' = 0$. With f_i given by (30), the resultant expressions for the impedance elements (26) are

$$\begin{aligned} (Z^{tt}_n)_{ij} &= \int_0^N dt' \int_0^N dt \Big[j\omega\mu T(t'-i)\, T(t-j) \Big(\sin v \sin v' \frac{g_{n+1} + g_{n-1}}{2} + \\ &+ \cos v \cos v'\, g_n \Big) + \frac{1}{j\omega\varepsilon} T'(t'-i)\, T'(t-j) g_n \Big] \\ (Z^{t\phi}_n)_{ij} &= \int_0^N dt' \int_0^N dt \Big[-\omega\mu T(t'-i)\, T(t-j) \sin v' \frac{g_{n+1} - g_{n-1}}{2} + \\ &+ \frac{n}{\omega\varepsilon\rho} T'(t'-i)\, T(t-j)\, g_n \Big] \\ (Z^{\phi t}_n)_{ij} &= \int_0^N dt' \int_0^N dt \Big[+\omega\mu T(t'-i)\, T(t-j) \sin v \frac{g_{n+1} - g_{n-1}}{2} - \\ &- \frac{n}{\omega\varepsilon\rho'} T(t'-i)\, T'(t-j)\, g_n \Big] \end{aligned} \tag{39}$$

$$(Z_n^{\phi\phi})_{ij} = \int_0^N \mathrm{d}t' \int_0^N \mathrm{d}t \Big[\mathrm{j}\omega\mu T(t'-i)\, T(t-j)\, \frac{g_{n+1}+g_{n-1}}{2} +$$

$$+ \frac{n^2}{\mathrm{j}\omega\varepsilon\rho\rho'} T(t'-i)\, T(t-j)\, g_n \Big].$$

Here T' is the derivative of a triangle function

$$T'(t) = \begin{cases} 1, & -1 < t < 0 \\ -1, & 0 < t < 1 \\ 0, & |t| > 1 \end{cases} \tag{40}$$

which is a pulse doublet. Further evaluation of (39) is done by numerical methods.

The integrations of (39) involve many different integrands, and to reduce the number of integrations the following approximations are made. For the t integration, the T function is approximated by four pulses of amplitudes $\frac{1}{4}, \frac{3}{4}, \frac{3}{4}, \frac{1}{4}$, as shown in Fig. 2(a). The derivative of T is represented exactly by four pulses of amplitude $1, 1, -1, -1$, as shown in Fig. 2(b). The functions ρ, $\sin v$, and $\cos v$ are assumed constant over each pulse, equal to their values at the midpoint of the pulse. For the t' integration, the T function is approximated by four impulse functions, of strengths $\frac{1}{8}, \frac{3}{8}, \frac{3}{8}, \frac{1}{8}$, as shown in Fig. 2(a). The derivative of T is approximated by four impulse functions of strengths $\frac{1}{2}, \frac{1}{2}, -\frac{1}{2}, -\frac{1}{2}$, as shown in Fig. 2(b). We now define the numbers

$$\begin{aligned} T_1 &= \tfrac{1}{8} & T_1' &= \tfrac{1}{2} \\ T_2 &= \tfrac{3}{8} & T_2' &= \tfrac{1}{2} \\ T_3 &= \tfrac{3}{8} & T_3' &= -\tfrac{1}{2} \\ T_4 &= \tfrac{1}{8} & T_4' &= -\tfrac{1}{2}, \end{aligned} \tag{41}$$

the midpoints of the pulses

$$t_p = i + \frac{p-2.5}{2}, \quad t_q = j + \frac{q-2.5}{2} \tag{42}$$

and the pulse Green functions

$$G_n = 2 \int_{j+(q-3)/2}^{j+(q-2)/2} \mathrm{d}t \int_0^{\pi} \mathrm{d}\phi \frac{\mathrm{e}^{-\mathrm{j}kR_p}}{R_p} \cos n\phi \tag{43}$$

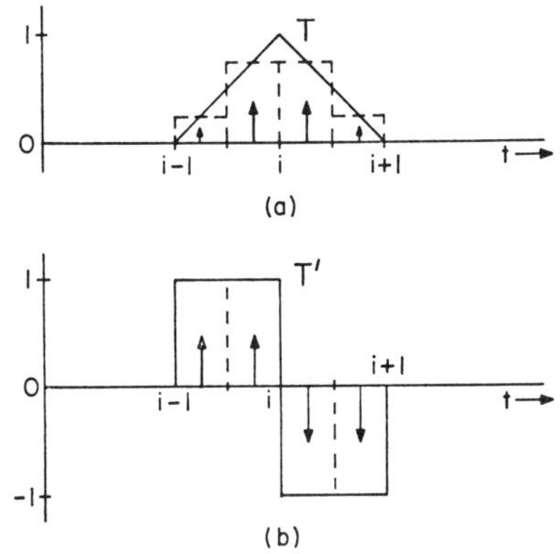

Fig. 2. (a) Triangle function (solid), four pulse approximation (dashed), impulse approximation (arrows).
(b) Derivative of triangle function (solid), four pulse representation (dashed), impulse approximation (arrows).

$$R_p = \{\rho^2 + \rho_p^2 - 2\rho\rho_p \cos\phi + (z - z_p)^2\}^{\frac{1}{2}}. \tag{44}$$

In terms of these definitions and approximations, the matrix elements (39) reduce to

$$(Z_n^{tt})_{ij} = \sum_{p=1}^{4} \sum_{q=1}^{4} \left[j\omega\mu T_p T_q \left(\sin v_p \sin v_q \frac{G_{n+1} + G_{n-1}}{2} + \right.\right.$$
$$\left.\left. + \cos v_p \cos v_q G_n \right) + \frac{1}{j\omega\varepsilon} T'_p T'_q G_n \right]$$

$$(Z_n^{t\phi})_{ij} = \sum_{p=1}^{4} \sum_{q=1}^{4} \left[-\omega\mu T_p T_q \sin v_p \frac{G_{n+1} - G_{n-1}}{2} + \frac{n}{\omega\varepsilon} T'_p \frac{T_q}{\rho_q} G_n \right]$$

(45)

$$(Z_n^{\phi t})_{ij} = \sum_{p=1}^{4} \sum_{q=1}^{4} \left[+ \omega\mu T_p T_q \sin v_q \frac{G_{n+1} - G_{n-1}}{2} - \frac{n}{\omega\varepsilon} \frac{T_p}{\rho_p} T'_q G_n \right]$$

$$(Z_n^{\phi\phi})_{ij} = \sum_{p=1}^{4} \sum_{q=1}^{4} \left[j\omega\mu T_p T_q \frac{G_{n+1} + G_{n-1}}{2} + \frac{n^2}{j\omega\varepsilon} \frac{T_p}{\rho_p} \frac{T_q}{\rho_q} G_n \right].$$

Here ρ_p, v_p, ρ_q, v_q are the ρ and v evaluated at t_p and t_q, respectively.

Finally, we evaluate the G_n of (43) by a combination of analytical and numerical techniques as follows. The interval $0 < \phi < \pi$ is divided into M equal intervals, and (43) approximated by

$$G_n = \frac{2\pi}{M} \sum_{m=1}^{M} \cos n\phi_m \int_{j+(q-3)/2}^{j+(q-2)/2} \mathrm{d}t \frac{\mathrm{e}^{-\mathrm{j}kR_p}}{R_p} \tag{46}$$

where $\phi_m = (m - \frac{1}{2})\pi/M$. The R_p of (46) is evaluated at $\phi = \phi_m$. Define

$$R_{pq} = \{\rho_p^2 + \rho_q^2 - 2\rho_p\rho_q \cos\phi_m + (z_p - z_q)^2\}^{\frac{1}{2}}. \tag{47}$$

Then $|R_p - R_{pq}| \leq \frac{1}{4}$. If the distance between peaks of the triangle functions is a tenth of a wavelength, then $k|R_p - R_{pq}| \leq \pi/20$, and $\exp(-\mathrm{j}kR_p)$ may be represented by two terms of its Taylor expansion about R_{pq}. This gives

$$G_n = \frac{2\pi}{M} \sum_{m=1}^{M} \cos n\phi_m \, \mathrm{e}^{-\mathrm{j}kR_{pq}} \int_{j+(q-3)/2}^{j+(q-2)/2} \mathrm{d}t \frac{1 - \mathrm{j}k(R_p - R_{pq})}{R_p}. \tag{48}$$

For the final integration, we approximate t by a straight line between $t = j + (q-3)/2$ and $t = j + (q-2)/2$. Then z and ρ are linear in t, and the integrand of (48) can be simplified by completing the square. This gives

$$G_n = \frac{2\pi}{M} \sum_{m=1}^{M} \cos n\phi_m \, \mathrm{e}^{-\mathrm{j}kR_{pq}} \int_{t_0-\frac{1}{4}}^{t_0+\frac{1}{4}} \mathrm{d}t \frac{1 - \mathrm{j}k\{(t^2 + d^2)^{\frac{1}{2}} - R_{pq}\}}{(t^2 + d^2)^{\frac{1}{2}}} \tag{49}$$

where

$$t_0 = |(z_p - z_q)\cos v_q + (\rho_p \cos\phi_m - \rho_q)\sin v_q| \tag{50}$$

$$d^2 = R_{pq}^2 - t_0^2. \tag{51}$$

The integration of (49) is now straightforward, and one obtains [8]

$$G_n = \frac{\pi}{M} \sum_{m=1}^{M} \cos n\phi_m \, f(\phi_m) \tag{52}$$

where

$$f(\phi_m) = \mathrm{e}^{-\mathrm{j}kR_{pq}} \left[2(1 + \mathrm{j}kR_{pq}) \ln \frac{t_2 + (t_2^2 + d^2)^{\frac{1}{2}}}{t_1 + (t_1^2 + d^2)^{\frac{1}{2}}} - \mathrm{j}k \right] \tag{53}$$

and

$$t_1 = t_0 - \tfrac{1}{4} \qquad t_2 = t_0 + \tfrac{1}{4}. \tag{54}$$

This completes the evaluation of the impedances.

Some computational details will now be discussed. In (54), if t_0 is larger than $\frac{1}{4}$ both t_1 and t_2 are positive, but if t_0 is less than $\frac{1}{4}$,

t_1 is negative. In this case the ln term of (53) should be replaced by

$$\ln \frac{\{t_2 + (t_2^2 + d^2)^{\frac{1}{2}}\}\{-t_1 + (t_1^2 + d^2)^{\frac{1}{2}}\}}{d^2}. \qquad (55)$$

The quantity G_n is the scalar potential due to a charge distribution on the lateral surface of a frustum. The approximation (46) supposes line charges at ϕ_m. If the source point is very close to the field point, the granular nature of the line charges may be too restrictive. When the source and field points coincide on the first interval $0 \leq \phi \leq \pi/M$, the line charge at $\phi = \pi/2M$ is spread out into a uniform surface charge. Furthermore, if the surface of this first interval is approximated by a plane rectangle of dimension $\frac{1}{2}$ by $\rho_q\pi/M$,

$$G_n = \frac{4}{\rho_q} \int_0^{\frac{1}{4}} \mathrm{d}y \int_0^{\rho_q\pi/M} \mathrm{d}x \frac{1 - \mathrm{j}k(x^2 + y^2)^{\frac{1}{2}}}{(x^2 + y^2)^{\frac{1}{2}}} + \sum_{m=2}^{M} \cos n\phi_m \, f(\phi_m). \qquad (56)$$

Now, since

$$\frac{\partial^2}{\partial x \partial y} [x \ln \{y + (x^2 + y^2)^{\frac{1}{2}}\} + + y \ln \{x + (y^2 + x^2)^{\frac{1}{2}}\}] = (x^2 + y^2)^{-\frac{1}{2}}, \qquad (57)$$

one can reduce (56) to

$$G_n = \frac{4}{\rho_q} \left[x_0 \ln \left(\frac{y_0 + (y_0^2 + x_0^2)^{\frac{1}{2}}}{x_0} \right) + y_0 \ln \left(\frac{x_0 + (y_0^2 + x_0^2)^{\frac{1}{2}}}{y_0} \right) \right] \times \times \frac{-\mathrm{j}kx_0}{\rho_q} + \sum_{m=2}^{M} \cos n\phi_m \, f(\phi_m) \qquad (58)$$

where

$$x_0 = \frac{\rho_q \pi}{M}, \qquad y_0 = \tfrac{1}{4}. \qquad (59)$$

A more elaborate method involving a localized increase in the density of line charges was attempted but was abandoned when the results did not differ appreciably from (58).

To check the accuracy of the program, the impedance for current on a sphere has been evaluated by spherical mode expansions. The details of the solution are given in Mautz's thesis [9]. The final result for any mode with $\exp(\mathrm{j}m\phi)$ variation on a sphere of radius a is

$$
\begin{aligned}
(Z^{tt}_m)_{ij} &= \sum_{n=|m|}^{\infty} \frac{1}{D_{mn}} [\lambda_n^{\mathrm{TE}} b_{ni} b_{nj} + \lambda_n^{\mathrm{TM}} c_{ni} c_{nj}] \\
(Z^{\phi t}_m)_{ij} &= \mathrm{j} \sum_{n=|m|}^{\infty} \frac{1}{D_{mn}} [\lambda_n^{\mathrm{TE}} c_{ni} b_{nj} + \lambda_n^{\mathrm{TM}} b_{ni} c_{nj}] \\
(Z^{t\phi}_m)_{ij} &= -(Z^{\phi t}_m)_{ji} \\
(Z^{\phi\phi}_m)_{ij} &= \sum_{n=|m|}^{\infty} \frac{1}{D_{mn}} [\lambda_n^{\mathrm{TE}} c_{ni} c_{nj} + \lambda_n^{\mathrm{TM}} b_{ni} b_{nj}],
\end{aligned}
\tag{60}
$$

where

$$
D_{mn} = \frac{4\pi a^2 (n)(n+1)(n+m)!}{(2n+1)(n-m)!}. \tag{61}
$$

The eigenvalues are given in terms of careted spherical Bessel functions as

$$
\begin{aligned}
\lambda_n^{\mathrm{TE}} &= \eta \hat{J}_n(ka)\, \hat{H}_n^{(2)}(ka) \\
\lambda_n^{\mathrm{TM}} &= \eta \hat{J}_n'(ka)\, \hat{H}_n^{(2)\prime}(ka).
\end{aligned}
\tag{62}
$$

The b and c coefficients are

$$
b_{nj} = -2\pi a^2 m \int_0^{\pi} P_n^m(\cos\theta)\, f_j(t)\, \mathrm{d}\theta \tag{63}
$$

$$
c_{nj} = 2\pi a^2 \int_0^{\pi} \frac{\partial P_n^m(\cos\theta)}{\partial\theta} \sin\theta\, f_j(t)\, \mathrm{d}\theta,
$$

where $f_j(t)$ is the j-th expansion function as defined in general by (22) and in particular by (30). The P_n^m in (63) are the associated Legendre polynomials. The evaluation of (63) was done numerically.

Table I shows representative computations for a sphere of radius $a = 0.2\lambda$, using 9 equispaced triangles. In other words, the expansion functions are (30) with the $t_i = 1, 2, \ldots, 9$, and the z-axis cuts the sphere at $t_i = 0$ and 10. The M appearing in (52) is twenty. The mode chosen is $n = 1$ in (45), and correspondingly, $m = 1$ in (60). Column 1 of Table I gives the computation using the general solution (45) of Section 3. Column 2 gives the solution using 18 terms of the infinite series (60). Column 3 uses 9 terms of the series (60). We notice that the real parts of Z^{tt} and $Z^{\phi\phi}$ and the imaginary part of $Z^{t\phi}$ are identical for the 18 term and the 9 term solution of (60) and are only slightly different from those of (45). The imaginary parts of Z^{tt} and $Z^{\phi\phi}$ and the real parts of

TABLE I

Impedance element	Formula (45) of Section 3	18 TE and TM modes in (60)	9 TE and TM modes in (60)
Z^{tt}_{11}	53.46 − j2477.	53.36 − j1908.	53.36 − j616.5
Z^{tt}_{55}	26.80 − j768.9	26.83 − j701.0	26.83 − j567.9
Z^{tt}_{15}	10.26 + j52.28	10.08 + j52.36	10.08 + j151.8
$Z^{\phi\phi}_{11}$	57.67 − j2641	57.36 − j2236.	57.36 − j1754
$Z^{\phi\phi}_{55}$	32.90 + j9.467	32.93 + j11.05	32.93 + j9.903
$Z^{\phi\phi}_{15}$	29.26 − j34.36	29.27 − j34.32	29.27 − j40.57
$Z^{t\phi}_{11}$	1569. − j54.88	1100. − j55.08	321.6 − j55.08
$Z^{t\phi}_{55}$	$(-1.68 + j0.343) \times 10^{-4}$	$(-2.12 + j0.405) \times 10^{-5}$	$(-2.48 + j0.405) \times 10^{-5}$
$Z^{t\phi}_{15}$	−37.15 − j27.17	−37.13 − j27.25	−27.87 − j27.25

Comparison of impedance elements on a sphere of radius 0.2λ, using 9 expansion triangles, $n = 1$ mode.

$Z^{\phi t}$ and $Z^{t\phi}$ of (60) are converging very slowly in the modal solution, but are converging towards the corresponding parts of the general solution; $Z^{t\phi}_{55}$ is supposed to be zero, and the nonzero values of $Z^{t\phi}_{55}$ in Table I are so small that all the significant figures are probably meaningless yet. On the basis of comparisons such as Table I, it appears that our general program is sufficiently accurate for most purposes.

§ 4. Measurement matrices

Any linear measurement of the field from a current $\boldsymbol{J}$ on a body S can be expressed as a linear functional of $\boldsymbol{J}$, that is

$$\text{measurement} = \iint_S \boldsymbol{E}^r \cdot \boldsymbol{J} \, \mathrm{d}s \tag{64}$$

where $\boldsymbol{E}^r$ is a known function. If the excitation-measurement system is viewed as a two port system, it can be shown that $\boldsymbol{E}^r$ in (64) is the field on S when the measurement port is excited [3]. For a moment solution, the current is given by a superposition $\boldsymbol{J} = \sum I_j \boldsymbol{J}_j$, and (64) reduces to

$$\text{measurement} = [R][I] \tag{65}$$

where $[I]$ is the matrix (15) and $[R]$ is a measurement *row* matrix

$$[R] = [\langle \boldsymbol{J}_j, \boldsymbol{E}^r \rangle]. \tag{66}$$

We note the similarity of $[R]$ to the excitation matrix defined by (14). If the matrix solution (17) is now substituted into (65), one has

$$\text{measurement} = [R][Y][V]. \tag{67}$$

The symmetry of (67) with respect to $[R]$ and $[V]$ reflects the reciprocity theorem of electromagnetic theory.

For bodies of revolution, the expansion for $\boldsymbol{J}$ can be separated into t- and ϕ-directed components, according to (23). It is then convenient to partition $[R]$ into t and ϕ component terms as

$$\begin{aligned} (R_n^t)_i &= \langle \boldsymbol{J}_{ni}^t, \boldsymbol{E}^r \rangle \\ (R_n^\phi)_i &= \langle \boldsymbol{J}_{ni}^\phi, \boldsymbol{E}^r \rangle. \end{aligned} \tag{68}$$

The analogous partition for excitation $[V]$ is given by (27). Now one can rewrite (67) in partitioned form as

$$\text{measurement} = [[R_n^t] \ [R_n^\phi]] \begin{bmatrix} [Y_n^{tt}] & [Y_n^{t\phi}] \\ [Y_n^{\phi t}] & [Y_n^{\phi\phi}] \end{bmatrix} \begin{bmatrix} [V_n^t] \\ [V_n^\phi] \end{bmatrix}, \tag{69}$$

where the submatrices Y are obtained after the Z matrix is inverted, and are not the inverses of the corresponding Z submatrices.

An important special case is that of radiation field measurement. It has been shown that the radiation field from currents $\boldsymbol{J}$ on S is given by [10]

$$\boldsymbol{E} \cdot \boldsymbol{u} = \frac{-\mathrm{j}\omega\mu}{4\pi r} \, \mathrm{e}^{-\mathrm{j}kr} [R][I] \tag{70}$$

where the elements of $[R]$ are given by (66) with

$$\boldsymbol{E}^r = \boldsymbol{u} \, \mathrm{e}^{-\mathrm{j}\boldsymbol{k}\cdot\boldsymbol{r}}. \tag{71}$$

This is a unit plane wave with polarization vector $\boldsymbol{u}$ and propagation vector $\boldsymbol{k}$. An arbitrary plane wave is a superposition of two orthogonal components, say $\boldsymbol{E}_\theta$ and $\boldsymbol{E}_\phi$. Hence, one can treat the general case as two applications of (71), one for $\boldsymbol{u} = \boldsymbol{u}_\theta$ and the other for $\boldsymbol{u} = \boldsymbol{u}_\phi$. To distinguish between the two cases let

$$\begin{aligned} (R_n^{t\theta})_i &= \langle \boldsymbol{J}_{ni}^t, \boldsymbol{E}_\theta^r \rangle \\ (R_n^{\phi\theta})_i &= \langle \boldsymbol{J}_{ni}^\phi, \boldsymbol{E}_\theta^r \rangle \end{aligned} \tag{72}$$

for the θ-polarized case, and

$$(R_n^{t\phi})_i = \langle \boldsymbol{J}_{ni}^t, \boldsymbol{E}_\phi^r \rangle$$
$$(R_n^{\phi\phi})_i = \langle \boldsymbol{J}_{ni}^\phi, \boldsymbol{E}_\phi^r \rangle \tag{73}$$

for the ϕ-polarized case.

The excitation matrices can now be evaluated as follows. Let

$$\boldsymbol{E}_\theta^r = \boldsymbol{u}_\theta^r \mathrm{e}^{\,\mathrm{j}k(\rho \sin \theta_r \cos \phi + z \cos \theta_r)}, \tag{74}$$

where θ_r and $\phi_r = 0$ are the angles to the field point of measurement. The dot products required in (72) are given by

$$\boldsymbol{u}_t \cdot \boldsymbol{u}_\theta^r = \cos \theta_r \, \sin v \, \cos \phi - \sin \theta_r \, \cos v$$
$$\boldsymbol{u}_\phi \cdot \boldsymbol{u}_\theta^r = -\cos \theta_r \, \sin \phi. \tag{75}$$

Using the integral formula for Bessel functions

$$J_n(\rho) = \frac{\mathrm{j}^n}{2\pi} \int_0^{2\pi} \mathrm{e}^{-\mathrm{j}\rho \cos \phi} \, \mathrm{e}^{-\mathrm{j}n\phi} \, \mathrm{d}\phi \tag{76}$$

one can now evaluate the ϕ integrations in (72), obtaining

$$(R_n^{t\theta})_i = 2\pi \mathrm{j}^{n+1} \int_0^N \rho f_i(t) \, \mathrm{e}^{\mathrm{j}kz \cos \theta_r} \times$$
$$\times \left[\cos \theta_r \sin v \, \frac{J_{n+1} - J_{n-1}}{2} + \mathrm{j} \sin \theta_r \cos v \, J_n \right] \mathrm{d}t \tag{77}$$
$$(R_n^{\phi\theta})_i = -2\pi \mathrm{j}^{n+1} \int_0^N \rho f_i(t) \, \mathrm{e}^{\mathrm{j}kz \cos \theta_r} \cos \theta_r \, \frac{J_{n+1} + J_{n-1}}{2\mathrm{j}} \, \mathrm{d}t,$$

where

$$J_n = J_n(k\rho \sin \theta_r). \tag{78}$$

Similarly, to evaluate (73) let

$$\boldsymbol{E}_\phi^r = \boldsymbol{u}_\phi^r \, \mathrm{e}^{\mathrm{j}k(\rho \sin \theta_r \cos \phi + z \cos \theta_r)}. \tag{79}$$

The required dot products are then

$$\boldsymbol{u}_t \cdot \boldsymbol{u}_\phi^r = \sin v \, \sin \phi$$
$$\boldsymbol{u}_\phi \cdot \boldsymbol{u}_\phi^r = \cos \phi \tag{80}$$

and the ϕ integrations in (73) can be performed. The result is

$$(R_n^{t\phi})_i = 2\pi j^{n+1} \int_0^N \rho f_i(t)\, e^{jkz\cos\theta_r} \sin v\, \frac{J_{n+1} + J_{n-1}}{2j}\, dt$$

$$(R_n^{\phi\phi})_i = 2\pi j^{n+1} \int_0^N \rho f_i(t)\, e^{jkz\cos\theta_r} \frac{J_{n+1} - J_{n-1}}{2}\, dt, \tag{81}$$

where (78) again applies. For computation, the $\rho f_i(t)$ in (77) and (81) were the triangle functions (31). The remaining integrals must be evaluated numerically for bodies of arbitrary contour. In general, $R_n^{t\theta}$ and $R_n^{\phi\phi}$ are even in n, while $R_n^{\phi\theta}$ and $R_n^{t\phi}$ are odd in n. As shown by (22) and (24), the excitation matrix $[V]$ differs from the measurement matrix $[R]$ only by the sign of n. Hence, for plane wave excitation of the body,

$$(V_n^{pq})_i = (R_{-n}^{pq})_i, \tag{82}$$

where pq represents $t\theta$, $\phi\theta$, $t\phi$, or $\phi\phi$. Equation (82) means that the V_i are given by (77) and (81) with n replaced by $-n$ and θ_r by θ_t, and does not imply equality of excitation and measurement.

For axial incidence of plane waves, (77) and (81) reduce considerably. In this case $\theta_r = \pi$ and all $R_n = 0$ except

$$\begin{aligned}
(R_1^{t\theta})_i &= (R_{-1}^{t\theta})_i = -\pi \int_0^N \rho f_i(t)\, e^{-jkz} \sin v\, dt \\
(R_1^{\phi\theta})_i &= -(R_{-1}^{\phi\theta})_i = j\pi \int_0^N \rho f_i(t)\, e^{-jkz}\, dt \\
(R_1^{t\phi})_i &= -(R_{-1}^{t\phi})_i = -j(R_1^{t\theta})_i \\
(R_1^{\phi\phi})_i &= (R_{-1}^{\phi\phi})_i = -j(R_1^{\phi\theta})_i.
\end{aligned} \tag{83}$$

Hence, only the $n = -1$ and $n = +1$ modes are excited by axially incident plane waves.

§ 5. Aperture antennas

An aperture antenna is a radiating system consisting of a conducting body with apertures through which electromagnetic energy is supplied. For analysis purposes, it is assumed that the tangential electric field $\boldsymbol{E}_t$ is known over the aperture. It is, of course, zero over the conducting body. This known field $\boldsymbol{E}_t$ corresponds to $\boldsymbol{E}_t^{s} = -\boldsymbol{E}_t^{i}$ in the general analysis of Section 1. Fig. 3 represents

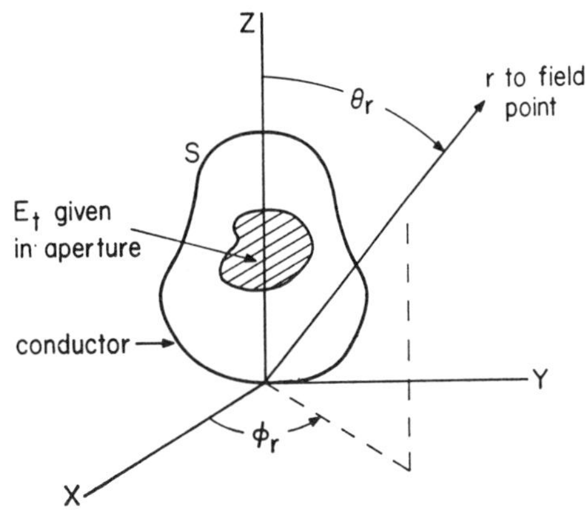

Fig. 3. An aperture in a conducting body of revolution.

the general problem, showing the body S, the aperture, and the radius vector to a distant field point.

In terms of a moment solution, the excitation is the known $\boldsymbol{E}_t$ in the aperture, from which the excitation matrix may be computed by (14) by setting $\boldsymbol{E}^{\mathrm{i}} = -\boldsymbol{E}_t$. In general, this results in an excitation of many $\exp(\mathrm{j}n\phi)$ modes, each of which can be treated separately. The current on S for each mode is found by inverting (25), and the total current is the sum of the modal currents, that is

$$\boldsymbol{J} = \sum_n [\boldsymbol{J}_n][Y_n][V_n], \tag{84}$$

where n denotes the mode. Similarly, the radiation field for each mode is found from (70), and the total radiation field is the sum of the modal fields, that is

$$E_u = \frac{-\mathrm{j}\omega\mu}{4\pi r} \mathrm{e}^{-\mathrm{j}kr} \sum_n [R_n][Y_n][V_n], \tag{85}$$

where $u = \theta$ or ϕ and n again denotes the mode.

For simplicity, illustrative computations have been made solely for excitation independent of ϕ, which excites the $n = 0$ mode only. This results in a further simplification in that $[Z_0^{t\phi}] = [Z_0^{\phi t}] = 0$, that is, there is no coupling between the t-directed currents and the ϕ-directed currents. From symmetry considerations, an excitation having only an E_t component produces a J_t current only, and an excitation having only an E_ϕ component produces a J_ϕ

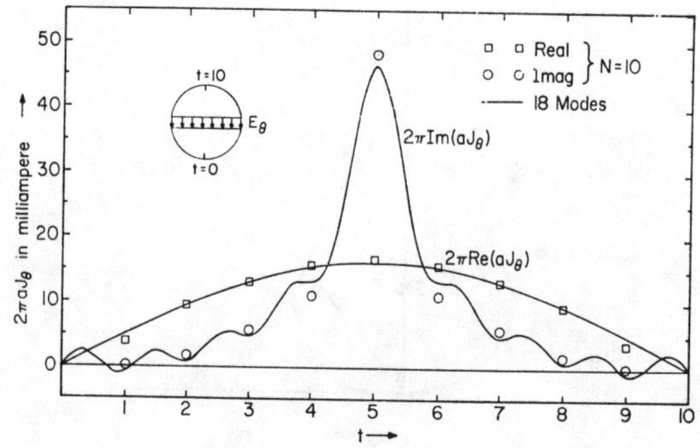

Fig. 4. Current density on a conducting sphere of radius 0.2λ, excited by 1 V across an equatorial slot.

current only. Hence, the problem reduces to two independent ones,

$$\begin{aligned} J_t &= [J_0^t][Y_0^{tt}][V_0^t] \\ J_\phi &= [J_0^\phi][Y_0^{\phi\phi}][V_0^\phi] \end{aligned} \tag{86}$$

for the current, and

$$\begin{aligned} E_\theta &= \frac{-j\omega\mu}{4\pi r}\, e^{-jkr}\, [R_0^{t\theta}][Y_0^{tt}][V_0^t] \\ E_\phi &= \frac{-j\omega\mu}{4\pi r}\, e^{-jkr}\, [R_0^{\phi\phi}][Y_0^{\phi\phi}][V_0^\phi] \end{aligned} \tag{87}$$

for the radiation field. The $[Y_0^{tt}]$ and $[Y_0^{\phi\phi}]$ in (86) and (87) are the inverses of $[Z_0^{tt}]$ and $[Z_0^{\phi\phi}]$, respectively.

To check the program, radiation from an equatorially slotted conducting sphere was computed both by the classical eigenfunction expansion and by the general program. Fig. 4 shows the current density on a sphere of radius $a = 0.2\lambda$ excited by an $E_\theta = \delta(\theta - \pi/2)/a$ in the equatorial slot. This gives rise to $V_{0j}^{t\theta} = 2\pi$ if t_j is on the equator, and zero otherwise. The eigenfunction solution was computed using 18 terms of the Fourier series. The real part of the current converged very rapidly, and can be considered as exact. The imaginary part of the current does not converge at the source, and converges slowly elsewhere, and hence the imaginary part computed from 18 terms of the Fourier series is not exact. As can be seen in Fig. 4, the $N = 10$ solution using the general program is in good agreement with the eigenfunction solution. Computations of radiation field patterns for the two solutions

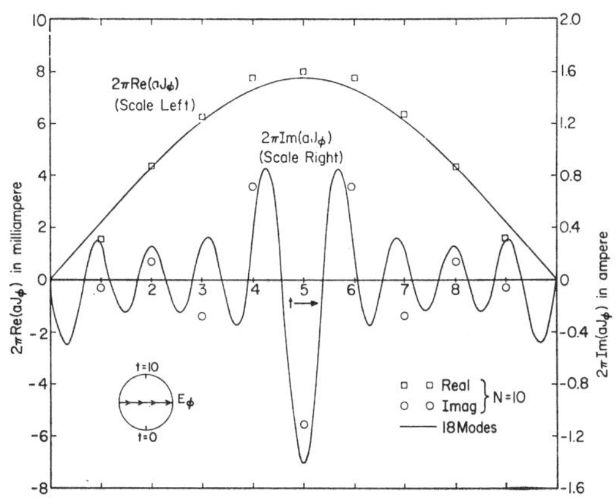

Fig. 5. Current density on a conducting sphere of radius 0.2λ, excited by an impulsive E_ϕ in an equatorial slot.

agree to within less than one percent accuracy, and hence are not shown.

Fig. 5 shows the current density on a sphere of radius $a = 0.2\lambda$ excited by an $E_\phi = -\delta(\theta - \pi/2)/a$ in the equatorial slot. This gives rise to $V_{0j}^{\phi\phi} = 2\pi$ when t_j is on the equator, and zero otherwise. Again the eigenfunction solution was computed using 18 terms of the Fourier series, and the real part of the current converged rapidly. However, the imaginary part of the current does not converge. The real part of the current computed from the $N = 10$ case of the general program again agrees well with the eigenfunction solution. It is difficult to say anything about the imaginary part because of the divergent nature of the solution. Computations of the radiation field patterns for the two solutions again agree to within a few percent. Evidently the radiation pattern does not depend strongly on the nonconverging imaginary part of the current.

As an example of a body of more complicated shape, computations were made for aperture excitation of a cone-sphere. This body is formed by wedging a conducting sphere into a conducting cone, resulting in a shape similar to that of an ice cream cone. The radius of the sphere is $a = 0.2\lambda$, and the cone half angle is 10 deg. The aperture is a narrow slot near the cone-to-sphere junction. Fig. 6 shows the current density when the excitation is an impulsive E_t

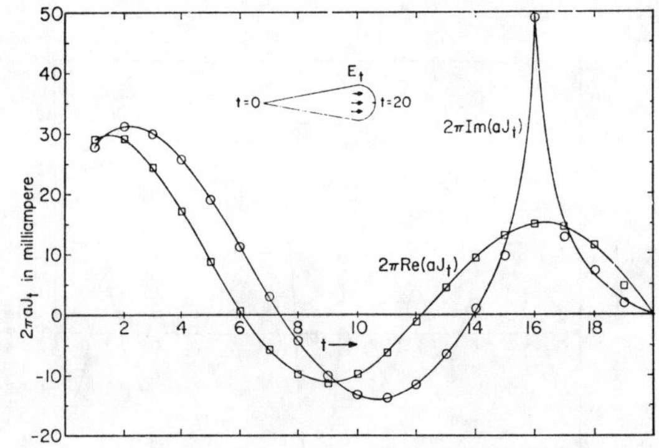

Fig. 6. Current density on a conducting cone-sphere excited by 1 V across a slot at the cone-to-sphere junction.

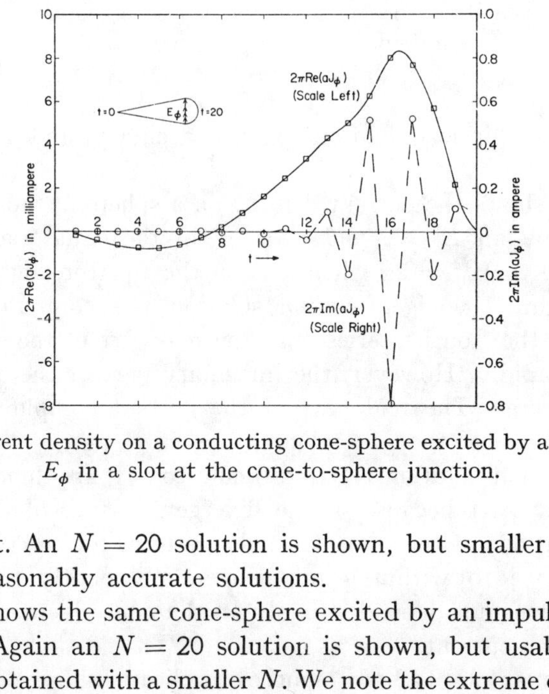

Fig. 7. Current density on a conducting cone-sphere excited by an impulsive E_ϕ in a slot at the cone-to-sphere junction.

in the slot. An $N = 20$ solution is shown, but smaller values of N give reasonably accurate solutions.

Fig. 7 shows the same cone-sphere excited by an impulsive E_ϕ in the slot. Again an $N = 20$ solution is shown, but usable results could be obtained with a smaller N. We note the extreme oscillatory nature of the imaginary part of the current. Theoretically, the imaginary part does not converge, as discussed previously for the sphere.

Figs. 8 and 9 show the power gain patterns for the two cone-sphere excitations discussed above. For a given polarization, the

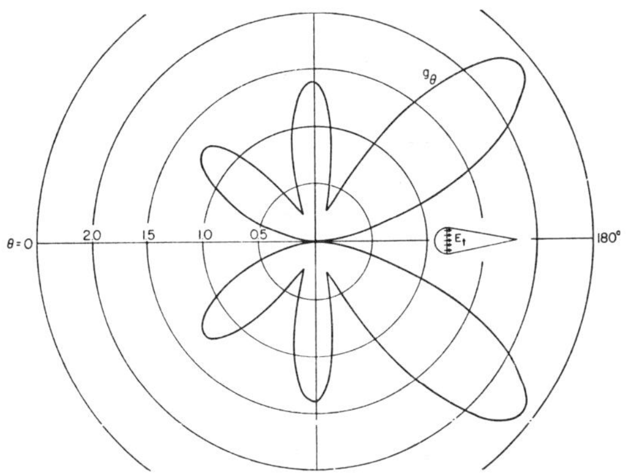

Fig. 8. Power gain pattern for radiation from a conducting cone-sphere excited by 1 V across a slot at the cone-to-sphere junction.

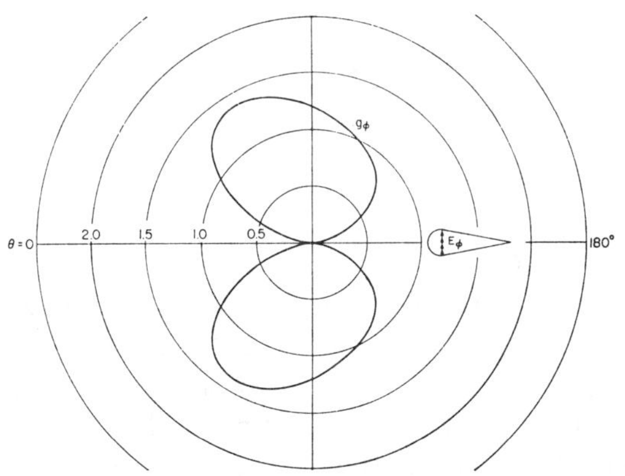

Fig. 9. Power gain pattern for radiation from a conducting cone-sphere excited by an impulsive E_θ in a slot at the cone-to-sphere junction.

power gain normalized to omnidirectional radiation is [11]

$$g_u = \frac{4\pi r^2 |E_u|^2}{\eta P_{\text{rad}}}, \tag{88}$$

where $\eta = (\mu/\varepsilon)^{\frac{1}{2}}$ and P_{rad} is the power radiated. From (70) and (17), the numerator of (88) can be expressed as

$$4\pi r^2 |E_u|^2 = \frac{1}{4\pi} |k\eta[R][Y][V]|^2. \tag{89}$$

In general, the power radiated is given by [12]

$$P_{\text{rad}} = -\operatorname{Re} \iint_S \boldsymbol{E}^* \cdot \boldsymbol{J} \, \mathrm{d}s = \operatorname{Re}[\hat{V}][Y][V], \tag{90}$$

where the elements of the row matrix $[\hat{V}]$ are

$$\hat{V}_i = -\iint_S \boldsymbol{E}^* \cdot \boldsymbol{J}_i \, \mathrm{d}s. \tag{91}$$

For the excitations used in Figs. 8 and 9, only one V_i is nonzero, and $V_i = \hat{V}_i = 2\pi$. Hence (90) reduces to a single element of the $[Y]$ matrix. In particular, for Fig. 8,

$$g_\theta = \frac{\eta k^2 \left| \sum_{j=1}^{N-1} (R_n^{t\theta})_j \, (Y_n^{tt})_{ji} \right|^2}{4\pi \operatorname{Re}(Y_n^{\theta\theta})_{ii}} \tag{92}$$

where t_i is the point of excitation. Similarly, for Fig. 9,

$$g_\phi = \frac{\eta k^2 \left| \sum_{j=1}^{N-1} (R_n^{\phi\phi})_j \, (Y_n^{\phi\phi})_{ji} \right|^2}{4\pi \operatorname{Re}(Y_n^{\phi\phi})_{ii}} \tag{93}$$

where again t_i is the point of excitation.

§ 6. Plane wave scattering

The radar scattering problem consists of a plane wave incident on a scattering body, plus measurement of the far zone scattered field. Fig. 10 illustrates the geometry of the problem for conducting bodies of revolution. In general, the incident wave $\boldsymbol{E}$ can be expressed as a superposition of the two orthogonal components $\boldsymbol{E}_\theta^{\mathrm{i}}$ and $\boldsymbol{E}_\phi^{\mathrm{i}}$, and similarly the far zone scattered wave $\boldsymbol{E}^{\mathrm{s}}$ as the superposition of $\boldsymbol{E}_\theta^{\mathrm{s}}$ and $\boldsymbol{E}_\phi^{\mathrm{s}}$. These are related by the scattering matrix $[s]$ of the

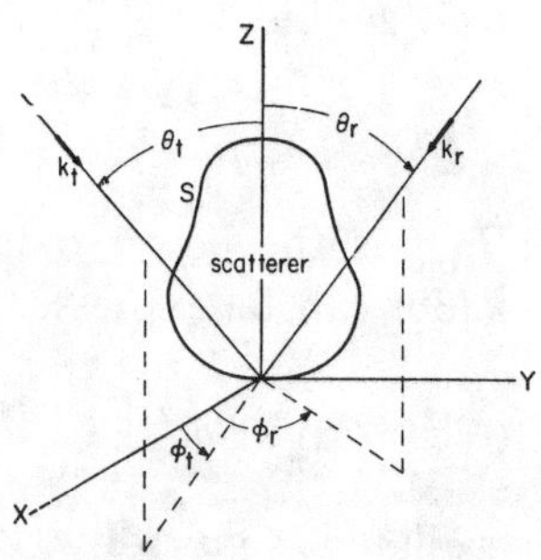

Fig. 10. Plane wave scattering by a conducting body of revolution.

body according to

$$\begin{bmatrix} E_\theta^s \\ E_\phi^s \end{bmatrix} = \frac{e^{-jkr}}{r} \begin{bmatrix} s^{\theta\theta} & s^{\theta\phi} \\ s^{\phi\theta} & s^{\phi\phi} \end{bmatrix} \begin{bmatrix} E_\theta^i \\ E_\phi^i \end{bmatrix}. \tag{94}$$

The elements of $[s]$ can be expressed as a summation over the modal components

$$s^{pq} = \sum_n s_n^{pq}, \tag{95}$$

where pq denotes $\theta\theta$, $\theta\phi$, $\phi\theta$, or $\phi\phi$. The scattered field is given by (70), and hence

$$s_n^{pq} = \frac{-j\omega\mu}{4\pi} [[R_n^{tp}]\ [R_n^{\phi p}]] \begin{bmatrix} [Y_n^{tt}] & [Y_n^{t\phi}] \\ [Y_n^{\phi t}] & [Y_n^{\phi\phi}] \end{bmatrix} \begin{bmatrix} [V_n^{tq}] \\ [V_n^{\phi q}] \end{bmatrix}, \tag{96}$$

where the elements R_n are given by (77) or (81), and the elements V_n by the same formulas with n replaced by $-n$ and θ_r by θ_t.

Radar scattering data are often presented in terms of the radar cross section, defined as

$$\sigma^{pq} = 4\pi r^2 \left| \frac{E_p^s}{E_q^i} \right|^2. \tag{97}$$

As indicated, σ depends both on the received polarization p and the incident polarization q. It follows from (94) and (97) that for a given polarization,

$$\sigma^{pq} = 4\pi |s^{pq}|^2. \tag{98}$$

An incident wave of arbitrary polarization can be expressed as the superposition

$$\boldsymbol{E}^i = (\boldsymbol{u}_\theta \alpha_\theta + \boldsymbol{u}_\phi \alpha_\phi) E^i \tag{99}$$

where $|\alpha_\theta|^2 + |\alpha_\phi|^2 = 1$. The matrix

$$[\alpha] = \begin{bmatrix} \alpha_\theta \\ \alpha_\phi \end{bmatrix} \tag{100}$$

is called the polarization matrix. Let $[\alpha^t]$ be the polarization matrix of the incident wave when the transmitting antenna is excited, and $[\alpha^r]$ be the transpose of the polarization matrix of the incident wave when the receiving antenna is excited. Then the component of scattered field that is polarization-matched to the receiver is

$$E^s = \frac{e^{-jkr}}{r} [\alpha^r][s][\alpha^t] E^i \tag{101}$$

and the radar cross section is

$$\sigma = 4\pi \, |[\alpha^{\mathrm{r}}][s][\alpha^{\mathrm{t}}]|^2, \tag{102}$$

where $[s]$ is given by (94). Equation (98) is a special case of (102).

For simplicity, only excitation by axially incident plane waves is considered for computation. Let

$$\boldsymbol{E}^{\mathrm{i}} = -\boldsymbol{u}_{\mathrm{x}} \, \mathrm{e}^{-\mathrm{j}kz} \tag{103}$$

which can be considered a θ-polarized plane wave from a transmitter at coordinates $\theta_t = \pi$, $\phi_t = 0$. The excitation matrix elements are thus given by the first two equations of (83) with R replaced by V, n by $-n$, and θ_r by θ_t. Only the $n = \pm 1$ modes are present in the excitation, giving rise to only the same two modes in the current. The symmetry relationships (29) and (82) can now be used to obtain the simpler form

$$\boldsymbol{J} = \boldsymbol{u}_t J^t \cos\phi + \boldsymbol{u}_\phi J^\phi \sin\phi \tag{104}$$

where

$$J^t = 2 \sum_{i=1}^{N-1} (I_1^t)_i \, f_i(\rho)$$
$$J^\phi = 2\mathrm{j} \sum_{i=1}^{N-1} (I_1^\phi)_i \, f_i(\rho) \tag{105}$$

and $f_i(\rho) = T(t - t_i)/\rho$. Hence, the current is completely specified by its $n = 1$ mode coefficients.

To check the program, the current on a sphere excited by the

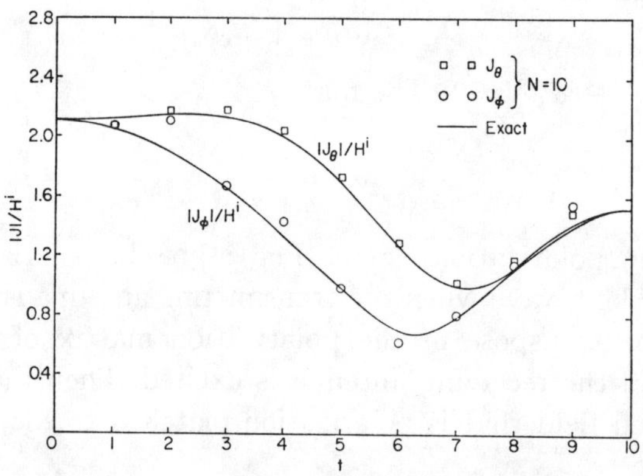

Fig. 11. Current density on a conducting sphere of radius 0.2λ excited by a uniform plane wave.

plane wave (103) was computed both by the classical eigenfunction method and by the general computer program. Fig. 11 shows the magnitudes of J^t and J^ϕ of (104) for a sphere of radius $a = 0.2\lambda$, using $N = 10$ for the general program. The eigenfunction expansion converges very rapidly and can be considered as exact. The current density is normalized with respect to the incident magnetic field intensity, giving a dimensionless quantity which is 2 for a plane conductor covering the $z = 0$ plane. The radar scattering patterns

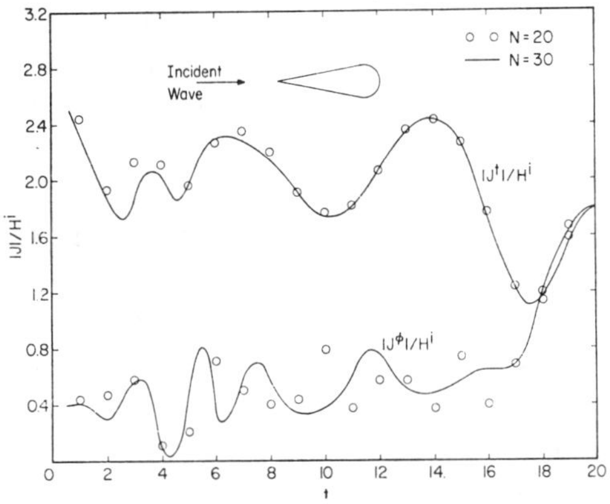

Fig. 12. Current density on a conducting cone-sphere (cone angle 20 deg, sphere radius 0.2λ) excited by a uniform plane wave axially incident on the tip.

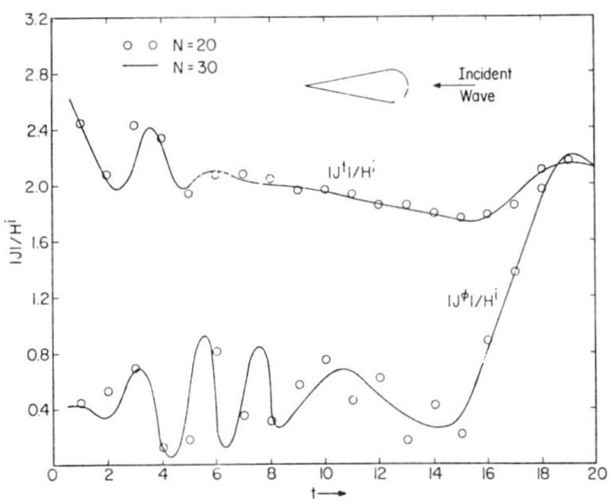

Fig. 13. Current density on a conducting cone-sphere (cone angle 20 deg, sphere radius 0.2λ) excited by a uniform plane wave axially incident on the sphere.

computed by the eigenfunction method and by the $N = 10$ general solution are almost identical, and are not shown.

Fig. 12 shows the current on a cone-sphere excited by a plane wave axially incident on the tip. The radius of the sphere is again 0.2λ, which is the same as used in the antenna examples. Again J^t and J^ϕ of (104), normalized with respect to H^i, are plotted. To illustrate convergence, both the $N = 20$ (circles) and the $N = 30$ (solid line) solutions are plotted. The J^t current converges fairly rapidly, but the J^ϕ current converges more slowly. Even the $N = 30$ solution cannot be considered accurate for J^ϕ. Apparently the rate of convergence is affected by the sharp tip of the cone. For better convergence one could include a term in the current expansion to properly represent the singularity at the tip. However, radar scattering patterns are insensitive to small oscillations in the currents, and more accurate solutions are not needed in most cases.

Fig. 13 shows the current on the same cone-sphere excited by a plane wave axially incident on the spherical end, that is, the end opposite to the point. The above comments concerning convergence apply also to this excitation.

Figs. 14 and 15 show radar cross section patterns for the cone-sphere excitations corresponding to Figs. 12 and 13, respectively. The plots were made using the $N = 20$ solution, but the corresponding $N = 30$ solution gave almost identical results. The only difference occurred in the vicinity of forward scattering where a change in σ

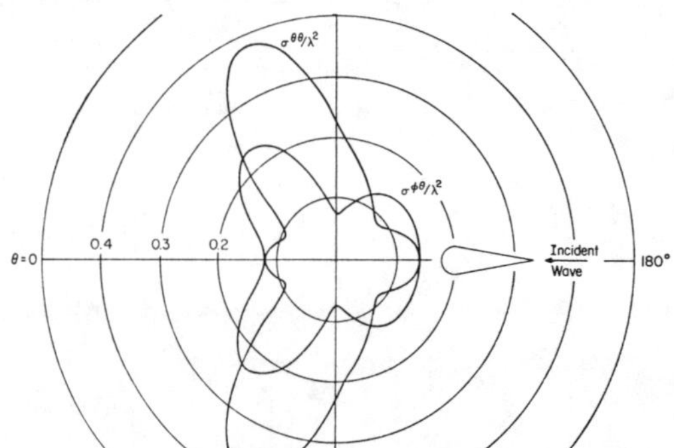

Fig. 14. Bistatic scattering cross section for conducting cone-sphere (cone angle 20 deg, sphere radius 0.2λ) excited by a uniform plane wave axially incident on the tip; $\sigma^{\theta\theta}$ is in the plane $\phi = 0$, and $\sigma^{\phi\theta}$ is in the plane $\phi = \pi/2$.

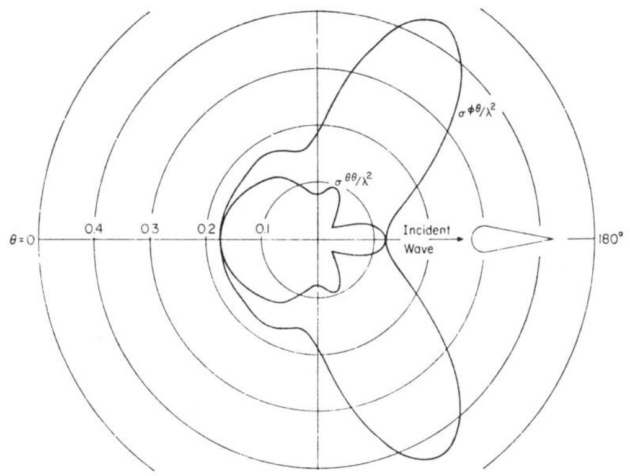

Fig. 15. Bistatic scattering cross section for a conducting cone-sphere (cone angle 20 deg, sphere radius 0.2λ) excited by a uniform plane wave axially incident on the sphere; $\sigma^{\theta\theta}$ is in the plane $\phi = 0$, and $\sigma^{\phi\theta}$ is in the plane $\phi = \pi/2$.

of the order of a few percent was obtained. This region is most sensitive to inaccuracies because all parts tend to radiate in phase in the forward direction. All plots are normalized to λ^2 to make them dependent only on length/wavelength, not on absolute dimensions. The curves labeled $\sigma^{\theta\theta}/\lambda^2$ are in the $\phi = 0$ plane, and are the radar cross sections measured by a θ-polarized receiver. The curves labelled $\sigma^{\phi\theta}/\lambda^2$ are in the $\phi = \pi/2$ plane, and are the radar cross sections measured by a ϕ-polarized receiver.

Because only the $n = 1$ and $n = -1$ modes are present in the excitation, these are the only modes present in the scattered field. From (95) and the symmetry relationships (29) and (82) one can show that

$$\begin{aligned} s^{\theta\theta} &= 2s_1^{\theta\theta} \cos \phi_r \\ s^{\phi\theta} &= 2\mathrm{j}\; s_1^{\phi\theta} \sin \phi_r, \end{aligned} \tag{106}$$

where $s_1^{\theta\theta}$ and $s_1^{\phi\theta}$ are the $n = 1$ modal solutions evaluated at $\phi_r = 0$. Hence, from (98) and (106) it follows that the radar cross sections are given by

$$\begin{aligned} \sigma^{\theta\theta} &= 16\pi\,|s_1^{\theta\theta}|^2 \cos^2 \phi_r \\ \sigma^{\phi\theta} &= 16\pi\,|s_1^{\phi\theta}|^2 \sin^2 \phi_r. \end{aligned} \tag{107}$$

Hence, the graphs of Figs. 14 and 15 are proportional to the coefficients of $\cos^2 \phi_r$ and $\sin^2 \phi_r$ in (107).

§ 7. Discussion

This report develops a solution for the electromagnetic behavior of conducting bodies of revolution with arbitrary excitation in terms of generalized network parameters [3]. Basically, this involves the application of a moment method to the superposition integral representation of the problem. Because of the rotational symmetry, a Fourier series expansion on ϕ is used. To facilitate the handling of arbitrary body contours, a subsectional expansion on t is used. The simplest subsectional function is the pulse function, but it is not in the domain of the original operator because derivatives must be taken. The derivative can be replaced by a finite difference approximation, thereby approximating the operator. However, it was found that difference approximations did not yield as good accuracy as desired. The next simplest subsectional function is the triangle function, and this was used for the general program. Its derivative is a doublet pulse, and hence no approximation of the operator is necessary.

Three types of singularities are encountered in the general solution: (a) a singularity in the coordinate system along the z-axis, (b) singularities of curvature in the body contour, such as at a cone tip, and (c) singularities in the excitation. The first type introduces difficulties in the numerical evaluation of the impedance elements, but can be overcome by careful analysis. The second type could be taken care of by special subroutines at points of discontinuity in body curvature, but would be difficult to implement. No attempt has been made to do this in the general program. The third type of singularity is unpractical to include in a general solution, because then the impedance matrix would depend on the excitation instead of being a function of the body geometry alone. It could, however, be incorporated into particular solutions. For example, the singularity in the current at the feed point of a slot-fed antenna can be accounted for by including the proper singularity in the expansion functions.

If only the radiation and scattering patterns are wanted, but not the current, then less accuracy is required in the solution. This is because such patterns are continuous linear functionals of the current, and, as shown by the calculus of variations, are accurate to the second order when the currents are accurate to the first order. For example, if the current has an error of the order of ten percent, the radiation field has an error of the order of only one percent. Furthermore, the radiation depends primarily on the current moment ρJ instead of the current itself. Hence, inaccuracies at the poles ($\rho = 0$) of the body have little effect on the radiation pattern. Finally, rapid oscillations in the current about the correct value also have little effect on the radiation field. They contribute mainly to energy stored in the near field.

The generalized network parameters of a body are basically a matrix approximation to the functional operator equation for the problem. Hence, as long as the computational approximations are valid, one can compute the response of the body to arbitrary electromagnetic excitation. Aperture antennas (Section 5) and plane wave scattering (Section 6) are but two examples. Another problem of interest is excitation by current dipoles in the vicinity of the body. Still more generally, both the excitation and the measurement can be taken in the near zone if desired. Radiation by loaded aperture antennas and scattering by loaded scattering antennas can also be treated by known methods [4, 13].

Other problems can be treated by modifications or extensions of the theory. For example, the internal resonances of a cavity of revolution can be found by a procedure similar to that used for waveguides of arbitrary cross section [14]. The problem of scattering by a homogeneous dielectric or magnetic body of revolution can be treated in terms of equivalent electric and magnetic surface currents. Hence, an extension to include both electric and magnetic currents could be used to solve such problems.

Acknowledgements

The research reported in this paper was sponsored by the Air Force Cambridge Research Laboratories, Office of Aerospace Research, U.S.A., under contract No. F-19628-67-C-0233, but this does not necessarily reflect endorsement by the sponsor.

Received 28 May 1968

REFERENCES

[1] HARRINGTON, R. F., Proc. IEEE **55** (1967) 136.
[2] HARRINGTON, R. F., Field Computation by Moment Methods, Macmillan, New York 1968.
[3] HARRINGTON, R. F., Generalized Network Parameters in Field Theory, p. 51, Proc. Sym. on Generalized Networks, MRIS Ser. XVI, Polytechnic Press, Brooklyn (N.Y.) 1966.
[4] HARRINGTON, R. F., Proc. IEE (London) **111** (1964) 617.
[5] HARRINGTON, R. F. and J. R. MAUTZ, IEEE Trans. **AP-15** (1967) 502.
[6] ANDREASEN, M. G., IEEE Trans. **AP-13** (1965) 303.
[7] MAUTZ, J. R. and R. F. HARRINGTON, Generalized Network Parameters for Bodies of Revolution, Scientific Report, Air Force Cambridge Research Laboratories, Bedford (Mass.) May 1968.
[8] DWIGHT, H. B., Tables of Integrals and Other Mathematical Data, p. 50, Eq. 200.01, Macmillan Company, New York 1961.
[9] MAUTZ, J. R., Radiation and Scattering from Bodies of Revolution, Ph.D. thesis, Appendix B, Syracuse University, Syracuse (N.Y.) May 1968.
[10] See [2], p. 70, Eq. 4-36.
[11] See [2], p. 70, Eq. 4-37.
[12] See [2], p. 90, Eq. 5-39.
[13] See [2], Chapter 6.
[14] See [2], Section 8-8.

H-Field, E-Field, and Combined-Field Solutions for Conducting Bodies of Revolution

by Joseph R. Mautz * and Roger F. Harrington *

The H-field, E-field, and combined-field integral equations have been solved by the method of moments for the surface electric currents and the scattered far fields of a body of revolution. The combined-field equation is a linear combination of the H-field and E-field integral equations. Exact solutions to the H-field and E-field equations are not unique at frequencies corresponding to interior resonances of the conducting surface. Exact solutions to the combined-field equations are unique. Numerical solutions to the H-field and E-field equations degenerate in the vicinity of interior resonant frequencies, but numerical solutions to the combined-field equations do not degenerate. A general computer program has been written, and sample computations are given for a sphere, a finite cylinder, and a flat back cone.

H-Feld, E-Feld- und kombinierte Feldlösungen für leitende Rotationskörper

Die Integralgleichungen mit dem H-Feld, dem E-Feld und mit beiden Feldern werden mit der Momentenmethode für die Oberflächenströme und das gestreute Fernfeld bei einem Rotationskörper gelöst. Die Gleichung mit beiden Feldern bildet eine Linearkombination der Integralgleichungen mit H-Feld und E-Feld. Bei Resonanzfrequenzen im Inneren der leitenden Oberfläche sind die exakten Lösungen nicht eindeutig. Exakte Lösungen der kombinierten Feldgleichungen sind eindeutig. Numerische Lösungen der H- und E-Feldgleichungen entarten in der Nähe der inneren Resonanzen, numerische Lösungen der kombinierten Feldgleichungen entarten dagegen nicht. Von einem allgemeinen Rechnerprogramm werden Musterrechnungen angegeben für die Kugel, den endlichen Zylinder und den Kegel mit ebener Basis.

1. Introduction

A general discussion of the H-field, E-field, and combined-field integral equations is given by Poggio and Miller in [1]. Solutions to the H-field equation for bodies of revolution have been given by Oshiro et al. [2] and by Uslenghi [3]. Solutions to the E-field equation for bodies of revolution have been given by Mautz and Harrington [4]. Solutions to the combined-field equation for a sphere have been given by Oshiro et al. [1], [5], but no general solution for bodies of revolution was reported. In this paper we use a modification of the solution [3] for the H-field equation and a modification of the solution [4] for the E-field equation. The combined-field equation is a linear combination of the H-field and E-field equations, and it is solved using a linear combination of the matrices obtained for the H-field and E-field solutions. Detailed developments of our solutions are given in the research report [6], and the documented computer program is given in the research report [7]. These reports are available from the authors.

It is shown theoretically that neither the H-field equation nor the E-field equation has unique solutions for the current on a conducting body at frequencies corresponding to resonant frequencies of the region enclosed by the conducting surface, but the combined-field equation does have a unique solution. The H-field equation does not have a unique solution for the scattered field at these resonant frequencies, but the E-field and the combined-field equations do have unique solutions. Our numerical solutions of the H-field and the E-field equations degenerate in the vicinity of internal resonant frequencies for both the current and the scattered field. Our numerical solution of the combined-field equation does not degenerate at or in the vicinity of the internal resonant frequencies.

2. Formulation of the Solutions

We seek the electric surface current and the far scattered field of the perfectly conducting body of revolution of Fig. 1 excited by an incident plane wave. In Fig. 1, ϱ, ϕ, z are cylindrical coordinates,

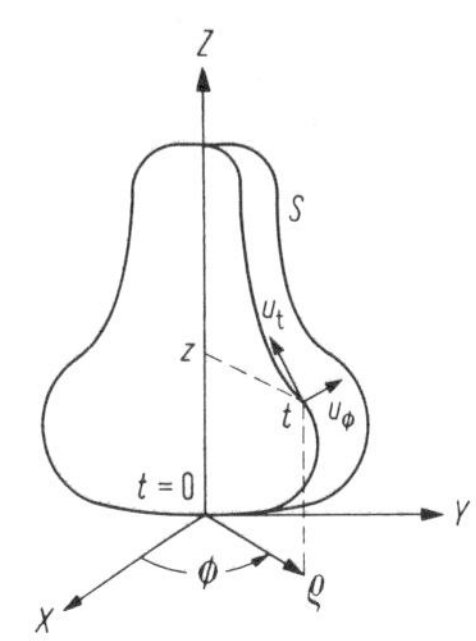

Fig. 1. Body of revolution and coordinate system.

and t, ϕ form an orthogonal curvilinear coordinate system on the surface S of the body of revolution. Also, $\boldsymbol{u}_t$ and $\boldsymbol{u}_\phi$ are orthogonal unit vectors in the t

* Dr. J. R. Mautz, Professor R. F. Harrington, Department of Electrical and Computer Engineering, Syracuse University, 111 Link Hall, Syracuse, N.Y. 13210, U.S.A.

and ϕ directions, respectively. The coordinate origin is on the axis of the body of revolution but not necessarily at the lower pole as in Fig. 1. Fig. 2 defines the propagation vector $\boldsymbol{k}_t$ of the incident plane wave, the transmitter coordinate θ_t, the coordinates θ_r, ϕ_r (receiver coordinates) at which the

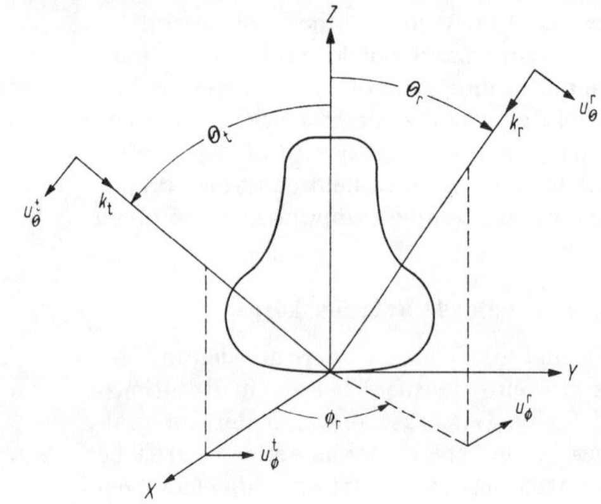

Fig. 2. Plane wave scattering by a conducting body of revolution.

far scattered field is observed, and the propagation vector $\boldsymbol{k}_r$ of a hypothetical measurement plane wave which travels from the receiver location (θ_r, ϕ_r) toward the origin. In Fig. 2, $\boldsymbol{u}_\theta^t$, $\boldsymbol{u}_\phi^t$, $\boldsymbol{u}_\theta^r$, and $\boldsymbol{u}_\phi^r$ are unit vectors in the θ_t, y, θ_r, and ϕ_r directions respectively.

We consider separately a θ polarized incident plane wave defined by

$$\begin{aligned} \boldsymbol{E}^i &= \boldsymbol{u}_\theta^t k \eta \exp(-\mathrm{j}\, \boldsymbol{k}_t \cdot \boldsymbol{r}), \\ \boldsymbol{H}^i &= (\boldsymbol{k}_t \times \boldsymbol{u}_\theta^t) \exp(-\mathrm{j}\, \boldsymbol{k}_t \cdot \boldsymbol{r}) \end{aligned} \tag{1}$$

and a ϕ polarized incident plane wave defined by

$$\begin{aligned} \boldsymbol{E}^i &= \boldsymbol{u}_\phi^t k \eta \exp(-\mathrm{j}\, \boldsymbol{k}_t \cdot \boldsymbol{r}), \\ \boldsymbol{H}^i &= (\boldsymbol{k}_t \times \boldsymbol{u}_\phi^t) \exp(-\mathrm{j}\, \boldsymbol{k}_t \cdot \boldsymbol{r}) \end{aligned} \tag{2}$$

where $\boldsymbol{E}^i$ and $\boldsymbol{H}^i$ denote incident electric and magnetic fields respectively, $\boldsymbol{r}$ is the radius vector from the origin, k is the propagation constant, and η is the intrinsic impedance. Either plane wave gives rise to t and ϕ directed electric surface currents on S and θ_r and ϕ_r directed far scattered fields.

A) H-field solution

The H-field integral equation is derived by setting the component tangential to S of the total magr etic field equal to zero just inside S, that is

$$-\boldsymbol{n} \times \boldsymbol{H}^s = \boldsymbol{n} \times \boldsymbol{H}^i \quad \text{just inside } S \tag{3}$$

where $\boldsymbol{n}$ is the unit outward normal vector to S, $\boldsymbol{H}^s$ is the magnetic field due to the electric surface current on S and $\boldsymbol{H}^i$ is the incident magnetic field given by either eq. (1) or eq. (2). By expressing $\boldsymbol{H}^s$ in terms of the magnetic vector potential integral, and taking the limit as the field point approaches the surface from the interior of S, we obtain the H-field equation [1]

$$L_H(\boldsymbol{J}) = \boldsymbol{n} \times \boldsymbol{H}^i \tag{4}$$

where

$$L_H(\boldsymbol{J}) = \tfrac{1}{2}\, \boldsymbol{J}(\boldsymbol{r}) - \iint_S \boldsymbol{n} \times \boldsymbol{\nabla} \times \left[\boldsymbol{J}(\boldsymbol{r}') \frac{\exp(-\mathrm{j}\, k\, |\boldsymbol{r} - \boldsymbol{r}'|)}{4\pi\, |\boldsymbol{r} - \boldsymbol{r}'|} \right] \mathrm{d}s' . \tag{5}$$

Here $\boldsymbol{J}$ is the surface current on S, $\boldsymbol{r}'$ is a source point on S, and $\boldsymbol{r}$ is the field point also on S. The improper integral in eq. (5) is convergent.

Our solution to eq. (4) is obtained by the method of moments [8]. For this, we let

$$\boldsymbol{J}(\boldsymbol{r}) = \sum_{n,j} [I_{nj}^t \boldsymbol{J}_{nj}^t(t, \phi) + I_{nj}^\phi \boldsymbol{J}_{nj}^\phi(t, \phi)] \tag{6}$$

where $\boldsymbol{J}_{nj}^t(t, \phi)$ and $\boldsymbol{J}_{nj}^\phi(t, \phi)$ are expansion functions defined by

$$\boldsymbol{J}_{nj}^p = \boldsymbol{u}_p f_j(t)\, \mathrm{e}^{\mathrm{j} n \phi} \tag{7}$$

where p is t or ϕ. Expansion functions whose t and ϕ components are proportional to $\mathrm{e}^{\mathrm{j} n \phi}$ are especially suitable because they make the t and ϕ components of the left-hand side of eq. (4) proportional to $\mathrm{e}^{\mathrm{j} n \phi}$. The appropriate symmetric product for the problem is

$$\langle \boldsymbol{A}, \boldsymbol{B} \rangle = \iint_S \boldsymbol{A} \cdot \boldsymbol{B}\, \mathrm{d}s \tag{8}$$

where $\boldsymbol{A}$ and $\boldsymbol{B}$ are tangential vectors on S. We next substitute eq. (6) into eq. (4), and take the symmetric product with respect to each testing function $\boldsymbol{J}_{mi}^{t*}$ and $\boldsymbol{J}_{mi}^{\phi*}$, where the asterisk denotes complex conjugate. The result is a system of equations which, in matrix form, can be represented as

$$\begin{pmatrix} Y_n^{tt} & Y_n^{t\phi} \\ Y_n^{\phi t} & Y_n^{\phi\phi} \end{pmatrix} \begin{pmatrix} \vec{I}_n^t \\ \vec{I}_n^\phi \end{pmatrix} = \begin{pmatrix} \vec{\tilde{I}}_n^t \\ \vec{\tilde{I}}_n^\phi \end{pmatrix}, \qquad n = 0, \pm 1, \pm 2, \ldots . \tag{9}$$

Here $\vec{I}_n^t$ and $\vec{I}_n^\phi$ are column vectors whose jth elements are I_{nj}^t and I_{nj}^ϕ, respectively. Also, $\vec{\tilde{I}}_n^t$ and $\vec{\tilde{I}}_n^\phi$ are column vectors whose ith elements are given by

$$\tilde{I}_{ni}^p = \langle \boldsymbol{J}_{ni}^{p*}, \boldsymbol{n} \times \boldsymbol{H}^i \rangle \tag{10}$$

where p is t or ϕ. Finally, the Y_n^{tt}, $Y_n^{\phi t}$, $Y_n^{t\phi}$, and $Y_n^{\phi\phi}$ are square matrices whose ijth elements are given by

$$(Y_n^{pq})_{ij} = \langle \boldsymbol{J}_{ni}^{p*}, L_H(\boldsymbol{J}_{nj}^q) \rangle \tag{11}$$

where p and q are permutations of t and ϕ. Formulas (11) represent complicated expressions which are given in detail in [6]. Note that there is a set of equations (9) for each $\mathrm{e}^{\mathrm{j} n \phi}$ mode. For a rotationally symmetric problem, only the $n = 0$ mode would be required. For a plane wave axially incident on the body of revolution, only the $n = \pm 1$ modes would be required. In general, as many modes must be taken as needed to obtain a good representation of the solution as a Fourier series in ϕ.

To evaluate the elements (11), we use expansion functions $f_j(t)$ equal to four impulse approximations to triangle functions divided by the cylindrical radius, as described in [4]. This is in contrast to Uslenghi's solution [3], which uses impulse functions. This difference was chosen not to make the solution more efficient, but to make it compatible with our E-field solution when used for the combined-field solution. For testing functions, we use a four impulse approximation to triangle functions divided by radius, again as described in [4]. For the ϕ integration, we use Gaussian quadrature instead of Simpson's rule for integration. Numerical tests have shown this to be more efficient for a given accuracy. The details of this solution are given in [6].

B) E-field solution

The E-field equation is derived by setting the component tangential to S of the total electric field equal to zero on S, that is,

$$-\frac{1}{\eta}\boldsymbol{E}^{\mathrm{s}}_{\tan} = \frac{1}{\eta}\boldsymbol{E}^{\mathrm{i}}_{\tan} \quad \text{on } S. \tag{12}$$

Here $\boldsymbol{E}^{\mathrm{s}}$ is the electric field due to the electric surface current on S, $\boldsymbol{E}^{\mathrm{i}}$ is the incident electric field given by either eq. (1) or eq. (2) and η is the intrinsic impedance. The subscript tan denotes tangential components on S. The $1/\eta$ factor is included in eq. (12) to give it the dimensions of current. By expressing $\boldsymbol{E}^{\mathrm{s}}$ in terms of the magnetic vector potential and electric scalar potential, we obtain a common form of the E-field equation [8]

$$\frac{1}{\eta}L_E(\boldsymbol{J}) = \frac{1}{\eta}\boldsymbol{E}^{\mathrm{i}}_{\tan} \tag{13}$$

where

$$L_E = (\mathrm{j}\,\omega\,\boldsymbol{A} + \boldsymbol{\nabla}\Phi)_{\tan}. \tag{14}$$

$\boldsymbol{A}$ is the magnetic vector potential integral and Φ is the electric scalar potential integral.

Following the method of moments, we again write $\boldsymbol{J}$ as in eq. (6), expansion functions as in eq. (7), and testing functions equal to the complex conjugate of the expansion functions. The symmetric product is again given by eq. (8). When eq. (6) is substituted into eq. (13), and the symmetric product taken with respect to each testing function, there result systems of equations

$$\begin{pmatrix} Z_n^{tt} & Z_n^{t\phi} \\ Z_n^{\phi t} & Z_n^{\phi\phi} \end{pmatrix}\begin{pmatrix} \vec{I}_n^{\,t} \\ \vec{I}_n^{\,\phi} \end{pmatrix} = \begin{pmatrix} \vec{V}_n^{\,t} \\ \vec{V}_n^{\,\phi} \end{pmatrix}, \qquad n = 0, \pm 1, \pm 2, \ldots \tag{15}$$

Again the $\vec{I}_n^{\,t}$ and $\vec{I}_n^{\,\phi}$ are column vectors with elements I_{nj}^t and I_{nj}^ϕ, respectively. The $\vec{V}_n^{\,t}$ and $\vec{V}_n^{\,\phi}$ are column vectors whose ith elements are given by

$$V_{ni}^p = \frac{1}{\eta}\iint_S \boldsymbol{J}_{ni}^{p*}\cdot\boldsymbol{E}^{\mathrm{i}}\,\mathrm{d}s \tag{16}$$

where p is either t or ϕ. Finally, the Z_n^{tt}, $Z_n^{\phi t}$, $Z_n^{t\phi}$, $Z_n^{\phi\phi}$ are square matrices whose ijth elements are given by

$$(Z_n^{pq})_{ij} = \left\langle \boldsymbol{J}_{ni}^{p*}, \frac{1}{\eta}L_E(\boldsymbol{J}_{nj}^q)\right\rangle \tag{17}$$

where p and q are permutations of t and ϕ. Again, formulas (17) represent complicated expressions.

A numerical evaluation of the elements (17) is described in [4]. For the present work we have made some minor modifications which improve the speed of computation about threefold. The details of this modified solution are given in [6].

C) Combined-field solution

The true solution to the problem must satisfy both conditions (3) and (12), that is, both $\boldsymbol{E}_{\tan}$ and $\boldsymbol{H}_{\tan}$ must be zero just inside S. It must also satisfy a linear combination of eqs. (3) and (12), which can be written as [1], [5]

$$-\boldsymbol{n}\times\boldsymbol{H}^{\mathrm{s}}(\boldsymbol{J}) - \frac{\alpha}{\eta}\boldsymbol{E}^{\mathrm{s}}_{\tan}(\boldsymbol{J}) = \boldsymbol{n}\times\boldsymbol{H}^{\mathrm{i}} + \frac{\alpha}{\eta}\boldsymbol{E}^{\mathrm{i}}_{\tan} \quad \text{just inside } S \tag{18}$$

where, for now, we can view α as an arbitrary constant. The representation of $\boldsymbol{n}\times\boldsymbol{H}^{\mathrm{s}}$ as an integral operator is the same as in part A above, and the representation of $\boldsymbol{E}^{\mathrm{s}}_{\tan}$ as an integral operator is the same as in part B above. The same expansion and testing functions were used to solve the H-field equation as were used to solve the E-field equation. Therefore, the moment solution to the combined field equation is just a linear combination of eqs. (9) and (15) with the same constant α, or

$$\left(\begin{pmatrix} Y_n^{tt} & Y_n^{t\phi} \\ Y_n^{\phi t} & Y_n^{\phi\phi} \end{pmatrix} + \alpha\begin{pmatrix} Z_n^{tt} & Z_n^{t\phi} \\ Z_n^{\phi t} & Z_n^{\phi\phi} \end{pmatrix}\right)\begin{pmatrix} \vec{I}_n^{\,t} \\ \vec{I}_n^{\,\phi} \end{pmatrix} = \begin{pmatrix} \vec{\bar{I}}_n^{\,t} \\ \vec{\bar{I}}_n^{\,\phi} \end{pmatrix} + \alpha\begin{pmatrix} \vec{V}_n^{\,t} \\ \vec{V}_n^{\,\phi} \end{pmatrix}, \qquad n = 0, \pm 1, \pm 2, \ldots \tag{19}$$

Here all matrices have the same meanings as in parts A and B.

3. Uniqueness of Solutions

The solution to a linear operator equation is not unique if the corresponding homogeneous equation has a nonzero solution. For the E-field equation (13), the solution therefore will not be unique if $L_E(\boldsymbol{J}) = 0$ has a nonzero solution. In terms of differential equations this means that the solution is not unique whenever there is a nonzero solution to

$$\boldsymbol{\nabla}\times\boldsymbol{\nabla}\times\boldsymbol{E}^{\mathrm{s}}(\boldsymbol{J}) = k^2\boldsymbol{E}^{\mathrm{s}}(\boldsymbol{J}) \quad \text{within } S, \tag{20}$$

$$\boldsymbol{n}\times\boldsymbol{E}^{\mathrm{s}}(\boldsymbol{J}) = 0 \quad \text{on } S. \tag{21}$$

But these are the equations for the modes of a resonator formed by a perfect electric conductor covering S. It is known that these modes exist at specific resonant frequencies, and that $\boldsymbol{J}$ is nonzero on S for these modes. Therefore the E-field equation does not have a unique solution whenever the frequency is such that S, when covered by a perfect electric conductor, forms a resonant cavity.

At an interior resonant frequency the solution for $\boldsymbol{J}$ is not unique in that it contains an undetermined amount of the resonant mode current. However, the resonant mode current produces no field external to S, and therefore the field $\boldsymbol{E}^{\mathrm{s}}$ external to S determined from the E-field equation theoretically should be unique. Unfortunately, when numerically computing the solution, the matrix representing L_E becomes ill conditioned and the solution for $\boldsymbol{E}^{\mathrm{s}}$ external to S may degenerate at and in the vicinity of a modal resonance.

For the H-field equation (4), the solution will not be unique if $L_H(\boldsymbol{J}) = 0$ has a nonzero solution. In terms of differential equations this means that the solution will not be unique whenever there is a nonzero solution to

$$\nabla \times \nabla \times \boldsymbol{H}^s(\boldsymbol{J}) = k^2 \boldsymbol{H}^s(\boldsymbol{J}) \quad \text{within } S, \tag{22}$$

$$\boldsymbol{n} \times \boldsymbol{H}^s(\boldsymbol{J}) = 0 \quad \text{just inside } S\,. \tag{23}$$

These are the equations for the modes of a resonator formed by a perfect magnetic conductor covering S. However, eqs. (22) and (23) are mathematically the same form as eqs. (20) and (21), and have the same resonant frequencies. Therefore the frequencies for which the solution to the H-field equation is not unique are the same frequencies at which the solution to the E-field equation is not unique.

The current associated with the $\boldsymbol{H}^s$ of eqs. (22) and (23) is obtained from

$$\boldsymbol{J} = \boldsymbol{n} \times (\boldsymbol{H}^s_{\text{ext}} - \boldsymbol{H}^s_{\text{int}}) \tag{24}$$

where $\boldsymbol{H}^s_{\text{ext}}$ denotes $\boldsymbol{H}^s$ just outside S and $\boldsymbol{H}^s_{\text{int}}$ denotes $\boldsymbol{H}^s$ just inside S. Since $\boldsymbol{n} \times \boldsymbol{H}^s_{\text{int}}$ is zero, $\boldsymbol{n} \times \boldsymbol{H}^s_{\text{ext}}$ is not zero and hence there will be a field $\boldsymbol{H}^s$ external to S at the interior resonant frequencies. Thus, in contrast to the E-field equation, the H-field equation theoretically does not have a unique solution for the field external to S at interior resonant frequencies.

We now show that the solution to the combined-field formulation (18) is unique whenever α is a positive real number. The solution to eq. (18) is unique if

$$-\boldsymbol{n} \times \boldsymbol{H}^s(\boldsymbol{J}) - \frac{\alpha}{\eta} \boldsymbol{E}^s_{\tan}(\boldsymbol{J}) = 0 \tag{25}$$

implies that $\boldsymbol{J} = 0$. Scalar multiplication of eq. (25) by its complex conjugate and integration over S lead to

$$\iint_S \left[|\boldsymbol{H}^s_{\tan}(\boldsymbol{J})|^2 + \frac{\alpha^2}{\eta^2} |\boldsymbol{E}^s_{\tan}(\boldsymbol{J})|^2 \right] \mathrm{d}s + \tag{26}$$

$$+ \frac{2\alpha}{\eta} \{\text{Real} \iint_S [\boldsymbol{E}^s(\boldsymbol{J}) \times \boldsymbol{H}^{s*}(\boldsymbol{J})] \cdot (-\boldsymbol{n})\, \mathrm{d}s\} = 0\,.$$

The bracketed quantity in eq. (26) is the real power flowing inside S and hence is either zero if the media is loss-free or greater than zero if the media is lossy. Hence

$$\boldsymbol{H}^s_{\tan}(\boldsymbol{J}) = 0 \quad \text{just inside } S\,, \tag{27}$$

$$\boldsymbol{E}^s_{\tan}(\boldsymbol{J}) = 0 \quad \text{on } S \tag{28}$$

if α is real and positive. Eq. (28) implies that $\boldsymbol{H}^s_{\tan}(\boldsymbol{J}) = 0$ just outside S. Using this fact and eq. (27) in eq. (24) shows that $\boldsymbol{J} = 0$, and the solution to the combined-field equation is unique at all frequencies.

If the media is loss-free, α could be chosen real and less than zero, but there appears to be no advantage in doing so. In fact, if $\alpha < 0$ were chosen, the right-hand (excitation) terms of eq. (18) would tend to be smaller on the illuminated portion of the surface than in the shadow zone, contrary to our physical intuition.

4. Examples

The solution for bodies of revolution has been implemented by a general computer program for the case of plane-wave incidence [7]. This program computes the current on the body and the scattered far field. A parameter of interest is the scattering cross section σ, which is the area for which the incident wave contains enough power to produce by omnidirectional radiation the scattered power density at the far field point. Specializing σ to the four different polarizations, we obtain

$$\frac{\sigma^{pq}}{\lambda^2} = \frac{1}{4\pi^3} \left| \frac{2\pi r_r E^s_{pq}(\boldsymbol{J})}{\eta} \right|^2 \tag{29}$$

where p is either θ or ϕ and q is either θ or ϕ. In eq. (29), p is the receiver polarization, q is the transmitter polarization, $\lambda = 2\pi/k$ is the wavelength, and $E^s_{pq}(\boldsymbol{J})$ is the received component of the scattered far field.

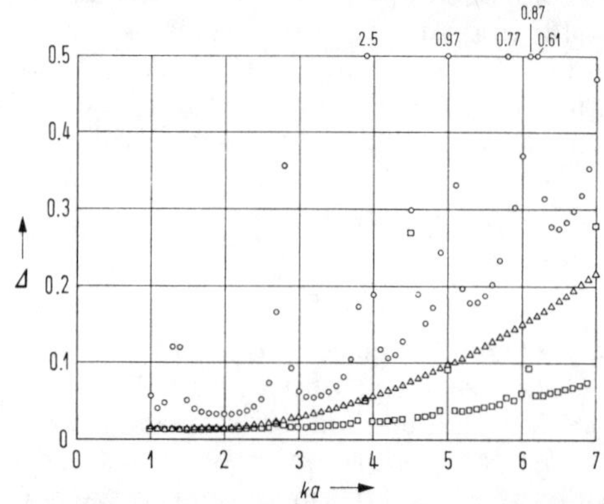

Fig. 3. Electric current error Δ versus ka for a conducting sphere of radius a excited by an incident plane wave. Squares denote H-field solution, circles denote E-field solution, and triangles denote combined-field solution. Values of $\Delta > 0.5$ are plotted at the top.

We compare computed approximations to the electric current induced on the surface of a conducting sphere by an axial incident plane wave with the known exact solution. Fig. 3 is a plot of Δ versus ka where k is the propagation constant, a is the radius of the sphere, and Δ is defined by

$$\Delta = \sqrt{\frac{\iint_S |\boldsymbol{J} - \boldsymbol{J}^c|^2 \mathrm{d}s}{|\boldsymbol{H}^i|^2 \iint_S \mathrm{d}s}}\,. \tag{30}$$

Here, S is the surface of the sphere and $\boldsymbol{H}^i$ is the incident magnetic field. Also, $\boldsymbol{J}^c$ is the exact electric current given by (6-103) of [9] and $\boldsymbol{J}$ is the computed approximation to $\boldsymbol{J}^c$. The squares, circles, and triangles in Fig. 3 tell from which field solution (H-field, E-field, or combined field with $\alpha = 1$) $\boldsymbol{J}$ is obtained.

The number of data points on the generating curve is 31 for all computations made for the sphere. These data points, equally spaced from the lower pole to the upper pole of the sphere, give 14 expansion functions for the t directed electric current and 14 expansion functions for the ϕ directed current. Both surface integrals in eq. (30) are evaluated by integrating analytically in ϕ and by sampling in t at 30 points.

In Fig. 3, the H-field solution is somewhat better than the E-field solution as far as $\boldsymbol{J}$ is concerned. For both H-field and E-field solutions, the error in $\boldsymbol{J}$ is large at resonances of the spherical cavity which are tabulated on page 270 of [9]. The combined-field solution is not affected by these resonances. Note that the error in $\boldsymbol{J}$ for the E-field solution also has a peak around $ka = 1.35$ which is far from any resonance. This peak disappeared when the number of data points was reduced from 31 to 21.

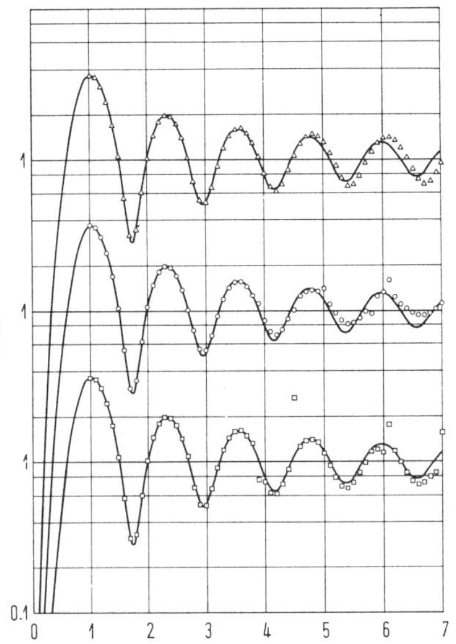

Fig. 4. Normalized radar cross section $\sigma/\pi a^2$ versus ka for a conducting sphere of radius a. Solid line denotes exact solution, squares denote H-field solution, circles denote E-field solution, and triangles denote combined field solution.

Fig. 4 compares the normalized radar cross section $\sigma/\pi a^2$ in the backscattering direction obtained from the H-field solution, the E-field solution, and the combined-field solution with the exact $\sigma/\pi a^2$. The exact $\sigma/\pi a^2$ is calculated from (6-105) of [9] in which A_e is our σ. In Fig. 4, the squares denote the H-field solution, the circles denote the E-field solution, the triangles denote the combined-field solution, and the solid line denotes the exact solution. The H-field, E-field and combined-field solutions for $\sigma/\pi a^2$ are beginning to deteriorate for the larger values of ka in Fig. 4. At $ka = 6$, there are only five expansion functions per wavelength.

Fig. 5a and b shows the curves of Figs. 3 and 4, respectively, in more detail the vicinity of the first resonance, which occurs at $ka = 2.744$. The disturbance in the H-field solution occurs quite close to the resonant frequency, but the disturbance in the E-field solution occurs at a slightly higher frequency. Although the error in $\boldsymbol{J}$ for the E-field solution is large, the error in $\sigma/\pi a^2$ for the E-field solution is small except at two or three points. The

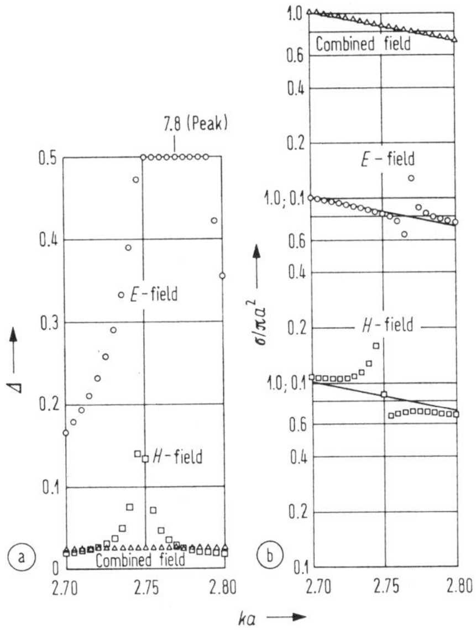

Fig. 5. Solution for the conducting sphere of radius a in the vicinity of the first resonance ($ka = 2.744$) excited by an incident plane wave; (a) electric current error Δ versus ka, (b) normalized radar cross section versus ka.

following explanation is offered. The E-field solution does not determine how much of the electric cavity mode current is contained in $\boldsymbol{J}$. Hence, our numerical E-field solution probably does not contain the right amount of the electric cavity mode. If this is true, then the radar cross section can still be quite accurate because the electric cavity mode does not radiate any external field. This is in contrast to the H-field solution, for which an external field does exist at magnetic cavity resonances. Fig. 5a and b shows that the combined-field solution is much better than either the H-field solution or the E-field solution in the vicinity of the first resonance.

Oshiro et al. [1], [5] conclude from their plots of mean error versus α for $0 \leq \alpha \leq 1$ that an α value on the order of 0.2 is best. To give some additional information on this choice, Table 1 gives the RMS error in current (Δ) and the backscattering ($\sigma/\pi a^2$) for a conducting sphere of radius a. The three values $ka = 1.50$, 2.75, and 4.00 were chosen for ka, and the parameter α was varied from 0.2 to 1.0. Also shown are results for the H-field solution ($\alpha = 0$) and for the E-field solution ($\alpha = \infty$). The exact values of $\sigma/\pi a^2$ are shown in the last row, computed from the spherical mode solution [10]. As is evident from the table, the value $\alpha = 0.2$ always gave good results, but the error was small for any value of α from 0.2 to 1.0.

The problem of scattering by a conducting right circular cylinder of cross sectional radius a and height $2a$ illustrates further the effect of internal resonance on the various solutions. The excitation is a plane wave axially incident on the flat end of the cylinder. The first internal resonance varying as $\cos\phi$ occurs at $ka = 2.42$ approximately. Fig. 6a shows the normalized backscattering cross section (σ/λ^2) in the vicinity of the first internal resonance.

Table 1. RMS error in current (Δ) and backscattering ($\sigma/\pi a^2$) for various solutions for a conducting sphere of radius a. H-field solution corresponds to $\alpha = 0$. E-field solution corresponds to $\alpha = \infty$.

	RMS error (Δ)			backscattering ($\sigma/\pi a^2$)		
$ka =$	1.50	2.75	4.00	1.50	2.75	4.00
H-field	0.0132	0.1342	0.0243	1.076	0.8702	0.7274
$\alpha = 0.2$	0.0118	0.0164	0.0274	1.075	0.8645	0.7842
$\alpha = 0.4$	0.0117	0.0187	0.0362	1.070	0.8648	0.7821
$\alpha = 0.6$	0.0127	0.0211	0.0447	1.065	0.8647	0.7912
$\alpha = 0.8$	0.0141	0.0233	0.0516	1.060	0.8645	0.8040
$\alpha = 1.0$	0.0154	0.0254	0.0571	1.057	0.8637	0.8159
E-field	0.0514	0.5991	0.1894	1.043	0.8246	0.8586
exact	0	0	0	1.076	0.8552	0.7853

The range of ka is from 2.37 to 2.47 in increments of 0.005. Note that the internal resonance has no observable effect on the combined-field solution with $\alpha = 0.3$ (triangles), a small effect on the E-field solution (circles), and a larger effect on the H-field solution (squares).

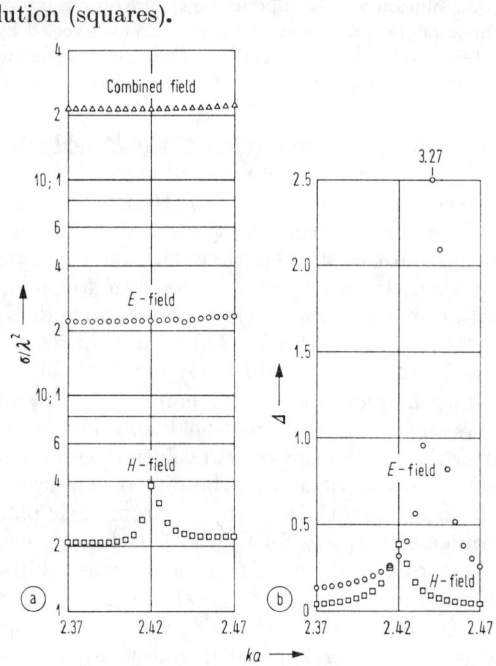

Fig. 6. Solution for a closed conducting cylinder of radius a and height $2a$ in the vicinity of first resonance ($ka = 2.42$); (a) normalized radar cross section σ/λ^2 versus ka, (b) RMS current error Δ versus ka. Excitation is by an axially incident plane wave.

To estimate the RMS error in the current, the combined-field solution is now considered to be the exact solution. The RMS error is then defined by eq. (30) where $\boldsymbol{J}^c$ is the combined field solution and the integration is over the surface of the scatterer. Fig. 6b shows Δ for the right circular cylinder versus ka in the vicinity of the first internal resonance. Note that the greatest RMS error occurs in the E-field solution (circles), with a somewhat smaller error in the H-field solution (squares). However, since the H-field eigencurrents radiate external to the cylinder, and the E-field eigencurrents do not (at least theoretically), the H-field backscattering cross section suffers a greater error at resonance than does the E-field backscattering cross section. This is shown by Fig. 6a.

In the E-field solution a resonance phenomenon sometimes occurs that is unrelated to any internal resonance. For example, in the case of a sphere there was a small peak in Δ at about $ka = 1.3$ (Fig. 3). In that case the resonance effect was too small to show up in the backscattering cross section (Fig. 4). This E-field "false resonance" effect was also found to occur in the case of scattering by a flat-back cone. The cone geometry is shown in the insert in Fig. 7. The excitation is a plane wave axially incident on the tip of the cone. Fig. 7a

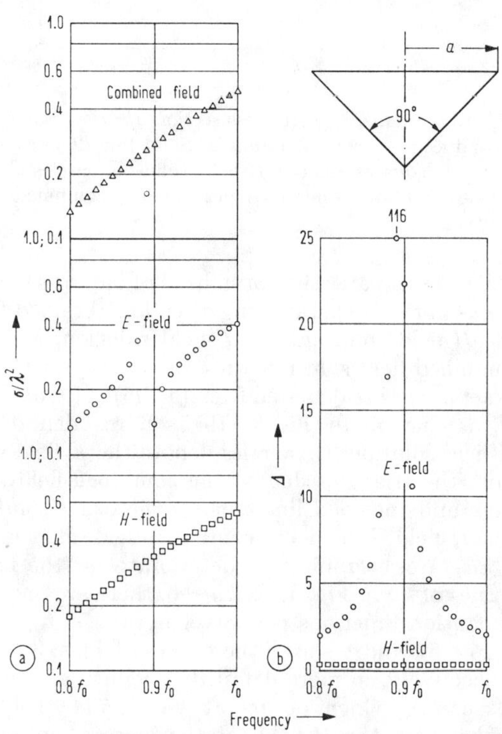

Fig. 7. Solution for a flat-back cone of radius a and $90°$ cone angle (see insert) in the vicinity of an E-field solution false resonance ($a = 3\lambda/16$ at f_0); (a) normalized radar cross section σ/λ^2 versus frequency, (b) RMS current error Δ versus frequency. The excitation is a plane wave axially incident on the cone tip.

shows the normalized cross section (σ/λ^2) versus frequency in the vicinity of an E-field false resonance, which occurred at $0.9\,f_0$, where f_0 is the frequency at which $a = 3\lambda/16$. Note that the E-field solution (circles) suffers a considerable perturbation at this false resonance, while the H-field solution (squares) and the combined-field solution (triangles) are unaffected by it. Again α is taken to be 0.3 in the combined-field solution. It should also be noted that the H-field solution differs from the combined-field solution by an appreciable amount in this range of frequencies.

Fig. 7b shows the RMS error in current for the H-field and E-field solutions in the vicinity of the false resonance for the flatback cone. Again the combined-field solution with $\alpha = 0.3$ is taken to be the exact solution, with Δ computed according to eq. (30). Note that there is no resonance effect apparent in the H-field solution, even though there is a large effect in the E-field solution. It is felt that this false resonance is a consequence of energy storage due to the current being approximated by triangular expansion functions. In support of this conjecture, we have noted that a false resonance either disappears or occurs at a different frequency when the number of expansion functions is changed.

5. Discussion

The H-field solution is a modification and generalization to oblique incidence of Uslenghi's impulse solution [3]. Instead of impulse expansion functions and Simpson's rule for integration with respect to the azimuth ϕ, we use four impulse approximations to triangle functions and Gaussian quadrature integration. The E-field solution is that of [4] with impulse Green's functions obtained from Gaussian quadrature integration in the azimuth ϕ instead of pulse Green's functions obtained from equal interval, equal weight sampling. The combined-field formulation is a linear combination of the equations for the H-field and E-field solutions. Computer program subroutines which calculate the square matrices for the H-field and E-field solutions and the plane wave measurement vectors are given in a research report [7]. The subroutine which calculates the square matrix for the E-field solution executes appreciably faster and requires considerably less storage to obtain the same kind of accuracy as our previous solution [11].

The H-field and E-field solutions deteriorate in the vicinity of cavity resonances because their homogeneous equations admit nontrivial solutions at these resonances. It is shown that the homogeneous equation associated with the combined field formulation has no nontrivial solution if the relative weight α of the E-field equation is real and, if the inside media is lossy, positive. For this reason the combined field solution is much better than either the H-field or E-field solutions in the vicinity of cavity resonances.

For the examples of the sphere, the computations were made with the relative weight α of the E-field equation in the combined-field formulation equal to unity. This put the H-field and E-field equations on a more or less equal footing because the magnitude of the excitation due to the H-field equation is then that due to the E-field equation rotated 90° in space. According to Fig. 4, the H-field solution for $\sigma/\pi a^2$ is generally a bit more accurate than the E-field solution away from the cavity resonances. This suggests that one weight the E-field equation less than the H-field equation in the combined field formulation. In general, α should be chosen in the range $0.2 \leqq \alpha \leqq 1$. At internal resonances, the RMS error in current is greater for the E-field solution than for the H-field solution. However, the scattering cross section error is usually smaller for the E-field solution than for the H-field solution. No effects due to internal resonances have ever been observed in the combined-field solution. Finally, the E-field solution sometimes exhibits a false resonance effect, giving incorrect solutions at some frequency which does not correspond to an internal resonance.

It is worth noting that there is still another method for obtaining unique solutions at interior resonant frequencies, called a "combined-source" formulation. This is an extension to electromagnetic fields of the scalar combined-source formulation of Brakhage and Werner [12]. Solutions to two-dimensional electromagnetic problems, which are reducible to scalar problems, have been obtained with this formulation by Bolomey and Tabbara [13]. The extension of this method to three-dimensional electromagnetic fields is currently being investigated by us.

Acknowledgement

This work was supported by the Rome Air Development Center, Deputy for Electronic Technology, Hanscom Air Force Base, U.S.A., under Contract No. F19628-76-C-0300.

(Received December 7th, 1977.)

References

[1] Poggio, A. J. and Miller, E. K., Integral equation solutions of three-dimensional scattering problems. In: R. Mittra (ed.), Computer techniques for electromagnetics, chap. 4. Pergamon Press, London 1973.

[2] Oshiro, F. K. and Mitzner, K. M., Digital computer solution of three-dimensional scattering problems. Symposium Digest, IEEE Internat. Symp. on Antennas and Propagation, Ann Arbor, Mich., Oct. 1967, pp. 257–263.

[3] Uslenghi, P. L. E., Computation of surface currents on bodies of revolution. Alta Frequenza **39** [1970], No. 8, 1–12.

[4] Mautz, J. R. and Harrington, R. F., Radiation and scattering from bodies of revolution. Appl. Sci. Res. **20** [1969], 405–435.

[5] Oshiro, F. K. et al., Calculation of radar cross section. Air Force Avionics Lab. Tech. Rep. AFAL-TR-70-21, Part II, April 1970.

[6] Mautz, J. R. and Harrington, R. F., H-field, E-field, and combined field solutions for bodies of revolution. Interim Tech. Rep. RADC-TR-77-109, Rome Air Development Center, Griffiss Air Force Base, N.Y., March 1977.

[7] Mautz, J. R. and Harrington, R. F., Computer programs for *H*-field, *E*-field, and combined field solutions for bodies of revolution. Interim Tech. Rep. RADC-TR-77-215, Rome Air Development Center, Griffiss Air Force Base, N.Y., June 1977.

[8] Harrington, R. F., Field computation by moment methods. Macmillan Co., New York 1968.

[9] Harrington, R. F., Time-harmonic electromagnetic fields. McGraw-Hill Book Co., New York 1961.

[10] Adler, S. B. and Johnson, R. S., New backscattering computation and tables for dielectric and metal spheres. Appl. Optics **1** [1962], 655–660.

[11] Harrington, R. F. and Mautz, J. R., Radiation and scattering from bodies of revolution. Rep. AFCRL-69-0305, Contract No. F19628-67-C-0233 between Syracuse University and Air Force Cambridge Research Laboratories, July 1969.

[12] Brakhage, H. and Werner, P., Über das Dirichletsche Aussenraumproblem für die Helmholtzsche Schwingungsgleichung. Arch. Math. **16** [1965], 325–329.

[13] Bolomey, J. C. and Tabbara, W., Numerical aspects on coupling between complementary boundary value problems. Transact. IEEE AP-**21** [1973], 356–363.

Scattering from Axisymmetric Obstacles Embedded in Axisymmetric Dielectrics: The Method of Moments Solution★

L. N. Medgyesi-Mitschang and *C. Eftimiu*★★

McDonnell Douglas Research Laboratories, McDonnell Douglas Corp., St. Louis, MO 63166, USA

Received 1 December 1978/Accepted 24 February 1979

Abstract. The electromagnetic scattering from axisymmetric conducting or nonconducting (dielectric) obstacles, embedded in an axisymmetric dielectric body is treated. A surface integral equation formulation, consisting of coupled Fredholm equations of the first kind for the electric and magnetic fields, is solved by the method of moments. The outer surfaces of the internal obstacles and the embedding dielectrics can be nonconcentric, depart significantly from a spherical shape, but must be rotationally symmetric about a common axis. The embedding dielectric can be multilayered. Computer implementable expressions are given for the scattering cross sections for any desired polarization and for both backscatter (monostatic) and bistatic illumination. Comparisons are made with the extended boundary condition method for homogeneous dielectric bodies and the Mie theory extended for dielectrically clad conducting spheres. The generality of the present formulation is demonstrated for several other cladded scatterer configurations.

PACS: 42.10

Numerous formulations have been devised to study the scattering and diffraction from three-dimensional conducting or non-conducting (dielectric) obstacles. These investigations are motivated by a plethora of diverse applications ranging from radar meteorology to biological sciences. The suitability of a given method is determined by the scatterer geometry, composition, and size as a function of wavelength λ. At the extrema, namely scattering from objects very small or very large relative to λ, the Rayleigh approximation [1] and the optic-derived methods have been found effective [2, 3]. The intermediate region between these regimes, on the other hand, has defied a unified treatment. Scattering from penetrable and coated bodies in the resonance region, i.e., $\sim\lambda$ in size, has been treated, depending on the geometry, by point matching [4], perturbation [5], and other approximation [6] techniques, as well as generalizations of the Mie theory [7, 8]. The Extended Boundary Condition Method (EBCM), originally due to Waterman [9, 10] has likewise been applied to treat the scattering from axisymmetric homogeneous dielectric bodies [11, 12], from spherical conductors embedded in axisymmetric dielectrics [13], and from multilayered scatterers [14]. Recently, the unimoment method [15] and the Method of Moments (MoM) [16, 17] have been used for these problems, with the latter applied to homogeneous axisymmetric scatterers. The principles involved in generalizing the MoM concepts to layered dielectric scatterers are discussed in [18].

An approximate MoM formulation proposed by Harrington and Mautz [19] treated the scattering from very thin ($\ll\lambda$) axisymmetric dielectric shells. In that analysis, the normal components of the currents in the dielectric region are assumed to be negligible compared with the tangential components.

The present work generalizes the original MoM formulation for axisymmetric conducting bodies [20] to

★ This research was conducted under the McDonnell Douglas Independent Research and Development Program.

★★ Department of Physics, University of Missouri, St. Louis, MO 63166, USA.

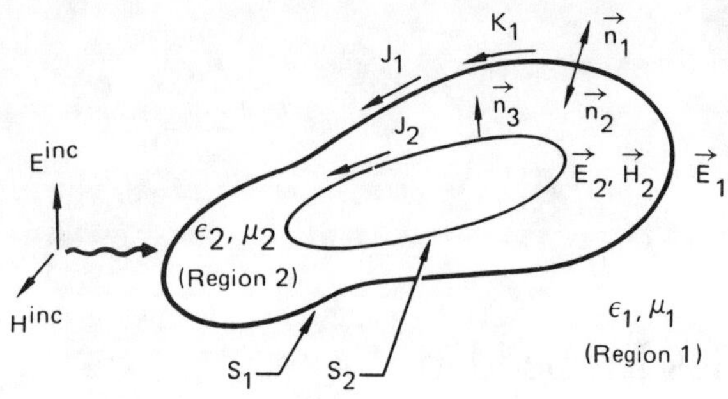

Fig. 1. Representation of general dielectrically clad body

rigorously treat the scattering from an axisymmetric conductor surrounded by an arbitrarily thick axisymmetric dielectric region (shell).

Basic Theory – Single Layer Case

The case of the scattered electromagnetic fields associated with a conducting body embedded in a single homogeneous dielectric layer is considered first. Later the extension to an arbitrary number of layers is discussed. Generalizing the expressions for a penetrable body given in [21, Eq. (4.15)], and referring to Fig. 1, the surface integral equations relating the **E** and **H** fields in free space (region 1) to the incident electric field $\mathbf{E}^{\mathrm{inc}}$ are

$$\theta(\mathbf{r})\mathbf{n}_1\times\mathbf{E}_1=\mathbf{n}_1\times\mathbf{E}^{\mathrm{inc}}-\mathbf{n}_1\times\int_{S_1}[\mathrm{j}\omega\mu_1\Phi_1(\mathbf{n}_1'\times\mathbf{H}_1)-(\mathbf{n}_1'\times\mathbf{E}_1)\times\nabla'\Phi_1-(\mathbf{n}_1'\cdot\mathbf{E}_1)\nabla'\Phi_1]ds', \quad (1)$$

$$\theta(\mathbf{r})E_2=-\int_{S_1}[\mathrm{j}\omega\mu_2\Phi_2(\mathbf{n}_2'\times\mathbf{H}_2)-(\mathbf{n}_2'\times\mathbf{E}_2)\times\nabla'\Phi_2-(\mathbf{n}_2'\cdot\mathbf{E}_2)\nabla'\Phi_2]ds'-\int_{S_2}[\mathrm{j}\omega\mu_2\Phi_2(\mathbf{n}_3'\times\mathbf{H}_2)-(\mathbf{n}_3'\times\mathbf{E}_2)\times\nabla'\Phi_2-(\mathbf{n}_3'\cdot\mathbf{E}_2)\nabla'\Phi_2]ds', \quad (2)$$

where S_1 defines the surface of the dielectric covering of the body with outward and inward normals of $\mathbf{n}_1$ and $\mathbf{n}_2$, respectively; S_2 defines the surface of the conducting body with normal $\mathbf{n}_3$. Similarly, ~~in the dielectric (region 2)~~ the governing equations in terms of an incident magnetic field $\mathbf{H}^{\mathrm{inc}}$ are

$$\theta(\mathbf{r})\mathbf{n}_1\times\mathbf{H}_1=\mathbf{n}_1\times\mathbf{H}^{\mathrm{inc}}+\mathbf{n}_1\times\int_{S_1}[\mathrm{j}\omega\varepsilon_1\Phi_1(\mathbf{n}_1'\times\mathbf{E}_1)+(\mathbf{n}_1'\times\mathbf{H}_1)\times\nabla'\Phi_1+(\mathbf{n}_1'\cdot\mathbf{H}_1)\nabla'\Phi_1]ds', \quad (3)$$

$$\theta(\mathbf{r})\mathbf{H}_2=\int_{S_1}[\mathrm{j}\omega\varepsilon_2\Phi_2(\mathbf{n}_2'\times\mathbf{E}_2)+(\mathbf{n}_2'\times\mathbf{H}_2)\times\nabla'\Phi_2+(\mathbf{n}_2'\cdot\mathbf{H}_2)\nabla'\Phi_2]ds'+\int_{S_2}[\mathrm{j}\omega\varepsilon_2\Phi_2(\mathbf{n}_3'\times\mathbf{E}_2)+(\mathbf{n}_3'\times\mathbf{H}_2)\times\nabla'\Phi_2+(\mathbf{n}_3'\cdot\mathbf{H}_2)\nabla'\Phi_2]ds', \quad (4)$$

where the Green's functions in regions 1 and 2 are

$$\Phi_i=\frac{1}{4\pi|\mathbf{r}-\mathbf{r}'|}\mathrm{e}^{-\mathrm{j}k_i|\mathbf{r}-\mathbf{r}'|};\qquad k_i=\frac{2\pi}{\lambda}\sqrt{\varepsilon_i\mu_i};\qquad i=1,2$$

and $\mu_i=\mu_0\mu_{ir}$; and $\varepsilon_i=\varepsilon_0\varepsilon_{ir}$; μ_0 and ε_0 are the permeability and permittivity of free space; the subscripts r denote the relative permittivity and permeability; in general, these can be complex; and

$$\theta(\mathbf{r})=\begin{cases}1, & \text{for } \mathbf{r}\in\text{free space}\\ 1/2; & \text{for } \mathbf{r}\in S_1 \text{ or } S_2\\ 0; & \text{otherwise.}\end{cases}$$

The boundary conditions on the fields on S_1 and S_2 are

$$\mathbf{n}_1\times\mathbf{E}_1|_{S_1}=\mathbf{n}_1\times\mathbf{E}_2|_{S_1}=-\mathbf{K}_1,$$

$$\mathbf{n}_1\times\mathbf{H}_1|_{S_1}=\mathbf{n}_1\times\mathbf{H}_2|_{S_1}=\mathbf{J}_1,$$

$$\mathbf{n}_1\cdot\mathbf{E}_1|_{S_1}=\frac{\varepsilon_2}{\varepsilon_1}\mathbf{n}_1\cdot\mathbf{E}_2|_{S_1}=-\frac{1}{\mathrm{j}\omega\varepsilon_1}\nabla_s\cdot(\mathbf{J}_1),$$

$$\mathbf{n}_1\cdot\mathbf{H}_1|_{S_1}=\frac{\mu_2}{\mu_1}\mathbf{n}_1\cdot\mathbf{H}_2|_{S_1}=-\frac{1}{\mathrm{j}\omega\mu_1}\nabla_s\cdot(\mathbf{K}_1),$$

and

$$\mathbf{n}_3\times\mathbf{E}_2|_{S_2}=0,$$

where $\mathbf{K}_1$ and $\mathbf{J}_1$ are the unknown magnetic and electric currents on S_1. Similarly $\mathbf{J}_2$ is the unknown electric current on S_2. Applying the foregoing boundary conditions yields a set of integro-differential equations summarized in matrix form as

$$\begin{bmatrix}L_{11} & L_{12} & L_{13}\\ L_{21} & L_{22} & L_{23}\\ L_{31} & L_{32} & L_{33}\end{bmatrix}\begin{bmatrix}\mathbf{J}_1\\ \mathbf{K}_1\\ \mathbf{J}_2\end{bmatrix}\begin{bmatrix}\mathbf{n}_1\times\mathbf{E}^{\mathrm{inc}}\\ \mathbf{n}_1\times\mathbf{H}^{\mathrm{inc}}\\ \mathbf{0}\end{bmatrix}, \quad (5)$$

where the $L_{ij}(\cdot)$ operators are defined as

$$L_{11}(\mathbf{X})=\mathbf{n}_1\times\fint_{S_1}\left[\mathrm{j}\omega(\mu_1\Phi_1+\mu_2\Phi_2)\mathbf{X}+\frac{1}{\mathrm{j}\omega}(\nabla_S\cdot\mathbf{X})\mathrm{grad}'\left(\frac{\Phi_1}{\varepsilon_1}+\frac{\Phi_2}{\varepsilon_2}\right)\right]\Bigg|_{S_1}ds', \quad (6)$$

$$L_{12}(\mathbf{X})=\mathbf{n}_1\times\fint_{S_1}\mathbf{X}\times\mathrm{grad}'(\Phi_1+\Phi_2)\Big|_{S_1}ds'=-L_{21}(\mathbf{X}), \quad (7)$$

$$L_{13}(\mathbf{X})=-\mathbf{n}_1\times\int_{S_2}\left[\mathrm{j}\omega\mu_2\Phi_2\mathbf{X}+\frac{1}{\mathrm{j}\omega\varepsilon_2}(\nabla_s\cdot\mathbf{X})\mathrm{grad}'\Phi_2\right]\Bigg|_{S_1}ds', \quad (8)$$

$$L_{23}(\mathbf{X})=\mathbf{n}_1\times\int_{S_2}[\mathbf{X}\times\mathrm{grad}'\Phi_2]\Big|_{S_1}ds', \quad (9)$$

and

$$L_{33}(\mathbf{X})=-\mathbf{n}_3\times\fint_{S_2}\left[\mathrm{j}\omega\mu_2\Phi_2\mathbf{X}+\frac{1}{\mathrm{j}\omega\varepsilon_2}(\nabla_s\cdot\mathbf{X})\mathrm{grad}'\Phi_2\right]\Bigg|_{S_2}ds', \quad (10)$$

where $⨍$ is the Cauchy principal value integral with the singularities deleted. The remaining operators can be obtained from (6)–(10). Thus $L_{22}(\mathbf{X})=L_{11}(\mathbf{X})$ with ε_i and μ_i $(i=1,2)$ interchanged; $L_{31}(\mathbf{X})=L_{13}(\mathbf{X})$ with S_1 and S_2 interchanged and $\mathbf{n}_1\to-\mathbf{n}_3$. Similarly, $L_{32}(\mathbf{X})=L_{23}(\mathbf{X})$ with S_1 and S_2 interchanged and $\mathbf{n}_1\to\mathbf{n}_3$. The operators in (6) and (7) have been given previously in [Ref. 17, pp. 94–95] and [18] for the totally penetrable body. $L_{33}(\cdot)$ corresponds to the operator associated with the conducting body [20].

Method of Moments

To solve (5) for the unknown currents on S_1, explicit expressions must be obtained for $L_{ij}(\cdot)$ in terms of body geometry. The subsequent analysis is specialized for an arbitrary axisymmetric configuration [i.e., body of revolution (BOR)] shown in Fig. 2. The surfaces S_1 and S_2 need not be concentric or similar. Using the MoM technique, the surfaces are segmented axially into annuli on which the unknown electric and magnetic currents are expressed in a finite series expansion. Thus on the i-th segment of S_1,

$$\mathbf{J}_i=\sum_n (a^t_{ni}\mathbf{J}^t_{ni}+a^\phi_{ni}\mathbf{J}^\phi_{ni})$$
$$=\sum_n (\mathbf{u}_t a^t_{ni} f^t_i(t)+\mathbf{u}_\phi a^\phi_{ni} f^\phi_i(t))e^{jn\phi}. \tag{11}$$

Similarly, $\mathbf{K}_1$ and $\mathbf{J}_2$ can be expanded, replacing $a^\alpha_{ni}(\alpha=t$ or $\phi)$ by b^α_{ni} and c^α_{ni}, respectively. In (11) the circumferential current modes on S_1 and S_2 are denoted by n, and $f_i(t)$ are expansion functions along the generating curve t of S_1 or S_2. Substituting the expansions for $\mathbf{J}_1, \mathbf{K}_1$, and $\mathbf{J}_2$ into (5) and forming the inner products over S_1 and S_2 with respect to testing functions $\mathbf{W}^\alpha_{ni}=\mathbf{u}^\alpha f^\alpha_i(t)\exp(-jn\phi)$ yields the MoM matrix equation for the n-th mode, written in compact form as

$$\begin{bmatrix} \mathbf{S}_{EE} & \mathbf{S}_{EH} & -\mathbf{M}_E \\ \mathbf{S}_{HE} & \mathbf{S}_{HH} & -\mathbf{M}_H \\ -\tilde{\mathbf{M}}_E & \tilde{\mathbf{M}}_H & \mathbf{Z}_0 \end{bmatrix}_n \begin{bmatrix} \mathbf{A}_n \\ \mathbf{B}_n \\ \mathbf{C}_n \end{bmatrix} = \begin{bmatrix} \bar{\mathbf{E}} \\ \bar{\mathbf{H}} \\ \mathbf{0} \end{bmatrix}_n \tag{12}$$

where $\mathbf{A}_n$, $\mathbf{B}_n$, $\mathbf{C}_n$ are column vectors comprising the unknown expansion coefficients $\mathbf{a}^\alpha_{ni}$, $\mathbf{b}^\alpha_{ni}$, and $\mathbf{c}^\alpha_{ni}$. The elements of the submatrices in (12) are expressed in terms of the "inner" product defined as, for example

$$(\mathbf{S}_{EE})_{ij}=\langle \mathbf{W}_{ni}, L_{11}(\mathbf{J}_{nj})\rangle=\int_0^{2\pi} d\phi \int_i \mathbf{W}_{ni}\cdot L_{11}(\mathbf{J}_{nj})$$

where the index i and j run over all the segments of S_1. (For the sake of notational simplicity, $\mathbf{S}_{EE}$ is not partitioned explicitly in terms of the t and ϕ components.) Note that $\mathbf{Z}_0$ is the impedance matrix for the conducting core, previously given [Ref. 20, Eq. (39)].

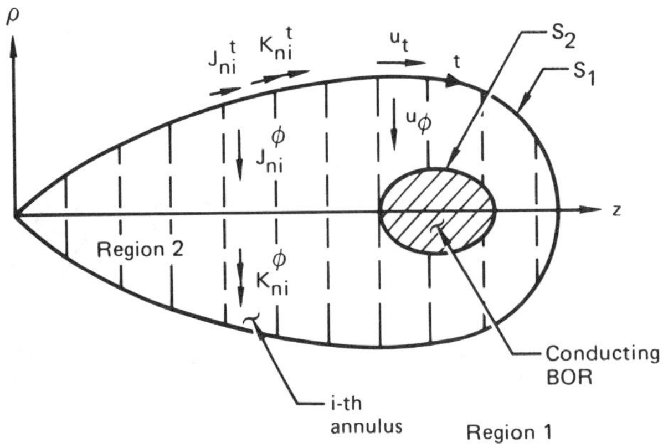

Fig. 2. Segmented axisymmetric scatterer geometry

For computational purposes, it is convenient to express each of the partitioned submatrices in (12) in terms of two operators $\mathscr{L}_{ij}(\cdot)$ and $\mathscr{K}_{ij}(\cdot)$, defined explicitly in Appendix A in terms of computer implementable expressions. Thus

$$(\mathbf{S}^{\alpha\beta}_{EEij})_n=\mathscr{L}^{\alpha\beta}_{ij}\left[1,1;\mu_1\Phi_1+\mu_2\Phi_2;\frac{\partial}{\partial t}\left(\frac{\Phi_1}{\varepsilon_1}+\frac{\Phi_2}{\varepsilon_2}\right);n\right], \tag{13}$$

$$(\mathbf{S}^{\alpha\beta}_{HHij})_n=\mathscr{L}^{\alpha\beta}_{ij}\left[1,1;\varepsilon_1\Phi_1+\varepsilon_2\Phi_2;\frac{\partial}{\partial t}\left(\frac{\Phi_1}{\mu_1}+\frac{\Phi_2}{\mu_2}\right);n\right], \tag{14}$$

$$(\mathbf{M}^{\alpha\beta}_{Eij})_n=\mathscr{L}^{\alpha\beta}_{ij}\left[1,2;\mu_2\Phi_2;\frac{\partial}{\partial t}\left(\frac{\Phi_2}{\varepsilon_2}\right);n\right], \tag{15}$$

$$(\tilde{\mathbf{M}}^{\alpha\beta}_{Eij})_n=\mathscr{L}^{\alpha\beta}_{ij}\left[2,1;\mu_2\Phi_2;\frac{\partial}{\partial t}\left(\frac{\Phi_2}{\varepsilon_2}\right);n\right], \tag{16}$$

$$(\mathbf{Z}^{\alpha\beta}_{ij})_n=\mathscr{L}^{\alpha\beta}_{ij}\left[2,2;\mu_2\Phi_2;\frac{\partial}{\partial t}\left(\frac{\Phi_2}{\varepsilon_2}\right);n\right], \tag{17}$$

$$(\mathbf{M}^{\alpha\beta}_{Hij})_n=\mathscr{K}^{\alpha\beta}_{ij}(1,2;\Phi_2;n), \tag{18}$$

$$(\tilde{\mathbf{M}}^{\alpha\beta}_{Hij})_n=\mathscr{K}^{\alpha\beta}_{ij}(2,1;\Phi_2;n), \tag{19}$$

$$(\mathbf{S}^{\alpha\beta}_{EHij})_n=\mathscr{K}^{\alpha\beta}_{ij}(1,1;\Phi_1+\Phi_2;n), \tag{20}$$

and

$$(\mathbf{S}^{\alpha\beta}_{HEij})_n=-(\mathbf{S}^{\alpha\beta}_{EHij})_n. \tag{21}$$

In the preceding operator notation, α or β denotes t or ϕ; the first two parameters within the parenthesis indicate the surfaces S_1 or S_2 over which the integrations are carried out. The parameter n denotes the n-th circumferential mode. Note that $\tilde{\mathbf{M}}$ is identical to $\mathbf{M}$, except that the surfaces of integration S_1 and S_2 are interchanged. Matrices $\mathbf{S}_{EE}$ and $\mathbf{S}_{HH}$ give the interactions of the electric surface currents on S_1 and those for the magnetic surface currents, respectively. The off-diagonal matrices, $\mathbf{S}_{EH}$ and $\mathbf{S}_{HE}$, give the interaction between the electric and magnetic currents on surface S_1. Matrices $\mathbf{M}_E$ and $\tilde{\mathbf{M}}_E$ represent the

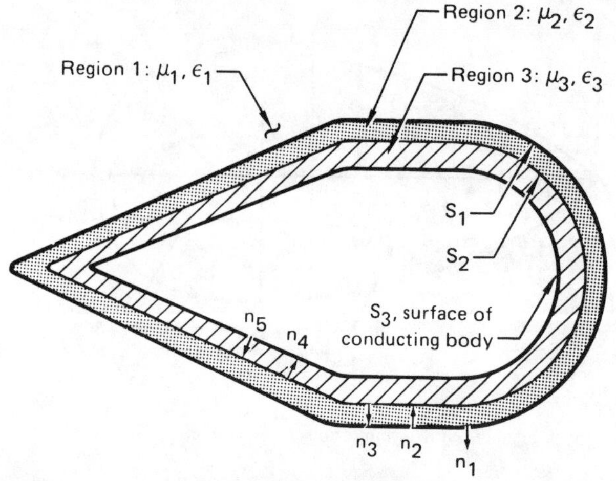

Fig. 3. Representation of a multilayered dielectrically clad axisymmetric body

electric field interactions on S_1 and S_2 due to the electric currents on S_1 and S_2, respectively. A similar interpretation holds for matrices $\mathbf{M}_H$ and $\tilde{\mathbf{M}}_H$.

To compute the scattered fields from the coated body, the currents on the outer surface S_1 are obtained in terms of the currents on S_2. Thus solving (12) yields

$$\begin{bmatrix}\mathbf{A}_1\\ \mathbf{B}_1\end{bmatrix}_n = [\mathbf{Z}_1]_n^{-1}\begin{bmatrix}\bar{\mathbf{E}}\\ \bar{\mathbf{H}}\end{bmatrix}_n, \tag{22}$$

where $\bar{\mathbf{E}}$ and $\bar{\mathbf{H}}$ are dependent upon the incident electromagnetic fields illuminating the body. $[\mathbf{Z}_1]_n$ is the composite MoM matrix defining all electromagnetic interactions on the dielectrically clad body, defined by

$$[\mathbf{Z}_1]_n = [\mathscr{S}]_n - {}_t[\mathbf{M}_E \mid \mathbf{M}_H]_n [\mathbf{Z}_0]_n^{-1} [\tilde{\mathbf{M}}_E \mid -\tilde{\mathbf{M}}_H]_n, \tag{23}$$

where

$$[\mathscr{S}]_n = \begin{bmatrix}\mathbf{S}_{EE} & \mathbf{S}_{EH}\\ \mathbf{S}_{HE} & \mathbf{S}_{HH}\end{bmatrix}_n$$

and t denotes the matrix transpose. Subsequently, the scattering cross section will be computed using $[\mathbf{Z}_1]_n$.

Multi-Layered Case

Next the formulation is extended for a conducting body embedded within two layers of dielectric (Fig. 3). The extension developed here is immediately applicable to the case where S_3 bounds a homogeneous or possibly layered dielectric region. Note that the latter can always be reformulated as a sequence of two layer problems.

The formulation for the case in Fig. 3 proceeds by adding to the original integral equations (1)–(4) a pair of equations for the fields $\mathbf{E}_3$ and $\mathbf{H}_3$ in region 3. Formally these expressions are identical to (2) and (4), with $\mathbf{n}_2 \rightarrow \mathbf{n}_4$, and $\mathbf{n}_3 \rightarrow \mathbf{n}_5$ and all other subscripts increased by one. The present discussion is again specialized to the case where the surfaces are not necessarily concentric but have a common axis of revolution. The boundary conditions are applied on S_1, S_2, and S_3. Expanding the unknown currents in a finite series of functions spanning these surfaces, and forming the inner product on these surfaces, the following MoM matrix equation is obtained:

$$\left[\begin{array}{c|c|c}\mathscr{S} & -\mathscr{M} & 0\\ \hline -\tilde{\mathscr{M}} & \mathscr{S}' & \begin{matrix}-\mathbf{M}'_E\\ -\mathbf{M}'_H\end{matrix}\\ \hline 0 & -\tilde{\mathbf{M}}'_E \quad \tilde{\mathbf{M}}'_H & \mathbf{Z}'_0\end{array}\right]_n \left[\begin{array}{c}\mathbf{A}_1\\ \mathbf{B}_1\\ \hline \mathbf{A}_2\\ \mathbf{B}_2\\ \hline \mathbf{A}_3\end{array}\right]_n = \left[\begin{array}{c}\bar{\mathbf{E}}\\ \bar{\mathbf{H}}\\ \hline 0\\ 0\\ \hline 0\end{array}\right]_n \tag{24}$$

where

$$\eta_2 = \sqrt{\frac{\mu_2}{\varepsilon_2}}$$

and

$$\mathscr{M} = \begin{bmatrix}\mathbf{M}_E & -\mathbf{M}_H\\ \mathbf{M}_H & \eta_2^{-2}\mathbf{M}_E\end{bmatrix} \quad \text{and similarly for } \tilde{\mathscr{M}}.$$

The submatrices are defined as before in (12), the primed matrices being associated with region 3. Thus, for example,

$$(\mathbf{M}^{\alpha\beta}_{Hij})'_n = \mathscr{K}^{\alpha\beta}_{ij}(2, 3; \Phi_3; n)$$

and similarly for the others. Solving (24) for the currents on the outermost surface yields

$$\begin{bmatrix}\mathbf{A}_1\\ \mathbf{B}_1\end{bmatrix} = [\mathbf{Z}_2]_n^{-1}\begin{bmatrix}\bar{\mathbf{E}}\\ \bar{\mathbf{H}}\end{bmatrix} \tag{25}$$

where

$$[\mathbf{Z}_2]_n = [\mathscr{S}]_n - [\mathscr{M}]_n [\mathbf{Z}_1]_n^{-1} [\tilde{\mathscr{M}}]_n \tag{26}$$

and $[\mathbf{Z}_1]_n$ is the composite, MoM matrix for the single layer case (23), with all matrices primed. If region 3 does not enclose any conducting body, then $[\mathbf{Z}_1]_n = [\mathscr{S}']_n$ and (25) is similar in form to [18, Eq. 4].

Computation of Scattering Cross Sections

In general, the scattering cross section for a dielectrically cladded conducting body or a multi-layered penetrable body for a p and q polarized scattered and incident fields, respectively, is given as

$$\sigma^{pq} = 4\pi |S^{pq}(\theta_i, \phi_i, \theta_s, \phi_s)|^2; \ (p, q = \theta \text{ or } \phi) \tag{27}$$

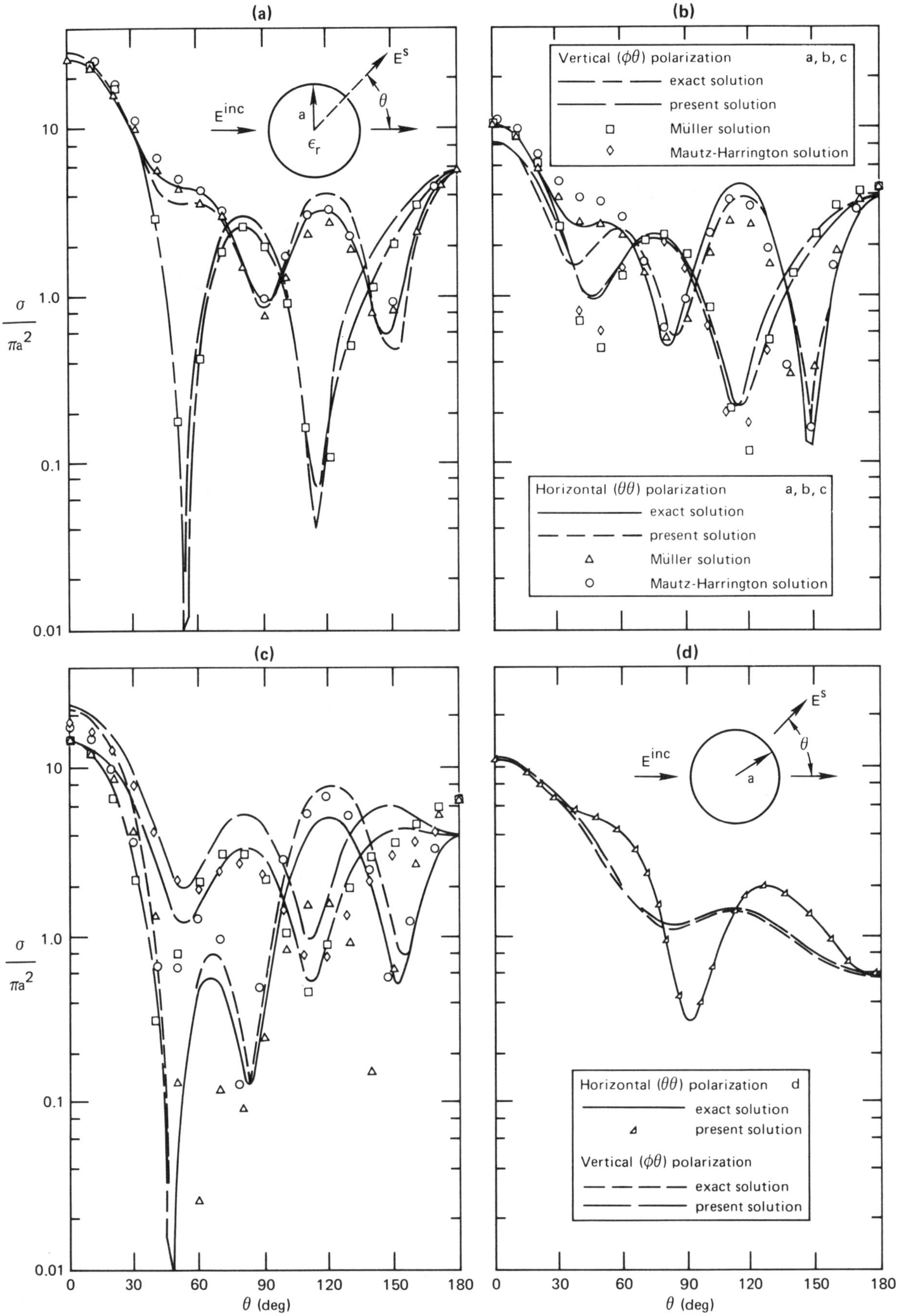

Fig. 4a–d. Scattering cross section for a dielectric sphere: (a) $ka=3.0$, $\varepsilon_r=4$; (b) $ka=3.0$, $\varepsilon_r=10$; (c) $ka=3.0$, $\varepsilon_r=20$; and (d) scattering cross section for conducting sphere, $ka=3.0$

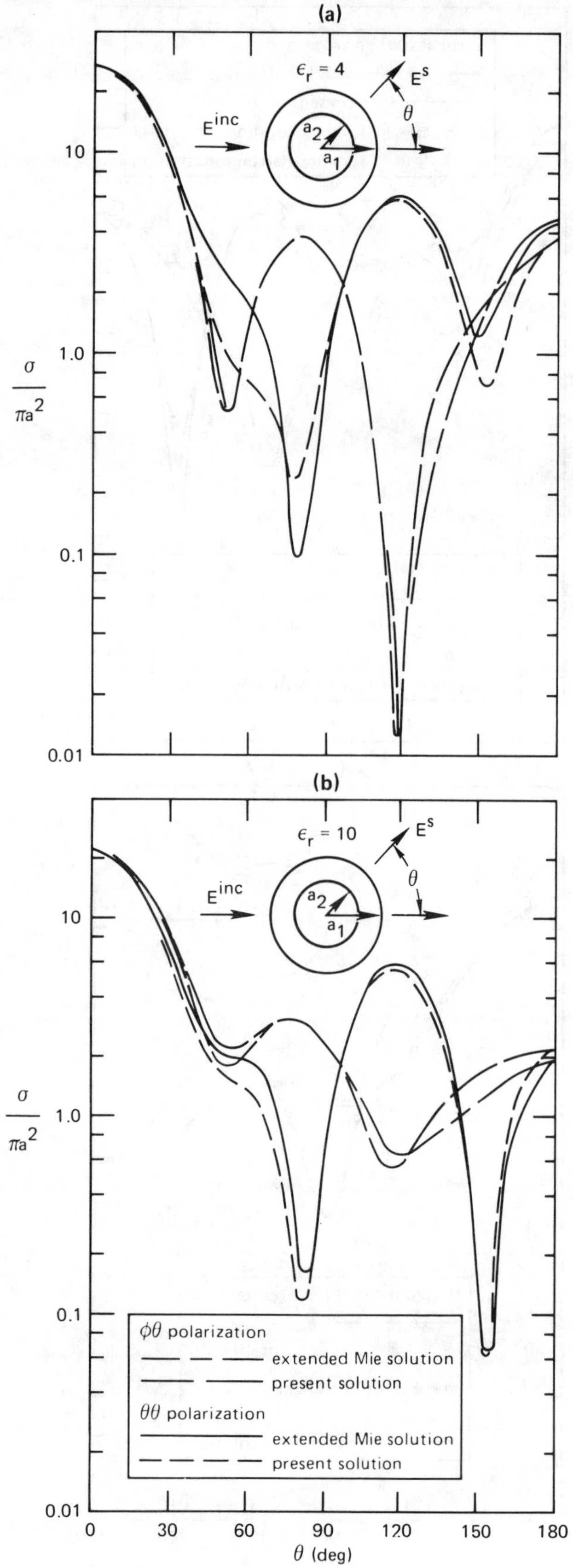

Fig. 5a and b. Scattering cross section for coated sphere: $ka_1 = 3.0$, $a_2/a_1 = 0.25$, (a) $\varepsilon_r = 4$, and (b) $\varepsilon_r = 10$

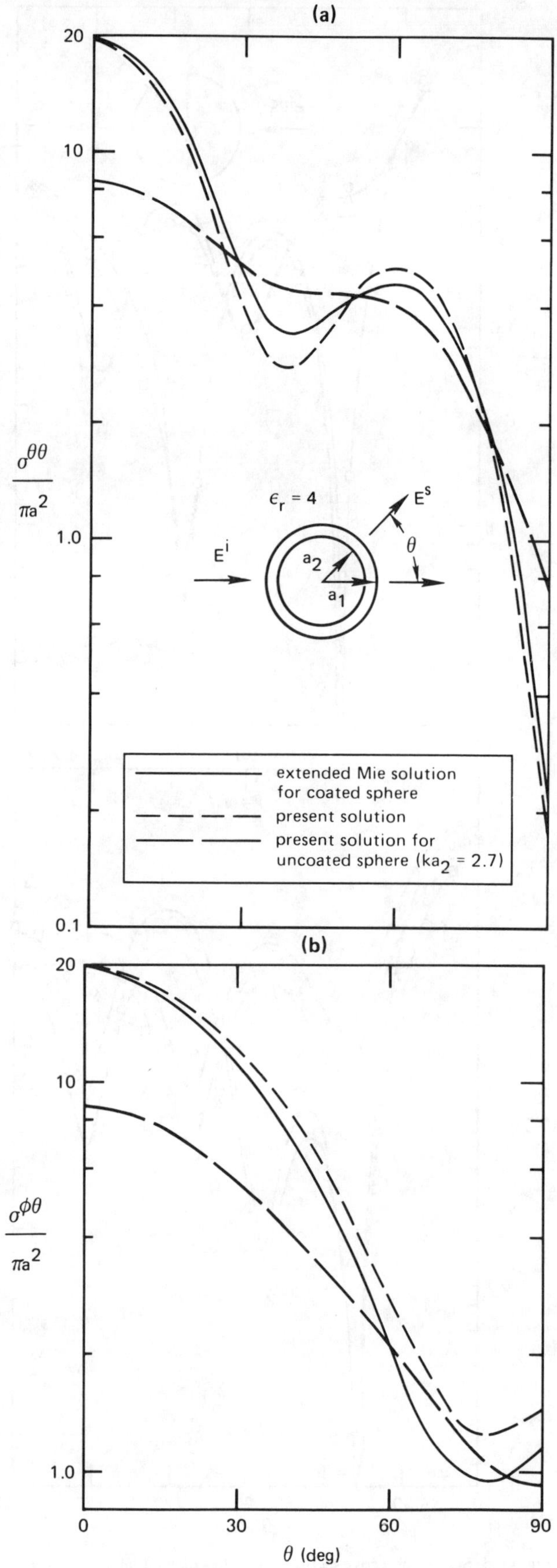

Fig. 6a and b. Scattering cross section for thinly coated sphere: $ka_1 = 3.0$, $a_2/a_1 = 0.9$, $\varepsilon_r = 4$

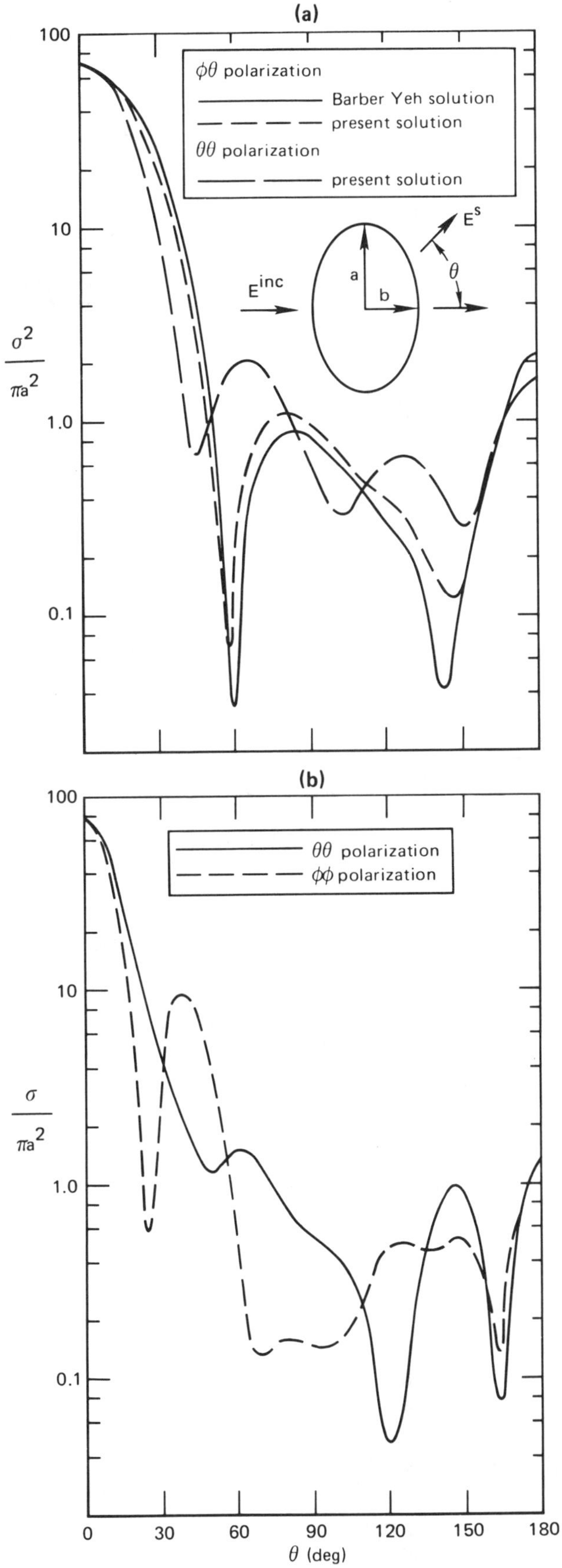

Fig. 7a and b. Scattering cross section for dielectric oblate spheroid: (a) $ka=4.932$, $a/b=3$, $\varepsilon_r=5$, and (b) $ka=7.114$, $a/b=3$; $\varepsilon_r=5$

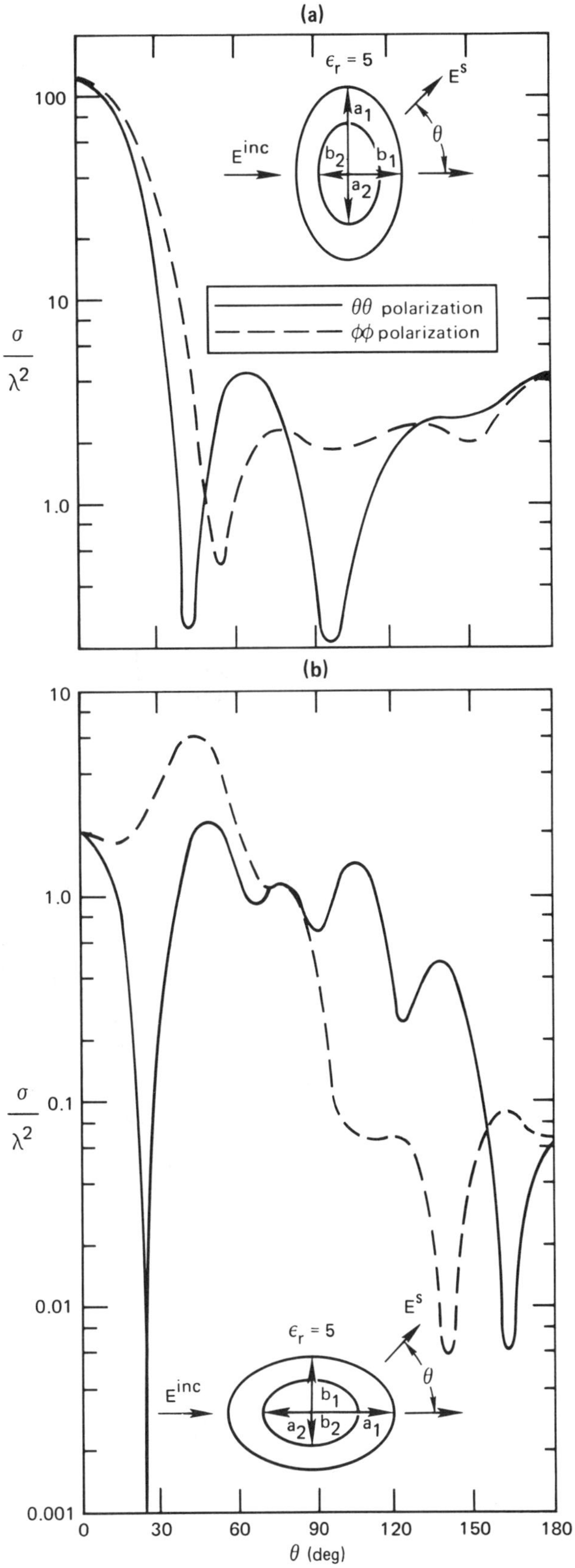

Fig. 8. (a) Scattering cross section for coated oblate spheroid; $ka_1=4.932$, $a_1/b_1=a_2/b_2=3$, $a_2/a_1=0.223$, $\varepsilon_r=5$. (b) Scattering cross section for coated prolate spheroid: $ka_1=7.114$, $a_1/b_1=a_2/b_2=3$, $a_2/a_1=0.223$, $\varepsilon_r=5$

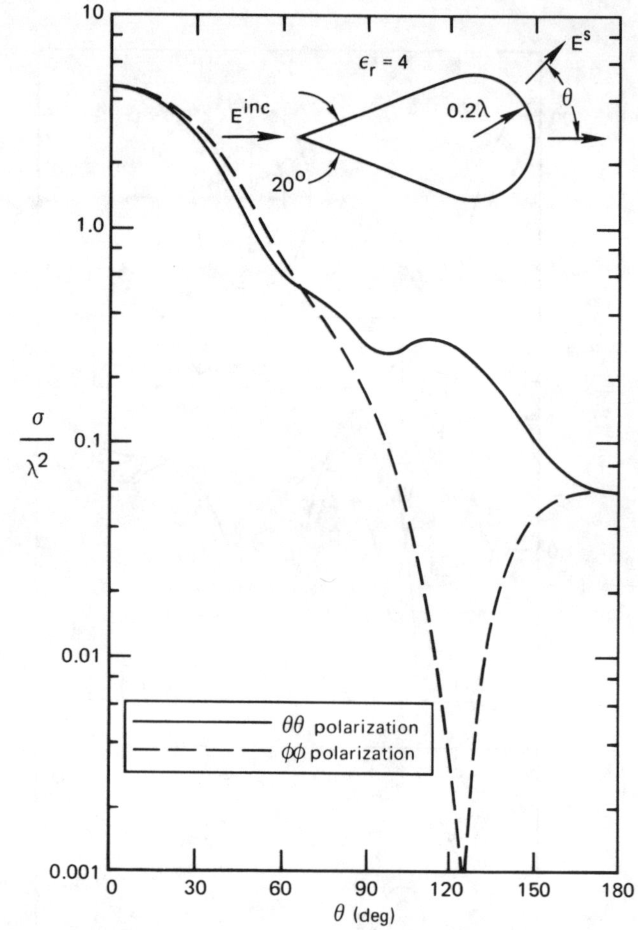

Fig. 9. Scattering cross section for dielectric conesphere: 10° half angle, $a=0.2\lambda$, $\varepsilon_r=4$

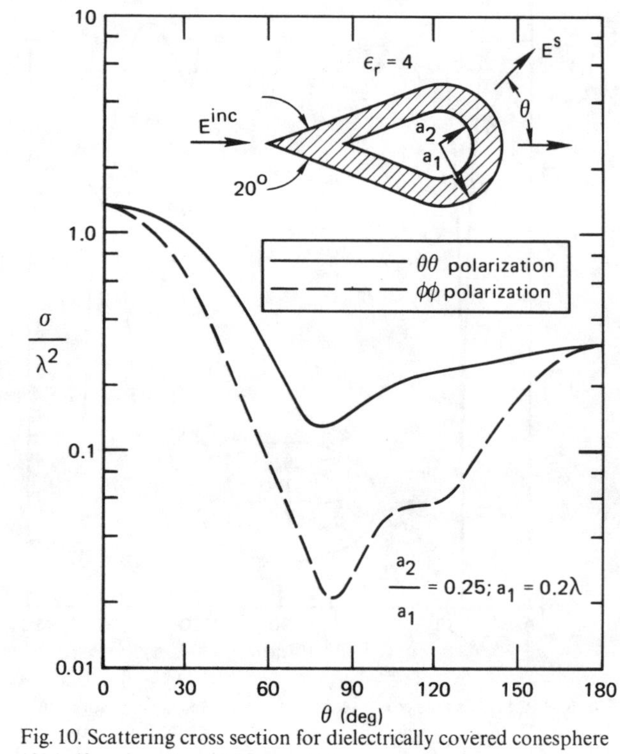

Fig. 10. Scattering cross section for dielectrically covered conesphere 10° half angle, $a_2/a_1=0.25$, $a_1=0.2\lambda$, $\varepsilon_r=4$

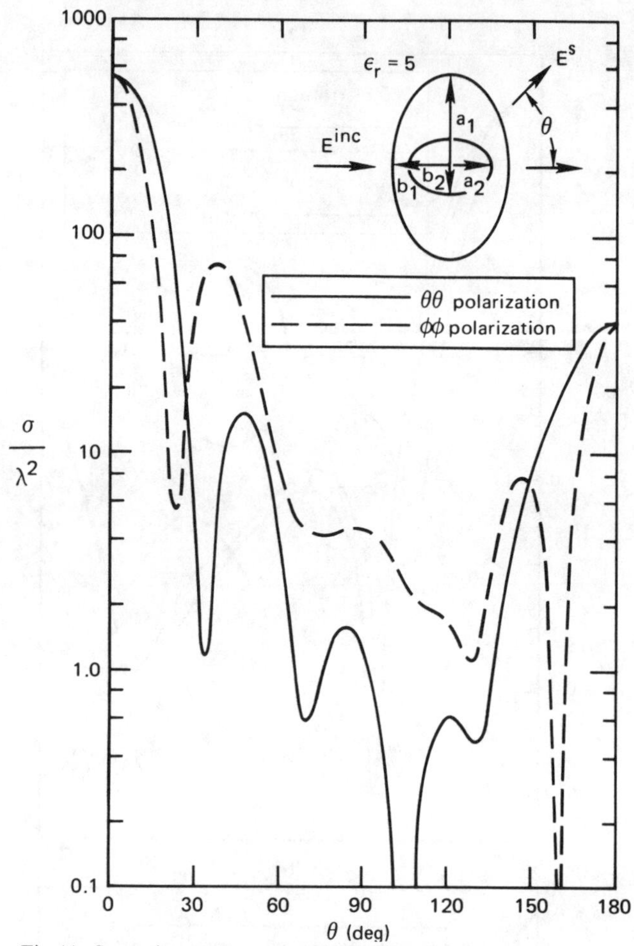

Fig. 11. Scattering cross section for conducting prolate spheroid ($ka_2=1.59$), embedded in a dielectric oblate spheroid ($ka_1=7.114$), $a_1/b_1=a_2/b_2=3.0$, $\varepsilon_r=5$

where

$$S^{pq}=-\frac{j\omega\mu}{4\pi}\sum_{n=0,\pm1\ldots}{}_t\mathscr{R}_n^p(\theta_s,\phi_s)[\mathbf{Z}]_n^{-1}\,\mathscr{R}_{-n}^q(\theta_i,\phi_i) \quad (28)$$

and where the column vector $\mathscr{R}_n^p(\cdot)$ is expressed in terms of the transfer matrices $\mathbf{R}$ [20, 22]. Thus, for example,

$$\mathscr{R}_n^\theta(\theta_s,\phi_s)=\begin{bmatrix}\mathbf{R}_n^{t\theta}\\ \mathbf{R}_n^{\phi\theta}\\ \dfrac{-1}{\eta_0}\mathbf{R}_n^{t\phi}\\ \dfrac{-1}{\eta_0}\mathbf{R}_n^{\phi\phi}\end{bmatrix}e^{\mp jn\phi_s},$$

$$\mathscr{R}_n^\phi(\theta_s,\phi_s)=\begin{bmatrix}\mathbf{R}_n^{t\phi}\\ \mathbf{R}_n^{\phi\phi}\\ \dfrac{+1}{\eta_0}\mathbf{R}_n^{t\theta}\\ \dfrac{+1}{\eta_0}\mathbf{R}_n^{\phi\theta}\end{bmatrix}e^{\mp jn\phi_s};\eta_0=377. \quad (29)$$

In the above expressions, ϕ_s and θ_s, ϕ_i and θ_i denote the angle of the scattered and incident fields, respectively. The cross section for the single layered case (Fig. 2) is computed with $\mathbf{Z}=\mathbf{Z}_\phi$; for the case in Fig. 3, $\mathbf{Z}=\mathbf{Z}_2$.

Using the present formulation, the scattering cross sections $\sigma^{\theta\theta}$, $\sigma^{\phi\theta}$, and $\sigma^{\phi\phi}$ were computed for various axisymmetric penetrable (dielectric) or conducting bodies embedded in an axisymmetric dielectric region. Comparisons were made with published data. Here, $\sigma^{\theta\theta}$ refers to the θ (horizontally) polarized patterns measured in the plane of scattering ($\phi=0$). Similarly, $\sigma^{\phi\phi}$ denotes the ϕ polarized patterns in the plane of scattering, and $\sigma^{\phi\theta}$ is the ϕ (vertically) polarized pattern normal to the plane of scattering ($\phi=90°$). The case of spherical homogeneous, penetrable scatterers is considered first. This is an important subcase of the present formulation. The $\theta\theta$ and $\phi\theta$ polarized cross sections for a homogeneous dielectric sphere with a $ka=3.0$ and $\varepsilon_r=4, 10$, and 20, are depicted in Figs. 4a–c, respectively. The scattering angle θ is measured from the incident field direction. Note $\theta=0°$ and 180° are the forward and backscatter directions, respectively. The present analysis is in good agreement with the exact (Mie) solution. For additional comparison, the results of a MoM based solution for homogeneous axisymmetric dielectric bodies using the Mautz-Harrington and the Müller formulations are given [16]. (For vertical polarization the former solution overlays the Mie results in Fig. 4a.) Correlation of the foregoing results for the dielectric spheres with the cross section for a conducting sphere of $ka=3.0$ (Fig. 4d) shows that strong nulls develop in the patterns due to internal wave cancellations in the dielectric material. These nulls are particularly pronounced in the vertical polarization. Also the focusing effect of the dielectric is evident in the enhancement of the forward scatter, expecially for $\varepsilon_r=4$ (Fig. 4a).

Application of the single-layer analysis is illustrated in Fig. 5. The cross sections for $\theta\theta$ and $\phi\theta$ polarizations are given for conducting spheres of $ka_2=0.75$, embedded in dielectric coatings of $\varepsilon_r=4$ and 10. The ratio of radii of the conducting surface (S_2) to that of the outer surface of the dielectric shell (S_1) is a $a_2/a_1=0.25$. Comparisons with the extended Mie solution [7,8] are also given. In this case, the shell thickness is 0.358 λ_0, where λ_0 is the free-space wavelength. For a wave penetrating the dielectric coating, the effective shell thicknesses appear to be 0.715 λ_0 and 1.13 λ_0 for Figs. 5a and b, respectively. The results for a very thin shell (0.047 λ_0 thickness) are shown in Fig. 6 for a conducting sphere of $ka_2=2.7$, $a_2/a_1=0.9$ and $\varepsilon_r=4$ in the coating. Also shown are the cross sections for the uncoated conducting sphere ($ka_2=2.7$). Again, the dielectric coating provides strong enhancement of the scattered fields in the forward direction.

The application of the present formulation to nonspherical scatterers is demonstrated in Figs. 7–11. The case of spheroids has been treated by Barber and Yeh [11] using the EBCM solution. Comparison of $\sigma^{\phi\theta}$ for an oblate spheroid with $a/b=3$ and $ka=4.932$, predicted by [11] and the present formulation is shown in Fig. 7a, as well as results for $\sigma^{\theta\theta}$. The cross sections $\sigma^{\theta\theta}$ and $\sigma^{\phi\phi}$ for an oblate spheroid with $a/b=3$ and $ka=7.114$ are depicted in Fig. 7b. The horizontal plane cross sections for a conducting oblate spheroid with $ka_2=1.099$ embedded concentrically in a dielectric oblate spheroid of $\varepsilon_r=5$ and $ka_1=4.932$ are shown in Fig. 8a. A similar result for a conducting prolate spheroid with $ka_2=1.59$ embedded in a dielectric prolate spheroid of $\varepsilon_r=5$ and $ka_1=7.114$ is given in Fig. 8b.

The present formulation allows any axisymmetric scatterer to be considered, including those whose shapes depart radically from a sphere. An example of such a configuration is the dielectric ($\varepsilon_r=4$) 10° (half-angle) cone-sphere shown in Fig. 9. The $\sigma^{\theta\theta}$ and $\sigma^{\phi\phi}$ cross sections are given for axial (nose-on) illumination. The corresponding cross sections for a dielectrically clad 10° (half-angle) cone-sphere is shown in Fig. 10 for axial illumination. The ratio of the radii of the spherical caps for the inner and outer surfaces is $a_2/a_1=0.25$. In addition to coated nonspherical scatterers, the present formulation also treats nonconcentric scatterers that contain penetrable or conducting bodies. An example of the latter is a conducting prolate spheroid embedded in an oblate spheroidal dielectric region, with $\varepsilon_r=4$. The prolate and the oblate spheroids have $ka_2=1.59$ and $ka_1=7.114$, respectively, and the ratio of their major to minor axes is 3. The $\phi\phi$ and $\theta\theta$ polarized cross sections for this configuration are shown in Fig. 11. Comparison of these results with those of Fig. 7b (which must be multplied by 4.03 to obtain σ/λ^2) shows that the presence of the internal scatterer enhances the forward-scattered cross section and diminishes the backscatter cross section. Furthermore, the patterns as a function of θ are greatly altered. Scattering calculations were also carried out for concentric layered dielectric spheres and spheroids using the foregoing multilayer analysis. The results were in agreement with those predicted by the Mie theory and provide additional validation of the present formulation.

In all of the above cases, the computations were carried out by segmenting the surfaces into annular rings not wider than 0.15 λ, where λ is the wavelength measured in the medium in which a given surface is embedded. For computing the bistatic cross section for axial illumination, only the $n=1$ mode is needed, since the $\mathbf{R}_n$ matrices in (29) vanish for the other modes. For backscatter (monostatic) calculations at an arbitrary angle, (ϕ_s, θ_s), the highest mode number required in $n \sim d\pi/\lambda$, where d is the largest diameter of the axisymmetric scattering geometry. This condition is identical to that for perfectly conducting axisymmetric scatterers.

Summary

A moment method formulation for treating rigorously the scattering from axisymmetric bodies embedded in axisymmetric dielectrics was presented. This analysis is applicable to dielectric coatings (shells) of arbitrary thickness. The extension to the multilayered case was discussed. The analytical formulation was expressed in a computer implementable form. Representative cases were given to illustrate the flexibility of the analysis for a wide variety of scatterer configurations. Comparisons were made with published data for spheres and spheroids. Applicability of the formulation to more complex scatterer geometries was demonstrated.

Acknowledgements. The authors thank Dr. A. C. Lind for making available his computer program for the Mie-Güttler theory and Mr. P. W. Bierdeman for assistance with the computer coding and computations.

References

1. M. Kerker: *The Scattering of Light and Other Electromagnetic Radiation.* (Academic Press, New York 1969)
2. R.C. Kouyoumjian: Proc. IEEE **53**, 864–876 (1965)
3. J.B. Keller: J. Opt. Soc. Am. **52**, 116–130 (1962)
4. D.R. Wilton, R. Mittra: IEEE Trans. AP-**20**, 310–317 (1972)
5. C. Yeh: Phys. Rev. **135**, A1193–A1201 (1964)
6. W.G. Swarner, L. Peters: IEEE Trans. AP-**11**, 558–569 (1963)
7. A. Güttler: Ann. Physik **6**, (11) 65–98 (1952)
8. G.R. Ruck (ed.); *Radar Cross Section Handbook* (Plenum Press, New York 1970), Vol. I, Chapt. 3
9. P.L. Waterman: Proc. IEEE **53**, 805–812 (1965)
10. P.L. Waterman, Phys. Rev. D **3**, 825–839 (1971)
11. P. Barber, C. Yeh: J. Appl. Opt. **14**, 2864–2872 (1975)
12. P.W. Barber: IEEE Trans. MTT-**25**, 373–381 (1977)
13. V.N. Bringi, T.A. Seliga: IEEE Trans. AP-**25**, 575–580 (1977)
14. B. Peterson, S. Ström: Phys. Rev. D **10**, 2670–2684 (1974)
15. S.K. Chang, K.K. Mei: IEEE Trans. AP-**24**, 35–42 (1976)
16. J.R. Mautz, R.F. Harrington: Electromagnetic Scattering from a Homogeneous Body of Revolution, Report TR-77-10, Department of Electrical and Computer Engineering, Syracuse University, Syracuse, NY (1977)
    See also R. Mittra (ed.): *Numerical and Asymptotic Techniques in Electromagnetics.* Topics Appl. Phys. **3** (Springer, Berlin, Heidelberg, New York 1975)
17. T.K. Wu, L.L. Tsai: Radio Sci. **12**, 709–718 (1977)
18. R.J. Pogorzelski: IEEE Trans. AP-**26**, 616–618 (1978)
19. R.F. Harrington, J.R. Mautz: IEEE Trans. AP-**23**, 531–534 (1975)
20. J.R. Mautz, R.F. Harrington: Appl. Sci. Res. **20**, 405–435 (1969)
21. R. Mittra (ed.): *Computer Techniques for Electromagnetics* (Pergamon Press, Oxford 1973) Chapt. 4
22. L.N. Medgyesi-Mitschang, J.H. Mullen: IEEE Trans. AP-**24**, 90–93 (1976)

A.1. J.R. Mautz, R.F. Harrington: J. Appl. Sci. Res. **20**, 405–435 (1969)

A.2. T.K. Wu: Electromagnetic Scattering from Arbitrarily Shaped Lossy Dielectric Bodies. Ph. D. Dissertation, University of Mississippi, (May 1976)

Appendix A

Definition of the $\mathcal{L}(\cdot)$ and $\mathcal{K}(\cdot)$ Operators

Explicit forms for $\mathcal{L}(\cdot)$ and $\mathcal{K}(\cdot)$, suitable for computer implementation, are summarized. The expressions for $\mathcal{L}(\cdot)$ embrace as a special case the impedance equations derived in [A.1] for a perfectly conducting BOR. The equations for $\mathcal{K}(\cdot)$ are a generalization of the results in [A.2], which were derived for certain specific cases of homogeneous dielectric BOR geometries. Here the expressions for $\mathcal{K}(\cdot)$ are applicable to any BOR configuration. Using the notational scheme introduced for axisymmetric bodies in [A.1], the $\mathcal{L}_{ij}^{\alpha\beta}[a,b;u_\gamma\Phi_\gamma;(\partial/\partial t)(\Phi_\gamma/w_\gamma);n]$ operators with $\alpha,\beta=t$ or ϕ evaluated on surface S_a and S_b are

$$\mathcal{L}_{ij}^{tt}(\cdot)=\int_{S_a} ds' \int_{S_b} ds\Big[\mathrm{j}\omega\Psi f_i f_j(\sin v \sin v' \cos\varphi + \cos v \cos v') - \frac{1}{\mathrm{j}\omega}\frac{1}{\varrho'}\frac{\partial}{\partial t'}(\varrho' f_j) f_i \frac{\partial\Psi}{\partial t}\Big]\mathrm{e}^{\mathrm{j}n\varphi},$$

where u_γ and w_γ denote the permeability or permittivity, and where $\varphi=\phi'-\phi$ and for notational simplicity $\Psi\equiv u_\gamma\Phi_\gamma$ and $\partial\Psi/\partial t\equiv(\partial/\partial t)(\Phi_\gamma/w_\gamma)$; v' and v are the angles subtended between a tangent to the generating curves of S_a or S_b and the z-axis. Correspondingly

$$\mathcal{L}_{ij}^{t\phi}(\cdot)=\int_{S_a} ds' \int_{S_b} ds\left(-j\omega\Psi f_i f_j \sin v' \sin\varphi - \frac{n}{\omega}\frac{1}{\varrho'} f_j f_i \frac{\partial\Psi}{\partial t}\right)\mathrm{e}^{\mathrm{i}n\varphi} \tag{A.2}$$

$$\mathcal{L}_{ij}^{\phi t}(\cdot)=-\mathcal{L}_{ij}^{t\phi}(\cdot) \tag{A.3}$$

with i and j, and primed and unprimed quantities interchanged in (A.2).

$$\mathcal{L}_{ij}^{\phi\phi}(\cdot)=\int_{S_a} ds' \int_{S_b} ds\left(\mathrm{j}\omega\Psi\cos\varphi - \frac{n}{\omega\varrho'\varrho}\frac{\partial\Psi}{\partial\phi}\right) f_i f_j \mathrm{e}^{\mathrm{j}n\varphi}, \tag{A.4}$$

Explicit expressions suitable for computer implementation can be written for the foregoing equations. Choosing the basis functions f_i to be triangle functions, i.e.,

$$f_i=\begin{cases}\dfrac{1}{\varrho}(1-|t-t_i|), & |t-t_i|\leqq 1\,,\\ 0, & |t-t_i|>1\end{cases}$$

and approximating these with pulses as in (A.1), yields

$$\mathcal{L}_{ij}^{tt}(\cdot)=\sum_{p,q=1}^{4}\left\{\mathrm{j}\omega T_p T_q\left(u_\gamma \sin v_p \sin v_q\left[\frac{G_{n+1}^{(\gamma)}+G_{n-1}^{(\gamma)}}{2}\right] + \cos v_p \cos v_q u_\gamma G_n^{(\gamma)}\right) - \frac{\mathrm{j}}{\omega}\dot{T}_p\dot{T}_q\frac{G_n^{(\gamma)}}{w_\gamma}\right\}, \tag{A.5}$$

$$\mathcal{L}_{ij}^{t\phi}(\cdot)=\sum_{p,q=1}^{4}\left\{-\omega u_\gamma \sin v_p T_p T_q\left[\frac{G_{n+1}^{(\gamma)}-G_{n-1}^{(\gamma)}}{2}\right] + \frac{n}{\omega\varrho_q}\dot{T}_p T_q\frac{G_n^{(\gamma)}}{w_\gamma}\right\}, \tag{A.6}$$

$$\mathcal{L}_{ij}^{\phi t}(\cdot)=-\mathcal{L}_{ji}^{t\phi}(\cdot) \tag{A.7}$$

with p and q interchanged, and

$$\mathcal{L}_{ij}^{\phi\phi}(\cdot)=\sum_{p,q=1}^{4}\left[\mathrm{j}\omega u_\gamma T_p T_q\left(\frac{G_{n+1}^{(\gamma)}+G_{n-1}^{(\gamma)}}{2}\right) + \frac{n^2}{\mathrm{j}\omega}\frac{T_p}{\varrho_p}\frac{T_q}{\varrho_q}\frac{G_n^{(\gamma)}}{w_\gamma}\right], \tag{A.8}$$

where ϱ_p and ϱ_q are evaluated at points corresponding to the i-th and j-th annular segments on surface S_a and S_b into which they are subdivided.

The terms T_p and $\dot{T}_p$ are the pulse approximations for the triangle functions and their derivatives, i.e., $T_p=\{1/4,\ 3/4,\ 3/4,\ 1/4\}$ and $\dot{T}_p=\{1, 1, -1, -1\}$. $G_n^{(\gamma)}$ is the integral of the Green's function Ψ, and is given in [A.1] as

$$G_n^{(\gamma)} = \frac{\pi}{M} \sum_{m=1}^{M} \cos n\phi_m f(\phi_m), \tag{A.9}$$

where

$$f(\phi_m) = e^{-jk_\gamma R_{pq}} \left\{ 2(1 + jk_\gamma R_{pq}) \ln \left[\frac{t_2 + \sqrt{t_2^2 + d^2}}{t_1 + \sqrt{t_1^2 + d^2}} \right] - jk_\gamma \right\},$$

and

$$\mathscr{K}_{ij}^{\phi\phi}(\cdot) = \int_{S_a} ds' \int_{S_b} ds \left(-\mathbf{u}_{\phi'} \cdot \mathbf{n} \frac{\partial \Psi}{\partial t} - \sin v \sin \varphi \frac{\partial \Psi}{\partial n} \right) f_i f_j e^{jn\varphi}, \tag{A.13}$$

where $\partial\Psi/\partial n$ is the derivative of the Green's function with respect to the outward normal $\mathbf{n}$ on surface, S_a. Written explicitly for triangle functions as a basis set, the computer implementable expressions for $\mathscr{K}(\cdot)$ are

$$\mathscr{K}_{ij}^{tt}(\cdot) = \sum_{p,q=1}^{4} T_p T_q \left\{ \frac{j}{\varrho_p} \left(\cos v_p \sin v_q \left[\frac{G_{n+1}^{(\gamma)} + G_{n-1}^{(\gamma)}}{2} \right] - \sin v_p \cos v_q G_n^{(\gamma)} \right) \right.$$
$$\left. + (\varrho_p \cos v_p - (z_p - z_q) \sin v_p) \sin v_q B_{n+1}^{(\gamma)} - \tfrac{1}{2} \varrho_q \cos v_p \sin v_q B_{n+2}^{(\gamma)} \right\}, \tag{A.14}$$

$$\mathscr{K}_{ij}^{t\phi}(\cdot) = \sum_{p,q=1}^{4} T_p T_q \left\{ -\frac{j}{\varrho_p} \cos v_p \left[\frac{G_{n+1}^{(\gamma)} - G_{n-1}^{(\gamma)}}{2} \right] + j(\varrho_p \cos v_p - (z_p - z_q) \sin v_p) A_{n+1}^{(\gamma)} - j \frac{\varrho_q}{2} \cos v_p D_n^{(\gamma)} \right\}, \tag{A.15}$$

$$\mathscr{K}_{ij}^{\phi t}(\cdot) = \sum_{p,q=1}^{4} \left\{ \dot{T}_p T_q \left(\cos v_p \sin v_q \left[\frac{G_{n+1}^{(\gamma)} + G_{n-1}^{(\gamma)}}{2} \right] - \sin v_p \cos v_q G_n^{(\gamma)} \right) \right.$$
$$- jT_p T_q [\cos v_p \cos v_q (\varrho_p \cos v_p - (z_p - z_q) \sin v_p)(k_2 H_n^{(\gamma)} - jK_n^{(\gamma)})$$
$$\left. + (\sin v_p \sin v_q (\varrho_p \cos v_p - (z_p - z_q) \sin v_p) - \varrho_q \cos^2 v_p \cos v_q) A_{n+1}^{(\gamma)} - \tfrac{1}{2} \varrho_q \cos v_p \sin v_p \sin v_q D_n^{(\gamma)}] \right\}, \tag{A.16}$$

and

$$\left.\begin{matrix} t_1 \\ t_2 \end{matrix}\right\} = t_0 \mp \frac{1}{4}; \quad d^2 = R_{pq}^2 - t_0^2; \quad \phi_m = \left(m - \frac{1}{2} \right) \pi / M$$

$$t_0 = |(z_p - z_q) \cos v_q + (\varrho_p \cos \phi_m - \varrho_q) \sin v_q|$$

and R_{pq} is the distance between the p-th and q-th annuli on the BOR surfaces. If the $\mathscr{L}(\cdot)$ operators are defined over the same surfaces (i.e., $S_a = S_b$), then a self-term exists when $R_{pq} = 0$, and (A.9) is modified so that

$$G_n^{(\gamma)} = \frac{4}{\varrho_q} \left[x_0 \ln \left(\frac{y_0 + \sqrt{y_0^2 + x_0^2}}{x_0} \right) + y_0 \ln \left(\frac{x_0 + \sqrt{y_0^2 + x_0^2}}{y_0} \right) \right]$$
$$- \frac{jk_\gamma x_0}{\varrho_q} + \sum_{m=2}^{M} \cos n\phi_m f(\phi_m),$$

where

$$x_0 = \frac{\varrho_q \pi}{M}; \quad y_0 = 1/4.$$

If $a = b = 1$ so that S_1 defines a perfectly conducting BOR, and $\gamma = 1$, $u = \mu$, and $w = \varepsilon$ then (A.5–8) are identical to the impedance expressions given in [Ref. A.1, Eq. (45)]. Computational experience has shown that for stable numerical results, M should be $> 2\pi^2(\varrho_{max}/\lambda)$, where ϱ_{max} is the largest radius of surface S_a or S_b. The corresponding expressions for $\mathscr{K}_{ij}^{\alpha\beta}[a, b; \Phi_\gamma; n]$ are

$$\mathscr{K}_{ij}^{tt}(\cdot) = \int_{S_a} ds' \int_{S_b} ds \left(\mathbf{u}_{t'} \cdot \mathbf{n} \frac{1}{\varrho} \frac{\partial \Psi}{\partial \varphi} + \sin v' \sin \varphi \frac{\partial \Psi}{\partial n} \right) f_i f_j e^{jn\varphi}, \tag{A.10}$$

where

$$\varphi = \phi' - \phi; \quad \Psi \equiv \Phi_\gamma, \frac{\partial \Psi}{\partial t} \equiv \frac{\partial \Phi_\gamma}{\partial t},$$

and similarly for all the other derivatives,

$$\mathscr{K}_{ij}^{t\phi}(\cdot) = \int_{S_a} ds' \int_{S_b} ds \left(\mathbf{u}_{\phi'} \cdot \mathbf{n} \frac{1}{\varrho} \frac{\partial \Psi}{\partial \phi} - \cos \varphi \frac{\partial \Psi}{\partial n} \right) f_i f_j e^{jn\varphi}, \tag{A.11}$$

$$\mathscr{K}_{ij}^{\phi t}(\cdot) = \int_{S_a} ds' \int_{S_b} ds \left[-\mathbf{u}_{t'} \cdot \mathbf{n} \frac{\partial \Psi}{\partial t} + (\sin v \sin v' \cos \varphi \right.$$
$$\left. + \cos v \cos v') \frac{\partial \Psi}{\partial n} \right] f_i f_j e^{jn\varphi}, \tag{A.12}$$

$$\mathscr{K}_{ij}^{\phi\phi}(\cdot) = \sum_{p,q=1}^{4} \left\{ j\dot{T}_p T_q \cos v_p \left[\frac{G_{n+1}^{(\gamma)} - G_{n-1}^{(\gamma)}}{2} \right] \right.$$
$$+ T_p T_q \left(\varrho_p \cos v_p \sin v_p - (z_p - z_q) \sin^2 v_p \right) B_{n+1}^{(\gamma)}$$
$$\left. - \tfrac{1}{2} T_p T_q \varrho_q \cos v_p \sin v_p B_{n+2}^{(\gamma)} \right\}, \tag{A.17}$$

where

$$A_{n+l}^{(\gamma)} = \tfrac{1}{2}(k_\gamma H_{n+l}^{(\gamma)} - jK_{n+l}^{(\gamma)}) + \tfrac{1}{2}(k_\gamma H_{n-l}^{(\gamma)} - jK_{n-l}^{(\gamma)}),$$
$$B_{n+l}^{(\gamma)} = \tfrac{1}{2}(k_\gamma H_{n+l}^{(\gamma)} - jK_{n+l}^{(\gamma)}) - \tfrac{1}{2}(k_\gamma H_{n-l}^{(\gamma)} - jK_{n-l}^{(\gamma)}),$$
$$D_n^{(\gamma)} = (k_\gamma H_n^{(\gamma)} - jK_n^{(\gamma)}) + A_{n+2}^{(\gamma)},$$

and

$\gamma = 1, 2,$ and $l = 1, 2.$

G_n is defined as before in (A.9), and

$$H_n^{(\gamma)} = \frac{\pi}{M} \sum_{m=1}^{M} h(\phi_m) \cos n\phi_m, \tag{A.18}$$

where in terms of the previously defined parameters,

$$h(\phi_m) = 2e^{-jk_\gamma R_{pq}} \left\{ \frac{1 + jk_\gamma R_{pq}}{d} \left[\tan^{-1}\left(\frac{t_2}{d}\right) - \tan^{-1}\left(\frac{t_1}{d}\right) \right] \right.$$
$$\left. - jk_\gamma \ln \left[\frac{t_2 + \sqrt{t_2^2 + d^2}}{t_1 + \sqrt{t_1^2 + d^2}} \right] \right\} \tag{A.18a}$$

and similarly,

$$K_n^{(\gamma)} = \frac{\pi}{M} \sum_{m=1}^{M} k(\phi_m) \cos n\phi_m, \tag{A.19}$$

where

$$k(\phi_m) = 2e^{-jk_\gamma R_{pq}} \left\{ \frac{1 + jk_\gamma R_{pq}}{d^2} \left(\frac{t_2}{\sqrt{t_2^2 + d^2}} - \frac{t_1}{\sqrt{t_1^2 + d^2}} \right) \right.$$
$$\left. - \frac{jk_\gamma}{d} \left[\tan^{-1}\left(\frac{t_2}{d}\right) - \tan^{-1}\left(\frac{t_1}{d}\right) \right] \right\}. \tag{A.19a}$$

Errata:

In (A.7) replace $-\mathscr{L}_{ij}^{t\phi}(\cdot)$ by $-\mathscr{L}_{ji}^{t\phi}(\cdot)$

In (A.14) and (A.15) replace $G_{n\pm1}^{(v)}$ by $(n\pm1)\,G_{n\pm1}^{(v)}$

PART 3:
Conducting and Dielectric Bodies

A surface solution using a quadratic basis with point matching and harmonic azimuthal terms for a finite tubular cylinder is found in a relatively early paper by Kao (1970). Triangular patches, which allow almost any surface to be covered with piecewise triangular functions, are used by Rao, Wilton, and Glisson (1982). Another excellent approach uses quadrilateral patches with piecewise sinusoidal functions; see Newman and Tulyathan (1982). Greatly improved convergence results from bases incorporating a proper edge singularity, as developed by Wilton and Govind (1977) for a strip. Ricoy, Kilberg, and Volakis (1989) reduce the number of equations through the use of equivalent currents. Dielectric strip scattering is treated by Richmond (1985) using plane and surface wave bases.

Many wire antennas are mounted on metallic bodies, with or without dielectric coatings. Albertsen, Hansen, and Jensen (1974) use surface patches with quadratic bases, and wires with pulse bases. An attachment mode to handle the singularity in current at the wire-surface join is contributed by Newman and Pozar (1978). Combined field integral equations and other techniques used to subdue the effects of internal cavity resonances are listed in the Bibliography.

In the medical electronics field, calculation of energy deposition is important. Block models are usually employed: Chen and Guru (1977) and Hagmann and Gandhi (1979) offer examples of these calculations. These methods are also useful for thick dielectric layer geometries.

BIBLIOGRAPHY

Wire Grid Modeling

J.L. Lin, W.L. Curtis and M.C. Vincent, "Radar Cross Section of a Rectangular Conducting Plate by Wire Mesh Modeling," *Trans. IEEE,* Vol. AP-22, September 1974, pp. 718–720.

Y.T. Lin and J.H. Richmond, "EM Modeling of Aircraft at Low Frequencies," *Trans. IEEE,* Vol. AP-23, January 1975, pp. 53–56.

K.S.H. Lee, L. Marin and J.P. Castillo, "Limitations of Wire Grid Modeling of a Closed Surface," *Trans. IEEE,* Vol. EMC-18, August 1976, pp. 123–129.

P.K. Agrawal, "A Hybrid Technique for Wire Antennas in a Cavity," *Trans. IEEE,* Vol. AP-26, May 1978, pp. 434–438.

A.C. Ludwig, "Wire Grid Modeling of Surfaces," *Trans. IEEE,* Vol. AP-35, September 1987, pp. 1045–1048.

Surface Patches

D.L. Knepp and J. Goldhirsh, "Numerical Analysis of Electromagnetic Radiation Properties of Smooth Conducting Bodies of Arbitrary Shape," *Trans. IEEE,* Vol. AP-20, May 1972, pp. 383–388.

L.L. Tsai, D.G. Dudley and D.R. Wilton, "Electromagnetic Scattering by a Three-dimensional Conducting Rectangular Box," *J. Appl. Phys.,* Vol. 45, October 1974, pp. 4393–4400.

N.N. Wang, J.H. Richmond and M.C. Gilreath, "Sinusoidal Reaction Formulation for Radiation and Scattering from Conducting Surfaces," *Trans. IEEE,* Vol. AP-23, May 1975, pp. 376–382.

J.J.H. Wang, "Numerical Analysis of Three Dimensional Arbitrarily-Shaped Conducting Scatterers by Trilateral Surface Cell Modeling," *Radio Science,* Vol. 13, November-December 1978, pp. 947–952.

J. Singh and A.T. Adams, "A Non-Rectangular Patch Model for Scattering from Surfaces," *Trans. IEEE,* Vol. AP-27, July 1979, pp. 531–535.

A.W. Glisson and D.R. Wilton, "Simple and Efficient Numerical Methods for Problems of Electromagnetic Radiation and Scattering from Surfaces," *Trans. IEEE,* Vol. AP-28, September 1980, pp. 593–603.

J.J.H. Wang and C.J. Drane, "Numerical Analysis of Arbitrarily Shaped Bodies Modeled by Surface Patches," *Trans. IEEE,* Vol. MTT-30, August 1982, pp. 1167–1173.

D.R. Wilton, S.M. Rao, A.W. Glisson, D.H. Schaubert, O.M. Al-Bundak and C.M. Butler, "Potential Integrals for Uniform and Linear Source Distributions on Polygonal and Polyhedral Domains," *Trans. IEEE,* Vol. AP-32, March 1984, pp. 276–281.

E.H. Newman, P. Alexandropoulos and E.K. Walton, "Polygonal Plate Modeling of Realistic Structures," *Trans. IEEE,* Vol. AP-32, July 1984, pp. 742–747.

L.M. Medgyesi-Mitschang and J.M. Putnam, "Electromagnetic Scattering from Extended Wires and Two- and Three-dimensional Surfaces," *Trans. IEEE,* Vol. AP-33, October 1985, pp. 1090–1100.

R. Janaswamy, "A Simplified Expression for the Self/Mutual Impedance Between Coplanar and Parallel Surface Monopoles," *Trans. IEEE,* Vol. AP-35, October 1987, pp. 1174–1176.

E. Alanen, "Pyramidical Basis Functions in the Method of Moments," *Microw. & Opt. Tech. Letters,* Vol. 2, July 1989, pp. 265–268.

Conducting Cylinders

R.A. Ross, H.R. Witt and J.R. Graham, "Radar Cross-Section of a Finite Rectangular Cylinder," *Electronics Letters,* Vol. 3, 1967, p. 336.

D.M. Bolle and D.M. Fye, "Application of Point-Matching Method to Scattering from Quadrilateral Cylinders," *Electronics Letters,* Vol. 7, 1971, p. 577.

P.L. Huddleston, "Scattering from Conducting Finite Cylinders with Thin Coatings," *Trans. IEEE,* Vol. AP-35, October 1987, pp. 1128–1136.

E. Arvas, Y. Qian, T.K. Sarkar and F. Aslan, "TE Scattering from a Conducting Cylinder of Arbitrary Cross-Section Covered by Multiple Layers of Lossy Dielectric," *Proc. IEE,* Vol. 136H, December 1989, pp. 425–430.

J.L. Tsalamengas, "Electromagnetic Scattering from Conducting Circular Cylinders in the Presence of a Stratified Anisotropic Medium—TM Case," *Trans. IEEE,* Vol. AP-37, December 1989, pp. 1582–1590.

A.G. Papayiannakis and E.E. Kriezis, "Scattering from a Dielectric Cylinder of Finite Length," *Trans. IEEE,* Vol. AP-31, September 1983, pp. 725–731.

L.L. Tsai, T.K. Wu, R.D. Darone and G.L. Brown, "Some Aspects of Valid EMC Testing of Missiles," *Trans. IEEE,* Vol. EMC-20, May 1978, pp. 306–313.

Reflectors

J.A. Kinzel, "Large Reflector Antenna Pattern Computation Using Moment Methods," *Trans. IEEE,* Vol. AP-22, January 1974, pp. 116–118.

K.R.G. Bailey, T.S.M. Maclean and G. Morris, "Field-Contour Plots in Parabolic Cylinder by Method of Moments," *Electronics Letters,* Vol. 13, March 31, 1977, pp. 207–208.

B. Azarbar and L. Shafai, "Application of Moment Method to Large Cylindrical Reflector Antennas," *Trans. IEEE,* Vol. AP-26, May 1978, pp. 500–502.

R. Neri and T.S.M. Maclean, "On-Axis Field of Small Grid Paraboloid by Moment Method," *Proc. IEE-MOA,* Vol. 2, July 1978, pp. 123–128.

Other Surfaces

R. Mittra, Y. Rahmat-Samii, D.V. Jamnejad, and W.A. Davis, "A New Look at the Thin-Plate Scattering Problem," *Radio Science,* Vol. 8, October 1973, pp. 869–875.

D.R. Wilton and R. Mittra, "A New Numerical Approach to the Calculation of EM Scattering Properties of Two-dimensional Bodies of Arbitrary Cross Section," *Trans. IEEE,* Vol. AP-20, May 1972, pp. 310–317.

A.C. Ludwig, "A Comparison of Spherical Wave Boundary Value Matching Versus Integral Equation Scattering Solutions for a Perfectly Conducting Body," *Trans. IEEE,* Vol. AP-34, July 1986, pp. 857–865.

R.F. Harrington and J.R. Mautz, "An Impedance Sheet Approximation for Thin Dielectric Shells," *Trans. IEEE,* Vol. AP-23, July 1975, pp. 531–534.

L.N. Medgyesi-Mitschang and J.M. Putnam, "Integral Equation Formulations for Imperfectly Conducting Scatterers," *Trans. IEEE,* Vol. AP-33, February 1985, pp. 206–214.

D.J. Heedy and W.D. Burnside, "An Aperture-Matched Compact Range Feed Horn Design," *Trans. IEEE,* Vol. AP-33, November 1985, pp. 1249–1255.

A.J. Terzuoli, J.H. Richmond and L. Peters, "The VSWR of E-Plane Dihedral Horns," *Trans. IEEE,* Vol. AP-26, March 1978, pp. 236–239.

A.J. Terzuoli and L. Peters, "VSWR Properties of E-Plane Dihedral Corrugated Horns," *Trans. IEEE,* Vol. AP-26, March 1978, pp. 239–243.

R. Janaswamy, "An Accurate Moment Method Model for the Tapered Slot Antenna," *Trans. IEEE,* Vol. AP-37, December 1989, pp. 1523–1528.

L.N. Medgyesi-Mitschang, and J.M. Putnam, "Electromagnetic Scattering from Ducts with Irregular Edges: Part I—Circular Case," *Trans. IEEE,* Vol. AP-36, March 1988, pp. 383–397.

J.M. Putnam and L.N. Medgyesi-Mitschang, "Electromagnetic Scattering from Ducts with Irregular Edges; Part II—Noncircular Case," *Trans. IEEE,* Vol. AP-36, March 1988, pp. 398–404.

H.K. Chen, N.H. Sun and M.Y. Li, "A Study of the Effect of Buildings on the Radiation Patterns of Simple Antenna Arrays," *Trans. IEEE,* Vol. AP-31, July 1983, pp. 689–692.

A. Boag, Y. Leviatan and A. Boag, "Analysis of Two-dimensional Electromagnetic Scattering from Nonplanar Periodic Surfaces Using a Strip Current Model," *Trans. IEEE,* Vol. AP-37, November 1989, pp. 1437–1446.

Wires Connected to Surfaces

F.M. Tesche and A.R. Neureuther, "Radiation Patterns for Two Monopoles on a Perfectly Conducting Sphere," *Trans. IEEE,* Vol. AP-18, September 1970, pp. 692–694.

J. Goldhirsch, D.L. Knepp and R.J. Doviak, "Radiation from a Dipole Near a Conducting Cylinder of Finite Length," *Trans. IEEE,* Vol. EMC-12, August 1970, pp. 96–105.

E.K. Miller and J.B. Morton, "The RCS of a Metal Plate with a Resonant Slot," *Trans. IEEE,* Vol. AP-18, March 1970, pp. 290–292.

A.H. Cherin and J. Goldhirsh, "Impedance and Far Field Characteristics of a Linear Antenna Near a Conducting Cylinder," *Trans. IEEE,* Vol. EMC-15, August 1973, pp. 110–117.

E.H. Newman and D.M. Pozar, "Considerations for Efficient Wire/Surface Modeling," *Trans. IEEE,* Vol. AP-28, January 1980, pp. 121–125.

E.K. Yung and C.M. Butler, "Coaxial-Line Driven Monopole on an Electrically Thick Conducting Cylinder Over a Ground Plane," *Proc. IEE-MOA,* Vol. 131, February 1984, pp. 54–60.

J. Perini and D.J. Buchanan, "Assessment of MOM Techniques for Shipboard Applications," *Trans. IEEE,* Vol. EMC-24, February 1982, pp. 32–39.

Internal Resonances

W.A. Davis and R. Mittra, "A New Approach to the Thin Scatterer Problem Using the Hybrid Equations," *Trans. IEEE,* Vol. AP-25, May 1977, pp. 402–406.

J.R. Mautz and R.F. Harrington, "A Combined-Source Solution for Radiation and Scattering from a Perfectly Conducting Body," *Trans. IEEE,* Vol. AP-27, July 1979, pp. 445–454.

A.D. Yaghjian, "Augmented Electric- and Magnetic-Field Integral Equations," *Radio Science,* Vol. 16, November-December 1981, pp. 987–1001.

C.A. Klein and R. Mittra, "An Application of the 'Condition Number' Concept to the Solution of Scattering Problems in the Presence of Interior Resonant Frequencies," *Trans. IEEE,* Vol. AP-23, May 1975, pp. 431–435.

N. Morita, "Resonant Solutions Involved in the Integral Equation Approach to Scattering from Conducting and Dielectric Cylinders," *Trans. IEEE,* Vol. AP-27, November 1979, pp. 869–871.

F.X. Canning, "Singular Value Decomposition of Integral Equations of EM and Applications to the Cavity Resonance Problem," *Trans. IEEE,* Vol. AP-37, September 1989, pp. 1156–1163.

W.V.T. Rusch and R.J. Pogorzelski, "A Mixed-Field Solution for Scattering from Composite Bodies," *Trans. IEEE,* Vol. AP-34, July 1986, pp. 955–958.

R. Naishadham and L.W. Pearson, "Numerical Evaluation of Complex Resonances of an Elliptic Cylinder," *Trans. IEEE,* Vol. AP-33, June 1985, pp. 674–676.

S.M. Rao, E. Arvas and T.K. Sarkar, "Combined Field Solution for TM Scattering from Multiple Conducting and Dielectric Cylinders of Arbitrary Cross Section," *Trans. IEEE,* Vol. AP-35, April 1987, pp. 447–451.

J.R. Parkinson and M.J. Mehler, "Uniqueness of Numerical Solutions to Electric-, Magnetic- and Combined-field Integral Equations for Open-ended Circular Waveguides," *Proc. IEE,* Vol. 136H, June 1989, pp. 269–275.

Dielectric Cylinders

A.F. Peterson and P.W. Klock, "An Improved MFIE Formulation for TE-Wave Scattering from Lossy, Inhomogeneous Dielectric Cylinders," *Trans. IEEE,* Vol. AP-36, January 1988, pp. 45–49.

R.G. Rojas, "Scattering by an Inhomogeneous Dielectric/Ferrite Cylinder of Arbitrary Cross-Section Shape—Oblique Incidence Case," *Trans. IEEE,* Vol. AP-36, February 1988, pp. 238–246.

A.F. Peterson, "A Comparison of Integral, Differential and Hybrid Methods for TE-Wave Scattering from Inhomogeneous Dielectric Cylinders," *J. Electromag. Waves and Appl.,* Vol. 3, No. 2, 1989, pp. 87–106.

Y. Leviatan and A. Boag, "Analysis of Electromagnetic Scattering from Dielectric Cylinders Using a Multifilament Current Model," *Trans. IEEE,* Vol. AP-35, October 1987, pp. 1119–1127.

Y. Leviatan, A. Boag and A. Boag, "Analysis of TE Scattering from Dielectric Cylinders Using a Multifilament Magnetic Current Model," *Trans. IEEE,* Vol. AP-36, July 1988, pp. 1026–1031.

L.N. Medgyesi-Mitschang and J.M. Putnam, "Electromagnetic Scattering from Electrically Large Coated Flat and Curved Strips: Entire Domain Galerkin Formulation," *Trans. IEEE,* Vol. AP-35, July 1987, pp. 790–801.

L.N. Medgyesi-Mitschang and J.M. Putnam, "Scattering from Composite Laminate Strips," *Trans. IEEE,* Vol. AP-37, November 1989, pp. 1427–1436.

Dielectric Bodies

R. Ruppin, "Calculation of Electromagnetic Energy Absorption in Prolate Spheroids by the Point Matching Method," *Trans. IEEE,* Vol. MTT-26, February 1978, pp. 87–90.

K. Umashankar, A. Taflove and S.M. Rao, "Electromagnetic Scattering by Arbitrary Shaped Three-dimensional Homogeneous Lossy Dielectric Objects," *Trans. IEEE,* Vol. AP-34, June 1986, pp. 758–766.

B. Beker, K.R. Umashankar and A. Taflove, "Numerical Analysis and Validation of the Combined Field Surface Integral Equations for Electromagnetic Scattering by Arbitrary Shaped Two-dimensional Anisotropic Objects," *Trans. IEEE,* Vol. AP-37, December 1989, pp. 1573–1581.

R.J. Pogorzelski, "On the Numerical Computation of Scattering from Inhomogeneous Penetrable Objects," *Trans. IEEE,* Vol. AP-26, July 1978, pp. 616–618.

M.A. Ricoy and J.L. Volakis, "Integral Equations with Reduced Unknowns for the Simulation of Two-dimensional Composite Structures," *Trans. IEEE,* Vol. AP-37, March 1989, pp. 362–372.

J.M. Jin and V.V. Liepa, "Simple Moment Method Program for Computing Scattering from Complex Cylindrical Obstacles," *Proc. IEE,* Vol. 136H. August 1989, pp. 321–329.

R.D. Graglia and P.L.E. Uslenghi, "Electromagnetic Scattering from Anisotropic Materials, Part I: General Theory," *Trans. IEEE,* Vol. AP-32, August 1984, pp. 867–869.

R.D. Graglia and P.L.E. Uslenghi, "Electromagnetic Scattering from Anisotropic Materials, Part II: Computer Code and Numerical Results," *Trans. IEEE,* Vol. AP-35, February 1987, pp. 225–232.

J.M. Jin, J.L. Volakis and V.V. Liepa, 'A Moment Method Solution of a Volume—Surface Integral Equation Using Isoparametric Elements and Point Matching," *Trans. IEEE,* Vol. MTT-37, October 1989, pp. 1641–1645.

R.D. Graglia, P.L.E. Uslenghi and R.S. Zich, "Moment Method with Isoparametric Elements for Three-dimensional Anisotropic Scatterers," *Proc. IEEE,* Vol. 77, May 1989, pp. 750–560.

M.F. Sultan and R. Mittra, "An Iterative Moment Method for Analyzing the Electromagnetic Field Distribution Inside Inhomogeneous Lossy Dielectric Objects," *Trans. IEEE,* Vol. MTT-33, February 1985, 163–168.

R.W.C. van der Veeken, "Extended Least Squares Iterative Techniques in the Solution of Electromagnetic Scattering by Penetrable Objects," *J. Electromag. Waves and Appl.* Vol. 2, 1988, pp. 503–522.

Block Models of Humans

D.E. Livesay and K.M. Chen, "Electromagnetic Field Induced Inside Arbitrary Shaped Biological Bodies," *Trans. IEEE,* Vol. MTT-22, December 1974, pp. 1273–1280.

K.M. Chen, D.E. Livesay and B.S. Guru, "Induced Current in and Scattered Field from a Finite Cylinder with Arbitrary Conductivity and Permittivity," *Trans. IEEE,* Vol. AP-24, May 1976, pp. 330–336.

B.S. Guru and K.M. Chen, "Experimental and Theoretical Studies on Electromagnetic Fields Induced Inside Finite Biological Bodies," *Trans. IEEE,* Vol. MTT-24, July 1976, pp. 433–440.

K.M. Chen and B.S. Guru, "Internal EM Field and Absorbed Power Density in Human Torsos Induced by 1–500 MHz EM Waves," *Trans. IEEE,* Vol. MTT-25, September 1977, pp. 746–756.

M.J. Hagmann, O.P. Gandhi and C.H. Durney, "Upper Bound on Cell Size for Moment-Method Solutions," *Trans. IEEE,* Vol. MTT-25, October 1977, pp. 831–832.

D.P. Nyquist, K.M. Chen and B.S. Guru, "Coupling Between Small Thin-Wire Antennas and a Biological Body," *Trans. IEEE,* Vol. AP-25, November 1977, pp. 863–866.

E.H. Newman and P. Tulyathan, "Wire Antennas in the Presence of a Dielectric/Ferrite Inhomogeneity," *Trans. IEEE,* Vol. AP-26, July 1978, pp. 586–587.

M.J. Hagmann, O.P. Gandhi and C.H. Durney, "Procedures for Improving Convergence of Moment-Method Solutions in Electromagnetics," *Trans. IEEE,* Vol. AP-26, September 1978, pp. 743–748.

M.J. Hagmann, O.P. Gandhi and C.H. Durney, "Improvement of Convergence in Moment-Method Solutions by the Use of Interpolants," *Trans. IEEE,* Vol. MTT-26, November 1978, pp. 904–908.

C.H. Durney, "Electromagnetic Dosimetry for Models of Humans and Animals: A Review of Theoretical and Numerical Techniques," *Proc. IEEE,* Vol. 68, January 1980, pp. 33–40.

K. Karimullah, K.M. Chen and D.P. Nyquist, "Electromagnetic Coupling Between a Thin-Wire Antenna and a Neighboring Biological Body: Theory and Experiment," *Trans. IEEE,* Vol. MTT-28, November 1980, pp. 1218–1225.

M.J. Hagmann, "Convergence Testing; Scattering by Dielectric Slab," *Trans. IEEE,* Vol. AP-29, May 1981, pp. 547–550.

M.J. Hagmann, I. Chatterjee and O.P. Gandhi, "Dependence of Electromagnetic Energy Deposition Upon Angle of Incidence for an Inhomogeneous Block Model of Man Under Plane-Wave Irradiation," *Trans. IEEE,* Vol. MTT-29, March 1982, pp. 252–255.

J.F. DeFord, O.P. Gandhi and M.J. Hagmann, "Moment-Method Solutions and SAR Calculations for Inhomogeneous Models of Man with Large Number of Cells," *Trans. IEEE,* Vol. MTT-31, October 1983, pp. 848–851.

H. Massoudi, C.H. Durney and M.F. Iskander, "Limitations of the Cubical Block Model of Man in Calculating SAR Distributions," *Trans. IEEE,* Vol. MTT-32, August 1984, pp. 746–752.

M.J. Hagmann, "Comments on 'Limitations of the Cubical Block Model of Man in Calculating SAR Distributions'," *Trans. IEEE,* Vol. MTT-33, April 1985, pp. 347–350.

M.J. Hagmann and R.L. Levin, "Convergence of Local and Average Values in Three-dimensional Moment-Method Solutions," *Trans. IEEE,* Vol. MTT-33, July 1985, pp. 649–654.

C.T. Tsai, H. Massoudi, C.H. Durney and M.F. Iskander, "A Procedure for Calculating Fields Inside Arbitrarily Shaped, Inhomogeneous Dielectric Bodies Using Linear Basis Functions with the Moment Method," *Trans. IEEE,* Vol. MTT-34, November 1986, pp. 1131–1139.

M.J. Hagmann and R.L. Levin, "Criteria for Accurate Usage of Block Models," *J. Microwave Power,* 1987, pp. 19–27.

M.J. Hagmann, "Numerical Solution of Two-dimensional Electromagnetic Problems Based on the Integral Definition of the Curl Operator," *J. Microwave Power,* 1987, pp. 103–105.

Electromagnetic scattering from a finite tubular cylinder: Numerical solutions

Cheng C. Kao

Gordon McKay Laboratory, Harvard University
Cambridge, Massachusetts 02138

(Received June 13, 1969; revised September 29, 1969.)

The method of approximate product integration is applied to solve numerically the integral equations of the total currents on the surfaces of the finite tubular cylinder with complete generality both in radius and length. Procedures have been devised to take care of the square-root singularity of the transverse total current at the end of the cylinder. Both total currents and scattering cross sections are presented for three special cases with $ka = 1.0$ and $kh = 0.5\pi$, 1.0π, and 1.5π. A complete set of the computer program in Fortran 4 has been prepared for the IBM 7094 at Harvard Computing Center. It can solve simultaneously for cylinders of the same radius but discrete and different lengths.

Electromagnetic scattering from an infinite cylinder is a two-dimensional problem, and the solutions have been established for decades [*King and Wu*, 1959]. In practical applications, however, one always encounters finite structures rather than infinite ones. Because of the inherent, complicated coupling between the induced axial and transverse components of the total current, previously related researches have been somewhat restricted for limiting cases. For example, *Ufimtsev* [1962], *Kieburtz* [1963], and *Fialkovskii* [1967] were interested in thin cylinders subject to the condition of $ka \ll 1$; *Adey* [1958], though dealing with $ka \cong 1$, was interested in rather long cylinders; *Williams* [1956] solved the scalar diffraction for sound wave; and *Vainshtein* [1963] solved for the static case.

A rigorous theory has recently been developed by the author [*Kao*, 1969*a*, *b*] for three-dimensional electromagnetic scattering from a thin-walled, perfectly conducting circular tube of finite length with complete generality both in radius and length. An electromagnetic plane wave in either E or H polarization is normally incident on this finite tubular cylinder. A pair of integral equations has been obtained for the total currents; these have been decoupled in a special sense permitting the solution of the transverse total current to be obtained first, followed by that of the axial total current. All the remaining quantities pertinent to the scattering problem, such as the near fields, the far-field patterns, and the scattering cross sections have been derived in various integral forms in terms of the total currents.

Since the integral equations derived are rather involved, no explicit solution has been attempted except a numerical one by using a high-speed electronic computer. It is the purpose of this paper to present the numerical procedures that have been specially devised to take care of the square-root singularity of the transverse total current at the end of the cylinder. Results of three representative cases are presented in graphical forms. Numerical data can be found in *Kao* [1969*c*].

The complete set of computer source programs for solving the problem has been prepared in Fortran 4 for the IBM 7094 at the Harvard Computing Center. It has been designed to solve simultaneously for cylinders of the same radius but discrete and different lengths.

INTEGRAL EQUATIONS

The center of the finite tubular cylinder with radius a and half-length h is at the origin of the cylindrical coordinates (r, θ, z), and the axis aligns with the z axis. The electromagnetic plane wave in either E or H polarization is normally incident on the cylinder from the negative x axis $(\theta = \pi)$. The problem is first transformed into determining an infinity of Fourier components by expanding each function of θ into a Fourier series. The functions $J_z^t(z|n)$ and $J_\theta^t(z|n)$ represent, respectively, the nth order Fourier component of the total currents $J_z^t(\theta, z)$ and $J_\theta^t(\theta, z)$ on the surfaces of the cylinder. For each n, a pair of integral equations have been obtained for $0 \leq u \leq kh$ [*Kao*, 1969*b*] as follows:

$$\int_0^{kh} G_\theta{}^t(u, u' \mid n) J_\theta{}^t(u' \mid n)\, du' = \tfrac{1}{2}[\pm ink^{-1}a^{-1}C \,{}^{\sin}_{\cos}\, u + C_\theta(n)] \tag{1}$$

$$\int_0^{kh} G_z{}^t(u, u' \mid n) J_z{}^t(u' \mid n)\, du' + nk^{-1}a^{-1} \cdot \int_0^{kh} G_{z\theta}{}^t(u, u' \mid n) J_\theta{}^t(u' \mid n)\, du' = \tfrac{1}{2}[C \,{}^{\cos}_{\sin}\, u + C_z(n)] \tag{2}$$

where $u = kz$, k being the propagation constant. The upper line in a double notation refers to the E polarization and the lower line to the H polarization. The unknown constant C is determined by the condition that $J_z{}^t(u|n)$ vanish at $u = kh$. The expansion parameters $C_z(n)$ and $C_\theta(n)$ are, for E polarization,

$$C_z(n) = -4\alpha_n i^n(\zeta_0 ka)^{-1} J_n(ka) \tag{3a}$$

$$C_\theta(n) = 0 \tag{3b}$$

and for H polarization

$$C_\theta(n) = -4\alpha_n i^{n-1}(\zeta_0 ka)^{-1} J_n'(ka) \tag{4a}$$

$$C_z(n) = 0 \tag{4b}$$

where $J_n(ka)$ is the Bessel function with argument ka and order n; $J_n'(ka)$ is its first derivative with respect to the argument; ζ_0 is the free-space wave impedance; and $i = (-1)^{1/2}$. One of the two expansion parameters is always zero. The parameter α_n is defined as 1 if $n = 0$, and 2 if $n \neq 0$.

The kernels are found to be:

$$G_\theta{}^t(u, u' \mid n) = i \int_0^\infty \{ J_n'[ka(1 + x^2)^{1/2}]\}^2 [e^{-|u-u'|x} \mp e^{-|u+u'|x}]\, dx - \int_0^1 \{ J_n'[ka(1 - x^2)^{1/2}]\}^2 [e^{i|u-u'|x} \mp e^{i|u+u'|x}]\, dx - \tfrac{1}{2} n(ka)^{-2}[e^{i|u-u'|} \mp e^{i|u+u'|}] \tag{5}$$

$$G_z{}^t(u, u' \mid n) = i \int_0^\infty \{ J_n[ka(1 + x^2)^{1/2}]\}^2 [e^{-|u-u'|x} \pm e^{-|u+u'|x}]\, dx - \int_0^1 \{ J_n[ka(1 - x^2)^{1/2}]\}^2 [e^{i|u-u'|x} \pm e^{i|u+u'|x}]\, dx \tag{6}$$

$$G_{z\theta}{}^t(u, u' \mid n) = 0 \quad \text{if} \quad n = 0 \tag{7a}$$

$$G_{z\theta}{}^t(u, u' \mid n) = \int_0^\infty x(1 + x^2)^{-1}\{ J_n[ka(1 + x^2)^{1/2}]\}^2 \cdot [\operatorname{sgn}(u - u')e^{-|u-u'|x} \mp \operatorname{sgn}(u + u')e^{-|u+u'|x}]\, dx + \int_0^1 x(1 - x^2)^{-1}\{ J_n[ka(1 - x^2)^{1/2}]\}^2 \cdot [\operatorname{sgn}(u - u')e^{i|u-u'|x} \mp \operatorname{sgn}(u + u')e^{i|u+u'|x}]\, dx - \tfrac{1}{2}n^{-1}[\operatorname{sgn}(u - u')e^{i|u-u'|} \mp \operatorname{sgn}(u + u')e^{i|u+u'|}] \quad \text{if} \quad n \neq 0 \tag{7b}$$

with the function sgn (z) defined as 1 if $z > 0$; 0 if $z = 0$; and -1 if $z < 0$.

The complete solutions for the total currents can be written into the following forms:

$$J_e{}^t(\theta, u) = \sum_{n=0}^{\infty} C_p(n) J_e{}^t(u \mid n) \cos n\theta \tag{8}$$

$$J_o{}^t(\theta, u) = i \sum_{n=1}^{\infty} C_p(n) J_o{}^t(u \mid n) \sin n\theta \tag{9}$$

where the subscripts p, e are z or θ, and the subscript o is θ or z for E or H polarization, respectively. In (8) and (9) the nth-order Fourier component of the total currents are the normalized solutions of (1) and (2) with the nonzero expansion parameter set equal to unity.

APPROXIMATE PRODUCT INTEGRATION

Numerical methods and high-speed electronic computers have been very useful tools for solving extremely complex mathematical problems. A concise introduction was given by *Tanner and Andreason* [1967]. The numerical methods for solving integral equations can be of many different forms, depending on the preference of each individual researcher. The essence is nothing but the moment methods as presented by *Harrington* [1968]. Many important references can also be found in Harrington's book. However, the method known as the approximate product integration proposed by *Young* [1954*a*, *b*] has been followed here.

Let the following integral equation be considered:

$$\int_0^{kh} G(u, u') J(u')\, du' = \tfrac{1}{2} F(u) \tag{10}$$

where $J(u)$ is unknown, but $G(u, u')$ and $F(u)$ are assumed to be given. The range of the integration is evenly divided into M sections as shown in Figure 1. The unknown function $J(u)$ is expanded at the center of each section into a quadratic form. The total number of unknown points amounts to $2M +$

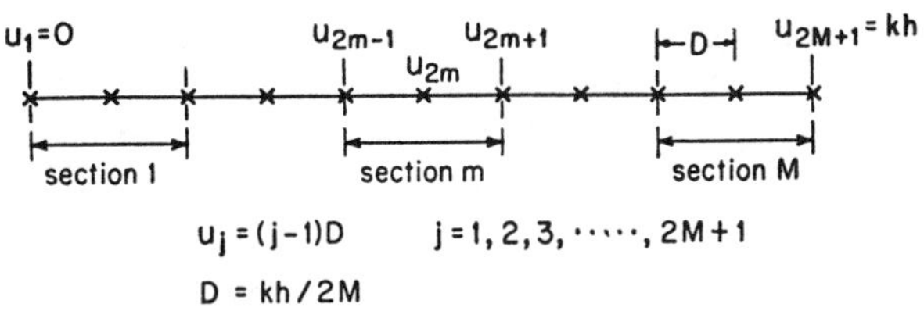

Fig. 1. Subdividing the finite cylinder.

1; they are located at $u_j = (j-1)D$ with $j = 1, 2, 3, \cdots, 2M+1$. The half-length of each subsection is $D = (kh/2M)$.

It can be shown that (10) can be approximated by

$$\sum_{j=1}^{2M+1} A_j(u)\, J(u_j) = F(u) \tag{11}$$

with the weighting function $A_j(u)$ defined by

$$A_j(u) = -\mu(1, 1, u) + \mu(2, 1, \mu) \qquad j = 1 \tag{12a}$$

$$= 2[\mu(0, \tfrac{1}{2}j, u) - \mu(2, \tfrac{1}{2}j, u)] \quad j \text{ even} \tag{12b}$$

$$= \mu(1, \tfrac{1}{2}j - \tfrac{1}{2}, u) + \mu(2, \tfrac{1}{2}j - \tfrac{1}{2}, u) - \mu(1, \tfrac{1}{2}j + \tfrac{1}{2}, u) + \mu(2, \tfrac{1}{2}j + \tfrac{1}{2}, u) \quad j \text{ odd} \tag{12c}$$

$$= \mu(1, M, u) + \mu(2, M, u) \qquad j = 2M + 1 \tag{12d}$$

where

$$\mu(\beta, m, u) = D^{-\beta} \int_{u_{2m-1}}^{u_{2m+1}} (u' - u_{2m})^{\beta} G(u, u')\, du' \tag{13}$$

for $\beta = 0, 1,$ and 2; and $m = 1, 2, 3, \cdots, M$.

The matrix equation corresponding to (10) is thus:

$$[A]\{J\} = \{F\} \tag{14}$$

The elements of the matrices are defined for $i, j = 1, 2, 3, \cdots, 2M+1$ as:

$$A_{ij} = A_j(u_i) \tag{15}$$

$$J_j = J(u_j) \tag{16}$$

$$F_i = F(u_i) \tag{17}$$

In (14), $[A]$ is a $2M+1$ by $2M+1$ square matrix; and $\{J\}$ and $\{F\}$ are both $2M+1$ by 1 column matrices.

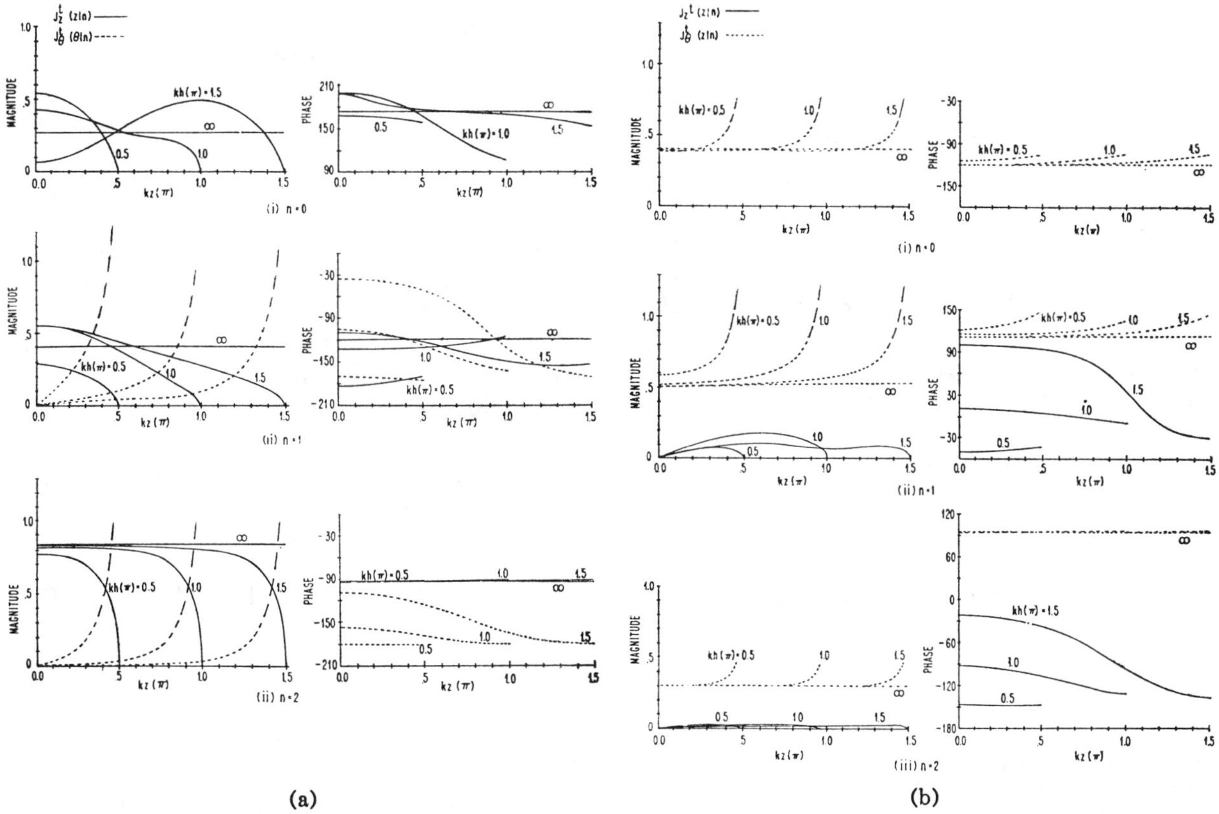

Fig. 2. Normalized total current $J_z^t(z|n)$ and $J_\theta^t(z|n)$ as functions of z. The radius of the cylinders is $ka = 1.0$; the half-lengths are $kh = 0.5\pi$, 1.0π, and 1.5π. *Left*, E polarization; *right*, H polarization.

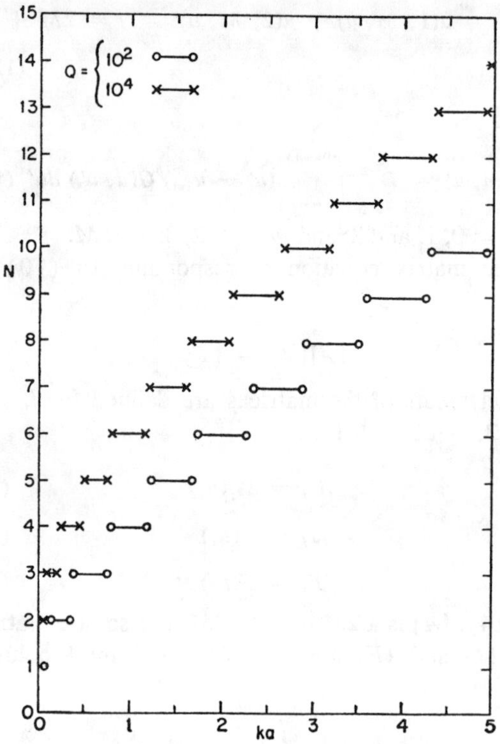

Fig. 3. Truncation of infinite Fourier series at $n = N$ as a function of ka. Only the terms with $n \leq N$ are kept for which $|H_n^{(1)}(ka)| \leq Q$.

TOTAL CURRENT

$J_z^t(z|n)$ and $J_\theta^t(z|n)$

In order to take care of the square-root singularity of the transverse total current at the end of the cylinder, a special function $\phi(u)$ is defined to be unity all over the cylinder except at the last (Mth) subsection, where it behaves oppositely to $J_\theta^t(u|n)$, such as

$$\phi(u) = 1 \qquad 0 \leq u \leq 2(M-1)D \tag{18a}$$

$$= [(2MD - u)/2D]^{1/2} \qquad 2(M-1)D \leq u \leq 2MD \tag{18b}$$

Accordingly, the modified, transverse total current defined by

$$J_\phi^t(u \mid n) = J_\theta^t(u \mid n)\phi(u) \tag{19}$$

will be finite at the end of the cylinder, and the modified kernels defined by

$$G_\phi^t(u, u' \mid n) = G_\theta^t(u, u' \mid n)/\phi(u') \tag{20a}$$

$$G_{z\phi}^t(u, u' \mid n) = G_{z\theta}^t(u, u' \mid n)/\phi(u') \tag{20b}$$

will remain integrable. Note that the subscript ϕ replaces θ for the modified quantities.

From (18) and (19) the values of $J_\theta^t(u|n)$ and $J_\phi^t(u|n)$ are equal at every point except at the last two. At the $2M$th point the actual value is $2^{1/2}$ times that of the modified one. At the end of the cylinder the magnitude of $J_\phi^t(u|n)$ is not interesting, since the infinite value of $J_\theta^t(u|n)$ at this last point is always understood. However, its phase does provide useful information to check the numerical computation.

With the method of the approximate product integration, it is possible to transform (1) and (2) into:

$$[A_\phi^t]\{J_\phi^t\} = \pm ink^{-1}a^{-1}C\{F_\theta\} + C_\theta(n)\{1\} \tag{21}$$

$$[A_z^t]\{J_z^t\} + nk^{-1}a^{-1}[A_{z\phi}^t]\{J_\phi^t\} = C\{F_z\} + C_z(n)\{1\} \tag{22}$$

where $\{1\}$ is a unit column matrix. The elements of $\{F_\theta\}$ or $\{F_z\}$ are, respectively, $\begin{smallmatrix}\sin\\ \cos\end{smallmatrix}\, u_i$ or $\begin{smallmatrix}\cos\\ \sin\end{smallmatrix}\, u_i$. It is trivial, although tedious, to derive the analytic expressions for the weighting matrices A in (21) and (22) [*Kao*, 1969*c*]. They are neglected here, however, owing to extreme complexity.

The procedures to solve (21) and (22) are different for $n = 0$ and $n \neq 0$: For $n = 0$, it is known that the axial and the transverse total currents are decoupled. For E polarization the constant $C_\theta(n)$ vanishes; thus, $\{J_\phi^t\}$ vanishes from (21) and $\{J_z^t\}$ is determined completely from (22). To solve (22), the last column of $[A_z^t]$ is replaced by $\{F_z\}$, and thereby the last element of $\{J_z^t\}$, which is zero, is replaced by the integration constant C. For H polarization, the constant $C_z(n)$ vanishes, so that both $\{J_z^t\}$ and C are zero, because of (22), and the modified transverse total current $\{J_\phi^t\}$ is determined completely from (21).

For $n \neq 0$, both components of the total currents exist. With the properties of linearity, (21) can be solved to obtain

$$\{J_\phi^t\} = C\{J_{\phi 1}^t\} + \{J_{\phi 2}^t\} \tag{23}$$

where both terms exist for H polarization; but the second one vanishes for E polarization. With (22) and (23) one can obtain $\{J_z^t\}$ and C by means of the procedure described for E polarization with $n = 0$. After the integration constant C is obtained, the solution of $\{J_\phi^t\}$ is then completed by (23).

The numerical results for three finite cylinders with $ka = 1.0$ and $kh = 0.5\pi$, 1.0π, and 1.5π are shown in Figure 2 for E and H polarization. Only three different values of n (0, 1, 2) are included. Both the magnitude and the phase are plotted. Because of the normalization, the curves for the magnitude are approximately of the same order.

The corresponding (normalized) values for an infinite cylinder are:

For E polarization

$$J_z{}^t(n) = -[2\pi J_n(ka) H_n{}^{(1)}(ka)]^{-1} \tag{24}$$

For H polarization

$$J_\theta{}^t(n) = -[2\pi J_n{}'(ka) H_n{}^{(1)\prime}(ka)]^{-1} \tag{25}$$

These are included in Figure 2 as horizontal lines. A lengthwise observation along the finite cylinder shows that the transverse total current for H polarization is very close to that of an infinite cylinder. The agreement is good except at the end of the cylinder, where the three-dimensional solution is infinite. But the axial total current for E polarization differs very much from its corresponding two-dimensional value, especially for the lower orders of n. The agreement is not apparent until $n = 2$.

$J_z{}^t(\theta, z)$ and $J_\theta{}^t(\theta, z)$

The complete solutions for $J_z{}^t(\theta, z)$ and $J_\theta{}^t(\theta, z)$ are given in (8) and (9), where the infinite Fourier series must be truncated for practical purposes. It is observed from Figure 2 that the corresponding two- and three-dimensional solutions are of the same order of magnitude. The infinite series for a three-dimensional problem can, therefore, be truncated in a manner familiar in the two-dimensional one.

The total current for a two-dimensional problem is as follows [*King and Wu*, 1959]:

For E polarization

$$J_z{}^t(\theta) = 2(\pi\zeta_0 ka)^{-1} \sum_{n=0}^{\infty} \alpha_n i^n [H_n{}^{(1)}(ka)]^{-1} \cos n\theta \tag{26}$$

For H polarization

$$J_\theta{}^t(\theta) = 2(\pi\zeta_0 ka)^{-1} \sum_{n=0}^{\infty} \alpha_n i^{n-1} [H_n{}^{(1)\prime}(ka)]^{-1} \cos n\theta \tag{27}$$

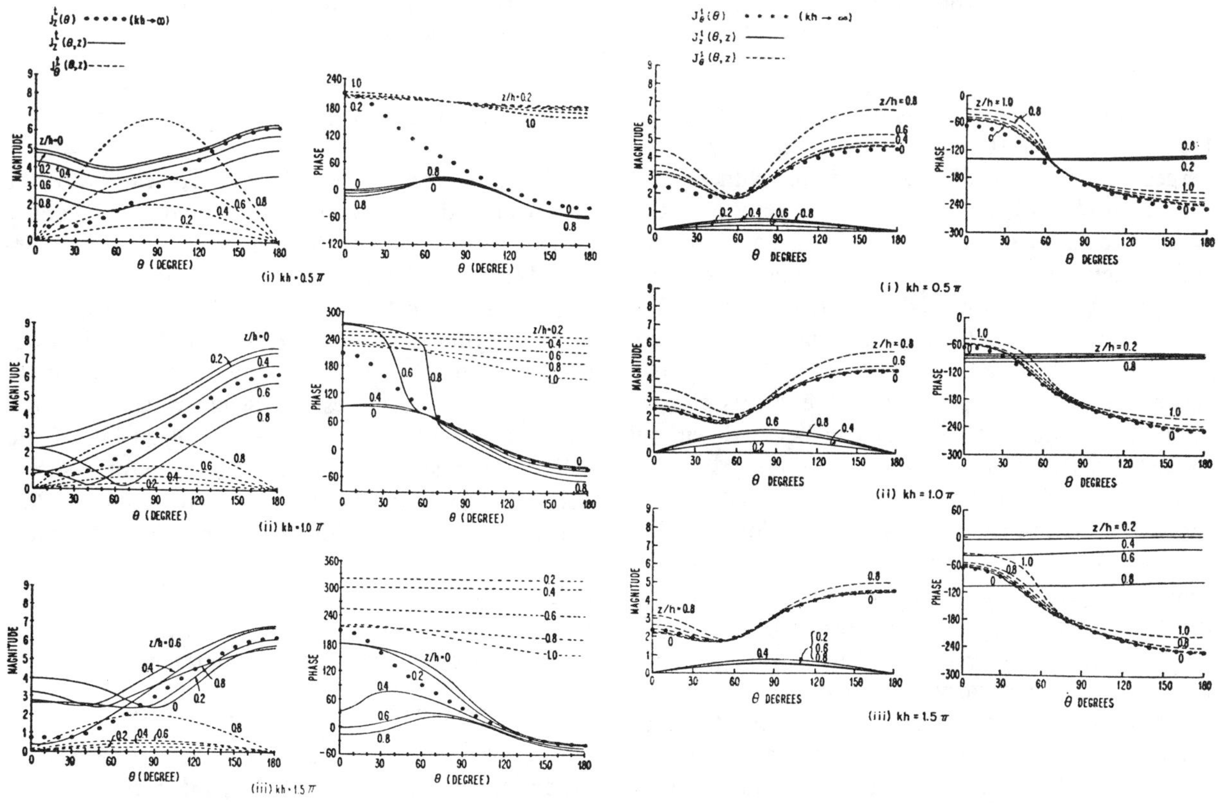

(a) (b)

Fig. 4. **Total currents** $J_3{}^t(\theta, z)$ and $J_\theta{}^t(\theta, z)$ as functions of θ at z/h = 0(0.2)1.0. The radius of the cylinders is ka = 1.0; the half-lengths are kh = 0.5π, 1.0π, and 1.5π. *Left*, E polarization; *right*, H polarization.

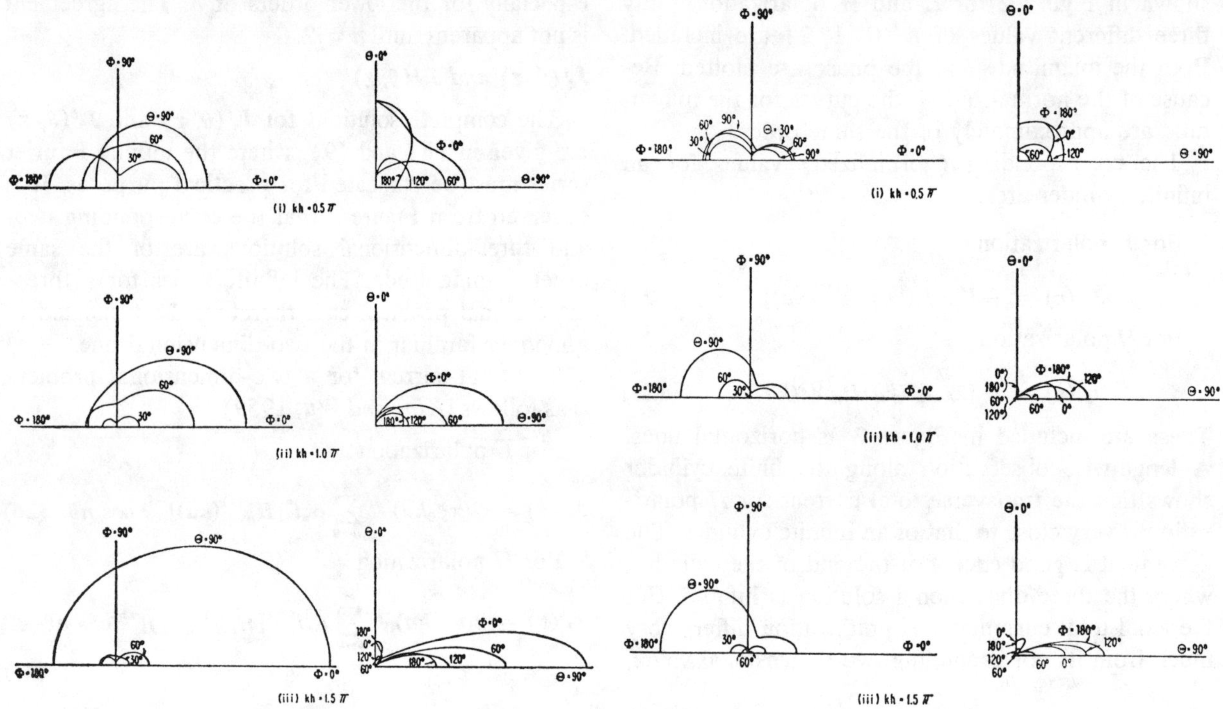

Fig. 5. Normalized scattering cross section $\sigma(\Theta, \Phi)$ $4ah$ as a function of Θ and Φ. The radius of the cylinders is $ka = 1.0$; the half-lengths are $kh = 0.5\pi$, 1.0π, and 1.5π. *Left*, E polarization; *right*, H polarization.

The convergence depends on the behavior of the Hankel function. It can be shown that (27) converges faster than (26). Let the terms with $n \leq N$ for which $H_n^{(1)}(ka) \leq Q$ in (26) be kept; a plot of N as a function of ka will then indicate where the series ought to be truncated for a cylinder of radius ka in order to satisfy a predetermined accuracy. The accuracy depends on the chosen value of Q^{-1}. Figure 3 shows two broken curves with $Q = 10^2$ and 10^4, respectively. At the points where the curves are discontinuous, either value of N can be used without much difference of accuracy.

For the cylinder with $ka \leq 0.05$, a series of at most three terms ($N = 2$) is needed. For a thicker cylinder with $ka = 1.0$, seven terms ($N = 6$) are required to give an accuracy of approximately four digits; five terms ($N = 4$) are required for roughly an accuracy of two digits. The curve with $Q = 10^2$ is, in general, satisfactory.

The numerical results of $J_z^t(\theta, z)$ and $J_\theta^t(\theta, z)$ for three representative cylinders of $ka = 1.0$ and $kh = 0.5\pi$, 1.0π, and 1.5π are shown in Figure 4 for E and H polarization. Both magnitude and phase are presented as functions of θ with $z/h = 0(0.2)1.0$. For comparison, the corresponding results of a two-dimensional problem are superimposed in dotted curves.

For H polarization a finite cylinder can be better approximated by an infinite cylinder. Generally speaking, for E polarization the transverse total current $J_\theta^t(\theta, z)$, which vanishes on an infinite cylinder, is comparable to the axial total current $J_z^t(\theta, z)$ on a finite cylinder. But for H polarization the axial total current $J_z^t(\theta, z)$ which vanishes on an infinite cylinder, is more or less negligible when compared with the transverse total current $J_\theta^t(\theta, z)$ on a finite cylinder. Consideration of the functions of θ shows that, for E polarization, the results for a finite cylinder are comparable with those for an infinite cylinder in the illuminated region ($\theta \cong \pi$) but are significantly different in the shadow region ($\theta \cong 0$). For H polarization, on the contrary, they are nearly the same around the entire circumference.

SCATTERING CROSS SECTIONS

The spherical coordinates (R, Θ, Φ) have been used in the expressions of the general scattering cross section $\sigma(\Theta, \Phi)$. It has been derived in terms of the far-field patterns which, in turn, are derived in integral forms with respect to the total currents [*Kao,*

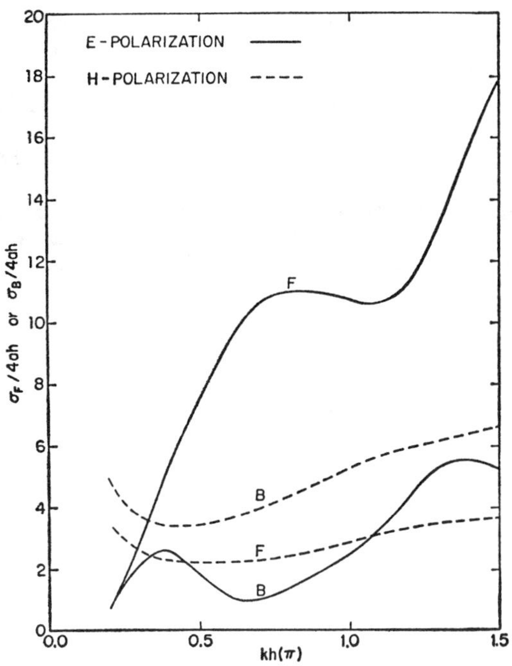

Fig. 6. Normalized forward- and backward-scattering cross sections as functions of kh. The radius of the cylinders is $ka = 1.0$.

1969*b*]. After obtaining the total currents, it is only a routine job to compute the scattering cross sections by using the numerical data of the total currents. Sample results for the scattering cross section normalized with respect to the geometrical cross section of the cylinder are given in Figure 5 for E and H polarization.

The forward- and backward-scattering cross sections are plotted as functions of kh in Figure 6. The forward direction is defined for $\Theta = \pi/2$ and $\Phi = 0$, and the backward for $\Theta = \pi/2$ and $\Phi = \pi$. For E polarization, both normalized forward- and backward-scattering cross sections oscillate and increase rather rapidly as kh increases, the forward increasing faster than the backward. This indicates that relatively more power is scattered in the forward direction than in the backward direction.

For H polarization both scattering cross sections are nearly constant for different values of kh, and the ratio between them remains roughly constant. Therefore, the ratio of the powers scattered in these two directions does not differ very much for cylinders of different lengths. But, what makes the H polarization different from the E polarization is the fact that the scattered power in the backward direction is greater than that in the forward direction rather than vice versa.

CONCLUSIONS

The integral equations for the three-dimensional electromagnetic scattering from a circular tube of finite length have been solved numerically by the method known as the approximate product integration. In general, scattering from a finite cylinder is significantly different from scattering from an infinite cylinder for E polarization, but is approximately equivalent for H polarization.

The theory itself is general for cylinders with any radius or length. The limit of using this theory depends on a consideration of the computer time involved and the amount of storage locations available. As is shown in Figure 3, it is clear that a thicker cylinder requires more terms in the series and is, therefore, more expensive to solve. For the computer programs prepared at Harvard, there is no limit on the thickness of the cylinder; however, the largest number of subsections M that divide the half-length of the cylinder is set at 50. Therefore, depending on the half-length of the subsection (D), this program does have an upper limit on the length of the cylinder. For example, it can solve for a cylinder as long as $kh = 10\pi$ if D is 0.1π.

Acknowledgments. I thank Professors R. W. P. King and T. T. Wu for their patient guidance.

This research was supported in part by the U. S. Air Force.

REFERENCES

Adey, A. W. (1958), Scattering of electromagnetic waves by long cylinders, *Electronic Radio Eng., 1958,* 149–158.

Fialkovskii, A. T. (1967), Scattering of plane electromagnetic waves by a thin, cylindrical conductor of finite length, *Soviet Phys.-Tech. Phys., 11,* 1300–1304.

Harrington, R. F. (1968), *Field Computation by Moment Methods,* MacMillan, New York.

Kao, C. C. (1969*a*), On the electromagnetic scattering from a finite cylinder, Ph.D. thesis, Division of Engineering and Applied Physics, Harvard University.

Kao, C. C. (1969*b*), Three-dimensional electromagnetic scattering from a circular tube of finite length, *J. Appl. Phys., 40*(12), 4732–4740.

Kao, C. C. (1969*c*), Electromagnetic scattering from a finite tubular cylinder: numerical solutions and data, 1, Development of theory, 2, Tables. *Cruft Sci. Rep. 6,* Division Eng. Appl. Phys., Harvard University, Cambridge, Mass.

Kieburtz, R. B. (1963), Scattering by a finite cylinder, *Electromagnetic Theory and Antennas,* part 1, edited by E. C. Jordon, Pergamon Press, pp. 145–156.

King, R. W. P., and T. T. Wu (1959), *The Scattering and Diffraction of Waves,* Harvard University Press, Cambridge.

Tanner, R. L., and M. G. Andreason (1967), Numerical solution of electromagnetic problems, *IEEE Spectrum, 4*(9), 53–61.

Ufimtsev, P. Y. (1962), Diffraction of plane electromagnetic waves by a thin cylindrical conductor, *Radioteck. Elekchtron., 7,* 241–249.

Vainshtein, L. A. (1963), Static boundary problems for a hollow cylinder of finite length, 2, Numerical results, and 3, Approximate formula, *Soviet Phys.-Tech. Phys., 7,* 855–866.

Williams, W. E. (1956), Diffraction by a cylinder of finite length, *Proc. Cambridge Phil. Soc., 52,* 322–335.

Young, A. (1954*a*), Approximate product-integration, *Proc. Roy. Soc. London, A, 224,* 552–561.

Young, A. (1954*b*), The application of approximate product-integration to the numerical solution of integral equations, *Proc. Roy. Soc. London, A, 224,* 561–573.

Electromagnetic Scattering by Surfaces of Arbitrary Shape

SADASIVA M. RAO, DONALD R. WILTON, SENIOR MEMBER, IEEE, AND ALLEN W. GLISSON, MEMBER, IEEE

Abstract—**The electric field integral equation (EFIE) is used with the moment method to develop a simple and efficient numerical procedure for treating problems of scattering by arbitrarily shaped objects. For numerical purposes, the objects are modeled using planar triangular surfaces patches. Because the EFIE formulation is used, the procedure is applicable to both open and closed surfaces. Crucial to the numerical formulation is the development of a set of special subdomain-type basis functions which are defined on pairs of adjacent triangular patches and yield a current representation free of line or point charges at subdomain boundaries. The method is applied to the scattering problems of a plane wave illuminated flat square plate, bent square plate, circular disk, and sphere. Excellent correspondence between the surface current computed via the present method and that obtained via earlier approaches or exact formulations is demonstrated in each case.**

I. INTRODUCTION

ENGINEERS AND researchers in electromagnetics have been quick to take advantage of the expanding capabilities of digital computers over the past two decades by developing effective numerical techniques applicable to a wide variety of practical electromagnetic radiation and scattering problems. As new computer developments dramatically increase computational capabilities, however, it becomes less cost effective to develop highly efficient but specialized codes for treating certain classes of geometries than to use less efficient but existing general purpose codes that can handle a wide variety of problems. For these reasons there has been a growing interest in the use and development of computer codes for treating scattering by arbitrarily shaped conducting bodies.

To date, the most notable approaches for treating such problems have used integral equation formulations in conjunction with the method of moments. The body surface in these approaches is generally modeled either as a wire mesh—the so-called wire-grid model—or as a surface partitioned into smooth or piecewise-smooth patches—the so-called surface patch model.

The wire-grid modeling approach has been remarkably successful in treating many problems, particularly in those requiring the prediction of far-field quantities such as radiation patterns and radar cross sections [1]. Not only is the connectivity of a wire-grid model easily specified for computer input, but the approach also has the advantage that all numerically computed integrals in the moment matrix are one dimensional.

Manuscript received May 28, 1980; revised August 6, 1981. This work was supported by the Rome Air Development Center, Griffiss AFB, NY, under Contract No. F30602-78-C-0148.

S. M. Rao was with Syracuse University, Syracuse, NY, on leave from the Department of Electrical Engineering, University of Mississippi, University, MS 38677. He is now with the Department of Electrical Engineering, Rochester Institute of Technology, Rochester, NY 14623.

D. R. Wilton was with Syracuse University, Syracuse, NY, on leave from the Department of Electrical Engineering, University of Mississippi, University, MS 38677.

A. W. Glisson is with the Department of Electrical Engineering, University of Mississippi, University, MS 38677.

However, the approach is not well suited for calculating near-field and surface quantities such as surface current and input impedance. Some of the problems encountered include the presence of fictitious loop currents in the solution, ill-conditioned moment matrices and incorrect currents at the cavity resonant frequencies of the scatterer [2], and difficulties in interpreting computed wire currents and relating them to equivalent surface currents. The accuracy of wire-grid modeling has also been questioned on theoretical grounds [3]. Most of these difficulties can be either wholly or partially overcome by surface patch approaches, however, which account for much of the recent activity in this area.

Several approaches to surface patch modeling have been reported in the literature. Knepp and Goldhirsh [4] partitioned a conducting surface into nonplanar quadrilateral patches and employed the magnetic field integral equation (MFIE) to solve the electromagnetic scattering problem. Albertsen *et al.* [5] solved for the current and computed radiation patterns for satellite structures with attached wire antennas, booms, and solar panels. They employed a hybrid formulation in which the MFIE, with planar quadrilateral surface patches, was used to model the satellite, and the electric field integral equation (EFIE) was used to treat the wire antennas. Their approach also forms the basis for the arbitrary surface treatment of the widely used numerical electromagnetic code (NEC) developed at the Lawrence Livermore Laboratory [6]. Wang *et al.* [7] used an EFIE formulation and modeled relatively complex surfaces by means of planar rectangular patches. Newman and Pozar [8] extended the use of the well-known piecewise-sinusoidal basis functions of thin-wire theory to the treatment of surfaces in their EFIE formulation for surfaces with attached wires. Sankar and Tong [9] employed planar triangular patches to model a square plate and pointed out the possibility of extending their approach to arbitrary bodies. Their formulation, based on a variational formula for the current made stationary with respect to a set of trial functions, is equivalent to a Galerkin solution of the EFIE. Wang [10], [11] employed planar triangular patches in conjunction with the MFIE, but used basis functions containing the phase variation of the incident field in each patch, which unfortunately yield a moment matrix dependent on the incident field. Jeng and Wexler [12] suggested using the MFIE and nonplanar triangles to model arbitrary surfaces, while Singh and Adams [13] proposed the use of planar quadrilateral patches and sinusoidal basis functions in conjunction with the EFIE for the same purpose.

In arbitrary surface modeling the EFIE has the advantage of being applicable to both open and closed bodies, whereas the MFIE applies only to closed surfaces. On the other hand, for arbitrarily shaped objects the EFIE is considerably more difficult to apply than the MFIE. In fact, of the above authors, only Wang *et al.* and Newman and Pozar have actually applied the EFIE to nonplanar structures—and the use of rectangular patches limits their approaches to structures with curvature in one dimension only.

The difficulites with the EFIE stem primarily from the presence of derivatives appearing in conjunction with a singular kernel in the integral equation. For example, if the vector basis functions which represent the surface current are not constructed so that their normal components are continuous across surface edges, then the continuity equation demands the presence of line or point charges at such edges. These fictitious charges, when present, usually cause anomalies or inconsistencies in the solution. Although the approaches of Wang *et al.* and Newman and Pozar are free of these difficulties, their use of rectangular patches, with their consequent limitation to surfaces with curvature in one dimension only, is too restrictive for many applications.

For modeling arbitrarily shaped surfaces, planar triangular patch models, an example of which is shown in Fig. 1, are particularly appropriate. Some of the advantages of triangular patch surface modeling have been noted by Sankar and Tong [9], as well as by Wang [10], and are similar to those of wire-grid modeling. For example, triangular patches are capable of accurately conforming to any geometrical surface or boundary, the patch scheme is easily specified for computer input, and a varying patch density can be used according to the resolution required in the surface geometry or current. Although planar quadrilateral (nonrectangular) patches share some of these features, it is difficult to construct basis functions defined on them which are free of line charges. Furthermore, the vertices of planar quadrilaterals cannot be independently specified—a restriction that is a severe inconvenience to the modeler.

In this paper, a numerical solution of the problem of scattering by either open or closed arbitrarily shaped conducting bodies is presented. The approach combines the advantages of triangular patch modeling and the EFIE formulation, and results in an algorithm which is both simple and efficient. Crucial to the approach is the development of special basis functions defined on triangular patches which are free of fictitious line or point charges and which are analogous to the so-called "rooftop" functions used in rectangular patch models [14].

In the following section the EFIE formulation is presented, the special set of basis functions is developed, and the method of moments [15] is applied to obtain a linear system of equations for the surface current. In Section III, numerical results are presented for scattering by a flat square plate, a bent square plate, a circular disk, and a sphere. Section IV summarizes the contents of the paper.

II. ELECTRIC FIELD FORMULATION

In this section an integral equation for the surface current induced on a conducting scatterer is derived from boundary conditions on the electric field. To solve the integral equation by the method of moments, a set of expansion functions and a testing procedure are developed and used to derive the elements of the moment matrix. Finally, the numerical computation of the moment matrix elements is discussed.

Electric Field Integral Equation

Let S denote the surface of an open or closed perfectly conducting scatterer with unit normal $\hat{\mathbf{n}}$. An electric field $\mathbf{E}^i$, defined to be the field due to an impressed source in the absence of the scatterer, is incident on and induces surface currents $\mathbf{J}$ on S. If S is open, we regard $\mathbf{J}$ as the vector sum of the surface currents on opposite sides of S; hence the normal component of $\mathbf{J}$ must vanish on boundaries of S. The scattered electric field

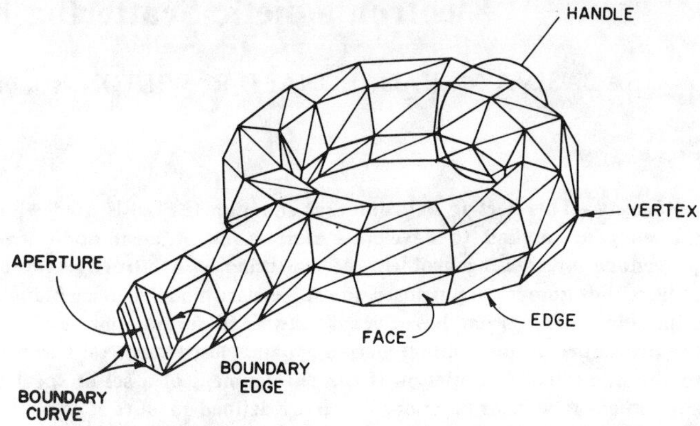

Fig. 1. Arbitrary surface modeled by triangular patches.

$\mathbf{E}^s$ can be computed from the surface current by

$$\mathbf{E}^s = -j\omega\mathbf{A} - \nabla\Phi \tag{1}$$

with the magnetic vector potential defined as

$$\mathbf{A}(\mathbf{r}) = \frac{\mu}{4\pi}\int_S \mathbf{J}\frac{e^{-jkR}}{R}\,dS' \tag{2}$$

and the scalar potential as

$$\Phi(\mathbf{r}) = \frac{1}{4\pi\epsilon}\int_S \sigma\frac{e^{-jkR}}{R}\,dS'. \tag{3}$$

A harmonic time dependence $\exp(j\omega t)$ is assumed and suppressed, and $k = \omega\sqrt{\mu\epsilon} = 2\pi/\lambda$, where λ is the wavelength. The permeability and permittivity of the surrounding medium are μ and ϵ, respectively, and $R = |\mathbf{r} - \mathbf{r}'|$ is the distance between an arbitrarily located observation point $\mathbf{r}$ and a source point $\mathbf{r}'$ on S. Both $\mathbf{r}$ and $\mathbf{r}'$ are defined with respect to a global coordinate origin O. The surface charge density σ is related to the surface divergence of $\mathbf{J}$ through the equation of continuity,

$$\nabla_s \cdot \mathbf{J} = -j\omega\sigma. \tag{4}$$

We derive an integrodifferential equation for $\mathbf{J}$ by enforcing the boundary condition $\hat{\mathbf{n}} \times (\mathbf{E}^i + \mathbf{E}^s) = \mathbf{0}$ on S, obtaining

$$-\mathbf{E}_{\tan}{}^i = (-j\omega\mathbf{A} - \nabla\Phi)_{\tan}, \quad \mathbf{r} \text{ on } S. \tag{5}$$

Equation (5), with (2)–(4), constitutes the so-called electric field integral equation. One notes that the presence of derivatives on the current in (4) and on the scalar potential in (5) suggests that care should be taken in selecting the expansion functions and testing procedure in the method of moments.

Development of Basis Functions

In this section we discuss a set of basis functions introduced by Glisson [16] which is suitable for use with the EFIE and triangular patch modeling. We assume that a suitable triangulation, defined in terms of an appropriate set of faces, edges, vertices, and boundary edges, as illustrated in Fig. 1, has been found to approximate S. (More detailed considerations concerning the mathematical representation and topological properties of triangular patch models may be found in [17].)

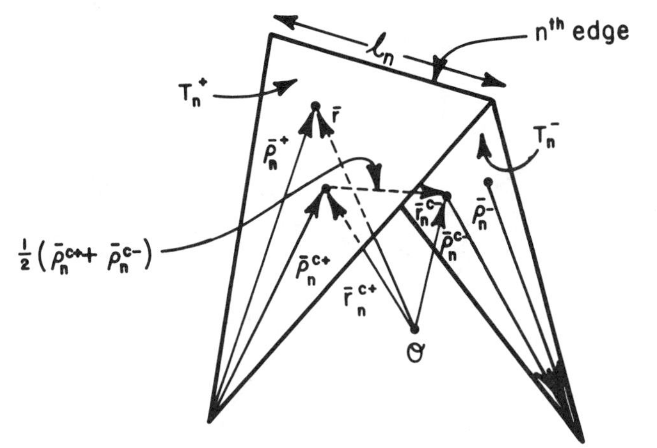

Fig. 2. Triangle pair and geometrical parameters associated with interior edge.

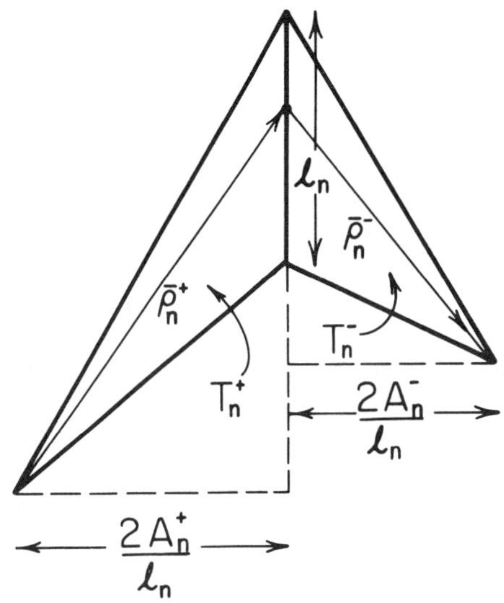

Fig. 3. Geometry for construction of component of basis function normal to edge.

It is convenient to start our development by noting that each basis function is to be associated with an *interior edge* (i.e., nonboundary edge (cf Fig. 1)) of the patch model and is to vanish everywhere on S except in the two triangles attached to that edge. Fig. 2 shows two such triangles, T_n^+ and T_n^-, corresponding to the nth edge of a triangulated surface modeling a scatterer. Points in T_n^+ may be designated either by the position vector $\mathbf{r}$ defined with respect to $\mathcal{O}$, or by the position vector $\boldsymbol{\rho}_n^+$ defined with respect to the free vertex of T_n^+. Similar remarks apply to the position vector $\boldsymbol{\rho}_n^-$ except that it is directed toward the free vertex of T_n^-. The plus or minus designation of the triangles is determined by the choice of a positive current reference direction for the nth edge, the reference for which is assumed to be from T_n^+ to T_n^-.[1] We define the vector basis function associated with the nth edge as

$$\mathbf{f}_n(\mathbf{r}) = \begin{cases} \dfrac{l_n}{2A_n^+}\boldsymbol{\rho}_n^+, & \mathbf{r} \text{ in } T_n^+ \\[2ex] \dfrac{l_n}{2A_n^-}\boldsymbol{\rho}_n^-, & \mathbf{r} \text{ in } T_n^- \\[2ex] \mathbf{0}, & \text{otherwise,} \end{cases} \tag{6}$$

where l_n is the length of the edge and $A_n^\pm$ is the area of triangle $T_n^\pm$. (Note that we use the convention, followed throughout the paper, that subscripts refer to edges while superscripts refer to faces.) The basis function $\mathbf{f}_n$ is used to approximately represent the surface current, and we list and discuss below some properties which make it uniquely suited to this role.

1) The current has no component normal to the boundary (which excludes the common edge) of the surface formed by the triangle pair T_n^+ and T_n^-, and hence no line charges exist along this boundary.
2) The component of current normal to the nth edge is constant and continuous across the edge as may be seen with the aid of Fig. 3, which shows that the normal component of $\boldsymbol{\rho}_n^\pm$ along edge n is just the height of triangle $T_n^\pm$ with edge n as the base and the height expressed as $(2A_n^\pm)/l_n$. This latter factor normalizes $\mathbf{f}_n$ in (6) such that its flux density normal to edge n is unity, ensuring continuity of current normal to the edge. This result, together with 1), implies that *all* edges of T_n^+ and T_n^- are free of line charges.
3) The surface divergence of $\mathbf{f}_n$, which is proportional to the surface charge density associated with the basis element, is

$$\nabla_s \cdot \mathbf{f}_n = \begin{cases} \dfrac{l_n}{A_n^+}, & \mathbf{r} \text{ in } T_n^+ \\[2ex] -\dfrac{l_n}{A_n^-}, & \mathbf{r} \text{ in } T_n^- \\[2ex] 0, & \text{otherwise,} \end{cases} \tag{7}$$

since the surface divergence in $T_n^\pm$ is $(\pm 1/\rho_n^\pm)\partial(\rho_n^\pm f_n)/\partial\rho_n^\pm$. The charge density is thus constant in each triangle, the total charge associated with the triangle pair T_n^+ and T_n^- is zero, and the basis functions for the charge evidently have the form of *pulse doublets* [14].
4) The *moment* of $\mathbf{f}_n$ is given by $(A_n^+ + A_n^-)\mathbf{f}_n^{\text{avg}}$ where

$$\begin{aligned} (A_n^+ + A_n^-)\mathbf{f}_n^{\text{avg}} &\equiv \int_{T_n^+ + T_n^-} \mathbf{f}_n \, dS \\ &= \frac{l_n}{2}(\boldsymbol{\rho}_n^{c+} + \boldsymbol{\rho}_n^{c-}) \\ &= l_n(\mathbf{r}_n^{c+} - \mathbf{r}_n^{c-}) \end{aligned} \tag{8}$$

and $\boldsymbol{\rho}_n^{c\pm}$ is the vector between the free vertex and the centroid of $T_n^\pm$ with $\boldsymbol{\rho}_n^{c-}$ directed toward and $\boldsymbol{\rho}_n^{c+}$ directed away from the vertex, as shown in Fig. 2, and $\mathbf{r}_n^{c\pm}$ is the vector from $\mathcal{O}$ to the centroid of $T_n^\pm$. Equation (8) may be most easily derived by expressing the

[1] For orientable surfaces, the current reference direction may be obtained from the connection matrix used to describe the triangulation scheme. This matrix merely lists the vertices linked by each edge, the order of appearance of the vertices for each edge effectively assigning an orientation to the edge. The direction of the cross product of this edge orientation vector with the surface normal in each adjacent triangle may be taken as the positive current reference direction in that triangle for the basis function associated with the edge.

integral in terms of area coordinates, to be discussed below.

The current on S may be approximated in terms of the $\mathbf{f}_n$ as

$$\mathbf{J} \cong \sum_{n=1}^{N} I_n \mathbf{f}_n(\mathbf{r}) \tag{9}$$

where N is the number of interior (nonboundary) edges. Since a basis function is associated with each nonboundary edge of the triangulated structure, up to three basis functions may have nonzero values within each triangular face. But at a given edge only the basis function associated with that edge has a current component *normal to the edge* since, according to 1), all other basis currents in adjacent faces are parallel to the edge. Furthermore, since the normal component of $\mathbf{f}_n$ at the nth edge is unity, *each coefficient I_n in (9) may be interpreted as the normal component of current density flowing past the nth edge.* Also, we see that the basis functions are independent in each triangle since the current normal to the nth edge, I_n in (9), is an independent quantity. At surface boundary edges, the sum of the normal components of current on opposite sides of the surface cancel because of current continuity. Therefore we neither define nor include in (9) contributions from basis functions associated with such edges.

Because of the considerable variation in the direction of the flow lines of $\mathbf{f}_n$ within a triangle, it is not at first obvious that a linear superposition of basis functions is capable of representing, say, a constant current flowing in an arbitrary direction within a triangle. That this is possible, however, can be seen with the aid of Fig. 4, which shows a triangle T^q with edges arbitrarily labeled 1, 2, and 3 (in effect, we employ here a "local indexing scheme," in contrast to the "global indexing scheme" used earlier). With the vectors ρ_1, ρ_2, and ρ_3 as shown, the basis functions in T^q are $\mathbf{f}_i = (l_i/2A^q)\rho_i$, $i = 1, 2, 3$, where A^q is the triangle area and where, for simplicity, the current reference directions are assumed to be out of the triangle for each edge. It is apparent from the figure and the definition of $\mathbf{f}_i$ that the linear combinations $l_2\mathbf{f}_1 - l_1\mathbf{f}_2$ and $l_3\mathbf{f}_1 - l_1\mathbf{f}_3$ are *constant* vectors for every point $\mathbf{r}$ in T^q and are parallel to sides 3 and 2, respectively. Since the two composite forms are linearly independent (i.e., nonparallel), a constant vector of arbitrary magnitude and direction within T^q may be synthesized by an appropriate linear combination of the two forms, as asserted.

Testing Procedure

The next step in the method of moments is to select a testing procedure. We choose as testing functions the expansion functions $\mathbf{f}_n$ developed in the previous section. With a symmetric product defined as

$$\langle \mathbf{f}, \mathbf{g} \rangle \equiv \int_S \mathbf{f} \cdot \mathbf{g} \, dS, \tag{10}$$

(5) is tested with $\mathbf{f}_m$, yielding

$$\langle \mathbf{E}^i, \mathbf{f}_m \rangle = j\omega \langle \mathbf{A}, \mathbf{f}_m \rangle + \langle \nabla\Phi, \mathbf{f}_m \rangle. \tag{11}$$

If one makes use of a surface vector calculus identity [18] and the properties of $\mathbf{f}_m$ at the edges of S, the last term in (11) can

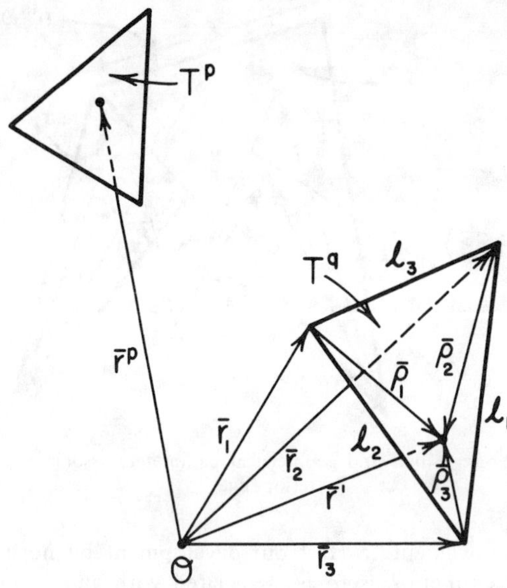

Fig. 4. Local coordinates and edges for source triangle T^q with observation point in triangle T^p.

be rewritten as

$$\langle \nabla\Phi, \mathbf{f}_m \rangle = -\int_S \Phi \nabla_s \cdot \mathbf{f}_m \, dS. \tag{12}$$

With (7), the integral in (12) may now be written and approximated as follows:

$$\begin{aligned} &\int_S \Phi \nabla_s \cdot \mathbf{f}_m \, dS \\ &= l_m \left(\frac{1}{A_m^{+}} \int_{T_m^{+}} \Phi \, dS - \frac{1}{A_m^{-}} \int_{T_m^{-}} \Phi \, dS \right) \\ &\cong l_m [\Phi(\mathbf{r}_m^{c+}) - \Phi(\mathbf{r}_m^{c-})]. \end{aligned} \tag{13}$$

In (13) the average of Φ over each triangle is approximated by the value of Φ at the triangle centroid. With similar approximations, the vector potential and incident field terms in (11) may be written as

$$\begin{aligned} &\left\langle \begin{Bmatrix} \mathbf{E}^i \\ \mathbf{A} \end{Bmatrix}, \mathbf{f}_m \right\rangle \\ &= l_m \left[\frac{1}{2A_m^{+}} \int_{T_m^{+}} \begin{Bmatrix} \mathbf{E}^i \\ \mathbf{A} \end{Bmatrix} \cdot \rho_m^{+} \, dS \right. \\ &\left. + \frac{1}{2A_m^{-}} \int_{T_m^{-}} \begin{Bmatrix} \mathbf{E}^i \\ \mathbf{A} \end{Bmatrix} \cdot \rho_m^{-} \, dS \right] \\ &\cong \frac{l_m}{2} \left[\begin{Bmatrix} \mathbf{E}^i(\mathbf{r}_m^{c+}) \\ \mathbf{A}(\mathbf{r}_m^{c+}) \end{Bmatrix} \cdot \rho_m^{c+} \right. \\ &\left. + \begin{Bmatrix} \mathbf{E}^i(\mathbf{r}_m^{c-}) \\ \mathbf{A}(\mathbf{r}_m^{c-}) \end{Bmatrix} \cdot \rho_m^{c-} \right], \end{aligned} \tag{14}$$

where the integral over each triangle is eliminated by approximating $\mathbf{E}^i$ (or $\mathbf{A}$) in each triangle by its value at the triangle centroid and carrying out integrations similar to those used to obtain (8). With (12)-(14), (11) now becomes

$$j\omega l_m \left[\mathbf{A}(\mathbf{r}_m{}^{c+}) \cdot \frac{\boldsymbol{\rho}_m{}^{c+}}{2} + \mathbf{A}(\mathbf{r}_m{}^{c-}) \cdot \frac{\boldsymbol{\rho}_m{}^{c-}}{2} \right] + l_m[\Phi(\mathbf{r}_m{}^{c-}) - \Phi(\mathbf{r}_m{}^{c+})] = l_m \left[\mathbf{E}^i(\mathbf{r}_m{}^{c+}) \cdot \frac{\boldsymbol{\rho}_m{}^{c+}}{2} + \mathbf{E}^i(\mathbf{r}_m{}^{c-}) \cdot \frac{\boldsymbol{\rho}_m{}^{c-}}{2} \right], \tag{15}$$

which is the equation enforced at each triangle edge, $m = 1, 2, \cdots, N$.

We remark that another interpretation of the testing procedure arriving at (15) is also possible. One may equate line integrals of the form $\int_{C_m} \mathbf{F} \cdot d\mathbf{r}$, where $\mathbf{F}$ represents the right and left sides of (5), and C_m is the piecewise linear path from the point $\mathbf{r}_m{}^{c+}$ to the midpoint of edge m and thence to $\mathbf{r}_m{}^{c-}$. $\mathbf{E}^i$ and $\mathbf{A}$ can be approximated along each portion of the path by their respective values at the triangle centroids. The resulting equality, apart from the factor l_m, is (15). Under either interpretation, the testing procedure reduces the differentiability requirement on Φ in (5) by integrating $\nabla\Phi$ first, the procedure having been constructed with this goal in mind. The purpose of approximations (13) and (14) is to eliminate surface integrals of the potential quantities, allowing a double surface integral to be approximated by a quantity involving a single surface integral in the numerical computation of the moment matrix elements. These approximations are justified by observing that the potentials are locally smooth within each subdomain, as follows from their integral definitions and the locally smooth nature of the source representation in terms of the basis functions [14].[2]

Matrix Equation Derivation

Substitution of the current expansion (9) into (15) yields an $N \times N$ system of linear equations which may be written in matrix form as

$$ZI = V \tag{16}$$

where $Z = [Z_{mn}]$ is an $N \times N$ matrix and $I = [I_n]$ and $V = [V_m]$ are column vectors of length N. Elements of Z and V are given by

$$Z_{mn} = l_m \left[j\omega \left(\mathbf{A}_{mn}{}^{+} \cdot \frac{\boldsymbol{\rho}_m{}^{c+}}{2} + \mathbf{A}_{mn}{}^{-} \cdot \frac{\boldsymbol{\rho}_m{}^{c-}}{2} \right) + \Phi_{mn}{}^{-} - \Phi_{mn}{}^{+} \right] \tag{17}$$

$$V_m = l_m \left(\mathbf{E}_m{}^{+} \cdot \frac{\boldsymbol{\rho}_m{}^{c+}}{2} + \mathbf{E}_m{}^{-} \cdot \frac{\boldsymbol{\rho}_m{}^{c-}}{2} \right) \tag{18}$$

where

$$\mathbf{A}_{mn}{}^{\pm} = \frac{\mu}{4\pi} \int_S \mathbf{f}_n(\mathbf{r}') \frac{e^{-jkR_m{}^{\pm}}}{R_m{}^{\pm}} \, dS', \tag{19}$$

$$\Phi_{mn}{}^{\pm} = -\frac{1}{4\pi j\omega\epsilon} \int_S \nabla_s{}' \cdot \mathbf{f}_n(\mathbf{r}') \frac{e^{-jkR_m{}^{\pm}}}{R_m{}^{\pm}} \, dS', \tag{20}$$

$$R_m{}^{\pm} = |\mathbf{r}_m{}^{c\pm} - \mathbf{r}'|$$

and

$$\mathbf{E}_m{}^{\pm} = \mathbf{E}^i(\mathbf{r}_m{}^{c\pm}). \tag{21}$$

For plane wave incidence, we set

$$\mathbf{E}^i(\mathbf{r}) = (E_\theta \hat{\boldsymbol{\theta}}_0 + E_\phi \hat{\boldsymbol{\phi}}_0) e^{j\mathbf{k}\cdot\mathbf{r}} \tag{22}$$

where the propagation vector $\mathbf{k}$ is

$$\mathbf{k} = k(\sin\theta_0 \cos\phi_0 \hat{\mathbf{x}} + \sin\theta_0 \sin\phi_0 \hat{\mathbf{y}} + \cos\theta_0 \hat{\mathbf{z}}) \tag{23}$$

and $(\hat{\theta}_0, \hat{\phi}_0)$ defines the angle of arrival of the plane wave in terms of the usual spherical coordinate convention. Unit vectors θ_0 and ϕ_0 are *constant* vectors which coincide with the usual spherical coordinate unit vectors only at points on the line from O in the direction of $\mathbf{k}$.

Once the elements of the moment matrix and the forcing vector V are determined, one may solve the resulting system of linear equations (16) for the unknown column vector I. The elements of Z in (16) may be evaluated by naively computing Z_{mn} directly by (17) (with aid of (19) and (20)) for each index combination m and n. However, as shown in the next section, this procedure is extremely inefficient since the integrals required for each combination of m and n are generally required for a number of other combinations as well.

Efficient Numerical Evaluation of Matrix Elements

Evaluation of each matrix element Z_{mn} associated with edges m and n involves integrations over triangles $T_n{}^{\pm}$ with observation points located at the centroids of triangles $T_m{}^{\pm}$. One is easily convinced that some of the same integrals required for an element Z_{mn} are also needed to compute an element Z_{rs}, $r \neq m$, $s \neq n$, if edge r happens to be an edge of $T_m{}^{+}$ or $T_m{}^{-}$ while edge s is an edge of $T_n{}^{+}$ or $T_n{}^{-}$. Indeed, if one focuses attention on a single pair of *faces* rather than on a pair of *edges*, one observes that the integrals evaluated for a source face with scalar and vector potentials observed at the centroid of another face are involved in all the elements Z_{mn} having edges n as (nonboundary) edges of the source triangle and edges m as the (nonboundary) edges of the observation triangle. Thus, the total number of matrix elements requiring evaluation of the same potential integrals can be as large as nine. Clearly then, it is far more efficient to compute the required potential integrals by face-pair combinations, rather

[2] Note that if the approximations (13) and (14) had not been made, the procedure leading to (15) would have been identical to Galerkin's method since the basis and testing functions chosen are identical [15]. For the EFIE, the matrix would then satisfy the symmetry property $Z_{mn} = Z_{nm}$, but this desirable property is lost due to the approximations made. Since Z_{mn} and Z_{nm} are different approximations to the same quantity, however, then their average also approximates the quantity, and one tempting possibility is to average Z with its transpose, thus restoring the symmetry property to the moment matrix. This approach, however, has not to date been tested.

than directly compute single elements of Z by edge-pair combinations. For each face-pair combination, the potential integrals may be multiplied by the appropriate coefficients (cf (17)) and their contributions accumulated in the appropriate elements of Z as they are computed.

In accordance with the above discussion, consider the evaluation of the vector and scalar potential integrals for a given source and observation face combination. Fig. 4 illustrates such a face pair with an observation point in face p and with source currents residing in face q. Each of the three basis functions which may exist simultaneously in T^q is proportional to one of the vectors ρ_1, ρ_2, or ρ_3 defined in the figure. Each vector ρ_i, $i = 1, 2, 3$, is shown directed away from its associated vertex in the figure, but would be directed toward the vertex if the current reference direction for the associated edge were into the triangle. Consequently,

$$\rho_i = \pm(\mathbf{r}' - \mathbf{r}_i), \qquad i = 1, 2, 3, \tag{24}$$

where the positive sign is used if the positive current reference direction is out of T^q and the negative sign is used otherwise. We wish to evaluate the magnetic vector potential,

$$\mathbf{A}_i^{pq} = \frac{\mu}{4\pi} \int_{T^q} \left(\frac{l_i}{2A^q}\right) \rho_i \frac{e^{-jkR^p}}{R^p} \, dS', \tag{25}$$

and the electric scalar potential,

$$\Phi_i^{pq} = -\frac{1}{4\pi j\omega\epsilon} \int_{T^q} \left(\frac{l_i}{A^q}\right) \frac{e^{-jkR^p}}{R^p} \, dS', \tag{26}$$

associated with the ith basis function on face q observed at the centroid of face p. In (25) and (26),

$$R^p = |\mathbf{r}^{cp} - \mathbf{r}'| \tag{27}$$

where $\mathbf{r}^{cp}$ is the position vector of the centroid of face p.

Integrals (25) and (26) are most conveniently evaluated by transforming from the global coordinate system to a local system of coordinates defined within T^q. To define these coordinates, note that the vectors ρ_i in Fig. 4 divide T^q into three subtriangles of areas A_1, A_2, and A_3, with l_1, l_2, and l_3, respectively, as one of their sides. The areas are not independent, however, since they must satisfy $A_1 + A_2 + A_3 = A^q$. We now introduce the so-called *normalized area coordinates* [19]

$$\xi = \frac{A_1}{A^q}, \quad \eta = \frac{A_2}{A^q}, \quad \zeta = \frac{A_3}{A^q}, \tag{28}$$

which, because of the area constraint, must satisfy

$$\xi + \eta + \zeta = 1. \tag{29}$$

Note that all three coordinates vary between zero and unity in T^q and that at the triangle corners $\mathbf{r}_1$, $\mathbf{r}_2$, and $\mathbf{r}_3$, the triplet (ξ, η, ζ) takes on the values (1, 0, 0), (0, 1, 0), and (0, 0, 1), respectively. The transformation from Cartesian to normalized area coordinates may be written in vector form as

$$\mathbf{r}' = \xi\mathbf{r}_1 + \eta\mathbf{r}_2 + \zeta\mathbf{r}_3, \tag{30}$$

where ξ, η, and ζ are subject to the constraint (29). It can easily be shown that surface integrals over T^q transform as follows:

$$\int_{T^q} g(\mathbf{r}) \, dS = 2A^q \int_0^1 \int_0^{1-\eta} g[\xi\mathbf{r}_1 + \eta\mathbf{r}_2 + (1 - \xi - \eta)\mathbf{r}_3] \, d\xi \, d\eta. \tag{31}$$

With (24), (27), (30), and (31), (25) and (26) may now be written as

$$\mathbf{A}_i^{pq} = \pm\frac{\mu l_i}{4\pi}(\mathbf{r}_1 I_\xi^{pq} + \mathbf{r}_2 I_\eta^{pq} + \mathbf{r}_3 I_\zeta^{pq} - \mathbf{r}_i I^{pq}) \tag{32}$$

and

$$\Phi_i^{pq} = \mp\frac{l_i}{j2\pi\omega\epsilon} I^{pq}, \tag{33}$$

where

$$I^{pq} = \int_0^1 \int_0^{1-\eta} \frac{e^{-jkR^p}}{R^p} \, d\xi \, d\eta, \tag{34a}$$

$$I_\xi^{pq} = \int_0^1 \int_0^{1-\eta} \xi \frac{e^{-jkR^p}}{R^p} \, d\xi \, d\eta, \tag{34b}$$

$$I_\eta^{pq} = \int_0^1 \int_0^{1-\eta} \eta \frac{e^{-jkR^p}}{R^p} \, d\xi \, d\eta, \tag{34c}$$

$$I_\zeta^{pq} = I^{pq} - I_\xi^{pq} - I_\eta^{pq}. \tag{34d}$$

Thus only three independent integrals, (34a)–(34c), must be numerically evaluated for each combination of face pairs p and q. The three integrals, in turn, contribute to up to nine elements of Z in (16). For a closed object with N edges the number of independent integrals computed is $4N^2/3$. By contrast, the edge-by-edge approach would require the evaluation of $12N^2$ integrals or nine times as many. Numerical evaluation of the integrals (34a)–(34c) may be accomplished by using numerical quadrature techniques specially developed for triangular domains [20]. However, for the terms in which $p = q$ the integrands are singular, and for these cases the singular portion of each integrand must be removed and integrated analytically [17].

III. NUMERICAL RESULTS

In this section numerical results are presented for surface current distributions induced on selected scatterers under plane wave illumination. Although one radar cross-section example is given, emphasis in the examples is on the calculation of current distributions, not only because of their practical value in problems such as electromagnetic compatibility and nuclear electromagnetic pulse (EMP) penetration, but also because we believe that the calculation of accurate surface currents is a much more stringent test of a numerical approach than is ultimate calculation of far-field quantities. The geometries considered are a conducting square plate, a bent plate, a circular disk, and a sphere. The plate and disk problems involve open surfaces and therefore are a test of the EFIE approach when edges are present. The disk also serves as an example of a structure with a curved boundary, while the sphere exemplifies both a closed surface and a doubly curved surface. Both problems are examples of surfaces not amenable to rectangular patch modeling. A summary of the computational resources required to calculate results for several of the examples is given at the end of the section.

Flat Plate

Figs. 5 and 6 show the dominant component current distributions along the two principal cuts on a square plate illuminated by a normally incident plane wave. For comparison, the solution of Glisson [16], obtained using rectangular patches, is also given. The number of patches listed in the figures refers to the number of *charge* patches in the earlier solution of Glisson and to the number of triangles (also equal to the number of charge patches) in the present solution. Note that these quantities play similar roles in the two approaches. No comparison of the convergence rate of the two approaches should be inferred from the figures since both solutions are already well converged for the number of unknowns used. Note also that the density of data points appearing in the figures for the triangular patch solution is not truly indicative of the linear density of the subdomains. This is because, in effect, we show data points only for every other edge, i.e., only for those edges where the current reference direction vector is parallel to the current component we wish to observe.

Fig. 5 shows the current induced on a plate 0.15 λ on each side. At this low frequency the current distribution is largely determined by the edge conditions and hence this case provides a good test of the technique's capacity for handling surface edges. We note the absence of any anomalies in the computed distribution near the plate edges. The elimination of such anomalies is attributed to the use of basis functions in which the expansion coefficients are not associated with current components parallel to plate edges and to a testing procedure in which potentials are not evaluated at edges [14].

Fig. 6 shows corresponding results for a 1.0 λ square plate. From the figure, one observes that the edge behavior of the current distribution is confined to a relatively smaller region near the edges than for the 0.15 λ plate and that the current on the interior portion of the plate has begun to exhibit the physical optics-plus standing wave distribution characteristic of the higher frequencies. Also shown for comparison are the corresponding results reported in [21].[3] Note that because of the placement of subdomains, the component of current normal to the plate edge in the solution of [21] vanishes prematurely in the subdomain nearest the edge.

Fig. 7 compares the calculated radar cross section (RCS) with the thin plate measurements of Kouyoumjian [22] and the computations of Rahmat-Samii and Mittra [23] for a square plate. In the latter case, we suspect that the premature vanishing of the current near the plate edge causes the underestimation in the RCS observed in the figure for lower frequencies. Kouyoumjian has also computed the RCS from a variational formula and we find no discernable difference between his computations and our results for square plates whose sides are smaller than 0.4 λ, the range for which his formula should be most accurate. Also shown in Fig. 7 is a plot of the

[3] The data reported in [21] is actually for the electric field in a square aperture in a ground plane but has been converted to an equivalent electric current on a square conducting plate via the duality of the plate and aperture problems.

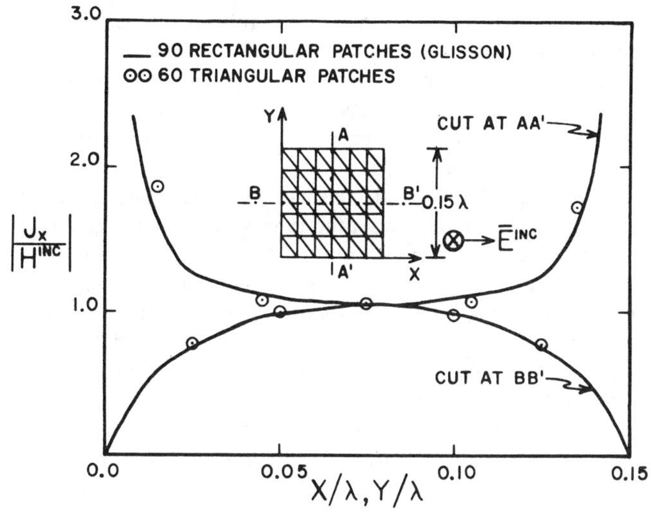

Fig. 5. Distribution of dominant component of current on 0.15 λ square flat plate.

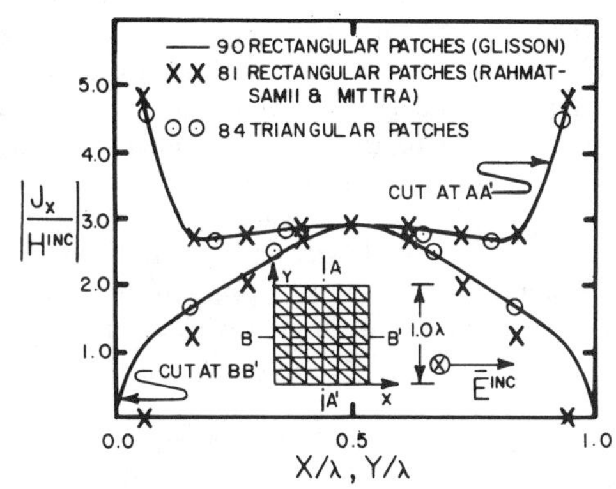

Fig. 6. Distribution of dominant component of current on 1.0 λ square flat plate.

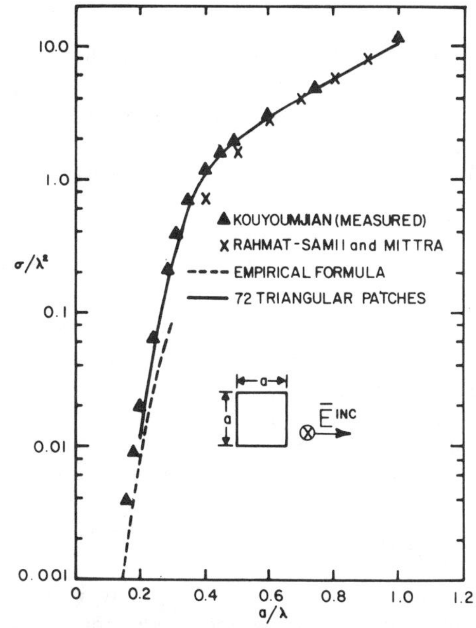

Fig. 7. Monostatic radar cross section versus frequency for square plate, normal incidence. Measurements are for plate of thickness 0.000127 λ.

RCS for a square plate given by an empirical formula derived from measurements [24].

While there is generally very good agreement between the various results reported in Figs. 6 and 7, we have not been able to obtain a favorable comparison with the corresponding current distribution and RCS calculations of Wang *et al.* [7]. There are several reasons to suspect their calculations. First, they employ a set of basis functions which would generally cause spurious oscillations in the current distribution along the direction of current flow, and which would hence be expected to adversely affect the solution for the current distribution and, ultimately, the RCS. Secondly, their model ignores the presence of cross-polarized components of surface current on the plate. Finally, in the RCS case, the excellent agreement which they obtain between computed and measured radar cross section is unfortunately based on a comparison to the thick plate rather than the thin plate measurements of Kouyoumjian [22] for which their theory more appropriately applies and which we have shown in Fig. 7.

Bent Plate

Fig. 8 shows the dominant component current distribution along a cut through the symmetry plane of a bent square plate. The bend is parallel to and located a distance of one-third the plate width from one edge, and a plane wave with the electric field polarized parallel to the bend is incident normal to the larger section of the bent plate. The smaller plate section is bent through an angle of 50° toward the shadow side of the plate. Other frequencies, polarizations, and angles of incidence have been examined, and the resulting current distributions show good correspondence with those of Glisson [16].

Circular Disk

Fig. 9 shows the computed current distribution on a circular disk illuminated by a normally incident plane wave. The component J_ϕ is shown along a diameter cut oriented perpendicular to the incident electric field vector. Also shown for comparison is the quasi-static solution valid at low frequencies [25].

Sphere

Fig. 10 shows the computed current distribution along the principal cuts on a 0.2 λ radius conducting sphere. The cases of axial and equatorial incidence are both considered in order to observe the influence of the triangulation scheme on the solution. Also shown for comparison is the exact eigenfunction solution. Results for both illuminations are in very good agreement with the exact solution.

Although we have not attempted to solve the sphere problem at frequencies near its cavity resonances, we expect the usual difficulties associated with the EFIE formulation at the internal resonance frequencies of closed bodies to arise. To alleviate the problem, one might make use of the present treatment of the EFIE operator in a formulation which combines it with the MFIE operator in such a way as to eliminate the singularity which is present in both operators alone [26].

Computational Aspects

Table I summarizes data on the number of triangular patches, number of unknowns, and the computation time required on a Univac 1100/83 computer to generate the results of Figs. 5, 6, 8, and 10. The timing data for Fig. 6 were not re-

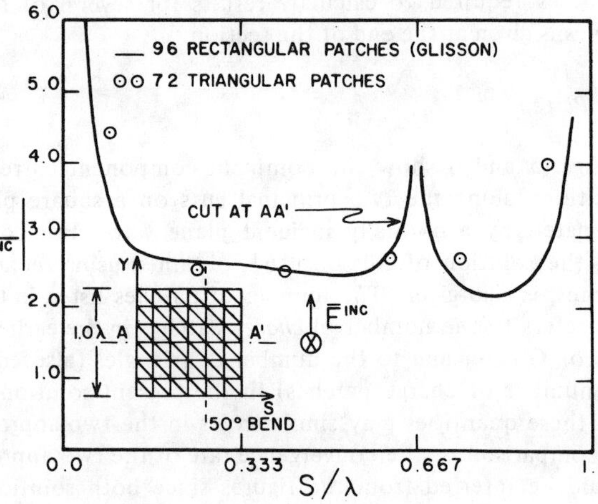

Fig. 8. Distribution of dominant component of current on 1.0 λ bent square plate.

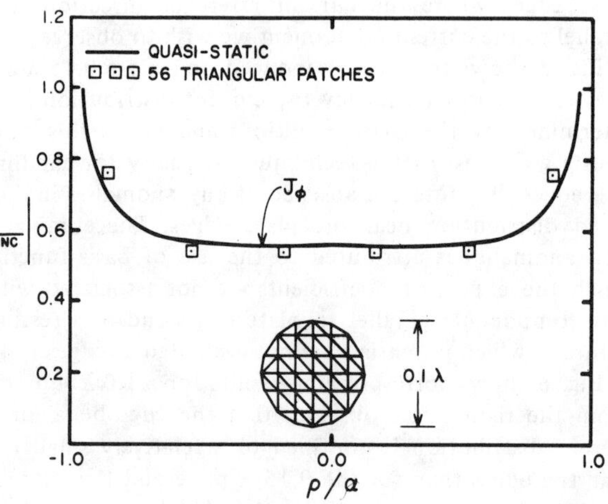

Fig. 9. Distribution of current on 0.05 λ radius circular disk.

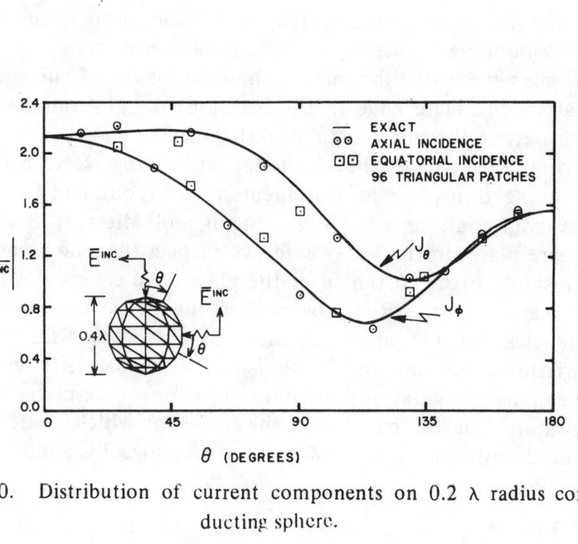

Fig. 10. Distribution of current components on 0.2 λ radius conducting sphere.

TABLE I
SUMMARY OF COMPUTATION DATA FOR SELECTED EXAMPLE PROBLEMS

Case (Figure Number)	Number of Patches N_f	Number of Boundary Edges N_b	Number of Unknowns N†	Matrix Fill Time (s)	Matrix Inversion Time (s)	Total CPU Time (s)
5	60	22	79	65	20	92
6	84	26	113	127*	57*	195*
8	72	24	96	94	35	139
10	96	0	144	163	118	294

* Estimated from convergence study.
† $N = (3N_f - N_b)/2$.

corded during computation but are estimated from the convergence study discussed below. The present version of the computer code has not been completely optimized with respect to computation speed, and it should be possible to increase this speed by at least a factor of two or three by 1) decreasing the order of integration used to calculate the potential integrals when R^p in (34) is large compared to source patch size, and 2) replacing the matrix inversion procedure, used primarily for diagnostic purposes, by a linear equation solving procedure.

A convergence study of the current distribution on the 1.0 λ square plate of Fig. 6 was carried out using 18, 32, 50, and 72 patches corresponding to 21, 40, 65, and 96 unknowns, respectively. It was found that, except for the nearly singular component of current nearest the plate edge, the computed currents in the sequence of computations differed from the results of Glisson (cf Fig. 6) by no more than 10 percent.

IV. SUMMARY

In this paper, the electric field integral equation (EFIE) is used with the method of moments to develop a simple and efficient numerical procedure for treating problems of scattering by arbitrarily shaped objects. For numerical purposes the objects are modeled by planar triangular surface patch models. Because the EFIE formulation is used, the procedure is applicable to both open and closed bodies. Crucial to the formulation is the development of a set of special subdomain basis functions which are defined on pairs of adjacent triangular patches and yield a current representation free of line or point charges at subdomain boundaries.

The approach is applied to the scattering problems of plane wave illumination of a flat square plate, a bent square plate, a circular disk, and a sphere. Comparisons of surface current density with previous computations or exact formulations show good correspondence in each case.

A listing of the computer code is available from the authors.

ACKNOWLEDGMENT

The authors are grateful to Mr. Robert Baxter and Dr. Deanne Pecora of the Lockheed Missiles and Space Company, Inc., for providing us with the results of their timing and convergence study summarized in Table I.

REFERENCES

[1] J. H. Richmond, "A wire-grid model for scattering by conducting bodies," *IEEE Trans. Antennas Propagat.*, vol. AP-14, no. 6, pp. 782–786, Nov. 1966.
[2] E. K. Miller and F. J. Deadrick, "Some computational aspects of thin-wire modeling," in *Numerical and Asymptotic Techniques in Electromagnetics*, R. Mittra, Ed. New York: Springer-Verlag, 1975, ch. 4.
[3] K. S. H. Lee, L. Marin, and J. P. Castillo, "Limitations of wire-grid modeling of a closed surface," *IEEE Trans. Electromagn. Compat.*, vol. EMC-18, no. 3, pp. 123–129, Aug. 1976.
[4] D. L. Knepp and J. Goldhirsh, "Numerical analysis of electromagnetic radiation properties of smooth conducting bodies of arbitrary shape," *IEEE Trans. Antennas Propagat.*, vol. AP-20, no. 3, pp. 383–388, May 1972.
[5] N. C. Albertsen, J. E. Hansen, and N. E. Jensen, "Computation of radiation from wire antennas on conducting bodies," *IEEE Trans. Antennas Propagat.*, vol. AP-22, no. 2, pp. 200–206, Mar. 1974.
[6] G. J. Burke and A. J. Poggio, "Numerical Electromagnetic Code (NEC) - method of moments," Naval Ocean Systems Center, San Diego, CA, Tech. Document 116, July 1977.
[7] N. N. Wang, J. H. Richmond, and M. C. Gilreath, "Sinusoidal reaction formulation for radiation and scattering from conducting surfaces," *IEEE Trans. Antennas Propagat.*, vol. AP-23, no. 3, pp. 376–382, May 1975.
[8] E. H. Newman and D. M. Pozar, "Electromagnetic modeling of composite wire and surface geometries," *IEEE Trans. Antennas Propagat.*, vol. AP-26, no. 6, pp. 784–789, Nov. 1978.
[9] A. Sankar and T. C. Tong, "Current computation on complex structures by finite element method," *Electron. Lett.*, vol. 11, no. 20, pp. 481–482, Oct. 1975.
[10] J. J. H. Wang, "Numerical analysis of three-dimensional arbitrarily-shaped conducting scatterers by trilateral surface cell modelling," *Radio Sci.*, vol. 13, no. 6, pp. 947–952, Nov.–Dec. 1978.
[11] J. J. H. Wang and C. Papanicolopulos, "Surface patch modeling of scatterers of arbitrary shapes," *Antennas Propagat. Soc. Int. Symp. Digest*, Univ. Washington, Seattle, WA, June 1979, pp. 159–162.
[12] G. Jeng and A. Wexler, "Finite element, boundary integral equation analysis of scattering problems," *URSI Symp. on Electromagnetic Wave Theory*, Stanford Univ., Stanford, CA, June 1977, pp. 179–181.
[13] J. Singh and A. T. Adams, "A non rectangular patch model for scattering from surfaces," *IEEE Trans. Antennas Propagat.*, vol. AP-27, no. 4, pp. 531–535, July 1979.
[14] A. W. Glisson and D. R. Wilton, "Simple and efficient numerical methods for problems of electromagnetic radiation and scattering from surfaces," *IEEE Trans. Antennas Propagat.*, vol. AP-28, no. 5, pp. 593–603, Sept. 1980.
[15] R. F. Harrington, *Field Computation by Moment Methods*. New York: Macmillan, 1968.
[16] A. W. Glisson, "On the development of numerical techniques for treating arbitrarily-shaped surfaces," Ph.D. dissertation, Univ. Mississippi, 1978.
[17] D. R. Wilton, S. S. M. Rao, and A. W. Glisson, "Electromagnetic scattering by arbitrary surfaces," Rome Air Development Center, Griffiss AFB, NY, Tech. Rep. RADC-TR-79-325, Mar. 1980.
[18] J. Van Bladel, *Electromagnetic Fields*. New York: McGraw-Hill, 1964, p. 502.
[19] O. C. Zienkiewicz, *The Finite Element Method in Engineering Science*. New York: McGraw-Hill, 1971.
[20] P. C. Hammer, O. P. Marlowe, and A. H. Stroud, "Numerical integration over simplexes and cones," *Math. Tables Aids Comp.*, vol. 10, pp. 130–137, 1956.

[21] C. M. Butler, Y. Rahmat-Samii, and R. Mittra, "Electromagnetic penetration through apertures in conducting surfaces," *IEEE Trans. Antennas Propagat.*, vol. AP-26, no. 1, pp. 82–93, Jan. 1978.

[22] R. G. Kouyoumjian, "The calculation of the echo areas of perfectly conducting objects by the variational method," Ph.D. dissertation, The Ohio State Univ., 1953.

[23] Y. Rahmat-Samii and R. Mittra, "Integral equation solution and RCS computation of a thin rectangular plate," *IEEE Trans. Antennas Propagat.*, vol. AP-22, no. 4, pp. 608–610, July 1974.

[24] C. T. Ruck, Ed., *Radar Cross Section Handbook.* New York: Plenum, 1970, p. 523.

[25] J. J. Bowman, T. B. A. Senior, and P. L. E. Uslenghi, *Electromagnetic and Acoustic Scattering by Simple Shapes.* Amsterdam: North-Holland, 1969, pp. 576–577.

[26] J. R. Mautz and R. F. Harrington, "*H*-field, *E*-field, and combined field solutions for conducting bodies of revolution," *AEÜ*, vol. 32, no. 4, pp. 157–164, Apr. 1978.

Sadasiva M. Rao received the bachelor's degree in electrical and communications engineering from Osmania University, Hyderbad, India, in 1974, the master's degree from the Indian Institute of Sciences, Bangalore, India, in 1976, and the Ph.D. degree from the University of Mississippi, University, in 1980.

He is currently a Visiting Assistant Professor at the Department of Electrical Engineering, Rochester Institute of Technology, Rochester, NY. His research interests are in the areas of electromagnetic theory and numerical methods applied to antennas and scattering.

Donald R. Wilton (S'63–M'65–M'70–SM'80), for a photograph and biography please see page 835 of the November 1981 issue of this TRANSACTION.

Allen W. Glisson (S'71–M'78) was born in Meridian, MS, on June 26, 1951. He received the B.S., M.S., and Ph.D. degrees in electrical engineering from the University of Mississippi, University, in 1973, 1975, and 1978, respectively.

From 1973 to 1978 he served as a Research Assistant in the Electrical Engineering Department of the University of Mississippi. In 1978, he joined the faculty of the University of Mississippi, where he is currently Assistant Professor of Electrical Engineering. His current research interests include the development and application of numerical techniques for treating electromagnetic radiation and scattering problems.

Dr. Glisson is a member of Sigma Xi, Tau Beta Pi, Phi Kappa Phi, Eta Kappa Nu, and an associate member of Commission B of the International Union of Radio Science.

A Surface Patch Model for Polygonal Plates

EDWARD H. NEWMAN, MEMBER, IEEE, AND PRAVIT TULYATHAN, MEMBER, IEEE

Abstract—**A method of moments surface patch model for polygon plates is presented. Three methods for segmenting polygon plates are considered. Also, the choice between full surface testing and point matching is discussed as it affects the efficiency and generality of the solution. Numerical results are compared with measurements and independent calculations for polygonal plates of three different shapes, including a wire to plant junction.**

I. INTRODUCTION

THIS PAPER will describe a moment method surface patch modeling technique, based upon the reaction integral equation, for modeling nonrectangular plates. A surface patch model is a model for a solid surface which uses surface current densities, as opposed to a wire grid model which uses a fine mesh or grid of crossing wires. In surface patch modeling, the unknown surface current density is expanded in a basis set, and the unknown coefficients are determined using the method of moments.

The early work in surface patch modeling was limited to rectangular plates [1]. For example, the authors have developed a surface patch technique and its associated user-oriented code for treating combinations of thin wires, rectangular plates, wire to plate and plate to plate junctions [2]. The problem with the rectangular plate codes is that they cannot easily model a general structure. For example, they cannot efficiently model the swept wing of an aircraft or the pointed bow of a ship.

Several techniques exist to permit the modeling of general or nonrectangular structures. These seem to be divided into two approaches. In the first approach, the structure is viewed as a general curved surface, and the nonrectangular surface patch modes make up a piecewise flat approximation to the surface current density on the surface [3]-[8]. In the second approach the structure is viewed as an interconnection of polygonal plates. The currents on the polygonal plates are then expanded in terms of nonrectangular surface patch modes. The work of Singh and Adams [9], as well as this work, are examples of the second approach. Although the first approach appears to be more general, we chose the second, since we feel that it will lead to a more user-oriented code. That is, the user need only define the coordinates of the corners of the several plates which approximate the structure in the second approach, rather than the coordinates of possibly several hundred modes which approximate the structure in the first approach.

Previous work [3]-[9] clearly shows that it is possible to model polygonal plates using nonrectangular surface patch modes. However, we feel that the practical use of these techniques requires additional effort in two areas, which will be considered in this paper. First, an automated scheme for segmenting the polygonal plates into surface patch modes is required. Requiring a code user to specify the coordinates of several hundred modes individually on a complex structure is felt to be so tedious, that it is impractical. The second problem involves the code efficiency and generality and specifically how they are impacted by the choice of full surface testing versus point matching. This problem will be discussed, and it will be demonstrated that reasonable run times are possible for a code based on full surface testing. Numerical examples, compared with measurements and other independent solutions, will illustrate the accuracy and versatility of the techniques.

Manuscript received April 1, 1981; revised July 13, 1981. This work was supported in part by Contract N00014-78-C-0049 and Grant DAAG29-76-C-0331 between the Joint Services Electronic Progam and the U.S. Army Research Office, and the Ohio State University Research Foundation.

The authors are with the ElectroScience Laboratory, Department of Electrical Engineering, The Ohio State University, 1320 Kinnear Road, Columbus, OH 43212.

II. BRIEF THEORY

Consider the impressed sources $(\mathbf{J}_i, \mathbf{M}_i)$ in the volume V, and radiating in the presence of a perfectly conducting obstacle defined by the surface S (open or closed). The reaction integral equation for the current $\mathbf{J}_s$, induced on S is

$$\iint_S \mathbf{J}_s \cdot \mathbf{E}_m \, ds = -\iiint_V (\mathbf{J}_i \cdot \mathbf{E}_m - \mathbf{M}_i \cdot \mathbf{H}_m) \, dv \tag{1}$$

where $(\mathbf{E}_m, \mathbf{H}_m)$ are the free space fields of an electric test source, $\mathbf{J}_m$, placed on S [2]. The $e^{j\omega t}$ time dependence of all sources and fields are suppressed. Equation (1) insures that, since the total tangential electric field vanishes on the conducting surface, there is zero reaction between a test mode $\mathbf{J}_m$ on the surface and the currents $\mathbf{J}_s$ and $(\mathbf{J}_i, \mathbf{M}_i)$ which generate the total field. Note that for open surfaces, $\mathbf{J}_s$ is the vector sum of the current on the inner and outer surfaces.

The integral equation (1) is solved by the moment method. The unknown current $\mathbf{J}_s$ is expanded in a set of N basis (expansion functions),

$$\mathbf{J}_s = \sum_{n=1}^{N} I_n \mathbf{J}_n$$

and (1) is enforced for N electric test sources placed in S. Thus, (1) reduces to the set of simultaneous linear equations

$$\sum_{n=1}^{N} I_n Z_{mn} = V_m, \qquad m = 1, 2, \cdots, N$$

where

$$Z_{mn} = -\iint_n \mathbf{J}_n \cdot \mathbf{E}_m \, ds \tag{2}$$

$$V_m = \iiint_V (\mathbf{J}_i \cdot \mathbf{E}_m - \mathbf{M}_m \cdot \mathbf{H}_m) \, dv \tag{3}$$

and the integral in (2) is over the surface of the nth expansion mode.

For expansion modes on a polygonal plate we choose a generalization of the rectangular piecewise sinusoidal surface patch dipole mode [2] and [10]. Fig. 1 shows a general quadrilateral piecewise sinusoidal surface patch dipole mode. It is composed of two monopoles. Consider some general point in monopole A (shown as the heavy dot in Fig. 1), which has terminal length V and end length U. Construct a line through the point which intersects the end and terminal such that $u/v = U/V$. Let L be the length of this line and l be the distance, along the line, from the end to the point. Then, the surface current density on monopole A is

$$\mathbf{J}_A = \frac{C \sin kl\hat{l}}{w(l/L)} \tag{4}$$

where $w(l/L)$ is the local width of monopole A, k is the free space wave number, and C is a normalizing constant such that the terminal current is 1 A. The surface current density on monopole B, $\mathbf{J}_B$, is defined in a similar way except that the current is in the $-\hat{l}$ direction. Two overlapping, and nearly orthogonal, arrays of surface patch dipole modes are placed on the polygon plate, as described in the next section.

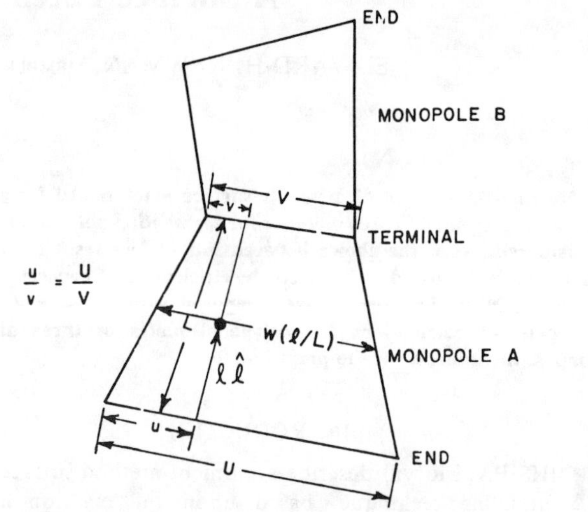

Fig. 1. General quadrilateral piecewise-sinusoidal surface patch dipole mode.

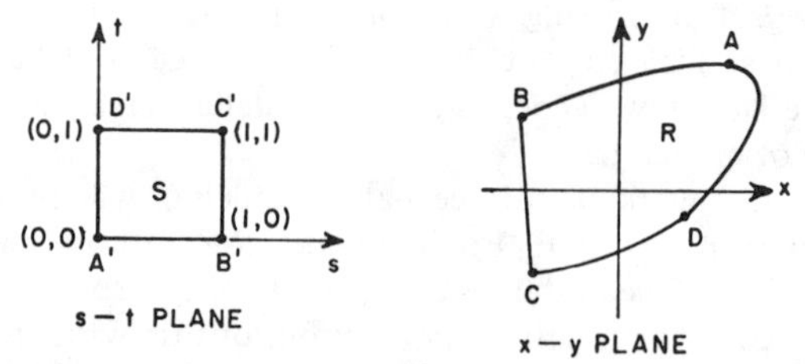

Fig. 2. Gordon–Hall domains.

III. PLATE SEGMENTATION TECHNIQUES

Probably the most important problem in the surface patch modeling of polygonal plates is the development of a technique (in the form of a computer algorithm) for segmenting the plate into surface patch modes. Segmenting a rectangular plate into rectangular modes, or a triangular plate into triangle modes is fairly straightforward. The problem is segmenting an arbitrary polygon into a given mode shape, not chosen to fit the polygon. The segmentation scheme must

1) completely fill (and not overlap) the area of the polygonal plate;
2) allow for two orthogonal components of current at each point on the plate. At edges the normal current should vanish and the parallel current should be finite.

In addition to these requirements, it is desirable

1) for a given maximum segment size, the technique should divide the plate into as few modes as possible;
2) as many as possible of the modes should be rectangular;
3) as many as possible of the modes should be parallel.

These will all tend to reduce run time [10], and 1) will of course reduce the required storage. Simpler formulas are available for the mutual impedance of parallel sinusoidal filaments, than for general skew filaments. Also, one integration can be eliminated in the mutual impedance of parallel and rectangular surface patch modes [10].

Three techniques for segmenting a polygonal plate will now be described. The first is a standard technique, known as the Gordon–Hall (G-H) procedure [11], and the last two, denoted Method 1 and 2, were developed by the authors. The descriptions presented here are intended to give the reader a basic feel for the methods. More details are available elsewhere [12].

Gordon-Hall Procedure

Referring to Fig. 2, the G-H technique maps the unit square $A'B'C'D'$ in the s–t plane (S domain) to the closed curve $ABCDA$ in the x–y plane (R domain or real space). The G-H procedure was used by Okon and Harrington [13] to compute the capacitance of arbitrarily shaped disks, and the description here follows their presentation. Here, the planar curve $ABCDA$ consists of up to four analytically defined arcs. The vector function $\mathbf{F}(s, t)$ maps the s–t plane onto the x–y plane, and in particular, maps $A'B'C'D'A'$ onto $ABCDA$. Essentially, a grid is drawn on the unit square in the s–t plane. The intersection points in this grid are mapped (by a transfinite mapping [11]) into the x–y plane.

Fig. 3 shows the G-H procedure applied to a triangle such that no segment exceeds a given maximum size W. The hypotenuse is split in half to make the triangle a four sided figure. The G-H procedure seems to work best on quadrilaterals with nearly equal sides. The arrows in Fig. 3 indicate the overlapping surface patch dipole modes. Note how the modes tend to bunch near the middle of the hypotenuse.

Segmentation Method 1

The general approach of Method 1 is to choose a side of the polygonal plate, usually the longest side, and divide the plate into strips which are parallel to this side. The width of the strips are chosen to be as large as possible, and yet less than a given maximum size W. The ends of each strip are a straight line segment which exactly fits the plate boundary. Each strip is segmented into the same number of monopoles or segments. The number of segments is chosen so that the longest segment is less than W.

Fig. 4 shows Method 1 applied to the same triangle which was segmented by the G-H procedure in Fig. 3. Note that breaking each strip into the same number of segments is ideal for the longest strip, adjacent to the hypotenuse, but is waste-

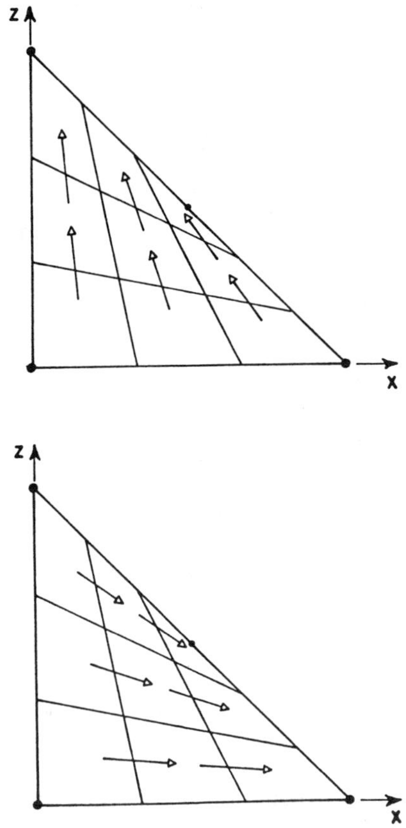

Fig. 3. Surface patch dipole modes generated by Gordon-Hall procedure for a triangle.

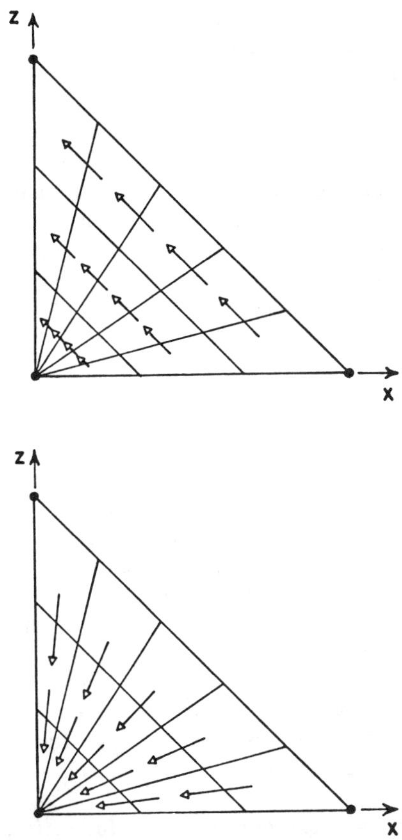

Fig. 4. Surface patch dipole modes generated by Method 1 for same triangle shown in Fig. 3.

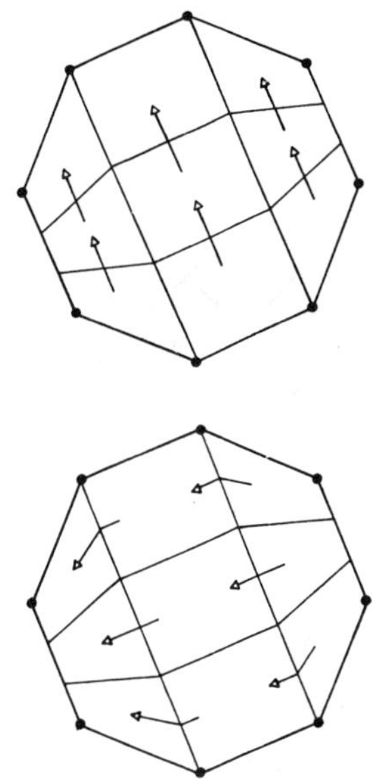

Fig. 5. Surface patch dipole modes generated by Method 1 for regular octagon.

ful for the last strip which is much shorter. Method 1 is best suited to a polygon that can be divided into strips of approximately equal lengths, such as the regular octagon of Fig. 5.

Segmentation Method 2

Segmentation Method 2 is designed to eliminate the inefficiency in Method 1 caused by breaking each strip into the same number of segments. In Method 2, each strip is broken into a different number of segments such that the longest segment is less than W. This is shown in Fig. 6(a) where the longer strips have more segments. In doing this note that the ends of the segments of adjacent strips do not line up. Thus, the grid in Fig. 6(a) is suitable for only one current polarization. The second polarization, shown in Fig. 6(b), is constructed from strips parallel to a line orthogonal to the hypotenuse. With the modes shown in Fig 6(a) and (b) there is a current parallel to the hypotenuse edge but not the remaining two sides. Thus Fig. 6(c) shows two final strips added parallel to these sides. Fig. 6 shows that Method 2 does a good job of satisfying the above stated criteria.

Computer Algorithm

A computer subroutine has been written which implements Methods 1 and 2. The subroutine contains logic which chooses Method 2 for plates with a relatively large difference in the strip lengths and Method 1 otherwise. The user specifies the coordinates of the corners of the polygon, and the maximum segment size W. The subroutine generates the coordinates of the N surface patch dipole modes on the polygon plate. Although we have illustrated in Figs. 4-6 how the subroutine segments a plate, the real test of the segmentation process is its accuracy and efficiency when implemented in a surface patch moment method solution. The attributes of the methods described will be shown by the examples in Section V.

IV. CHOICE OF TEST MODES

In this section we will discuss some considerations in the choice of the test modes. By this choice we do not mean pulse versus triangle versus sinusoids, etc., but rather the choice between point matching and full surface testing. As usual, the

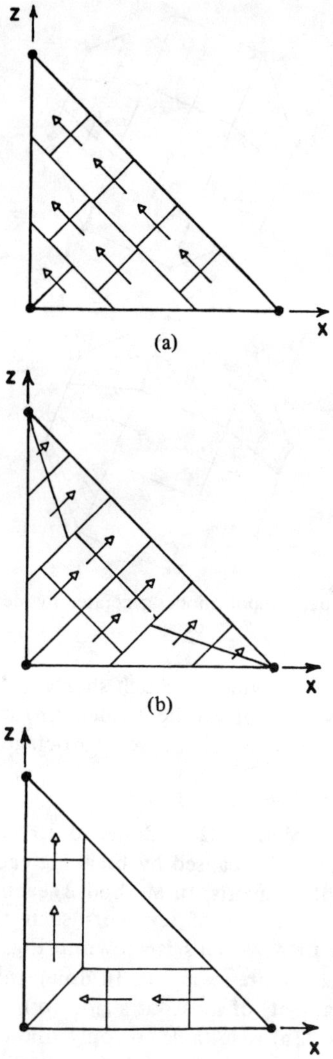

Fig. 6. Surface patch dipole modes generated by Method 2 for triangle shown in Figs. 3 and 4.

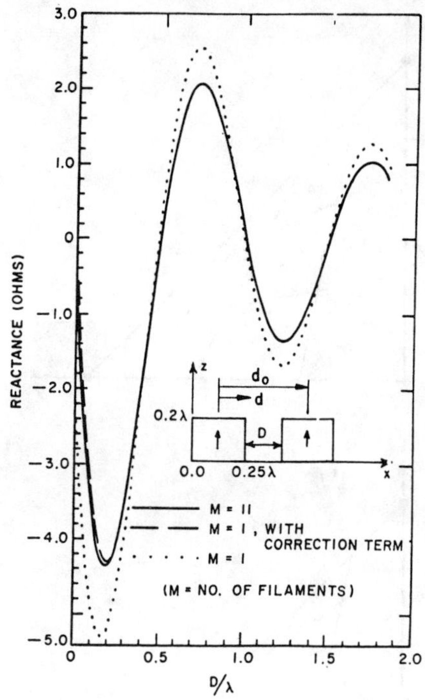

Fig. 7. Reactance of two rectangular piecewise-sinusoidal surface patch monopoles.

decision is not clear cut, but rather, involves trade-offs between computer run time, storage, and code generality.

In surface patch modeling, the matrix fill time usually far exceeds the matrix inversion time. The time to compute a single element in the impedance matrix is generally proportional to the number of integrations required. Full surface testing requires four integrations (two on the test mode and two on the expansion mode), while point matching requires only two integrations. The relatively short time to compute a single element using point matching is offset by the fact that point matching solutions often require more modes to achieve a given level of convergence.

Although a parameter study investigating time trade-offs for point matching and surface test solutions would be interesting, we did not make such a study. We chose surface testing since it was felt that it would be required when considering the problem of a wire attached to a plate. We felt that the point matching solution would be overly sensitive to the proximity of the attachment point and the match point. Our previous experience with rectangular modes [2] indicated that with full surface testing, the attachment point may be anywhere in the surface patch grid.

Having chosen full surface testing, it is essential that the computation time for each element in the impedance matrix be made as small as possible, and yet result in reasonably accurate results. Besides the advantages of the segmentation procedure just described and those described in [10], probably the most effective technique for reducing computational time is simply to reduce to a minimum the number of points used in numerical integrations. Thus, it is desirable to represent the mutual impedance between two surface patch modes as simply the mutual impedance between two filaments located along the center line of each patch. Fig. 7 shows a comparison of the mutual reactance between two surface patch monopoles versus their separation, for $M = 11$ (i.e., essentially the exact result) filaments per patch as the solid line, and $M = 1$ filament per patch as the dotted line. Note that, although generally small, the error in the one point integration is more than 10 per cent at $D = 1.25\lambda$ or $D = 1.75\lambda$. This result can be improved by using the simple approximation

$$Z(d) \approx Z(d_0)e^{-jk(d-d_0)}\frac{d_0}{d} \tag{5}$$

for d near d_0. Here $Z(d)$ is the mutual impedance for filaments separated by a distance d. Using a correction obtained with (5), Fig. 7 shows that the one point integration mutual reactance can be made essentially exact, even for small D/λ. Note that using the one point integration in conjunction with the interpolation formula of (5) is not an application of point matching. Rather, it is full surface testing which requires only one filament to filament impedance to be computed per patch to patch impedance. Being able to use a one point integration for all but touching patches, results in a substantial savings in time. The extension of (5) to skew patches is straightforward [12].

In summary, we view our solution as full surface expansion and testing, (i.e., Galerkin) using overlapping piecewise-sinusoidal surface patch test and expansion modes. For the

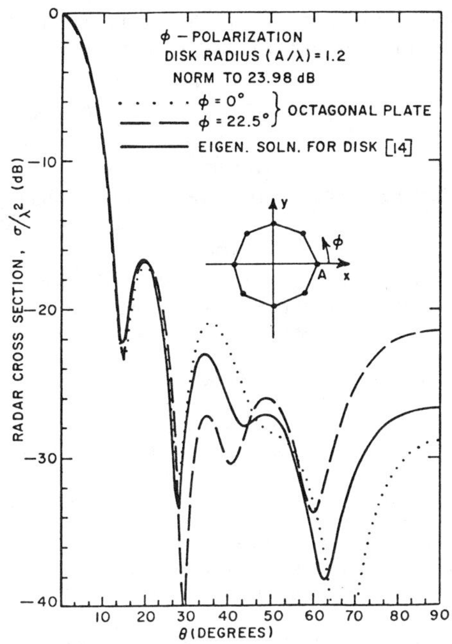

Fig. 8. Backscatter from disk with radius $A/\lambda = 1.2$

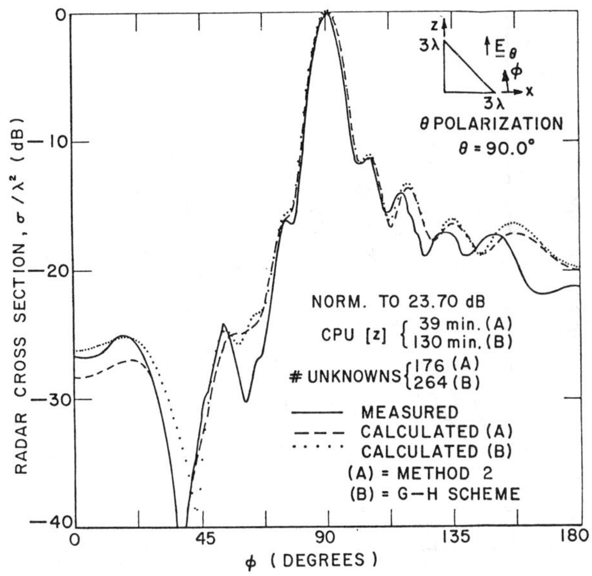

Fig. 9. Backscatter from 3λ right angle triangle (θ-polarization).

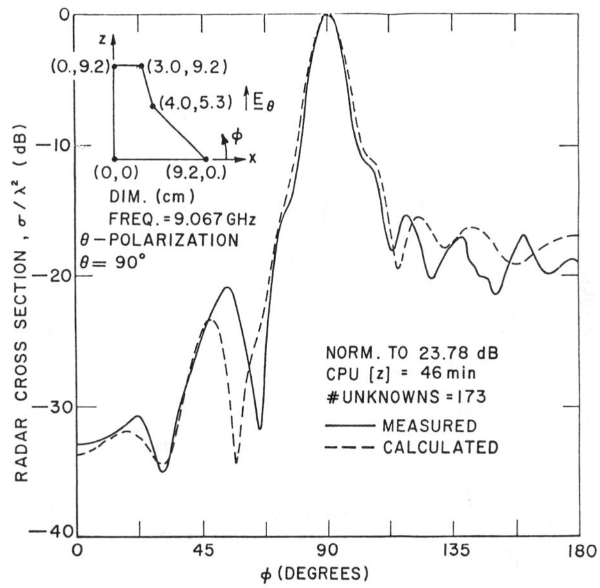

Fig. 10. Backscatter from five-sided plate with 3λ nominal size (θ-polarization).

purpose of numerical integrations, the surface patch modes are replaced by a number of $\hat{l}$ (see Fig. 1) directed filaments. To minimize run time, the number of filaments is kept to a minimum. If the modes are not touching, then the full surface testing can be done with only one filament to filament evaluation and using (5).

V. NUMERICAL RESULTS

In this section, numerical results will be presented that illustrate the ability of the technique to treat plates of different shape, as well as wire to plate junctions. In each case, results computed by the present technique will be compared with measurements or independent computations. In all moment method surface patch computations, the maximum segment size was 0.25λ. All run times shown are for the Digital Equipment Corporation VAX-11, which is about three to four times slower than the IBM-370/168. Using the usual spherical coordinates, θ will refer to the angle with respect to the z axis, and ϕ is the angle with respect to the x axis and in the x-y plane.

The insert in Fig. 8 shows an octagon plate approximating (equal areas) a circular disk of radius 1.2λ. Fig. 8 compares the exact or eigenfunction solution [14] of the disk backscatter radar cross section with the moment method solution for the octagonal plate. The backscatter pattern for the circular disk is, of course, independent of ϕ. However, the patterns for the octagon do depend on ϕ. Thus the cross section of the octagon plate is shown at $\phi = 0°$ and 22.5°. Note that the three curves are in essentially perfect agreement over the first 20 dB. Below 20 dB, the octagon approximation to the circle causes the curves to differ. Note that the disk solutions always fall between the $\phi = 0°$ and $\phi = 22.5°$ solutions for the octagonal plate. The type of segmentation for the octagonal plate is illustrated in Fig. 5.

Fig. 9 shows the backscatter radar cross section of a 3λ right triangle. Moment method solutions are shown for the Gordon–Hall and a Method 2 segmentation of the triangular plate. In each case, segment sizes were restricted to be less than or equal to 0.25λ. The Gordon–Hall technique, which tends to have many small modes near the center of the hypotenuse, required 264 modes, while the Method 2 segmentation required 176 modes. The authors' code required 130 min to compute the impedance matrix for the Gordon–Hall segmentation, but only 39 min for the Method 2 segmentation. Part of the reduced time for Method 2 results from the fact that it required fewer modes. However, a significant part results from the large number of parallel or rectangular modes formed by Method 2. Fig. 9 shows that both segmentation schemes result in a radar cross section in close agreement with measurements. Fig. 10 shows good agreement between computed and measured backscattered cross section of a five-sided plate.

The solid curve in Fig. 11 shows the input impedance of a 0.25λ monopole on a circular disk, computed versus disk radius [15]. The dashed and dotted curves show the input

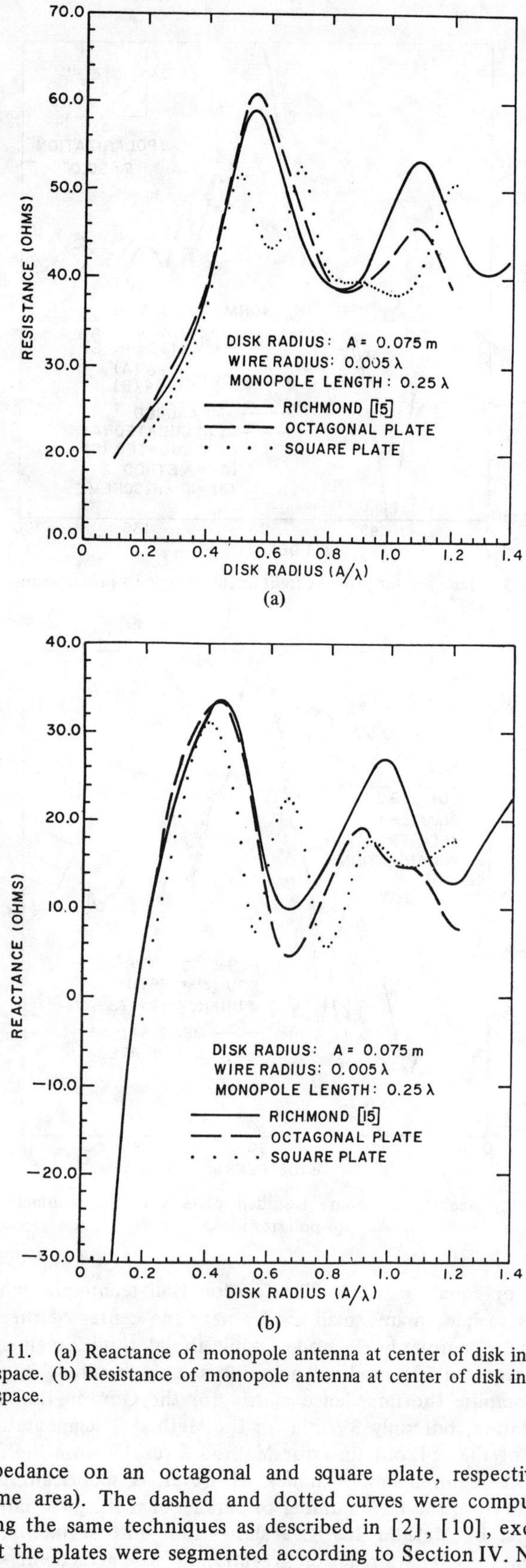

Fig. 11. (a) Reactance of monopole antenna at center of disk in free space. (b) Resistance of monopole antenna at center of disk in free space.

impedance on an octagonal and square plate, respectively (same area). The dashed and dotted curves were computed using the same techniques as described in [2], [10], except that the plates were segmented according to Section IV. Note that the impedance of the monopole over the square plate departs from that over the disk, for disk radii greater than about 0.5λ. The impedance of the monopole on the octagonal plate agrees well with that over the disk, for disk radii up to about 1.0λ.

VI. SUMMARY

This paper has considered the surface patch moment method solution for polygonal plates. Three different automated plate segmentation schemes were considered. Full surface testing was used so that the method would be applicable to wires attached to plates. Numerical results illustrated the accuracy and computational efficiency of the technique applied to three plates of distinctively different shapes.

REFERENCES

[1] N. N. Wang, J. H. Richmond, and M. C. Gilbreath, "Sinusoidal reaction formulation for radiation and scattering from conducting surfaces," *IEEE Trans. Antennas Propagat.*, vol. AP-23, pp. 376–382, May 1975.

[2] E. H. Newman and D. M. Pozar, "Electromagnetic modeling of composite wire and surfce geometries," *IEEE Trans. Antennas Propagat.*, vol. AP-26, pp. 784–789, Nov. 1978.

[3] D. L. Knepp and J. Goldhirsh, "Numerical analysis of electromagnetic radiation properties of smooth conducting bodies of arbitrary shape," *IEEE Trans. Antennas Propagat.*, vol. 20, pp. 383–388, May 1972.

[4] N. C. Albertsen, J. E. Hansen, and N. E. Jensen, "Computation of radiation from wire antennas on conducting bodies," *IEEE Trans. Antennas Propagat.*, vol. AP-22, pp. 200–206, March 1974.

[5] G. J. Burke and A. J. Poggio, "Numerical electromagnetic code (NEC)-method of moments," Naval Ocean Systems Center, Tech. Doc. 116, AFWL-TR-76-320, July 1977.

[6] A. Sankar and T. C. Tong, "Current computation on complex structures by finite-element method," *Electronics Lett.*, vol. 11, pp. 481–482, Oct. 1975.

[7] J. J. H. Wang, "Numerical analysis of three-dimensional arbitrarily-shaped conducting scatterers by trilateral surface cell modelling," *Radio Sci.*, vol. 13, pp. 947–952, Nov.-Dec. 1978.

[8] D. R. Wilton, S. S. M. Rao, and A. W. Glisson, "Electromagnetic scattering by surfaces of arbitrary shape," University of Mississippi, RADC-TK-79-325, March 1980.

[9] J. Singh and A. T. Adams, "A nonrectangular patch model for scattering from surfaces," *IEEE Trans. Antennas Propagat.*, vol. AP-27, pp. 531–535, July 1979.

[10] E. H. Newman and D. M. Pozar, "Considerations for efficient wire/surface modeling," *IEEE Trans. Antennas Propagat.*, vol. AP-28, pp. 121–125, Jan. 1980.

[11] W. J. Gordon and C. A. Hall, "Construction of Curvilinear coordinate systems and applications to mesh generation," *Int. J. Numerical Methods Eng.*, vol. 7, pp. 461–477, 1973.

[12] P. Tulyathan, "Moment method solutions for radiation and scattering from arbitrarily shaped surfaces," Ph.D. dissertation, Ohio State Univ., Columbus, OH March 1981.

[13] E. E. Okon and R. F. Harrington, "The capacitance of discs of arbitrary shape," Dept. Elec. Comput. Eng. Syracuse Univ. Tech. Rep. 10, April 1979, prepared under Contract N00014-76-C-0225 for the Office of Naval Research, Department of the Navy.

[14] D. B. Hodge, "Scattering by circular metallic disks," *IEEE Trans. Antennas Propagat.*, vol. AP-28, pp. 707–712, Sept. 1980.

[15] J. H. Richmond, "Monopole antenna on circular disk," ElectroScience Lab. Dept. Elec Eng., Ohio State Univ., Tech. Rep. 711639-1, July 1979, prepared under Contract MDA 904-79-C-0407 for Maryland Procurement Office, Fort George G. Meade, MD.

Incorporation of Edge Conditions in Moment Method Solutions

D. R. WILTON, MEMBER, IEEE, AND S. GOVIND, STUDENT MEMBER, IEEE

Abstract—The use of an edge condition in moment method solutions is investigated. Using as an example TM scattering by a strip, various numerical schemes for treating the edge behavior are examined for low frequencies where an analytical solution of the quasi-static integral equation is available for comparison. The same numerical techniques are then applied in the general dynamic case. It is found that failure to incorporate the correct edge behavior in the current expansion can result in erroneous currents and anomalous behavior of the solution near edges. The use of ordinary pulses or triangle expansion functions with appropriately singular pulses at the edges circumvents these difficulties. In the case of pulses, a simple correction factor is derived which can be used *a posteriori* to correct the calculated edge current.

Manuscript received December 22, 1975; revised August 17, 1976.

The authors are with the Department of Electrical Engineering, University of Mississippi, University, MS 38677.

I. INTRODUCTION

In electromagnetic scattering and radiation problems, it is well known that currents parallel to a conducting edge may be singular and must obey an edge condition [1]. In the classical eigenfunction solution of problems with edges, the edge condition is used explicitly to select the correct eigenfunctions. In an integral equation formulation with unknown currents, the kernel is such that only currents with the required edge behavior will satisfy the equation. In a moment method solution of the integral equation, however, one often chooses the simple nonsingular pulse or triangle functions to represent the current [2]. The resulting solution often appears to satisfy the singular edge behavior in all but the subdomains nearest the edge where the current often exhibits anomalies or is grossly in error. This poor behavior at edges should not be unexpected; for if the exact solution must have a certain edge behavior in order to maintain equality on both sides of the integral equation, then choosing a current representation that does not possess the correct edge behavior should certainly result in poor overall satisfaction of the boundary conditions.

To improve solutions arising from the choice of pulse or triangle expansions in such problems, it is necessary to explicitly build the edge behavior of the current into the expansion functions. To illustrate the procedure, scattering by an infinitely long conducting strip illuminated by a TM plane wave is considered. The procedure used is adaptable to any problem where the current has a known singularity. It may be mentioned here that other workers, such as Jansen [3] and Minor and Bolle [4], have incorporated the edge condition in representing currents with entire domain expansion functions. The discussion in this paper pertains essentially to the subdomain type of expansion functions.

In Section II a quasi-static approximation to the strip scattering problem for plane wave illumination is considered. The resulting approximate integral equation permits one to obtain a simple closed form solution for the unknown current against which numerical results can be easily compared. The improvement in the solution of the edge current when the edge behavior is incorporated is illustrated. A simple *a posteriori* correction factor for the edge current behavior is also obtained for the pulse expansion. Section III examines the procedure when applied to the exact integral equation where no simple analytical solution of the strip problem is available.

II. TM INTEGRAL EQUATION WITH QUASI-STATIC KERNEL

The geometry considered is shown in Fig. 1, wherein the strip is assumed to lie in the xz plane and to be of width $2a$. The integral equation for TM illumination is given by [2]

$$\int_{-a}^{a} J_s(x') H_0^{(2)}(k\,|x - x'|)\,dx' = \frac{4}{k\eta} E_z^{\text{inc}}(x), \tag{1}$$

where $J_s(x')$ is the unknown z-directed strip current. For low frequencies one may use the small argument approximation for the Hankel function,

$$H_0^{(2)}(z) \cong 1 - j\frac{2}{\pi}\ln\left(\frac{\gamma z}{2}\right), \tag{2}$$

where $\ln \gamma = 0.57722 \cdots$ is Euler's constant. Substituting (2)

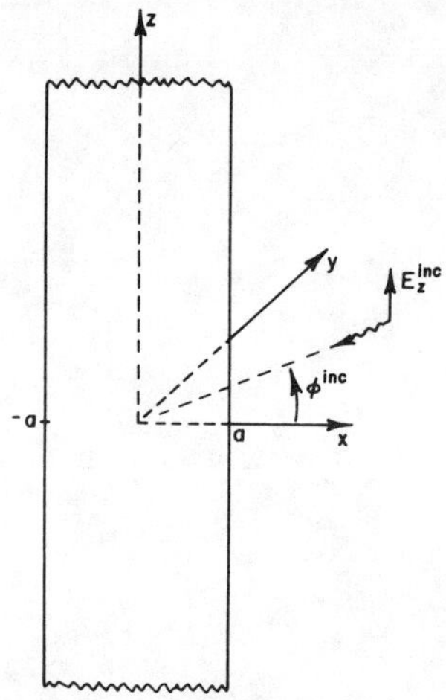

Fig. 1. Geometry of strip.

into (1) and rearranging, one obtains

$$\int_{-a}^{a} J_s(x') \ln|x - x'|\,dx' = \frac{j2\pi E_0}{k\eta} - \frac{j\pi}{2}\left(1 - j\frac{2}{\pi}\ln\frac{\gamma k}{2}\right)\int_{-a}^{a} J_s(x')\,dx', \tag{3}$$

where $E_z^{\text{inc}}(x) = E_0$, for a normally incident plane wave. One notes that the right side of (3) is independent of x and hence is constant for a given strip size. Using the integral

$$\int_{-1}^{1} (1 - x'^2)^{-1/2} \ln|x - x'|\,dx' = -\pi \ln 2, \tag{4}$$

it can easily be shown that the current $J_s(x')$ in (3) is given by

$$J_s(x') = \frac{j2E_0/\eta}{ka\left[\ln\left(\frac{\gamma ka}{4}\right) + j\frac{\pi}{2}\right]\sqrt{1 - \left(\frac{x'}{a}\right)^2}} \tag{5}$$

which is accurate provided that $ka \ll 1$. We note that the current contains the factor $1/\sqrt{1 - (x'/a)^2}$, which exhibits the correct singularity at the edges. Furthermore the kernel in (3) contains the basic logarithmic singularity which is common to almost all electric field integral equations and which determines the order of the current singularity. Thus this quasi-static integral equation displays explicitly the basic features of electric field type integral equations for structures with edges. By properly normalizing (5), it suffices to consider the normalized integral equation

$$\int_{-1}^{1} J_s(x') \ln|x - x'|\,dx' = -\pi \ln 2 \tag{6}$$

whose solution is obviously $1/\sqrt{1 - x'^2}$.

To solve for the current $J_s(x')$ by the moment method procedure, we examine both the pulse and triangle expansion schemes shown in Fig. 2 where the edge behavior is explicitly built into the end subdomains. The factor $\sqrt{\Delta/2}$ normalizes the singular expansion terms to a value of unity at the center of the end subdomains. Using either of the two expansion

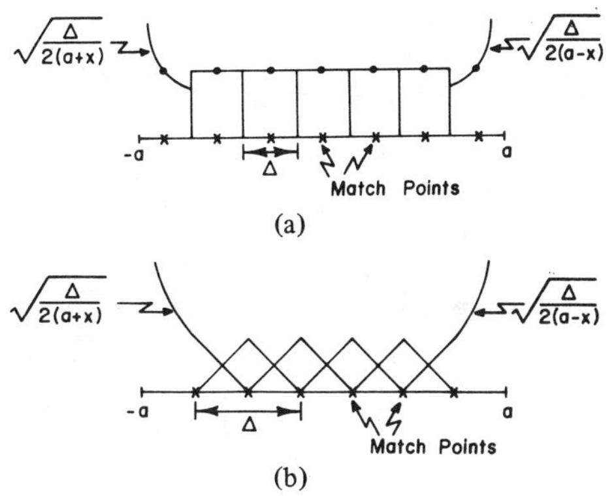

Fig. 2. Expansion schemes for (a) rectangular pulses with singular edge terms and (b) triangles with singular edge terms.

schemes in (6) and point matching at the center of the pulses and at the peaks of the triangles, one converts the integral equation (6) into a matrix equation and solves for the current $J_s(x')$. The results are shown in Fig. 3. Also shown are the currents obtained when only pulses or triangles are used everywhere along the strip as well as the exact $1/\sqrt{1-x^2}$ distribution. From Fig. 3(a), we note that the use of pulses all along the strip, including the end subsections, results in a computed current near the edges which is significantly higher than the correct value. This anomaly is corrected with the introduction of the singular terms in the current expansion scheme. In obtaining the current in Fig. 3(b) a triangle expansion has been used with a half triangle used in the end subsections. Match points have been placed at the triangle peaks and at the edges. One observes a sharp dip or "glitch" in the current distribution near the edges. This glitch is attributed to the inability of the half triangles at the edges to adequately model the singular current there. If one incorporates the edge behavior in the expansion functions, the nonphysical glitch is removed as seen in Fig. 3(b). One observes, however, that the edge current solution in this case is still not as good as was obtained for pulses with the singular edge term. This fact is attributed to placing the match points at slope discontinuities in the triangle expansion scheme where the fields are poorly behaved. It has also been verified that the boundary conditions are satisfied better for pulses with the singular edge term than for triangles with the singular edge term. Again, this is attributed to placement of the match points at slope discontinuities in the triangle expansion scheme (Fig. 2(b)).

Several variations on the above approach were also tried. First, a pulse solution using only half pulses on the end (that is, pulses of width $\Delta/2$) was computed using equally spaced match points and placing match points at each edge of the strip. As with triangles, a glitch was observed in the subdomains adjacent to the edges. Apparently, then, it is not the inherent shape of the triangles which causes the glitch noted in Fig. 3(b), but the fact that match points were located directly at the edges where the proper edge behavior is most critical. Indeed, when the half pulses were replaced by terms with the appropriate singularity in the end sections and match points again placed at the strip edges, results very similar to those of Fig. 3(a) which used full domain singular pulses were obtained. Hence we conclude that the anomalous glitch in the current with triangle expansions is essentially due to the placement of match points at the edges. Another investigator, Tsai [5], has reported using half triangle *testing* functions [2] together with half triangle expansion functions in the end subdomains, in an attempt to satisfy the boundary conditions in an average or integrated sense. This approach, however, did not seem to eliminate anomalous behavior of the current in the end sections. We also allowed the singular current in the end sections of Fig. 2(b) to be separately determined from that of the linear portion of the end expansion functions of Fig. 2(b), by adding match points at the ends. However, in this representation, the current is discontinuous with match points located at the discontinuities. The resulting currents were not as accurate as when the continuous expansion functions of Fig. 2(b) were used.

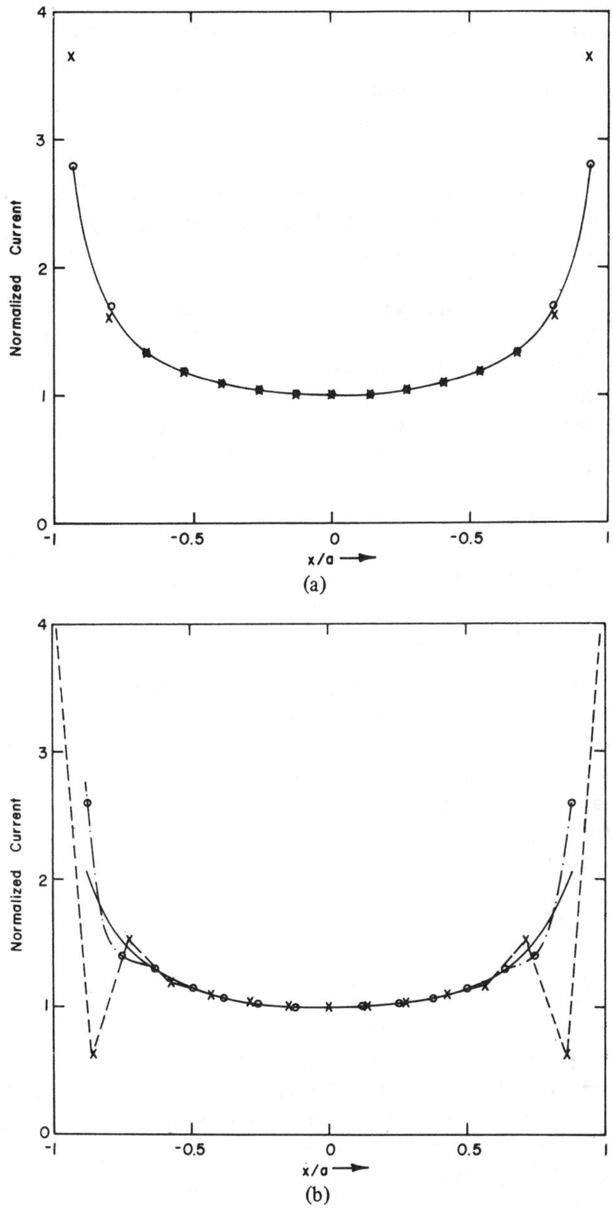

Fig. 3. Solutions of integral equation with quasi-static kernel for (a) pulses (xxx), pulses with singular edge terms (ooo); and (b) triangles (xxx), and triangles with singular edge terms (– – –). Exact solution is denoted by (—).

From the above discussion we are led to conclude the following.

1) In a current expansion scheme where match points are placed at edges, the resulting current exhibits anomalous behavior near the edges.

2) To eliminate anomalous behavior when match points are at the edge, or to improve the accuracy of the current representation near the edge, one must incorporate the correct singular form of the edge current into the expansion functions.

The anomalous behavior described above has also been observed in various other problems containing edges such as strips, plates, and apertures, both in our own data and in that of other researchers.

One further observation concerning pulse expansions is in order. From Fig. 3(a) one observes that the use of the singular pulse term essentially corrects only the currents at the edges. With this empirical observation, we derive an *a posteriori* correction factor for the end currents in the pulse solution. To do this we first write the matrix equation as

$$ZI = V \tag{7a}$$

for the pulse expansion solution, and

$$Z'I' = V' \tag{7b}$$

for the case of pulses with singular edge terms (Fig. 2(a)). In (7a) $Z = [z_{ij}]$ is the $N \times N$ moment matrix, $I = [I_j]$ is the current coefficient column vector, and $V = [v_i]$ is the incident field column vector. Similar definitions apply for primed quantities. Noting that $V = V'$ and assuming that the first and Nth elements of I, I', V, V' correspond to the end subdomains, we have

$$z_{11}I_1 + z_{12}I_2 + \cdots + z_{1N}I_N$$
$$= z_{11}'I_1' + z_{12}'I_2' + \cdots + z_{1N}'I_N'.$$

Noting that $z_{1i} = z_{1i}'$ and $I_i \approx I_i'$ for $i = 2, 3, \cdots, N-1$, we can approximately cancel the terms $z_{1i}I_i$ and $z_{1i}'I_i'$ for $i = 2, 3, \cdots, N-1$. Furthermore the terms $z_{1N}I_N$ and $z_{1N}'I_N'$ are small compared to the remaining terms $z_{11}I_1$ and $z_{11}'I_1'$. Hence we have

$$\frac{I_1}{I_1'} \approx \frac{z_{11}'}{z_{11}} = \frac{\sqrt{\dfrac{\Delta}{2}}\displaystyle\int_{-1}^{-1+\Delta} (1+x')^{-1/2} \ln\left|-1+\frac{\Delta}{2}-x'\right| dx'}{\displaystyle\int_{-1}^{-1+\Delta} \ln\left|-1+\frac{\Delta}{2}-x'\right| dx'}$$

$$= \sqrt{2}\left(1 + \frac{C}{\ln\left(\dfrac{\Delta}{2e}\right)}\right), \tag{8}$$

where

$$C = \frac{1}{\sqrt{2}} \ln\left|\frac{\sqrt{2}+1}{\sqrt{2}-1}\right| - 1 = 0.2465.$$

This ratio can be used to estimate the factor by which the numerically computed value of the current near the edges needs to be altered when only pulse functions are used over the whole strip. Table I compares the edge correction ratio one obtains from (8) with that obtained from the numerical solution of the integral equation (6). One notes immediately

TABLE I
COMPARISON OF NUMERICALLY OBTAINED EDGE CURRENT RATIO WITH THAT PREDICTED BY THE APPROXIMATE FORMULAS

Quasi-Static Integral Equation

Number of Unknowns	I_1/I_1' Our Formula	I_1/I_1' Numerical* Solution
9	1.3052	1.2871
15	1.3205	1.3082
21	1.3280	1.3278
31	1.3356	1.3353
49	1.3450	1.3427

Exact Integral Equation

Strip Width	Number of Unknowns	$\lvert I_1/I_1'\rvert$ Our Formula	$\lvert I_1/I_1'\rvert$ Numerical** Solution
0.2λ	15	1.3428	1.3635
0.5λ	31	1.3404	1.3584
1.5λ	31	1.3239	1.3337
2.0λ	45	1.3253	1.3358

*Pulse solution compared to exact solution.
**Pulse solution compared to pulse solution with edge condition built in.

the accuracy of the above formula. According to (8) the error *increases* as the subdomain size decreases and as Δ approaches zero, $I_1/I_1' \rightarrow \sqrt{2}$. Although the maximum possible error is thus about 41 percent, over a wide range in subdomain sizes the error predicted by (8) is closer to 30 percent. The appearance of the factor $\sqrt{2}$ is rather interesting because it is also the ratio of the areas of the singular end sections of Fig. 2(a) and a unit pulse of width Δ. Hence in calculations where the current *moments* of the subdomain expansion functions are of more importance than the actual *values* of the current at the center of the subdomains (as in pattern or scattering cross-section calculations), then the value by which the calculated pulse current exceeds the correct value of current at the center of the singular pulse almost compensates for the factor by which the area of the singular pulse exceeds that of the rectangular pulse.

III. TM INTEGRAL EQUATION WITH EXACT KERNEL

In the previous section, a quasi-static approximation to the kernel was employed to demonstrate the utility of a singular pulse in the expansion functions. In this section, the same expansion schemes of Fig. 2 are employed in the exact integral equation (1). Fig. 4 shows the comparison of currents obtained with and without the use of the singular pulse in the end sections for both the pulse and triangle expansion schemes for normal incidence and for a wave incident at an angle of 50° from the plane of the strip. One notes again the improvement in the edge current solution in the pulse expansion scheme and the removal of the nonphysical glitch in the

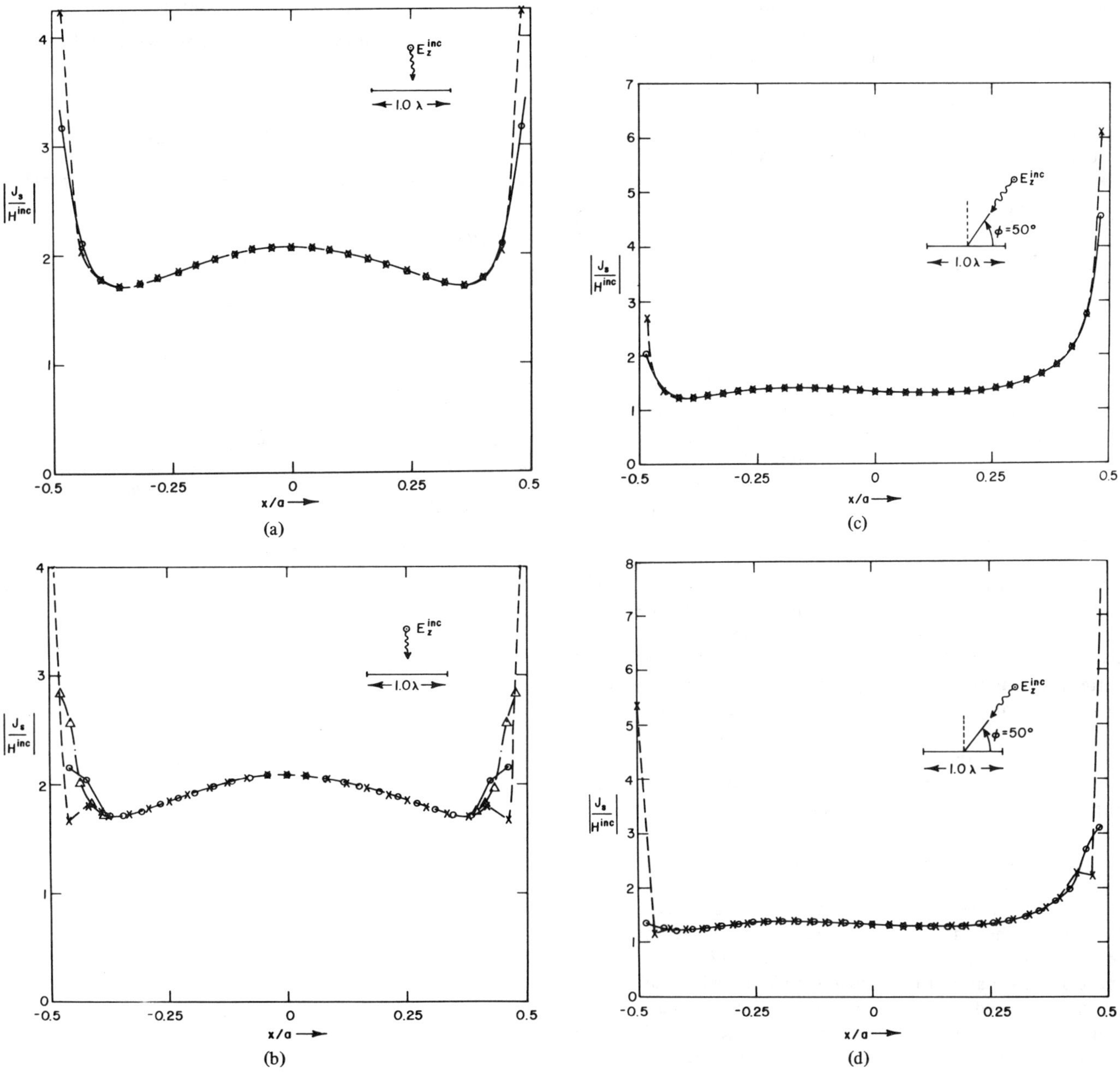

Fig. 4. Currents on a 1.0λ strip due to an incident TM plane wave for (a) pulses (xxx), pulses with singular edge terms (ooo), normal incidence, 25 unknowns; (b) triangles (xxx) with 25 unknowns, triangles with singular edge terms, 25 unknowns (ooo), triangles with singular edge terms, 45 unknowns (-Δ-·-Δ-), normal incidence; (c) pulses (xxx), pulses with singular edge terms (ooo), angle of incidence = 50°, 31 unknowns; (d) triangles (xxx), triangles with singular edge terms (ooo), angle of incidence = 50°, 31 unknowns.

current distribution associated with the triangle expansion scheme. Here also the rectangular pulse expansion scheme is superior to the triangle functions. Several factors associated with the frequency dependence of the current should be noted. As the frequency increases, the singularity becomes more and more confined to the strip edge while the current in the center of the strip approaches the physical optics limit. Hence at higher frequencies, one must take care not to choose so few subdomains that the singular pulse extends into the region where the current is relatively flat, thus artificially extending the region where the current tends to become singular. In such a case, a rectangular pulse in the end subdomains would actually give more accurate results. Similar care needs to be observed when the incident field is not normally incident, where, as seen in Figs. 4(c) and 4(d), the extent of the singularity is different on the leading and trailing edges. We also note in Figs. 4(a) and 4(b) the appearance of a dip in the current distribution near the edge just before the current tends to its singular behavior. This dip, which is a physical phenomenon, occurs even at high frequencies and for many choices of subdomain sizes approximately coincides with the nonphysical dip produced by the triangle expansions. (In Fig. 4(b), the two dips are distinct.) It is apparently this coincidence that has led many investigators to erroneously conclude that their results were correct when using triangle or similar expansion functions near edges.

A correction factor for the pulse expansion currents near the edges is obtained by paralleling the development in Section II. Using the same notation as in Section II, we have

$$\frac{I_1}{I_1'} \approx \frac{\sqrt{\frac{\Delta}{2}} \int_{-a}^{-a+\Delta} (a + x')^{-1/2} \left(1 - j\frac{2}{\pi} \ln\left|\frac{\gamma k}{2}\left(-1 + \frac{\Delta}{2} - x'\right)\right|\right) dx'}{\int_{-a}^{-a+\Delta} \left(1 - j\frac{2}{\pi} \ln\left|\frac{\gamma k}{2}\left(-1 + \frac{\Delta}{2} - x'\right)\right|\right) dx'} = \sqrt{2}\left(1 + \frac{C}{\ln\left(\frac{\gamma k \Delta}{4e}\right) + j\frac{\pi}{2}}\right), \tag{9}$$

where the small argument approximation for the Hankel function has been used and γ and C are defined as before. Table I gives a comparison of the ratio predicted by (9) with that obtained from the two numerical solutions. The correspondence is again good and hence (9) provides a simple formula for correcting the edge current when one wishes to avoid the complication of using a singular pulse.

Our primary concern has been to accurately represent the currents on TM strips. We note that great accuracy in the surface currents are not usually required in the calculation of far field patterns or radar cross sections except for the very low sidelobes of the scattered fields. Numerical results of pattern calculations on strips of medium sizes (typically 1λ–5λ) indicate that the difference in the sidelobe levels with and without the singular pulses is on the order of 0.1 to 0.2 dB. Apparently these minor differences are partially due to the current moment being approximately equal with or without the singular pulses as discussed in Section II.

IV. CONCLUSIONS

The importance of incorporating the edge behavior into the moment method solution procedure has been demonstrated for TM plane wave scattering by a conducting strip. A quasistatic approximation to the integral equation whose solution is known in closed form was used to demonstrate the improvement resulting from incorporating the edge condition. The method was then generalized to the exact integral equation. The appearance of a glitch in the current distribution near an edge is attributed to placement of a match point at the edge of the strip where the current is singular. This happens, for example, when half triangles are used to represent the edge currents. This anomaly, however, does not appear for the pulse expansion where the match point is not normally at the ends. However, for pulses, the current near the edges is typically too large in magnitude. A simple correction factor is derived which permits an *a posteriori* correction of the current nearest the edge. Interestingly, the numerically calculated value, which is larger than the actual value for the edge pulse, almost compenthe effect of producing a compensating error in calculating, for example, far fields.

Finally, we mention that when the current flow is normal to an edge, as for the TE strip, the edge condition requires that the current approach zero at the edge in a certain manner [1]. However, numerical experiments have indicated that incorporating the correct behavior of the current at the edges for a TE strip resulted in little improvement in the accuracy of the currents near the edges, both for pulse and triangle expansions.

ACKNOWLEDGMENT

The authors acknowledge helpful discussions with Dr. L. L. Tsai.

REFERENCES

[1] D. S. Jones, *The Theory of Electromagnetism.* New York: Pergamon, 1964.

[2] R. F. Harrington, *Field Computation by Moment Methods.* New York: MacMillan, 1968.

[3] R. Jansen, "A moment method for covered microstrip dispersion," *Arch. Elek. Übertragung,* vol. 29, no. 1, pp. 17–20, Jan. 1975.

[4] J. C. Minor and D. M. Bolle, "Modes in the shielded microstrip on a ferrite substrate transversely magnetized in the plane of the substrate," *IEEE Trans. Microwave Theory Tech.*, MTT-19, no.7, pp. 570–577, July 1971.

[5] L. L. Tsai, Private communication.

Simple integral equations for two-dimensional scattering with further reduction in unknowns

M.A. Ricoy
S.M. Kilberg
J.L. Volakis

Indexing terms: Electromagnetic theory, Numerical analysis

Abstract: A simple set of integral equations with reduced unknowns and kernel singularity are derived for simulating arbitrarily-shaped two-dimensional inhomogeneous composite scatterers. By utilising a known equivalence between electric and magnetic currents, new equivalent currents are introduced with the resultant integral formulation exhibiting a volume integral in addition to a surface integral, each in terms of a single equivalent current component. A pulse basis-point matching moment method implementation of the reduced unknown integral equations is presented. Scattering patterns computed with the numerical code are compared with results obtained via alternative analytical techniques.

1 Introduction

The computation of the scattered electromagnetic field from two-dimensional inhomogeneous dielectric/magnetic structures has traditionally been accomplished via a numerical solution of the appropriate integral equations. Over the past twenty five years, a variety of volume and surface integral formulations have been developed and applied to cylinders of arbitrary cross-section and composition [1–5], as well as specialized cylindrical configurations [6 and 7]. A point of commonality among these techniques is the necessity of solving for three independent current components within a material body having a nontrivial permittivity and permeability profile. Thus, for a total of N cells comprising the discretised composite structure, the resultant total number of system unknowns is $3N$. As is well known, the expenditure of computer resources (CPU time and memory allocation) for numerical codes is proportional to $\sum \alpha_i M^{\alpha_i}$ where M denotes the total number of unknowns and, in general, $\alpha_i \leqslant 3$. For example, the required CPU time varies as $\alpha_0 + \alpha_1 M + \alpha_2 M^2 + \alpha_3 M^3$ in the case of nonoptimized moment method formulations, and thus any reduction in M will produce a significant improvement in the solution algorithm's computational efficiency and economy.

Recently, Ricoy and Volakis [8] presented an integral formulation for modelling inhomogeneous composite cylinders of arbitrary cross-section utilising only two current components across the volume of the scatterer. In this paper, a simple integral equation set which accomplishes a further reduction in unknowns is developed. By replacing the traditional polarisation currents with new equivalent currents, the resultant integral equation exhibits a volume integral in addition to a surface integral, each in terms of a single independent equivalent current component. Consequently, this implies a substantially reduced system with only $N^v + N^s$ unknowns, where N^v is the number of volumetric sampling points and N^s the number of surface sampling points. In contrast to the development in Reference 8, this reduction of unknowns is obtained without an increase in the singularity or complexity of the derived integral equations.

To illustrate the validity and applicability of the volume/surface integral formulation, a general numerical code was developed using a pulse basis-point matching moment method procedure. A variety of scattering patterns computed using this code are presented and compared to results obtained by alternative analytical techniques.

Paper 6756H (E11), first received 4th August 1988 and in revised form 22nd February 1989

The authors are with the Radiation Laboratory, Department of Electrical Engineering and Computer Science, The University of Michigan, Ann Arbor, Michigan 48109-2122, USA

2 Formulation

From Stratton [9], an $e^{-i\omega t}$ time dependence being assumed and suppressed, the electric and magnetic scattered fields can be expressed as

$$\bar{E}^s = \nabla\nabla \cdot \bar{\Pi}^e + k_o^2 \bar{\Pi}^e + ik_o Z_o \nabla \times \bar{\Pi}^m \tag{1}$$

$$\bar{H}^s = \nabla\nabla \cdot \bar{\Pi}^m + k_o^2 \bar{\Pi}^m - ik_o Y_o \nabla \times \bar{\Pi}^e \tag{2}$$

with

$$\bar{\Pi}^e = \frac{iZ_o}{k_o} \iint_{A'} \bar{J}(\bar{r}')G(\bar{r}, \bar{r}')\, dA' \tag{3}$$

$$\bar{\Pi}^m = \frac{iY_o}{k_o} \iint_{A'} \bar{M}(\bar{r}')G(\bar{r}, \bar{r}')\, dA' \tag{4}$$

In the above equations, $\bar{\Pi}^e$ and $\bar{\Pi}^m$ are the Hertz potentials, $\bar{J}$ and $\bar{M}$ denote the electric and magnetic currents generating the scattered fields, k_o is the free space wave number and $Z_o = 1/Y_o$ is the free space intrinsic impedance. Additionally,

$$G(\bar{r}, \bar{r}') = \frac{i}{4} H_o^{(1)}(k_o \rho) \tag{5}$$

is the two-dimensional Green's function with

$$\rho = |\bar{r} - \bar{r}'| \tag{6}$$

where $\bar{r}$ and $\bar{r}'$ are the observation and source position vectors as shown in Fig. 1, respectively. It is further assumed that the infinite dimension of the scatterer is along the z-direction and thus the direction of the ∇ operator is transverse to $\hat{z}$.

If we are interested in the scattered field by a composite structure due to a given excitation, then $\bar{J}$ and $\bar{M}$

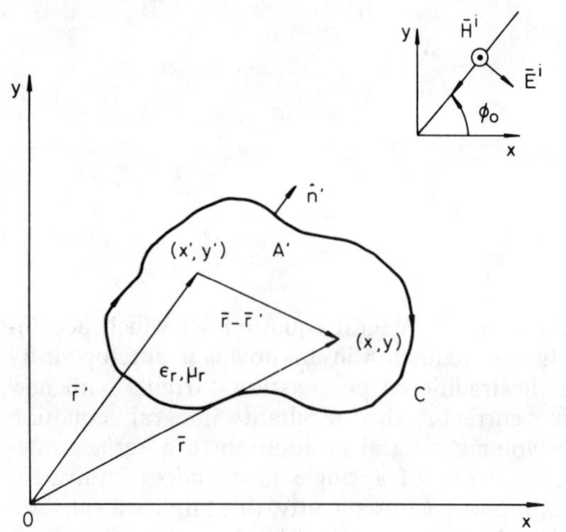

Fig. 1 *Arbitrary cylindrical structure in free space*

appearing in eqns. 3 and 4 become the traditional polarisation currents [10]

$$\bar{J} = -ik_o Y_o(\varepsilon_r - 1)\bar{E} \tag{7}$$

$$\bar{M} = -ik_o Z_o(\mu_r - 1)\bar{H} \tag{8}$$

where $(\bar{E}, \bar{H})$ represent the total fields within the scatterer. The relative permittivity and permeability of the composite medium are denoted by ε_r and μ_r, respectively, and henceforth these will refer to the observation point only. The integral equations for determining $\bar{J}$ and $\bar{M}$ can now be generated by employing the relations

$$\bar{E}^s + \bar{E}^i = \bar{E} \tag{9}$$

$$\bar{H}^s + \bar{H}^i = \bar{H} \tag{10}$$

within the scatterer, where $(\bar{E}^i, \bar{H}^i)$ denote the source or incident fields.

Since we are concerned with a two-dimensional problem we may now assume H_z-incidence (TE incidence) without loss of generality for generating the integral equations implied by eqns. 9 and 10. The corresponding integral expressions for E_z-incidence will then be obtained via duality.

For the case of H_z-incidence, the traditional polarisation currents take the form

$$\bar{J} = \hat{x}J_x + \hat{y}J_y, \quad \bar{M} = \hat{z}M_z \tag{11}$$

Thus, by substituting eqns. 1, 2 and 11 into eqn. 9 we obtain

$$\bar{E}^i(\bar{r}) = Z_o \frac{i}{k_o}\left[\frac{\varepsilon_r}{\varepsilon_r - 1}\right]\bar{J}(\bar{r}) - \nabla \times \left\{\frac{iZ_o}{k_o}\iint_{A'} \nabla \times [\bar{J}(\bar{r}')G(\bar{r}, \bar{r}')]\, dA'\right\} + \nabla \times \iint_{A'} \hat{z}M_z(\bar{r}')G(\bar{r}, \bar{r}')\, dA' \tag{12}$$

where we have also employed the identities

$$\nabla \times \nabla \times \bar{B}(\bar{r}) = \nabla\nabla \cdot \bar{B}(\bar{r}) - \nabla^2 \bar{B}(\bar{r}) \tag{13}$$

$$(\nabla^2 + k_o^2)G(\bar{r}, \bar{r}') = -\delta(\bar{r} - \bar{r}'), \quad \bar{r} \in A' \tag{14}$$

in the given sequence. In the above, $\delta(\bar{r})$ denotes the usual Dirac delta function. Proceeding with similar substitutions into eqn. 10 we obtain

$$-\frac{i}{k_o}\nabla \times \bar{E}^i = \frac{i}{k_o(\mu_r - 1)}\bar{M} - ik_o \iint_{A'} \bar{M}(\bar{r}')G(\bar{r}, \bar{r}')\, dA' - Z_o \nabla \times \iint_{A'} \bar{J}(\bar{r}')G(\bar{r}, \bar{r}')\, dA' \tag{15}$$

The coupled equations 12 and 15 represent one form of the standard integral equations employed in the solution of the scattered field by an inhomogeneous scatterer. As seen, in the case of a general two-dimensional composite scatterer they involve three unknown current components per scatterer location. Clearly, any reduction in the number of unknowns is very desirable when considering their numerical implementation. Also, of equal importance is the singularity associated with the kernel of the integrals to be evaluated in such an implementation. The purpose of the following sections is to address both of the above issues.

3 Reductions in unknowns

In this section we develop a new set of integral equations involving fewer unknown current components per scatterer location than those quoted above. This is accomplished via a manipulation of the coupled set of integral equations to allow the introduction of a new equivalent current component to replace two of those now appearing in eqns. 12 and 15. In proceeding to do so, let us first concentrate on eqn. 12.

By twice employing the identity

$$\nabla \times (\bar{B}f) = \nabla f \times \bar{B} + f\nabla \times \bar{B} \tag{16}$$

in the first integral of eqn. 12 we obtain

$$\bar{E}^i(\bar{r}) = Z_o \frac{i}{k_o}\left(\frac{\varepsilon_r}{\varepsilon_r - 1}\right)\bar{J}(\bar{r}) + \nabla \times \left\{\frac{iZ_o}{k_o}\iint_{A'} \nabla' \times [\bar{J}(\bar{r}')G(\bar{r}, \bar{r}')]\, dA'\right\} + \iint_{A'} \nabla \times \left\{\left[\hat{z}M_z - \frac{iZ_o}{k_o}\nabla' \times \bar{J}(\bar{r}')\right]G(\bar{r}, \bar{r}')\right\} dA' \tag{17}$$

It is now instructive to define the new equivalent magnetic current [11]

$$\tilde{M}_z(\bar{r}') = M_z(\bar{r}') - \frac{iZ_o}{k_o} z \cdot \nabla' \times \bar{J}(\bar{r}') \tag{18}$$

and seek a modification of eqns. 17 and 15 for a more global use of eqn. 18. With this in mind, we multiply eqn. 17 by $-(\varepsilon_r - 1/\varepsilon_r)$ and take its curl to obtain

$$-\nabla \times \left(\frac{\varepsilon_r - 1}{\varepsilon_r}\bar{E}^i(\bar{r})\right) = -\frac{iZ_o}{k_o}\nabla \times \bar{J}(\bar{r}) - \frac{iZ_o}{k_o}\nabla \times \int_{C'} \frac{\varepsilon_r - 1}{\varepsilon_r}\nabla[\hat{n}' \times \bar{J}(\bar{r}')G(\bar{r}, \bar{r}')]\, dl' - \nabla \times \iint_{A'} \frac{\varepsilon_r - 1}{\varepsilon_r}\nabla \times [\hat{z}\tilde{M}_z(\bar{r}')G(\bar{r}, \bar{r}')]\, dA' \tag{19}$$

after also applying the vector Stokes' theorem [12] in the first integral of eqn. 17. Accordingly, C' is the contour enclosing the area A' and $\hat{n}'$ denotes the outward unit normal to C' (Fig. 1).

To eliminate the presence of $\nabla \times \bar{J}$ in eqn. 19, we now return to eqn. 15. By employing eqn. 16 in the second integral of that equation and invoking the definition in eqn. 18 we find

$$-\frac{i}{k_o}\nabla \times \bar{E}^i(\bar{r}) = \hat{z}\frac{i}{k_o(\mu_r - 1)}M_z(\bar{r}) - \hat{z}ik_o\iint_{A'}\tilde{M}_z(\bar{r}')G(\bar{r},\bar{r}')\,dA' + Z_o\iint_{A'}\nabla' \times [\bar{J}(\bar{r})G(\bar{r},\bar{r}')]\,dA' \quad (20)$$

Further, applying the vector Stokes' theorem as before in the last integral of eqn. 20 we deduce

$$-(\mu_r - 1)\nabla \times \bar{E}^i(\bar{r}) = \hat{z}M_z(\bar{r}) - \hat{z}k_o^2(\mu_r - 1) \times \iint_{A'}\tilde{M}_z(\bar{r}')G(\bar{r},\bar{r}')\,dA' - iZ_o k_o(\mu_r - 1) \times \int_{C'}[n' \times \bar{J}(\bar{r}')]G(\bar{r},\bar{r}')\,dl' \quad (21)$$

By adding eqns. 19 and 21 we finally obtain the scalar integral equation

$$\left[-\nabla \times \left(\frac{\varepsilon_r - 1}{\varepsilon_r}\bar{E}^i(\bar{r})\right) - (\mu_r - 1)\nabla \times \bar{E}^i(\bar{r})\right]\cdot\hat{z} = \tilde{M}_z(\bar{r}) - k_o^2\iint_{A'}\hat{z}\cdot\left[(\mu_r - 1) + \frac{1}{k_o^2}\nabla \times \left(\frac{\varepsilon_r - 1}{\varepsilon_r}\nabla x\right)\right]\cdot\hat{z}\tilde{M}_z(\bar{r}')G(\bar{r},\bar{r}')\,dA' - ik_o\int_{C'}\hat{z}\cdot\left[(\mu_r - 1) + \frac{1}{k_o^2}\nabla \times \left(\frac{\varepsilon_r - 1}{\varepsilon_r}\nabla x\right)\right]\cdot\hat{z}\tilde{J}_z(\bar{r}')G(\bar{r},\bar{r}')\,dl' \quad (22)$$

where we have set

$$\tilde{J}_z(\bar{r}') = Z_o\hat{z}\cdot[\hat{n}' \times \bar{J}(\bar{r}')] \quad (23)$$

in accordance with eqn. 11.

We observe that eqn. 22 involves only one unknown magnetic current density component ($\tilde{M}_z$) across the cross-section of the scatterer and another electric current density component ($\tilde{J}_z$) just over the boundary of the inhomogeneous scatterer. A similar conclusion could also be reached by examining the integral equation derived recently by Tai [13] using a different procedure. It should be emphasised, however, that $\tilde{J}_z$ is not a surface current sheet, but simply the associated current density at the locations coinciding with the scatterer's boundary. From eqns. 9, 21 and 23, the scattered field in terms of these new unknowns is given by

$$H_z^s = +ik_oY_o\iint_{A'}\tilde{M}_z(\bar{r}')G(\bar{r},\bar{r}')\,dA' - Y_o\int_{C'}\tilde{J}_z(\bar{r}')G(\bar{r},\bar{r}')\,dl' \quad (24)$$

Clearly, the above formulation implies a substantial reduction in unknowns when compared with the coupled set of eqns. 12 and 15. However, the reduction in unknowns appears to have been achieved at the expense of some complexity. As will be shown in the following Section, however, the kernel of the integrals in eqn. 22 can be substantially simplified leading to one of lower singularity and thus more attractive for numerical implementation.

4 Reduction of kernel singularity

The integral equation (eqn. 22) is attractive because it involves less unkowns. However, its numerical implementation could be difficult due to the complexity and high singularity associated with the kernels of the relevant integrals. The integrals in question are

$$\bar{I}_1 = \frac{-i}{k_o}\int_{C'}\nabla \times \left\{\left(\frac{\varepsilon_r - 1}{\varepsilon_r}\right)\nabla \times [\hat{z}\tilde{J}_z(\bar{r}')G(\bar{r},\bar{r}')]\right\}dl' \quad (25)$$

and

$$\bar{I}_2 = -\iint_{A'}\nabla \times \left\{\left(\frac{\varepsilon_r - 1}{\varepsilon_r}\right)\nabla \times [\hat{z}\tilde{M}_z(\bar{r},\bar{r}')G(\bar{r},\bar{r}')]\right\}dA' \quad (26)$$

and, because of their apparent similarity, it will suffice to consider the simplification of $\bar{I}_2$ only.

By employing eqn. 16, $\bar{I}_2$ can be written as

$$\bar{I}_2 = -\iint_{A'}\nabla\left(\frac{\varepsilon_r - 1}{\varepsilon_r}\right) \times \nabla \times [\hat{z}\tilde{M}_z(\bar{r}')G(\bar{r},\bar{r}')]\,dA' - \frac{(\varepsilon_r - 1)}{\varepsilon_r}\iint_{A'}\nabla \times \nabla \times [\hat{z}\tilde{M}_z(\bar{r}')G(\bar{r},\bar{r}')]\,dA' \quad (27)$$

Since ∇ is transverse to $\hat{z}$, the integrand in the first integral of eqn. 27 can be simplified as

$$\nabla\left(\frac{\varepsilon_r - 1}{\varepsilon_r}\right) \times \nabla \times [\hat{z}\tilde{M}_z(\bar{r}')G(\bar{r},\bar{r}')] = \nabla\left(\frac{\varepsilon_r - 1}{\varepsilon_r}\right) \times \nabla G(\bar{r},\bar{r}') \times \hat{z}\tilde{M}_z(\bar{r}') = -\left[\nabla\left(\frac{\varepsilon_r - 1}{\varepsilon_r}\right)\cdot\nabla G(\bar{r},\bar{r}')\right]\hat{z}\tilde{M}_z(\bar{r}') \quad (28)$$

after first employing eqn. 16 followed by the use of a standard vector identity. In the case of the second integral in eqn. 27 we find that

$$\nabla \times \nabla \times [\hat{z}\tilde{M}_z(\bar{r}')G(\bar{r},\bar{r}')] = -\nabla^2[\hat{z}\tilde{M}_z(\bar{r}')G(\bar{r},\bar{r}')] = \hat{z}\tilde{M}_z(\bar{r}')[k_o^2G(\bar{r},\bar{r}') + \delta(\bar{r} - \bar{r}')] \quad (29)$$

after first employing eqn. 13 followed by the use of eqn. 14. Substituting eqns. 28 and 29 into eqn. 27 we have

$$\bar{I}_2 = -\hat{z}\frac{(\varepsilon_r - 1)}{\varepsilon_r}\tilde{M}_z(r) + \hat{z}\iint_{A'}\tilde{M}_z(\bar{r}')\left[\nabla\left(\frac{\varepsilon_r - 1}{\varepsilon_r}\right)\cdot\nabla G(\bar{r},\bar{r}')\right] \times dA' - \frac{\hat{z}k_o^2(\varepsilon_r - 1)}{\varepsilon_r}\iint_{A'}\tilde{M}_z(\bar{r}')G(\bar{r},\bar{r}')\,dA' \quad (30)$$

Similarly, $\bar{I}_1$ can be rewritten as

$$\bar{I}_1 = \hat{z}\frac{i}{k_o}\int_{C'} \bar{J}_z(\bar{r}')\left[\nabla\left(\frac{\varepsilon_r - 1}{\varepsilon_r}\right)\cdot \nabla G(\bar{r}, \bar{r}')\right] dl'$$

$$-\frac{\hat{z}ik_o(\varepsilon_r - 1)}{\varepsilon_r}\int_{C'} \bar{J}_z(\bar{r}')G(\bar{r}, \bar{r}')\, dl' \quad (31)$$

since $\nabla^2 G(\bar{r}, \bar{r}') = -k_o^2 G(\bar{r}, \bar{r}')$ on C'.

Using eqns. 30 and 31, the integral equation (eqn. 22) can be written as

$$\left[-\nabla\left(\frac{\varepsilon_r - 1}{\varepsilon_r}\right) \times \bar{E}^i(\bar{r}) - \left(\frac{\varepsilon_r\mu_r - 1}{\varepsilon_r}\right)\nabla \times \bar{E}^i(\bar{r})\right]\cdot \hat{z}$$

$$= \left(\frac{1}{\varepsilon_r}\right)\tilde{M}_z(\bar{r}) - ik_o \iint_{A'} \tilde{M}_z(\bar{r}')K(\bar{r}, \bar{r}')\, dA'$$

$$+ \int_{C'} \bar{J}_z(\bar{r}')K(\bar{r}, \bar{r}')\, dl' \quad (32)$$

with

$$K(\bar{r}, \bar{r}') = \left[\frac{-ik_o(\varepsilon_r\mu_r - 1)}{\varepsilon_r} + \frac{i}{k_o}\nabla\left(\frac{\varepsilon_r - 1}{\varepsilon_r}\right)\cdot \nabla\right] G(\bar{r}, \bar{r}') \quad (33)$$

As seen, the singularity of $K(\bar{r}, \bar{r}')$ is only $O(1/\rho)$. In comparison, the singularity of the kernels associated with eqn. 22 was $O(1/\rho^2)$.

The numerical implementation of eqn. 32 is now a straightforward task. However, the condition of the resulting system matrix may be further improved by utilising an auxiliary integral equation when testing on C', i.e. the surface of the scatterer. One such integral equation can be obtained from eqn. 19 by removing the indicated curl operation. This gives

$$\bar{E}^i(r) = \frac{i\varepsilon_r Z_o}{k_o(\varepsilon_r - 1)}\bar{J}(\bar{r}) + \frac{i}{k_o}\int_{C'} \nabla \times [\hat{z}\bar{J}_z(\bar{r}')G(\bar{r}, \bar{r}')]\, dl'$$

$$+ \iint_{A'} \nabla \times [\hat{z}\tilde{M}_z(\bar{r}')G(\bar{r}, \bar{r}')]\, dA'$$

Alternatively, we may write

$$\hat{z}\cdot[\hat{n} \times \bar{E}^i(\bar{r})] = \frac{i\varepsilon_r}{k_o(\varepsilon_r - 1)}\bar{J}_z(\bar{r}) + \frac{i}{k_o}\int_{C'} \bar{J}_z(\bar{r}')$$

$$\times \{\hat{z}\cdot\hat{n} \times [\nabla G(\bar{r}, \bar{r}') \times \hat{z}]\}\, dl' + \iint_{A'} \tilde{M}_z(\bar{r}')$$

$$\times \{\hat{z}\cdot\hat{n} \times [\nabla G(\bar{r}, \bar{r}') \times \hat{z}]\}\, dA' \quad (34)$$

after employing eqn. 16 and the implied vector products.

The above integral equation can now be employed in place of eqn. 32 for generating the system equations corresponding to the testing points on the surface of the scatterer.

5 Numerical implementation

In proceeding with a moment method solution of the integral equations 32 and 34, the cross-section of the two-dimensional cylindrical scatterer is subdivided into a discrete number of rectangular and non-rectangular cells (Fig. 2) totalling N^v for the entire material body. Additionally, the exterior contour of the scatterer and any interior boundaries, where step-discontinuities in permittivity or permeability occur, are discretised to generate the required contour segments totalling N^s. Since neither integral expression involves derivatives of the unknown equivalent magnetic or electric currents, a pulse basis-point matching procedure is implemented by assuming the appropriate current density to be constant within

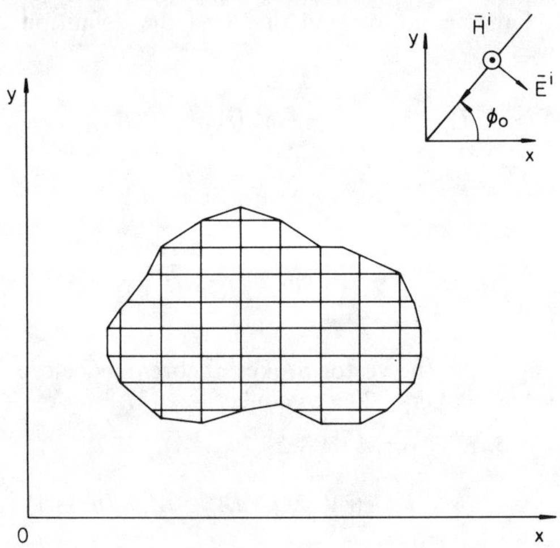

Fig. 2 *Discretisation of scatterer into rectangular and nonrectangular volume cells*

each cell/segment and by subsequently satisfying the derived integral equations at the centroid/centre of each volume cell/contour segment. The generated system of $N^v + N^s$ equations can be represented in matrix form as

$$\begin{bmatrix} V_l^v \\ V_l^s \end{bmatrix} = \begin{bmatrix} [F_{lm}^1] & [F_{lm}^2] \\ [F_{lm}^3] & [F_{lm}^4] \end{bmatrix}\begin{bmatrix} \tilde{M}_{zm} \\ \bar{J}_{zm} \end{bmatrix} \quad (35)$$

where $\tilde{M}_{zm}$, $m = 1, 2, \ldots, N^v$ denotes the value of the equivalent magnetic current density at the mth volume cell and $\bar{J}_{zm}$, $m = N^v + 1, \ldots, N^v + N^s$ denotes the equivalent electric current density at the mth contour segment. The excitation vector elements of the system (eqn. 35) are given by

$$V_l^v = -ik_o\left(\mu_r - \frac{1}{\varepsilon_r}\right)ZH_z^i(\bar{r}_l) + \left(\frac{\partial}{\partial s}\frac{1}{\varepsilon_r}\right)E_n^i(\bar{r}_l) - \left(\frac{\partial}{\partial n}\frac{1}{\varepsilon_r}\right)E_s^i(\bar{r}_l)$$

for $l = 1, 2, \ldots, N_v$

$$V_l^s = -E_s^i(\bar{r}_l)$$

for

$$l = N^v + 1, \ldots, N^v + N^s. \quad (36)$$

where $\bar{r}_l$ denotes the position vector associated with the centroid/centre of the observation element. Finally the four impedance matrix elements can be expressed as

$$F_{lm}^1 = \frac{1}{\varepsilon_r}\delta_{lm} - k_o^2 \iint_{A_m}\left\{\left(\mu_r - \frac{1}{\varepsilon_r}\right) + \frac{1}{k_o^2}\left[\left(\frac{\partial}{\partial s}\frac{1}{\varepsilon_r}\right)\frac{\partial}{\partial s}\right.\right.$$

$$\left.\left. + \left(\frac{\partial}{\partial n}\frac{1}{\varepsilon_r}\right)\frac{\partial}{\partial n}\right]\right\} G(\bar{r}_l, \bar{r}')\, ds'\, dn'$$

$$F_{lm}^2 = -ik_o \int_{C_m}\left\{\left(\mu_r - \frac{1}{\varepsilon_r}\right) + \frac{1}{k_o^2}\left[\left(\frac{\partial}{\partial s}\frac{1}{\varepsilon_r}\right)\frac{\partial}{\partial s}\right.\right.$$

$$\left.\left. + \left(\frac{\partial}{\partial n}\frac{1}{\varepsilon_r}\right)\frac{\partial}{\partial n}\right]\right\} G(\bar{r}_l, \bar{r}')\, ds'$$

$$F_{lm}^3 = -\iint_{A_m} \frac{\partial}{\partial n} G(\bar{r}_l, \bar{r}')\, ds'\, dn'$$

$$F_{lm}^4 = \frac{i}{k_o}\left(\frac{\varepsilon_r}{\varepsilon_r - 1}\right)\delta_{lm} - \frac{i}{k_o}\int_{C_m} \frac{\partial}{\partial n} G(\bar{r}_l, \bar{r}')\, ds' \quad (37)$$

where

$$\delta_{lm} = \begin{cases} 1 & l = m \\ 0 & \text{otherwise} \end{cases}$$

and $(\hat{s}, \hat{n}, \hat{z})$ represent a general rectilinear co-ordinate system oriented at the observation cell/segment as illustrated in Fig. 3.

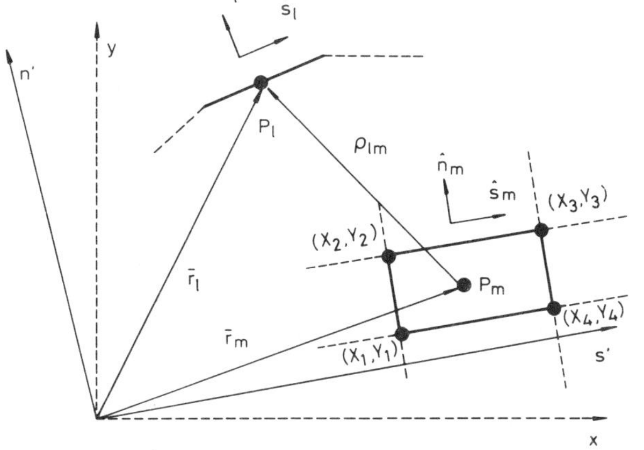

Fig. 3 *Co-ordinate frame used in the evaluation of impedance elements requiring integration over a rectangular cell*

To simplify the rather complex numerical evaluation of the smaller nonrectangular cells, we have chosen to replace them with square cells of equal area and perform an evaluation similar to that given by Richmond [2] under the premise that the generated inaccuracy is minimal due to the comparatively small size and number of these cells. The remaining integrals may be accurately evaluated by simple numerical means when their integrands are not singular or near-singular over the range of integration. When the latter is true, the impedance matrix element must then be evaluated analytically.

An analytical evaluation of the integrals (eqn. 37) can be obtained for the singular condition by invoking the small argument approximation of the Hankel function and subsequently integrating each of the resulting terms. This procedure results in a highly accurate evaluation of the diagonal and near-diagonal impedance matrix elements in the case of both the line and area integrals occurring in eqn. 37. To explicitly evaluate the analytical form of the impedance elements we first refer to Fig. 3. With the testing point P located at the centroid of the lth volume cell or alternately at the midpoint of the lth contour segment, the impedance elements can be expressed in terms of three generic integrals, defined as

$$U_0(\alpha, \beta) = \lim_{\varepsilon \to 0} \int_\varepsilon^\alpha \int_\varepsilon^\beta H_o^{(1)}(k_o\sqrt{(t^2 + u^2)})\, dt\, du$$

$$U_1(\alpha, \beta) = \lim_{\varepsilon \to 0} \int_\varepsilon^\alpha H_o^{(1)}(k_o\sqrt{(t^2 + \beta^2)})\, dt$$

$$U_2(\alpha, \beta) = \lim_{\varepsilon \to 0} \frac{\partial}{\partial \beta} \int_\varepsilon^\alpha H_o^{(1)}(k_o\sqrt{(t^2 + \beta^2)})\, dt \qquad (38)$$

For this specific implementation an $O(\rho^4, \rho^4 \ln \rho)$ small argument expansion for $H_o^1(\rho)$ was employed and the reader is referred to Reference 8, where explicit expressions are given for $U_i(\alpha, \beta)$ when α and β are small.

Since the details for the evaluation of the above F_{lm}^p are rather lengthy, we will simply state the results. With the introduction of the additional definitions

$$r_1 = \sqrt{(x_1^2 + y_1^2)}$$

$$r_2 = \sqrt{(x_2^2 + y_2^2)}$$

$$r_3 = \sqrt{(x_3^2 + y_3^2)}$$

$$r_4 = \sqrt{(x_4^2 + y_4^2)} \qquad (39)$$

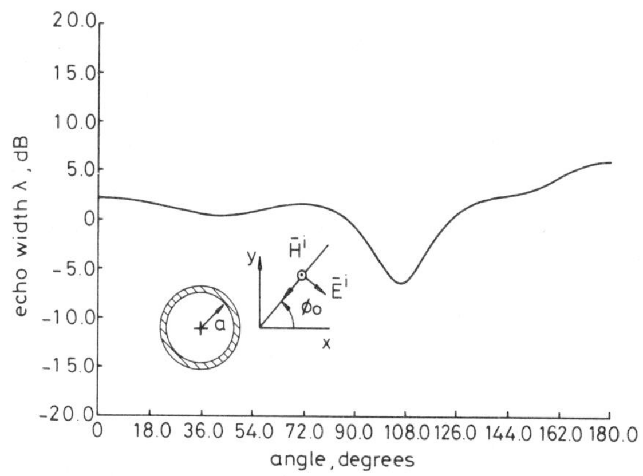

Fig. 4 *Comparison of computed H-polarisation bistatic echowidth patterns for a 1 λ diameter perfectly conducting circular cylinder*

——— reduced unknown formulation
- - - - - - traditional formulation

where (x_i, y_i) are defined in Fig. 3, the analytic expressions of the impedance elements take the form

$$F_{lm}^1 = \frac{1}{\varepsilon_r}\delta_{lm} - \frac{ik_o^2}{4}\left(\mu_r - \frac{1}{\varepsilon_r}\right)$$

$$\times [U_o(x_3, y_3) - U_o(x_2, y_2) - U_o(x_4, y_4) + U_o(x_1, y_1)]$$

$$+ \frac{i}{4}\left\{(\hat{s}_l \cdot \hat{n}_m)\left(\frac{\partial}{\partial s}\frac{1}{\varepsilon_r}\right) + (\hat{n}_l \cdot \hat{n}_m)\left(\frac{\partial}{\partial n}\frac{1}{\varepsilon_r}\right)\right\}$$

$$\times [U_1(x_3, y_3) - U_1(x_2, y_2) - U_1(x_4, x_4) + U_1(x_1, y_1)]$$

$$+ \frac{i}{4}\left\{(\hat{s}_l \cdot \hat{s}_m)\left(\frac{\partial}{\partial s}\frac{1}{\varepsilon_r}\right) + (\hat{n}_l \cdot \hat{s}_m)\left(\frac{\partial}{\partial n}\frac{1}{\varepsilon_r}\right)\right\}$$

$$\times [U_1(y_3, x_3) - U_1(y_2, x_2) - U_1(y_4, x_4) - U_1(y_1, x_1)]$$

$$F_{lm}^2 = \frac{k_o}{4}\left(\mu_r - \frac{1}{\varepsilon_r}\right)[U_1(x_2, y_2) - U_1(x_1, y_1)]$$

$$- \frac{1}{4k_o}\left\{(\hat{s}_l \cdot \hat{s}_m)\left(\frac{\partial}{\partial s}\frac{1}{\varepsilon_r}\right) + (\hat{n}_l \cdot \hat{s}_m)\left(\frac{\partial}{\partial n}\frac{1}{\varepsilon_r}\right)\right\}$$

$$\times [H_o^{(1)}(k_o r_2) - H_o^{(1)}(k_o r_1)]$$

$$- \frac{1}{4k_o}\left\{(\hat{s}_l \cdot \hat{n}_m)\left(\frac{\partial}{\partial s}\frac{1}{\varepsilon_r}\right) + (\hat{n}_l \cdot \hat{n}_m)\left(\frac{\partial}{\partial n}\frac{1}{\varepsilon_r}\right)\right\}$$

$$\times [U_2(x_2, y_2) - U_2(x_1, y_1)]$$

$$F_{lm}^3 = \frac{i}{4}(\hat{n}_l \cdot \hat{n}_m)[U_1(x_3, y_3) - U_1(x_2, y_2)$$

$$- U_1(x_4, y_4) + U_1(x_1, y_1)] + \frac{i}{4}(\hat{n}_l \cdot \hat{s}_m)$$

$$\times [U_1(y_3, x_3) - U_1(y_2, x_2) - U_1(y_4, x_4) + U_1(y_1, x_1)]$$

$$F^4_{lm} = \frac{i}{k_o}\left(\frac{\varepsilon_r}{\varepsilon_r - 1}\right)\delta_{lm} - \frac{\hat{n}_l \cdot \hat{s}_m}{4k_o}$$
$$\times [H_o^{(1)}(k_o r_2) - H_o^{(1)}(k_o r_1)]$$
$$- \frac{\hat{n}_l \cdot \hat{n}_m}{4k_o}[U_2(x_2, y_2) - U_2(x_1, y_1)] \quad (40)$$

We note that in the case of perfectly conducting scatterers or substructures, some of the current components vanish *a priori* and this can be exploited by an altered discretisation scheme to simplify and reduce the numerical computations. Specifically in the case of H-polarisation, there are no equivalent volumetric currents within the perfectly conducting scatterer and as a result the entire scatterer may be represented by only contour/surface currents. However, for E-polarisation, the contour/surface currents are not excited for purely dielectric scatterers. In that case, the perfectly conducting substructure may be modelled by a thin conducting shell conforming to the contour of the scatterer. The equivalent $\hat{z}$-directed volumetric currents within the shell are then sufficient to represent the E-Polarisation scattering.

6 Numerical results

In this section, we present scattering patterns for a variety of two-dimensional geometries that have been modelled using the reduced unknown integral equations. A comparison of the computational efficiencies associated with the numerical implementations based on the traditional and reduced integral equations is also presented.

Figs. 4–6 present the backscatter echo width patterns corresponding to a variety of composite configurations.

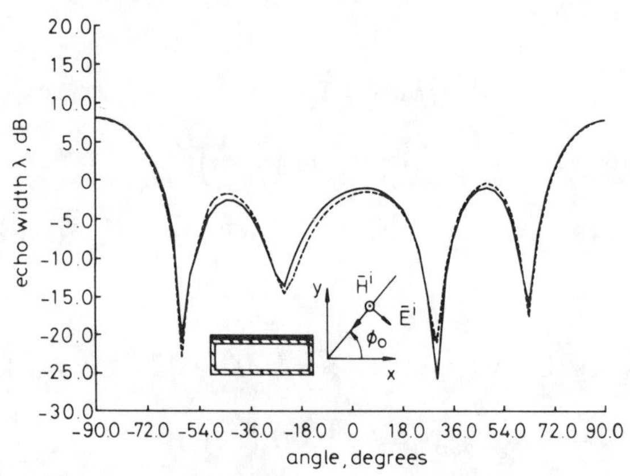

Fig. 5 *Comparison of computed H-polarisation backscatter echowidth for a .05 λ thick material ($\varepsilon_r = 5$, $\mu_r = 1$) layer upon a 1 λ × 0.25 λ perfectly conducting rectangular cylinder*

——— reduced unknown formulation
------ traditional formulation

These include a perfectly conducting circular cylinder, a partially-coated perfectly conducting rectangular cylinder, and an inhomogeneous square cylinder. The specific geometrical parameters associated with each of these configurations are included in the corresponding figure. As seen, the agreement between results obtained by the reduced unknown integral equations and those via alternate techniques [2, 3 and 8] is excelent. In modelling the illustrated test configurations a sampling interval of 0.1 λ was employed throughout the structure except near the corners (with 0.1 λ) of the perfect conductors where a higher sampling rate is required to accurately describe the currents in this region.

For practical purposes, it is of interest to compare the computational efficiency of numerical codes based on the traditional and reduced integral equations. Fig. 7 compares the CPU times in computing the scattering by a square dielectric cylinder and it is seen that the code based on the reduced unknown formulation is substantially more efficient in comparison with that based on the traditional formulation. As expected, when one of the dimensions of the dielectric cylinder is kept small, the advantage offered by the reduced unknown formulation diminishes analogously and, in the particular case of the strip, the new formulation is nearly as efficient as the traditional.

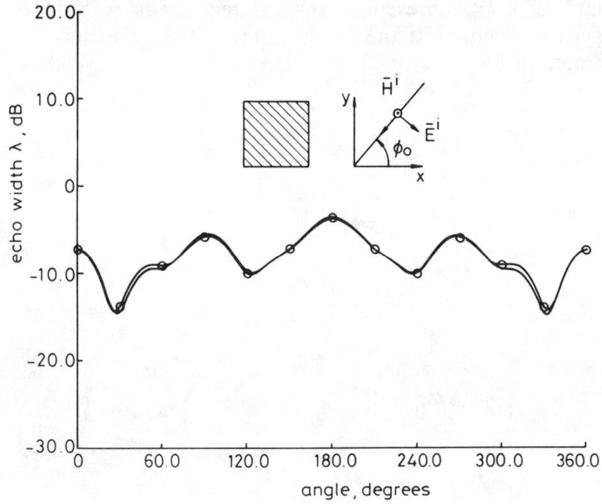

Fig. 6 *Comparison of computed H-polarisation backscatter echowidth patterns for a 0.5 λ inhomogeneous square dielectric cylinder*

$\mu_r = 1$, ε_r varies by a Blackman-Harris taper from 4 to 2 along the positive x direction

——— reduced unknown formulation
—○— traditional formulation

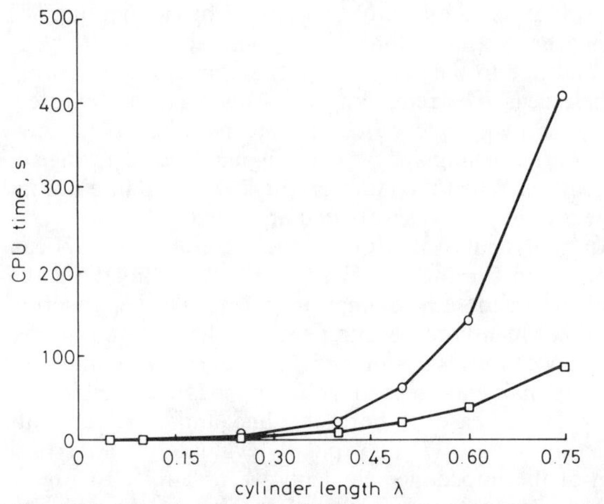

Fig. 7 *Comparison of H-polarisation backscatter CPU time between reduced unknown and traditional numerical codes for a uniformly sampled (0.05 λ × 0.05 λ cells) square cylinder*

$\varepsilon_r, \mu_r \neq 1$
—□— reduced
—○— traditional

7 Summary

A computationally efficient and simple set of integral equations has been presented for modelling two-dimensional scatterers of arbitrary composition and geometrical cross-section. The derived integral equations employ only one current component over the scatterer's cross section and another over its surface. In comparison to the three current components generally required with traditional formulations, this highly significant reduction in unknowns was obtained without the expense of computational complexity or increased singularity of the resulting equations. To verify the validity of the derived integral equations a pulse basis-point matching moment method implementation was considered. This procedure involves a highly accurate analytical evaluation of the diagonal and near-diagonal impedance matrix elements. A representative selection of scattering patterns produced by the developed computer code implementation has also been presented and, in all cases, an excellent agreement was observed with corresponding results obtained via alternate methods. Finally, a comparison of the computational efficiencies of codes based on the new and traditional formulations was presented. It was observed that, for larger scatterers of low eccentricity, the code based on the new formulation provided a significant improvement in computational efficiency over that based on the traditional.

8 References

1 GLISSON, A.W.: 'An integral equation for electromagnetic scattering from homogeneous dielectric bodies', *IEEE Trans.*, 1984, **AP-32,** pp. 173–175

2 RICHMOND, J.H.: 'Scattering by a dielectric cylinder of arbitrary cross section shape', *IEEE Trans.*, 1965, **AP-13,** pp. 334–341

3 RICHMOND, J.H.: 'TE-wave scattering by a dielectric cylinder of arbitrary cross section shape', *IEEE Trans.*, 1966, **AP-14,** pp. 460–464

4 IZADIAN, J.Z., PETERS, JR., L., and RICHMOND, J.H.: 'Computation of scattering from penetrable cylinders with improved numerical efficiency', *IEEE Trans.*, 1984, **GE-22,** pp. 52–61

5 LANGAN, D.K., and WILTON, D.R.: 'Numerical solution of TE scattering by inhomogeneous two-dimensional composite dielectric/metallic bodies of arbitrary cross section'. Abstracts of 1986 National Radio Science Meeting, Philadelphia, Pennsylvania, June 1986

6 NEWMAN, E.H.: 'TM scattering by a dielectric cylinder in the presence of a half plane', *IEEE Trans.*, 1985, **AP-33,** pp. 773–782

7 NEWMAN, E.H.: 'TM and TE scattering by a dielectric/ferrite cylinder in the presence of a half plane', *IEEE Trans.*, 1986, **AP-34,** pp. 804–813

8 RICOY, M.A., and VOLAKIS, J.L.: 'Integral equations with reduced unknowns for the simulation of two-dimensional composite structures', *IEEE Trans.*, 1989, **AP-37,** pp. 362–372

9 STRATTON, J.A.: 'Electromagnetic theory' (McGraw-Hill Company, New York, 1941), p. 394

10 HARRINGTON, R.F.: 'Time-harmonic electromagnetic fields' (McGraw-Hill Company, New York, 1961, p. 127

11 MAYES, P.E.: 'The equivalence of electric and magnetic sources', *IEEE Trans.*, 1958, **AP-6,** pp. 295–296

12 TAI, C.-T.: 'Dyadic Green's functions in electromagnetic theory' (International Textbook Company, San Francisco, 1971), p. 5

13 TAI, C.-T.: 'A note on the integral equations for the scattering of a plane wave by an electromagnetically permeable body', *Electromagnetics*, 1985, **5,** pp. 79–88

14 HARRIS, F.J.: 'On the use of windows for harmonic analysis with the discrete Fourier transform', *Proc. IEEE*, 1978, **66,** pp. 51–83

Computation of Radiation from Wire Antennas on Conducting Bodies

N. CHRISTIAN ALBERTSEN, JESPER E. HANSEN, AND NIELS E. JENSEN

***Abstract*—A theoretical formulation, in terms of combined magnetic and electric field integral equations, is presented for the class of electromagnetic problems in which one or more wire antennas are connected to a conducting body of arbitrary shape. The formulation is suitable for numerical computation provided that the overall dimensions of the structure are not large compared to the wavelength. A computer program is described, and test runs on various configurations involving a cylindrical body with one or more straight wires are presented. The results obtained agree well with experimental data.**

Manuscript received May 1, 1973; revised October 1, 1973. This work was supported by the European Space Research and Technology Centre, Noordwijk, The Netherlands, under Contract 1340-71.
N. C. Albertsen and J. E. Hansen are with the Laboratory of Electromagnetic Theory, The Technical University of Denmark, Lyngby, Denmark.
N. E. Jensen was with the Laboratory of Electromagnetic Theory, The Technical University of Denmark, Lyngby, Denmark. He is now with the European Space Research and Technology Centre, Noordwijk, The Netherlands.

I. Introduction

FOR RADIATING systems that are not large compared to the wavelength, several methods are available for the determination of radiation patterns. For antennas mounted on structures which closely approximate simple geometrical shapes, the problem may well be attacked analytically [1]–[3], or by combined analytical and numerical methods [4]. However, when more complex

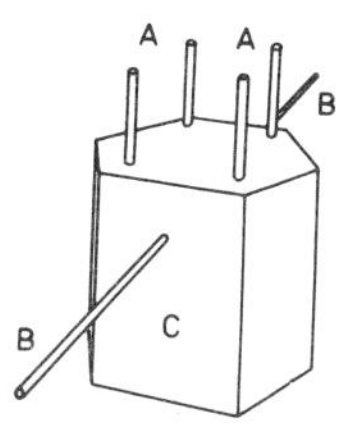

Fig. 1. Example of structure.

geometries are involved, as in the case of antennas on spacecrafts, resort has to be taken to purely numerical techniques, in which solutions to Maxwell's equations or integral equations derived from these, are obtained with the aid of a fast computer. A computer program for solving complicated radiation or scattering problems of the kind indicated is in many cases equally as accurate as experimental procedures, and more convenient. Various examples of such programs have been described in the past few years [5]–[10].

The present work deals with a specific class of radiating structures, the typical features of which are illustrated in Fig. 1. Here, a number of wire antennas A and a number of booms (passive antennas) B are attached to the surface of a metallic body C which may, at least in principle, be arbitrarily shaped. The complete structure is assumed to have a maximum linear dimension which is not large compared to the wavelength. Fig. 1 is actually an idealization of the shape of a small scientific satellite with four antennas operating in a turnstile mode, and two booms for support of measurement equipment. Although this particular interpretation of Fig. 1 formed the primary motivation of the work reported here, the geometry is common to other types of radiating systems of interest, e.g., antennas on ships, cars, etc.

For the class of structures described it seems a natural choice to attempt a formulation in terms of integral equations. Several forms of integral equations suitable for solid bodies and thin wires exist [6]. Although the electric-field integral equation (EFIE) has been used as a basis for the numerical solution of scattering or antenna problems involving (large) bodies of revolution [11], the magnetic-field integral equation (MFIE) is more attractive for generally shaped voluminous structures. For thin wires or plates, however, the MFIE is known to fail, and one is forced to use the EFIE equation in such cases, or to modify the MFIE equation in a suitable way [12], [13].

In view of the properties of the MFIE and the EFIE, the authors have used the MFIE for the solid body, whereas the EFIE in its thin-wire approximation has been applied to the wire parts of the structure. The mutual coupling between wires and body, and vice versa, is taken account of through the source terms in the integral equations [14]. A computer program based on this technique has been developed.

In Section II the important features of the MFIE and EFIE are summarized. In Section III it is explained how the coupling between body and wires may be computed using a surface current interpolation scheme. A formulation of the combined MFIE and EFIE equations particularly suitable for computer solution is given in Section IV. A short description of the structure of the computer program and examples of computations with experimental verifications are given in Section V.

II. Magnetic and Electric Field Integral Equations

From Maxwell's equations, the following MFIE for the surface current density $\bar{J}$ on the surface S of a perfectly conducting body in an incident field with magnetic field strength $\bar{H}^{\text{inc}}$ may be derived [6]:

$$-\tfrac{1}{2}\bar{J}(\xi,\eta) + \hat{n} \times \int_S^* \bar{J}(\xi',\eta') \times \nabla'\phi \, dS' = -\hat{n} \times \bar{H}^{\text{inc}}(\xi,\eta). \quad (2.1)$$

Here, $\hat{n}$ is an outward unit normal vector to the surface. Furthermore, $\phi = \exp(-jkR)/4\pi R$, where R is the distance between the observation point (ξ,η) and source point (ξ',η') on area element dS', and where k is the wavenumber. The asterisk denotes that the integration is carried out on the surface punctured at (ξ,η). The symbol ∇' denotes a gradient with respect to the primed coordinates. The time factor used is $\exp(j\omega t)$. In operator notation (2.1) may be written

$$L_M\bar{J} = -\hat{n} \times \bar{H}^{\text{inc}}. \quad (2.2)$$

For a perfectly conducting thin straight wire, the following EFIE may be derived [6]:

$$\frac{1}{j\omega\epsilon}\int_L \left(\frac{\partial^2\phi}{\partial s^2} + k^2\phi\right) I(s')\, ds' = -E_s^{\text{inc}}(s) \quad (2.3)$$

where it is assumed that the current I has an axial component only (thin wire approximation). Here, L denotes the interval for the coordinate s' of a source point on line element ds' along the wire, while

$$\phi = \exp\{-jk[(s-s')^2 + a^2]^{1/2}\}/[(s-s')^2 + a^2]^{1/2}$$

where a is the wire radius. E_s^{inc} is the component of the electric field strength of the incident field in the direction of the wire, and ϵ the dielectric constant. In operator notation we have

$$L_E I = -E_s^{\text{inc}}. \quad (2.4)$$

Using pulse functions for expansion and point-matching, the moment method formulations [15] of (2.2) and (2.4) become, respectively

$$[\langle\delta_p, L_M W_q\rangle][\bar{J}_q] = -[\langle\delta_p, \hat{n} \times \bar{H}^{\text{inc}}\rangle],$$
$$p = 1,2,\cdots,q_m; \quad q = 1,2,\cdots,q_m \quad (2.5)$$

and

$$[\langle\delta_i, L_E P_j\rangle][I_j] = -[\langle\delta_i, E_s^{\text{inc}}\rangle],$$
$$i = 1,2,\cdots,j_m; \quad j = 1,2,\cdots,j_m. \quad (2.6)$$

In these formulas δ_p and δ_i are Dirac delta functions on S and L, respectively. The centers of these (match points) are on the pth of the q_m patches into which the surface is divided, and on the ith of the j_m segments into which the wire is divided, respectively. W_q and P_j are pulse functions with amplitude unity on the qth patch, and on the jth segment, respectively, and zero elsewhere. Long-hand expansion of (2.6) may be found in [16].

While (2.5) is of a vector character, and may be split into two scalar systems of q_m linear equations in the $2q_m$ unknown components of the surface current density coefficients $\bar{J}_q$, (2.6) represents a system of j_m linear equations in the j_m unknown current coefficients I_j.

When implemented on a computer both the total number of surface patches q_m at the MFIE and the total number of wire segments j_m at the EFIE must be chosen large enough in order to obtain sufficiently accurate solutions [6].

III. Coupling Between Body and Wire

In order to take the wire–body coupling into account, as described in Section I, we need formulas for the magnetic and electric fields from known current distributions. These may be found in standard texts, e.g., [17]. At a patch with unit normal vector $\hat{n}$ and a segment with unit tangential vector $\hat{s}$ we get the following contributions to the tangential magnetic and electric fields, respectively

$$\hat{n} \times \bar{H}(I) = \hat{n} \times \int_L I(s')\hat{s}' \times \nabla'\phi \, ds' \doteq L_M'I \quad (3.1)$$

and

$$E_s(\bar{J}) = \frac{1}{j\omega\epsilon} \hat{s} \cdot \int_S \nabla \times (\nabla \times (\bar{J}\phi)) \, dS \doteq L_E'\bar{J} \quad (3.2)$$

where we have introduced an operator notation along the same lines as above.

After having expanded the currents in (3.1) and (3.2) into pulse functions with coefficients I_j and $\bar{J}_q$, in a similar manner as in Section II, we obtain the following matrix equation for the unknown currents in the mixed problem

$$\left[\begin{array}{c|c} \langle \delta_p, L_M W_q \rangle & \langle \delta_p, L_M' P_j \rangle \\ \hline \langle \delta_i, L_E' W_q \rangle & \langle \delta_i, L_E P_j \rangle \end{array}\right] \left[\begin{array}{c} \bar{J}_q \\ \hline I_j \end{array}\right] = -\left[\begin{array}{c} \langle \delta_p, \hat{n} \times \bar{H}^{\text{inc}} \rangle \\ \hline \langle \delta_i, E_s^{\text{inc}} \rangle \end{array}\right] \quad (3.3)$$

with $p = 1,2,\cdots,q_m$; $q = 1,2,\cdots,q_m$; $i = 1,2,\cdots,j_m$; and $j = 1,2,\cdots,j_m$.

This equation is of mixed scalar and vector form, and may be expanded into scalar form alone by introducing the proper decompositions of the vector quantities.

In order to compute the off-diagonal (coupling) coefficients in (3.3) a careful treatment of the currents close to the wire attachment area is necessary. The operator L_M' is well behaved, hence it suffices to use pulse functions in (3.1) for all segments including the attachment segment.

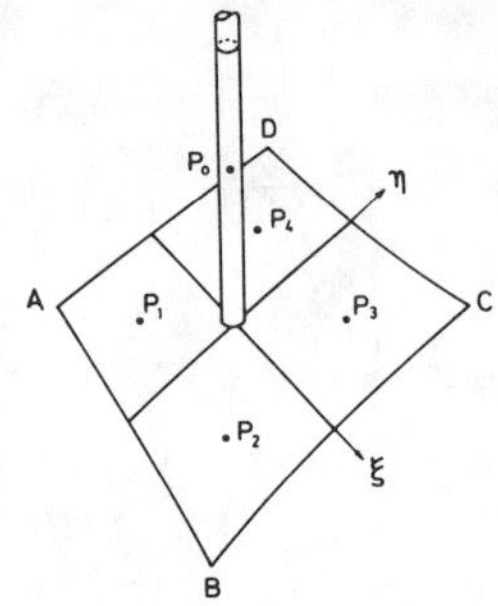

Fig. 2. Details of wire attachment area.

The operator L_E', on the other hand, has a singularity of third order, which makes it necessary to use at patches on which wires are attached, a more accurate expression for the surface current density $\bar{J}$ than pulse functions. The technique used is described below for the case of quadrilateral patches.

In Fig. 2, $ABCD$ is a quadrilateral patch in a body segmentation scheme set up before the wires are brought into the picture. In order to introduce an interpolation of the surface current density on $ABCD$ which takes into account the attachment of a wire, $ABCD$ is divided into four patches as shown. A local coordinate system (ξ,η) is introduced in which A, B, C, and D have the coordinates (ξ_1,η_1), (ξ_2,η_2), (ξ_3,η_3), and (ξ_4,η_4), respectively. The match points of the four patches are denoted $P_1,\cdots,P_4$, and the unknown surface current densities in these points $\bar{J}_1,\cdots,\bar{J}_4$. The match point on the first wire segment is P_0 and the corresponding unknown current I_0. The coordinates for $P_1,\cdots,P_4$ are $(\xi_1/2,\eta_1/2),\cdots,(\xi_4/2,\eta_4/2)$, respectively.

A suitable expansion of the surface current density $\bar{J}(\xi,\eta)$ over $ABCD$ should meet the following requirements:

$$\bar{J}(\xi_i/2,\eta_i/2) = \bar{J}_i, \qquad i = 1,\cdots,4 \quad (3.4)$$

and

$$\nabla_s \cdot \bar{J}(\xi,\eta) = J_0(\xi,\eta) - I_0\delta(\xi,\eta) \quad (3.5)$$

where $\nabla_s\cdot$ denotes surface divergence, and where $J_0(\xi,\eta)$ is continuous across $ABCD$. One such expansion is

$$\bar{J}(\xi,\eta) = I_0 \frac{\xi\hat{\xi} + \eta\hat{\eta}}{2\pi(\xi^2 + \eta^2)} + \sum_{i=1}^{4} \bar{J}_i' g_i(\xi,\eta). \quad (3.6)$$

The interpolation functions, $g_1(\xi,\eta),\cdots,g_4(\xi,\eta)$, are required to have the following properties across $ABCD$: 1) $g_i(\xi,\eta)$ is differentiable; 2) $g_i(\xi_j,\eta_j) = \delta_{ij}$, for $i = 1,\cdots,4$ and $j = 1,\cdots,4$, where δ_{ij} is the Kronecker delta; and 3) $g_1(\xi,\eta) + \cdots + g_4(\xi,\eta) = 1$.

These requirements are fulfilled by the hyperbolic paraboloids used in this study. Hyperbolic paraboloids are particularly well suited for quadrilateral surface patches, and are merely specific examples of the more general "wedge functions" introduced by Wachspress [18] as interpolation functions for surface patches of a general shape.

The fields due to the currents on the attachment patches and segment may be expressed in terms of $\bar{J}_1',\cdots,\bar{J}_4'$, and I_0 as the moments

$$[\langle \delta_p, \hat{n} \times \bar{H} \rangle] = [B][\bar{J}'] + [W_B]I_0 \quad (3.7)$$

$$\langle \delta_0, E_s \rangle = [B_W][\bar{J}'] + WI_0 \quad (3.8)$$

where $[B]$ is a matrix derived from (2.5), W is a coefficient derived from (2.6) and $[W_B]$ and $[B_W]$ are matrices derived as inner products between (3.1) and δ_p, and between (3.2) and δ_0, respectively. Here δ_p and δ_0 are delta functions defined in Section II. In the computations $[B]$, $[W_B]$, and W are evaluated using the simple technique used for all other elements of (3.3), i.e., the unknown current densities are approximated by pulse functions, while $[B_W]$ is computed as the moments of δ_0 (at P_0) and (3.2) with (3.6) inserted.

If (3.6) is inserted into (3.4), a set of linear equations is obtained which may be solved for $\bar{J}_1', \cdots, \bar{J}_4'$. In this manner the fields at $P_0, P_1, \cdots, P_4$ due to currents on the attachment patches and segment are expressed in terms of the unknowns $\bar{J}_1, \cdots, \bar{J}_4$ and I_0 occurring in (3.3).

It is noted that the interpolation technique described permits continuity of the current at the attachment point of the wire without introduction of new unknown quantities.

IV. Formulation for Computer Solution

In this section we shall show how (3.3) may be rewritten in order to cut down computer storage requirements. From (3.3) we have by multiplying out formally

$$[\langle \delta_p, L_M W_q \rangle][\bar{J}_q] + [\langle \delta_p, L_M' P_j \rangle][I_j] = -[\langle \delta_p, \hat{n} \times \bar{H}^{\text{inc}} \rangle] \quad (4.1)$$

and

$$[\langle \delta_i, L_E' W_q \rangle][\bar{J}_q] + [\langle \delta_i, L_E P_j \rangle][I_j] = -[\langle \delta_i, E_s^{\text{inc}} \rangle]. \quad (4.2)$$

Equation (4.1) may now be solved for $[\bar{J}_q]$.

$$[\bar{J}_q] = -[\langle \delta_p, L_M W_q \rangle]^{-1}[\langle \delta_p, \hat{n} \times \bar{H}^{\text{inc}} \rangle] - [\langle \delta_p, L_M W_q \rangle]^{-1}[\langle \delta_p, L_M' P_j \rangle][I_j] \quad (4.3)$$

which by insertion in (4.2) yields

$$(-[\langle \delta_i, L_E' W_q \rangle][\langle \delta_p, L_M W_q \rangle]^{-1}[\langle \delta_p, L_M' P_j \rangle] + [\langle \delta_i, L_E P_j \rangle])[I_j] = [\langle \delta_i, L_E' W_q \rangle][\langle \delta_p, L_M W_q \rangle]^{-1}[\langle \delta_p, \hat{n} \times \bar{H}^{\text{inc}} \rangle] - [\langle \delta_i, E_s^{\text{inc}} \rangle]. \quad (4.4)$$

When this equation has been solved for $[I_j]$, $[\bar{J}_q]$ may be found from (4.3).

Equation (4.4) is of a form similar to the moment solution (2.6) of the EFIE for a wire in free space. However, extra terms occur on both sides of the equality sign. These precisely account for the presence of the conducting body, and (4.4) may be considered to be the moment method formulation of an electric field integral equation similar to (2.3), with the (free space) Green's function ϕ modified to take the conducting body into account. This is analogous to the integral equation formulation given in [4] for the particular case of a monopole on a sphere, where the modified Green's function is available in analytical form.

An equation for $[\bar{J}_q]$ analogous to (4.4) may also be derived. The result is

$$([\langle \delta_p, L_M' P_j \rangle][\langle \delta_i, L_E P_j \rangle]^{-1}[\langle \delta_i, L_E' W_q \rangle] + [\langle \delta_p, L_M W_q \rangle])[\bar{J}_q] = [\langle \delta_p, L_M' P_j \rangle][\langle \delta_i, L_E P_j \rangle]^{-1} [\langle \delta_i, E_s^{\text{inc}} \rangle] - [\langle \delta_p, \hat{n} \times \bar{H}^{\text{inc}} \rangle]. \quad (4.5)$$

This equation may be considered to be the moment method formulation of an MFIE similar to (2.1), with a kernel that takes the conducting wire into account.

For a problem involving j_m segments and q_m patches the original matrix in (3.3) has $(2q_m + j_m)^2$ elements. With the preceding approach totally no more than two matrices with a total number of elements not exceeding $2q_m(2q_m + j_m)$ or $j_m(j_m + 2q_m)$, any of which is the largest figure, must be in core storage in the computer, simultaneously.

In the present work, (4.4) forms the basis of computations. In the computer program, the operators L_M and L_E have been evaluated using rectangular rule for the integration. The coupling between the attachment segment and the attachment patch is computed using Romberg integration in L_M' and a rectangular rule with 100 subdivisions in L_E'. All other couplings are computed using a simple rectangular rule in both L_M' and L_E'.

The computation of the radiation pattern from $[I_j]$ and $[\bar{J}_q]$ is a standard procedure, which may be found in textbooks, e.g., [17]. For this reason details are omitted.

V. Computations and Comparisons with Experiments

The computer program constructed for the present investigation consists of a main program and 34 Fortran subroutines. The sequence of computations is divided into four distinct parts.

First, the complete geometry is specified. This includes all coordinates of the structure as well as the segmentation parameters for body, antennas, and passive booms. In the second part of the computer program, the left-hand side matrix in (4.4) is formed and stored in an array. In the third part of the program, (4.4) is solved for the currents on all antennas and other wires attached to the body. Equation (4.3) then yields the surface current densities. Finally, the radiation pattern is computed. A number of test runs of the program and comparisons with experiments have been made for a circular cylindrical body with straight antennas and passive booms.

The experimental model is shown in Fig. 3. The height and diameter of the cylindrical body are 220 mm and 200 mm, respectively. Inside the model, a battery-powered transmitter for the frequency range 0.5–1.0 GHz is placed. On the surface of the model, monopole antennas and passive booms may be mounted in a variety of positions.

Fig. 3. Cylindrical model used for experiments.

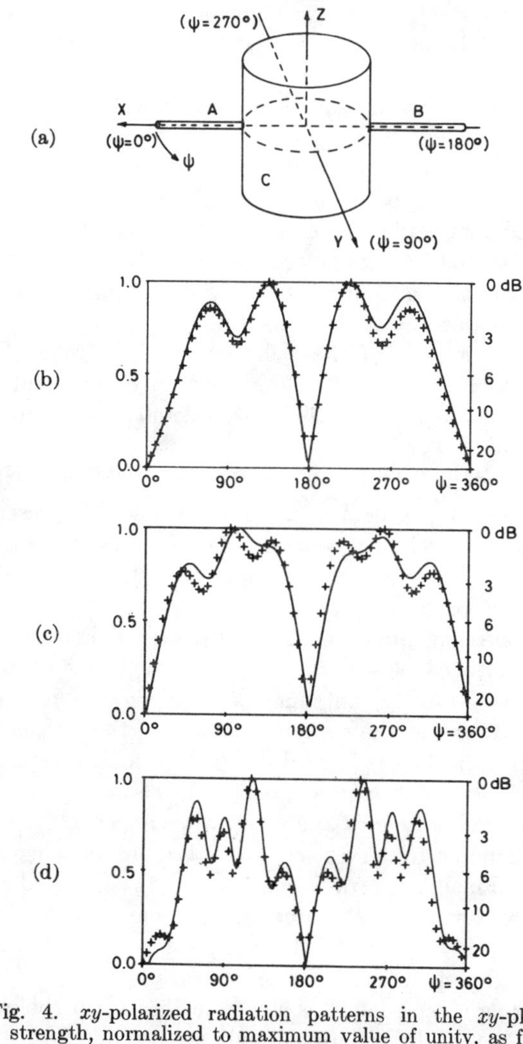

Fig. 4. xy-polarized radiation patterns in the xy-plane (field strength, normalized to maximum value of unity, as function of angle ψ) for monopole antenna (A) on metallic cylinder (C) with and without straight boom (B). Cylinder height = 0.22 m, cylinder diameter = 0.20 m, wavelength = 0.36 m. (a) Configuration, xyz-coordinate system (centered at cylinder midpoint), and definition of angular variable ψ. (b) Radiation pattern without boom; monopole length = 0.12 m. (c) Radiation pattern for boom length of 0.09 m; monopole length = 0.08 m. (d) Radiation pattern for boom length of 0.44 m; monopole length = 0.08 m.

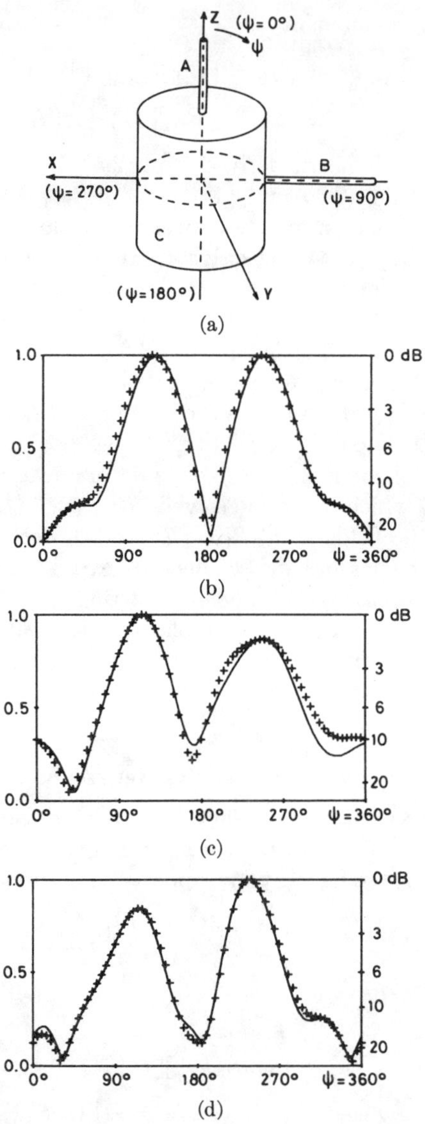

Fig. 5. zx-polarized patterns in the zx plane (field strength, normalized to maximum value of unity, as function of the angle ψ) for monopole antenna (A) on a metallic cylinder (C) with and without a straight boom (B). Cylinder height = 0.22 m, cylinder diameter = 0.20 m, monopole length = 0.12 m; wavelength = 0.48 m. (a) Configuration, xyz-coordinate system (centered at cylinder midpoint), and definition of angular variable ψ. (b) Radiation pattern without boom. (c) Radiation pattern for boom length of 0.18 m. (d) Radiation pattern for boom length of 0.44 m.

The antenna is 2 mm in diameter while booms (in metallic connection with the body) are 4 mm in diameter.

During the measurements the received signals were recorded digitally on paper tape. These tapes were used later as input data for a plotting program. Measured patterns (solid line) for a number of representative configurations are shown together with the corresponding computer radiation patterns (crosses) in Figs. 4–6.

The small unsymmetries in the measured patterns in Figs. 4 and 6, where the configurations have complete symmetry with respect to the xz plane, give an impression of the role of the measurement inaccuracies. These were due to wall reflections in the Radio Anechoic Room where the patterns were taken.

In Fig. 4 the influence on the radiation pattern of

TABLE I

Configuration illustrated in figure	Number of surface patches	Number of wire segments	Run times (in seconds) on IBM 370/165				
			(T)[a]	(T)[b]	(T)[c]	(T)[d]	(T)[e]
4(a)	57	4	4	17	~0	21	10
4(b)	60	12	5	23	~0	28	19
4(c)	60	20	6	25	~0	31	22
5(a)	57	6	4	17	~0	21	10
5(b)	60	14	6	23	~0	29	18
5(c)	60	27	7	27	~0	34	25
6	66	24	8	34	~0	42	32

[a] Computation of L_M and L_M' for (4.1).
[b] Solution of (4.1).
[c] Computation of L_E and L_E' and solution of (4.4).
[d] Total time for computation of current distributions.
[e] Additional time for computation of radiation patterns.

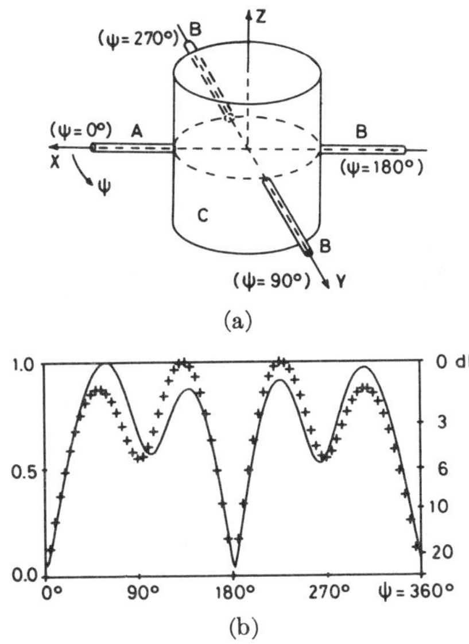

Fig. 6. xy-polarized radiation pattern in xy plane (field strength, normalized to maximum value of unity, as function of the angle ψ) for a monopole antenna (A) on metallic cylinder (C) with four straight booms (B). Cylinder height = 0.22 m, cylinder diameter = 0.20 m, monopole length = 0.12 m, boom length = 0.13 m, wavelength = 0.58 m. (a) Configuration, xyz-coordinate system (centered at cylinder midpoint), and definition of angular variable ψ. (b) Radiation pattern.

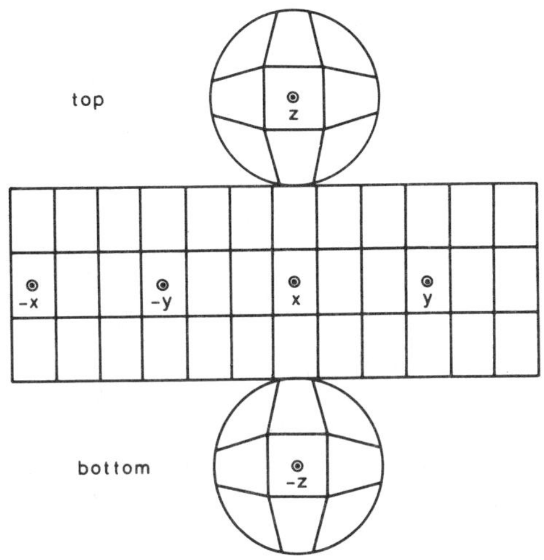

Fig. 7. Segmentation of cylinder.

adding booms diametrically opposite to an antenna on the curved surface of the cylinder, is shown. The undisturbed pattern is shown in Fig. 4(a), while patterns for booms of lengths $\lambda/4$ and $5\lambda/4$ are given in Figs. 4(b) and (c), respectively. It is seen that the radiation pattern in the xy plane is changed noticeably by the addition of booms. In all three cases the computations predict the measured behavior satisfactorily.

For the examples in Fig. 5 the antenna is placed at the top of the cylinder. The undisturbed pattern is shown in Fig. 5(a), while patterns for booms of lengths $\lambda/4$ and $3\lambda/4$ are given in Figs. 5(b) and (c), respectively. Also in these cases the measured results are satisfactorily predicted by the computations.

Fig. 6 demonstrates the numerical prediction of the radiation pattern of an antenna on a cylinder with three quarterwave booms.

No studies of the relative numerical convergence as a function of sampling density were made. The segmentation of body and wires were made with as many patches and segments as permitted by the available core storage in the computer (500k bytes) and the computations were verified by comparison with experiments. In all cases the number of surface patches was larger than the number used in [6] in an MFIE computation of surface currents on a sphere in a plane wave field.

The number of surface patches, wire segments, and run times for the seven examples described above were as given in Table I.

Other computational examples and a more detailed description of the computer program may be found in [19]. The location of booms and antennas in the cylinder version of the program is restricted to surface patches of quadrilateral shape. No boom or antenna should be placed directly on the edge of the cylinder or in the triangular shaped patches on the end surfaces of the cylinder shown in Fig. 7. No special treatment of the cylinder edges were attempted. Apparently, this is not of any practical importance for the computations, although the MFIE obviously is not valid for field points right on the edge.

VI. Conclusion

A theoretical formulation in terms of combined magnetic and electric field integral equations has been presented for antenna and scattering problems involving wires connected to a conducting body. Numerical computations using the method of moments for the solution have been made for a configuration consisting of a cylindrical body with straight wires attached to the surface. Comparison between the computations and experiments shows good agreement.

The formulation and computer program described in this paper seem suitable for solving electromagnetic problems in which thin wires are connected to a conducting body. Although the primary motivation for the work presented was the need for computerized prediction of radiation patterns for spacecraft VHF-monopole antennas, the technique may prove useful also for the examination of radiation from antennas on small aircrafts, portable radio sets, etc.

Acknowledgment

The authors wish to thank G. Paci of the European Space Research Organisation for suggesting the present investigation. The numerical computations were made at the NEUCC computing center at the Technical University of Denmark.

References

[1] F. V. Schultz, "The scattering of electromagnetic waves by perfectly reflecting objects of complex shape," School Elec. Eng., Purdue Univ., Indiana, Rep. TR-EE68-39, 1968.

[2] D. M. Bolle and M. D. Morganstern, "Monopole and conic antennas on spherical vehicles," *IEEE Trans. Antennas Propagat.*, vol. AP-17, pp. 477–84, July 1969.

[3] P. Bruscaglioni and A. Consortini, "Irraggiamento di un monopolo su una sfera conduttrice," *Alta Freq.*, vol. 36, pp. 224–26, 1967.

[4] F. M. Tesche and A. R. Neureuther, "Radiation patterns for two monopoles on a perfectly conducting sphere," *IEEE Trans. Antennas Propagat.*, vol. AP-18, pp. 692–94, Sept. 1970.

[5] E. K. Miller *et al.*, "Numerical analysis of aircraft antennas," in *Proc. Conf. Environmental Effects on Antenna Performance*, pp. 55–58, 1969.

[6] A. J. Poggio and E. K. Miller, "Integral equation solutions of three-dimensional scattering problems," in *Computer Techniques for Electromagnetics*, R. Mittra, Ed. Oxford: Pergamon, 1973, ch. 4.

[7] P. L. E. Uslenghi, "Computation of surface currents on bodies of revolution," *Alta Freq.*, vol. 39, pp. 213E–224E, Aug. 1970.

[8] G. A. Thiele and M. Travieso-Diaz, "Radiation of a monopole antenna on the base of a conical structure," in *Proc. Conf. Environmental Effects on Antenna Performance*, pp. 99–102, 1969.

[9] C. L. Bennett, A. M. Auckenthaler, and J. D. DeLorenzo, "Transient scattering by three-dimensional conducting surfaces with wires," in *1971 G-AP Int. Symp. Dig.*, pp. 349–351, Sept. 1971.

[10] J. E. Jones, J. H. Richmond, and T. G. Campbell, "A numerical method for obtaining radiation patterns of space shuttle annular slot antennas," *1972 G-AP Int. Symp. Dig.*, pp. 189–192, Dec. 1972.

[11] J. R. Mautz and R. F. Harrington, "Radiation and scattering from bodies of revolution," *Appl. Sci. Res.*, vol. 20, pp. 405–435, June 1969.

[12] R. Mittra, "How to use the H-integral equation for electrically thin antennas and scatterers," in *1972 USNC-URSI Spring Meeting Abst.*, 1972.

[13] R. Mittra, W. A. Davis, and D. V. Jamnejad, "An integral equation for plane wave scattering by thin plates," in *1972 USNC-URSI Fall Meeting Abst.*, pp. 93–94, Dec. 1972.

[14] N. C. Albertsen, J. E. Hansen, and N. E. Jensen, "Numerical prediction of radiation patterns for antennas mounted on spacecrafts," Conference on Aerospace Antennas, IEE Conference Publication Number 77, London, pp. 219–228, June 1971.

[15] R. F. Harrington, *Field Computation by Moment Methods.* New York: Macmillan, 1968.

[16] J. H. Richmond, "A wire-grid model for scattering by conducting bodies," *IEEE Trans. Antennas Propagat.*, vol. AP-14, pp. 782–86, Nov. 1966.

[17] R. E. Collin and F. J. Zucker, *Antenna Theory.* New York: McGraw-Hill, 1969, pt. I, ch. 2.

[18] E. L. Wachspress, "A rational basis for function approximation," in *Publ. Conf. Applications of Numerical Analysis.* Berlin, Germany: Springer-Verlag, 1971.

[19] N. C. Albertsen, J. E. Hansen, and N. E. Jensen, "Computation of spacecraft antenna radiation patterns," Lab. Electromag. Theory, Tech. Univ. Denmark, Lyngby, Denmark. Rep. R 108, June 1972; also Rep. ESRO CR-207, Sept. 1973.

Electromagnetic Modeling of Composite Wire and Surface Geometries

E. H. NEWMAN, MEMBER, IEEE, AND D. M. POZAR, STUDENT MEMBER, IEEE

Abstract—**A moment method solution to the problem of radiation or scattering from geometries consisting of open or closed surfaces, wires, and wire/surface junctions is presented. The method is based on the sinusoidal reaction formulation. Several examples of input impedance calculations illustrate the versatility, accuracy, and computational efficiency of the method.**

I. INTRODUCTION

THE METHOD of moments is perhaps the most widely used tool for the electromagnetic modeling of bodies which are not large in terms of a wavelength. The advantages of moment methods are accuracy, versatility, and the ability to compute near- as well as far-zone parameters.

The most widely used forms of the method of moments are the thin-wire computer programs [1]–[3]. These programs are ideal for modeling most wire antennas. By forming a wire-grid model they also can be used to model solid surfaces. Unfortunately, the size of the surface which can be modeled practically is severely limited since it requires many wires to accurately model a solid surface. Further, even a very fine wire mesh may not yield accurate near-zone parameters, such as current distribution or impedance.

More recently, surface-patch models have been developed for modeling solid surfaces. The advantage of the surface patch solution is that fewer unknowns are required per square wavelength of surface area. Wang *et al.* [4] have developed a model for wires and plates based on the sinusoidal reaction formulation. Albertsen *et al.* [5] used pulse expansion modes to model wires and plates, including the case where a wire touches a plate. Burke and Poggio have incorporated these results into a user-oriented computer code [3]. Parhami *et al.* have treated the problem of a wire surface junction using the finite difference technique [6].

The purpose of this paper is to present a technique for modeling wires and surfaces, including the case where the wires contact the surfaces, which is sufficiently accurate to compute impedance. The technique is based on the sinusoidal reaction formulation and is applicable to open as well as closed surfaces. Continuity of current is enforced on the wires, on the surfaces and at the wire/surface junctions. The accuracy and versatility of the model is demonstrated by several numerical examples. Attachment points are restricted to be somewhat removed from an edge.

II. THEORY

The Reaction Method

Consider a geometry of arbitrarily shaped scatterers in a homogeneous medium as shown in Fig. 1. S is the surface

Manuscript received October 14, 1977; revised February 23, 1978. This work was supported in part by the U.S. Army Research Office and in part by the Ohio State University Research Foundation, under Grant DAAG29-76-G-0331.

The authors are with the ElectroScience Laboratory, Department of Electrical Engineering, the Ohio State University, Columbus, OH 43212.

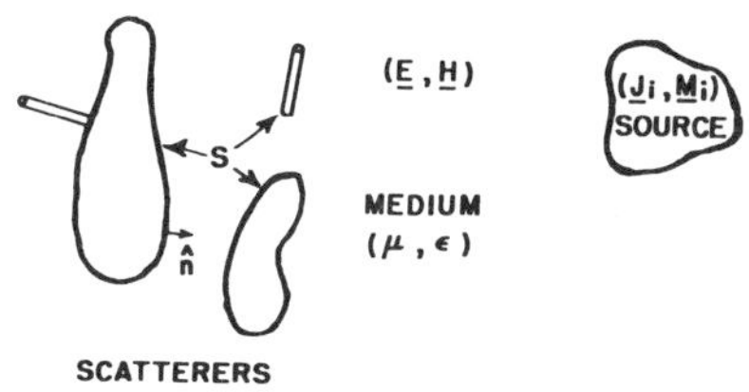

Fig. 1. General problem geometry.

enclosing the scatterers, and $\hat{n}$ is the unit outward normal to S. The scatterers may consist of solid surfaces and thin wires, and the wires may contact the surfaces. For now, the surfaces will be considered closed. The sources ($\mathbf{J}_i$, $\mathbf{M}_i$) generate the fields (E, H) in the presence of the scatterers. The $e^{j\omega t}$ time dependence is suppressed. From the surface-equivalence theorem the field interior to the surface S will vanish without changing the exterior fields if the surface current densities

$$\mathbf{J}_s = \hat{n} \times \mathbf{H}, \tag{1}$$

$$\mathbf{M}_s = \mathbf{E} \times \hat{n} \tag{2}$$

are introduced on the surface S. The scattering obstacles may then be replaced by the ambient medium without altering the field anywhere. The scattered field, radiated by ($\mathbf{J}_s$, $\mathbf{M}_s$) in the ambient medium, is defined as

$$\mathbf{E}_s = \mathbf{E} - \mathbf{E}_i \tag{3}$$

$$\mathbf{H}_s = \mathbf{H} - \mathbf{H}_i. \tag{4}$$

An electric test source, which radiates the fields ($\mathbf{E}_T$, $\mathbf{H}_T$) in the ambient medium, is now placed in the interior region of S. Noting that this test source has zero reaction with the sources ($\mathbf{J}_s$, $\mathbf{M}_s$) and ($\mathbf{J}_i$, $\mathbf{M}_i$), the reaction integral equation [4] is obtained:

$$\iint_S (\mathbf{J}_s \cdot \mathbf{E}_T - \mathbf{M}_s \cdot \mathbf{H}_T)\, ds + \iiint_V (\mathbf{J}_i \cdot \mathbf{E}_T - \mathbf{M}_i \cdot \mathbf{H}_T)\, dv = 0 \tag{5}$$

where the volume integral is over the source volume. The reaction equation (5) is used to determine the unknown surface currents ($\mathbf{J}_s$, $\mathbf{M}_s$). For conductors of finite conductivity the impedance boundary condition

$$\mathbf{M}_s = Z_s \mathbf{J}_s \times \hat{n} \tag{6}$$

is used where Z_s may be a function of position. For simplicity, only perfect conductors are considered here; thus $Z_s = \mathbf{M}_s = 0$.

In deriving (5) it was assumed that all surfaces were closed so that Schelkunoff's equivalence theorem could be used. However, it can be shown that (5) applies equally well for open surfaces. An open surface, such as a fictitious plate of zero thickness, can be considered to be a limiting case of a real plate of finite thicknes as the plate thickness goes to zero. In general, different currents exist on the top and bottom surfaces of the real plate. As the plate thickness goes to zero, the fields radiated by the top and bottom currents become identical to the fields radiated by a single surface current located on the plate center. This single surface current is the vector sum of the top and bottom surface currents and is the current which must be determined to treat open surfaces. If electric test sources are used in (5), then $\mathbf{J}_s$ will be the vector sum of the current on the top and bottom surfaces [7].

The integral equation (5) is solved by the moment method. The unknown current $\mathbf{J}_s$ is expanded in a set of N basis (expansion) functions,

$$\mathbf{J}_s = \sum_{n=1}^{N} I_n \mathbf{J}_n, \tag{7}$$

and (5) is enforced for N electric test sources placed in S. Thus (5) reduces to the set of simultaneous linear equations

$$\sum_{n=1}^{N} I_n Z_{mn} = V_m; \qquad m = 1, 2 \cdots N \tag{8}$$

where

$$Z_{mn} = -\iint_n \mathbf{J}_n \cdot \mathbf{E}_m \, ds \tag{9}$$

$$V_m = \iiint_V (\mathbf{J}_i \cdot \mathbf{E}_m - \mathbf{M}_i \cdot \mathbf{H}_m) \, dv \tag{10}$$

where $(\mathbf{E}_m, \mathbf{H}_m)$ are the fields of the mth test source radiating in the medium (μ, ϵ) and the integration in (9) is over the surface of the nth expansion mode.

The expansion and test modes used here are identical. Thus the method is a Galerkin method, and a symmetric impedance matrix $[Z]$ results. The form of the expansion/test functions will now be defined. It is this form which ultimately determines the accuracy and efficiency of the solution as well as the types of geometries which may be modeled.

Expansion/Testing Functions

Three basic types of modes are used: wire dipole modes, surface dipole modes, and a special attachment mode whenever a wire connects to a surface. With this choice of functions geometries consisting of flat surfaces, thin wires, and wire-surface connections may be modeled. A piecewise flat approximation can be made to model singly curved surfaces. Note that all of the modes will involve sinusoids with the free-space wavenumber. This allows all but one of the integrations required to find the $\mathbf{E}_m$ to be done in closed form, thus avoiding the very difficult $1/r^3$ singularity or second derivative associated with finding the extreme near-zone fields of an electric current source, as is required to evaluate self or overlapping impedance elements Z_{mn}. The one integration not available in closed form is associated with the disk component of the attachment mode, as discussed below.

1) Thin-Wire Mode: The wire mode used is identical to that used by Richmond [1]. It is a piecewise-sinusoidal V-dipole consisting of two sinusoidal monopoles. A V-dipole with a 180° internal angle lying on the z-axis is shown in Fig. 2(a).

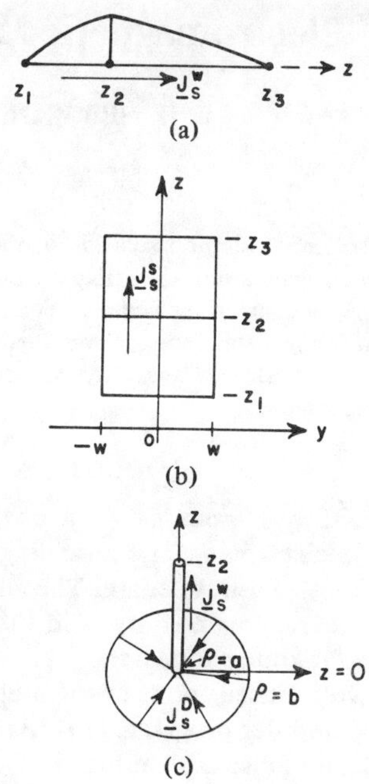

Fig. 2. Expansion and test modes. (a) Thin-wire V-dipole. (b) Sinusoidal surface V-dipole. (c) Attachment dipole.

Using the thin-wire approximation the current on this dipole is given by

$$\mathbf{J}_s{}^w = \frac{\hat{z}}{2\pi a}\left[P_1 \frac{\sin k(z - z_1)}{\sin k(z_2 - z_1)} + P_2 \frac{\sin k(z_3 - z)}{\sin k(z_3 - z_2)}\right] \tag{11}$$

where P_1 and P_2 represent pulse functions with unit value when $z_1 < z < z_2$ and $z_2 < z < z_3$, respectively, and are zero elsewhere. Also, a is the wire radius and $k = 2\pi/\lambda$. This choice of mode has the advantages that the near-zone fields and the Z_{mn} are known in closed form [8]. These modes are placed in an overlapping array on the wire ensuring continuity of current on the wire.

2) Surface-Patch Mode: The surface-patch mode is a surface V-dipole consisting of two sinusoidal surface monopoles. A surface V-dipole with interior angle of 180° is shown in Fig. 2(b). The current on this dipole is given by

$$\mathbf{J}_s{}^s = \hat{z}\frac{kP_1 \sin k(z - z_1) \cos ky}{2 \sin k(z_2 - z_1) \sin kw} + \hat{z}\frac{kP_2 \sin k(z_3 - z) \cos ky}{2 \sin k(z_3 - z_2) \sin kw} \tag{12}$$

where P_1 and P_2 represent unit pulse functions as before. This mode is similar to that used by Wang *et al.* [4] except that the sinusoidal variation is with the speed of light and a cosine variation is used transverse to $\mathbf{J}_s{}^s$ instead of the constant variation used by Wang *et al.* Because of this the fields due to the current of (12) may be found in closed form [9]. Two orthogonal and overlapping arrays of the surface-patch modes are placed on the surface, allowing a two-dimensional vector surface current density. Results presented in the next section indicate that reasonable accuracy can be obtained with $2w$ or $z_2 - z_2$ or $z_3 - z_2$ as large as 0.25 λ.

3) Attachment Mode: When a wire is attached to a surface a special attachment mode is introduced. The purpose of the attachment mode is twofold: first is to establish continuity of current at the wire/surface junction and second is to insure that in the immediate vicinity of the attachment the surface current density has the proper $\hat{\rho}$ polarization and $1/\rho$ dependence. This mode consists of two parts: a wire monopole and a disk monopole, as shown in Fig. 2(c). The wire monopole current density is similar to the thin-wire monopole mode:

$$\mathbf{J}_s{}^w = \frac{1}{2\pi a} \frac{\sin k(z_2 - z)}{\sin kz_2} \hat{z}. \tag{13}$$

Thus

$$\mathbf{J}_s{}^D = \frac{-\sin k(b-\rho)}{2\pi\rho \sin k(b-a)} \hat{\rho}, \qquad a \leqslant \rho \leqslant b \tag{14}$$

where a, b are the inner and outer radii of the annulus. Note that the total current on the disk at $\rho = a$ is equal to the total current on the wire segment at $z = 0$ insuring continuity of current at the attachment. Also observe that the $\sin k(b - \rho)$ function in the numerator of (14) forces the disk current to be zero at the disk edge, $\rho = b$. It is this property which allows the disk to be placed on the surface and still maintain continuity of current on the surface. The $\sin k(b - \rho)$ function was chosen (rather than, say, $b - \rho$ or $\sin \alpha(b - \rho)$) since this permits the fields of the disk to be obtained with only one numerical integration.

The attachment mode is applied by placing it directly on the surface-patch modes wherever the wire meets the surface. Attachment points are not restricted to be in the center of or at the corners of the surface patch modes. However, a more detailed treatment of the attachment mode, including the edge singularity, would be required for attachment points less than about 0.1 λ from an edge. The inner radius a of the annulus corresponds to the wire radius, and results presented below indicate that the outer radius b has little effect on the final result if it is chosen to be between 0.1 λ to 0.25 λ (see Fig. 3).

With these expansion functions, current continuity in the direction of $\mathbf{J}_s$ is always maintained, although the current may not be continuous in directions orthogonal to the direction of current flow. Thus $\nabla \cdot \mathbf{J}_s = -j\omega\rho_s$ is always finite, and no line charges appear on the surface.

For antennas, feeds may be placed at the endpoints of any wire segment, including the attachment segment. For example, a monopole on a ground plane may be modeled by inserting a generator at the base of the monopole between the attachment mode wire segment and disk. In the work presented here the delta-gap feed was used, although the magnetic frill model also could be used.

The scatterer or antenna geometry is thus modeled by N_w wire modes, N_s surface-patch modes, and N_A attachment modes. Then, $N = N_w + N_s + N_A$, and (8) is solved for the unknown current samples I_n by standard Gauss–Jordan elimination.

III. NUMERICAL EXAMPLES

Several numerical examples will now be presented which illustrate the accuracy and versatility of the solution presented in Section II. In each case, the number of modes used is summarized in Table I. Note that N_s is the total number of surface-patch dipoles. Thus there are $N_s/2$ dipoles in each of the two orthogonal directions on the surface. The data presented are input impedance or admittance, since this tends to be a sensitive indicator of overall accuracy.

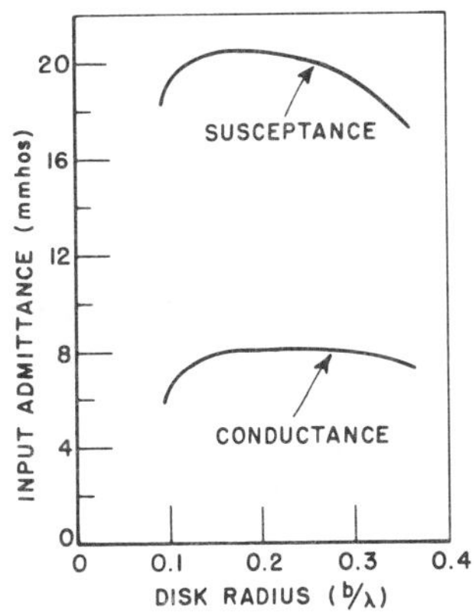

Fig. 3. Input admittance for geometry of Fig. 4, with N = 14, versus disk radius. f = 160 MHz.

TABLE I
PARAMETERS OF CALCULATIONS

Figure/Curve	N_W	N_S	N_A	N (Total Unknowns)
3/ — — —	2	40	1	43
3/ ———	1	12	1	14
3/ ----	121	-	-	121
5/• • •	2	12	1	15
6/• • •	4	47	1	52
7/ ———	4	12	1	17
8/ ———	4	12	2	18
9/ ———	7	77	3	87

In the experimental models the wires were brass rods, 0.0008 m in radius, with a conductivity of 15 × 10^6 mho/m. The finite conductivity of the wires was taken into account in the manner used by Richmond [10]. The metal plates were made of brass or aluminum, and their finite conductivity was not taken into account. In all cases the driven wire element was driven with a coaxial feed with outer radius 0.0047 m. All measurements were made by the authors.

The calculated results were all made using the same general purpose computer code. No advantage of inherent symmetries in the geometries was made. The resulting symmetries in the expansion mode currents offered a useful check on the calculations.

Fig. 4 presents the input admittance for a monopole on a square finite ground plane versus frequency in the vicinity of the first resonance. The calculation was done with N = 14 and with 43 unknowns and is compared with measured data and with a wire-grid model which used 121 unknowns. This example shows the accuracy and rapid convergence of the patch solution as well as a savings in storage over the wire-grid model. The run time for the N = 14 case was about 25 min on a Datacraft 6024/3, which is equivalent to about 2–3 min on an IBM 370-165.[1] In comparison the wire-grid model took about 40 min on the Datacraft 6024/3.

Fig. 3 shows the effect of the attachment disk radius b on the input admittance of the same geometry as Fig. 4. This graph shows that the disk radius is not critical and usually can be selected between 0.1 λ → 0.25 λ.

The impedance versus height of a monopole on a 1.4 λ square ground plane is shown in Fig. 5. These calculations are compared with the input impedance of a monopole of the same height on an infinite ground plane computed by image

[1] Recent improvements in the technique and computer code have reduced the run time by about a factor of 6, while increasing accuracy.

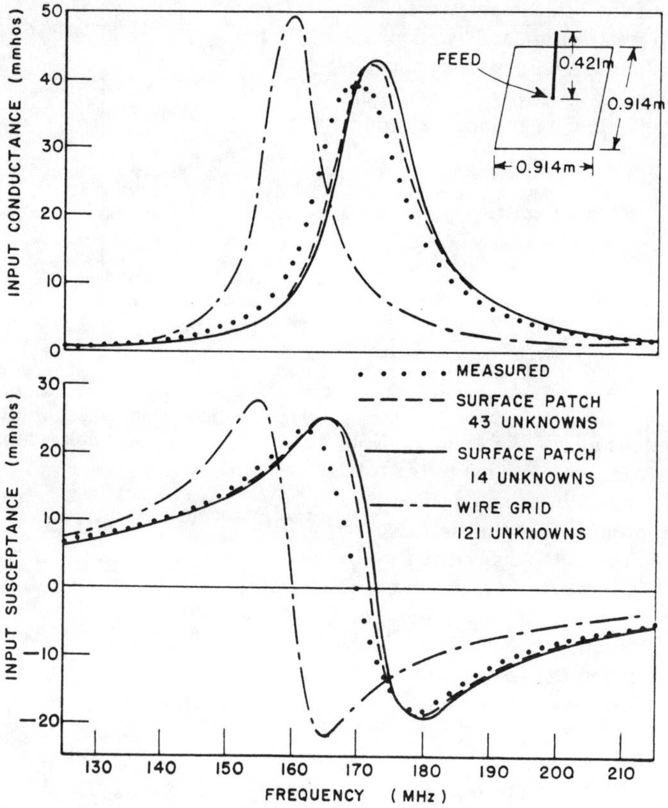

Fig. 4. Input admittance versus frequency for monopole (h = 0.421 m, a = 0.0008 m) on square ground plane (9.914 × 0.914 m), compared with measured data and wire-grid modeling. b = 0.2 λ.

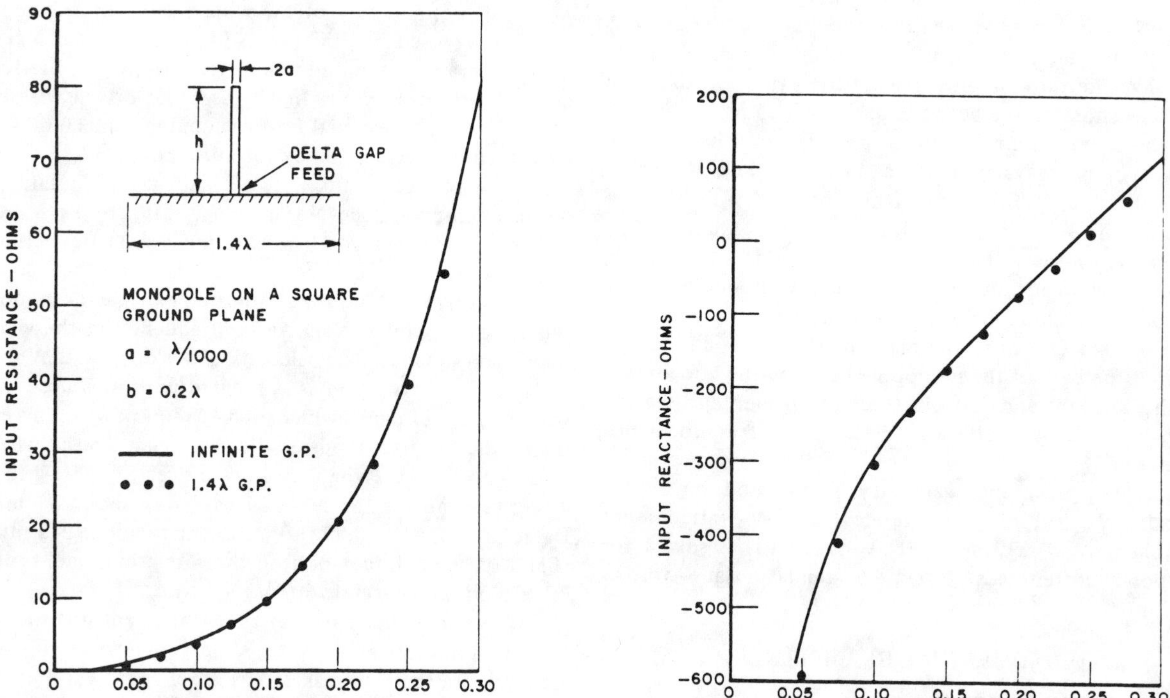

(a) (b)

Fig. 5. (a) Input resistance versus monopole height for monopole centered on 1.4 λ square ground plane compared with same monopole on infinite ground plane as computed by image theory. (b) Input reactance versus monopole height for monopole centered on 1.4 λ square gound plane compared with same monopole on infinite ground plane, as computed by image theory.

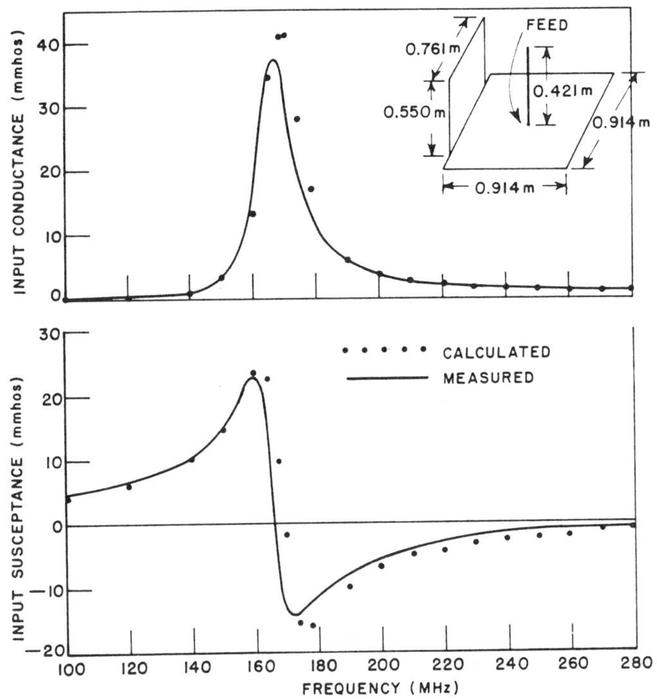

Fig. 6. Input admittance versus frequency for geometry of Fig. 4 with attached reflected plate compared with measured values.

theory. Since the 1.4 λ ground plane is fairly large, good agreement is obtained.

Fig. 6 shows the input admittance for a geometry identical to that of Fig. 3, except that a reflecting plate is attached to one side of the ground plane. Note that the width of the reflecting plate differs from that of the ground plane and that the theory and measurements are in close agreement. Overlap or hinge surface-patch dipoles are used to insure continuity of current at the plate to plate junction.

Fig. 7 shows the input admittance of the geometry of Fig. 3 with the addition of a parasitic wire element 1.5 cm above the ground plane and parallel to the fed monopole. The admittance is shown as a function of the separation d of the two wires. Fig. 8 shows a similar geometry except that the parasitic element now touches the ground plane.

Fig. 9 shows the measured and calculated input impedance of a T-bar fed, rectangular cavity-backed slot antenna opening into a finite ground plane.

IV. CONCLUSION

A moment method solution to the problem of radiation or scattering from geometries consisting of wires, open or closed surfaces, and wire/surface junctions has been presented. The technique is a Galerkin solution based on the sinusoidal reaction formulation. The expansion and test modes were chosen to obtain continuity of current on the composite structure, the proper $\hat{\rho}/\rho$ behavior of the surface current density at attachment points, and facilitate the numerical evaluation of the impedance matrix. Numerical examples presented illustrate the accuracy, versatility, stability, and computational efficiency of the method. Future work will center on further improving the speed and computational efficiency of the method and adding the presence of dielectrics.

ACKNOWLEDGMENT

The authors wish to acknowledge the assistance of J. H. Richmond for helpful discussions and computer subroutines.

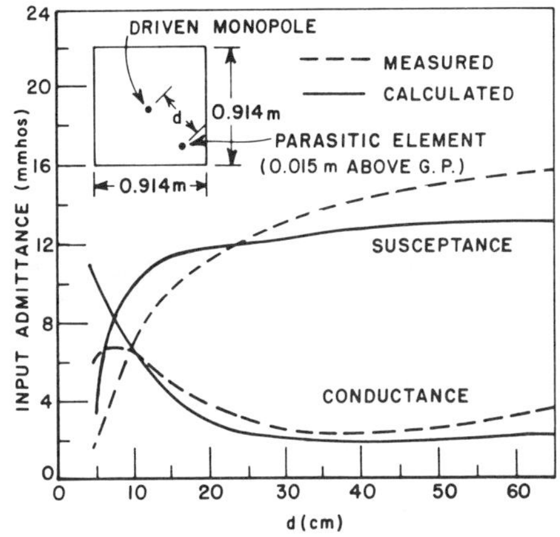

Fig. 7. Input admittance versus separation of driven monopole and parasitic wire element 0.015 m above the 0.0914 × 0.914 m ground plane. Driven monopole height = 0.421 m. Parasitic wire length = 0.842 m. f = 150 MHz.

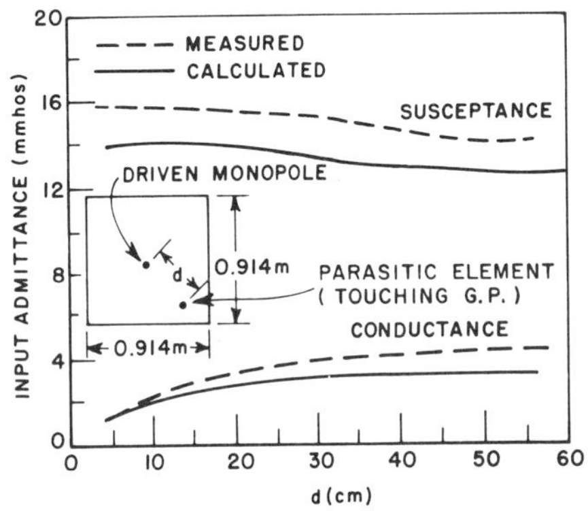

Fig. 8. Input admittance versus separation of driven monopole and parasitic wire element touching 0.914 × 0.914 m ground plane. Driven monopole height = 0.421 m. Parasitic wire height = 0.842 m. f = 150 MHz.

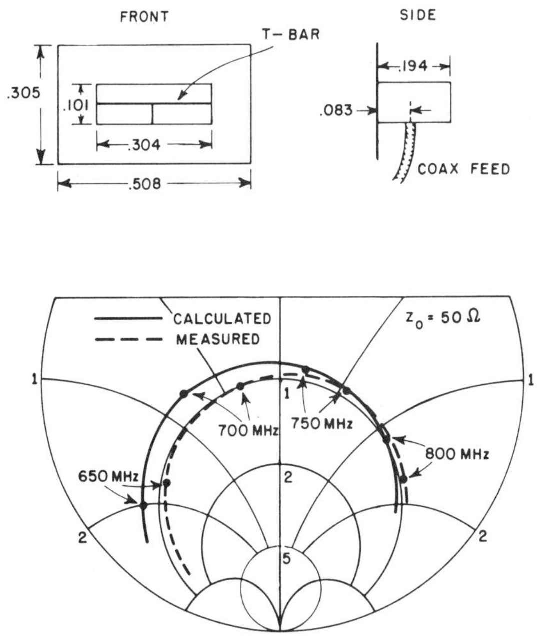

Fig. 9. Geometry and input impedance of T-bar fed slot antenna. Note: dimensions in meters.

REFERENCES

[1] J. H. Richmond, "A wire-grid model for scattering by conducting bodies," *IEEE Trans. Antennas Propagat.*, vol. AP-14, pp. 782–786, Nov. 1966.

[2] H. H. Chao and B. S. Strait, "Computer programs for radiation and scattering by arbitrary configurations of bent wires," Scientific Rep. 7 AFCRL-70-0374, Syracuse Univ., Syracuse, NY, Sept. 1970.

[3] G. J. Burke and A. J. Poggio, "Numerical electromagnetic code–Method of moments," Naval Ocean Systems Center tech. document 116, AFWL-TR-76-320, July 1977.

[4] N. N. Wang, J. H. Richmond, and M. C. Gilreath, "Sinusoidal reaction formulation for radiation and scattering from conducting surfaces," *IEEE Trans. Antennas and Propagat.*, vol. AP-23, pp. 376–382, May 1975.

[5] N. C. Albertsen, J. E. Hansen, and N. E. Jensen, "Computation of radiation from wire antennas on conducting bodies," *IEEE Trans. Antennas and Propagat.*, vol. AP-22, pp. 200–206, Mar. 1974.

[6] P. Parhami, Y. Rahmat-Samii, and R. Mittra, "Investigation of antennas on a finite ground plane," *AP-S Int. Symp.*, Univ. of Massachusetts Amherst, Oct. 11–15, 1976, pp. 511–514.

[7] W. A. Davis and R. Mittra, "A new approach to the thin scatterer problem using the hybrid equations," *IEEE Trans. Antennas and Propagat.*, vol. AP-25, May 1977.

[8] J. H. Richmond and N. H. Geary, "Mutual impedance of non-planar-skew sinusoidal dipoles," *IEEE Trans. Antennas and Propagat.*, vol. AP-23, May 1975.

[9] J. H. Richmond, D. M. Pozar, and E. H. Newman, "Rigorous near-zone field expressions for rectangular sinusoidal monopole," *IEEE Trans. Antennas and Propagat.*, vol. AP-26, p. 509, May 1978.

[10] J. H. Richmond, "Radiation and scattering by thin-wire structures in the complex frequency domain," Rep. 2902-10, Ohio State Univ. ElectroScience Laboratory, Dep. of Electrical Engineering, prepared under Grant NGL 35-008-138 for National Aeronautics and Space Administration, July 1973.

Scattering by Thin Dielectric Strips

JACK H. RICHMOND, FELLOW, IEEE

Abstract—**A plane wave incident on a thin dielectric strip with infinite length is considered, letting the incident electric field vector be parallel with the edges of the strip. The field is expanded in the dielectric region as the sum of three plane waves (the forced wave and two surface waves). The *x*-axis and *y*-axis propagation constants are known for each wave, and Galerkin's method is employed to determine the amplitudes of these waves. Finally, the far-zone scattered field is determined by considering the polarization currents radiating in free space. Numerical data are presented to illustrate the scattering properties of lossless and lossy dielectric strips as a function of the angle of incidence and the width of the strip. The calculations show excellent agreement with an earlier moment method using pulse bases and point matching.**

I. INTRODUCTION

A DIELECTRIC strip with infinite length, width w, and thickness d is illustrated in Fig. 1. A plane wave is incident on this strip, and the incident electric field vector is parallel with the edges of the strip. This geometry is of interest in connection with certain antenna design problems.

This problem was treated by Richmond [1] in 1965 with the moment method, but the technique becomes inefficient as the width of the strip increases. Burnside and Burgener [2] presented the geometrical theory of diffraction (GTD) analysis, but they neglected the surface waves which propagate on the strip. Kastner and Mittra [3] presented the spectral-iteration technique, but they encountered difficulties as the width of the strip increased. Pathak and Rojas-Teran [4] obtained excellent results with the Wiener–Hopf procedure, except for difficulties in the vicinity of grazing incidence.

In this paper we expand the field in the dielectric strip as the sum of three plane waves: the forced wave and two surface waves. The form is known for each wave, but the amplitude is unknown. We apply Galerkin's technique to develop three equations for the three unknown amplitudes. Finally, the far-zone scattered field is determined by considering the electric polarization current radiating in free space.

We present the theory in the next section. This is followed by numerical results for the scattering levels of thin dielectric strips versus the angle of incidence and the width.

II. THEORY

Consider a time-harmonic wave incident on a thin dielectric strip, as illustrated in Fig. 2. The time dependence $e^{j\omega t}$ is assumed and suppressed. In this two-dimensional problem, we let the electric field intensity E have only a z-component which is independent of z. At each point in space,

$$E = E^i + E^s \tag{1}$$

where E, E^i, and E^s denote the total field, incident field, and scattered field, respectively. Since the electric field has only a z-component, this scalar notation is suitable.

Manuscript received June 4, 1984. This work was supported in part by the Joint Services Electronics Program under Contract N00014-78-C-0049 with The Ohio State University Research Foundation.

The author is with the ElectroScience Laboratory, Department of Electrical Engineering, The Ohio State University, Columbus, OH 43212.

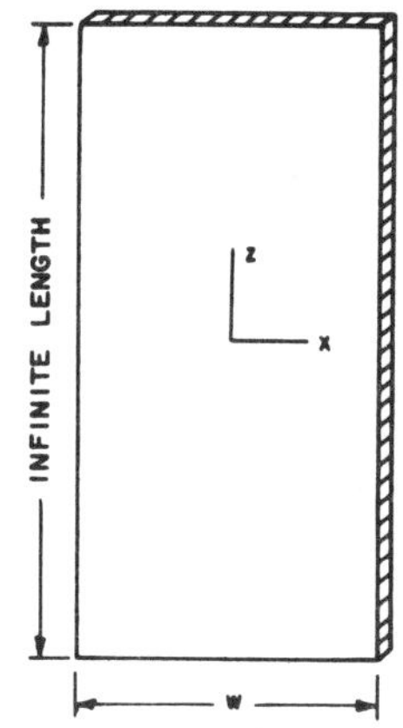

Fig. 1. Dielectric strip and coordinate system.

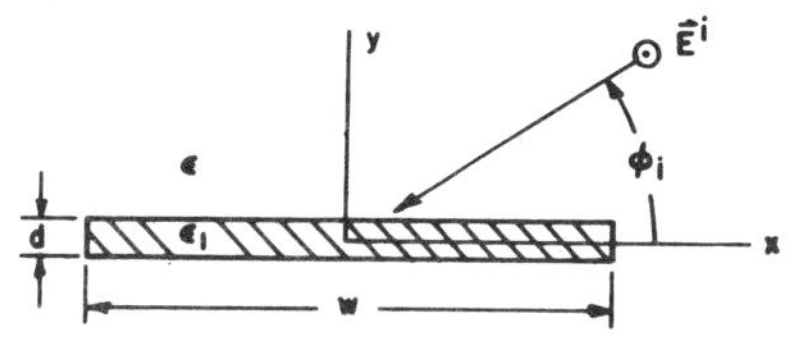

Fig. 2. Cross-sectional view of thin dielectric strip with incident plane wave.

The scattered field can be generated by the equivalent electric polarization-current density

$$J = j\omega(\epsilon_1 - \epsilon)E \tag{2}$$

radiating in the unbounded homogeneous ambient medium with parameters μ and ϵ. We consider the dielectric strip to be homogeneous, with permittivity ϵ_1 and permeability $\mu_1 = \mu$. (If μ_1 differed from μ, the scattered field would have another contribution from the magnetic polarization current.) We let ϵ_1 be complex to represent a lossy dielectric material. At each point in space, the scattered field is given by

$$E^s(x, y) = -(\omega\mu/4)\iint J(x', y')H_0(kR)\, dx'\, dy' \tag{3}$$

$$R^2 = (x' - x)^2 + (y' - y)^2 \tag{4}$$

$$k^2 = \omega^2\mu\epsilon. \tag{5}$$

In (3) the integration extends over the cross-sectional surface of the dielectric strip, and $H_0(kR)$ denotes the Hankel function of the second kind.

From (1), (2), and (3) we obtain the integral equation

$$E(x, y) + j(k^2/4)(\epsilon_r - 1)\iint E(x', y')H_0(kR)\, ds' = E^i(x, y) \tag{6}$$

where $\epsilon_r = \epsilon_1/\epsilon$ and $ds' = dx'dy'$. The integration extends over the cross-sectional surface of the dielectric strip.

In the moment method [5], the next step is to expand the unknown function $E(x, y)$ in a series of basis functions. We selected the following series for the field in the dielectric strip:

$$E(x, y) = \sum_{n=1}^{3} C_n \exp(f_n x) \cosh(g_n y) \tag{7}$$

$$f_n^2 + g_n^2 = -\omega^2 \mu \epsilon_1. \tag{8}$$

We choose the propagation constants f_n and g_n so that the wave $n = 1$ represents the forced wave, wave 2 is a surface wave propagating in the negative x direction, and wave 3 is a surface wave propagating in the positive x direction.

By "forced wave," we mean the field induced in an infinitely wide dielectric slab by an incident plane wave. Referring to Fig. 2, we see that the propagation constant of the forced wave is

$$f_1 = jk \cos \phi_i. \tag{9}$$

For the surface wave ($n = 2$), the propagation constant is determined from (8) and the transcendental equation

$$g_2 \tanh(g_2 t) = -\sqrt{-k^2 - f_2^2} \tag{10}$$

$$t = d/2. \tag{11}$$

The theory of the surface wave is presented in the Appendix. Since mode 3 is a surface wave propagating in the opposite direction, $f_3 = -f_2$.

For the surface wave on a thin dielectric slab, it turns out that E_z is an even function of y. For the forced mode, E_z has even and odd components but the even component dominates if the slab is thin and $\mu_1 = \mu$. For these reasons, we represent the field in the strip as an even function of y in (7).

The three modes in (7) form an incomplete set, and we may hope for success only if the true field in the dielectric strip is dominated by these three modes. If the dielectric strip were thick enough to support higher-order surface waves, one would have to include them in (7).

It may be helpful to compare our basis functions for the dielectric strip with the GTD modes on a perfectly conducting strip. The forced mode on the dielectric strip is analogous to the physical-optics current wave on the conducting strip. Furthermore, the surface waves on the dielectric strip are analogous to the creeping waves on the conducting strip. Thus, in our approach to the dielectric strip, we are using the moment method in combination with GTD basis functions.

From (6) and (7), we obtain the following equation which holds at each point within the dielectric strip:

$$\sum_{n=1}^{3} C_n \left[\exp(f_n x) \cosh(g_n y) + j(k^2/4)(\epsilon_r - 1) \iint \exp(f_n x') \cosh(g_n y') H_0(kR)\, ds' \right] = E^i(x, y). \tag{12}$$

In the moment method, the next step is to choose a set of testing or weighting functions $W_m(x, y)$. We selected the functions

$$W_m(x, y) = \exp(-f_m x) \cosh(g_m y), \qquad m = 1, 2, 3. \tag{13}$$

Except for a sign reversal in the exponent, these weighting functions are the same as the basis functions in (7). Each basis function represents a wave traveling across the dielectric strip, and the corresponding weighting function represents a wave traveling in the opposite direction across the strip. In Galerkin's method the weighting functions are the same as the basis functions or the complex conjugate of the basis functions. With the goal of obtaining an impedance matrix that is symmetric and diagonally dominant, we deviated slightly from Galerkin's method.

Multiplying both sides of (12) by $W_m(x, y)$ and integrating over the cross section of the strip, we obtain the following system of simultaneous linear equations:

$$\sum_{n=1}^{3} C_n Z_{mn} = V_m \qquad m = 1, 2, 3 \tag{14}$$

$$V_m = \iint \exp(-f_m x) \cosh(g_m y) E^i(x, y)\, ds \tag{15}$$

$$Z_{mn} = S_{mn} + T_{mn} \tag{16}$$

$$S_{mn} = j(k^2/4)(\epsilon_r - 1) \cdot \iiiint \cosh(g_m y) \cosh(g_n y') \cdot \exp(f_n x' - f_m x) H_0(kR)\, d_s'\, ds \tag{17}$$

$$T_{mn} = \iint \cosh(g_m y) \cosh(g_n y) \exp(f_n x - f_m x)\, ds. \tag{18}$$

It can be shown that the impedance matrix is symmetric, so that $Z_{mn} = Z_{nm}$. The incident plane-wave field is given by

$$E^i = E_0 \exp(jkx \cos \phi_i) \exp(jky \sin \phi_i). \tag{19}$$

In (15), the limits of integration are $: = \pm h$ and $y = \pm t$ where $h = w/2$. From (15) and (19), the excitation voltages are given by

$$V_m = E_0 X_m Y_m \tag{20}$$

$$X_m = w \operatorname{Sinh}[(f_1 - f_m)h] \tag{21}$$

$$Y_m = t[\operatorname{Sinh}[(g_0 + g_m)t] + \operatorname{Sinh}[(g_0 - g_m)t]] \tag{22}$$

$$h = w/2 \tag{23}$$

$$\operatorname{Sinh}(z) = \frac{\sinh(z)}{z} \tag{24}$$

where $g_0 = jk \sin \phi_i$ and $X_1 = w$.

The integrals in (18) are easy to evaluate, with the following result:

$$T_{mn} = X_{mn} Y_{mn} \tag{25}$$

$$X_{mn} = w \operatorname{Sinh}[(f_n - f_m)h] \tag{26}$$

$$Y_{mn} = t[\operatorname{Sinh}[(g_n + g_m)t] + \operatorname{Sinh}[(g_n - g_m)t]]. \tag{27}$$

In evaluating S_{mn} we use a change of variables (such as $v = x' - x$ and $s = y' - y$) to reduce the number of integrations required. The range of integration is split into three regions. When kR is small, we employ the small-argument form for the Hankel function. When kR is large, we use the large-argument form to express S_{mn} in terms of complex error functions. In the intermediate region, we employ numerical integration techniques.

The constants C_n are determined from (14), and the scattered field (from (2), (3), and (7)) is calculated as follows:

$$E^s(x, y) = -j(k^2/4)(\epsilon_r - 1) \sum_{n=1}^{3} C_n \iint \exp(f_n x') \cdot \cosh(g_n y') H_0(kR)\, ds'. \tag{28}$$

The far-zone scattered field is obtained from (28) and the large-argument form for the Hankel function [6], as follows:

$$E^s(\rho, \phi_s) = -j(k^2/4)(\epsilon_r - 1)\sqrt{\frac{2j}{\pi k \rho}}\, e^{-jk\rho} \sum_{n=1}^{3} C_n F_n(\phi_s) \tag{29}$$

$$F_n(\phi) = \iint \exp(f_n x + jkx \cos \phi) \exp(jky \sin \phi) \cdot \cosh(g_n y)\, dx\, dy \tag{30}$$

$$F_n(\phi) = X_n Y_n \tag{31}$$

$$X_n = w \operatorname{Sinh}[(f_n + f_0)h] \tag{32}$$

$$Y_n = t[\operatorname{Sinh}[(g_n + g_0)t] + \operatorname{Sinh}[(g_n - g_0)t]] \tag{33}$$

$$f_0 = jk \cos \phi \tag{34}$$

$$g_0 = jk \sin \phi. \tag{35}$$

III. NUMERICAL RESULTS

In this section we present some numerical results for a thin dielectric strip with thickness $d = 0.025\ \lambda_0$. The ambient medium is free space, and λ_0 denotes the wavelength in free space. The permittivity of the dielectric strip is $\epsilon_1 = 4\epsilon_0(1 - j \tan \delta)$. Data are included for lossless strips ($\tan \delta = 0$) and lossy strips with $\tan \delta = 0.1$. For a surface wave propagating on this strip, the electric field intensity in the dielectric region is $E = \exp(\pm fx) \cdot \cosh(gy)$. Letting $f = \alpha + j\beta$, we used the theory in the Appendix to determine that $\alpha = 0$ for the lossless strip and $\alpha = 0.0438$ Np/m for the lossy strip ($\tan \delta = 0.1$) at 300 MHz. Let c denote the velocity of light in free space, and let v denote the phase velocity of the surface wave. We found that $v/c = 0.9732$ and 0.9737 for the lossless and lossy strips, respectively. Now β and v are related as follows: $\beta = k_0 c/v$.

Fig. 3 illustrates the electric field distribution $E_z(x)$ induced in a thin dielectric strip by a plane wave with grazing incidence ($\phi_i = 0$). The incident field strength is 1 V/m. As shown in Fig. 2, x is measured from the center of the strip. As a check, we calculated the field distribution by two methods. In the old method (pulse bases), the dielectric strip was divided into 200 square cells and we generated a system of 200 simultaneous linear equations. In the new method (plane-wave expansion), the field in the dielectric strip was represented as the sum of three plane waves, and we generated a system of three simultaneous linear equations. It is evident in Fig. 3 that the two methods show good agreement. The amplitudes of the three plane waves are tabulated in Fig. 3.

Fig. 4 illustrates the backscatter pattern for the same dielectric strip that we considered in Fig. 3.

Fig. 5 illustrates forward scattering versus the width of the strip with grazing incidence. As the width increases, the pulse-basis technique requires ever increasing computational time and storage. When the width is equal to 25 wavelengths, there are 1000 unknowns and the central processing unit (CPU) time is 120 s on a VAX 11/780 computer. For this reason, the pulse-bases calculations were terminated at this point. Although the results are not shown in Fig. 5 for narrow strips, the calculations (with the plane-wave expansion and with the pulse bases) show excellent agreement for all widths from $w = 0$ to $w = 25\ \lambda_0$.

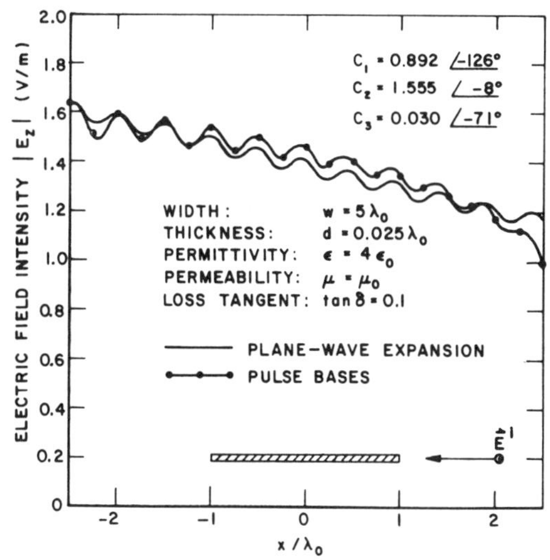

Fig. 3. Electric field distribution induced in thin dielectric strip by plane wave with grazing incidence.

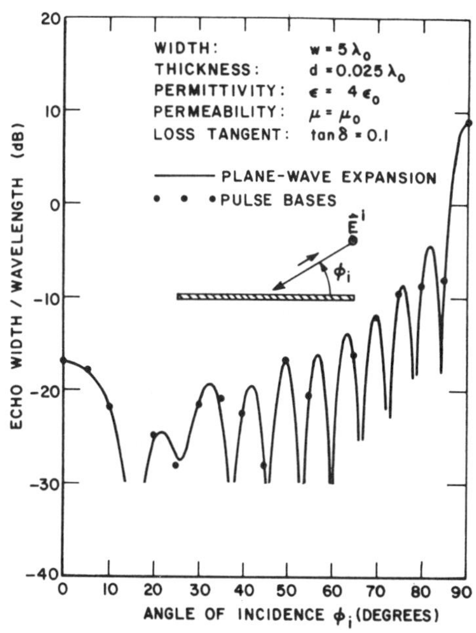

Fig. 4. Backscatter pattern for thin dielectric strip.

Fig. 6 illustrates backscatter versus width for dielectric strips with grazing incidence, as calculated with pulse bases. In Figs. 7 and 8, it may be noted that the new plane-wave expansion shows excellent agreement with pulse bases for lossless and lossy strips with grazing incidence.

All calculations were carried out on a VAX 11/780 computer. With the plane-wave expansion the computational time is approximately 0.3 s to solve for the backscatter or bistatic scatter from a lossless or lossy thin dielectric strip. For comparison, Fig. 9 illustrates the computational time with pulse bases when we take advantage of the Toeplitz nature of the impedance matrix.

The limitations of the new technique have not yet been established, but the method should not be relied on if the thickness d of the strip exceeds one skin depth or one tenth of the wavelength in the dielectric medium.

IV. SUMMARY AND CONCLUSION

In this report we consider a plane wave to be incident on a thin dielectric strip with infinite length, and the incident electric

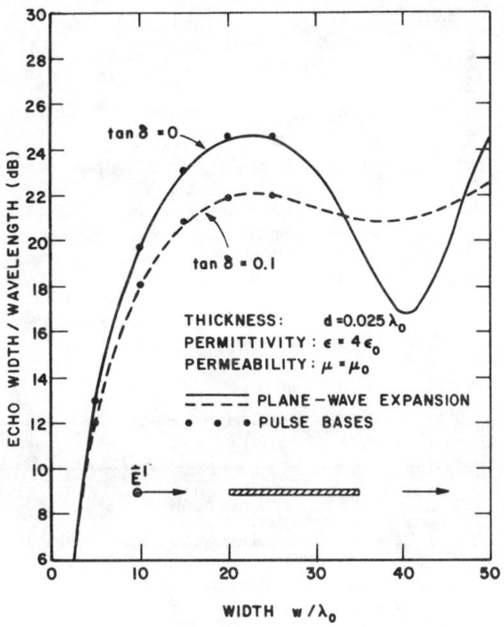

Fig. 5. Forward scatter versus width for thin dielectric strip with grazing incidence.

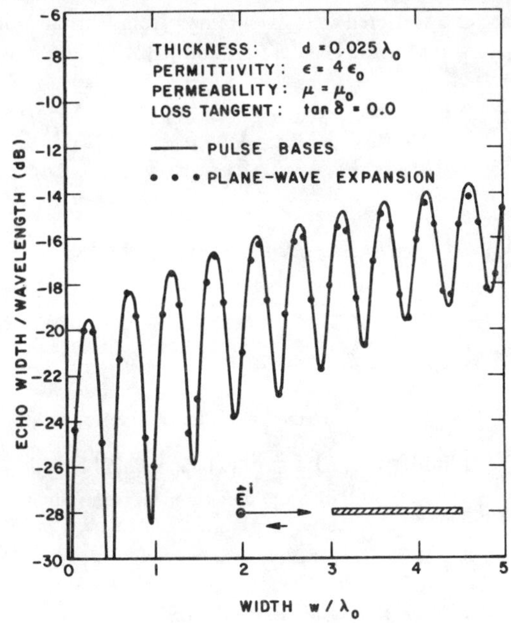

Fig. 7. Backscatter versus width for thin dielectric strip with grazing incidence.

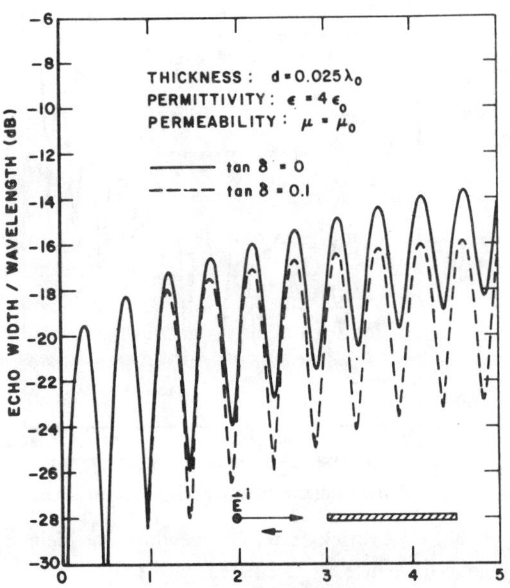

Fig. 6. Backscatter versus width for thin dielectric strip with grazing incidence, calculated with pulse basis.

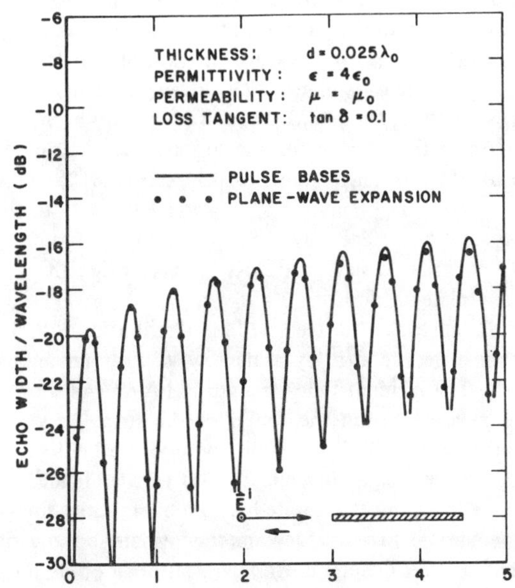

Fig. 8. Backscatter versus width for thin dielectric strip with grazing incidence.

field vector is parallel with the edges of the strip. The field in the dielectric region is expanded as the sum of three plane waves (the forced wave and two surface waves). The propagation constants are known for each wave, and Galerkin's method is employed to determine the amplitudes of these waves. Finally, the far-zone scattered field is determined by considering the polarization currents radiating in free space.

Numerical data are presented to illustrate the scattering properties of lossless and lossy dielectric strips as a function of the angle of incidence and the width of the strip. The calculations show excellent agreement with an earlier moment method using pulse bases and point matching.

APPENDIX
SURFACE WAVES

Let us consider a z-polarized surface wave propagating in the x direction on the lossy dielectric slab shown in Fig. 2. Let E^{I} denote the electric field intensity in region I (the dielectric slab) where the medium has parameters μ_1 and ϵ_1. Similarly, E^{II} denotes the electric field intensity in region II (outside the slab) where the ambient medium has parameters μ_2 and ϵ_2. For the mode that can propagate even on a very thin slab, the electric field intensity is given by

$$E_z^{\mathrm{I}} \doteq B \exp(-fx) \cosh(g_1 y) \tag{36}$$

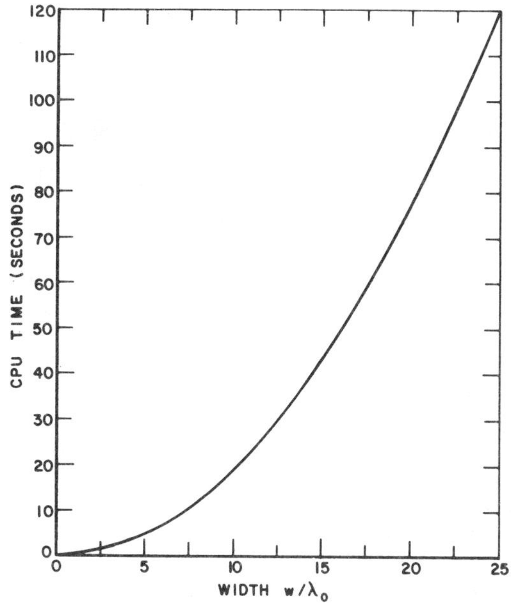

Fig. 9. Computer time versus width for thin dielectric strip with pulse bases.

$$E_z^{\mathrm{II}} = A \exp(-fx) \exp(-g_2 y). \tag{37}$$

The magnetic field intensity has x and y components, and the x component is given by

$$H_x^{\mathrm{I}} = -g_1 B \exp(-fx) \sinh(g_1 y)/(j\omega\mu_1) \tag{38}$$

$$H_x^{\mathrm{II}} = g_2 A \exp(-fx) \exp(-g_2 y)/(j\omega\mu_2). \tag{39}$$

Enforcing continuity of E_z and H_x across the boundary at $y = t$, we find that

$$A \exp(-g_2 t) = B \cosh(g_1 t) \tag{40}$$

$$\mu_1 g_2 A \exp(-g_2 t) = -\mu_2 g_1 B \sinh(g_1 t). \tag{41}$$

From (40) and (41),

$$P = g_1 \tanh(g_1 t) + \mu_r g_2 = 0 \tag{42}$$

where $\mu_r = \mu_1/\mu_2$. From the wave equation,

$$g_1^2 = -\omega^2 \mu_1 \epsilon_1 - f^2 \tag{43}$$

$$g_2^2 = -\omega^2 \mu_2 \epsilon_2 - f^2. \tag{44}$$

We may use (43) and (44) to eliminate g_1 and g_2 in (42), so that P is a function of f. Letting P' denote the derivative of P with respect to f, we find that

$$P' = -[g_1 g_2 t \operatorname{sech}^2(g_1 t) + g_2 \tanh(g_1 t) + \mu_r g_1] f/(g_1 g_2). \tag{45}$$

For the surface wave, the complex x-axis propagation constant can be written as

$$f = \alpha + j\beta \tag{46}$$

where α and β are real. Let c denote the velocity of light in free space, and let v denote the phase velocity of the surface wave. Then

$$\beta = \omega\sqrt{\mu_0 \epsilon_0}\, c/v. \tag{47}$$

The Newton-Raphson method is most convenient for seeking a root of P in (42), and thus determining the propagation constant f.

REFERENCES

[1] J. H. Richmond, "Scattering by a dielectric cylinder of arbitrary cross section shape," *IEEE Trans. Antennas Propagat.*, vol. AP-13, pp. 334–341, May 1965.

[2] W. D. Burnside and K. W. Burgener, "High frequency scattering by a thin lossless dielectric slab," *IEEE Trans. Antennas Propagat.*, vol. AP-31, pp. 104–110, Jan. 1983.

[3] R. Kastner and R. Mittra, "A spectral-iteration technique for analyzing scattering from arbitrary bodies, part I: Cylindrical scatterers with E-wave incidence," *IEEE Trans. Antennas Propagat.*, vol. AP-31, pp. 499–506, May 1983.

[4] P. H. Pathak and R. Rojas-Teran, private communication, July 1983.

[5] R. F. Harrington, *Field Computation by Moment Methods.* New York: Macmillan, 1968.

[6] M. Abramowitz and I. A. Stegun, *Handbook of Mathematical Functions.* Washington, D.C.: National Bureau of Standards, 1964.

 1977 IMPI, Edmonton, Canada. Reprinted with permission.

Induced EM Fields Inside Human Bodies Irradiated by EM Waves of up to 500 MHz*

Kun-Mu Chen and B. S. Guru†

ABSTRACT

In this paper, the internal EM field and the specific absorption rate of EM energy induced inside human bodies by EM waves of up to 500 MHz are theoretically quantified based on a tensor integral equation method. Numerical results for a realistic model of a man of 177 cm high irradiated by EM waves of various frequencies and of vertical and horizontal polarizations are presented. The resonance phenomenon and the effect of body heterogeneity on the induced field are studied. Some theoretical results are compared with existing experimental results.

INTRODUCTION

In view of possible health hazards due to nonionizing electromagnetic radiation and in medical applications utilizing EM radiation, it is important and desirable to qualify the internal EM field and the absorbed power density induced inside a human body by an EM field. Since a human body is an irregularly shaped heterogeneous conducting body with frequency-dependent permittivity and conductivity, the theoretical quantification of the induced EM field inside an irradiated human body becomes a difficult problem.

The existing methods commonly used to predict the induced EM field inside a biological body are based on simplified models of a plane slab [1, 2], the sphere [3, 4, 5], the cylinder [6], and spheroids [7, 8]. Although these simple models provide estimates of the internal EM fields, the results have limited applicability to biological bodies with irregular shapes illuminated by an EM wave with a frequency higher than the VHF range.

This paper presents some numerical results on the internal EM fields and the specific absorption rate (SAR) of EM energy induced inside a human body with a realistic model by EM waves of 30 to 500 MHz and of both vertical and horizontal polarizations. The resonance phenomenon at discrete frequencies is discussed. The effect of body heterogeneity on the induced EM field is also studied. The results in this paper are considered to be more accurate than that reported in a previous paper [9] by the present authors based on a simpler, homegeneous model. Although a part of the numerical results has been presented in a meeting [10], a more complete set of results is reported in this paper with a comparison of experimental results.

Numerical results are obtained based on the tensor integral equation method [11]. The accuracy of this method has been verified theoretically by convergence

---

* Manuscript received March 17, 1977; in revised form July 6, 1977. This research was supported by Army Research Office under Grant DAAG 29-76-G-0201 and in part by NSF under Grant ENG74-12603.

† Department of Electrical Engineering and Systems Science, Michigan State University, E. Lansing, Michigan 48824 U.S.A.

tests [11] and experimentally by a study conducted in scaled models containing saline solution [12].

The tensor integral equation method is briefly outlined and the numerical technique is discussed in Section 2. Numerical results on the internal EM field and the SAR inside a typical adult body of 177 cm high induced by EM waves of various frequencies with different polarizations are presented in Section 3. In the same section, the resonance phenomenon and the effect of body heterogeneity are also discussed. Some theoretical results are compared with existing experimental results in Section 4.

THEORY AND NUMERICAL TECHNIQUE

If a finite biological body of arbitrary shape, with permittivity $\epsilon(\vec{r})$, conductivity $\sigma(\vec{r})$ and permeability μ_o, is illuminated in free space by an incident electromagnetic wave with an electric field $\vec{E}^i(\vec{r})$, the total induced electric field $\vec{E}(\vec{r})$ inside the body can be determined from the following tensor integral equation:

$$[1 + \frac{\tau(\vec{r})}{3j\omega\epsilon_0}]\vec{E}(\vec{r}) - \text{P.V.}\int_V \tau(\vec{r}\,')\vec{E}(\vec{r}\,')\cdot\overleftrightarrow{G}(\vec{r},\vec{r}\,')dV' = \vec{E}^i(r) \qquad (1)$$

where $\tau(\vec{r}) = \sigma(\vec{r}) + j\omega(\epsilon(\vec{r}) - \epsilon_o)$, ϵ_o is the free-space permittivity, the P.V. symbol means the principal value of the integral, $\overleftrightarrow{G}(\vec{r},\vec{r}\,')$ is the free-space tensor Green's function, and V is the body volume.

If the body is partitioned into N subvolumes or cells, and $\vec{E}(\vec{r})$ and $\tau(\vec{r})$ are assumed to be constant within each cell, equation (1) can be transformed into 3N simultaneous equations for E_x, E_y and E_z at the centers of N cells by the point matching method. These simultaneous equations can be written into matrix form as

$$\begin{matrix} [G_{xx}] & [G_{xy}] & [G_{xz}] \\ [G_{yx}] & [G_{yy}] & [G_{yz}] \\ [G_{zx}] & [G_{zy}] & [G_{zz}] \end{matrix} \quad \begin{matrix} [E_x] \\ [E_y] \\ [E_z] \end{matrix} = - \begin{matrix} [E_x^i] \\ [E_y^i] \\ [E_z^i] \end{matrix} \qquad (2)$$

the [G] matrix is a 3N × 3N matrix, while [E] and $[E^i]$ are 3N column matrices expressing the total electric field and the incident electric field at the centers of N cells. The elements of [G] matrix have been evaluated [11]. Therefore, with the known incident electric field $\vec{E}^i(\vec{r})$, the total induced electric field $\vec{E}(\vec{r})$ inside the body can be obtained from equation (2) by inverting the [G] matrix.

In the numerical analysis such as the present tensor integral equation method, it is important to establish an upper limit on the dimensions of the cells for the required accuracy. In Ref. [11], two convergence tests have been performed in a muscle cylinder and a muscle cube for this purpose. It was found from these tests that reliable numerical results can be obtained if the cell edge does not exceed 1/4 free-space wavelength ($\lambda/4$).

Other factors which dictate the selection of the cell size are the computer storage capacity and the size of the body. Our numerical computation was performed with a CDC 6500 computer which has the maximum capacity of investing a 162 × 162 matrix. This limits the maximum number of cells to 54. When the symmetry between the right-half and the left-half of the body is assumed, the maximum realistic shape can be constructed with 108 cubic cells of various sizes ranging from 5-centimeter cubes to 12-centimeter cubes as shown in Figure 1.

Since the largest cells used in the computation are 12-centimeter cubes, the highest frequency we can consider is 625 MHz without violating the criterion of the cell edge not exceeding $\lambda/4$. However, to provide a safe margin of accuracy, we consider a frequency range with its upper bound set at 500 MHz.

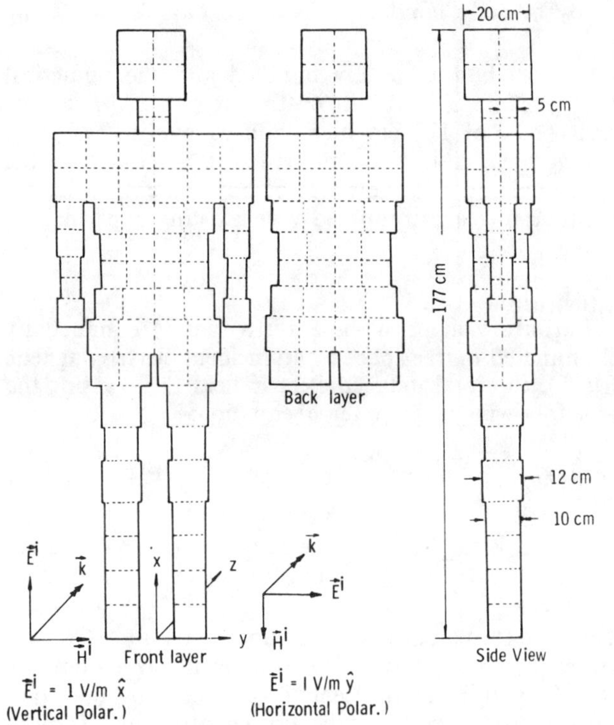

Figure 1 A model of a man 177 cm high used in the theoretical field quantification. The model is constructed with 108 cubic cells of various sizes ranging from 5 cm^3 to 12 cm^3. The model is irradiated either by a vertically or a horizontally polarized EM wave.

NUMERICAL RESULTS ON THE INDUCED ELECTRIC FIELD AND THE SPECIFIC ABSORPTION RATE

The model of a human body depicted in Figure 1 is used in the theoretical quantification. A vertically or horizontally polarized plane EM wave is assumed to be incident upon the body from front to back or from side-to-side. In the numerical calculation, the x-, y- and z-components of the induced electric field, E_x, E_y and E_z are computed at the centers of cells first, and the corresponding SAR is then determined based on the relation of $(\sigma/2)(E_x^2 + E_y^2 + E_z^2)$.

Figure 2 shows the x-component of the internal electric field induced by a vertically polarized EM wave of 80 MHz with an incident electric field of 1 V/m (max. value) at normal incidence (from front to back). The electrical properties of the body are assumed to be that of a high-water content tissue with $\epsilon = 80\epsilon_o$ and $\sigma = 0.84$ S/m at this frequency. Due to symmetry, the magnitudes of E_x are shown only in one half of the body. It is noted that the strongest induced electric field is about 0.441 times the incident electric field and is located in the thigh. Another local maximum of E_x is induced in the neck.

Figures 3 and 4 shows the y- and z-components of the induced electric field, respectively. It is noted that E_y and E_z are much smaller than E_x at this frequency.

Figure 5 shows the corresponding SAR. It is observed that "hot spots" are created in the thigh-leg region and another local hot spot is induced in the neck.

The case of 80 MHz was treated first because this is the major resonance frequency for a body 177 cm high irradiated by a vertically polarized EM wave as will be shown in a later figure. We have also computed the induced electric field and the SAR in the same body induced by various frequencies of both vertical and horizontal polarizations. Since it is not possible to present all the results, only the

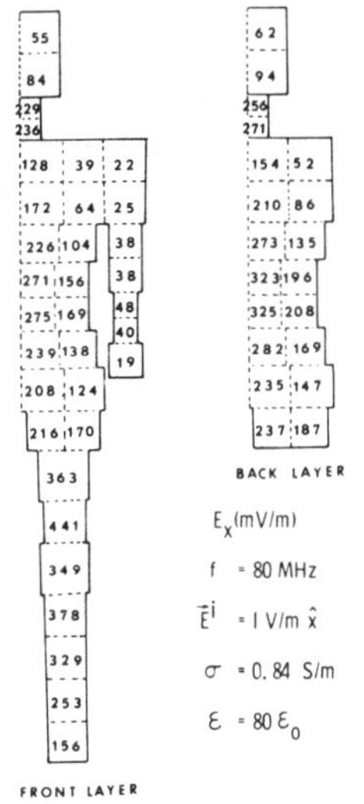

Figure 2 The x-component of induced electric field. Incident EM wave: vertical polarization, 80 MHz, $\vec{E}^i = 1$ V/m $\hat{x}$. Body parameters: $\sigma = 0.84$ S/m, $\epsilon = 80\ \epsilon_o$.

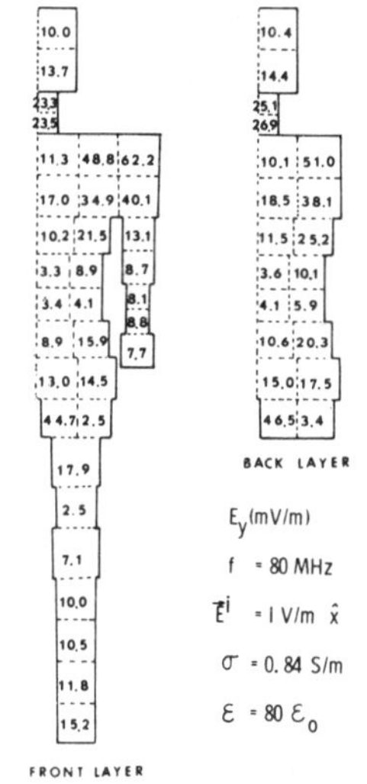

Figure 3 The y-component of induced electric field. Incident EM wave: vertical polarization, 80 MHz, $\vec{E}^i =$ 1 V/m $\hat{x}$. Body parameters: $\sigma = 0.84$ S/m, $\epsilon = 80\ \epsilon_o$.

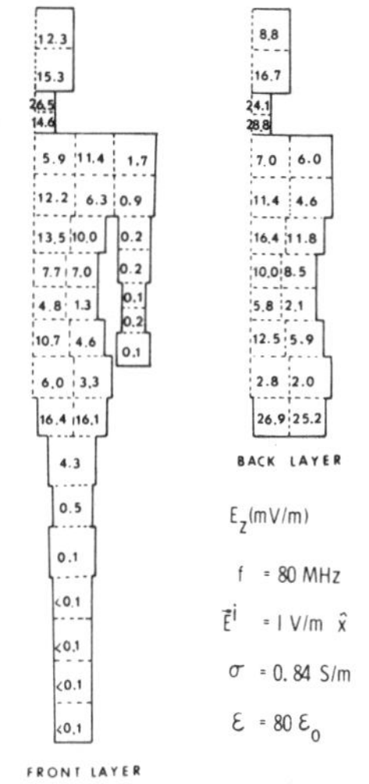

Figure 4 The z-component of induced electric field. Incident EM wave: vertical polarization, 80 MHz, $\vec{E}^i = 1$ V/m $\hat{x}$. Body parameters: $\sigma = 0.84$ S/m, $\epsilon = 80\ \epsilon_o$.

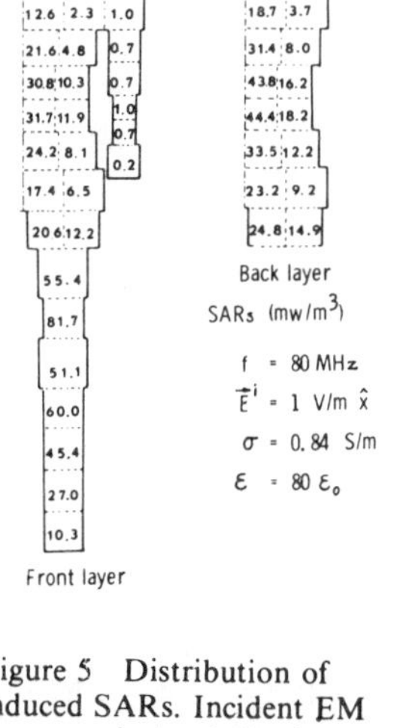

Figure 5 Distribution of induced SARs. Incident EM wave: vertical polarization, 80 MHz, $\vec{E}^i = 1$ V/m $\hat{x}$. Body parameters: $\sigma = 0.84$ S/m, $\epsilon = 80\ \epsilon_o$.

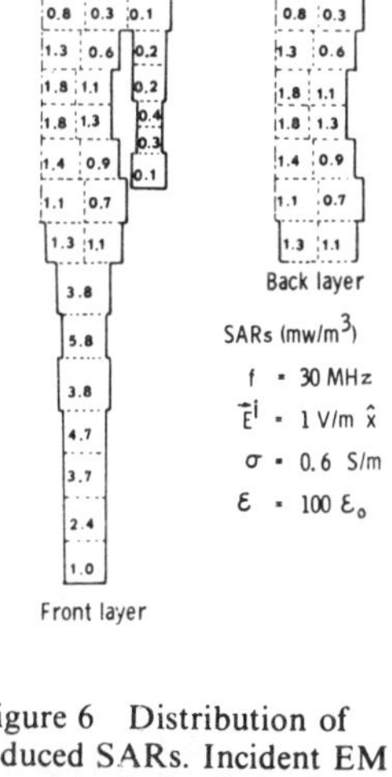

Figure 6 Distribution of induced SARs. Incident EM wave: vertical polarization, 30 MHz, $\vec{E}^i = 1$ V/m $\hat{x}$. Body parameters: $\sigma = 0.6$ S/m, $\epsilon = 100\ \epsilon_o$.

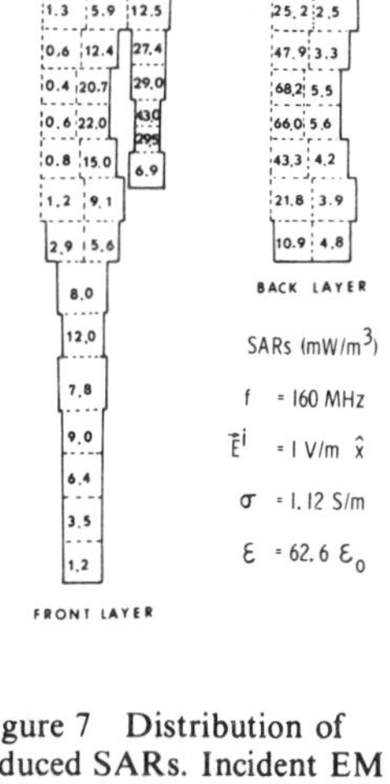

Figure 7 Distribution of induced SARs. Incident EM wave: vertical polarization, 160 MHz, $\vec{E}^i = 1$ V/m $\hat{x}$. Body parameters: $\sigma = 1.12$ S/m, $\epsilon = 62.6\ \epsilon_o$.

SARs induced by vertically polarized EM waves of 30, 80, 160, 200, 300, 400 and 500 MHz and horizontally polarized EM waves of 100, 200, 300 and 500 MHz are presented here.

Figure 6 shows the SARs in the same body induced by a vertically polarized EM wave of 30 MHz with an incident electric field of 1 V/m. The electrical properties of the body are assumed to be $\epsilon = 100\ \epsilon_o$ and $\sigma = 0.6$ S/m at this frequency. In this case, the hot spots are induced in the thigh-leg region and in the neck. Figure 7 shows the SARs induced by a vertically polarized EM wave of 160 MHz with an incident electric field of 1 V/m. The electrical properties of the body at this frequency are shown in the figure. At 160 MHz, the hot spots are created in the chest-stomach region in the back layer of the torso. The distribution of the SARs induced by a vertically polarized EM wave of 200 MHz is shown in Figure 8. At this frequency, the hot spots are created in the arms and in the neck. Figure 9 shows the SARs induced by a vertically polarized EM wave of 300 MHz. The hot spots are observed in the arms and in the neck at this frequency. Figure 10 shows the SARs induced by a vertically polarized EM wave of 400 MHz. At this frequency, a hot spot is created in the neck. The case of 500 MHz is shown in Figure 11, where a strong hot spot is induced in the neck.

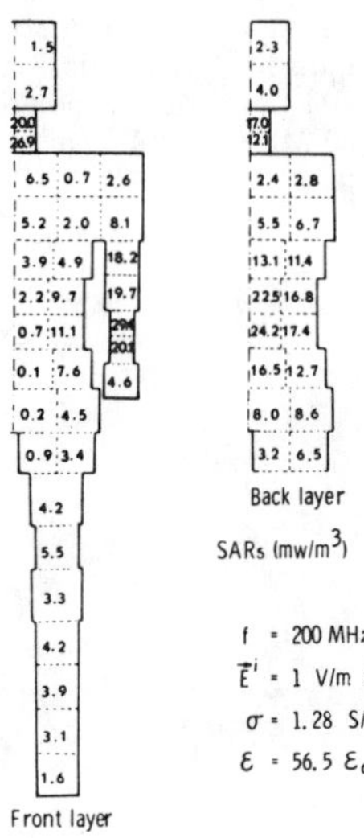

Figure 8 Distribution of induced SARs. Incident EM wave: vertical polarization, 200 MHz, $\vec{E}^i = 1$ V/m $\hat{x}$. Body parameters: $\sigma = 1.28$ S/m, $\epsilon = 56.5\ \epsilon_o$.

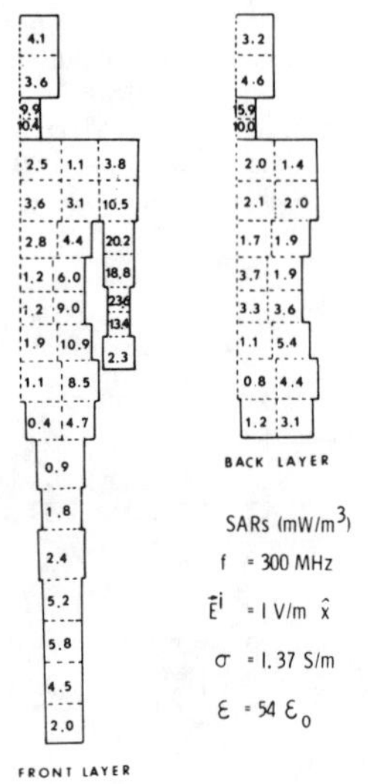

Figure 9 Distribution of induced SARs. Incident EM wave: vertical polarization, 300 MHz, $\vec{E}^i = 1$ V/m $\hat{x}$. Body parameters: $\sigma = 1.37$ S/m, $\epsilon = 54\ \epsilon_o$.

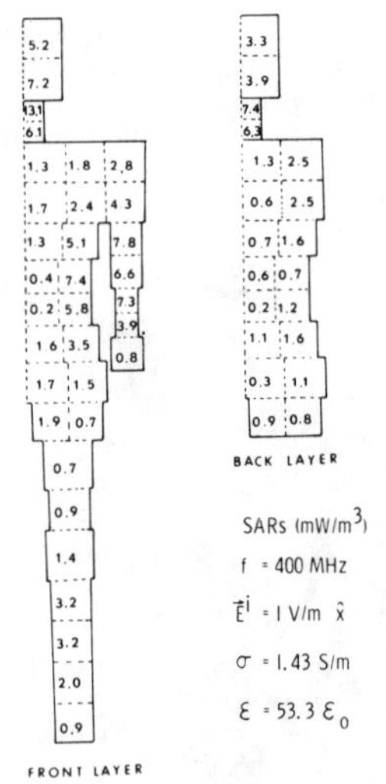

Figure 10 Distribution of induced SARs. Incident EM wave: vertical polarization, 400 MHz, $\vec{E}^i = 1$ V/m $\hat{x}$. Body parameters: $\sigma = 1.43$ S/m, $\epsilon = 53.3\ \epsilon_o$.

The case of the horizontal polarization are discussed in Figures 12 to 15. Figure 12 shows the distribution of the SARs in the same body induced by a horizontally polarized EM wave of 100 MHz with an incident electric field of 1 V/m. At this frequency and for this polarization, the hot spots are created in the mid-section of the body. Figure 13 shows the SARs induced by a horizontally polarized EM wave of 200 MHz. At this frequency, the hot spots are induced in the chest region in the back layer of the body. It is noted that for horizontally polarized incident EM waves, this body resonates near 200 MHz. The distribution of the SARs induced by

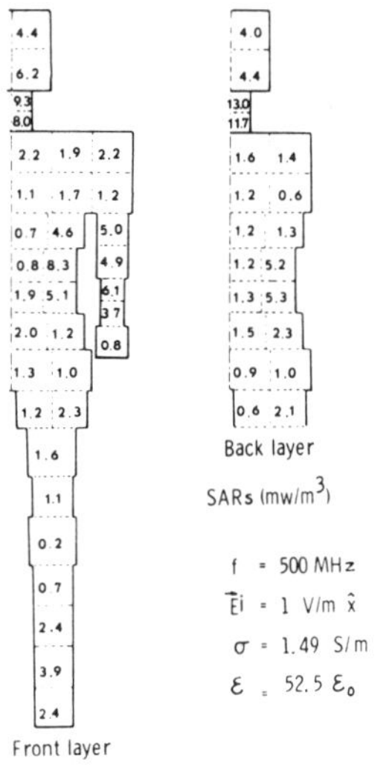

Figure 11 Distribution of induced SARs. Incident EM wave: vertical polarization, 500 MHz, $\vec{E}^i = 1$ V/m $\hat{x}$. Body parameters: $\sigma = 1.49$ S/m, $\epsilon = 52.5\ \epsilon_0$.

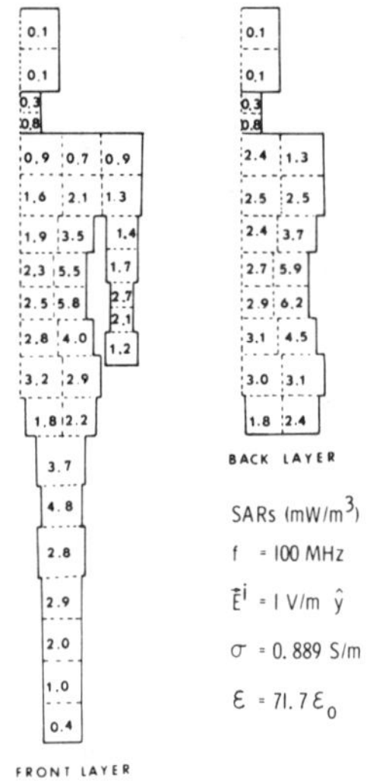

Figure 12 Distribution of induced SARs. Incident EM wave: horizontal polarization, 100 MHz, $\vec{E}^i = 1$ V/m $\hat{y}$. Body parameters: $\sigma = 0.889$ S/m, $\epsilon = 71.7\ \epsilon_0$.

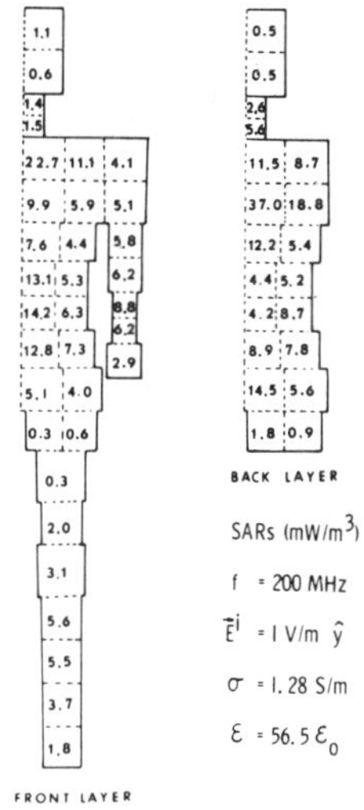

Figure 13 Distribution of induced SARs. Incident EM wave: horizontal polarization, 200 MHz, $\vec{E}^i = 1$ V/m $\hat{y}$. Body parameters: $\sigma = 1.28$ S/m, $\epsilon = 56.5\ \epsilon_0$.

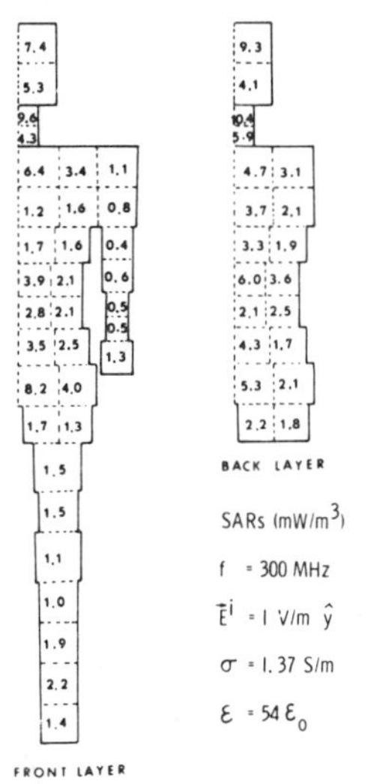

Figure 14 Distribution of induced SARs. Incident EM wave: horizontal polarization, 300 MHz, $\vec{E}^i = 1$ V/m $\hat{y}$. Body parameters: $\sigma = 1.37$ S/m, $\epsilon = 54\ \epsilon_0$.

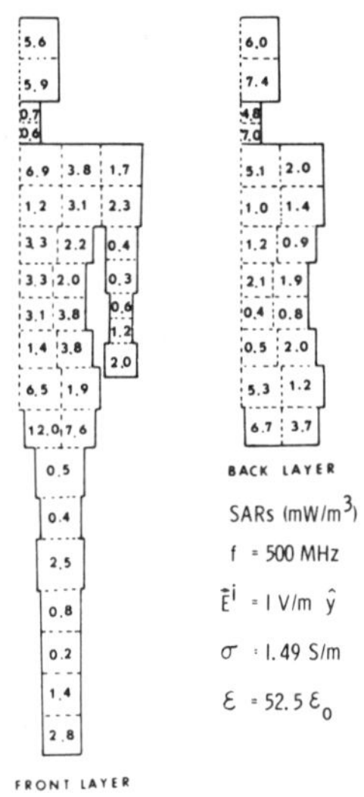

Figure 15 Distribution of induced SARs. Incident EM wave: horizontal polarization, 500 MHz, $\vec{E}^i = 1$ V/m $\hat{y}$. Body parameters: $\sigma = 1.49$ S/m, $\epsilon = 52.5\ \epsilon_0$.

a horizontally polarized EM wave of 300 MHz is shown in Figure 14, where the hot spot is induced in the neck. For the case of a 500 MHz horizontally polarized EM wave, the hot spot is observed in the groin region as shown in Figure 15.

From the numerical results of Figures 5-15, it is observed that the magnitude and the distribution of the SARs in a body of fixed dimensions are strongly dependent on the frequency and polarization of the incident EM wave. The hot spots are seen to move around as the frequency and polarization are varied. However, it is generally true that the hot spots are more often created in the regions with small cross-sectional areas such as the neck, the arms and the thigh-leg regions.

As indicated earlier, we have computed the induced electric field and the SARs inside the model of a man 177 cm high induced by EM waves of 1 to 500 MHz and of both vertical and horizontal polarizations. These results are summarized and expressed as functions of the frequency in Figure 16. The total absorbed power in the body is expressed in terms of the relative absorption area A_r. A_r is defined as: (total absorbed power/incident power density) divided by (total body surface area), because the heat dissipation capability of the body is essentially determined by the total surface area of the body. The maximum induced electric field at a certain location of the body is represented by its value normalized by the incident electric field such as $|E_{max}| / |E^i|$. In Figure 16, A_r and $|E_{max}| / |E^i|$ are shown as functions of the frequency for the cases of vertical and horizontal polarizations. For the vertical polarization, a major resonance is clearly observed at about 80 MHz and weaker resonances at 160 MHz and 240 MHz. This implies that the first major resonance occurs when the body height is about 0.472 times the free-space wavelength of the incident wave. For the horizontal polarization, only a weak resonance is observed near 200 MHz. At 200 MHz, the maximum horizontal dimension of the body is about 0.4 times the free-space wavelength.

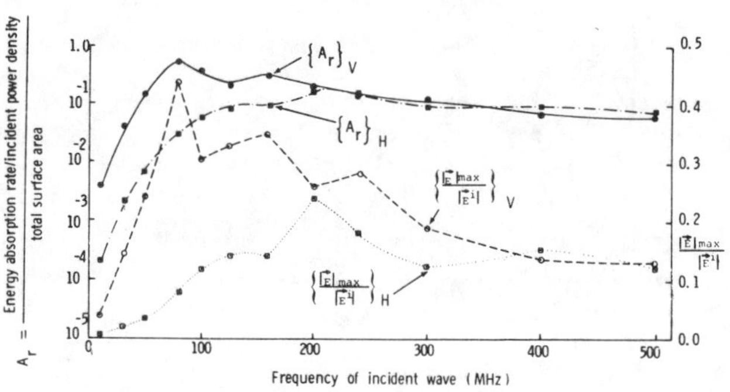

Figure 16 Relative absorption area and relative maximum induced electric field in a model of a man 177 cm high as functions of the frequency of vertically and horizontally polarized EM waves.

The effect of the body heterogeneity on the SAR induced inside the body is shown in Figure 17. In this figure, the distribution of the SARs induced by a vertically polarized EM wave of 80 MHz inside a heterogeneous body is shown. A region of low conductivity and permittivity with $\sigma = 0.045$ S/m and $\epsilon = 9.83\ \epsilon_o$, such as lungs indicated by a shaded area, is assumed to be imbedded in a body with $\sigma = 0.84$ S/m and $\epsilon = 80\ \epsilon_o$. The results indicate a greatly enhanced SAR in the low conductivity and permittivity region. The SAR was 31.4 mW/m^3 in the same region in a homogeneous body (see Figure 5), but its value is now increased to 117 mW/m^3 because the conductivity and permittivity of the region are reduced. This example indicates the fact that if the heterogeneity of the human body is considered, the induced EM field inside the body may be considerably more complicated than that inside a homogeneous body.

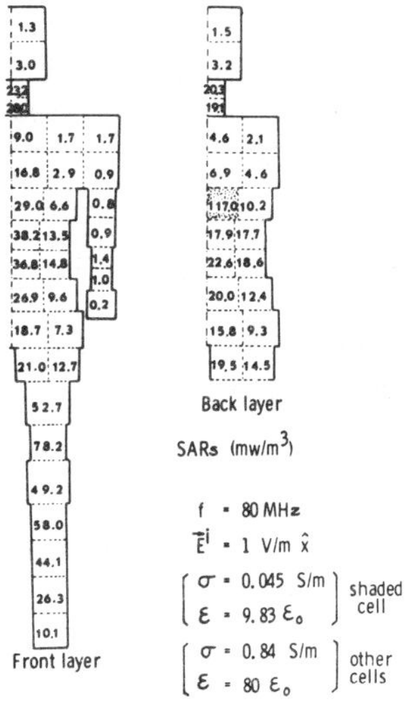

Figure 17 Distribution of induced SARs in a heterogeneous body with a region of low conductivity and permittivity (shaded region). Incident EM wave: vertical polarization, 80 MHz, $\vec{E}^i = 1$ V/m $\hat{x}$. Body parameters: $\sigma = 0.045$ S/m and $\epsilon = 9.83\ \epsilon_0$ for the region of low conductivity and permittivity, and $\sigma = 0.84$ S/m and $\epsilon = 80\ \epsilon_0$ for the rest of the body.

COMPARISON OF NUMERICAL RESULTS WITH EXPERIMENTAL RESULTS OF GANDHI ET AL.

To verify the accuracy of the theoretical results presented in the preceding section, some existing experimental results have been used to compare with the corresponding theoretical results.

Gandhi and his coworkers [14] have measured the distributions of power disposition in a phantom model of a human placed in front of a 90° corner reflector. Figure 18 shows the distribution of relative power disposition when a phantom model of 0.417 λ high was placed 1.5 λ away from the corner of the reflector, and irradiated by a vertically polarized EM wave of 71.5 MHz with a power density of 10 mW/cm^2 incident upon the model from the left to the right side of the figure. It is noted that in the presence of the corner reflector, the EM wave at the model was not a pure plane wave and its intensity was greatly enchanced. Thus, the actual

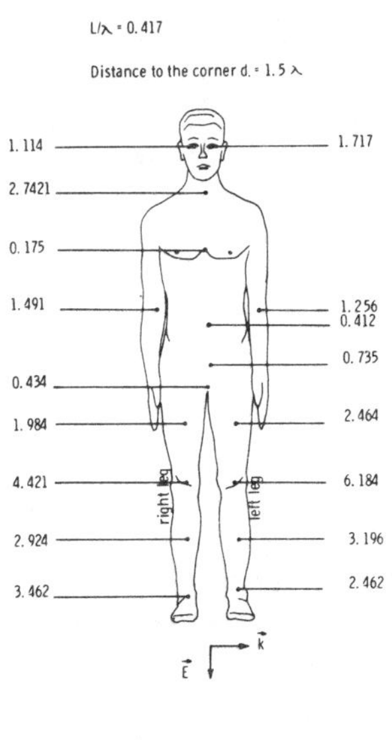

Figure 18 Distribution of relative SARs in a phantom model of a man placed 1.5λ in front of a 90° corner reflector, and irradiated by a 10 mW/cm^2 EM wave with the side-on incidence. (Measured by Gandhi, et al.)

incident power density at the model should be much higher than 10 mW/m^3. The enhancement factor for the energy coupling due to the reflector is estimated to be at least in the order of 10. We will compare our theoretical distribution of SARs in our theoretical model with the distribution of the relative SARs as given in Figure 18, measured by Gandhi et al. in their experimental model using the liquid crystal fiberoptic temperature probe.

To compare with these experimental results, we have computed the distribution of SARs in our theoretical model (177 cm high) induced by a 70 MHz, vertically polarized plane wave with a power density of 10 mW/cm^2, incident upon the model from the left to the right side of Figure 19. It is noted that this power density of 10 mW/cm^2 in free-space is much lower than that in the presence of a corner reflector and it is not intended to stimulate the actual power density at the experimental model. The height of the theoretical model is 0.41 λ. The electrical properties of the theoretical model are assumed to be $\epsilon = 60\ \epsilon_0$ and $\sigma = 0.62$ S/m at 70 MHz. This selection of parameters corresponds to the assumption of 65% muscle and 35% bone-fat as used by Gandhi and his coworkers in their experimental models. The theoretical results are shown in Figure 19.

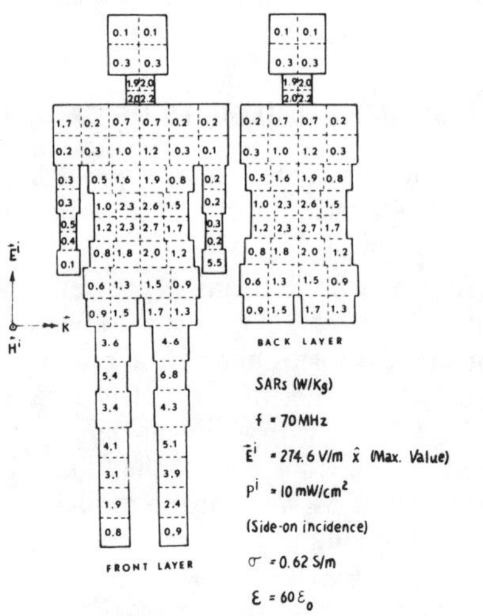

Figure 19 Distribution of induced SARs in the model of a man depicted in Figure 1. Incident EM wave: vertical polarization, side-on incidence, 70 MHz, $\vec{E}^i = 274.6$ V/m $\hat{x}$, P^i = 10 mW/cm^2. Body parameters: $\sigma = 0.62$ S/m, $\epsilon = 60\ \epsilon_0$.

Since the amplitudes of the incident power densities at the experimental and theoretical models are different, only experimental and theoretical distributions of relative SARs given in Figures 18 and 19 are compared. The comparison leads to the following observations.

(1) Both theoretically and experimentally, the "hot spots" are observed in the legs and the neck region.

(2) Both theory and experiment indicate that the absorbed power density is higher in the left leg (in the right side of the figure) than the right leg. This finding is not trivial because the EM wave is incident from the left side of the figure and, based on a simple intuition, the right leg should absorb more power. This phenomenon can be explained as follows: the total induced electric field in the body is the sum of the induced electric field due to the incident electric field and that due to the incident magnetic field. The former is linear in nature and of the same phase in both legs, while the latter is circulatory and of opposite phases in the two legs. When the body is near resonant to the incident EM wave as the

conditions of Figures 18 and 19, the induced electric fields of electric and magnetic modes are such that their sum in the left leg exceeds that in the right leg.

(3) Theoretical and experimental values of the SARs in the legs and the neck regions as given in Figures 18 and 19 are compared in Table 1. A good qualitative agreement is obtained between theory and experiment here.

TABLE 1

	Right Leg	Left Leg	Neck
Exp. Value	6.184 ~ 2.462	4.421 ~ 1.984	2.742
Theo. Value	6.82 ~ 2.36	5.38 ~ 1.89	2.2

(4) A major disagreement between theory and experiment is observed in the abdomen region where the theory predicts a higher SAR. This may be due to the fact that heat dissipation in this broader region of the model might be significant in the experiment, while the heat dissipation was neglected in the theory.

(5) The theoretical result in the head is lower than the experimental value because the theoretical head has a much bigger volume than the experimental head. The difference between theoretical and experimental values in the arms may be due to the difference in theoretical and experimental arm lengths and the fact that the latter has a direct contact with the body, while the former does not.

To summarize, the theory agrees quite well with the experiment over the regions of high power absorption. The agreement deteriorates over the regions of low absorbed power density. Since the regions of high power absorption are more important, the numerical results presented in this paper are considered useful. It is also noted that the agreement between Figures 18 and 19 is quite satisfactory in view of the fact that theoretical and experimental models were not identical, the heat dissipation was ignored in the theory, and the EM wave in the experiment was not a pure plane wave.

Figures 18 and 19 can only provide the comparison of the distributions of experimental and theoretical SARs, because the field intensity of the incident wave in the experiment was highly enhanced and its actual value was unknown because of the presence of the corner reflector. To check the absolute amplitude of the theoretical SAR, another set of experimental results obtained by Gandhi et al. [15] in saline-filled figurines will be used. They made projections for the average SAR in the human body at respective resonance frequencies for 10 mW/cm^2 incident EM waves. Due to limited theoretical results available, we will compare only two cases here. The first case is for the plane wave of 10 mW/cm^2 power density incident upon the body from arm-to-arm with its electric field parallel to the body height. The second case is for the same wave incidence but with its magnetic field parallel to the body height. Theoretical results were obtained, first by locating the resonance frequencies for the respective polarizations, and then computing the SAR with the assumption of a plane wave of resonance frequency incident side-on (or arm-to-arm) upon the body. Table 2 shows the comparison of experimental and theoretical projections of the resonance frequency and the average SAR at resonance in an adult body for two respective polarizations specified above. The difference between experimental and theoretical results is expected because the resonance frequency is quite sensitive to the body geometry and experimental and theoretical models are by no means identical.

TABLE 2

	E parallel to body height		H parallel to body height	
	Experiment (Gandhi et al)	Theory	Experiment (Gandhi et al)	Theory
Average SAR at resonance (W/kg)	2.18	1.88	0.37	0.302
Resonance freq. (MHz)	67.9	80	143 ~ 171	200

CONCLUSION

Extensive numerical computations were conducted to quantify the induced EM field and the SAR in a realistic model of a man irradiated by EM waves of up to 500 MHz and of both vertical and horizontal polarizations. Many other numerical results not reported in this paper are included in a report [13] which can be requested from the authors. The accuracy of the numerical results appears to be satisfactory as verified by existing experimental results. An experiment is planned to directly measure the induced electric field inside phantom models of human bodies.

REFERENCES

1 Schwan, H. P., "Radiation biology, medical applications, and radiation hazards," in *Microwave Power Engineering, 2*, E. C. Okress, Ed., New York: Academic Press, pp. 215-532, 1968.

2 Lehmann, J. F., Guy, A. W., Johnston, V. C., Brunner, G. D. and Bell, J. W., "Comparison to relative heating patterns produced in tissues by exposure to microwave energy at frequencies of 2450 and 900 megacycles," Arch. Phys. Med. Rehabil., *43*, pp. 69-76, February 1962.

3 Shapiro, A. R., Lutomirski, R. F. and Yura, H. T., "Induced fields and heating within a cranial structure irradiated by an electromagnetic plane wave," IEEE Trans. on Microwave Theory and Technique, *MTT-19*(2), pp. 187-196, February 1971.

4 Kritikos, H. N. and Schwan, H. P., "Hot spots generated in conducting spheres by electromagnetic waves and biological implications," IEEE Trans. on Biomedical Engineering, *BME-19*(1), pp. 53-58, January 1972.

5 Lin, J. C., Guy, A. W. and Johnson, C. C., "Power deposition in a spherical model of man exposed to 1-20 MHz electromagnetic field," IEEE Trans. on Microwave Theory and Techniques, *MTT-21*(12), pp. 791-797, December 1973.

6 Ho, H. S., Guy, A. W., Sigelmann, R. A. and Lehmann, J. F., "Electromagnetic heating patterns in circular cylindrical models of human tissue," in Proc. 8th Ann. Conf. Medical and Biological Engineering, pp. 27, (Chicago, Ill.) July 1969.

7 Durney, C. H., Johnson, C. C. and Massoudi, H., "Long-wavelength analysis of plane wave irradiation of a prolate spheroid model of man," IEEE Trans. on Microwave Theory and Techniques, *MTT-23*, pp. 246-253, February 1975.

8 Johnson, C. C., Durney, C. H. and Massoudi, H., "Long-wavelength electromagnetic power absorption in prolate spheroidal models of man and animals," IEEE Trans. on Microwave Theory and Techniques, *MTT-23*, pp. 739-747, September 1975.

9 Chen, Kun-Mu and Guru, B. S., "Induced EM field and absorbed power density inside a human torso," Proceedings of IEEE, *64*(9), pp. 1450-1453.

10 Chen, Kun-Mu and Guru, B. S., "Internal EM field induced by VHF and UHF EM waves in a human torso (realistic model)," Presented at 1976 Annual meeting of International Union of Radio Science, Oct. 11-15, Univ. of Mass. Amherst, Mass..

11 Livesay, D. E. and Chen, Kun-Mu, "Electromagnetic fields induced inside arbitrarily shaped biological bodies," IEEE Trans. on Microwave Theory and Techniques, *MTT-22*(12), pp. 1273-1280, Dec. 1974.

12 Guru, B. S. and Chen, Kun-Mu, "Experimental and theoretical studies on electromagnetic fields induced inside finite biological bodies," IEEE Trans. on Microwave Theory and Techniques, *MTT-24*(7), pp. 433-440, July 1976.

13 Chen, Kun-Mu and Guru, B. S., "Internal EM field and absorbed power density inside human torsos with realistic models induced by EM waves of up to 500 MHz," Technical Report, Division of Engineering Research, Michigan State University, E. Lansing, Michigan 1976.

14 Gandhi, O. P. Private Communication, 1976.

15 Gandhi, O. P., Sedigh, K., Beck, G. S. and Hunt, E. L., "Distribution of electromagnetic energy deposition in models of man with frequencies near resonance," Presented at 1975 Annual meeting of International Union of Radio Science, Oct. 20-23, Univ. of Colorado.

Numerical calculation of electromagnetic energy deposition in models of man with grounding and reflector effects

M. J. Hagmann and O. P. Gandhi

Departments of Electrical Engineering and Bioengineering, University of Utah, Salt Lake City, Utah 84112

(Received November 4, 1977.)

Image theory has been used to obtain moment-method solutions of the deposition of electromagnetic energy in standard man as a function of grounding and reflector effects. The calculated values are in good agreement with experimental data. For the $\mathbf{E} \| \hat{\mathbf{L}}$ orientation, the resonant frequency of standard man shifts from 77 MHz in free space to 47 MHz when standing on a ground plane. The dependence of reflector effects on spacing and frequency are in agreement with the gain enhancement calculated for dipoles by antenna theory.

1. INTRODUCTION

Experimental observations of grounding and reflector effects have been reported previously, but numerical methods have not been available to confirm the observations [*Gandhi et al.,* 1977b]. We have used image theory to reduce the problem of man either grounded or in front of reflectors to that of a multibody target in free space. Moment-method solutions of the electric-field integral equation have a pulse-function basis and delta functions for testing. A 180-cell model of the 50th-percentile standard man was used in the calculations [*Hagmann et al.,* 1977].

2. APPLICATION OF IMAGE THEORY WITH COMPRESSION

If all reflectors and ground planes are assumed to be perfect conductors and infinite in extent, then they may be removed if suitable images of the scattering body are included. Since polarization currents in the images contribute to the scattering, it is necessary to treat all parts of the scattering body and the images as discrete cells. If pulse functions are used as a basis, it is necessary that each subvolume be small enough that the electric field that it contains may be assumed constant.

A moment-method solution with pulse functions will relate the column matrix E, which contains the unknown electric-field values in each cell, to the column matrix E^i, which contains values of the incident electric field in each cell, through a square matrix A by the equation

$$AE = E^i \tag{1}$$

Equation 1 may be solved directly, but a reduction of matrix size will allow the solution to be found with greater economy; or alternately, more cells may be used for greater accuracy. A reduction in matrix size is possible by the use of "compression," which is described next.

The geometry involved when images are used causes the fields to have symmetry. Let m be the number of unique values of E that are repeated n times with possible sign changes. We may order the cells so that the indices from 1 to m correspond to the unique E values, and for any integer $1 \le \ell \le m$, then $E_{\ell+m}$, $E_{\ell+2m}$, ..., $E_{\ell+(n-1)m}$ correspond to values related to E_ℓ by symmetry.

The relationship of all $m \times n$ of the E values to the m-unique E values may be expressed in matrix form:

$$\begin{bmatrix} E_1 \\ E_2 \\ \\ \cdots \\ E_{m\times n} \end{bmatrix} = \begin{bmatrix} I_1 \\ I_2 \\ \\ \cdots \\ I_n \end{bmatrix} \begin{bmatrix} E_1 \\ E_2 \\ \\ \cdots \\ E_m \end{bmatrix} \tag{2}$$

where I_2, I_3, ..., I_n are m by m matrices with $\pm$ 1 on the diagonal and zero elsewhere, and I_1 is the m by m identity matrix.

Compression may be obtained by noting that only the first m simultaneous equations need to be used in equation (1); substituting (2) in (1) will give a series of m simultaneous equations in the m-unknown unique values of E.

$$A \begin{bmatrix} I_1 \\ I_2 \\ \\ \cdots \\ I_n \end{bmatrix} \begin{bmatrix} E_1 \\ E_2 \\ \\ \cdots \\ E_m \end{bmatrix} = \begin{bmatrix} E_1^i \\ E_2^i \\ \\ \cdots \\ E_m^i \end{bmatrix} \tag{3}$$

$$A' \begin{bmatrix} E_1 \\ E_2 \\ \\ \cdots \\ E_m \end{bmatrix} = \begin{bmatrix} E_1{}^i \\ E_2{}^i \\ \\ \cdots \\ E_m{}^i \end{bmatrix} \qquad (4)$$

where A' is an m by m matrix formally defined by A' = first m rows of:

$$A \begin{bmatrix} I_1 \\ I_2 \\ \\ \cdots \\ I_n \end{bmatrix} \qquad (5)$$

In implementing compression, the computation of matrix A' is made by making n modified passes through the loops that would normally be used to accumulate the scattering matrix.

We have used compression with image theory to treat a body with one plane of symmetry above a ground plane, as is shown in Figure 1. The field relationships used for a body with one plane of symmetry in front of a flat reflector are shown in Figure 2. It is also possible to use compression with image theory to treat a body in a corner reflector, if the corner reflector has a dihedral angle equal to 180°/N

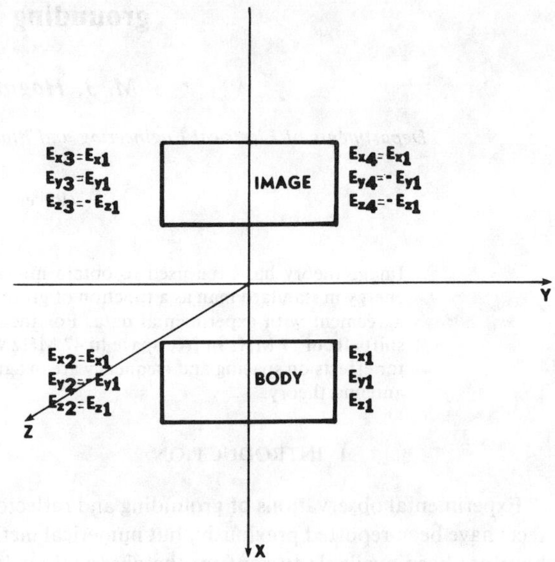

Fig. 2. Field relationships for a body with one plane of symmetry in front of a flat reflector; **E**‖$\hat{z}$, **k**‖$\hat{x}$.

where N is an integer. Field relationships used for a body with one plane of symmetry in front of a 90° corner reflector are shown in Figure 3.

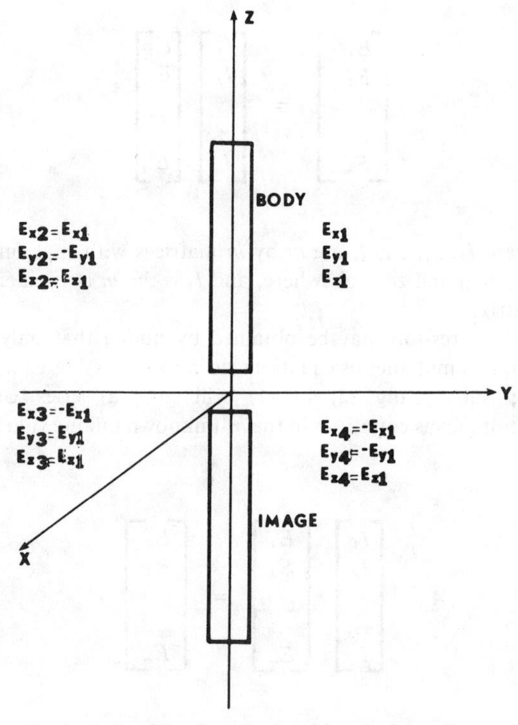

Fig. 1. Field relationships for a body with one plane of symmetry above a ground plane; **E**‖$\hat{z}$, **k**‖$\hat{x}$, x-y plane is the ground plane.

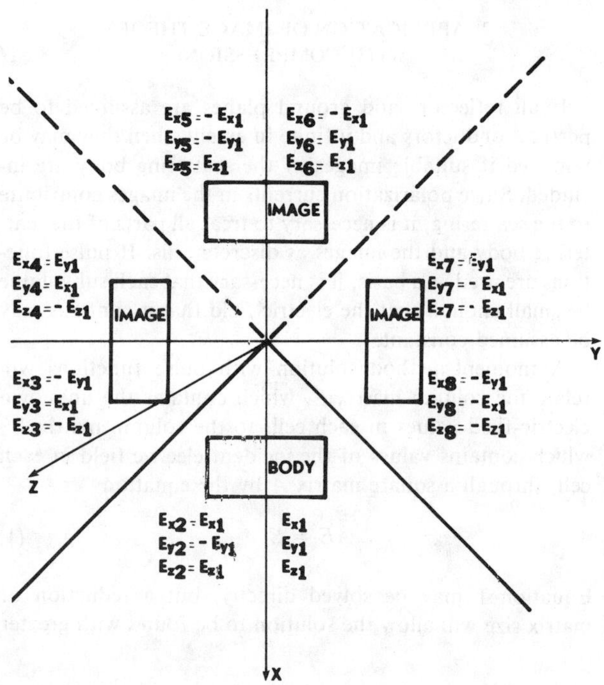

Fig. 3. Field relationships for a body with one plane of symmetry in front of a 90° corner reflector; **E**‖$\hat{z}$, **k**‖$\hat{x}$.

There are two types of symmetry relationships used for **E** values in the three figures: (1) Two points located symmetrically in respect to a plane of symmetry of the body that contains both **k** and $\mathbf{E}^i$ will necessarily have **E** values that are mirror images so that the two halves are indistinguishable under reflection. (2) Two points located symmetrically in respect to a plane of symmetry of the body that contains **k** and is perpendicular to $\mathbf{E}^i$ will necessarily have **E** values that are negated mirror images, so that if the values are exchanged, they will correspond to the solution for oppositely directed $\mathbf{E}^i$. This relationship is also found for two points located symmetrically in respect to a ground plane or to a flat reflector as required for symmetrical motion of a charge and its image.

An alternative to the use of image theory with compression is to use special Green's functions in place of the free-space Green's function. Special Green's functions are available that could be used to account for scattering objects located near targets having various shapes and even finite conductivities [*Tai,* 1971; *Baños,* 1966]. We have not yet pursued the use of such Green's functions.

The results of calculations for grounding and reflector effects are described in the next sections.

3. GROUND EFFECTS

All calculations of grounding effects have assumed that a man is standing on or above a perfectly conducting ground plane that is infinite in extent. The incident field is vertically polarized with **k** directed from the vental to the dorsal aspect of the man. Internal fields in the model are the same as those in one-half of a double-man, which consists of the model and its image in free space. A 180-cell model of the 50th-percentile standard man has been used for the calculations [*Hagmann et al.,* 1977].

The calculated resonant frequency of man standing on a ground plane is 47 MHz as compared to 77 MHz for man in free space. The specific absorption rate (SAR) of man on the ground plane at the reduced resonant frequency is 32.5 percent greater than the SAR for man in free space at the free-space resonant frequency. At lower frequencies, the SAR of man standing on a ground plane may be increased considerably from that of the same model in free space. This increase results from these frequencies being closer to resonance for the grounded man. For example, at 10 MHz the SAR of man standing on a ground plane is 0.0163 W $\mathrm{kg^{-1}}$ per mW $\mathrm{cm^{-2}}$, which is about seven times that found for the same model in free space.

Figures 4 and 5 illustrate, respectively, the distribution of absorbed energy in man at 10 MHz in free space and in contact with the ground plane. The enhancement of local energy deposition due to the grounding effect is approximately a factor of 60 in the area of the heel.

Figure 6 shows the part-body and whole-body SAR for the model of man with different spacings from the ground plane at 10 MHz. Figure 6 is consistent with the experimental observation on models that a small separation that breaks electrical contact with the ground plane is sufficient to eliminate much of the grounding effect [*Gandhi et al.,* 1975]. Several calculations made for the grounded resonant frequency of 47 MHz show a fall-off in magnitude of grounding effects with increasing distance from the ground plane, which is similar to the results at 10 MHz (Figure 6). A model composed of 292 cells was made by dividing each of the six lowest cells in each leg of the 180-cell model into eight identical cubical cells. Six solutions made for the 292-cell model and duplicated with the 180-cell model suggest that the fall-off of the grounding effect with increasing distance from the ground plane may be slightly slower than is found with the 180-cell model.

Figure 7 shows part- and whole-body SARs for the model of man standing on a ground plane at various frequencies. The part-body resonances shown in Figure 7 have also been observed in calculations for the same model of man in free space [*Gandhi et al.,* 1977a]. The enhancement in absorption of energy due to the presence of a ground plane is greatest in the leg and least in the head. Figure 8 shows whole-body SARs for the model of man both in free space and when standing on a ground plane at various frequencies. Grounding effects are most prominent at low frequen-

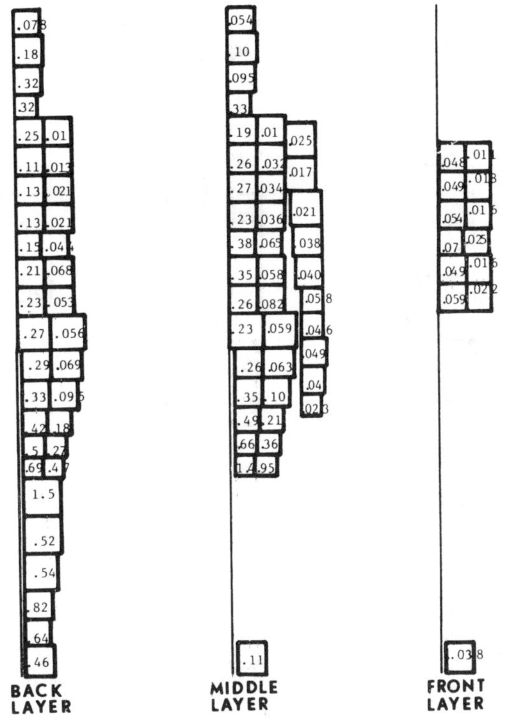

Fig. 4. Local SAR values (W $\mathrm{kg^{-1}}$ per mW $\mathrm{cm^{-2}}$) × 100 for inhomogeneous model of man in free space at 10 MHz.

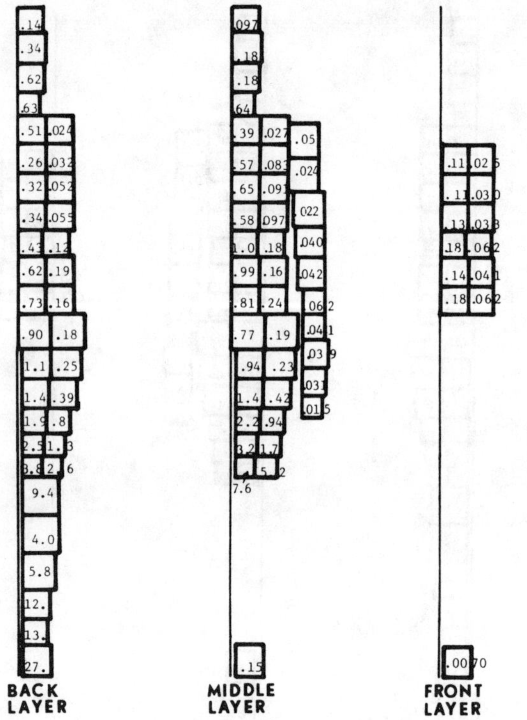

Fig. 5. Local SAR values (W kg^{-1} per mW cm^{-2}) × 100 for inhomogeneous model of man standing on a ground plane at 10 MHz.

cies. Above a frequency of about 200 MHz, part- and whole-body SAR values show little dependence on the presence of a ground plane.

4. REFLECTOR EFFECTS

All calculations of reflector effects have assumed that a man is standing in front of a reflector that is perfectly conducting and infinite in extent. The incident field is vertically polarized with **k** directed from the ventral to the dorsal aspect of the man. One image is required for a flat reflector and three are needed for a 90° corner reflector.

The enhancement of SAR due to reflector effects is approximately the same as the ratio of effective area of a half-wave dipole with a reflector to that of an isolated dipole in free space [*Gandhi et al.,* 1977b]. Antenna theory has been used to perform numerical calculations made for the model of man. Standard procedures have been used for computation of the required values of mutual impedance of dipoles [*Baker and LaGrone,* 1962].

Figures 9-12 show the enhancement of energy absorption as calculated for the model of man with reflector effects and the corresponding gain-enhancement factors calculated for thin dipoles from antenna theory. Values of enhancement for the whole-body average SAR have been used in preparing Figures 9-12, but the curves for part-body energy absorption are nearly identical, which indicates that the distribution of energy through the model is almost independent of the presence of a reflector. Experimental values in Figures 9-12 are for models of man and were reported earlier [*Gandhi et al.,* 1977b].

Figures 9 and 10 show the variation of enhancement of energy absorption as a function of spacing from a reflector at the free-space resonant frequency of 77 MHz. In Figure 9 the numerical values for the model of man are markedly below the curve for thin dipoles at small values of separa-

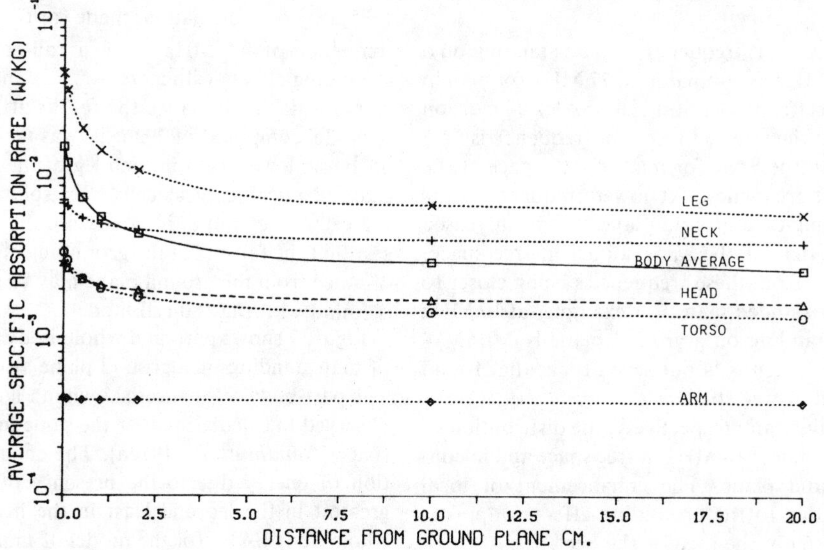

Fig. 6. Whole-body SARs in man for different spacings from the ground at 10 MHz; E∥L̂, **k** ventral to dorsal, incident power density of 1 mW cm^{-2}.

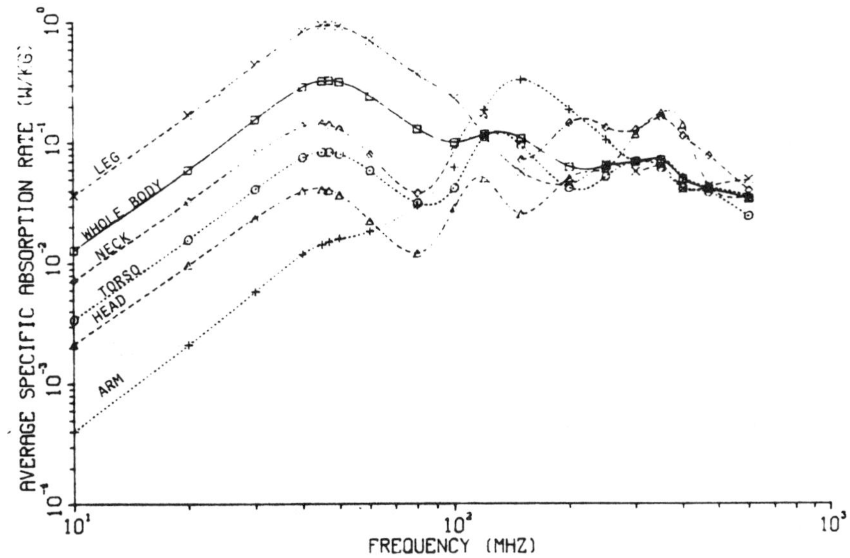

Fig. 7. Part-body SARs for a homogeneous model of man on a ground plane; **E**∥$\hat{\mathbf{L}}$, **k** ventral to dorsal, incident power density of 1 mW cm^{-2}.

tion. Antenna theory requires that the gain be small when the spacing is not much greater than the thickness of the dipole, but such variation is missed in the thin dipole approximation.

Figures 11 and 12 show the variation of enhancement of energy absorption as a function of frequency for fixed ratios of spacing to wavelength. The antenna-theory calculations were made for a thin dipole with a length of 1.75 meters, which is identical to the height of the model. Figures 11 and 12 show that the reflector effects are found over a wide range of frequencies. Values for the model of man have a pronounced roll-off at higher frequencies, which is not seen in calculations for the thin dipoles. In Figures 11 and 12 it can be seen that the rate of roll-off of the SAR at high frequencies is consistent with experimental results that are obtained from models of man. Antenna theory requires that the gain decrease at higher frequencies when the effective area of the dipole images is blocked by the physical cross section of the dipole, but such variation is missed in the thin-dipole approximation.

All gain calculations have been based on zero-order antenna theory; that is, they have no correction for non-sinusoidal current distributions. Significant first- and second-order corrections are required when the antennae are not very thin and the half-length is much different from λ/4 [*King*, 1956]. The sizable lateral extent of the model of man is thought to be responsible for the difference between

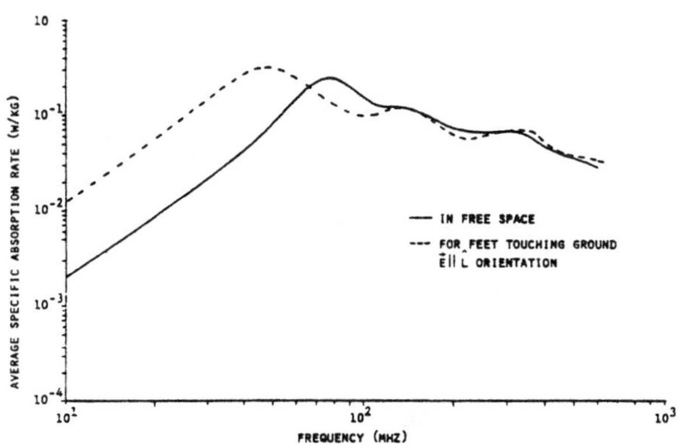

Fig. 8. Whole-body SARs for homogeneous model of man; **E**∥$\hat{\mathbf{L}}$, **k** ventral to dorsal, incident power density of 1 mW cm^{-2}.

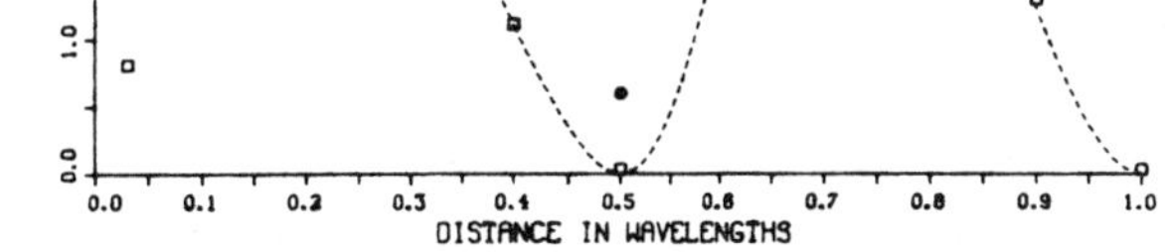

Fig. 9. Variation of gain enhancement with spacing for a half-wave dipole in front of a flat reflector. Squares represent numerical solutions for enhancement of SARs in man at 77 MHz (L/λ = 0.45). Solid circles represent experimental values for anatomically scaled figurines at L/λ = 0.417.

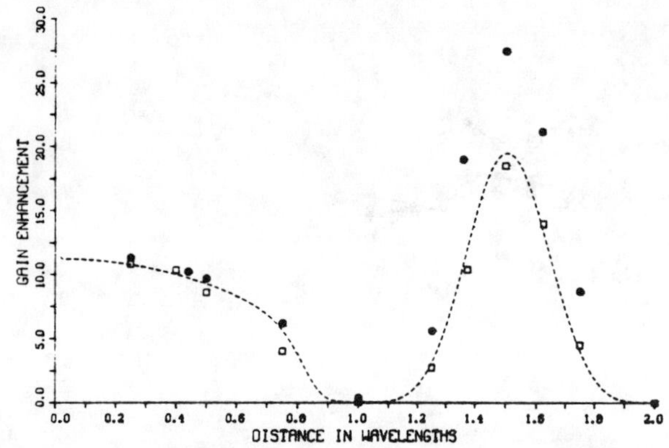

Fig. 10. Variation of gain enhancement with spacing for a half-wave dipole in front of a 90° corner reflector. Squares represent numerical solutions for enhancement of the SAR in man at 77 MHz (L/λ = 0.45). Solid circles represent experimental values for anatomically scaled figurines at L/λ = 0.417.

numerical solutions and antenna theory at low frequencies in Figure 12.

The numerical solutions indicate that for frequencies near resonance, the enhancement in SAR due to a reflector is approximately equal to the enhancement in gain of a half-wave dipole with the same reflector configuration. Experimental studies on models of man have also shown that the enhancement of SAR is equal to the enhancement in gain of a dipole, but the gain must be calculated for the finite experimental reflector rather than for an infinite reflector having the same dihedral angle. For certain combinations of length and width of the reflecting sheets, a finite corner reflector may exhibit supergain -- a gain significantly greater than that for infinite reflecting planes [*Cottany and Wilson*, 1958; *Schelkunoff and Friis*, 1952].

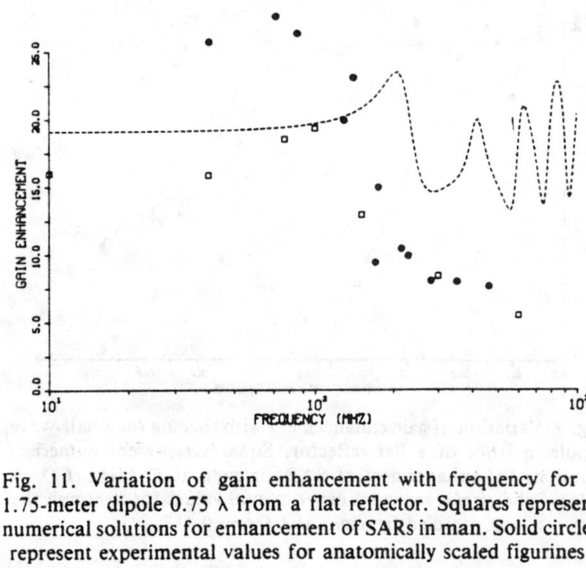

Fig. 11. Variation of gain enhancement with frequency for a 1.75-meter dipole 0.75 λ from a flat reflector. Squares represent numerical solutions for enhancement of SARs in man. Solid circles represent experimental values for anatomically scaled figurines.

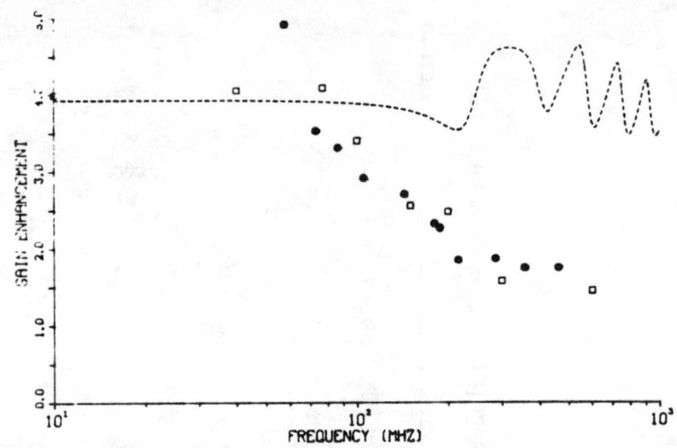

Fig. 12. Variation of gain enhancement with frequency for a 1.75-meter dipole 1.5 λ from a 90° corner reflector. Squares represent numerical solutions for enhancement of SARs in man. Solid circles represent experimental values for anatomically scaled figurines.

Examples of the effect of such supergain combinations are seen in the experimental values in Figures 10-12.

Figures 13 and 14 were copied from an experimental study of finite-size corner reflectors [*Cottany and Wilson*, 1958]. In both figures the angle of the aperture was adjusted for maximum gain at a particular distance of the dipole from the reflector. For reflector lengths greater than about 0.5 λ (Figure 13), the aperture angle is 90° and distance of the dipole from the reflector is 1.5 λ. For reflector lengths greater than 1.0 λ (Figure 14), the aperture angle

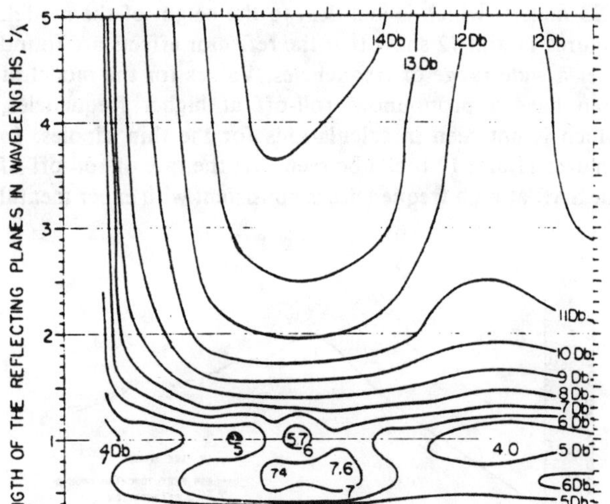

Fig. 13. Contours of constant gain for a corner-reflector antenna with the dipole in third position for various reflector sizes. For lengths greater than ~ 0.5 λ, the aperture angle is 90° and the distance of the dipole from the reflector is 1.5 λ [*Cottany and Wilson*, 1958].

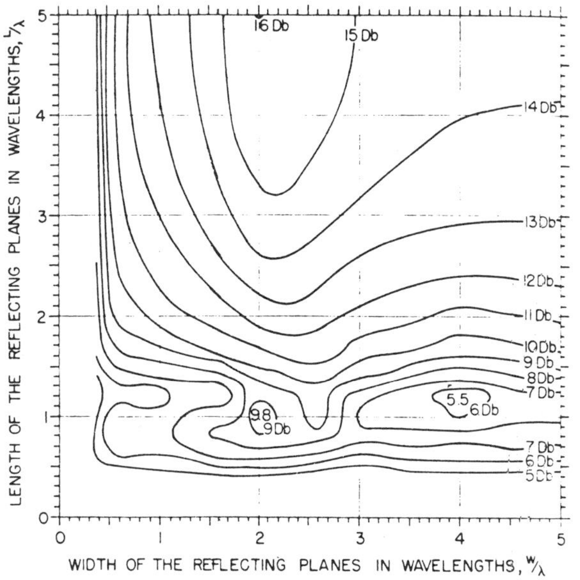

Fig. 14. Contours of constant gain for a corner-reflector antenna with the dipole in second position for various reflector sizes. For lengths greater than 1.0 λ, the aperture angle is 65° and the distance of the dipole from the reflector is 1.2 λ [*Cottany and Wilson,* 1958].

is 65° and distance of the dipole from the reflector is 1.2 λ.

Note in Figure 13 that for proper combinations of length and widths of the reflecting planes, the enhancement in SAR over the free-space value may be 30 to 45 percent greater than that for infinite reflectors. For such supergain combinations the SAR may be enhanced (see Figure 13) by as much as 14 to 14.5 dB (a factor of 25 to 28) rather than 12.9 dB (19.5) anticipated for infinite reflectors, or the factor of 16 that is projected from consideration of the local E fields in the standing-wave pattern of a 90° corner reflector. As is shown in Figure 13, the dimensions 2 λ × 3.5 λ of the reflectors used for the experimental results of Figures 10 and 12 correspond to one of the supergain combinations.

Supergain combinations of length and width are also present for a 65° corner reflector (see Figure 14). Antenna calculations with image theory are restricted to corner reflectors having dihedral angles of $180°/N$, where N is an integer. As seen in Figure 14, gain enhancement is not limited to such special angles, and we expect that significant enhancements of the SAR will occur for a wide variety of reflectors.

5. CONCLUSIONS

Image theory has been used to treat grounding and reflector effects so that all calculations are limited to a perfectly conducting ground plane, or to reflectors with an infinite extent. Experimental results suggest that reflectors of finite size may provide significantly more or less enhancement in energy deposition than that calculated for infinite reflectors. The effect of finite conductivity on grounding or reflector effects with models of man has not yet been evaluated either theoretically or experimentally.

The decrease in resonant frequency and alteration in energy deposition due to grounding may be explained by considering a double-length man in free space. If the model of man had a cross section independent of height and a complex permittivity independent of frequency, the grounded resonant frequency would be one-half the free-space resonant frequency, but a ratio of 0.610 is obtained for the model that has been used.

The dependence of reflector effects on spacing and frequency are in agreement with the gain enhancement calculated for dipoles by antenna theory. The enhancement in energy deposition due to reflector effects becomes small for small values of separation or for high frequencies, which effects agree with antenna theory only if finite width of the dipole is considered.

Acknowledgment. This work was supported by US Army Medical Research and Development Command, Washington, D. C., under Contract DAMD 17-74-C-4092.

REFERENCES

Baker, H. C., and A. H. LaGrone (1962), Digital computation of the mutual impedance between thin dipoles, *IRE Trans. Antenna Propagat., AP-10,* 172-178.

Baños, A., Jr. (1966), *Dipole Radiation in the Presence of a Conducting Half-Space,* 245 pp., Pergamon Press, Oxford.

Cottany, H. V., and A. C. Wilson (1958), Gains of finite-size corner reflector antennas, *IRE Trans. Antenna Propagat., AP-6,* 366-369.

Gandhi, O. P., J.A. D'Andrea, B. K. Jenkins, J. L. Lords, J. R. Mijanovich, and K. Sedigh (1975), Behavioral and biological effects of resonant electromagnetic absorption in rats, *Tech. Rep. UTEC MD-75-174* under Contract No. DAMD 17-74-C-4092, 29 pp., Microwave Device and Physical Electronics Laboratory, University of Utah, Salt Lake City, UT 84112.

Gandhi, O. P., M. J. Hagmann, and J. A. D'Andrea (1977a), Some recent results on deposition of electromagnetic energy in animals and models of man, paper presented at the 1977 USNC/URSI meeting, Airlie, VA.

Gandhi, O. P., E. L. Hunt, and J. A. D'Andrea (1977b), Deposition of electromagnetic energy in animals and in models of man with and without grounding and reflector effects, *Radio Sci., 12*(6S), 39-47.

Hagmann, M. J., O. P. Gandhi, and C. H. Durney (1977), Numerical calculation of electromagnetic energy deposition for a realistic model of man, paper presented at the 1977 USNC/URSI meeting, Airlie, VA, accepted for publication in *IEEE Trans. Microwave Theory and Tech.*

King, R. W. P. (1956), *The Theory of Linear Antennas,* 944 pp., Harvard University Press, Cambridge, MA.

Schelkunoff, S. A., and H. T. Friis (1952), *Antennas, Theory and Practice,* 639 pp., John Wiley and Sons, New York.

Tai, C. T. (1971), *Dyadic Green's Functions in Electromagnetic Theory,* 246 pp., Intext Educational Publishers, Scranton, PA.

PART 4:
Printed Circuit and Earth-Based Antennas

Microwave printed circuit antennas, such as microstrip patches and dipoles, have become important primarily due to their low fabrication cost and lower weight. At the other end of the frequency spectrum, HF and VHF antennas over the earth are widely used for various communications systems. These two disparate situations have much in common: both can support surface waves, and a Sommerfeld type integral is needed to model accurately the effects of the lossy dielectric half-space, or the dielectric slab with ground plane. Thus, key papers on these are grouped here.

The dielectric is represented by a Green's function integral; Rana and Alexopoulos (1981) use Galerkin piecewise sinusoidal bases for printed circuit dipoles. Dielectric-covered slots, where the dielectric represents a radome, are analyzed by Katehi (1989). Major improvements in the Sommerfeld integral calculation are realized by extracting the singularity: Pozar (1983), Johnson and Dudley (1983), and Michalski (1985). In two papers, Miller, Poggio, Burke, and Selden (1972) compare a simple reflection coefficient approximate solution with the exact one; Part I is included here, with Part II listed in the Bibliography. The choice of basis functions is important when the reflection coefficient approximation is used; Richmond and Newman (1979) show that piecewise sinusoidal bases give much better accuracy than point matching for that case.

BIBLIOGRAPHY

Microstrip Patches

E.H. Newman and P. Tulyathan, "Analysis of Microstrip Antennas Using Moment Methods," *Trans. IEEE,* Vol. AP-29, January 1981, pp. 47–53.

M.D. Deshpande and M.C. Bailey, "Input Impedance of Microstrip Antennas," *Trans. IEEE,* Vol. AP-30, July 1982, pp. 645–650.

M.C. Bailey and M.D. Deshpande, "Integral Equation Formulation of Microstrip Antennas," *Trans. IEEE,* Vol. AP-30, July 1982, pp. 651–656.

D.M. Pozar, "Input Impedance and Mutual Coupling of Rectangular Microstrip Antennas," *Trans. IEEE,* Vol. AP-30, November 1982, pp. 1191–1196.

D.M. Pozar, "Considerations for Millimeter Wave Printed Antennas," *Trans. IEEE,* Vol. AP-31, September 1983, pp. 740–747.

E.H. Newman, J.H. Richmond and B.W. Kwan, "Mutual Impedance Computation between Microstrip Antennas," *Trans. IEEE,* Vol. MTT-31, November 1983, pp. 941–945.

E.H. Newman, P. Alexandropoulous and E.K. Walton, "Polygonal Plate Modeling of Realistic Structures," *Trans. IEEE,* Vol. AP-32, July 1984, pp. 742–747.

M.C. Bailey and M.D. Deshpande, "Analysis of Elliptical and Circular Microstrip Antennas Using Moment Method," *Trans. IEEE,* Vol. AP-33, September 1985, pp. 954–959.

Y. Lin and L. Shafai, "Moment-Method Solution of the Near-Field Distribution and Far-Field Patterns of Microstrip Antennas," *Proc. IEE,* Vol. 132H, October 1985, pp. 369–374.

J.R. Mosig and F.E. Gardiol, "General Integral Equation Formulation for Microstrip Antennas and Scatterers," *Proc. IEE,* Vol. 132H, December 1985, pp. 424–432.

D.M. Pozar, "A Reciprocity Method of Analysis for Printed Slot and Slot-Coupled Microstrip Antennas," *Trans. IEEE,* Vol. AP-34, December 1986, pp. 1439–1446.

D.M. Pozar and S.M. Voda, "A Rigorous Analysis of a Microstripline Fed Patch Antenna," *Trans. IEEE,* Vol. AP-35, December 1987, pp. 1343–1350.

D.M. Pozar, "Radiation and Scattering from a Microstrip Patch on a Uniaxial Substrate," *Trans. IEEE,* Vol. AP-35, June 1987, pp. 613–621.

J.P. Damiano, "Computation of Input Impedance in Microstrip Antenna. Graphic Representation and Numerical Integration of Oscillating Functions," *Proc. IEE,* Vol. 134H, October 1987, pp. 456–466.

M. Marin, S. Barkeshli and P.H. Pathak, "Efficient Analysis of Planar Microstrip Geometries Using a Closed-Form Asymptotic Representation of the Grounded Dielectric Slab Green's Function," *Trans. IEEE,* Vol. MTT-37, April 1989, pp. 669–679.

R.C. Hall and J.R. Mosig, "The Analysis of Coaxially Fed Microstrip Antennas with Electrically Thick Substrates," *Electromagnetics,* Vol. 9, No. 4, 1989, pp. 367–384.

T.G. Livernois and P.B. Katehi, "A Generalized Method for Deriving the Space-Domain Green's Function in a Shielded, Multilayer Substrate Structure with Applications to MIS Slow-Wave Transmission Lines," *Trans. IEEE,* Vol. MTT-37, November 1989, pp. 1761–1767.

S. Pinhas, S. Shtrikman and D. Treves, "Moment-Method Solution of the Center-Fed Microstrip Disk Antenna Invoking Feed and Edge Current Singularities," *Trans. IEEE,* Vol. AP-37, December 1989, pp. 1516–1522.

J.T. Aberle and D.M. Pozar, "A Full-Wave Solution for the Mutual Coupling Between Circular Microstrip Patch Antennas," *Microw. & Opt. Tech. Letters,* Vol. 2, April 1989, pp. 130–132.

Microstrip Dipoles and Stubs

N.K. Uzunoglu, N.G. Alexopoulos and J.G. Fikioris, "Radiation Properties of Microstrip Dipoles," *Trans. IEEE,* Vol. AP-27, November 1979, pp. 853–858.

I.E. Rana and N.G. Alexopoulos, "Current Distribution and Input Impedance of Printed Dipoles," *Trans. IEEE,* Vol. AP-29, January 1981, pp. 99–105, Correction, July 1982, p. 822.

N.G. Alexopoulos and I.E. Rana, "Mutual Impedance Computation Between Printed Dipoles," *Trans. IEEE,* Vol. AP-29, January 1981, pp. 106–111.

K.A. Michalski and C.M. Butler, "Determination of Current Induced on a Conducting Strip Embedded in a Dielectric Slab," *Radio Science,* Vol. 18, November-December 1983, pp. 1195–1206.

U. Mahr, "Cubic Splines as Expansion Functions for the Current Distribution of Microstrip Stub Antennas," *Trans. IEEE,* Vol. AP-33, September 1985, pp. 1033–1036.

D.R. Jackson and N.G. Alexopoulos, "Analysis of Planar Strip Geometries in a Substrate—Superstrate Configuration," *Trans. IEEE,* Vol. AP-34, December 1986, pp. 1430–1439.

H. Nakano, S.R. Kerner and N.G. Alexopoulos, "The Moment Method Solution for Printed Wire Antennas of Arbitrary Configuration," *Trans. IEEE,* Vol. AP-36, December 1988, pp. 1667–1674.

A. Nakatani, N.G. Alexopoulos, N.K. Uzunoglu and P.L.E. Uslenghi, "Accurate Green's Function Computation for Printed Circuit Antennas on Cylindrical Substrates," *Electromagnetics,* Vol. 6, No. 3, 1986, pp. 243–254.

A. Nakatani and N.G. Alexopoulos, "Microstrip Elements on Cylindrical Substrates—General Algorithm and Numerical Results," *Electromagnetics,* Vol. 9, No. 4, 1989, pp. 405–426.

Earthed Antennas

E.K. Miller, A.J. Poggio, G.J. Burke and E.S. Selden, "Analysis of Wire Antennas in the Presence of a Conducting Half-space: Part II—The Horizontal Antenna in Free Space," *Canadian Journal of Physics,* Vol. 50, 1972, pp. 2614–2627.

W.A. Johnson and D.G. Dudley, "The Green's Function Moment Method," *Appl. Phys.,* Vol. 16, 1978, pp. 79–85.

J.N. Brittingham, E.K. Miller and J.T. Okada, "Bivariate Interpolation Approach for Efficiently and Accurately Modeling Antennas near a Half-Space," *Electronics Letters,* Vol. 13, November 10, 1977, pp. 690–691.

E.K. Miller, J.N. Brittingham and J.T. Okada, "Explicit Modeling of Antennas With Sparse Ground Screens," *Electronics Letters,* Vol. 14, September 14, 1978, pp. 627–629.

G.J. Burke, E.K. Miller, J.N. Brittingham, D.L. Lager and R.J. Lytle, "Computer Modeling of Antennas Near the Ground," *Electromagnetics,* Vol. 1 January-March 1981, pp. 29–49.

W.A. Johnson, "Analysis of a Vertical, Tubular Cylinder which Penetrates an Air-Dielectric Interface and which is Excited by an Azimuthally Symmetric Source," *Radio Science,* Vol. 18, November-December 1983, pp. 1273–1281.

G.J. Burke and E.K. Miller, "Modeling Antennas Near to and Penetrating a Lossy Interface," *Trans. IEEE,* Vol. AP-32, October 1984, pp. 1040–1049.

T.K. Sarkar and R.F. Harrington, "Radar Cross Sections of Conducting Bodies over a Lossy Half Space," *Radio Science,* Vol. 15, May-June 1980, pp. 518–858.

A. Karwowski, "Low-Frequency Approach to the Problem of a Horizontal Wire Antenna Above an Imperfect Ground," *Proc. IEE-MOA,* Vol. 131, June 1984, pp. 214–216.

K.A. Michalski and C.M. Butler, "Evaluation of Sommerfeld Integrals Arising in the Ground Stake Antenna Problem," *Proc. IEE,* Vol. 134H, February 1987, pp. 93–97.

P. Parhami and R. Mittra, "Wire Antennas over Lossy Half-Space," *Trans. IEEE,* Vol. AP-28, May 1980, pp. 397–403.

J.H. Richmond, "Monopole Antenna on Circular Disk Over Flat Earth," *Trans. IEEE,* Vol. AP-33, June 1985, pp. 633–637.

R.C. Hall and J.R. Mosig, "Vertical Monopoles Embedded in a Dielectric Substrate," *Proc. IEE,* Vol. 136H, December 1989, pp. 462–468.

Current Distribution and Input Impedance of Printed Dipoles

INAM E. RANA AND NICOLAOS G. ALEXOPOULOS, MEMBER, IEEE

Abstract—**Printed dipole antennas find increasing use in microwave as well as far infrared frequencies. The current distribution, input impedance, and radiation pattern are computed for wire antennas printed on a dielectric substrate. The current distribution is obtained by solving Pocklington's equation by moment methods. The Green's function pertinent to the problem involves improper Sommerfeld-type integrals. These integrals are computed by a real-axis integration technique which involves analytical and numerical steps. The effect of surface modes is carefully taken into account.**

I. INTRODUCTION

PRINTED CIRCUIT antennas find increasing use in microwave frequencies as well as in the far infrared where planar-integrated monolithic circuits is the ultimate goal. In the microwave regime these antennas are popular due to their inherent advantages such as conformality to a given surface, low cost, negligible volume, reproducibility, etc. In the submillimeter regime such antennas find use as an integral part of far infrared detectors. The analysis of printed antennas involves Sommerfeld-type integrals which are difficult to deal with when the source and field points are on the substrate, as is the case when one needs to solve Pocklington's equation for the printed antenna current distribution. For this reason the existing analytical techniques to study such antennas, as for example the microstrip patch element in microwave frequencies [1]–[5] or the printed dipole in the microwave regime [6], [7] as well as in the far infared [8], [9], are more of an empirical nature, and they do not provide adequate design criteria. In addition, heretofore published techniques do not account for the effect of the dielectric substrate on the antenna characteristics such as input and mutual impedance. Although for microwave printed antennas the substrate thickness is selected to be much smaller than the wavelength, the fundamental surface wave mode has zero cutoff, and therefore it is always launched in the substrate. Since surface modes decay in the substrate in a manner inversely proportional to the square root of distance (measured from the antenna), even an inefficiently launched surface mode may have an important effect on mutual coupling in a phased-array design. In addition, since the bandwidth of integrated antennas improves with increasing substrate thickness, surface mode effects become significant for the evaluation of a substrate-thickness bandwidth trade-off. On the other hand, in the far infrared, the printed antenna elements which have been tested to date [7], [8] are deposited on substrates which are several wavelengths thick. The presence of several surface modes in these designs provides for phase reversals of the printed dipole current distribution along the dipole length, thus resulting in undesired multiple beam radiation patterns [8].

Manuscript received November 30, 1979; revised July 23, 1980. This work was supported by the U.S. Army under Contract DAAG 29-79-C-0050.

I. E. Rana is with Suparco, P.O. Box 3209, Karachi 28, Pakistan.
N. G. Alexopoulos is with the Electrical Sciences and Engineering Department, University of California, Los Angeles, CA, 90024.

In this paper a theoretical approach to printed circuit antennas is presented with an application to dipoles printed on a grounded dielectric substrate. Although Pocklington's equation has been developed for printed patch antennas, in this paper attention is confined to printed dipoles with $\hat{x}$ orientation. Previously, the printed dipole problem was treated with an assumed sinusoidal current distribution [10]–[11] or at resonance [12], or as a current distribution derived in the absence of the substrate [13]. The assumption that the printed dipole current distribution is sinusoidal is valid provided that the substrate thickness is exceedingly small compared to the wavelength on the dielectric, or that the printed dipole is of resonant length [11]. In order to solve Pocklington's equation, the appropriate Green's function is obtained by considering a Hertzian dipole printed on the substrate. The resulting Green's function is in terms of Sommerfeld-type integrals [13]–[17], which are very slowly convergent and their derivatives are improper integrals, when the source and field points are on the substrate. An efficient numeric-analytical integration technique has been developed which computes the Green's function for all distances from the source. The solution to Pocklington's equation is then obtained by applying moment methods, yielding in this manner the antenna current distribution, input impedance, substrate thickness, and relative dielectric constant.

II. POCKLINGTON'S INTEGRAL EQUATION FOR A PRINTED WIRE

The printed wire dipole considered here is of length L and diameter $2a$, and it is shown in Fig. 1. It is assumed that the radius of the wire is very small as compared to the free-space wavelength ($a \ll \lambda$), and therefore the thin-wire approximation can be used. The current on the wire does not have circumferential symmetry, due to the presence of the grounded substrate. The antenna current distribution may have therefore, in addition to the axial component, a circumferentially nonsymmetric component; yet for a thin wire only the axial component is of significance, e.g., it has been shown [18] that only the axial component is dominant in the determination of the radiation and impedance characteristics. Pocklington's equation for printed wires can be written in the form

$$E_\alpha(x, y, B) = \int_{L_\alpha} I(\alpha') \left\{ k^2 \Pi_\alpha + \frac{\partial^2 \Pi_\alpha}{\partial \alpha^2} + \frac{\partial^2 \Pi_z}{\partial \alpha \partial z} \right\} d\alpha', \quad (1)$$

where $\alpha = x$ or y depending on whether the wire is in the x or y direction, respectively. For this case

$$\Pi_\alpha = \lim_{z \to B} 2K \int_0^\infty J_0(\lambda\rho) e^{-\mu(z-B)} \frac{\lambda d\lambda}{D_e(\lambda)} \quad (2)$$

and

$$\Pi_z = \lim_{z \to B} 2K(1 - \epsilon_r) \cdot \int_0^\infty J_1(\lambda\rho) e^{-\mu(z-B)} \frac{\lambda^2 \Phi_\alpha d\lambda}{D_e(\lambda) D_m(\lambda)}. \quad (3)$$

In the equations above the following parameters need to be defined:

$$D_e(\lambda) = \mu + \mu_e \coth \mu_e B \quad (4)$$

$$D_m(\lambda) = \mu\epsilon_r + \mu_e \tanh \mu_e B, \quad (5)$$

where

$$\mu = [\lambda^2 - k^2]^{1/2}, \qquad \mu_e = [\lambda^2 - \epsilon_r k^2]^{1/2},$$

$$K = -\frac{j}{4\pi\epsilon_0 \omega},$$

$$\rho = [(x - x')^2 + (y - y')^2]^{1/2},$$

and $\Phi_x = \cos\phi$ while $\Phi_y = \sin\phi$. In addition it must be mentioned that in order to obtain the Hertz potentials given by (2) and (3), the canonical problem of a horizontal Hertzian dipole at $z = B$ is solved with a dipole moment p_x. The time dependence $e^{-j\omega t}$ is omitted throughout the paper.

By considering henceforth the printed wire along the x direction, since the wire is unloaded at the ends, the unknown antenna current distribution $I(x')$ must satisfy the end boundary conditions

$$I(0) = I(L) = 0. \quad (6)$$

In order to solve (1) for the antenna current $I(\alpha')$, the method

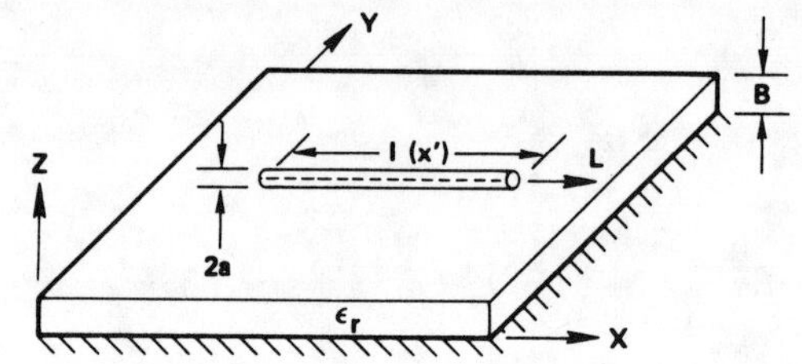

Fig. 1. Printed wire dipole geometry.

of moments is used. Accordingly, the wire is divided into N equal segments. As it is usually, the proper choice of current expansion functions must be made. Sinusoidal expansion functions are employed because they offer many advantages, such as 1) the boundary conditions (6) for the antenna current $I(\alpha')$ are automatically satisfied, 2) a closed-form expression for the fields is obtained, and as will be shown later, and 3) the improperness in the Green's function integrals is removed as $z \to B$. The form of the ith expansion function centered about x_i is simply

$$I(x') = \begin{cases} I_i \left[\dfrac{\sin k(d - |x' - x_i|)}{\sin kd} \right], & |x - x_i| \leq d \\ 0, & \text{otherwise,} \end{cases} \quad (7)$$

where d is the length of each segment given by $d = L/N$. By substituting the $N - 1$ term current expansion series into (1), and by making use of the identity $(\partial/\partial_x) = -(\partial/\partial_x')$, it can be shown that the x component of the electric field E is given at any point (x, y, B) on the substrate by

$$E_x(x, y, B) = \frac{k}{\sin kd} \sum_{i=2}^{N} I_x \Big[\Pi_x|_{x_{i-1}} + \Pi_x|_{x_{i+1}} - 2\cos kd \Pi_x|_{x_i} + 2\cos kd \Pi|_{x_i} - \Pi|_{x_{i-1}} - \Pi|_{x_{i+1}} + k \Big[\int_{x_{i-1}}^{x_i} \Pi \, dx' \sin k(x' - x_{i-1}) + \int_{x_i}^{x_{i+1}} \Pi \, dx' \sin k(x_{i+1} - x') \Big] \Big],$$

where Π_x is given by (2) and

$$\Pi = 2(\epsilon_r - 1)K \lim_{z \to B} \int_0^\infty J_0(\lambda\rho) \cdot e^{-\mu(z-B)} \frac{\lambda\mu \, d\lambda}{D_e(\lambda) \cdot D_m(\lambda)}$$

with

$$\rho = \sqrt{(x - x')^2 + y^2}.$$

Note that (8) does not involve derivatives of Π_x and Π_z. These derivatives yield improper integrals for $z = B$. Furthermore, the expression for E_x is in closed form, as it involves integrals which are convergent in nature.

For numerical convergence, sinusoidal functions have also been chosen as testing functions. Therefore each side of (8) is multiplied by the testing functions given by (7), with the subscript i replaced by j and the variable x' by x. The integration with respect to x reduces the problem to the matrix form

$$[V] = [I][Z],$$

where I is related to the currents on the subsections and V to the electromagnetic excitation column. Depending upon the feeding point, the corresponding excitation voltage is set to unity in the excitation column.

Matrix inversion is a simple process for high-speed computers, and hence the current distribution can be determined once a well-conditioned impedance matrix $[Z]$ is obtained. The ease and speed of evaluation of the matrix elements is of considerable importance. Evaluation of the matrix elements involves calculation of the following integrals:

$$I_1 = \int_{x_{j-1}}^{x_j} \Pi_x \sin\left[k(x - x_{j-1})\right] dx + \int_{x_j}^{x_{j+1}} \Pi_x \sin\left[k(x_{j+1} - x)\right] dx$$

and

$$I_2 = \int_{x_{j-1}}^{x_j} \sin\left[k(x - x_{j-1})\right] dx \cdot \left[\int_{x_{i+1}}^{x_i} \Pi \sin\left[k(x' - x_{i-1})\right] dx' + \int_{x_i}^{x_{i+1}} \Pi \sin\left[k(x_{i+1} - x')\right] dx'\right]$$
$$+ \int_{x_j}^{x_{j+1}} \sin\left[k(x_{j+1} - x)\right] dx \cdot \left[\int_{x_{i-1}}^{x_i} \Pi \sin k(x' - x_{i-1})\, dx' + \int_{x_i}^{x_{i+1}} \Pi \sin\left[k(x_{i+1} - x')\right] dx'\right].$$

The double integration with respect to x and x' is performed very efficiently by a numerical technique which makes use of the overlapping nature of the Green's function and removes redundancy in the calculation of the integrand. The integrals (2) and (9) are of the Sommerfeld-type; they require special treatment which is described in the next section.

III. EVALUATION OF THE SOMMERFELD-TYPE INTEGRALS

The Sommerfeld-type integrals given by (2) and (9) contain the two radicals given by μ and μ_e which are double-valued functions of the complex variable λ. However, the sign of the radical represented by μ_e does not affect the single-valuedness of the integrands since the terms involving μ_e are even functions of λ. This imples then that only the branch points contributed by μ need be considered. In order that the integrals represent a proper field solution, the integrand must be single-valued all along the path of integration. This is accomplished by drawing branch cuts through the branch points $\lambda = \pm k$. The choice of the branch cuts is arbitrary as far as single-valuedness is concerned; however, the convergence of the integrals is another matter. The convergence of the integrals and outgoing wave propagating character require Re $\lambda > 0$ and Im $\lambda < 0$, respectively. The choice of the branch cuts and the path of integration are shown in Fig. 2.

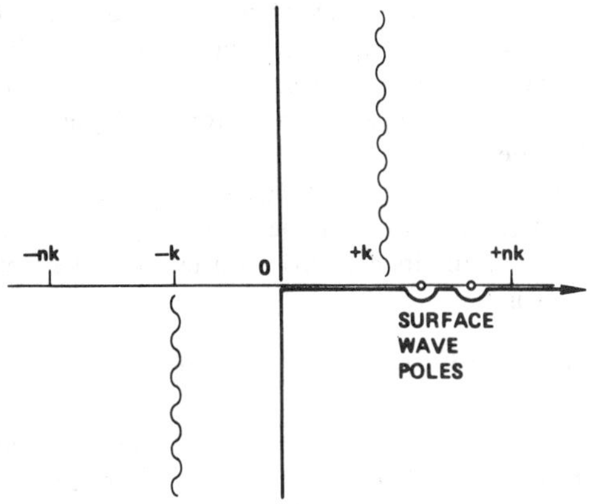

Fig. 2. Path of integration.

A numerical integration of the integrals given by (2) and (9) has been performed by real-axis integration over the $0 \leqslant \lambda < \infty$ interval. The convergence of these integrals can be enhanced due to the presence of the factor $\exp(-\mu(z - B))$ in the integrands. When the antenna is printed on the substrate $z = B$, this factor tends to unity. For this case the integrals given by (2) and (9) are proper but very slowly convergent. The numerical integratioin is carried out by employing Simpson's rule up to $\lambda = Q$, where Q is chosen so that the relation coth $(BQ) \cong 1.0$ is satisfied. When $\lambda > Q$ the integral can be evaluated analytically. The integration over the range $0 \leqslant \lambda < \infty$ is divided into the following subintervals.

1) $0 < \lambda < k$: Integration over this interval is performed numerically. The integrals contribute to the radiation resistance and to the reactive part of the input impedance.

2) $k < \lambda < (\epsilon_r)^{1/2} k$: The integrands have poles in this interval. The poles correspond to the surface wave modes excited by the antenna. The number of modes excited depends upon the thickness B of the substrate, its dielectric constant ϵ_r, and the frequency of operation. The modes are zeros of the factors

$$D_e(\lambda) = \mu + \mu_e \coth \mu_e B, \qquad \text{TE modes}$$

$$D_m(\lambda) = \epsilon_r \mu + \mu_e \tanh \mu_e B, \qquad \text{TM modes.}$$

By using the singularity extraction technique, the integral can be written as

$$\int_k^{nk} \frac{f(\lambda)}{(\lambda - \lambda_0)}\, d\lambda = \int_k^{nk} \frac{f(\lambda) - f(\lambda_0)}{(\lambda - \lambda_0)}\, d\lambda + \int_k^{nk} \frac{f(\lambda_0)}{(\lambda - \lambda_0)}\, d\lambda, \tag{10}$$

where $n = (\epsilon_r)^{1/2}$ is the index of refraction of the substrate. The first integral on the right-hand side of (10) can be evaluated numerically, while the second integral can be integrated analytically. This technique can be extended depending on the number of modes.

3) $nk < \lambda < Q$: Numerical integration is again invoked here. The choice of the number Q depends mainly on the thickness of the substrate B and its dielectric constant ϵ_r. The factor Q is such that it satisfies

$$\coth\left(B\sqrt{\lambda^2 - n^2k^2}\right) \cong 1.0, \qquad \text{for all } \lambda > Q. \tag{11}$$

4) $Q < \lambda < \infty$: The special nature of the integrand is involved in this interval. The use of (11) simplifies the integrand to a form which involves the radicals μ, μ_e. Since $\lambda > k$ and nk, after application of the binomial series expansion to these radicals the tail part of the integrals can be written as

$$2\int_Q^\infty \frac{J_0(\lambda\rho)}{\mu + \mu_e}\, d\lambda$$

$$= \int_Q^\infty d\lambda J_0(\lambda\rho)\left[1 + \frac{XL}{\lambda^2} + \frac{XT}{\lambda^4} + \frac{XC}{\lambda^6} + \cdots\right],$$

where $XL = Lk^2$, $XT = (L^2 + M)k^4$, $XC = (L^3 + 2LM + N)k^6$ with $L = (\epsilon_r + 1)/4$, $M = (\epsilon_r^2 + 1)/16$, and $N = (\epsilon_r^3 + 1)/32$. Thus the integrand has been reduced to a highly convergent series, and therefore term by term integration can be performed analytically.

IV. INPUT IMPEDANCE AND RADIATION PATTERN

The problem for the current distribution of the printed dipole was formulated in the previous section. The solution described uses sinusoidal subsection currents and Galerkin's method, while the dipole is assumed to be excited by an idealized source, the delta gap generator. In air it has been found that the input impedance is not critically dependent upon the excitation gap length as long as it is $< 0.1\ \lambda_0$. Therefore, neglecting the effect of the incoming excitation line, the input impedance for a small gap between the input terminals will be as seen by the delta gap generator. In the computations of current distribution, the circumferential variation of the antenna has been neglected. This is due to the fact that whatever device is used to measure the input impedance, it is insensitive to the circumferential variation of the current, provided the wire is very thin. The input impedance of the printed wire dipoles for a 1.0-V input excitation is given by

$$Z_{in} = 1.0/I_{in},$$

where I_{in} is the current at the input terminals of the antenna, obtained from knowledge of the current distribution.

In order to compute the radiation pattern, the stationary phase method is used [10]. The far-zone electric field components for $\theta < 90°$ are given by

$$E_\theta(\theta, \phi) = 2k^2\Pi_\theta \tag{13}$$

and

$$E_\phi(\theta, \phi) = 2k^2\Pi_\phi, \tag{14}$$

where Π_θ and Π_ϕ are expressed as

$$\Pi_\theta = K\frac{e^{jkr}}{r} \cdot \left[\frac{e^{-jkB\cos\theta}}{D_e(\theta,\phi)}\cos\theta + \frac{n^2-1}{n^2}\,\frac{\tan\theta\sin\theta}{D_m(\theta,\phi)}\cos\phi\right] I_x \tag{15}$$

and

$$\Pi_\phi = K\frac{e^{jkr}}{r}\,\frac{e^{-jkB\cos\theta}}{D_e(\theta,\phi)}\sin\phi I_x.$$

The parameter I_x is in this case

$$I_x = \int_{L_x} I(x')e^{-jkx\sin\theta\cos\phi}\, dx', \tag{17}$$

and it is computed numerically.

V. NUMERICAL RESULTS

Based on the previously developed analysis numerical computations have been performed for a very thin wire dipole with a typically chosen radius of $a = 0.0001\ \lambda_0$. The dipoles are center fed by an in-phase delta gap generator, and all pertinent dimensions are normalized to the free-space wavelength λ_0.

The computations have been performed so that the effect of one- or two-surface modes to the antenna current distribution is clarified. This is accomplished by considering a quartz-like substrate ($\epsilon_r \cong 3.25$) [11] for a single-surface mode and alumina-like substrate ($\epsilon_r \cong 8.5$) for the two-surface modes. Fig. 3 shows the Re $[I]$ and Im $[I]$ components of the dipoles current distribution for a dipole length of 0.5 λ_0, substrate thickness of $B = 0.1016\ \lambda_0$, and $\epsilon_r = 3.25$. The current distribution presents a dip at the feeding point because of the fact that the presence of the substrate increases the effective electrical length of the dipole [11]. When $\epsilon_r = 1.0$ (air substrate) the current distribution is about ten times as large in magnitude over the printed dipole, as indicated by a comparison between Fig. 3 and 4. An increase in the substrate thickness to $B = 0.127\ \lambda_0$ is seen to change the current level but to have a minor influence on the current distribution (see Fig. 5). This indicates that the input impedance is sensitive to changes in B. A substantial increase in B or a moderate increase in B to 0.15 λ_0, and a change of the substrate dielectric constant to $\epsilon_r = 8.5$, allows two propagating surface wave modes. The current distribution for this latter case is shown in Fig. 6. The figure indicates that both Re $[I]$ and Im $[I]$ undergo sign changes, and more importantly that there is a substantial increase in the effective electrical length of the dipole.

It can be inferred from the above discussion that the impedance level of the printed dipole is higher than the dipole, over the grounded plane. This higher impedance level is attributed to the fact that the printed dipole radiates power as space waves and launches a surface wave in the dielectric surface guide. However, this surface wave contributes only to the real part of the input impedance. The higher order non-

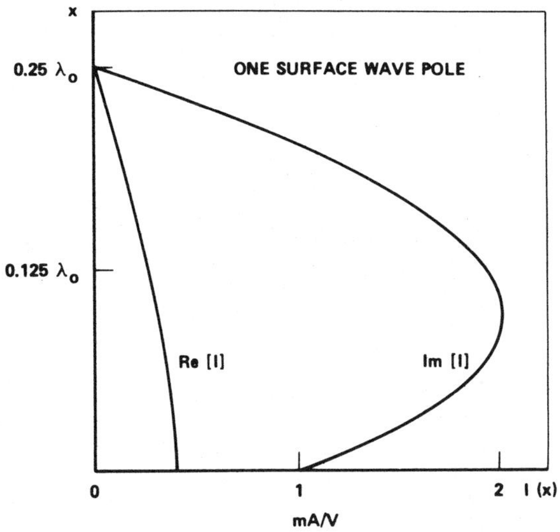

Fig. 3. Current distribution on a printed wire dipole Re[I] and Im[I] versus x (in λ_0), (B = 0.1016 λ_0, ϵ_r = 3.25, a = 0.00005 λ_0).

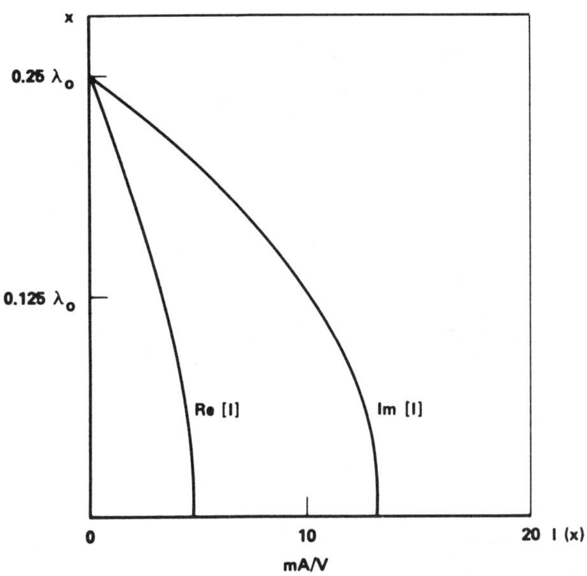

Fig. 4. Current distribution on a printed wire dipole Re[I] and Im[I] versus x (in λ_0), (B = 0.1016 λ_0, ϵ_r = 1.0, a = 0.00005 λ_0).

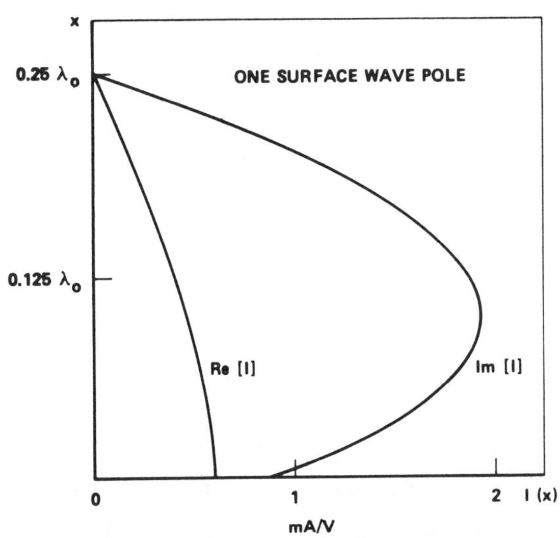

Fig. 5. Current distribution on a printed wire dipole Re[I] and Im[I] versus x (in λ_0), (B = 0.127 λ_0, ϵ_r = 3.25, a = 0.00005 λ_0).

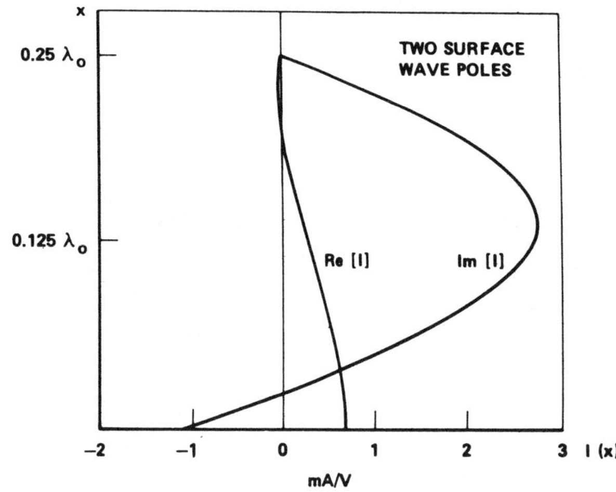

Fig. 6. Current distribution on a printed wire dipole Re[I] and Im[I] versus x (in λ_0), (B = 0.15 λ_0, ϵ_r = 8.5, a = 0.00005 λ_0).

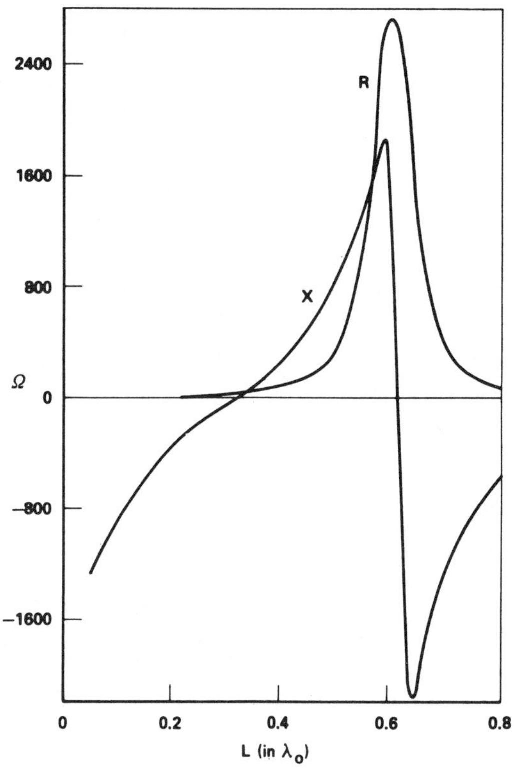

Fig. 7. Input impedance of a printed wire dipole versus L (in λ_0), (B = 0.1016 λ_0, ϵ_r = 3.25, a = 0.00005 λ_0).

propagating modes contribute to the reactive part of the input impedance. Figs. 7–9 show the input impedance of the printed dipole versus its length L, for the different substrate thicknesses and dielectric constants. Comparison of the resonant length L_r and the input impedance Z_{in} for all the above-mentioned cases is given in Table I. The real part of the Z_{in} is the sum of two terms, namely, the radiation resistance and the surface wave launching resistance. By increasing the substrate thickness from 0.1016 to 0.127 λ_0, the resonant impedance increases from 34.5 to 60 Ω. This increase in value of the resonant input impedance is essentially due to the fact that an increase in B (still referring to a single surface wave mode) increases both the surface wave launching as well as radiation efficiencies.

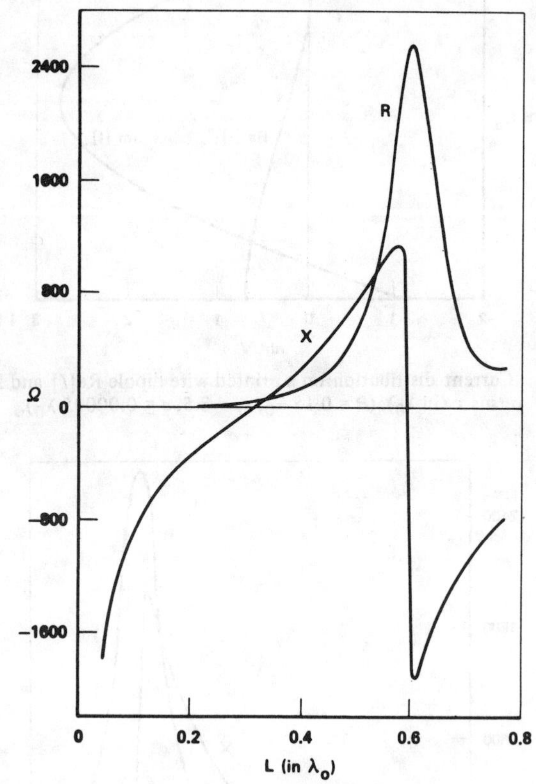

Fig. 8. Input impedance versus antenna length L (in λ_0), (B = 0.127 λ_0, ϵ_r = 3.25, a = 0.00005 λ_0).

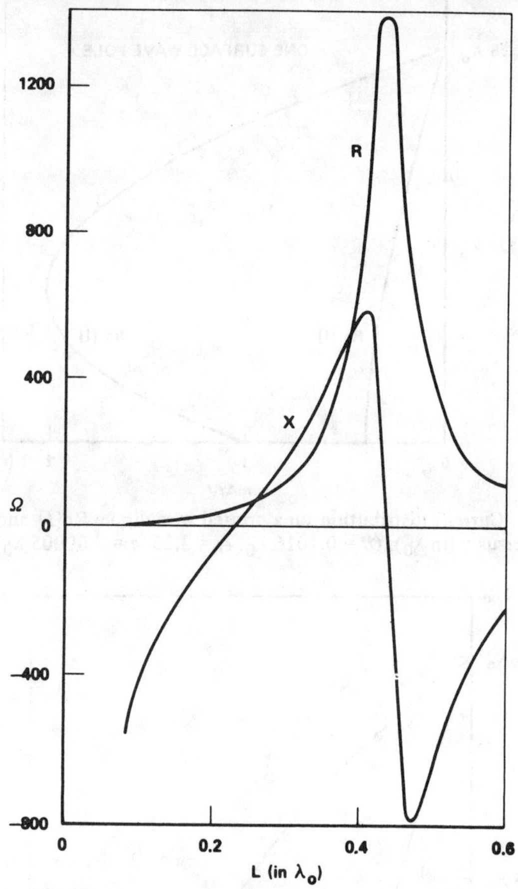

Fig. 9. Input impedance of a printed wire dipole versus L (in λ_0), (B = 0.15 λ_0), ϵ_r = 8.5, a = 0.000025 λ_0).

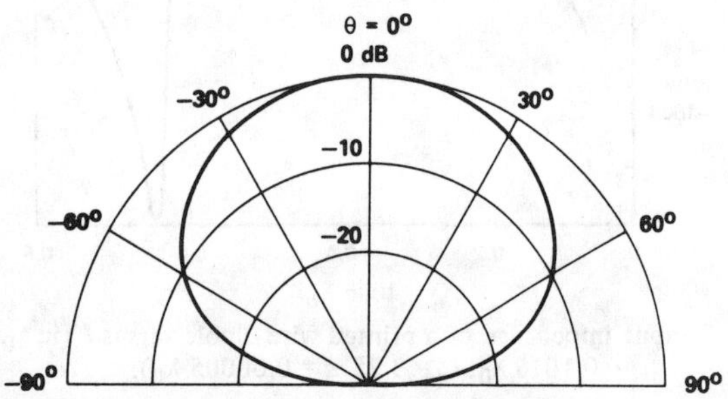

Fig. 10. E-plane radiation plot of a printed wire dipole (L = 0.333 λ_0, ϵ_r = 3.25).

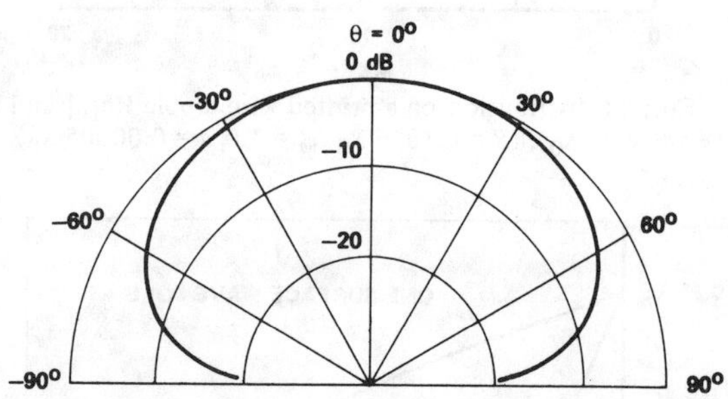

Fig. 11. H-plane radiation plot of a printed wire dipole (L = 0.333 λ_0, B = 0.1016 λ_0, ϵ_r = 3.25).

TABLE I
PRINTED DIPOLE PARAMETERS FOR DIFFERENT VALUES OF B AND ϵ_r

Number of modes	ϵ_r	B	L_r	Z_{in} (resonant)	Z_{in} (for L = 0.5 λ_0)
1	3.25	0.1016 λ_0	0.317 λ_0	34.5 + j0 Ω	330 − j880 Ω
1	3.25	0.127 λ_0	0.315 λ_0	60.0 + j0 Ω	535 − j788 Ω
2	8.50	0.150 λ_0	0.230 λ_0	50.0 + j0 Ω	418 + j664 Ω

The two propagating surface wave modes also effect the impedance characteristics of the printed dipole. The increased dielectric constant ($\epsilon_r \doteq 8.5$) reduces the radiation resistance (efficiency) of the printed dipole [11].

Lastly, E- and H-plane radiation patterns of the printed dipole are shown in Figs. 10 and 11. Near the air dielectric interface, i.e., $\theta = \pm 90°$, the radiation plots are not exact since the stationary phase method used in the development of the formula for the radiated fields is not valid when $\theta = \pm 90°$.

VIII. CONCLUSION

The radiation characteristics of wires printed on a grounded substrate have been investigated. The current distribution has been obtained by employing moment methods to solve Pocklington's equation. The improperness of the Sommerfeld-type integrals, when the field and source points are on the substrate, has been circumvented by employing a correct limiting process and sinusoidal expansion as well as testing functions for the numerical integration along the real axis. In addition, the effect of surface wave modes on the current distribution and input impedance was determined.

REFERENCES

[1] R. E. Munson, "Conformal microstrip antennas and microstrip phased arrays," *IEEE Trans. Antennas Propagat.*, vol. AP-22, pp. 74–78, Jan. 1974.

[2] J. E. Howell, "Microstrip antennas," *IEEE Trans. Antennas Propagat.*, vol. AP-23, pp. 90–93, Jan. 1975.

[3] A. G. Derneryd, "A theoretical investigation of the rectangular microstrip antenna element," *IEEE Trans. Antennas Propagat.*, vol. AP-26, pp. 532–535, July 1978.

[4] Y. T. Lo, D. Solomon, and W. F. Richards, "Theory and experiment on microstrip antennas," *IEEE Trans. Antennas Propagat.*, vol. AP-27, pp. 137–145, Mar. 1979.

[5] P. K. Agarwal and M. C. Bailey, "An analysis technique for microstrip antennas," *IEEE Trans. Antennas Propagat.*, vol. AP-25, pp. 756–759, Nov. 1977.

[6] D. A. Huebner, "An electrically small microstrip dipole planar array," in *Proc. Workshop on Printed Circuit Antenna Technol.*, Las Cruces, NM, Oct. 1979, pp. 17.1–17.16.

[7] H. G. Oltman and D. A. Huebner, "Electromagnetically coupled microstrip dipoles, *IEEE Trans. Antennas Propagat.*, vol. AP-29, pp. 152–158, Jan. 1981, this issue.

[8] D. B. Rutledge, S. E. Schwarz, and A. T. Adams, "Infrared and submillimeter antennas," *Infrared Phys.*, vol. 18, pp. 713–729, Dec. 1978.

[9] K. Mizuno, Y. Daiku, and S. Ono, "Design of printed resonant antennas for monolithic-diode detectors," *IEEE Trans. Microwave Theory Tech.*, vol. MTT-25, pp. 470–476, June 1977.

[10] G. N. Tsandoulas, "Excitation of a grounded dielectric slab by a horizontal dipole," *IEEE Trans. Antennas Propagat.*, vol. AP-17, pp. 156–161, Mar. 1969.

[11] N. K. Uzunoglu, N. G. Alexopoulos, and J. G. Fikioris, "Radiation properties of microstrip dipoles," *IEEE Trans. Antennas Propagat.*, vol. AP-27, pp. 853–858, Nov. 1979.

[12] G. Dubost, "Theory and experiments of a broadband short-circuited microstrip dipole at resonance," in *Proc. Workshop on Printed Circuit Antenna Technol.*, Las Cruces, NM, Oct. 17–19, 1979.

[13] A. Sommerfeld, *Partial Differential Equations in Physics*. New York: Academic, 1949, vol. VI.

[14] J. R. Wait, "Characteristics of antennas over lossy earth," in *Antenna Theory, Part 2*. New York: McGraw-Hill, 1969, ch. 23.

[15] P. Parhami, Y. Rahmat-Samii, and R. Mittra, "An efficient approach for evaluating the Sommerfeld integrals encountered in the current element radiating over lossy ground," *IEEE Trans. Antennas Propagat.*, vol. AP-28, pp. 100–104, Jan. 1980.

[16] J. A. Kong, "Electromagnetic fields due to dipole antennas over stratified anisotropic media," *Geophys.*, vol. 37, pp. 985–996, Dec. 1972.

[17] G. J. Burke, E. K. Miller, J. N. Brittingham, D. L. Lager, R. J. Lytle, and J. T. Okada, "Computer modeling of antennas near the ground," Lawrence Livermore Lab. Rep. UCID-18626, May 13, 1980.

[18] P. Tulyanthan and P. K. Newman, "The circumferential variation of the axial component of current in closely space thin wire antennas," *IEEE Trans. Antennas Propagat.*, vol. AP-27, Jan. 1979.

Inam E. Rana was born December 14, 1949, in Pasrar, Pakistan. He obtained the B.S.E.E. degree in 1973 from the University of Engineering and Technology Lahore, Pakistan. He also obtained the M.S.E.E. and Ph.D. degrees from the University of California, Los Angeles, CA in 1977 and 1979, respectively.

He is currently working for Suparco in Karachi, Pakistan. His current research interests are in the area of electromagnetic theory with applications to integrated microwave circuits and microstrip antennas.

Nicolaos G. Alexopoulos (S'68–M'69) was born in Athens, Greece, March 30, 1942. He obtained the B.S.E.E. degree in 1964, the M.S.E.E. degree in 1967 and the Ph.D. degree in 1968 at the University of Michigan, Ann Arbor, MI.

Since the Fall 1969 he has been with the Electrical Sciences and Engineering Department, University of California, Los Angeles. His research interests are in the area of applied electromagnetic theory with current emphasis on integrated microwave circuits, antennas, and scattering. He is the author of over 40 publications in scientific literature.

A space domain integral equation approach in the analysis of dielectric-covered slots

Pisti B. Katehi

Electrical Engineering and Computer Science Department, University of Michigan, Ann Arbor

(Received July 13, 1987; revised October 21, 1988; accepted October 25, 1988.)

Coupling through dielectric-covered slots is becoming a very important excitation mechanism for antennas and arrays in microwave and millimeter wave frequencies. The characteristics of these antennas can be analyzed by solving appropriate integral equations. A space domain method of moments applied for solving these equations may complicate the existing Sommerfeld integrals even more. In this paper a combination of analytical and numerical techniques has been developed which reduces the complexity of the integrals and improves efficiency and accuracy in their computation.

INTRODUCTION

Electromagnetic coupling through apertures is encountered in many applications. A few examples are the microstrip-to-microstrip coupling [*Pozar*, 1986], waveguide-to-waveguide coupling, and waveguide–to–exterior space coupling [*Elliott and O'Loughlin*, 1986; *Robertson and Elliott*, 1987; *Yang and Alexopoulos*, 1987]. When these apertures are in the proximity of the interface of electrically dissimilar materials, the resulting electromagnetic fields involve Sommerfeld type integrals [*Pozar*, 1986; *Katehi*, 1987].

Coupling of electromagnetic fields through dielectric-covered slots is used widely in the excitation of microwave and millimeter wave antennas. The development of models for such radiating structures necessitates the consideration of special techniques to solve the resulting Sommerfeld type integrals. In order to evaluate the electrical properties of the above antennas the electric field distribution on the slot must be obtained by solving an appropriate integral equation. Because of the existence of the dielectric layer the integrals appearing are slowly convergent and require special attention.

Rigorous solutions to Sommerfeld integrals have been reported by many authors and are divided into real axis [*Tsandoulas*, 1969; *Uzunoglu et al.*, 1979; *Rana and Alexopoulos*, 1981; *Pozar*, 1982] and imaginary axis [*Mosig and Gardiol*, 1985] integration techniques.

The real axis integration techniques used an integral representation for the microstrip Green's function in the form of improper Sommerfeld integrals. These integrals are computed numerically and take into account the existence of poles which for the case of zero dielectric or conductor losses lie along the real axis. In a similar manner the imaginary axis techniques employ an alternative integral representation which is obtained by deforming the original Sommerfeld type integration contour to one along the imaginary axis.

When the microstrip Green's function is used in a space or Fourier domain integral equation formulation, the resulting expressions involve multiple integrations of the above Sommerfeld type. As a result, solutions based on the previously described techniques suffer from excessive computation time and sometimes loss of accuracy. In order to preserve accuracy and improve efficiency the Sommerfeld integration cannot be considered alone but must be considered as part of a more complicated multiple integration scheme.

In this paper, techniques are developed to address this need. The validity of these techniques is more general and can be applied to any problem involving multiple dielectric interfaces. To demonstrate the developed analytical and numerical techniques, the problem of scattering by a slot on a dielectric-covered plane conductor will be considered (Figure 1).

For the solution of the resulting integral equation the method of moments is applied to reduce it into a system of linear equations. The unknown vector is the desired electric field distribution, while the known vector is derived from the electromagnetic field exciting the slot. The matrix to be inverted in-

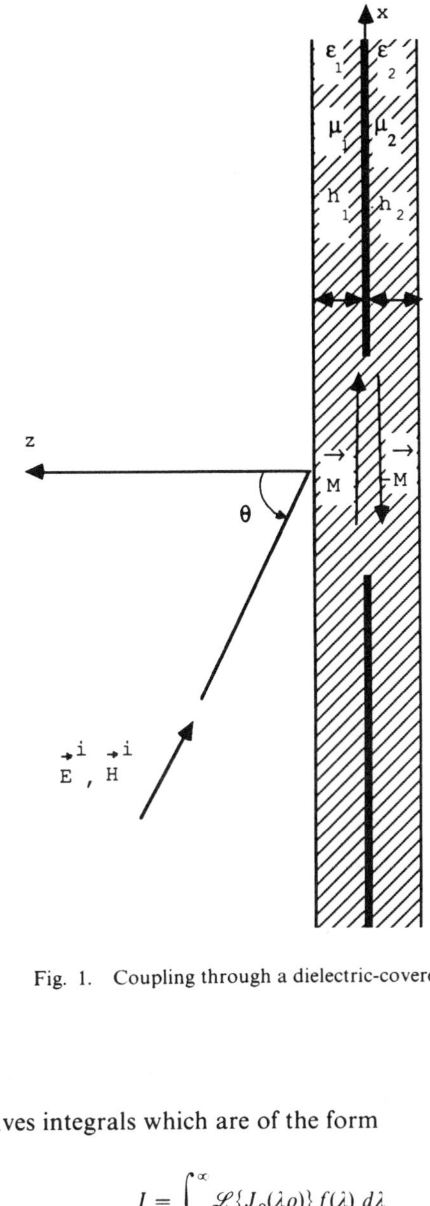

Fig. 1. Coupling through a dielectric-covered slot.

volves integrals which are of the form

$$I = \int_0^\infty \mathscr{L}\{J_0(\lambda\rho)\} f(\lambda)\, d\lambda \qquad (1)$$

In (1), $\mathscr{L}$ is a self-adjoint operator which involves integration with respect to spatial coordinates x, y, while $f(\lambda)$ is a transcendental function with discrete singularities. The number of these singularities (poles) depends on the electrical characteristics of the two dielectric covers [*Katehi and Alexopoulos*, 1983]. The difficulties encountered in the evaluation of the above integrals are resolved by applying a combination of analytical and numerical techniques. This procedure introduces an error which is of the order of 10^{-4} of the correct value of the integrals.

THEORY

Integral equation

Consider a slot on a dielectric-covered infinite conducting plate which is illuminated by an incident plane wave, as shown in Figure 1. The equivalence principle is used to divide the problem into two equivalent problems (Figure 2). In region b the field is produced by the incident field $\mathbf{E}^i$, $\mathbf{H}^i$ in this region and the equivalent magnetic current.

$$\mathbf{M} = \hat{n} \times \mathbf{E}^b \qquad (2)$$

over the slot region, with the slot covered by an electric conductor. In region c the field is produced by the equivalent magnetic current $-\mathbf{M}$ over the same slot covered by an electric conductor. The fact that the equivalent current in region b is the negative of that in region c ensures that the tangential component of the electric field is continuous across the aperture. The remaining boundary condition to be applied is continuity of the tangential component of the magnetic field across the aperture.

The tangential component $\mathbf{H}_t^b$ of the magnetic field in region b over the slot is the sum of that due to the incident plane wave $\mathbf{H}_t^i$ plus that due to the equivalent source $\mathbf{M}$, denoted by $\mathbf{H}_t^b(\mathbf{M})$. Therefore

$$\hat{n} \times [\mathbf{H}^b(\mathbf{M}) + \mathbf{H}^c(\mathbf{M})] = \hat{n} \times \mathbf{H}^i \qquad (3)$$

The magnetic fields $\mathbf{H}^b$ and $\mathbf{H}^c$ are given in terms of the magnetic current, as shown below [*Katehi*, 1987]:

$$\mathbf{H}^{b,c}(\mathbf{M}) = j\omega \iint_{\text{slot}} \left[(\mathsf{F}^{b,c}(\mathbf{r}/\mathbf{r}')) + \nabla\nabla\left(\frac{\mathsf{F}^{b,c}(\mathbf{r}/\mathbf{r}')}{k_{b,c}{}^2}\right)\right] \cdot \mathbf{M}(\mathbf{r}')\, ds' \qquad (4)$$

In (4), $k_{b,c}$ and $\mathsf{F}^{b,c}$ are the wave numbers and the dyadic Green's functions of the dielectric slabs, respectively, in regions b and c. For a thin slot the transverse component of the electric field is much larger than the longitudinal component ($E_y \gg E_x$). This results in one-directional magnetic current and therefore a single integral equation

$$\iint_{\text{slot}} K_{xx}(\mathbf{r}/\mathbf{r}') M_x(\mathbf{r}')\, ds' = -H_x^i \qquad z = 0 \qquad (5)$$

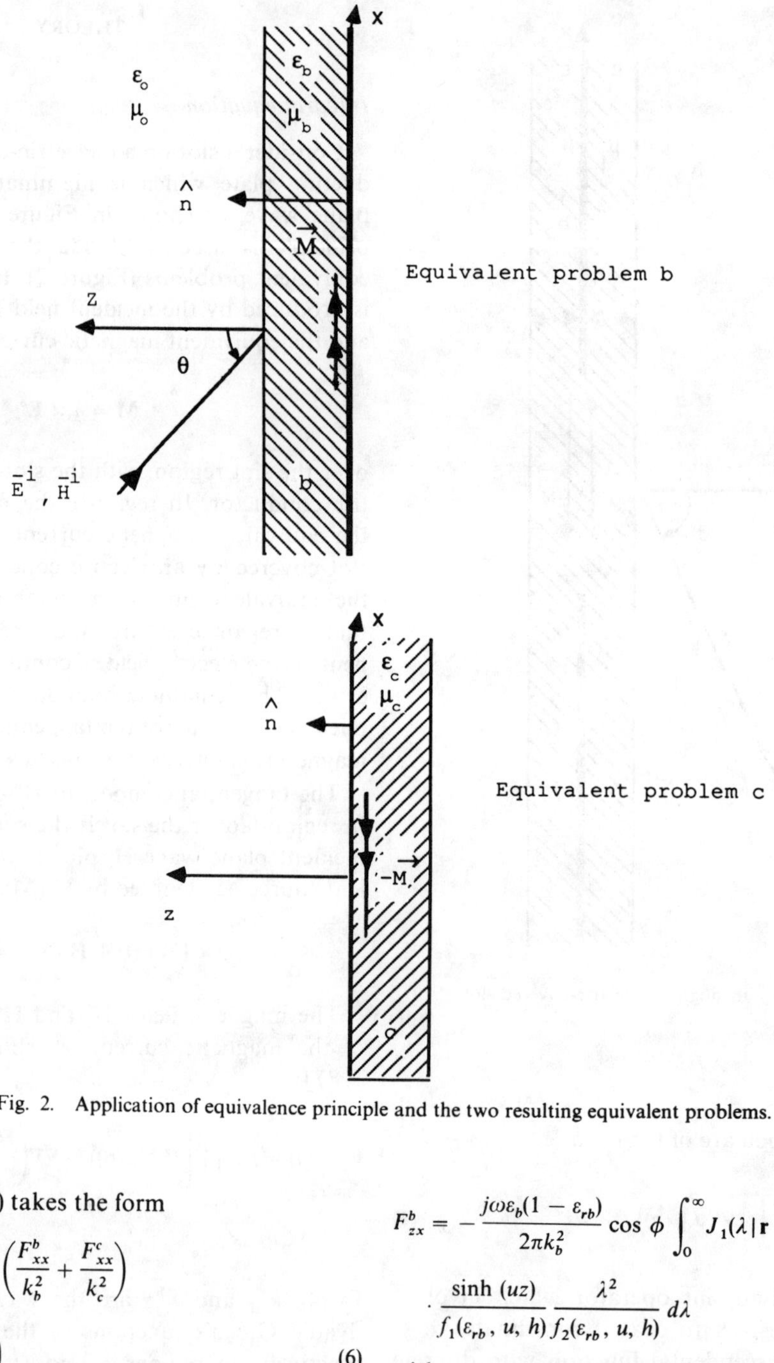

Fig. 2. Application of equivalence principle and the two resulting equivalent problems.

In this case, $K_{xx}(\mathbf{r}/\mathbf{r}')$ takes the form

$$K_{xx}=(F^b_{xx}+F^c_{xx})+\frac{\partial^2}{\partial x^2}\left(\frac{F^b_{xx}}{k_b^2}+\frac{F^c_{xx}}{k_c^2}\right)+\frac{\partial^2}{\partial x\,\partial z}\left(\frac{F^b_{zx}}{k_b^2}+\frac{F^c_{zx}}{k_c^2}\right) \tag{6}$$

with F^b_{xx} and F^b_{zx} given by [*Katehi*, 1987]

$$F^b_{xx}=-\frac{j\omega\varepsilon_b}{2\pi k_b^2}\int_0^\infty J_0(\lambda|\mathbf{r}-\mathbf{r}'|)\cdot\frac{\lambda}{u}\,\frac{u\cosh[u(h+z)]-\varepsilon_{rb}u_0\sinh[u(h+z)]}{f_1(\varepsilon_{rb},u,h)}\,d\lambda \tag{7}$$

$$F^b_{zx}=-\frac{j\omega\varepsilon_b(1-\varepsilon_{rb})}{2\pi k_b^2}\cos\phi\int_0^\infty J_1(\lambda|\mathbf{r}-\mathbf{r}'|)\cdot\frac{\sinh(uz)}{f_1(\varepsilon_{rb},u,h)}\,\frac{\lambda^2}{f_2(\varepsilon_{rb},u,h)}\,d\lambda \tag{8}$$

with

$$f_1(\varepsilon_{rb},u,h)=\varepsilon_{rb}u_0\cosh(uh)+u\sinh(uh) \tag{9}$$

$$f_2(\varepsilon_{rb},u,h)=u\cosh(uh)+u_0\sinh(uh) \tag{10}$$

$$u_0=(\lambda^2-k_0^2)^{1/2}\qquad u=(\lambda^2-k^2)^{1/2} \tag{11}$$

In (7) and (8) the functions $f_1(\varepsilon_{rb}, u, h)$ and $f_2(\varepsilon_{rb}, u, h)$ are analytic with discrete zeros. The contribution from these zeros in the Sommerfeld integration gives the power propagating in the substrate in the form of transverse electric (TE) and transverse magnetic (TM) surface waves. The functions F_{xx}^{c} and F_{zx}^{c} are given by relations similar to (7) and (8) with ε_c, ε_{rc}, and k_c substituting for ε_b, ε_{rb}, and k_b, respectively.

To obtain the unknown magnetic current $M_x(\mathbf{r}')$ in (5), the method of moments is applied. The unknown current is assumed to be in the form

$$M_x(\mathbf{r}') = \sum_n V_n f_n(x')g(y') \tag{12}$$

where $f_n(x')$ are piecewise sinusoidal functions and $g(y')$ has been chosen so that the edge condition is satisfied (see appendix).

The application of the method of moments reduces (5) into the following matrix equation:

$$[V_n] = [Y_{mn}]^{-1}[I_m] \tag{13}$$

where $[V_n]$ is the vector of the unknown coefficients and $[I_m]$ is the excitation vector given by

$$I_m = \langle -H_x^i, f_m \rangle \tag{14}$$

The admittance matrix $[Y_{mn}]$ can be split into two parts:

$$[Y_{mn}] = [Y_{mn}^b] + [Y_{mn}^c] \tag{15}$$

with

$$Y_{mn}^{b,c} = \langle f_m\, \delta(y),\ K_{xx}^{b,c}, f_n g \rangle \tag{16}$$

In (16), $\langle f_m\, \delta(y),\ K_{xx}^{b,c},\ f_n g\rangle$ indicates a double inner product of the form

$$\langle f_m\, \delta(y),\ K_{xx}^{b,c}, f_n g \rangle = \iint_{\text{slot}} dx\, dy \iint_{\text{slot}} dx'\, dy' \cdot f_m(x)\delta(y)K_{xx}^{b,c}(\mathbf{r}/\mathbf{r}')f_n(x')g(y') \tag{17}$$

In addition, $K_{xx}^{b,c}(\mathbf{r}/\mathbf{r}')$ is an integrodifferential operator operating on the zeroth-order, first-kind Bessel function $J_0(\lambda\rho)$ as

$$K_{xx}^{b,c}\{J_0(\lambda\rho)\} = \int_0^\infty d\lambda \left(f_{xx}^{b,c} + \frac{1}{k_{b,c}^2} f_{zx}^{b,c} \frac{\partial^2}{\partial x^2} \right) J_0(\lambda\rho) \tag{18}$$

The functions $f_{xx}^{b,c}$ and $f_{zx}^{b,c}$ are derived from $F_{xx}^{b,c}$ and $F_{zx}^{b,x}$, as shown by *Katehi* [1984].

In view of the above, (16) can take the form

$$Y_{mn}^{b,c} = \mathscr{K}_{xx}^{b,c}\{\langle f_m\, \delta(y),\ J_0(\lambda\rho), f_n g\rangle\} + \mathscr{K}_{zx}^{b,c}\left\{\left\langle f_m\, \delta(y),\ \frac{\partial^2}{\partial x^2} J_0(\lambda\rho), f_n g \right\rangle\right\} \tag{19}$$

where

$$\mathscr{K}_{xx}^{b,c} = \int_0^\infty d\lambda\, f_{xx}^{b,c} \tag{20}$$

$$\mathscr{K}_{zx}^{b,c} = \frac{1}{k_{b,c}^2} \int_0^\infty d\lambda\, f_{zx}^{b,c} \tag{21}$$

$$\rho = [(x - x')^2 + (y - y')^2]^{1/2} \tag{22}$$

For the solution of (13), most of the computation effort is spent on the evaluation of the elements of the admittance matrix. The complexity in these computations comes from the Sommerfeld integration as it is combined with multiple space integrals. As a result, these integrals need special treatment which consists of numerical and analytical techniques.

Evaluation of the elements of the generalized admittance matrix

It can be shown analytically that the semi-infinite integrodifferential operator in (7) has an integrand which exhibits multivalued behavior due to the radicals u, u_0 and has simple poles on the complex plane $\lambda = \lambda' + j\lambda''$ (Figure 3). For the case of lossless dielectric slabs these poles lie on the real axis in the interval (k_0, k), and their contribution corresponds to TM and TE surface waves propagating in the substrate [*Katehi*, 1984].

In order to simplify the computation of the elements of the admittance matrix the λ integration is split into two parts: The first one extends from 0 to A, where A satisfies the relationship $\tanh\,[(A^2 - k^2)^{1/2}h] \doteq 1$, and the second part extends from A to ∞.

As a result, the elements of the admittance matrix are considered as consisting of two parts:

$$Y_{mn}^{b,c} = Y_{mn}^{b,c}(A) + Y_{mn}^{b,c}(\infty) \tag{23}$$

Evaluation of $Y_{mn}^{b,c}(A)$

If we consider (17)–(23), the expression for $Y_{mn}^{b,c}(A)$ may be written as follows:

$$Y_{mn}^{b,c}(A) = \mathscr{K}_{xx}^{b,c}(A)\{\langle f_m\, \delta(y),\ J_0(\lambda\rho), f_n g\rangle\} + \mathscr{K}_{zx}^{b,c}(A)\left\{\left\langle f_m\, \delta(y),\ \frac{\partial^2}{\partial x^2} J_0(\lambda\rho), f_n g \right\rangle\right\} \tag{24}$$

where $\mathscr{K}_{xx}^{b,c}(A)$ and $\mathscr{K}_{zx}^{b,c}(A)$ are given by (20) and (21) with the λ integration extending over the interval $[0, A]$. The double inner products in (24) involve multi-

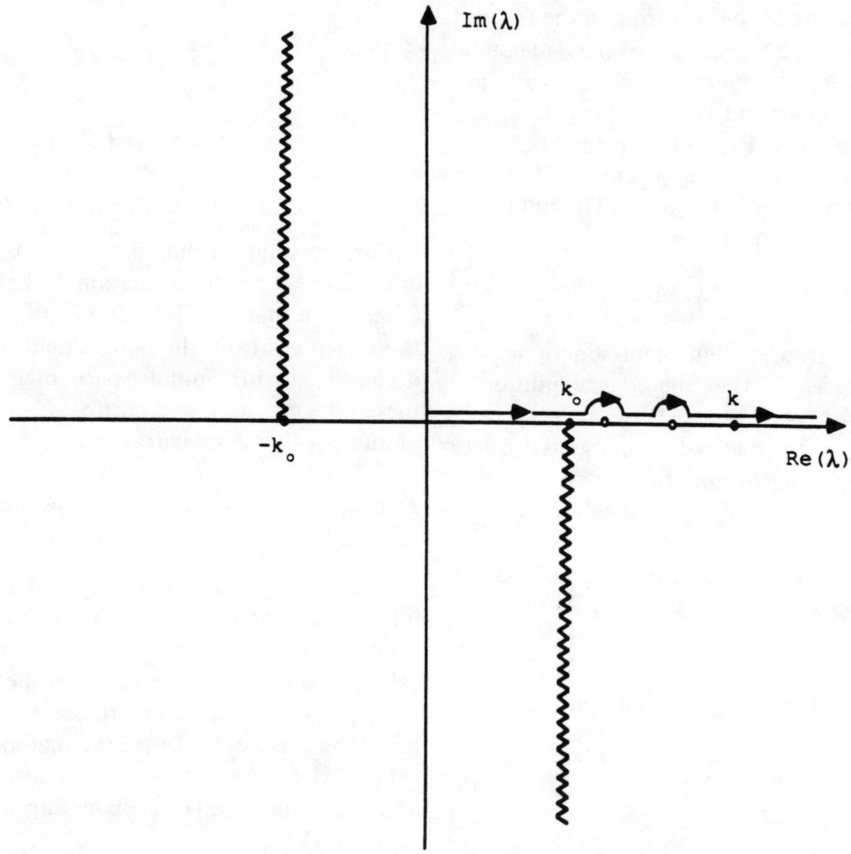

Fig. 3. Path of integration.

ple integrals. Numerical evaluation of this inner product for each pair (m, n) and for every $\lambda \varepsilon [0, A]$ can increase the computer time to prohibitively high levels restricting the applicability of the computer codes. One way to overcome this difficulty is to reduce the integrals from multiple to single ones.

From the known integral representations of $J_0(\lambda\rho)$ and $(\partial^2/\partial x^2)J_0(\lambda\rho)$,

$$J_0(\lambda\rho) = \frac{1}{\pi}\int_0^{\pi} e^{j\lambda(x-x')\cos\phi}\cos[\lambda(y-y')\sin\phi]\,d\phi \tag{25}$$

$$\frac{\partial^2}{\partial x^2}J_0(\lambda\rho) = \frac{1}{\pi}\int_0^{\pi}(i\lambda\cos\phi)^2 e^{j\lambda(x-x')\cos\phi}\cos[\lambda(y-y')\sin\phi]\,d\phi \tag{26}$$

the two double inner products can be reduced to the following simpler expressions:

$$\langle f_m\delta(y), J_0(\lambda\rho), f_n g\rangle = \sum_{v=0}^{\infty} C_v \int_0^{\pi} d\theta \cdot \left\{\frac{d^{2v}}{d|x|^{2v}} J_0\left[\lambda\left(x^2+\left(\frac{w}{2}\cos\theta\right)^2\right)^{1/2}\right]\right\}_{x=x_m-x_n} \tag{27}$$

$$\left\langle f_m\delta(y), \frac{\partial^2}{\partial x^2}J_0(\lambda\rho), f_n g\right\rangle = \sum_{v=0}^{\infty} C_v \int_0^{\pi} d\theta \cdot \left\{\frac{d^{2(v+1)}}{d|x|^{2(v+1)}} J_0\left[\lambda\left(x^2+\left(\frac{w}{2}\cos\theta\right)^2\right)^{1/2}\right]\right\}_{x=x_m-x_n} \tag{28}$$

where C_v is given in the appendix. The summations given by (27) and (28) are very fast converging and result in a very efficient and accurate computation of $Y_{mn}^{b,c}(A)$.

Evaluation of $Y_{mn}^{b,c}(\infty)$

The expression for this part of the elements of the generalized admittance matrix is given by

$$Y_{mn}^{b,c}(\infty) = \langle f_m \delta(y), K_{xx}^{b,c}(\infty), f_n g \rangle \tag{29}$$

where

$$K_{xx}^{b,c}(\infty) = \int_A^{\infty} d\lambda \left(f_{xx}^{b,c} + \frac{1}{k_{b,c}^2} f_{zx}^{b,c} \frac{\partial^2}{\partial x^2} \right) J_0(\lambda\rho) \tag{30}$$

Because of the assumption $\tanh\,[h(A^2 - k^2)^{1/2}] \doteq 1$, equation (4) can be simplified to the following form:

$$K_{xx}^{b,c}(\infty) = G_{xx}^{b,c}(\infty) - G_{xx}^{b,c}(A) \tag{31}$$

with

$$G_{xx}^{b,c}(\infty) = a_{xx}^{b,c}\left(\frac{1}{\rho} - \frac{\partial^2}{\partial x^2}\frac{1}{\rho}\right) + a_{zx}^{b,c}\frac{\partial^2}{\partial x^2}\left[\rho^2 + \left(\frac{2h}{b}\right)^2\right]^{-1/2} \tag{32}$$

$$G_{xx}^{b,c}(A) = \int_0^A d\lambda \left[a_{xx}^{b,c} + \left(a_{xx}^{b,c} + a_{zx}^{b,c} e^{-\lambda(2h/b)} \frac{\partial^2}{\partial x^2} \right) \right] J_0(\lambda\rho) \tag{33}$$

In (32) and (33), $a_{xx}^{b,c}$, $a_{zx}^{b,c}$, and b are constants which depend on A, k_b, or k_c only.

In view of (31), (29) gives

$$Y_{mn}^{b,c}(\infty) = \langle f_m \delta(y), G_{xx}^{b,c}(\infty), f_n g \rangle - \langle f_m \delta(y), G_{xx}^{b,c}(A), f_n g \rangle \tag{34}$$

The second part of (34) resembles $Y_{mn}^{b,c}(A)$ and therefore is evaluated using exactly the same techniques. In the first part the double inner products involved are evaluated using a combination of numerical and analytic integrations, as has been described extensively by *Katehi and Alexopoulos* [1983].

APPLICATIONS

Using the previously described method, the electric field distribution on the slot of Figure 1 is computed, and results are shown in Figures 4*a* and 5*a*. The difference in the field values comes mainly from the difference in the values of the dielectric constant of the layer in region *c*.

Also, the *E* and *H* plane patterns of the scattered fields are plotted in Figures 4*b* and 5*b*. It is expected that an appropriate combination of dielectric materials can easily modify the field distribution on the slot to desired values and the patterns of the scattered field to desired shapes.

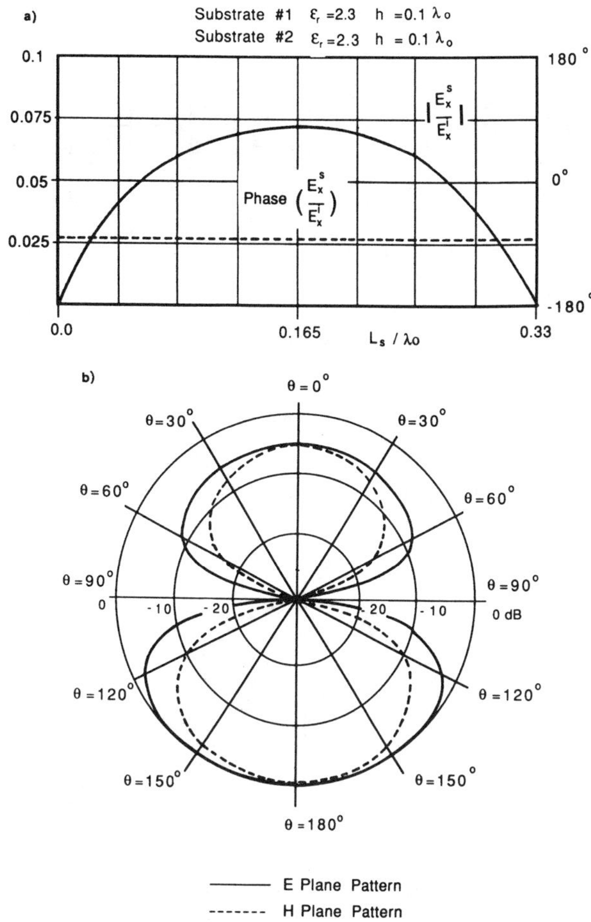

Fig. 4. (*a*) Electric field distribution on the slot ($\varepsilon_{r1} = \varepsilon_{r2} = 2.35$, $h_1 = h_2 = 0.1\lambda_0$, and $w = 0.05\lambda_0$). (*b*) Scattered field normalized with respect to its maximum value.

The computation of the above results was very fast owing to the special treatment of the Sommerfeld integrals. The amount of computer time required for the above computation was less than 10 CPU seconds.

CONCLUSIONS

In order to evaluate the electromagnetic coupling through dielectric-covered slots an integral equation which involves Sommerfeld type integrals has to be solved. These integrals include all the physical phenomena of the problem such as radiated space waves, evanescent waves, and surface wave modes propagating in the substrate. Because the resulting integrals are very slowly converging, special techniques are necessary for their proper evaluation.

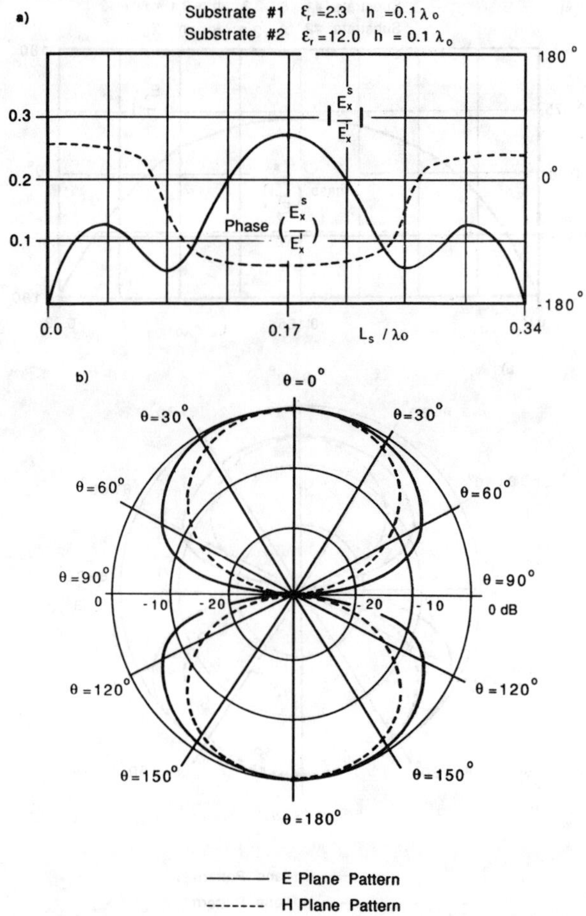

Fig. 5. (*a*) Electric field distribution on the slot for $\varepsilon_{r1} = 2.35$, $h_1 = 0.1\lambda_0$, $\varepsilon_{r2} = 12$, $h_2 = 0.1$, and $w = 0.05\lambda_0$. (*b*) Scattered field normalized with respect to its maximum value.

This paper has presented a systematic methodology for the accurate and efficient evaluation of the above integrals based on a real axis integration technique. The Sommerfeld integrals were part of a multiple integration scheme and as such were treated not alone but together with the other integrations, resulting in very good accuracy and short computer times.

APPENDIX

Evaluation of the double inner products

To obtain the unknown field distribution on the slot, the equivalent magnetic current M_x has to be computed. The integral equation (5) is solved numerically using the method of moments. The unknown current is assumed to be in the form

$$M_x(\mathbf{r}') = \sum_n V_n f_n(x')g(y') \tag{A1}$$

where

$$g(y') = (2/\pi w)[1 - (2y'/w)^2]^{-1/2} \tag{A2}$$

and $\{f_n(x'),\ n = 1, 2, \cdots, N\}$ is a set of expansion functions overlapping in nature. The expansion functions are given by

$$f_n(x') = \frac{\sin [k(x' - x_{n-1})]}{\sin (kl_x)} \qquad x_{n-1} \le x' \le x_n$$

$$f_n(x') = \frac{\sin [k(x_{n+1} - x')]}{\sin (kl_x)} \qquad x_n \le x' \le x_{n+1} \tag{A3}$$

$$f_n(x') = 0 \qquad \text{elsewhere}$$

where l_x is the subsection length and k the wave number in the dielectric.

In view of (25), (26), and (A1)–(A3) the double inner products of (24) take the form

$$\langle f_m \delta(y), J_0(\lambda\rho), f_n g\rangle = \int_{-w/2}^{w/2} dy \int_{-w/2}^{w/2} dy' \int_{x_{m-1}}^{x_{m+1}} dx \int_{x_{n-1}}^{x_{n+1}} dx'$$

$$\delta(y)g(y')f_m(x)f_n(x')J_0(\lambda\rho)$$

$$= \frac{1}{\pi}\int_0^{\pi} d\phi\ e^{j\lambda(x_m - x_n)\cos\phi}\{R(\lambda \cos \phi)\}^2$$

$$\int_0^{\pi} d\theta \cos\left[\left(\frac{\lambda w}{2} \cos \theta\right) \sin \phi\right] \tag{A4}$$

In (A4), R is a function of $\lambda \cos \phi$ given by

$$R(\lambda \cos \phi) = \sum_{\nu=0}^{\infty} A_{2\nu}(\lambda \cos \phi)^{2\nu} \tag{A5}$$

where

$$A_{2\nu} = \frac{2}{k^{2\nu+1}}\left\{[1 - \cos (kl_x)] + \sum_{\mu=1}^{\nu} (-1)^{\mu} \frac{(kl_x)^{2\mu}}{(2\mu)!}\right\} \tag{A6}$$

Similarly,

$$\left\langle f_m \delta(y), \frac{\partial^2}{\partial x^2} J_0(\lambda\rho), f_n g\right\rangle$$

$$= \frac{1}{\pi}\int_0^{\pi} d\phi\ e^{j\lambda(x_m - x_n)\cos\phi}(i\lambda \cos \phi)^2\{R(\lambda \cos \phi)\}^2$$

$$\int_0^{\pi} d\theta \cos\left[\left(\frac{\lambda w}{2} \cos \theta \sin \phi\right)\right] \tag{A7}$$

In (A4) and (A7) the second power of R can be written in the form of a simple summation:

$$\{R(\lambda \cos \phi)\}^2 = \sum_{v=0}^{\infty} A'_v(\lambda \cos \phi)^{2v} \tag{A8}$$

where

$$A'_v = A_v^2 + 2 \sum_{\mu=0,2,\cdots}^{v/2} A_\mu A_{2v-\mu} \qquad v \text{ even}$$

$$A'_v = 2 \sum_{\mu=0,2}^{v-1} A_0 A_{2v-\mu} \qquad v \text{ odd} \tag{A9}$$

Finally, from (A4)–(A9) we can conclude that

$$\langle f_m \delta(y), J_0(\lambda\rho), f_n g\rangle = \sum_{v=0}^{\infty} C_v \int_0^{\pi} d\theta \cdot \left\{\frac{d^{2v}}{d|x|^{2v}} J_0\left[\lambda\left(x^2 + \left(\frac{w}{2}\cos\theta\right)^2\right)^{1/2}\right]\right\}_{x=x_m-x_n} \tag{A10}$$

$$\left\langle f_m \delta(y), \frac{\partial^2}{\partial x^2} J_0(\lambda\rho), f_n g\right\rangle = \sum_{v=0}^{\infty} C_v \int_0^{\pi} d\theta \cdot \left\{\frac{d^{2(v+1)}}{d|x|^{2(v+1)}} J_0\left[\lambda\left(x^2 + \left(\frac{w}{2}\cos\theta\right)^2\right)^{1/2}\right]\right\}_{x=x_m-x_n} \tag{A11}$$

with

$$C_v = (-1)^v A'_v \tag{A12}$$

Considering that $k \sim 2\pi(\varepsilon_r)^{1/2}$ and $kl_x \ll 1$, the coefficients of the above summation converge as fast as $(1/k)^{2(v+1)}$.

Acknowledgment. This work was supported in part by Ericsson Radar Electronics AB and by Phraxos R and D.

REFERENCES

Elliott, R, S., and W. R. O'Loughlin, The design of slot arrays including internal mutual coupling, *IEEE Trans. Antennas Propag., AP-34*, 1149–1154, 1986.

Katehi, P. B., A generalized solution to a class of printed circuit antennas, Ph.D. dissertation, Department of Electr. Eng., Univ. of Calif., Los Angeles, June 1984.

Katehi, P. B., The Green's function for a slot on the ground of a dielectric slab, *Radiat. Lab. Rep. RL-841*, Electr. Eng. and Comput. Sci. Dep., Univ. of Mich., Ann Arbor, June 1987.

Katehi, P. B., and N. G. Alexopoulos, Real axis integration of Sommerfeld integrals with applications to printed circuit antennas, *J. Math. Phys., 24*(3), 527–533, 1983.

Mosig, J. R., and F. E. Gardiol, General integral equation formulation for microstrip antennas and scatterers, *IEE Proc., Part H, Microwaves Opt. Antennas, 132*, 175–182, 1985.

Pozar, D. M., Input impedance and mutual coupling of rectangular microstrip antennas, *IEEE Trans. Antennas Propag., AP-30*, 1191–1196, 1982.

Pozar, D. M., A reciprocity method of analysis for printed slots and slot-coupled microstrip antennas, *IEEE Trans. Antennas Propag., AP-34*, 1439–1446, 1986.

Rana, I. E., and N. G. Alexopoulos, Current distribution and input impedance of printed dipoles, *IEEE Trans. Antennas Propag., AP-29*, 99–105, 1981.

Robertson, R. S., and R. S. Elliott, The design of transverse slot arrays fed by the meandering strip of boxed stripline, *IEEE Trans. Antennas Propag., AP-35*, 252–257, 1987.

Tsandoulas, G. N., Excitation of a ground dielectric slab by a horizontal dipole, *IEEE Trans. Antennas Propag., AP-17*, 156–161, 1969.

Uzunoglu, N. K., N. G. Alexopoulos, and J. G. Fikioris, Radiation properties of microstrip dipole, *IEEE Trans. Antennas Propag., AP-27*, 853–858, 1979.

Yang, H.-Y., and N. G. Alexopoulos, Analysis of an aperture coupled dipole antenna, a microstrip fed slot and a slotline fed dipole, paper presented at IEEE AP-S International Symposium, Inst. of Electr. and Electron. Eng., Blacksburg, Va., June 1987.

P. B. Katehi, Radiation Laboratory, Department of Electrical Engineering and Computer Science, University of Michigan, 1301 Beal Avenue, Ann Arbor, MI 48109.

IMPROVED COMPUTATIONAL EFFICIENCY FOR THE MOMENT METHOD SOLUTION OF PRINTED DIPOLES AND PATCHES

David M. Pozar, *Department of Electrical and Computer Engineering, University of Massachusetts, Amherst, MA 01003*

This work was supported by Grants NAG-1-163 and NAG-1-279 between the National Aeronautics and Space Administration, Langley Research Center, Hampton, VA, and the University of Massachusetts, Amherst, MA 01003.

1. INTRODUCTION

Moment method solutions for printed dipoles and microstrip patches on grounded dielectric substrates have recently been proposed by a number of authors[1],[2],[3],[4],[5],[6],[7],[8]. These solutions use the rigorous Green's function for a grounded dielectric slab, so surface wave effects and mutual coupling can be treated as well as input impedance and radiation patterns. In addition, these methods can handle substrates which have high dielectric constants and/or substrates which are not electrically thin. Thus these solutions are proving to be of particular value in the analysis and design of millimeter wave printed antennas and arrays.

This versatility is not without a price, however: the use of the dielectric slab Green's function involves the numerical evaluation of Sommerfeld-type integrals, which results in a computationally inefficient solution. The purpose of this paper is to show the source of this slow convergence and a method of rewriting the Green's function as a sum of a closed-form expression and a quickly converging integral, resulting in a significantly more efficient computation. Computer times can be reduced by a factor of 6 to 20, or more, depending on the configuration under consideration. The improved method is particularly helpful for mutual impedance calculations, since the integrals in the original formulation have the undesirable property of converging more slowly as the distance between modes increases.

Briefly, the method involves subtraction of the source and image singularities from the Green's function. The source and image singularities, however, can be shown to be identical to the singularities arising from the same source in a grounded homogeneous medium of relative permittivity $\varepsilon_e=(\varepsilon_r+1)/2$, where ε_r is the relative permittivity of the original dielectric slab. The fields from this source and image can be evaluated in closed-form and the remaining integral will be better behaved than the original. Furthermore, if piecewise-sinusoidal (PWS) expansion and testing modes are used with a wavenumber based on ε_e, then the impedance contribution from the homogeneous medium terms can be efficiently evaluated using existing routines for thin-wire moment method solutions[9].

Section 2 presents the basic theory and equations for the Green's function moment method and shows the asymptotic behavior of the integrals. Section 3 discusses the source and image singularity removal technique and its physical interpretation. Section 4 discusses efficient evaluation of the separated source and image contributions, including effects of full surface testing/expansion and the derivation of an equivalent radius for a flat strip dipole. Some numerical examples of convergence are given in Section 5. For the interested reader, Section 6 presents some of the unsuccessful attempts at improving the computational efficiency of this Green's function solution, and why they did not work.

2. GREEN'S FUNCTION/MOMENT METHOD

The details of the Green's function and moment method formulation have appeared in [6], but for clarity it is worthwhile to summarize them here. Details will be given for the impedance between test and expansion modes with currents in the same direction, but the method applies to general impedance elements, as well as the excitation voltage vector.

The basic Green's function component, E_{xx}, representing the $\hat{x}$-electric field at (x,y,d) from an $\hat{x}$-directed infinitesimal dipole of unit strength located at (x_0,y_0,d) on a grounded dielectric slab of thickness d and relative permittivity ε_r (see Figure 1) is

$$E_{xx} = \frac{-jZ_0}{4\pi^2 k_0} \iint_{-\infty}^{\infty} Q(k_x,k_y) e^{jk_x(x-x_0)} e^{jk_y(y-y_0)} dk_x dk_y , \quad (1)$$

where

$$Q(k_x,k_y) = \frac{(\varepsilon_r k_0^2-k_x^2)k_2 \cos k_1 d + jk_1(k_0^2-k_x^2)\sin k_1 d}{T_e T_m} \sin k_1 d , \quad (2)$$

$$\begin{aligned} T_e &= k_1 \cos k_1 d + jk_2 \sin k_1 d , \\ T_m &= \varepsilon_r k_2 \cos k_2 d + jk_1 \sin k_1 d , \\ k_1^2 &= \varepsilon_r k_0^2 - \beta^2 , \quad (\mathrm{Im}k_1 < 0) \\ k_2^2 &= k_0^2 - \beta^2 , \quad (\mathrm{Im}k_2 < 0) \\ \beta^2 &= k_x^2 + k_y^2 , \\ Z_0 &= \sqrt{\mu_0/\varepsilon_0}. \end{aligned} \quad (3)$$

The poles of the integrand Q come from zeroes of the T_e, T_m functions, and represent TE and TM surface waves, respectively.

Expansion modes $J_m(x_0,y_0)$ are used to expand the unknown electric surface current density on the printed dipole or patch, and test modes $J_n(x,y)$ are used to reduce the electric field integral equation to a set of algebraic equations to be solved for the expansion mode coefficients. A typical moment method impedance matrix element is then computed as

$$Z_{mn} = \frac{jZ_0}{4\pi^2 k_0} \int_x \int_y \int_{x_0} \int_{y_0} \int_{k_x} \int_{k_y} J_m(x_0,y_0) J_n(x,y) Q(k_x,k_y) e^{jk_x(x-x_0)} e^{jk_y(y-y_0)} \cdot$$

$$\cdot dk_y dk_x dy_0 dx_0 dy dx. \quad (4)$$

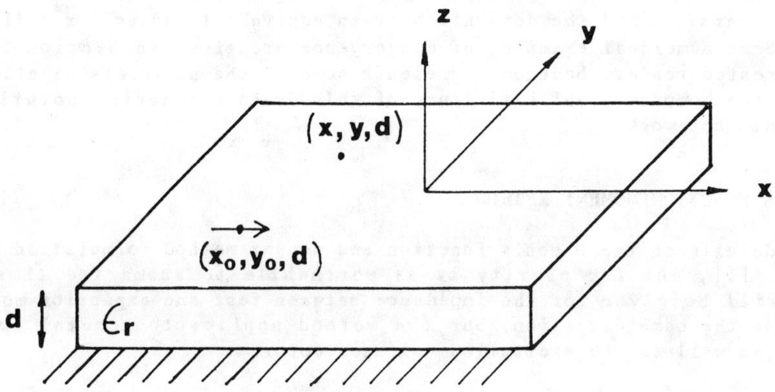

Figure 1. Geometry of an infinitesimal dipole and field point on a grounded dielectric slab.

Thus, a typical matrix element involves a six-fold integration. As discussed in [6], it is most expedient to do the four space-domain integrations in closed-form (amounting to the Fourier transforms of J_m and J_n), and convert the spectral domain integrations to polar form, involving one finite and one infinite integration:

$$Z_{mn} = \frac{jZ_0}{4\pi^2 k_0} \int_{\beta=0}^{\infty} \int_{\alpha=0}^{\pi/2} Q(k_x,k_y) F_n(k_x,k_y) F_m^*(k_x,k_y) \beta d\beta d\alpha \;, \tag{5}$$

where

$$F_k(k_x,k_y) = \int_x \int_y J_k(x,y) e^{jk_x x} e^{jk_y y} \, dxdy, \tag{6}$$

and $k_x = \beta\cos\alpha$, $k_y = \beta\sin\alpha$.
Equation (5) is then evaluated numerically with the symmetry properties of the integrand used to reduce the α-integration to the range 0 to $\pi/2$ [6].
This is basically the form used in [4],[5],[6],[7], and [8]; see Section V for alternative ways of evaluating (4).

The β-integration in (5) converges rather slowly; a typical upper limit for β when evaluating a self-impedance term is $200k_0$, while for mutual impedance terms an upper limit of $400k_0$ or more may be necessary. In addition, for large values of β, the integrand oscillates rapidly, requiring more sample points for the α-integration. Observe that the Green's function (1) does not contain an explicit 1/R dependence for the decay of the fields from the source; this range dependence (and source singularity) must be synthesized by the continuous spectrum of plane waves. Such is the nature of the spectral representation.

If PWS test and expansion modes of the form

$$J_k(x,y) = \begin{cases} \dfrac{\sin k_e[a-|x-x_k|]}{W \sin k_e a} & \text{for } |x-x_k|<a \\ 0 & \text{otherwise} \end{cases} , \tag{7}$$

(where k_e is the expansion mode wavenumber to be chosen), are used, then the asymptotic form of F_k for large β is

$$F_k \sim \begin{cases} 1/\beta^2 & \text{if } \alpha=0 \\ 1/\beta & \text{if } \alpha=\pi/2 \end{cases} .$$

The asymptotic form of Q for large β is

$$Q \sim \frac{k_0^2}{2\beta} - \frac{k_x^2}{\beta(\varepsilon_r+1)} \sim \begin{cases} \beta & \text{if } \alpha=0 \\ 1/\beta & \text{if } \alpha=\pi/2 \end{cases} . \tag{8}$$

The total integrand then is asymptotic to $1/\beta^2$ for $\alpha=0$ or $\pi/2$, and is no worse than this for other values of α. The voltage vector elements V_i, defined in [6], have a form similar to (5), and a similar asymptotic behavior.

3. SOURCE AND IMAGE SINGULARITY REMOVAL

Now consider the $\hat{x}$-electric field at (x,y,d) from an $\hat{x}$-directed infinitesimal dipole of unit strength at (x_0,y_0,d) in a grounded homogeneous medium of relative permittivity ε_e. See Figure 2. Image theory allows a simple closed-form expression as

$$E_{xx}^h = \frac{-jZ_0}{4\pi k_0 \varepsilon_e} \left(\varepsilon_e k_0^2 + \frac{\partial^2}{\partial x^2}\right) \left[\frac{e^{-jkR_s}}{R_s} - \frac{e^{-jkR_i}}{R_i} \right] , \tag{9}$$

where $R_s^2 = (x-x_0)^2+(y-y_0)^2$, $R_i^2 = (x-x_0)^2+(y-y_0)^2+(2d)^2$.

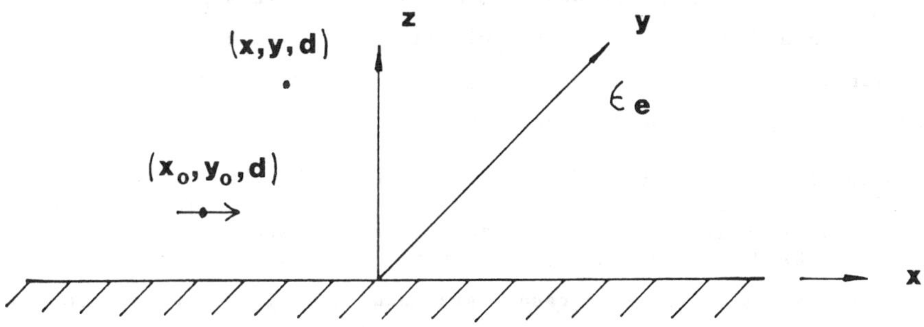

Figure 2. Geometry of an infinitesimal dipole and field point in a grounded homogeneous dielectric medium.

Standard identities can be used to rewrite (9) in a spectral representation:

$$E^h_{xx} = \frac{-jZ_0}{4\pi^2 k_0} \iint_{-\infty}^{\infty} Q^h(k_x,k_y) e^{jk_x(x-x_0)} e^{jk_y(y-y_0)} dk_x dk_y , \tag{10}$$

where

$$Q^h(k_x,k_y) = \frac{(\varepsilon_e k_0^2 - k_x^2)}{2j\varepsilon_e k_{1e}} (1-e^{-2jk_{1e}d}) , \tag{11}$$

and $k_{1e}^2 = \varepsilon_e k_0^2 - \beta^2$. (12)

For large β , Q^h behaves asymptotically as

$$Q^h \sim \frac{k_0^2}{2\beta} - \frac{k_x^2}{2\beta\varepsilon_e} . \tag{13}$$

Now, if the relative permittivity of the homogeneous medium is selected as $\varepsilon_e = (\varepsilon_r+1)/2$, then Q^h will have the same asymptotic form as Q for the grounded dielectric slab (8). Thus, Q^h of (11) can be subtracted from the integrand Q of (2), yielding a more quickly converging integral, and reinserted by adding the closed-form expression (9) for the fields of a source and image in a homogeneous medium of permittivity $\varepsilon_e=(\varepsilon_r+1)/2$:

$$E_{xx} = E^h_{xx} + \frac{-jZ_0}{4\pi^2 k_0} \iint_{-\infty}^{\infty} [Q-Q^h] e^{jk_x(x-x_0)} e^{jk_y(y-y_0)} dk_x dk_y . \tag{14}$$

This result also applies to the impedance element Z_{mn} , that is,

$$Z_{mn} = Z^h_{mn} + \frac{jZ_0}{4\pi^2 k_0} \int_\beta \int_\alpha [Q-Q^h] F_n(k_x,k_y) F_m^*(k_x,k_y) \beta d\beta d\alpha , \tag{15}$$

where Z^h_{mn} is the impedance between the modes m and n in a grounded homogeneous medium of relative permittivity $\varepsilon_e=(\varepsilon_r+1)/2$. Note that (9) contains an explicit range dependence and source singularities. Also, as $\varepsilon_r \to 1$, $\varepsilon_e \to 1$ and the integral terms in (14) and (15) go to zero.

Thus, rewriting the slab Green's function in the above manner allows the critical range dependence and source singularity to be separated off in closed-form, leaving a relatively well-behaved integral to account for the particular effects of the finite-thickness dielectric slab, such as surface waves. The surface wave poles in the integrand Q are unchanged by the subtraction of Q^h , and Q^h contributes no new poles.

Physically, the procedure amounts to the removal of terms representing the solution of the antenna geometry in an 'effective' homogeneous medium of $\varepsilon_e=(\varepsilon_r+1)/2$. It is interesting to note that this effective permittivity is close to the approximate values used for microstrip line calculations. The above procedure, however, involves no approximations.

This same basic result also applies to impedances between modes with currents in orthogonal directions as well as the excitation voltage vector. For example, a given voltage vector element can be expressed as a contribution from the source in a grounded homogeneous medium of relative permittivity $\varepsilon_e=(\varepsilon_r+1)/2$ plus a relatively well-behaved integral. The method is still valid for complex ε_r, so dielectric loss can be included.

4. EVALUATION OF IMPEDANCE IN A GROUNDED HOMOGENEOUS MEDIUM

It has been shown that the calculation of terms for the moment method solution of antennas printed on a grounded dielectric slab can be facillitated by the evaluation and subtraction of terms describing the same antenna geometry in a homogenous medium. The purpose of this section is to discuss the efficient evaluation of these impedance terms, Z^h_{xx}. The calculation of Z^h_{xx} is basically the same as the evaluation of surface patch impedances in the context of surface patch moment method formulations [10],[11], and so can use some of the efficient methods developed for these formulations.

Piecewise sinusoidal (PWS) modes have been chosen for expansion and testing in the moment method procedure, so if the wavenumber k_e in (7) is chosen as the wavenumber in the homogeneous medium ($k_e=\omega\sqrt{\mu_0\varepsilon_0(\varepsilon_r+1)/2}$), then the impedance between two filamentary PWS dipoles can be evaluated in closed-form [9] (actually, in terms of exponential integrals). The impedance between actual rectangular PWS modes, described by (7), on a dipole or patch in the grounded homogeneous medium can be found by image theory and the integration of filamentary PWS modes across the width of the dipole or patch. This procedure will be referred to as full surface testing and expansion. Two problems arise in practice, however. First, the impedance between PWS filaments contains a logarithmic singularity which hinders a simple, direct numerical integration. Second, for thin substrates ($d<0.02\lambda_0$) the presence of an image mode results in the subtraction of two numbers very close to each other, leading to a loss of accuracy. These two problems are discussed in more detail below. A technique for avoiding full surface testing and expansion (elimination of the integration across the width of the mode) for a thin dipole is also discussed.

The use of k_e as the wavenumber for the PWS modes, however, is sometimes a disadvantage in terms of the overall convergence of the moment method solution for microstrip patches. As discussed in [6] and [8], the number of expansion modes needed for convergence of the input impedance of patches can be significantly reduced if an expansion mode wavenumber is chosen to correspond to the effective dielectric constant of the microstrip medium. This wavenumber is basically the same as k_e for thin substrates, but may differ for thick substrates, in which case the moment method solution using k_e will require a larger number of expansion modes for good results. In other words, the number of expansion modes needed for a given accuracy depends on the choice of the expansion mode wavenumber, and k_e is not always the optimum choice.

4.1 Treatment of the Logarithmic Singularity

As discussed in [11], the impedance between two parallel filamentary PWS dipoles at x and x_0 is of the form

$$z(x-x_0) = C_1 + C_2 \ln|x-x_0| \ , \tag{16}$$

for small $|x-x_0|$. Thus, a direct numerical integration of this singularity is slow and expensive. In [11], however, a method for the integration of this singularity is described, and the same method has the time-saving feature of reducing the two-dimensional integration on x and x_0 to an integration which is effectively one-dimensional. The same technique can be applied to the computation of the voltage vector excitation terms. Details can be found in [11].

4.2 Thin Substrates

When the substrate is electrically very thin (say, $d<0.02\lambda_0$), the difference between the self-impedance of the source filament and the mutual impedance between the source and image filaments is small, leading to numerical difficulties. This problem is not severe for printed dipoles (especially when filamentary testing and expansion is used; see next sub-section), but can be a problem for patches on thin substrates, due to the numerical integration required. A given accuracy requires more and more integration points across the width of the patch, as the substrate gets thinner. Thus, for $d<0.02\lambda_0$, the new method may not be more efficient than the old(direct integration of (5)) for self-impedance, although mutual impedance calculations will still be speeded up by the new method, since the old method converges much slower as the patches get further apart.

This problem is the main disadvantage of the method; a way of calculating directly the impedance between patches close to a ground plane would be a useful result. Some work has been done on a Taylor series expansion of the impedance for small d, but generally useful results have not yet been obtained.

4.3 Filamentary Testing and Expansion for Thin Dipoles

Full surface and expansion across the width of a thin dipole for self-impedance is a tedious and unecessary task since, as will be shown, the same result can be obtained by filamentary testing and expansion with an appropriate separation. This result is analogous to thin-wire moment method solutions where a test filament is placed on the wire axis and an expansion filament on the wire surface. It has been shown [12] that this procedure is equivalent to full surface testing and expansion, when the separation between the filaments is selected as the wire radius. A corresponding result can be obtained for a thin flat strip.

The self-impedance of a thin flat dipole of width W, with uniform current across the width, can be computed using full surface testing and expansion as

$$Z_{xx} = \int_{x=-W/2}^{W/2} \int_{x_0=-W/2}^{W/2} z(x-x_0)\, dx dx_0 \ , \tag{17}$$

where $z(x-x_0)$ is the impedance between two parallel filamentary PWS dipoles at x and x_0, and uniform current is assumed across the width of the flat dipole. The double integration in (17) can be reduced to a single integration:

$$Z_{xx} = \frac{1}{W^2}\int_{-W}^{W} (W-|\xi|)\ z(\xi)\ d\xi \ . \tag{18}$$

Substituting the small-argument form of z given by (16) into (18) and integrating gives

$$Z_{xx} = C_1 + C_2\ (\ln W - 3/2). \tag{19}$$

If filamentary testing and expansion is used with a separation of a, the impedance will be

$$Z_{xx} = C_1 + C_2 \ln a. \tag{20}$$

Equating (19) and (20) gives an equivalent value for the separation of a = 0.223W.

A quasi-static approximation [13] gives an equivalent separation of a = 0.25W, and [14] incorporates the edge singularity in the current and also arrives at a = 0.25W, but to be self-consistent with the uniform current assumed here, a = 0.223W is the best choice.

This value has been validated by comparison with the full surface testing and expansion integration. Small changes in a can cause significant changes in the reactive part of the impedance, especially near resonance.

5. EXAMPLES OF CONVERGENCE

A few examples of the improvement in convergence are presented in this section. The selected examples show the convergence of the imaginary part of the moment method impedance matrix elements as computed by the direct method (4) and the new method (15). The real parts of the matrix elements converge quickly with both methods, unless perhaps the two modes are rather widely separated.

Figure 3 shows the convergence of the imaginary part of the self-impedance of a microstrip patch mode, vs. β_m/k_0 , the upper limit on the β-integration of (4) or (15). The direct method of (4) is seen to require β_m/k_0=100 or more, while the new technique of (15) has definitely converged for β_m/k_0=20.

Figure 4 shows the convergence of the imaginary part of the mutual impedance between two printed dipoles. Here convergence of (4) requires β_m/k_0=400 or more, while (15) has converged for β_m/k_0=50.

6. OTHER APPROACHES TO IMPROVED COMPUTATIONAL EFFICIENCY

A number of other attempts at improving the computational efficiency of the Green's function moment method solution were made, unsuccessfully, before the present approach was found. A brief discussion of some of these ideas may be of interest to some readers.

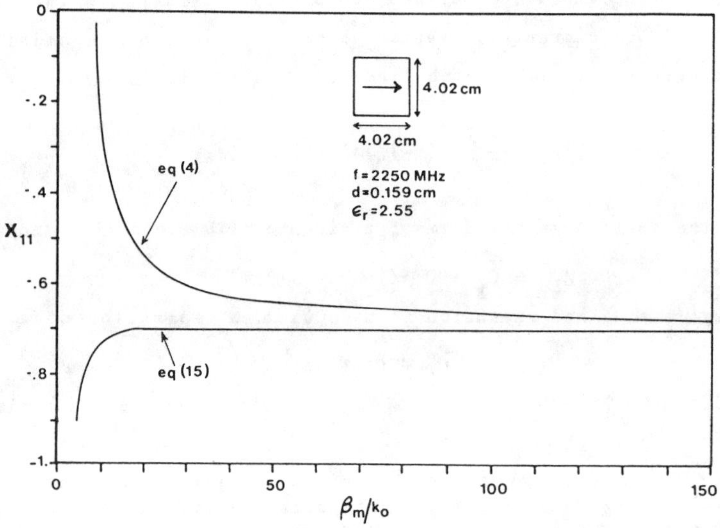

Figure 3. Convergence of the imaginary part of the self-impedance of a microstrip patch mode. R_{11}=0.167Ω.

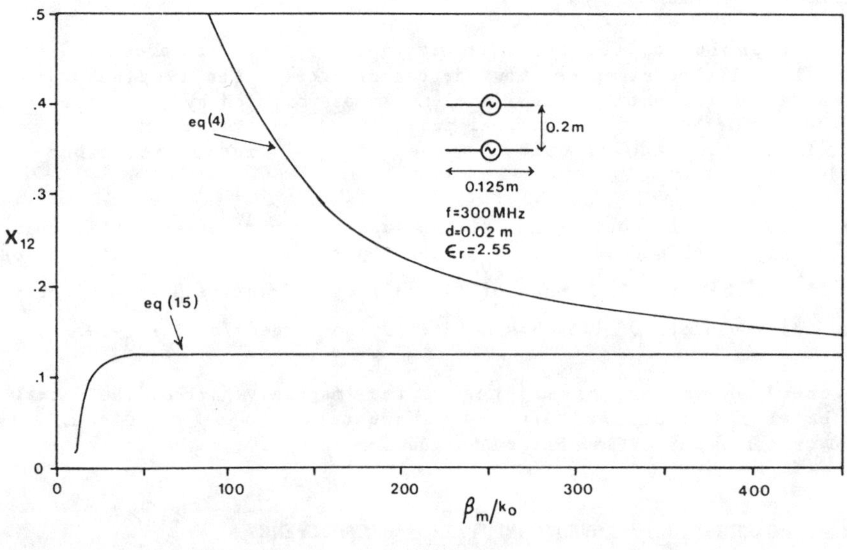

Figure 4. Convergence of the imaginary part of the mutual impedance between two printed dipoles. R_{12}=0.061Ω.

6.1 α-Integration in Closed-form

Instead of doing the x, y, x_0, y_0 integrations of (4) in closed-form, the k_x, k_y integrations can be converted to polar (β,α) form, and the α-integration done in closed-form, resulting in a Bessel function of the x, y, x_0, y_0 coordinates. The five remaining integrations then must be done numerically. This is obviously not an improvement in general, since the integrand involves a Bessel function and one of the integrations is infinite. If only dipoles are considered, however, the y, y_0 integrations are unnecessary, leaving three numerical integrations. This is the approach taken in [2] and [3], along with an asymptotic approximation of the Bessel function for large arguments. For dipoles the efficiency of this procedure is roughly the same as the direct numerical integration of (5), but for patches the efficiency is much less.

6.2 Integration Along the Branch Cut

The limits of integration in (5) can be changed to $\beta=-\infty$ to ∞ and $\alpha=0$ to π. Branch cuts exist in the complex-β plane, and a contour integral can be formed for the β-integration by closing the contour and deforming it on both sides of the branch cut. The infinite portion of the integrand then occurs for imaginary β, thus exponential convergence will result from the exponential terms in (4) or (6).

Unfortunately, the choice of whether to close the contour in the upper-half plane or the lower-half plane depends on the particular values of x, x_0, y, y_0 and α, and changes during the course of the α-integration(numerical). The number of possible combinations and the bookkeeping required renders this method impractical.

6.3 Asymptotic Form of Integrand for Large β

An attempt was made to derive a simple asymptotic expression for the integrand in (5) to permit closed-form integration on β after a certain (large) value of β had been reached by numerical integration. A very good asymptotic form for Q can be found, but the function F_k does not have any useful asymptotic form because the variables k_x and k_y take on all values from 0 to $\pm\beta$ as α changes from 0 to π.

7. CONCLUSION

A method for improving the computational efficiency of moment method solutions using the dielectric slab Green's function has been presented. This technique should permit wider use of these formulations, which previously had been hindered by exceptionally slow computer run times. The method can perhaps be applied to other Green's function geometries.

References

[1] N.K. Uzunoglu, N.G. Alexopoulos, and J.G. Fikioris, 'Radiation Properties of Microstrip Dipoles', IEEE Trans. Antennas and Prop., vol. AP-27, pp.853-858, Nov. 1979.

[2] I.E. Rana and N.G. Alexopoulos, 'Current Distribution and Input Impedance of Printed Dipoles', IEEE Trans. Antennas and Prop., vol. AP-29, pp.99-105, Jan. 1981.

[3] N.G. Alexopoulos and I.E. Rana, 'Mutual Impedance Computation between Printed Dipoles', IEEE Trans. Antennas and Prop., vol. AP-29, pp.106-111, Jan. 1981.

[4] M.D. Deshpande and M.C. Bailey, 'Input Impedance of Microstrip Antennas', IEEE Trans. Antennas and Prop., vol.AP-30, pp.645-650, July 1982.

[5] M.C. Bailey and M.D. Deshpande, 'Integral Equation Formulation of Microstrip Antennas', IEEE Trans. Antennas and Prop., vol.AP-30, pp.651-656, July 1982.

[6] D.M. Pozar, 'Input Impedance and Mutual Coupling of Rectangular Microstrip Antennas', IEEE Trans. Antennas and Prop., vol. AP-30, pp.1191-1196, Nov. 1982.

[7] D.M. Pozar, 'Considerations for Millimeter Wave Printed Antennas', IEEE Trans. Antennas and Prop., IEEE Trans. Antennas and Prop., vol. AP-31, pp.740-747, September 1983.

[8] E.H. Newman, J.H. Richmond, and B.W. Kwan, 'Mutual Impedance Computation between Microstrip Antennas', IEEE Trans. Microwave Theory and Techniques, accepted for publication.

[9] J.H. Richmond, 'Radiation and Scattering by Thin-wire Structures in the Complex Frequency Domain', Report 2902-10, Ohio State Univ. ElectroScience Laboratory, Dept. of Electrical Engineering, prepared under Grant NGL 35-008-138 for the National Aeronautics and Space Administration, July 1973.

[10] E.H. Newman and D.M. Pozar, 'Electromagnetic Modelling of Composite Wire and Surface Geometries', IEEE Trans. Antennas and Prop., vol. AP-26, pp.784-789, Nov. 1978.

[11] E.H. Newman and D.M. Pozar, 'Considerations for Efficient Wire/Surface Modelling', IEEE Trans. Antennas and Prop., vol. AP-28, pp.121-125, Jan. 1980.

[12] W.A. Imbriale and P.G. Ingerson, 'On Numerical Convergence of Moment Solutions of Moderately Thick Wire Antennas using Sinusoidal Basis Functions', IEEE Trans. Antennas and Prop., vol. AP-21, pp.363-366, May 1973.

[13] H. Jasik, Ed., <u>Antenna Engineering Handbook</u>, New York: McGraw-Hill, 1961, p.3-7.

[14] C.M. Butler, 'The Equivalent Radius of a Narrow Conducting Strip', IEEE Trans. Antennas and Prop. vol. AP-30, pp.755-758, July 1982.

Real axis integration of Sommerfeld integrals: Source and observation points in air

W. A. Johnson

Electronics Engineering Department, Lawrence Livermore National Laboratory, Livermore, California 94550

D. G. Dudley

Department of Electrical Engineering, University of Arizona, Tucson, Arizona 85721

(Received April 16, 1982; revised September 27, 1982; accepted September 27, 1982.)

Change of variables and subtraction of asymptotic terms are used to aid in the integration of Sommerfeld integrands along the real axis. The change of variables smooths oscillations which would otherwise become densely packed at the associated branch point. Subtraction of asymptotic terms is used to speed up the truncation process of an otherwise infinite integration interval. Interest is restricted to practical earth parameters and frequencies above 2 MHz. Comparison with a complex contour technique shows the present method to be particularly well suited for the analysis of a vertical, thin-wire antenna in the presence of a lossy earth.

1. INTRODUCTION

Conventional techniques for the analysis of scattering and radiation by antennas in the presence of a lossy earth require numerous evaluations of Sommerfeld integrals. With this motive to obtain efficient, accurate methods of evaluating Sommerfeld integrals, a vast literature has been produced [*Rahmat-Samii et al.*, 1981].

Direct numerical integration techniques have the advantage of a wide range of validity in medium, frequency, and spatial parameters. The major limitation of these techniques is computation time.

Siegel and King [1970] numerically integrated Sommerfeld integrands along the real axis. Their technique is applicable to both vertical and horizontal dipoles with arbitrary source and observation point locations. To avoid difficulties associated with integrating highly oscillatory integrands, Romberg integration was performed between the zeros of Bessel functions which occur in the Sommerfeld integrands. Modifications of the Romberg scheme were made to integrate carefully over regions where the integrand becomes singular. To accelerate the process of truncating the infinite integration interval, a Euler summation method was used. *Bubenik* [1977] improved this truncation process by use of Shank's summation technique, when he evaluated the magnetic field due to a quasi-static vertical electric dipole.

Integration along contours in the complex plane has been used to avoid singularities and reduce oscillations of the integrand. *Burke et al.* [1981] used Romberg-Shanks integration along complex contours to compute electric fields of vertical and horizontal electric dipoles when both source and observation points were located above a lossy ground. *Parhami et al.* [1980] evaluated Sommerfeld integrals, associated with the vector potential of a horizontal dipole with source and observer in air, by integrating along the steepest descent path with a Gauss-Laguerre quadrature.

To avoid integration along complex contours and the associated effort of computing Bessel functions with complex arguments, we use analytical techniques to remove singularities and reduce oscillations that occur when Sommerfeld integrands are evaluated along the real axis. The computation of electric fields due to vertical and horizontal electric dipoles is considered. For brevity, source and observation points are restricted to lie in air. Figure 1 illustrates the problem geometries for both vertical and horizontal electric dipoles. Of course, limitations on com-

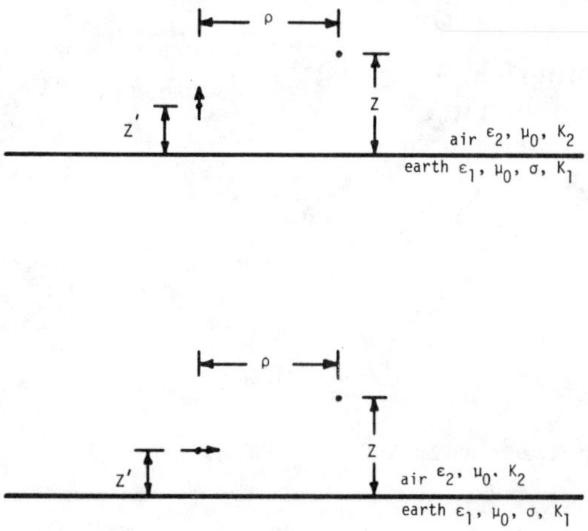

Fig. 1. Problem geometries for vertical and horizontal electric dipoles. The air parameters are an electric permittivity ε_2, a magnetic permeability μ_0, and a wave number K_2. The ground has a permittivity ε_1, a permeability μ_0, a conductivity σ, and a complex wave number K_1.

puter time will restrict the distances $z + z'$ and ρ (Figure 1) from becoming too large in terms of the free-space wavelength.

The present integration technique will use local change of variables to remove singularities, as well as rapidly varying oscillations, that occur along the real axis in the vicinity of a branch point. A compound Simpson's rule [*Davis and Rabinowitz*, 1975] is chosen to integrate over half-cycles of the integrand's oscillations. Subtraction of asymptotic terms, a technique with similarities to the work of *Rana and Alexopoulos* [1981], is used when necessary to speed up the process of truncating the infinite integration interval. Although our particular interest is in practical earth parameters, i.e., a dielectric constant less than 16 and a conductivity less than 0.01 mhos per meter, the technique is valid over a wider range of parameters. The major limitation of the technique is that when the magnitude of K_1^2/K_2^2 is larger than 90, subtraction of asymptotic terms becomes inefficient. The change of variable technique remains applicable for ratios of K_1^2/K_2^2 with magnitude larger than 600,000. Graphical illustrations show the effects of both the change of variables and subtraction of asymptotic terms for typical integrands. Comparison of results shows the present real axis integration technique has definite advantage over the complex contour methods of *Burke et al.* [1981] when the radial separation ρ between source and observation point is small with respect to the sum of the source and observation heights $z + z'$.

2. FORMULATION

The electric field of vertical and horizontal electric dipoles with source and observation points in air may be expressed as the sum of free-space and image Green's functions plus 5 Sommerfeld integrals [*Banos*, 1966]. These integrals are $((\partial^2/\partial z^2) + K_2^2)V_{22}$, U_{22}, $(\partial^2/\partial\rho\,\partial z)V_{22}$, $(-\partial/\partial\rho)V_{22}$, and $(\partial^2/\partial\rho^2)V_{22}$, where

$$U_{22} = 2\int_0^\infty \frac{e^{-\gamma_2|z+z'|}}{\gamma_1 + \gamma_2}\,\lambda J_0(\lambda\rho)\,d\lambda \tag{1}$$

$$V_{22} = 2\int_0^\infty \frac{e^{-\gamma_2|z+z'|}}{K_2^2\gamma_1 + K_1^2\gamma_2}\,\lambda J_0(\lambda\rho)\,d\lambda \tag{2}$$

$$K_i = \omega\left[\mu_0\left(\varepsilon_i - \frac{j\sigma_i}{\omega}\right)\right]^{1/2}$$
$$\mathrm{Re}\,(K_i) > 0 \qquad \mathrm{Im}\,(K_i) \le 0 \qquad i = 1, 2 \tag{3}$$

and

$$\gamma_i = (\lambda^2 - K_i^2)^{1/2}$$
$$\mathrm{Re}\,(\gamma_i) \ge 0 \qquad \mathrm{Im}\,(\gamma_i) \le 0 \qquad i = 1, 2 \tag{4}$$

Since

$$\frac{\partial^2}{\partial\rho^2}V_{22} = -\frac{1}{\rho}\frac{\partial}{\partial\rho}V_{22} - \left(\frac{\partial^2}{\partial z^2} + K_2^2\right)V_{22} \tag{5}$$

it is sufficient to consider the numerical evaluation of the first 4 Sommerfeld integrals. It is convenient to express these quantities as

$$\begin{bmatrix} \left(\dfrac{\partial^2}{\partial z^2} + K_2^2\right)V_{22} \\ U_{22} \\ \dfrac{\partial^2}{\partial\rho\,\partial z}V_{22} \\ -\dfrac{\partial}{\partial\rho}V_{22} \end{bmatrix} = 2\int_0^\infty \bar{I}(\lambda, |z + z'|, \rho)\,d\lambda \tag{6}$$

with

$$\bar{I}(\lambda, w, \rho) = \begin{bmatrix} \mathrm{I1} \\ \mathrm{I2} \\ \mathrm{I3} \\ \mathrm{I4} \end{bmatrix} = e^{-\gamma_2 w}\begin{bmatrix} \dfrac{\lambda^3}{K_1^2\gamma_2 + K_2^2\gamma_1}J_0(\lambda\rho) \\ \dfrac{\lambda}{\gamma_2 + \gamma_1}J_0(\lambda\rho) \\ \dfrac{\gamma_2\lambda^2}{K_1^2\gamma_2 + K_2^2\gamma_1}J_1(\lambda\rho) \\ \dfrac{\lambda^2}{K_1^2\gamma_2 + K_2^2\gamma_1}J_1(\lambda\rho) \end{bmatrix} \tag{7}$$

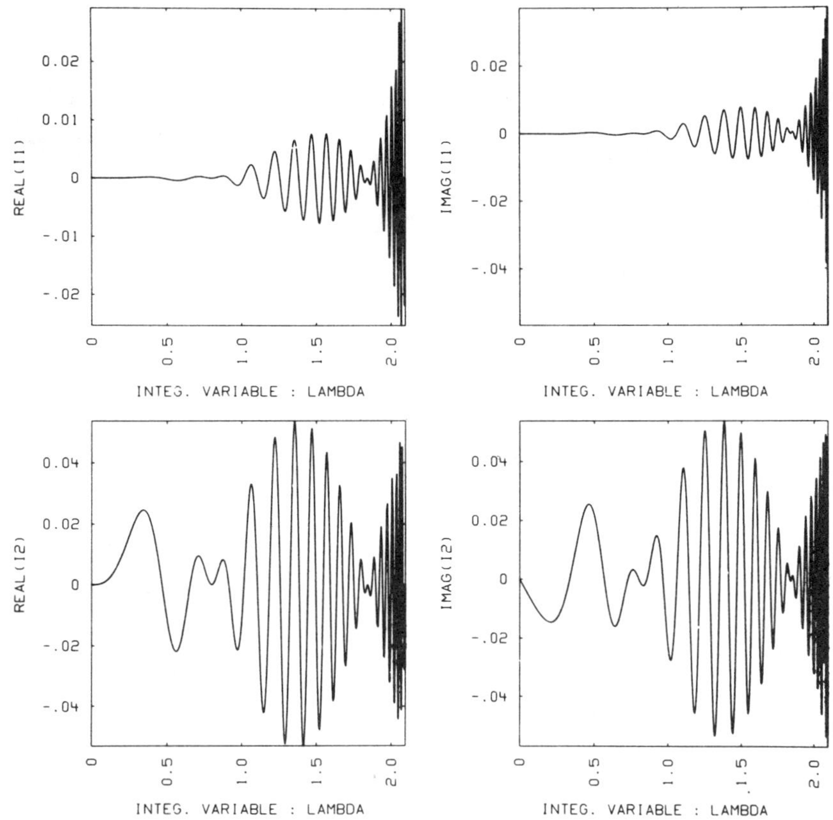

Fig. 2. Integrand behavior in the λ interval $[0, K_2]$. Parameters: $f = 10^8$ Hz, $\varepsilon_r = 16$, $\sigma = 10^{-4}$ mhos/m, $\rho = 1\lambda_0$, and $z + z' = 20\lambda_0$.

and $\gamma_i (i = 1, 2)$ given by (44). It is noted that if $|z + z'|$ is set to zero in (6), the integral becomes undefined. This problem will be remedied in the next section.

3. NUMERICAL CONSIDERATIONS

Obstacles to efficient evaluation of the Sommerfeld integrals (equations (6) and (7)) are now considered. A means of overcoming each obstacle is presented, and a brief description of a numerical integration procedure is given. Graphical results are chosen to illustrate typical integrand behavior before and after each obstacle has been overcome. Although these results are for fixed dielectric constant and conductivity, they represent typical integrand behavior throughout the range of interest.

The number of oscillations in the interval $\lambda \in [0, K_2]$ which are due to the exponential term $\exp(-(\lambda^2 - K_2^2)^{1/2}|z + z'|)$ is given by $|z + z'|$ divided by the free-space wavelength. As $|z + z'|$ becomes large in terms of free-space wavelengths, these oscillations become rapidly varying near the branch point at λ equal to K_2. Integration of these oscillations is further complicated by the peaked behavior of the integrand that occurs near this branch point. Figure 2 illustrates this behavior for the first two components of (7). Since the remaining components (equation (7)) behave similarly, attention will be focused on these first two components throughout this paper. In Figure 2, ρ equals 1 free-space wavelength, $z + z'$ equals 20 free-space wavelengths, the frequency f is 100 MHz, and the earth's relative dielectric constant and conductivity are 16 and 0.0001 mhos/m, respectively.

To make these densely packed oscillations more evenly spaced and to remove the peaked behavior of the integrand at branch point, the transformation

$$\lambda = K_2 \sin\theta \qquad \theta \in \left[0, \frac{\pi}{2}\right] \tag{8}$$

is employed over the λ interval $[0, K_2]$. The branch in the exponent

$$\exp[-(\lambda^2 - K_2^2)^{1/2}|z + z'|] = \exp[-iK_2|z + z'|\cos\theta] \tag{9}$$

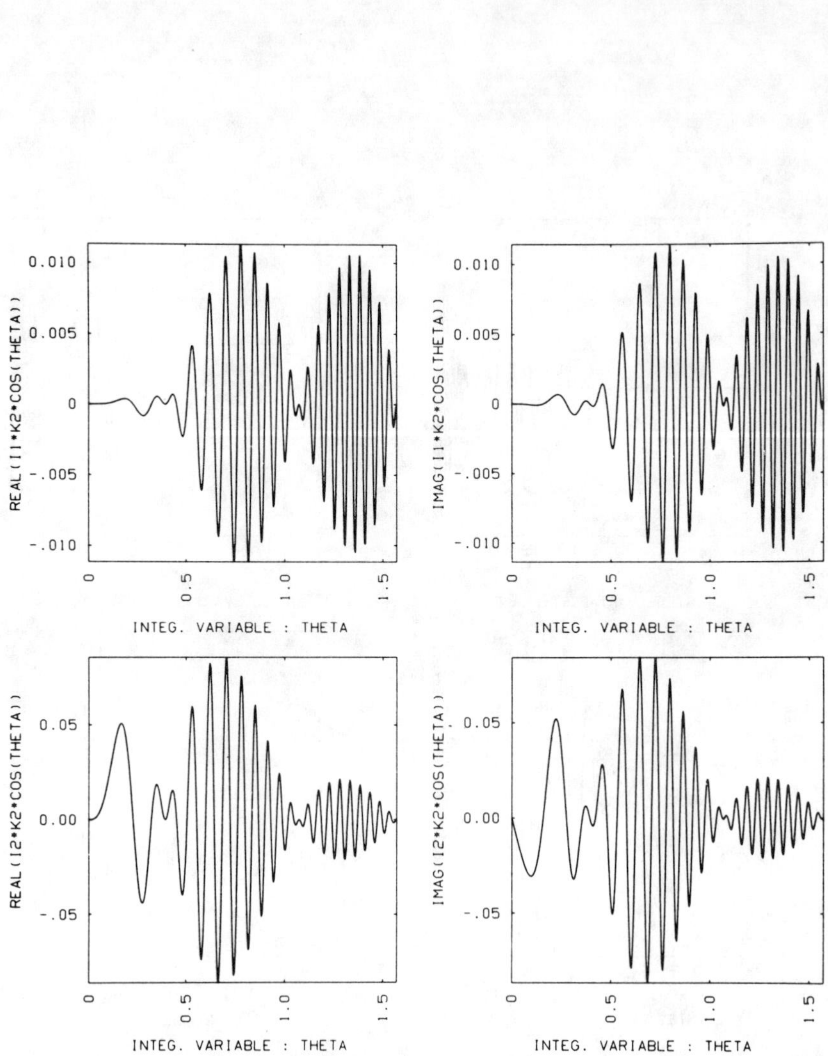

Fig. 3. Behavior of the transformed integrands in the θ interval $[0, \pi/2]$. Parameters: $f = 10^8$ Hz, $\varepsilon_r = 16$, $\sigma = 10^{-4}$ mhos/m, $\rho = 1\lambda_0$, and $z + z' = 20\lambda_0$.

is removed by this transformation. Since

$$d\lambda = K_2 \cos\theta \, d\theta \tag{10}$$

a typical denominator (equation (7)) multiplied by the transformation's Jacobian equals

$$D_1 = \frac{K_2 \cos\theta}{K_1^2 \gamma_2 + K_2^2 \gamma_1} = \frac{\cos\theta}{K_1^2 \cos\theta + K_2(K_2^2 \sin^2\theta - K_1^2)^{1/2}} \tag{11}$$

and remains nonsingular even when K_1 approaches K_2. The results of this transformation on the integrands I1 and I2 are shown in Figure 3. In contrast to Figure 2, the oscillations are more evenly spaced and the peaked behavior of the integrand no longer occurs.

A similar transformation may be used in the λ interval $[K_2, 2K_2]$. This transformation is given by

$$\lambda = K_2 \sec\theta \qquad \theta \in [0, \cos^{-1}(0.5)] \tag{12}$$

Again the branch in the exponent

$$\exp\left[-(\lambda^2 - K_2^2)^{1/2} |z + z'|\right] = \exp\left[-K_2 |z + z'| \tan\theta\right] \tag{13}$$

is no longer present. Also,

$$d\lambda = K_2 \sec\theta \tan\theta \, d\theta \tag{14}$$

so that a typical denominator (equation (7)) multiplied by the transformation's Jacobian equals

$$D_2 = \frac{\sec\theta \tan\theta}{K_1^2 \tan\theta + K_2(K_2^2 \sec^2\theta - K_1^2)^{1/2}} \tag{15}$$

which is nonsingular even when K_1 equals K_2.

When $|z + z'|$ is large, the rapid decay of the exponential $\exp(-(\lambda^2 - K_2^2)^{1/2} |z + z'|)$ facilitates integration for $\lambda > K_2$. This is illustrated in Figure 4, which has the same parameters as Figures 2 and 3. In particular, $z + z'$ equals 20 wavelengths while ρ equals 1 wavelength. As discussed previously, the transformation in (12) aids integration in the λ inter-

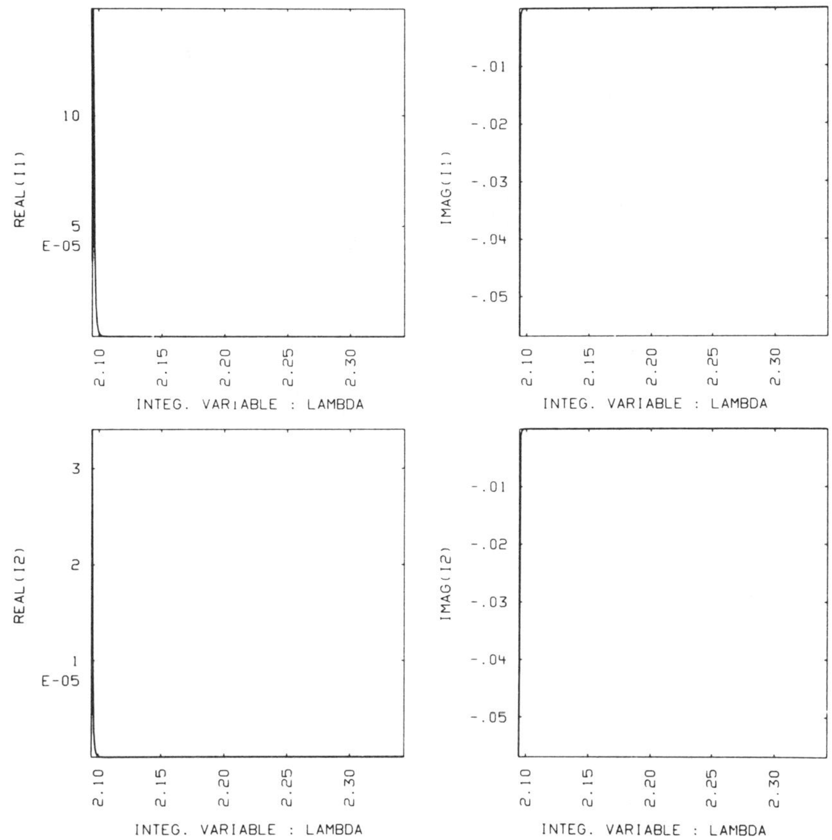

Fig. 4. Integrand behavior in the λ interval $[K_2, \infty]$. Parameters: $f = 10^8$ Hz, $\varepsilon_r = 16$, $\sigma = 10^{-4}$ mhos/m, $\rho = 1\lambda_0$, and $z + z' = 20\lambda_0$.

val $[K_2, 2K_2]$. In the interval $[2K_2, \infty)$, no transformation is necessary.

As $z + z'$ becomes small relative to ρ, integration over the interval $[2K_2, \infty)$ becomes difficult due to oscillations of the Bessel functions and slow exponential decay. Figure 5 shows the behavior of the integrands I1 and I2. In this figure, $z + z'$ and ρ have been set to 0.01 and 2 free-space wavelengths, respectively. All other parameters remain unchanged.

To accomplish the integration under these conditions, it is convenient to express (6) as

$$\int_0^\infty \bar{I}(\lambda, |z+z'|, \rho)\, d\lambda = \int_0^\infty [\bar{I}(\lambda, |z+z'|, \rho) - \overline{\text{ASYM}}(\lambda, |z+z'|, \rho)]\, d\lambda + \int_0^\infty \overline{\text{ASYM}}(\lambda, |z+z'|, \rho)\, d\lambda \tag{16}$$

where $\overline{\text{ASYM}}$ is the asymptotic approximation to $\bar{I}$ given by

$$\overline{\text{ASYM}}(\lambda, w, \rho) = \begin{bmatrix} \dfrac{e^{-\lambda w} J_0(\lambda\rho)}{K_2^2 + K_1^2}\left(\lambda^2 + \dfrac{K_2^2 w}{2}\lambda + \dfrac{K_2^4 w^2}{8} + \dfrac{K_1^2 K_2^2}{K_1^2 + K_2^2}\right) \\ \cdot 5e^{-\lambda w} J_0(\lambda\rho) \\ \dfrac{e^{-\lambda w} J_1(\lambda\rho)}{K_2^2 + K_1^2}\left(\lambda^2 + \dfrac{K_2^2 w}{2}\lambda + \dfrac{K_2^4 w^2}{8} - \dfrac{K_2^2}{2} + \dfrac{K_1^2 K_2^2}{K_1^2 + K_2^2}\right) \\ \dfrac{e^{-\lambda w} J_1(\lambda\rho)}{K_2^2 + K_1^2}\left(\lambda + \dfrac{K_2^2 w}{2}\right) \end{bmatrix} \tag{17}$$

These asymptotic forms have been derived in Appendix A. The second integral on the right of (16) has been obtained in closed form (Appendix A).

Figure 6 illustrates the behavior of the integrand when the asymptotic terms (Equation (17)) are subtracted from $\bar{I}$ (Equations (6) and (7)). The results are

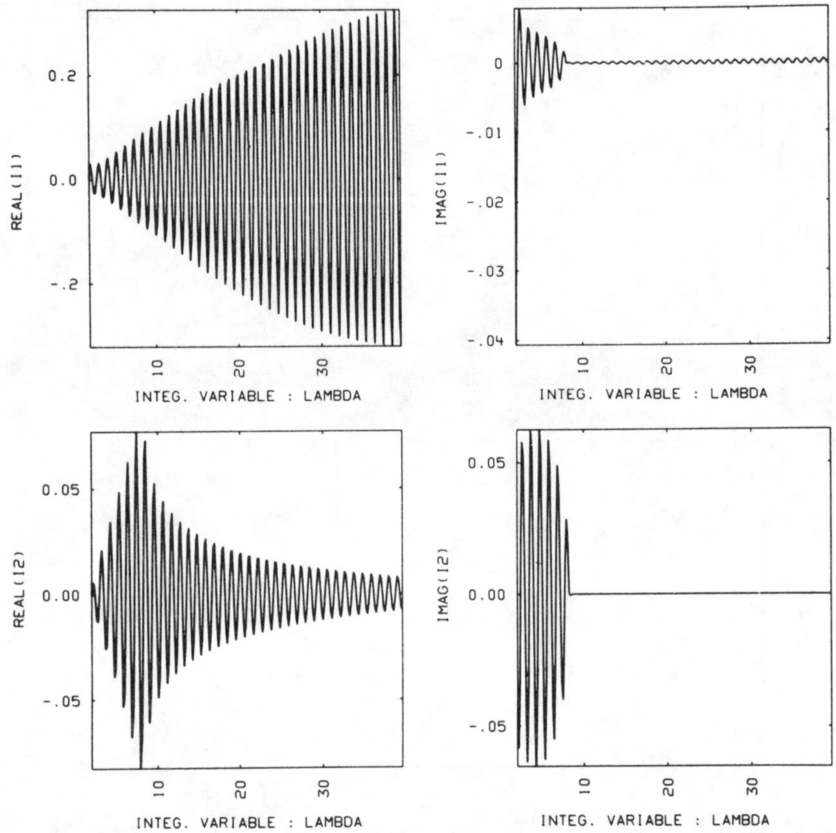

Fig. 5. Small exponential decay requires subtraction of asymptotic terms from the integrands in the λ interval $[K_2, \infty]$. Parameters: $f = 10^8$ Hz, $\varepsilon_r = 16$, $\sigma = 10^{-4}$, $\rho = 2\lambda_0$, and $z + z' = 0.01\lambda_0$.

shown for the first two components only, since the behavior of the remaining two components is similar. Clearly, these integrands are much more amenable to numerical integration than those of Figure 5 because of their faster rates of decay. The implementation of the analysis of this section by algorithmic means is left to Appendices B and C.

4. SAMPLE NUMERICAL RESULTS

In this section the real axis integration technique is compared with the complex contour methods of *Burke et al.* [1981]. Both programs were run on a CDC-7600 computer, and the computer run times were recorded. In each case both the Hankel and Bessel contours of *Burke et al.* [1981] were used. The better of these results were recorded. In Tables 1–4 the first two components of the integrals in (6) multiplied by the free-space wavelength λ_0 are recorded as (a, b), with a and b the real and imaginary parts of the result, respectively. Although brevity considerations allow only the first two components to be shown, computation times are given for the evaluation of all four components of (6).

In Tables 1–3 the frequency is 10^8 Hz, and the earth's relative dielectric constant and conductivity are 16 and 10^{-4} mhos/m, respectively. The number of points in the compound Simpson's rule (Appendix B) is 2^{IPTS}, where IPTS equals 3. The truncation criteria CONV (Appendix C) is 10^{-4}, 10^{-3}, and 10^{-4} for Tables 1, 2, and 3, respectively.

In Table 1, $z + z'$ is 10 free-space wavelengths, while ρ decreases from 5 to 0.1 free-space wavelengths. Subtraction of asymptotic terms was not needed for each real axis integration. For large ρ values, the complex Hankel contour is more efficient. However, as ρ decreases, the real axis integration becomes significantly more efficient. The Bessel contour has not been shown, since the Hankel results were faster. Similar results are shown in Table 2, where $z + z'$ is set to 3 free-space wavelengths and ρ varies from 3 to 0.01 free-space wavelengths. Again,

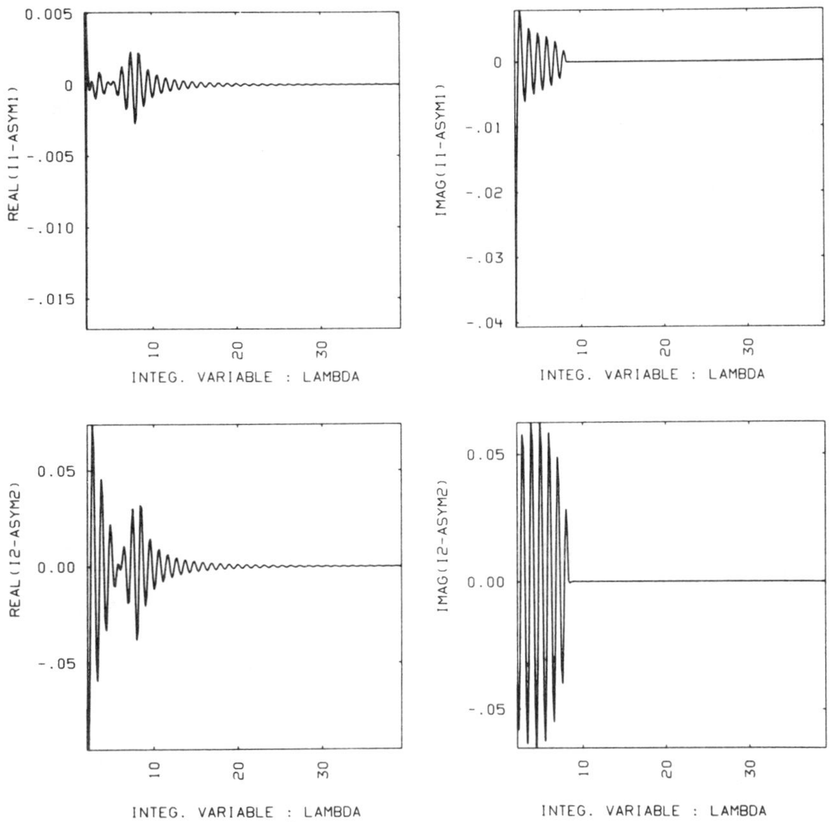

Fig. 6. Behavior of integrands for Figure 5 after asymptotic terms are subtracted. Parameters: $f = 10^8$ Hz, $\varepsilon_r = 16$, $\sigma = 10^{-4}$ mhos/m, $\rho = 2\lambda_0$, and $z + z' = 0.01\lambda_0$.

as ρ decreases, real axis integration becomes advantageous. In Table 3, ρ is set to 1 wavelength and $z + z'$ decreases from 1 wavelength to zero. Throughout, the real axis times are comparable to the Hankel contour, but computer experiments indicate the Hankel contour to be more accurate for $z + z'$ near zero and ρ equal to 1 free-space wavelength.

Table 4 shows sample calculations at a frequency of 2×10^6 Hz and earth's conductivity and relative dielectric constants of 0.01 mhos/m and 2, respec-

TABLE 1. A Comparison of Numerical Results as ρ Varies and $z + z' = 10\lambda_0$

	Distance in Wavelengths			
	$z + z' = 10\lambda_0$, $\rho = 5\lambda_0$	$z + z' = 10\lambda_0$, $\rho = 2.5\lambda_0$	$z + z' = 10\lambda_0$, $\rho = 1\lambda_0$	$z + z' = 10\lambda_0$, $\rho = 0.1\lambda_0$
Real axis	no asymptotics	no asymptotics	no asymptotics	no asymptotics
$\lambda_0((\partial^2/\partial z^2) + K_2^2)V_{22}$	(8.880E-4, −1.521E-3)	(4.615E-5, −6.308E-4	(1.951E-4, 2.625E-4)	(8.628E-6, 3.201E-4)
$\lambda_0 U_{22}$	(1.359E-2, −2.992E-2)	(1.387E-2, −3.531E-4)	(3.758E-2, −1.266E-2)	(3.999E-2, −5.870E-4)
CDC 7600 computer time, s	6.7E-2	5.8E-2	5.2E-2	4.2E-2
Complex contour	Hankel	Hankel	Hankel	Hankel
$\lambda_0((\partial^2/\partial z^2) + K_2^2)V_{22}$	(8.880E-4, −1.521E-3)	(4.612E-5, −6.306E-4)	(1.952E-4, 2.620E-4)	(8.625E-6, 3.183E-4)
$\lambda_0 U_{22}$	(1.359E-2, −2.992E-2)	(1.387E-2, −3.531E-2)	(3.759E-2, −1.266E-2)	(4.000E-2, −5.851E-4)
CDC 7600 computers time, s	4.8E-2	6.3E-2	8.7E-2	1.0E-1

Parameters: $f = 10^8$ Hz, $\sigma = 10^{-4}$ mhos/m, $\varepsilon_r = 16$, IPTS = 3, and CONV = 10^{-4}.

TABLE 2. A Comparison of Numerical Results as ρ Varies and $z + z' = 1\lambda_0$

	Distance in Wavelengths			
	$z + z' = 3\lambda_0, \rho = 3\lambda_0$	$z + z' = 3\lambda_0, \rho = 1\lambda_0$	$z + z' = 3\lambda_0, \rho = 0.1\lambda_0$	$z + z' = 3\lambda_0, \rho = 0.01\lambda_0$
Real axis	no asymptotics	no asymptotics	no asymptotics	no asymptotics
$\lambda_0((\partial^2/\partial z^2) + K_2^2)V_{22}$	(5.496-4, −1.097E-2)	(3.947E-3, −1.476E-3)	(3.251E-4, 3.529E-3)	(2.526E-4, 3.544E-3)
$\lambda_0 U_{22}$	(7.147E-4, −7.185E-2)	(5.969E-2, −1.062E-1)	(1.333E-1, −6.643E-3)	(1.335E-1, −5.266-3)
CDC 7600 time, s	3.5E-2	2.8E-2	2.5E-2	2.54E-2
Complex contour	Hankel	Hankel	Hankel	Bessel
$\lambda_0((\partial^2/\partial z^2) + K_2^2)V_{22}$	(5.495E-4, −1.097E-2)	(3.948E-3, −1.476E-3)	(3.270E-4, 3.526E-3)	(2.546E-4, 3.540E-3)
$\lambda_0 U_{22}$	(7.152E-4, −7.185E-2)	(5.968E-2, −1.061E-1)	(1.333E-1, −6.631E-3)	(1.335E-1, −5.254E-3)
CDC 7600 time, s	6.4E-2	1.0E-1	8.3E-2	9.4E-2

Parameters: $f = 10^8$ Hz, $\sigma = 10^{-4}$ mhos/m, $\varepsilon_r = 16$, IPTS = 3, and CONV = 10^{-3}.

tively. Results of the real axis integration seem to agree with the complex contours to 3 or 4 significant digits throughout. Again, for ρ small, the real axis integration has a distinct advantage.

5. CONCLUSIONS AND RECOMMENDATIONS FOR FUTURE WORK

The real axis integration technique presented in this paper is particularly well suited for the analysis of thin wires vertical to the air-earth interface. This analysis is not suitable for a long horizontal antenna with $z + z'$ equal to zero. To cover efficiently a wide range of spatial parameters, it is prudent to combine real axis and complex contour integration techniques. For example, if one desires to compute Sommerfeld integrals in the region $((z + z')^2 + \rho^2)^{1/2}$ less than or equal to 1 wavelength as done by *Burke et al.* [1981], it is efficient to use real axis techniques for $z + z' \geq \rho$ and complex contour techniques in the remaining region. Since the region in which the real axis integration is used does not require subtraction of asymptotic terms, the technique is valid for a wide range of frequencies and medium parameters.

Since extension of real axis integration of Sommerfeld integrands to the case where one or both of source and observer lies below the ground is straightforward, but tedious, results have not been given. It may be possible to extend results to lower frequencies by subtraction of higher order asymptotic terms. Use of integration routines other than Simpson's rule may be more efficient but would not change the basic conclusions of this paper. Horizontal rings of $\hat{\rho}$- and $\hat{z}$-directed current with aximuth symmetry may also be treated by this technique. The problem of a vertical cylinder which penetrates the air-earth interface and is excited by an aximuthally symmetric voltagé source will be treated with these techniques in a future paper.

TABLE 3. A Comparison of Numerical Results With $z + z'$ Variable and ρ Fixed

	Distance in Wavelengths			
	$z + z' = 1\lambda_0, \rho = 1\lambda_0$	$z + z' = 0.1\lambda_0, \rho = 1\lambda_0$	$z + z' = 0.01\lambda_0, \rho = 1\lambda_0$	$z + z' = 0, \rho = 1\lambda_0$
Real axis	no asymptotics	with asymptotics	with asymptotics	with asymptotics
$\lambda_0((\partial^2/\partial z^2) + K_2^2)V_{22}$	(−2.898E-2, −1.809E-2)	(6.451E-2, −4.516E-2)	(6.047E-2, −4.724E-2)	(5.984E-2, −4.701E-2)
$\lambda_0 U_{22}$	(−1.976E-1, −9.064E-2)	(4.521E-2, −2.192E-2)	(3.560E-2, 4.369E-2)	(5.185E-4, 6.245E-2)
CDC 7600 computer time, s	3.0E-2	5.4E-2	1.1E-1	1.1E-1
Complex contour	Hankel	Hankel	Hankel	Hankel
$\lambda_0((\partial^2/\partial z^2) + K_2^2)V_{22}$	(−2.900E-2, −1.808E-2)	(6.452E-2, −4.516E-2)	(6.045E-2, −4.725E-2)	(5.980E-2, −4.701E-2)
$\lambda_0 U_{22}$	(−1.976E-1, −9.059E-2)	(4.526E-2, −2.195E-2)	(3.376E-3, 4.369E-2)	(−6.417E-5, 6.245E-2)
CDC 7600 computer time, s	6.7E-2	1.1E-2	1.E-1	1.1E-1

Parameters: $f = 10^8$ Hz, $\sigma = 10^{-4}$ mhos/m, $\varepsilon_r = 16$, IPTS = 3, and CONV = 10^{-4}.

TABLE 4. Typical Numerical Results at 2 Megahertz

	Distance in Wavelengths			
	$z + z' = 0,\ \rho = 0.5\lambda_0$	$z + z' = 0,\ \rho = 0.01$	$z + z' = 3\lambda_0,\ \rho = 3\lambda_0$	$z + z' = 3\lambda_0,\ \rho = 0.1\lambda_0$
Real axis	with asymptotics	with asymptotics	no asymptotics	no asymptotics
$\lambda_0((\partial^2/\partial z^2) + K_2^2)V_{22}$	(−2.557E-2, −3.798E-2)	(−1.835E-1, −5.611E-2)	(2.393E-3, −1.103E-4)	(−7.244E-4, 1.253E-4)
$\lambda_0 U_{22}$	(−1.420E-2, 4.568E-3)	(7.308E-1, −2.079E-1)	(2.241E-2, 2.472E-2)	(5.188E-2, 3.915E-2)
CDC 7600 time, s	1.6E-1	7.0E-2	3.5E-2	2.5E-2
Complex contour	Hankel	Hankel	Hankel	Bessel
$\lambda_0((\partial^2/\partial z^2) + K_2^2)V_{22}$	(−2.562E-2, −3.801E-2)	(−1.840E-1, −5.618E-2)	(2.393E-3, −1.103E-4)	(−7.240E-4, 1.256E-4)
$\lambda_0 U_{22}$	(−1.413E-2, 4.666E-3)	(7.309E-1, −2.0778E-1)	(2.240E-2, −2.472E-2)	(5.189E-2, 3.917E-2)
CDC 7600 time, s	9.E-2	1.2E-1	3.6E-2	9.1E-2

Parameters: $\sigma = 10^{-2}$ mhos/m, $\varepsilon_r = 2$, IPTS = 3, and CONV = 10^{-5}.

APPENDIX A: ASYMPTOTIC FORMS OF THE SOMMERFELD INTEGRANDS

To overcome difficulties associated with the slow exponential decay of the integrands (equation (6)) as $|z + z'|$ becomes small, section 3 has made use of their asymptotic forms. These asymptotic forms are now obtained along with closed form expressions for their integrals over the definite interval $\lambda \in [0, \infty)$.

To obtain the desired result, some preliminary groundwork is necessary. For large λ,

$$(\lambda^2 - K_i^2)^{1/2} \approx \lambda - \frac{K_i^2}{2\lambda} \qquad i = 1, 2 \tag{A1}$$

$$e^{-\lambda_2 w} \approx e^{-\gamma w}\left(1 + \frac{K_2^2 w}{2\lambda} + \frac{K_2^4 w^2}{8\lambda^2}\right) \tag{A2}$$

$$\frac{1}{K_2^2\gamma_1 + K_1^2\gamma_2} \approx \frac{1}{K_2^2 + K_1^2}\frac{1}{\lambda}\left(1 + \frac{K_2^2 K_1^2}{K_2^2 + K_1^2}\frac{1}{\lambda^2}\right) \tag{A3}$$

and

$$\frac{1}{\gamma_2 + \gamma_1} \approx \frac{1}{2\lambda}\left(1 + \frac{K_1^2 + K_2^2}{4\lambda^2}\right) \tag{A4}$$

From these expressions an asymptotic approximation, $\overline{\text{ASYM}}$, to $\bar{I}$ (equation (7)) may be obtained. The result is

$$\text{ASYM1} = \frac{e^{-\lambda w}J_0(\lambda\rho)}{K_1^2 + K_2^2}\left(\lambda^2 + \frac{K_2^2 w}{2}\lambda + \frac{K_2^4 w^2}{8} + \frac{K_1^2 K_2^2}{K_1^2 + K_2^2}\right) \tag{A5}$$

$$\text{ASYM2} = 0.5e^{-\lambda w}J_0(\lambda\rho) \tag{A6}$$

$$\text{ASYM3} = \frac{e^{-\lambda w}J_0(\lambda\rho)}{K_2^2 + K_1^2}\left(\lambda^2 + \frac{K_2^2 w}{2}\lambda + \frac{K_2^4 w^2}{8} - \frac{K_2^2}{2} + \frac{K_1^2 K_2^2}{K_1^2 + K_2^2}\right) \tag{A7}$$

and

$$\text{ASYM4} = \frac{e^{-\lambda w}}{K_2^2 + K_1^2}J_1(\lambda\rho)\left(\lambda + \frac{K_2^2 w}{2}\right) \tag{A8}$$

In the above asymptotic expressions, terms which decrease faster than $e^{-\lambda w}J_0(\lambda\rho)$ have not been included so that the expressions may be integrated from zero to infinity analytically.

To obtain a closed form expression for the integral

$$\bar{Z} = \int_0^\infty \overline{\text{ASYM}}(\lambda)\, d\lambda \tag{A9}$$

recall [*Abramowitz and Stegun*, 1970]

$$T0 = \int_0^\infty e^{-\lambda w}J_0(\lambda\rho)\, d\lambda = \frac{1}{(w^2 + \rho^2)^{1/2}} \tag{A10}$$

$$T1 = \int_0^\infty e^{-\lambda w}J_0(\lambda\rho)\, \lambda d\lambda = -\frac{\partial}{\partial w}\frac{1}{(w^2 + \rho^2)^{1/2}} = \frac{w}{(w^2 + \rho^2)^{3/2}} \tag{A11}$$

$$T2 = \int_{0ZS\infty} e^{-\lambda w}\lambda^2 J_0(\lambda\rho)\, d\lambda = \frac{\partial^2}{\partial w^2}\frac{1}{(w^2 + \rho^2)^{1/2}} = \frac{2w^2 - \rho^2}{(w^2 + \rho^2)^{5/2}} \tag{A12}$$

Similarly,

$$U0 = \int_0^\infty e^{-\lambda w}J_1(\lambda\rho)\, d\lambda = \frac{1}{\rho}\frac{(w^2 + \rho^2)^{1/2} - w}{(w^2 + \rho^2)^{1/2}} \tag{A13}$$

$$U1 = \int_0^\infty e^{-\lambda w}\lambda J_1(\lambda\rho)\, d\lambda = -\frac{\partial}{\partial\rho}\int_0^\infty e^{-\lambda w}J_0(\lambda\rho)\, d\lambda = \frac{\partial}{\partial\rho}\frac{1}{(\rho^2 + w^2)^{1/2}} = \frac{\rho}{(\rho^2 + w^2)^{3/2}} \tag{A14}$$

and

$$U2 = \int_0^\infty e^{-\lambda w} \lambda^2 J_1(\lambda\rho)\, d\lambda = \frac{\partial^2}{\partial w\, \partial \rho} \int_0^\infty e^{-\lambda w} J_0(\lambda\rho)\, d\lambda$$

$$= \frac{\partial^2}{\partial w\, \partial \rho} \frac{1}{(\rho^2 + w^2)^{1/2}} = \frac{3\rho w}{(\rho^2 + w^2)^{5/2}} \qquad \text{(A15)}$$

From equations (A5)–(A15) it follows that

$$\bar{Z} = \begin{bmatrix} \dfrac{1}{K_2^2 + K_1^2}\left(T2 + \dfrac{K_2^2 w}{2} T1 + \dfrac{K_2^4 w^2}{8} + \dfrac{K_1^2 K_2^2}{K_1^2 + K_2^2} T0\right) \\ 0.5T0 \\ \dfrac{1}{K_2^2 + K_1^2}\left(U2 + \dfrac{K_2^2 w}{2} U1 + \dfrac{K_2^2 w}{8} - \dfrac{K_2^2}{2} + \dfrac{K_1^2 K_2^2}{K_1^2 + K_2^2} U0\right) \\ \dfrac{1}{K_2^2 + K_1^2}\left(U1 + \dfrac{K_2^2 w}{2} U0\right) \end{bmatrix} \qquad \text{(A16)}$$

Thus a closed form expression for equation (A9) has been obtained.

APPENDIX B: NUMERICAL IMPLEMENTATION

The analysis in section 3 has been implemented by means of a FORTRAN computer program. A brief description of how the algorithm works is now given. Attention is first focused on the case where the exponential term $\exp(-(\lambda^2 - K_2^2)^{1/2} |z + z'|)$ decays rapidly enough so that subtraction of asymptotic terms is not necessary. Integration of (6) with respect to λ over the three intervals $[0, K_2]$, $[K_2, T]$, and $[T, \infty)$ is considered separately, where

$$T = \frac{K_2 + (\mathrm{Re}\,(K_1^2))^{1/2}}{2} \qquad \text{(B1)}$$

Finally, subtraction of the asymptotic terms is briefly discussed.

Equation (8) is used to transform the λ integration over $[0, K_2]$ to an integration with respect to θ over the interval $[0, \pi/2]$. The peaks and nulls of the real and imaginary parts of the exponent (equation (9)) are given by

$$\theta_n = \cos^{-1}\left(\frac{n\pi}{2K_2|z + z'|}\right)$$

$$n = 0, 1, 2, \cdots, n_{\max} \qquad \text{(B2)}$$

where

$$n_{\max} = \mathrm{int}\left(\frac{2K_2|z + z'|}{\pi}\right)$$

The symbol int (x) denotes the largest integer less than or equal to x. The first 60 zeros of both $J_0(u)$ and $J_1(u)$ are interlaced and stored in increasing order in an array Z_i $(i = 1,\ 120)$. It follows from equations (8) and (6) that the θ values corresponding to these zeros are given by

$$\theta_i = \sin^{-1} \frac{Z_i}{K_2 \rho}$$

$$i = 1, \cdots, i_{\max} \qquad Z_{i_{\max}} \leq K_2 \rho \qquad \text{(B3)}$$

If $|K_1|/K_2 > 5$ the θ value

$$\theta = \cos^{-1} |4K_2/K_1| \qquad \text{(B4)}$$

is added to track the variation of equation (11). All of these θ values are combined and stored in increasing order. Integration between successive θ values, $[\theta_i, \theta_{i+1}]$, is carried out by a compound Simpson's rule with 2^{IPTS} points [*Davis and Rabinowitz*, 1975]. IPTS is an as yet undetermined input parameter.

Equation (12) is used to transform the λ interval $[K_2, T]$ into the θ interval $[0, \cos^{-1}(K_2/T)]$. To track the rapidly decaying exponential when $z + z'$ is large, the λ interval is subdivided by the points

$$\lambda = K_2, K_2 + \Delta, K_2 + 2\Delta, \cdots, K_2 + n\Delta, T \qquad \text{(B5)}$$

where

$$\Delta = \min\left(\frac{0.4\pi}{K_2|z + z'|}, K_2\right) \qquad \text{(B6)}$$

These values are easily converted into their corresponding θ values through equation (12). Next, the θ values corresponding to the zeros of the Bessel functions are computed from (6) and (7) along with the values of the Bessel function zeros that have been stored in an array. Use of this array may be avoided by replacing the zeros of the Bessel functions by their asymptotic values. If $|K_1| > 5K_2$, the θ value

$$\theta_0 = \tan^{-1}(4K_2/|K_1|) \qquad \text{(B7)}$$

is added to the other θ values to insure accurate tracking of (15). Again, all of the θ values are stored in an array of increasing order, and a compound Simpson's rule is used to integrate between the successive values.

Integration of (6) over the interval $[2K_2, \infty)$ is accomplished without use of any variable transformations. To ensure adequate tracking of the denominators of (7), this interval is broken into two subintervals $[T, \mathrm{RTRK1S}]$ and $[\mathrm{RTRK1S}, \infty)$, where

$$\mathrm{RTRK1S} = (\mathrm{Re}\,(K_1^2))^{1/2} \tag{B8}$$

The first subinterval is further subdivided by the points

$$\lambda_0 = 2K_2 \qquad \lambda_1 = \lambda_0 + \mathrm{STEP}, \cdots,$$
$$\lambda_{no-1} = \lambda_{no-2} + \mathrm{STEP} \qquad \lambda_{no} = \mathrm{RTRK1S} \tag{B9}$$

where STEP is set equal to Δ (equation (B6)) until λ is greater than $10K_2$, and then STEP is doubled each successive step. The second subinterval is also subdivided

$$\lambda_{no} = \mathrm{RTRK1S} \qquad \lambda_{no+1} = \lambda_{no} + \mathrm{STEP}$$
$$\lambda_{no+2} = \lambda_{no+1} + \mathrm{STEP}, \cdots \tag{B10}$$

where STEP has been defined previously.

The λ values associated with the zeros of the Bessel functions are computed from the stored values of the Bessel functions' zero crossings and equation (7). Again, all of these λ values are stored in an array, and a compound Simpson's rule with 2^{IPTS} points is used to integrate between successive values.

Techniques for implementing the analysis of section 3 have been presented under the assumption that subtraction of the asymptotic terms (equations (16) and (17)) were not necessary. The same techniques may be applied without change when these asymptotic terms are subtracted from the integrand. A further comment on the procedure for truncating the integration over the infinite interval is left to Appendix C.

APPENDIX C: TRUNCATION OF THE INFINITE INTEGRATION INTERVAL

In Appendix B, techniques were presented for implementing the analysis of section 3. A discussion of how integration over the infinite interval $[2K_2, \infty)$ is truncated was left for this appendix. Consideration is first given to the case when subtraction of asymptotic terms is not necessary.

In the case of significant exponential decay, say $z + z' \geq 0.5\rho$, the following truncation criterion has been found to be useful. As in Appendix B, the interval $[2K_2, \infty)$ is divided into subintervals. It is convenient to express this as

$$[2K_2, \infty) = \bigcup_{i=0}^{\infty} [\lambda_i, \lambda_{i+1}] \tag{C1}$$

Numerical integration is performed on each segment. After each segment is integrated, a check is made to determine whether

$$\frac{\left| \int_{[\lambda_m, \lambda_{m+1}]} \bar{I}(\lambda, |z + z'|, \rho)\, d\lambda \right|}{\left| \sum_{i=0}^{n} \int_{[\lambda_i, \lambda_{i+1}]} \bar{I}(\lambda, |z + z'|, \rho)\, d\lambda \right|} \leq \mathrm{CONV} \tag{C2}$$

where CONV is a convergence criterion that may be set by the user. If (C2) is satisfied IND consecutive times, where IND is a user-specified input parameter, the integration is truncated.

If $z + z' < 0.05\rho$, it has been found useful to use

$$\frac{\left| \sum_{j=0}^{3} \int_{[\lambda_{n+j}, \lambda_{n+1+j}]} \bar{I}(\lambda, |z + z'|, \rho)\, d\lambda \right|}{\left| \sum_{i=0}^{n+3} \int_{[\lambda_i, \lambda_{i+1}]} \bar{I}(\lambda, |z + z'|, \rho)\, d\lambda \right|} < \mathrm{CONV} \tag{C3}$$

as a convergence criteria. It is seen from (B10) that the partition points not associated with the zeros of the Bessel functions become more widely spaced as λ becomes large. Thus the partition points asymptotically become the zeros of the Bessel functions. Asymptotically for large λ, the nulls of J_0 become the peaks of J_1 and vice versa, and the nulls become uniformly spaced. Thus the sum of the intervals in the numerator of (C3) approaches the distance between the nulls of either $J_0(\lambda\rho)$ or $J_1(\lambda\rho)$. This sum tends to be much smaller than any of the individual elements due to the oscillation of the Bessel functions.

If subtraction of asymptotic terms is necessary, one merely replaces $\bar{I}$ by $\bar{I} - \overline{\mathrm{ASYM}}$ in (C2) and (C3) to obtain the truncation criteria. Similar criteria are used in the interval $[K_2, T]$.

Acknowledgments. This work was performed by the Lawrence Livermore National Laboratory under the auspices of the U.S. Department of Energy under contract W-7405-ENG-48 and was supported by CORADCOM of the U.S. Army. D. G. Dudley contributed to this work while a summer visitor at Lawrence Livermore National Laboratory. The authors wish to thank C. M. Butler, R. W. Ziolkowski, and G. J. Burke for valuable discussions throughout the course of this work. The contributions of the reviewers to the clarity of this work are gratefully acknowledged.

REFERENCES

Abramowitz, M., and I. A. Stegun (1970), *Handbook of Mathematical Functions*, p. 1024, Dover, New York.

Banos, A. (1966), *Dipole Radiation in the Presence of a Conducting Half-Space*, Pergamon, New York.

Bubenik, D. M. (1977), A practical method for the numerical

evaluation of Sommerfeld integrals, *IEEE Trans. Antennas Propag.*, *AP 25*(6), 904-906.

Burke, G. J., E. K. Miller, J. N. Brittingham, D. L. Lager, R. J. Lytle, and J. T. Okada (1981), Computer modeling of antennas near the ground, *Electromagnetics*, *1*, 29–44.

Davis, P. J., and P. Rabinowitz (1975), *Methods of Numerical Integration*, pp. 45–48, Academic, New York.

Parhami, P., Y. Rahmat-Samii, and R. Mittra (1980), An efficient approach for evaluating Sommerfeld integrals encountered in the problem of a current element radiating over a lossy ground, *IEEE Trans. Antennas Propag.*, *AP 28*(1), 100–104.

Rahmat-Samii, Y., R. Mittra, and P. Parhami (1981), Evaluation of Sommerfeld integrals for lossy half-space problems, *Electromagnetics*, *1*, 1–28.

Rana, I. E., and N. G. Alexopoulos (1981), Current distribution and input impedance of printed dipoles, *IEEE Trans. Antennas Propag.*, *AP 29*(1), 99–105.

Siegel, M., and R. W. P. King (1970), Electromagnetic fields in a dissipative half-space: A numerical approach, *J. Appl. Phys.*, *41*(6), 2415–2423.

On the efficient evaluation of integrals arising in the Sommerfeld halfspace problem

K.A. Michalski, Ph.D.

Indexing terms: *Mathematical techniques, Antennas*

Abstract: An efficient approach is described for the numerical evaluation of the Sommerfeld integrals which arise in analyses of antennas radiating over a dielectric halfspace. The main features of this approach can be summarised as follows. The contour deformation principle is employed to express the original integral as a sum of the steepest descent-path (SDP) integral and the possible branch-cut-path (BCP) integral, which are then evaluated numerically. If the branch point is captured, the SDP is allowed to go to the bottom sheet of the Riemann surface, resulting in a continuous behaviour of the integrand. In the last case, the BCP is deformed to the SDP emanating from the branch point, on which the integrand decays rapidly. Also, the Zenneck pole is subtracted, resulting in a smooth behaviour of the integrand on the SDP.

1 Introduction

A variety of methods exist for the evaluation of the integrals arising in the Sommerfeld halfspace problem [1]. For a comprehensive summary of these techniques, the reader is referred to the recent review papers [2–4]. These techniques can roughly be classified as analytic or numerical. The analytic methods are based on approximations, which inevitably apply for only a limited range of parameters. The numerical techniques are usually 'exact' in the sense that a numerical quadrature is applied to rigorous Sommerfeld integrals. These techniques, too, are only applicable in some range of parameters outside which the computational cost becomes prohibitive. Therefore, efforts continue to develop new, more efficient and more widely applicable methods for the evaluation of Sommerfeld integrals. The papers by Rahmat-Samii *et al.* [2, 5], Johnson and Dudley [6], and Lindell and Alanen [7] are examples of the more recent developments in this field.

The present paper elaborates upon the Rahmat-Samii *et al.* technique for the computation of the Sommerfeld integrals which arise in analyses of antennas radiating over a dielectric halfspace. In contrast to previous approaches [3], these authors deform the original path of integration to the steepest descent path (SDP) passing through the saddle point, as it is done in the asymptotic approximation of integrals [8, 9]. The Sommerfeld integral is then evaluated numerically along the SDP. It is quite obvious that this choice of the path of integration is superior to other approaches in the case when the distance between the source point and the observation point is large in terms of wavelengths. Rahmat-Samii *et al.* have found, however, that the integration along the SDP is very efficient even if the last condition is drastically violated. This surprising observation is the key to the success of their technique.

A complication in the Rahmat-Samii *et al.* method arises when a branch point is captured in the process of the contour deformation. In that case, in order to remain on the top sheet of the Riemann surface, the branch-cut-path (BCP) integral is added to the SDP integral. The BCP integral is only along the part of the fundamental branch cut which is captured by the SDP. This treatment of the branch point singularity has several disadvantages which can adversely affect the efficiency of the overall scheme. First, due to the fact that the SDP is kept on the top sheet, the integrand exhibits a discontinuity at the point where the SDP crosses the branch cut, which necessitates splitting the integration interval in two. Secondly, the limit of integration in the BCP integral must be found as a solution of a nonlinear equation, which is a rather unreliable process. Finally, the integrand along the BCP often varies rapidly, requiring a high-order quadrature.

Another difficulty in this scheme is due to the pole of the integrand (the Zenneck pole [8]), which, although never captured, can get arbitrarily close to the SDP when both the source point and the field point are close to the interface and the contrast between the two media is large. In that case, the integrand exhibits a sharp peak on the SDP in the vicinity of the pole, requiring a very careful integration.

In this paper, a modification of the Rahmat-Samii *et al.* approach is described, which does not suffer from the deficiencies stated above. The main features of the modified technique can be summarised as follows. If the branch point is captured, the SDP is allowed to go to the bottom sheet of the Riemann surface, resulting in a continuous behaviour of the integrand. To properly close the contour of integration, the SDP integral is augmented by an integral around the entire branch cut. The original branch cut is then deformed to the SDP emanating from the branch point, on which the integrand decays rapidly. Finally, the Zenneck pole is subtracted from the integrand along the SDP, making the integration efficient and reliable for a wide range of parameters. The integral of the subtracted term is then expressed in terms of the error function and is added to the modified integral to preserve its correct value.

2 Formulation

Consider a vertical electric dipole (VED) of unit strength located at height h above a planar interface separating two halfspaces (Fig. 1). The medium of the upper halfspace ($z > 0$) is taken to be air with permittivity ε_0, permeability μ_0 and wave number k_1. The lower halfspace ($z < 0$) is occupied by a dielectric characterised by ($\kappa\varepsilon_0$, μ_0) with $\kappa = \varepsilon - j\sigma/(\omega\varepsilon_0)$, where ε is the relative dielectric constant and σ is the conductivity of the medium (the exp ($j\omega t$) time convention is assumed). The Hertz vector due to the VED is given by $\mathbf{\Pi}_v = \hat{z}\Pi_{vz}$, where $\hat{z}$ is a unit vector in the z-direction, and where Π_{vz} can be expressed as [2]

$$\Pi_{vz} = (j\omega\varepsilon_0)^{-1}\{g(r_1) - g(r_2) + {}_0\Pi_{vz}\} \tag{1}$$

Paper 3965H (E11), received 13th July 1985

The author is with the Department of Electrical Engineering, University of Mississippi, Mississippi 38677, USA with

$$g(r) = \exp(-jk_1 r)/4\pi r \tag{2}$$

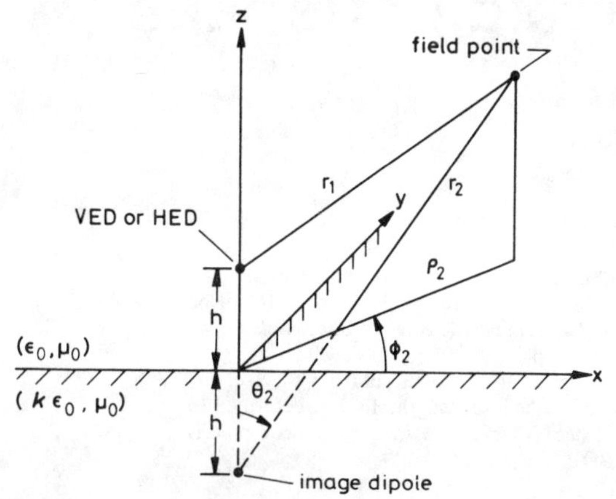

Fig. 1 *Elementary dipole radiating over a dielectric halfspace*

and

$$_0\Pi_{vz} = \frac{k_1\kappa}{4\pi j}\int_\Gamma \frac{\sin\xi\cos\xi}{\kappa\cos\xi + \sqrt{\kappa - \sin^2\xi}} \times H_0^{(2)}(k_1\rho_2\sin\xi)e^{-jk_1(z+h)\cos\xi}\,d\xi \tag{3}$$

In eqn. 3, $H_0^{(2)}$ is the Hankel function of the second kind and order zero, and Γ is the path of integration on the top sheet of the two-sheeted Riemann surface (Fig. 2). The top

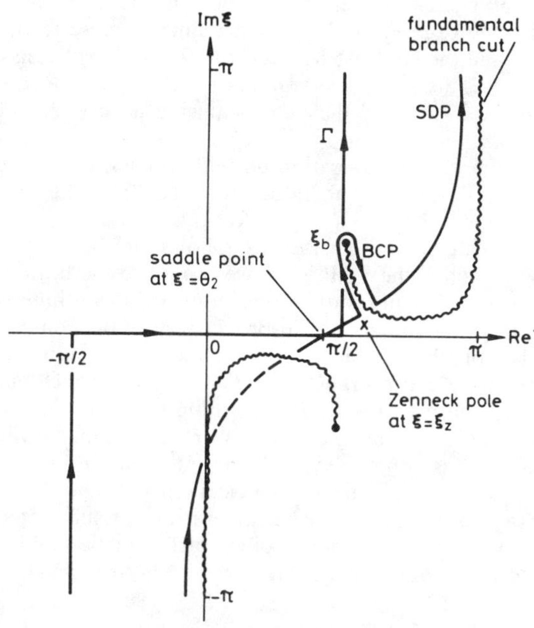

Fig. 2 *The original integration path Γ, the steepest-descent path (SDP), and the branch-cut path (BCP) in the complex ξ-plane*
f = 10 Mhz, ε = 10, $\sigma = 2 \cdot 10^{-4}$ S/m, and $\theta_2 = 78°$

and bottom sheets are defined by

$$\text{Im}\{\sqrt{\kappa - \sin^2\xi}\}\begin{cases} <0, \text{ on top sheet} & (4a) \\ =0, \text{ on fundamental branch cuts} & (4b) \\ >0, \text{ on bottom sheet} & (4c)\end{cases}$$

When the VED is replaced by a horizontal electric dipole (HED), which without loss of generality can be assumed to be along the x-axis, the Hertz vector is given by $\mathbf{\Pi}_h = \hat{x}\Pi_{hx} + \hat{z}\Pi_{hz}$ where Π_{hx} and Π_{hz} can be expressed as [2, 5]

$$\Pi_{hx} = (j\omega\varepsilon_0)^{-1}\{g(r_1) - g(r_2) + {}_0\Pi_{hx}\} \tag{5}$$

$$\Pi_{hz} = (j\omega\varepsilon_0)^{-1}\,{}_0\Pi_{hz} \tag{6}$$

with

$$_0\Pi_{hx} = \frac{k_1}{4\pi j}\int_\Gamma \frac{\sin\xi\cos\xi}{\cos\xi + \sqrt{\kappa - \sin^2\xi}} \times H_0^{(2)}(k_1\rho_2\sin\xi)e^{-jk_1(z+h)\cos\xi}\,d\xi \tag{7}$$

and

$$_0\Pi_{hz} = -\frac{k_1}{4\pi}\cos\phi_2\int_\Gamma \sin^2\xi\cos\xi\,\frac{\cos\xi - \sqrt{\kappa - \sin^2\xi}}{\kappa\cos\xi + \sqrt{\kappa - \sin^2\xi}} \cdot H_1^{(2)}(k_1\rho_2\sin\xi)e^{-jk_1(z+h)\cos\xi}\,d\xi \tag{8}$$

It is worth mentioning, before leaving this Section, that there are other ways of expressing Π_{vz} and Π_{hx} than given above. For example, eqn. 1 can be rewritten as

$$\Pi_{vz} = (j\omega\varepsilon_0)^{-1}\{g(r_1) + g(r_2) + {}_0\Pi'_{vz}\} \tag{9}$$

with

$$_0\Pi'_{vz} = -\frac{k_1}{4\pi j}\int_\Gamma \frac{\sin\xi\sqrt{\kappa - \sin^2\xi}}{\kappa\cos\xi + \sqrt{\kappa - \sin^2\xi}} \times H_0^{(2)}(k_1\rho_2\sin\xi)e^{-jk_1(z+h)\cos\xi}\,d\xi \tag{10}$$

When the lower medium is replaced by a perfect conductor, Π_{vz} and Π_{hx} are given by the first two terms of eqns. 9 and 5, respectively (Π_{hz} is zero in that case). Although eqn. 9 is more consistent with eqn. 5, the representation, eqn. 1, is used throughout this paper, as it was employed in the original formulation [2].

The remainder of this paper is concerned with the evaluation of the integrals incorporated in ${}_0\Pi_{vz}$, ${}_0\Pi_{hx}$, and ${}_0\Pi_{hz}$, which are known as the Sommerfeld integrals. The integral contained in ${}_0\Pi_{vz}$ is used in the numerical examples included in the next Sections, as it incorporates all the important features of all the Sommerfeld integrals.

3 Evaluation of the Sommerfeld integrals [2, 5]

Rahmat-Samii *et al.* evaluate the Sommerfeld integrals incorporated in eqns. 3 and 7–8 by deforming Γ to the steepest descent path (SDP) in the complex ξ-plane (Fig. 2) and by making the substitution

$$\cos(\xi - \theta_2) = 1 - js^2 \tag{11}$$

where s is a real variable ranging between $-\infty$ and $+\infty$, with $s = 0$ corresponding to the saddle point at $\xi = \theta_2$. From eqn. 11, the SDP is given by [2, 5]

$$\xi_{SDP} = \theta_2 - j\,\text{Ln}\left((1 - js^2) - s(1-j)\sqrt{1 - j\frac{s^2}{2}}\right) \tag{12}$$

In deforming Γ to the SDP, the branch point $\xi_b = \xi'_b + j\xi''_b$ can be captured, which occurs when $\theta_2 > \theta_c$, where θ_c is the angle of capture given by [8]

$$\theta_c = \xi'_b - \arccos\left(\frac{1}{\cosh(\xi''_b)}\right) \tag{13}$$

with

$$\xi_b = \frac{\pi}{2} + j\,\text{Ln}(\sqrt{\kappa} + \sqrt{\kappa - 1}) \tag{14}$$

When this happens, the contour of integration must be properly modified to take into account the branch-cut contribution. Rahmat-Samii *et al.* let the SDP enter the bottom sheet when it crosses the lower branch cut (Fig. 2), as it always re-emerges on the top sheet. However, they do not allow the SDP to go to the bottom sheet when it encounters the upper branch cut and integrate around it instead. The path around the captured part of the branch cut is designated BCP in Fig. 2.

Invoking the unit step function

$$u(x) = \begin{cases} 1, & x > 0 \\ 0, & x < 0 \end{cases} \tag{15}$$

one can express each of the Hertz potentials as a sum of the SDP contribution, and the possible BCP contribution as

$${}_0\Pi_{vz} = {}_0\Pi_{vz}^{SDP} + u(\theta - \theta_c)_0\,\Pi_{vz}^{BCP} \tag{16}$$

$${}_0\Pi_{hx} = {}_0\Pi_{hx}^{SDP} + u(\theta - \theta_c)_0\,\Pi_{hx}^{BCP} \tag{17}$$

$${}_0\Pi_{hz} = {}_0\Pi_{hz}^{SDP} + u(\theta - \theta_c)_0\,\Pi_{hz}^{BCP} \tag{18}$$

The SDP terms in eqns. 16–18 are given by

$${}_0\Pi_{vz}^{SDP} = \frac{k_1\kappa}{4\pi j} e^{-jk_1r_2} \int_{-\infty}^{\infty} P(\xi_{SDP})\alpha(s) e^{-k_1r_2s^2}\, ds \tag{19}$$

$${}_0\Pi_{hx}^{SDP} = \frac{k_1}{4\pi j} e^{-jk_1r_2} \int_{-\infty}^{\infty} S(\xi_{SDP})\alpha(s) e^{-k_1r_2s^2}\, ds \tag{20}$$

and

$${}_0\Pi_{hz}^{SDP} = -\frac{k_1}{4\pi}\cos\phi_2\, e^{-jk_1r_2} \times \int_{-\infty}^{\infty} T(\xi_{SDP})\alpha(s) e^{-k_1r_2s^2}\, ds \tag{21}$$

where

$$P(\xi) = \frac{1}{\kappa\cos\xi + \sqrt{\kappa - \sin^2\xi}}\, Q_0(\xi) \tag{22}$$

$$S(\xi) = \frac{1}{\cos\xi + \sqrt{\kappa - \sin^2\xi}}\, Q_0(\xi) \tag{23}$$

$$T(\xi) = \sin\xi\, \frac{\cos\xi - \sqrt{\kappa - \sin^2\xi}}{\kappa\cos\xi + \sqrt{\kappa - \sin^2\xi}}\, Q_1(\xi) \tag{24}$$

with

$$Q_i(\xi) = \sin\xi\cos\xi H_i^{(2)}(k_1r_2\sin\theta_2\sin\xi) \times e^{jk_1r_2\sin\theta_2\sin\xi} \quad i = 0, 1 \tag{25}$$

and where

$$\alpha(s) \equiv \frac{d\xi}{ds} = \frac{1+j}{\sqrt{1 - j\dfrac{s^2}{2}}} \tag{26}$$

The condition 4*a* is enforced along the SDP in eqns. 22–24, except for the part of the path which is on the bottom sheet (denoted by a dashed line in Fig. 2), where condition 4*c* holds. The effective interval of integration is determined from the condition [2]

$$|s| \leqslant \frac{3}{\sqrt{k_1r_2}} \tag{27}$$

As an illustration of the typical behaviour of the Sommerfeld integrands, the integrand of ${}_0\Pi_{vz}^{SDP}$ is plotted for two different sets of parameters in Figs. 3*a* and 4*a*. These examples show the undesirable discontinuity of the integrands at the point where the SDP crosses the upper branch cut.

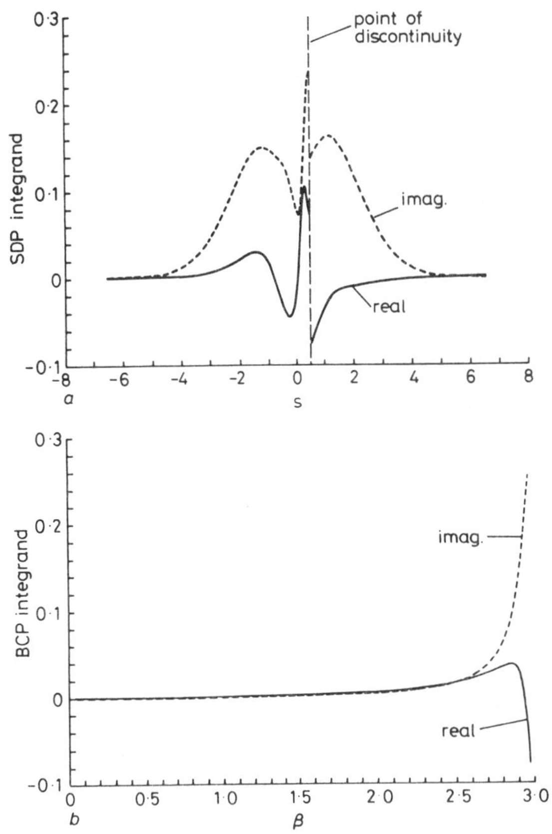

Fig. 3 *Plot of the ${}_0\Pi_{vz}$ integrand (Rahmat-Samii et al. [2, 5] approach)*
$f = 10$ MHz, $\varepsilon = 10$, $\sigma = 2 \cdot 10^{-4}$ S/m, $\theta_2 = 78°$, and $r_2 = 1$ m
a SDP integrand *b* BCP integrand

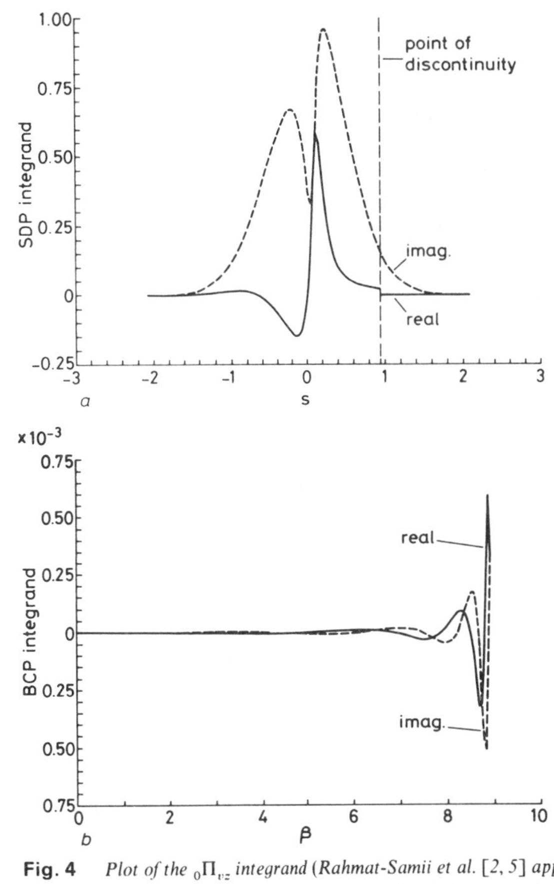

Fig. 4 *Plot of the ${}_0\Pi_{vz}$ integrand (Rahmat-Samii et al. [2, 5] approach)*
$f = 100$ MHz, $\varepsilon = 80$, $\sigma = 1 \cdot 10^{-2}$ S/m, $\theta_2 = 85°$, and $r_2 = 1$ m
a SDP integrand *b* BCP integrand

The BCP terms in eqns. 16–18 contain integrals around the captured part of the fundamental branch cut. Rahmat-Samii *et al.* express these integrals in terms of a real parameter β ranging between zero and some finite value β_1, which must be determined as a solution of a nonlinear equation. For more details, the reader is referred to References 2 and 5. To illustrate the typical behaviour of the integrands along the fundamental branch cut, the integrand of ${}_0\Pi_{vz}^{BCP}$ is plotted in Figs. 3*b* and 4*b* for the two cases. One observes a rapid variation of the integrands near the end of the interval, which necessitates a very careful integration.

4 Improved method

The discontinuity of the SDP integrand in the Rahmat-Samii *et al.* approach can be removed if one allows the SDP to continue on the bottom sheet when it encounters the upper fundamental branch cut. However, since the original path Γ must meet with the deformed contour on the same sheet at infinity, the SDP integral must be augmented in that case by an integral around the entire branch cut, as illustrated in Fig. 5*a*. Obviously, the necessity to integrate along this contour would nullify any gain of efficiency in the SDP integral. To remedy this, the fundamental branch cut is deformed to the steepest descent path from the branch point ξ_b (Fig. 5*b*), as is done in the asymptotic evaluation of branch-cut integrals [8, 9]. This is accomplished by making the substitution

$$\cos(\xi - \theta_2) = \cos(\xi_b - \theta_2) - jt^2 \quad (28)$$

where t is a real parameter ranging between $-\infty$ and $+\infty$. From eqn. 28, the deformed branch cut can be explicitly defined as

$$\xi_{BCP} = \theta_2 - j\,\mathrm{Ln}\,(\cos(\xi_b - \theta_2) - jt^2 + j\sqrt{t^4 + j2t^2\cos(\xi_b - \theta_2) + \sin^2(\xi_b - \theta_2)}) \quad (29)$$

No singularities are captured in the BCP deformation since the pole ξ_z (the Zenneck pole [8, 9]) exhibited by $P(\xi)$ and $T(\xi)$ lies below the original branch cut [9].

When eqn. 28 is used in eqns. 3 and 7–8, the BCP terms in eqns. 16–18 can be expressed as

$${}_0\Pi_{vz}^{BCP} = \frac{k_1\kappa}{4\pi j} e^{-jk_1r_2\cos(\xi_b - \theta_2)} \times \int_0^\infty P'(\xi_{BCP})\beta(t)e^{-k_1r_2t^2}\,dt \quad (30)$$

$${}_0\Pi_{hx}^{BCP} = \frac{k_1}{4\pi j} e^{-jk_1r_2\cos(\xi_b - \theta_2)} \times \int_0^\infty S'(\xi_{BCP})\beta(t)e^{-k_1r_2t^2}\,dt \quad (31)$$

and

$${}_0\Pi_{hz}^{BCP} = -\frac{k_1}{4\pi}\cos\phi_2\, e^{-jk_1r_2\cos(\xi_b - \theta_2)} \times \int_0^\infty T'(\xi_{BCP})\beta(t)e^{-k_1r_2t^2}\,dt \quad (32)$$

where

$$P'(\xi) - P^-(\xi) - P^+(\xi) = \frac{2\sqrt{\kappa - \sin^2\xi}}{(\kappa - 1)[(\kappa + 1)\cos^2\xi - 1]} Q_0(\xi) \quad (33)$$

$$S'(\xi) = S^-(\xi) - S^+(\xi) = -\frac{2\sqrt{\kappa - \sin^2\xi}}{(\kappa - 1)} Q_0(\xi) \quad (34)$$

$$T'(\xi) = T^-(\xi) - T^+(\xi) = \frac{2(\kappa + 1)\sqrt{\kappa - \sin^2\xi}}{(\kappa - 1)[(\kappa + 1)\cos^2\xi - 1]}\sin\xi\cos\xi Q_1(\xi) \quad (35)$$

and where

$$\beta(t) \equiv \frac{d\xi}{dt} = \frac{j2t}{\sqrt{1 - [\cos(\xi_b - \theta_2) - jt^2]^2}} \quad (36)$$

The minus or plus superscripts in the integrands in eqns. 33–35 signify that the branch of the square root is chosen

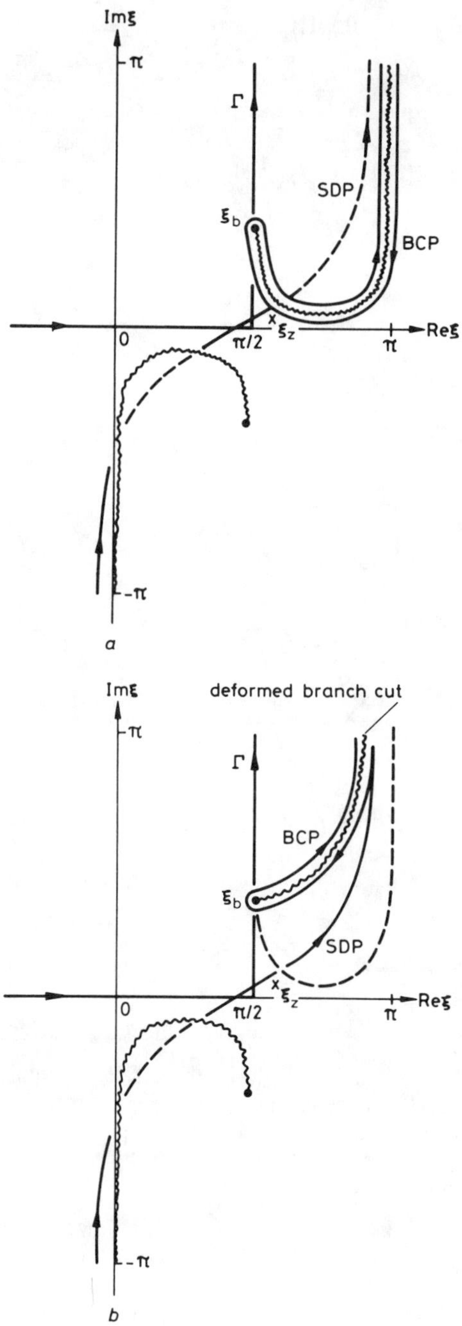

Fig. 5 *Proposed integration paths in the complex ξ-plane*
$f = 10$ MHz, $\varepsilon_r = 10$, $\sigma = 2.10^{-4}$ S/m, and $\theta_2 = 78°$
a Before branch-cut deformation
b After branch-cut deformation to the steepest-descent path from the branch point

according to eqn. 4*a* or 4*c*, respectively. The effective limits of integration in eqns. 30–32 are determined by the criterion, eqn. 27.

To illustrate the behaviour of the integrands in the modified method, the integrands of ${}_0\Pi_{vz}^{SDP}$ and ${}_0\Pi_{vz}^{BCP}$ are plotted in Figs. 6 and 7 for the cases of Figs. 3 and 4. It is

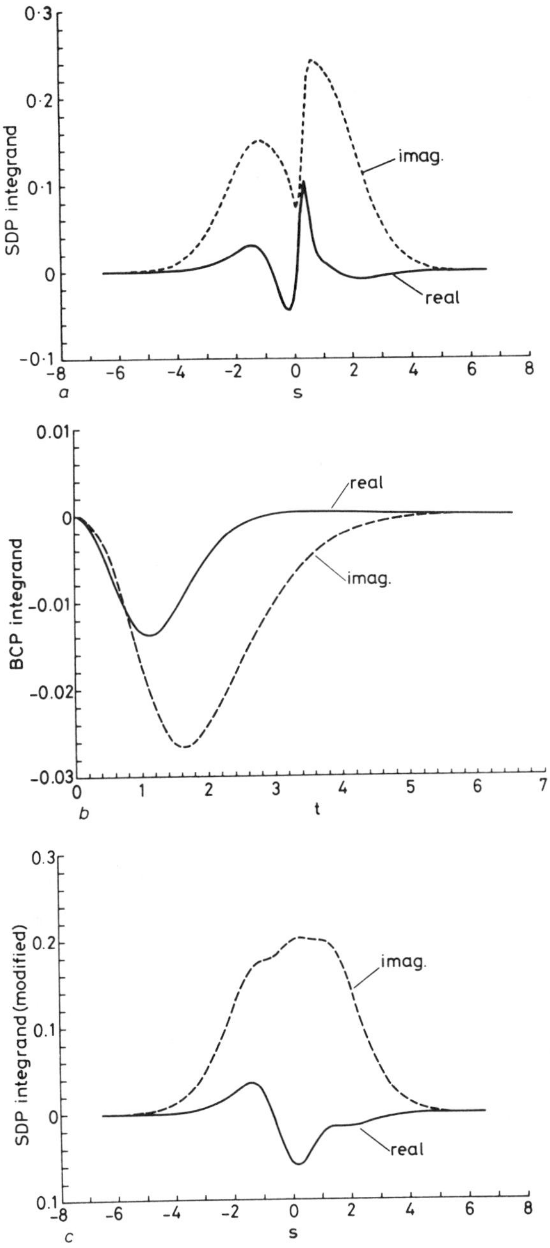

Fig. 6 *Plot of the ${}_0\Pi_{vz}$ integrand (improved approach) for the parameters of Fig. 3*

a SDP integrand
b BCP integrand
c Modified SDP integrand (Zenneck pole subtracted)

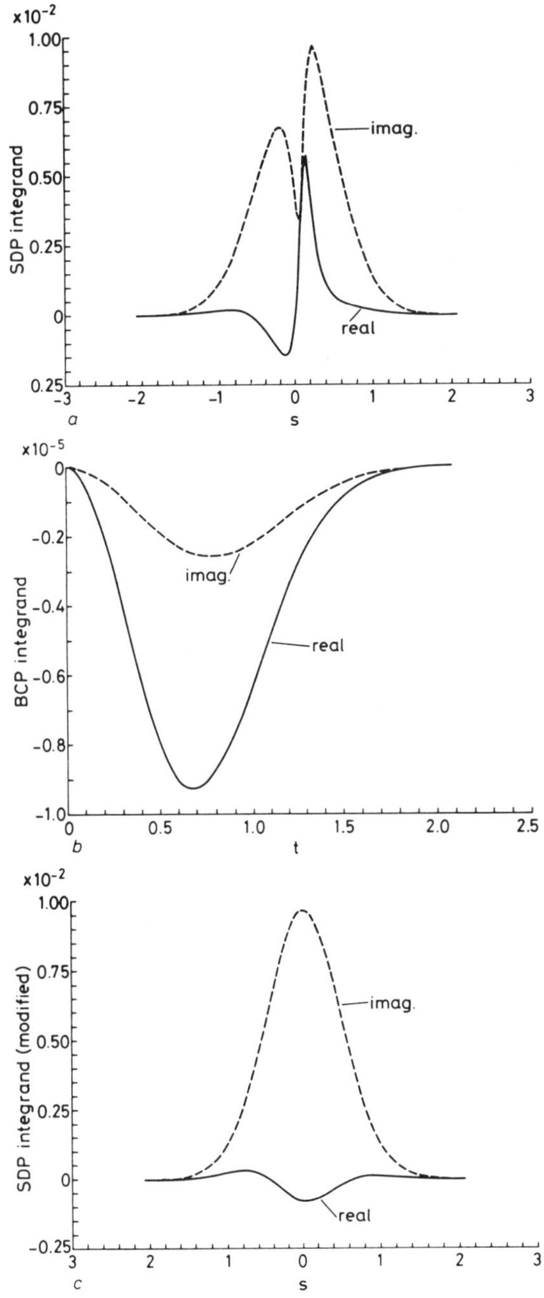

Fig. 7 *Plot of the ${}_0\Pi_{vz}$ integrand (improved approach) for the parameters of Fig. 4*

a SDP integrand
b BCP integrand
c Modified SDP integrand (Zenneck pole subtracted)

Table 1: Sample values of ${}_0\Pi_{vz}^{SDP}$, ${}_0\Pi_{vz}^{BCP}$, and ${}_0\Pi_{vz}$

		Rahmat-Samii *et al.* [2, 5]	Improved method
$f = 10$ MHz	SDP	$1.2587 \cdot 10^{-1} - j3.1269 \cdot 10^{-2}$	$1.4283 \cdot 10^{-1} - j4.4775 \cdot 10^{-2}$
$\varepsilon = 10$ $\sigma = 2 \cdot 10^{-4}$ S/m	BCP	$1.2524 \cdot 10^{-2} - j6.6071 \cdot 10^{-3}$	$-4.4418 \cdot 10^{-3} + j6.8982 \cdot 10^{-3}$
$\theta_2 = 78°$ $r_2 = 1$ m	Total	$1.3839 \cdot 10^{-1} - j3.7877 \cdot 10^{-2}$	$1.3839 \cdot 10^{-1} - j3.7877 \cdot 10^{-2}$
$f = 100$ MHz	SDP	$-8.4103 \cdot 10^{-2} - j1.1116 \cdot 10^{-1}$	$-8.4707 \cdot 10^{-2} - j1.1141 \cdot 10^{-1}$
$\varepsilon = 80$ $\sigma = 1 \cdot 10^{-2}$ S/m	BCP	$-6.1223 \cdot 10^{-4} - j2.2608 \cdot 10^{-4}$	$-7.5064 \cdot 10^{-6} + j1.5079 \cdot 10^{-5}$
$\theta_2 = 85°$ $r_2 = 1$ m	Total	$-8.4715 \cdot 10^{-2} - j1.1139 \cdot 10^{-1}$	$-8.4714 \cdot 10^{-2} - j1.1139 \cdot 10^{-1}$

noted that the SDP integrands are continuous and the BCP integrands are smooth and rapidly decaying. The values of the BCP and SDP integrals are given in Table 1 for the two examples. It is observed that, although the values of the particular integrals are different in the two techniques, the total values are the same, as expected (the small differences are due to numerical noise). These results also show that the values of the BCP integrals in the proposed method are one order of magnitude smaller than the corresponding values in the Rahmat-Samii *et al.* technique. Therefore, the branch-cut integrals are most likely to be negligible in comparison to the SDP integrals in the present method.

Although the SDP integrands in the modified approach are continuous, they still exhibit a rapid variation near $s = 0$, which is due to the Zenneck pole [2, 5]

$$\xi_z = \frac{\pi}{2} + j\{\mathrm{Ln}\,(\sqrt{\kappa} - j) - \mathrm{Ln}\,(\sqrt{\kappa + 1})\} \tag{37}$$

In the complex s-plane, which is defined by the transformation of eqn. 11, this pole is located at

$$s_z = e^{-j(\pi/4)} \sqrt{1 + \frac{\cos\theta_2}{\sqrt{\kappa + 1}} - \frac{\sqrt{\kappa}\sin\theta_2}{\sqrt{\kappa + 1}}} \tag{38}$$

Although the Zenneck pole is never captured in the path deformation, it can get very close to the SDP when both the source and the field point are close to the interface and when the contrast between the media is large (Fig. 5). To remedy this, the Zenneck pole is subtracted from the SDP integrands of ${}_0\Pi_{vz}$ and ${}_0\Pi_{hz}$, rendering them slowly varying. Subsequently, the integrals of the subtracted terms are expressed in terms of the error function and added to the modified integrals to preserve their correct value. As a result of these steps, which closely follow the modified saddle-point procedure [8–10], one obtains

$$ {}_0\Pi_{vz}^{SDP} = \frac{k_1\kappa}{4\pi j} e^{-jk_1r_2} \left\{ \int_{-\infty}^{\infty} \left(P(\xi_{SDP})\alpha(s) - \frac{R_P}{s - s_z} \right) \times e^{-k_1r_2s^2}\, ds - j\pi R_P\, w(-\sqrt{k_1r_2}\, s_z) \right\} \tag{39}$$

and

$$ {}_0\Pi_{hz}^{SDP} = -\frac{k_1}{4\pi} \cos\phi_2\, e^{-jk_1r_2} \times \left\{ \int_{-\infty}^{\infty} \left(T(\xi_{SDP})\alpha(s) - \frac{R_T}{s - s_z} \right) e^{-k_1r_2s^2}\, ds - j\pi R_T\, w(-\sqrt{k_1r_2}\, s_z) \right\} \tag{40}$$

where the residues R_P and R_T are given by

$$R_P = -\frac{\sqrt{\kappa}}{(\kappa - 1)\sqrt{\kappa + 1}} Q_0(\xi_z) \tag{41}$$

and

$$R_T = -\frac{\kappa}{(\kappa - 1)\sqrt{\kappa + 1}} Q_1(\xi_z) \tag{42}$$

with

$$Q_i(\xi_z) = -\frac{\sqrt{\kappa}}{(\kappa + 1)} H_i^{(2)}\left(k_1 r_2 \sqrt{\frac{\kappa}{\kappa + 1}} \sin\theta_2 \right) \times e^{jk_1r_2\sqrt{(\kappa/\kappa+1)}\sin\theta_2} \tag{43}$$

The function $w(\cdot)$ in eqns. 39–40 is the error function defined in Reference 11.

As an illustration, the modified integrand of eqn. 39 is plotted in Figs. 6*c* and 7*c* for the two cases considered before. It is observed that the sharp peaks present in Figs. 6*a* and 7*a* are completely removed. In Table 2 are shown the values of the modified integral in eqn. 39 and the values of the correction term containing the error function for the two examples (the term preceding the bracket in eqn. 39 has been included in these values).

Table 2: Sample values of the modified integral and the correction term in eqn. 39

Parameters		
$f = 10$ MHz, $\varepsilon = 10$, $\sigma = 2 \cdot 10^{-4}$ S/m, $\theta_2 = 78°$, $r_2 = 1$ m	Modified integral	$1.5844.10^{-1} - j3.1193 \cdot 10^{-2}$
	Correction	$1.5605 \cdot 10^{-2} + j1.3581 \cdot 10^{-2}$
	Total	$1.4283 \cdot 10^{-1} - j4.4775 \cdot 10^{-2}$
$f = 100$ MHz, $\varepsilon = 80$, $\sigma = 1 \cdot 10^{-2}$ S/m, $\theta_2 = 85°$, $r_2 = 1$ m	Modified integral	$-7.9890 \cdot 10^{-2} - j1.3813 \cdot 10^{-1}$
	Correction	$4.8167 \cdot 10^{-3} - j2.6720 \cdot 10^{-2}$
	Total	$-8.4707 \cdot 10^{-2} - j1.1141 \cdot 10^{-1}$

5 Summary and conclusions

A modification of the Rahmat-Samii *et al.* technique for the numerical evaluation of Sommerfeld integrals has been described. The modified method differs from the original one in a computationally more efficient treatment of both the steepest-descent path (SDP) and the branch-cut path (BCP) integrals. The gain in efficiency is achieved by (i) letting the SDP enter the bottom sheet, which removes the discontinuity of the integrands; (ii) deforming the fundamental branch cut to the steepest-descent path from the branch point, which results in smooth and rapidly decaying integrands; (iii) subtracting the Zenneck pole from the SDP integrands, which renders them slowly varying; and (iv) by eliminating the need of solving a nonlinear equation to determine the point of intersection of the SDP with the BCP. It should be pointed out, however, that the advantages of the improved technique are significant only in cases when the BCP integral is not negligible in comparison with the SDP integral (small losses), and when the Zenneck pole occurs close to the SDP (high media contrast, source and field points close to the interface).

As a result of the improvements mentioned above, the integrands on both the SDP and the BCP are well behaved and low-order Gaussian quadratures can be employed for a much broader range of parameters than in the Rahmat-Samii *et al.* approach. It is found that in most cases a 32-point rule is sufficient on the SDP, and a 16-point rule on the BCP. In contrast to some other techniques, the present approach is efficient for both the near-field and the far-field computation. The effective integration intervals contain relatively few oscillations of the integrands, which is an advantage over the approach given in Reference 6. Also, the integrands do not exhibit sharp peaks, as is the case in the method given in Reference 7.

Unlike the original technique [2, 5], the improved method is fully consistent with the asymptotic techniques for approximate evaluation of integrals in terms of a large parameter [8–10]. Therefore, for large values of $k_1 r_2$, the saddle-point method can be directly applied to eqns. 20

and 39–40, and the method of steepest descent can be used to approximately evaluate eqns. 30–32.

Finally, it is pointed out that this approach can, in principle, be applied in the case when the lower halfspace is replaced by a layered medium, such as the grounded dielectric slab, which is pertinent to microstrip antennas. However, the occurrence of the surface wave poles and the leaky wave poles, which may be exposed in the process of contour deformation, will make the application of this procedure much more difficult [12].

6 Acknowledgments

The author wishes to thank Professors Chalmers M. Butler and Charles E. Smith for their encouragement and interest in this work, and Dr. Yahya Rahmat-Samii for providing the subroutine for the computation of the error function of a complex argument.

This work was supported in part by the Committee on Faculty Research of the University of Mississippi.

7 References

1 SOMMERFELD, A.: 'Partial differential equations' (Academic Press, 1949), pp. 246–267

2 RAHMAT-SAMII, Y., MITTRA, R., and PARHAMI, P.: 'Evaluation of Sommerfeld integrals for lossy half-space problems', *Electromagnetics*, 1981, **1**, pp. 1–28

3 BURKE, G.J., MILLER, E.K., BRITTINGHAM, J.N., LAGER, D.L., LYTLE, R.J., and OKADA, J.T.: 'Computer modeling of antennas near the ground', *ibid.*, 1981, **1**, pp. 29–49

4 MOHSEN, A.: 'On the evaluation of Sommerfeld integrals', *IEE Proc. H, Microwaves, Opt. & Antennas*, 1982, **129**, pp. 177–182

5 PARHAMI, P., RAHMAT-SAMII, Y., and MITTRA, R.: 'An efficient approach for evaluating Sommerfeld integrals encountered in the problem of a current element radiating over lossy ground', *IEEE Trans.*, 1980, **AP-28**, pp. 100–104

6 JOHNSON, W.A., and DUDLEY, D.G.: 'Real axis integration of Sommerfeld integrals: source and observation points in air', *Radio Sci.*, 1983, **18**, pp. 175–186

7 LINDELL, I.V., and ALANEN, E.: 'Exact image theory for the Sommerfeld half-space problem, Part I: vertical magnetic dipole', *IEEE Trans.*, 1984, **AP-32**, pp. 126–133

8 FELSEN, L.B., and MARCUVITZ, N.: 'Radiation and scattering of waves' (Prentice-Hall, 1973)

9 BREKHOVSKIKH, L.M.: 'Waves in layered media' (Academic Press, 1980)

10 BERNARD, G.D., and ISHIMARU, A.: 'On complex waves', *Proc. IEE*, 1967, **114**, pp. 43–49

11 ABRAMOWITZ, M., and STEGUN, I.A. (Eds.): 'Handbook of mathematical functions' (Dover, 1965)

12 BLOK, H., VAN SPLUNTER, J.M., and JANSSEN, H.G.: 'Leaky-wave modes and their role in the numerical evaluation of the field excited by a line source in a non-symmetric, inhomogenously layered, slab waveguide'. Proc. International URSI Symposium, Universidad de Santiago de Compostela, Spain, 1983, pp. 523–526

Analysis of Wire Antennas in the Presence of a Conducting Half-Space. Part I. The Vertical Antenna in Free Space[1]

E. K. MILLER,[2] A. J. POGGIO,[3] G. J. BURKE,[4] AND E. S. SELDEN[4]

Received July 5, 1971[5]

An integral equation method (based upon the electric field) for the analysis of a vertical antenna located in free space near the interface with a lossy half-space is presented. Two methods are considered to account for the influence of the interface-reflected fields upon the antenna current distribution: (1) the rigorous Sommerfeld integral approach and (2) a plane-wave reflection coefficient approximation. Numerical results for antenna impedance from the two methods are compared. It is found that for the parameter variations studied, the reflection coefficient approximation provides results within 10% of those obtained from the rigorous approach at approximately 1% of the computer time. The reflection coefficient technique has the further advantage of being readily extended to arbitrary antenna geometries and orientations.

On présente une méthode d'équation intégrale pour l'analyse du champ électrique d'une antenne verticale placée dans le vide près de la surface d'un demi-espace dissipatif. Deux méthodes sont considérées pour tenir compte de l'influence des champs réfléchis à la surface de séparation sur la distribution de courant dans l'antenne: (1) le traitement rigoureux par l'intégrale de Sommerfeld et (2) une approximation utilisant le coefficient de réflexion d'une onde plane. Les résultats numériques pour l'impédance d'antenne calculés par ces deux méthodes sont comparés. On trouve que pour les valeurs des paramètres considérées, l'approximation du coefficient de réflexion donne des résultats compatibles à 10% près avec ceux qu'on obtient par le traitement rigoureux, avec seulement environ 1% du temps d'ordinateur. La technique du coefficient de réflexion a l'avantage d'être facilement généralisable à des géométries et des orientations d'antenne arbitraires.

Canadian Journal of Physics, 50, 879 (1972)

I. Introduction

A problem of continuing and widespread interest in electromagnetic theory is that of an antenna operated near the interface between two semi-infinite media of different electrical properties. Sommerfeld (1909, 1964) provided the basis for a formal solution to this problem by deriving the Hertz potentials of infinitesimal electric and magnetic sources arbitrarily located with respect to the interface. By properly combining the Sommerfeld integrals for vertically and horizontally oriented infinitesimal sources, it is in principle then possible to treat arbitrarily oriented antennas of arbitrary geometry in an essentially rigorous and self-consistent fashion.

A procedure for analyzing wire antennas in such an environment, based upon a version of the thin-wire approximation to the electric-field integral equation modified by inclusion of the Sommerfeld integrals in its kernel to account for the antenna–interface interaction, is described in this paper. Since the individual elements of the procedure have been well known for some time, no claim is made for the originality of the theoretical approach. However, the systematic and general numerical treatment of this problem which is undertaken below has not been previously reported. Use of the electric-field integral equation offers in the present problem the same advantage of geometrical flexibility as previously demonstrated for wire antennas in free space (see Poggio and Miller 1970) particularly relative to the vector potential (or Hallén) integral equation approach whose application to other than vertical antennas over the ground is not at all straightforward (Chang and Wait 1970a).

In addition to the rigorous Sommerfeld integral analysis of the antenna half-space problem, we present also an approximate analytical method for its treatment based upon the Fresnel or plane-wave reflection coefficients. Indeed, the primary contribution of the work reported here is, we feel, the systematic incorporation of the reflection coefficient approximation into an integral equation formulation

[1]Work prepared under the auspices of the U.S. Atomic Energy Commission.

[2]Lawrence Livermore Laboratory, University of California, Livermore, California.

[3]Cornell Aeronautical Laboratory, Buffalo, New York.

[4]M B Associates, San Ramon, California.

[5]Revision received November 23, 1971.

applicable to general wire structures to obtain an approach which, compared with the rigorous method, is reasonably accurate, yet vastly more efficient for the analysis of arbitrary geometry antennas located over a lossy ground. The analytical relationship between the rigorous and approximate approaches is demonstrated and numerical data derived from each are presented to verify their validity as well as to compare their results with independent analytical and experimental data.

It should be noted that reflection coefficient approximations in various forms have been previously mentioned by Krauss (1950) and Jasik (1961), and have been applied to a single vertical antenna by Ma and Walters (1969) and Chang and Wait (1970*a*, *b*). The approach taken here is a generalization of the basic idea, whose combination with the electric-field integral equation results in a methodology suitable for the treatment of antenna geometries considerably more complex than those previously considered.

The organization of the paper is as follows. In this, Part I, we discuss vertical antennas only. Section II of this paper contains the theoretical discussion, including the development of the rigorous and approximate integral equations and the numerical solution procedure. Numerical results are presented in Section III followed by the concluding Section IV. Consideration of the horizontally (and arbitrarily) oriented antennas is taken up in Part II. Further extensions of the analysis, both rigorous and approximate, to the treatment of buried antennas, is anticipated as Part III.

II. Theoretical Discussion

Integral equations of various kinds have been widely and successfully employed for the analysis of thin-wire antennas and scatterers. Since the details of their derivations and applicability are adequately discussed elsewhere (see Mei 1965 for example), we confine the discussion below to those primary aspects of the formulation peculiar to the antenna–ground interaction problem. It is convenient and conducive to an orderly presentation to begin with the case of the antenna located in free space, followed by the perfectly conducting half-space problem, thence to the finitely conducting half-space, for which both the rigorous Sommerfeld integral approach and the reflection coefficient method are discussed.

A. Free Space

The thin-wire approximation to the electric-field integral equation for a single straight wire of radius a and length L located in free space can be written

$$[1] \quad \frac{i\omega\mu_0}{4\pi}\int_L I(s')G(s,s')\,\mathrm{d}s' = -E^{\mathrm{A}}(s); \quad s\varepsilon L$$

where

$$G(s,s') = \left[1 + \frac{1}{k^2}\frac{\partial^2}{\partial s^2}\right]g$$

$$g = g_0 = \exp(-ikR)/R$$

$$R = [(s-s')^2 + a^2]^{1/2}$$

$k = \omega[\mu_0\varepsilon_0]^{1/2}$ is the wavenumber for free space, and the $\exp(i\omega t)$ time variation has been suppressed. The applied or exciting field is $E^{\mathrm{A}}(s)$, and $I(s')$ is the antenna current with s' and s denoting source and observation points respectively.

B. Perfect Ground

Consider the same antenna vertically oriented over a perfectly conducting ground. Then eq. 1 applies with the change

$$[2] \quad g = g_{\mathrm{pg}} = g_0 + \exp(-ikR')/R' = g_0 + g_i$$

where g_{pg} is the perfect ground Green's function,

$$R' = [(s+s')^2 + a^2]^{1/2}$$

and for a center height h, the s' integration range in eq. 1 is from $h - L/2$ to $h + L/2$. The added term, g_i, accounts for the image in the perfectly conducting ground, while g_0 of course is the direct free-space radiation field.

C. Lossy Ground

Sommerfeld Integral Approach

An extension of the integral equation to the finitely conducting or lossy ground can again be rigorously accomplished for the vertical antenna by a simple addition modification of g, which now becomes

$$[3] \quad g = g_{\mathrm{lg}} = g_0 + g_i - g_s$$

where

$$[4] \quad g_s = 2\int_0^\infty J_0(\lambda a)\, e^{-\mu(s+s')} \frac{\mu_E}{\varepsilon_E\mu + \mu_E} \frac{\lambda}{\mu} d\lambda$$

$$\mu^2 = \lambda^2 - k^2$$

$$\mu_E^2 = \lambda^2 - k_E^2$$

$$k_E^2 = \varepsilon_E k^2$$

and ε_E is the relative complex permittivity of the ground.

The lossy ground Green's function is denoted by g_{lg}, J_0 is the Bessel function of order zero, and the integral in eq. 4 is the Sommerfeld correction term. The integral equation obtained by inserting eq. 3 into eq. 1 is a rigorously correct one (within the restrictions imposed by the thin-wire approximation) for a wire antenna located in free space over a ground of arbitrary (*i.e.*, possibly complex) permittivity. The resulting integral equation thus has a portion which involves the double integration over s' and λ, a procedure that in general cannot be analytically carried out.

Reflection Coefficient Approximation

The reflection coefficient counterpart of eq. 3 for the lossy ground is only trivially different from the perfect ground Green's function, eq. 2, for the vertical antenna. We replace $g_i - g_s$ in eq. 3 by $R_M g_i$ where R_M is the Fresnel plane-wave reflection coefficient for the transverse magnetic (TM) polarization. Thus, the reflection coefficient Green's function for the lossy ground, $\tilde{g}_{lg}$, is written

$$[5] \quad g \cong \tilde{g}_{lg} = g_0 + R_M g_i$$

where

$$R_M = \frac{\varepsilon_E \cos\theta - [\varepsilon_E - \sin^2\theta]^{1/2}}{\varepsilon_E \cos\theta + [\varepsilon_E - \sin^2\theta]^{1/2}}$$

and θ is the angle of incidence with respect to the surface normal defined by a line (ray) from the image point to the observation point. This angle is analogous to the plane-wave incidence and reflection angles, and will be denoted below as the specular angle with the intersection point of the image ray to the interface called the specular point. Note that for a single vertical antenna, R_M is geometry independent (since θ is then zero) and is given by its normal incidence form

$$R_M = \frac{\varepsilon_E^{1/2} - 1}{\varepsilon_E^{1/2} + 1}$$

so that the approach discussed by Ma and Walters (1969) and Chang and Wait (1970*a*) is then recovered. In general, however, an aspect and polarization dependent reflection coefficient will be involved in the expression for $\tilde{g}_{lg}$.

Equation 5 is derived by expressing the reflected electric field as the product of the perfect ground image field and the TM reflection coefficient. This is discussed further below in comparing the rigorous and approximate integral equations in greater detail.

Extension of the integral equation to treat 2 or more vertical antennas is straightforward. Before proceeding to that case we will first discuss the moment method of integral equation solution, since some of the concepts which are introduced in the solution technique are useful for the subsequent discussion.

D. Numerical Solution

Subsectional collocation, a version of the moment method, has been used to solve the integral equations of concern here. This method, more fully described elsewhere (see Poggio and Miller (1970); also, for more general information, Harrington (1968) and Fenlon (1969)), involves segmenting the wire structure into a sequence of N segments on each of which the current is approximated by a basis function expansion involving a set of to-be-determined constants. For numerous reasons (Yeh and Mei 1967; Neureuther *et al.* 1968; Miller *et al.* 1971) we employ the so-called sinusoidal current expansion where

$$[6] \quad I_j(s') = A_j + B_j \sin k(s' - s_j) + C_j \cos k(s' - s_j)$$

with s_j the midpoint of segment j where $j = 1, \ldots, N$. The $3N$ unknown constants implied by eq. 6 are reduced to N unknown constants, the current values at the segment centers, by an extrapolation procedure which is applicable to singly or multiply connected segments (Gee *et al.* 1971).

Upon introducing the resulting segmental current expression into the integral equation, and point fitting it at the N segment centers (*i.e.*, delta function weights are used), a linear Nth-order system is obtained as an approximation to the original integral equation. The N^2 coefficients of this system represent the N^2 self and mutual impedances (impedance matrix) of the N structure segments, and the coefficients of the inverse matrix (admittance matrix) are the corresponding admittances, which when obtained allow calculation of the structure current due to any specified source field distribution.

Computation of the impedance matrix elements is relatively simple for the antenna in free space, and can be largely analytically performed when using eq. 6 for the current expansion. The Sommerfeld version of the integral equation may in general, however, require a double numerical integration, one integral over the current and the other over the Sommerfeld integral itself, although for the special case of the vertical antenna it is possible to perform the current integral analytically.

The Sommerfeld integral portion of eq. 1 when combined with eq. 3 has the form

$$[7a]\quad \int_L I(s')\left[k^2 + \frac{\partial^2}{\partial s^2}\right]\int_0^\infty J_0(\lambda a) \times e^{-\mu(s+s')}\frac{\mu_E}{\varepsilon_E\mu + \mu_E\,\mu}\,\frac{\lambda}{\mu}\,d\lambda\,ds'$$

so that the s' integral itself can be written

$$[7b]\quad \int_L I(s')\,e^{-\mu s'}\,ds'$$

Thus only the λ integration in eq. 7a must be performed numerically as eq. 7b is analytically integrable for the sinusoidal current expansion already adopted. This situation does not hold for nonvertical current segments, since then the source–observation point separation distance is dependent upon s' and consequently the Bessel function in eq. 7a must also be included in the s' integral eq. 7b. This point will be considered further in Part II.

Numerical evaluation of the λ integral remains. Because μ is present in the denominator of the integral, the integral has a singularity at $\lambda = k$. As a result, it is convenient to deform the integration (although other procedures could be used as well to handle the singularity) contour into the complex λ plane. Plots of the

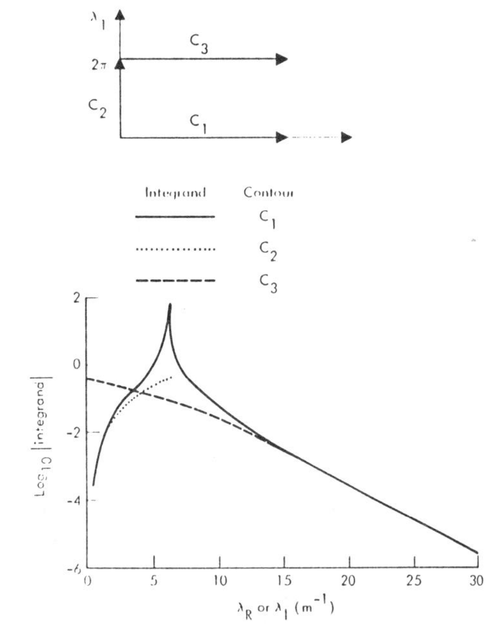

FIG. 1. Sommerfeld integrand for ground effect.

integrand magnitude are shown in Fig. 1 on the real λ axis and over a typical deformed contour. An adaptive numerical integration method was used to perform the integral. This same method is employed for the current integration in the free-space case; details of its use are described elsewhere (Miller 1970).

Once the impedance matrix coefficients have been obtained the remaining steps in the solution procedure involve well-known matrix operations. Consequently, there is no need to consider the numerical solution details beyond this point. Note, however, that the admittance matrix finally obtained numerically characterizes the complete electromagnetic properties of the structure for which it has been derived, as an antenna or scatterer, within the limitations imposed by the formulation and solution accuracy.

E. Multiple Antennas

The preceding discussion has specifically considered the single antenna case only. Application of the integral eq. 1 to 2 or more vertical antennas is straightforward. Consider an array of M vertical antennas of lengths L_m, radii a_m, and x–y coordinates x_m, y_m with center heights h_m, $m = 1, \ldots, M$. Then the rigorous and image integral equations both take the form

$$\sum_{m=1}^{M} \int_{h_m - L_m/2}^{h_m + L_m/2} I_m(s') G_{mn}(s, s')\, ds = -E_n^{\wedge}(s) \tag{8}$$

$$h_n - L_n/2 \leqslant s \leqslant h_n + L_n/2; \quad n = 1, \ldots, M$$

with I_i and $E_i^{\wedge}$ the current and source field respectively on antenna i. The new integral equation kernel, G_{mn}, is obtainable from the former Green's function with the changes

$$\begin{aligned} R \rightarrow R_{mn} &= [a_m^2 + (x_m - x_n)^2 + (y_m - y_n)^2 + (s' - s)^2]^{1/2} \\ R' \rightarrow R'_{mn} &= [a_m^2 + (x_m - x_n)^2 + (y_m - y_n)^2 + (s' + s)^2]^{1/2} \\ J_0(\lambda a) &\rightarrow J_0(\lambda\, [(x_m - x_n)^2 + (y_m - y_n)^2 + a^2]^{1/2}) \end{aligned} \tag{9}$$

where the a_m appears in R, R' and the argument of J_0 in accord with the thin-wire approximation.

In addition to the changes indicated, the reflection coefficient R_M can now be geometry dependent because of the possible horizontal separation of source and observation points. Consequently, the angle of incidence at the specular reflection point, determined by the source and observation point locations, is included in the calculation of the image-reflected field. Note, however, that the differential operation involved in determining G_{mn} from g_{mn} is not applied to R_M but only to the image Green's function it multiplies. The reason for this is that the reflection coefficient approximation which leads to eq. 5 for the Green's function is applied only to the reflected electric field. This field is of course obtained from a differential operation performed prior to multiplication by R_M, so that interchanging their order relative to g is permissible only if R_M is viewed as a constant.

It should be noted that the reflection coefficient approximation is, in our numerical approach, applied on a segmental basis. The total ground-reflected field at a given observation point is obtained as a summation over the N antenna segments. Each segment's field is multiplied by the reflection coefficient appropriate for the incidence angle made by a specularly reflected ray joining the centers of the source and observation segments (*i.e.*, the reflection coefficient is moved outside the current integral). A possible variation of the reflection coefficient with the source integration variable (the reflection coefficient is constant independent of position only for vertically aligned source and observation segments) within a segment is thus ignored, but allowance is made for the reflection coefficient variation from segment to segment.

In the case of vertical antennas, the latter variation occurs only through the changing specular reflection angle between source and operation segments and involves only the TM reflection coefficient. For the more general case of arbitrarily oriented antennas the ground-reflected field is then determined by both angle-dependent TM and transverse electric (TE) reflection coefficients and the relative orientations of the two segments. This procedure enables the reflection coefficient approximation to be applied in an approximate, yet reasonably accurate, piecewise sense on the structure in a manner consistent with the approximate expansion technique employed for the current. At the same time, the numerical solution time is not significantly increased over the perfect ground case since the reflection coefficient approximation introduces only a few additional multiplications per field calculation.

Note that the image method mentioned by Krauss and Jasik represents a coarser degree of approximation in that they suggest the use of a single reflection coefficient for the entire antenna. For a single antenna, this would be simply the normal incidence reflection coefficient. The mutual interaction of two or more antennas would presumably be accounted for by the reflection coefficient appropriate to the antenna feed points and their relative orientations.

F. Comparison of Sommerfeld and RC Approximation

Our discussion thus far has concentrated on the numerical solution of the integral equations applicable to a vertical antenna(s) over ground using the rigorous Sommerfeld integral formulation and the reflection coefficient approximations. In the following paragraphs, we briefly examine the specific electric-field expressions which arise from these separate approaches in order to more clearly demonstrate the relationship between them.

For purposes of simplicity, and with no loss of generality, we deal with the differential electric field, $\bar{e}$, due to a vertical Hertzian electric dipole located on the z axis a distance h above the ground. This electric field can be expressed in terms of the differential Hertz potential, $\mathrm{d}\bar{\pi}$, as

$$\bar{e} = (k^2 + \nabla\nabla_1)\,\mathrm{d}\bar{\pi}$$

where

$$\mathrm{d}\bar{\pi} = \frac{i\omega\mu_0}{4\pi k^2} I\,\mathrm{d}s\hat{z}g = P\hat{z}g$$

with g the appropriate Green's function. Since we are interested in only the ground-reflected field $\bar{e}^{\mathrm{R}}$ (the direct contributions are the same in both cases), we find for the rigorous case, with observation coordinates ρ, z,

$$\bar{e}^{\mathrm{R}} = P\left[\hat{z} + \frac{1}{k^2}\nabla\frac{\partial}{\partial z}\right](g_i - g_s)$$

from which

$$[10] \quad \begin{aligned} e_z^{\mathrm{R}} &= P\int_0^\infty \mathscr{R}_M \frac{\lambda^3}{\mu} J_0(\lambda\rho)\,\mathrm{e}^{-\mu(z+h)}\,\mathrm{d}\lambda \\ e_\rho^{\mathrm{R}} &= -P\int_0^\infty \mathscr{R}_M \lambda^2 J_0'(\lambda\rho)\,\mathrm{e}^{-\mu(z+h)}\,\mathrm{d}\lambda \end{aligned}$$

where $\mathscr{R}_M$ is the λ-dependent Fresnel TM plane-wave reflection coefficient given by

$$\mathscr{R}_M = \frac{\varepsilon_E\mu - \mu_E}{\varepsilon_E\mu + \mu_E}$$

and the prime denotes differentiation of J_0 with respect to argument.

Corresponding expressions for the reflection coefficient approximated field, $\tilde{\bar{e}}^{\mathrm{R}}$, are readily obtained. The vertical source (and its image) produces only radial and vertically directed electric fields, both of which lie in the vertical plane containing the source and observation points (or incidence plane). Thus the ground-reflected electric field of the vertical source is analogous to the TM polarization for plane-wave incidence, and we use only the TM plane-wave reflection coefficient. (For horizontal or arbitrary source orientations, there will in general be both in-plane and orthogonal electric fields relative to the incidence plane, in which case both the TM and TE reflection coefficients are required.) We consequently write $\bar{e}^{\mathrm{R}}$ for this case as

$$\tilde{\bar{e}} = R_M \bar{e}^{\mathrm{I}}$$

where $\bar{e}^{\mathrm{I}}$ is the image field given by

$$\bar{e}^{\mathrm{I}} = \left(1 + \frac{1}{k^2}\nabla\nabla\right) P\hat{z}g_i$$

We thus obtain, using the Fourier integral form for g_i,

$$[11] \quad \begin{aligned} \tilde{e}_z^{\mathrm{R}} &= PR_M \int_0^\infty \frac{\lambda^3}{\mu} J_0(\lambda\rho)\,\mathrm{e}^{-\mu(z+h)}\,\mathrm{d}\lambda \\ \tilde{e}_z^{\mathrm{R}} &= -PR_M \int_0^\infty \lambda^2 J_0'(\lambda\rho)\,\mathrm{e}^{-\mu(z+h)}\,\mathrm{d}\lambda \end{aligned}$$

A comparison of eqs. 10 and 11 shows them to differ only in the location of the reflection coefficient relative to the λ integral. The Sommerfeld integral is weighted by the λ-dependent reflection coefficient, whereas the approximate form has the reflection coefficient, evaluated at the specular angle, outside the integral. It is this difference which allows the approximate field integrals to be analytically evaluated in closed form while the Sommerfeld integrals cannot be so obtained. The relationship between the rigorous and approximate field expressions is, it should be mentioned (and as will be discussed in a subsequent paper), not in general so simple. That eq. 11 should reasonably approximate eq. 10 is not obvious, and is perhaps easiest to demonstrate via numerical computation.

Before concluding this section it is worthwhile to comment on the various equivalent forms into which the integral portion of the rigorous Green's function, g_{Ig}, may be cast. Some possible combinations are

$$[12a] \quad g_0 + g_i - (g_s) = g_0 + g_i - 2\int_0^\infty \frac{\mu_E}{\varepsilon_E\mu + \mu_E} F(\lambda)\,\mathrm{d}\lambda$$

$$[12b] \quad g_0 - g_i + (2g_i - g_s) = g_0 - g_i + 2\int_0^\infty \frac{\varepsilon_E\mu}{\varepsilon_E\mu + \mu_E} F(\lambda)\,\mathrm{d}\lambda$$

$$[12c] \quad g_0 + (g_i - g_s) = g_0 + \int_0^\infty \frac{\varepsilon_E\mu - \mu_E}{\varepsilon_E\mu + \mu_E} F(\lambda)\,\mathrm{d}\lambda$$

with

$$F(\lambda) = J_0(\lambda a)\frac{\lambda}{\mu}e^{-\mu(z+h)}$$

where the bracketed term is that which requires numerical integration. The first form is that used by Sommerfeld, and has the limits 0, g_i for $\varepsilon_E \to \infty$, 1 respectively. The second form is extensively employed by Baños (1966). It has no singularity at $\lambda = k$ as do the other two, but its limiting behavior requires it to approach $2g_i$, g_i for $\varepsilon_E \to \infty$, 1, respectively. The third expression, the form chosen for our numerical work (because of its obvious similarity to the reflection coefficient approximation), has a behavior the converse of eq. 12a, approaching g_i, 0 respectively for $\varepsilon_E \to \infty$, 1, so that its numerical evaluation is less demanding for the free-space case.

Generally speaking, the numerical accuracy required of the λ integral is least demanding when the contribution of this integral is small relative to the total Green's function, g_{lg}. Thus the three forms in eq. 12 are perhaps best suited to half-spaces having large, intermediate, and near unity values of ε_E, from an efficiency and accuracy viewpoint, in terms of the rigorous formulation. In this regard, it should be observed that the reflection coefficient approximation itself becomes rigorously correct in the limits $\varepsilon_E \to 1, \infty$, so that its least accurate region may be for intermediate values of ε_E. We take up the question of the reflection coefficient approximation accuracy in the next section, where numerical results for both methods are presented.

III. Numerical Results

Some representative numerical results for antenna resistance and radiation patterns are included here. Since the same 5-segment antenna model was used for both the rigorous Sommerfeld integral method and the approximate reflection coefficient approach, the feed region influence on the input susceptance is the same in each case, being a function of antenna parameters only. Consequently, a comparison either of admittance or impedance values will be equally useful. Several figures pertain to a single vertical antenna, followed by some results for two dipole array.

The input resistance of a half-wave dipole antenna is shown in Fig. 2 as a function of height

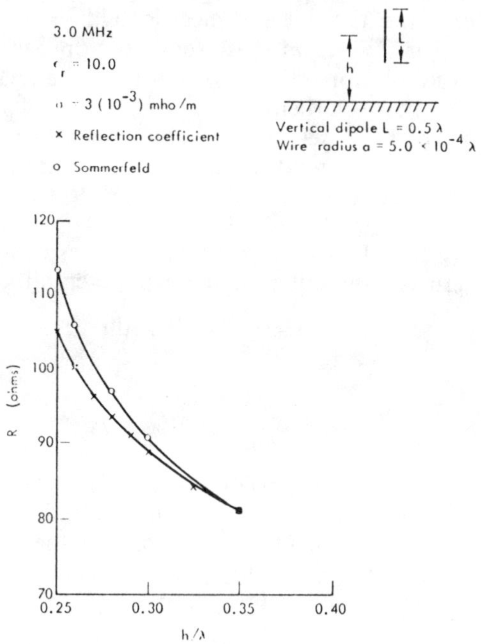

FIG. 2. Input resistance of a half-wave dipole over a lossy ground at 3.0 MHz as a function of center height.

above the ground for a relative ground permittivity (ε_r) of 10, a conductivity (σ) of 3×10^{-3} mhos/m ($\varepsilon_E = \varepsilon_r + \sigma/i\omega\varepsilon_0$), and a frequency of 3.0 MHz (note that $\sigma/\omega\varepsilon_0 \sim \varepsilon_r$). Agreement between the results obtained from the Sommerfeld and reflection coefficient approaches is maintained to within 10%, even when the lower end of the antenna touches the ground. Similar agreement is also found for the antenna reactance, which we do not include here.

Since the reflection coefficient approximation is rigorously correct in the limits $|\varepsilon_E| \to 1, \infty$, it is appropriate to examine its accuracy over an intermediate range of values where the errors which arise from its use might be expected to be greatest. We therefore present in Fig. 3 the half-wave dipole input resistance variation with σ for $\varepsilon_r = 10$ with the antenna height as a parameter. The discrepancy between the Sommerfeld and reflection coefficient data appears greatest in the range where $\sigma/\omega\varepsilon_0 \cong \varepsilon_r$, and is worst for the $h/\lambda = 0.25$ case, a not unexpected result.

Since the Sommerfeld integral derived results can be regarded as the correct ones (although some slight variation can be expected if the number of antenna segments is increased beyond

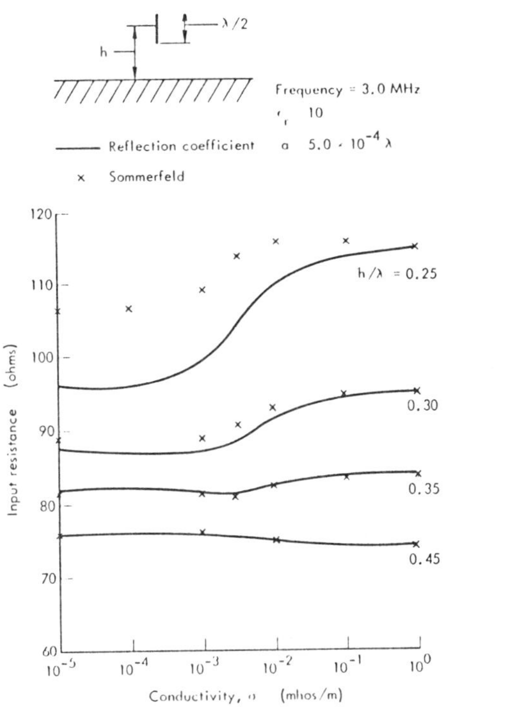

FIG. 3. Resistance variation with ground conductivity for vertical half-wave dipole over a lossy ground.

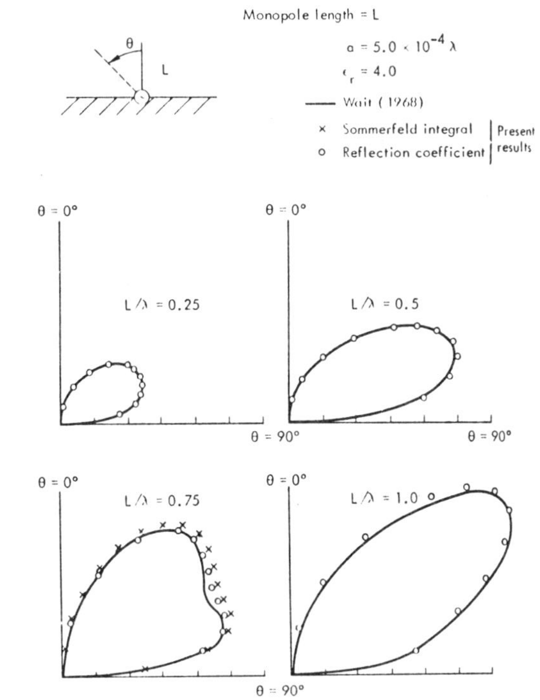

FIG. 4. Radiation patterns (linear voltage scale) of a monopole over a dielectric ground.

the 5 used here; see Miller *et al.* (1971))[6], we conclude that the reflection coefficient approximation yields values which are generally in error by no more than 10%. It is obvious from Fig. 3 that an error of this magnitude is unacceptable if, for example, one were to attempt a determination of the ground conductivity from a measured impedance value and calculation based upon the reflection coefficient approximation. An application of this kind would require the greater accuracy of the Sommerfeld integral approach and, incidentally, exemplary measurement techniques. However, for most engineering applications (antenna siting, for example) the 10% accuracy which the reflection coefficient approach appears capable of should be adequate.

Besides comparing the input impedance of an antenna as obtained from the two methods, it is also informative to compare their respective radiation patterns. Since the reflection coefficient approximation is an asymptotically correct form of the Sommerfeld integral in the far field (except near the interface) (see Baños 1966; Wait 1969), we evaluate the radiation pattern in the same fashion for both the rigorous and approximate approaches. We can consequently expect their absolute far-field behavior to agree at least as well as their input impedances, with the pattern shapes to be possibly even closer, considering the averaging effect of the current integration on the far-field variation. Illustrative of this expectation are the radiation patterns of a vertical monopole over a dielectric ground shown in Fig. 4, as obtained from the present

[6]For example, a 21-segment antenna model yields resistances corresponding to the $\sigma = 1$ mho/m case of Fig. 3 and h/λ values of 0.25, 0.3, 0.35, and 0.45 of 124, 99, 84, and 78 ohms respectively. By comparison Harrington *et al.* (1967), obtain a value of 125 ohms for the $h/\lambda = 0.25$ case and the perfectly conducting ground, a situation closely analogous to our $\sigma = 1$ results at this frequency. A 5-segment antenna model was used to perform the present computations in order to decrease the computer time required to obtain the Sommerfeld results (~ 20 s per data point on a CDC-6600 computer). Since both the rigorous and approximate calculations were performed using the same antenna model so that their respective resistance values can be expected to possess a similar degree of numerical convergence as a function of segmentation, a comparison of results obtained using a 5-segment model is meaningful.

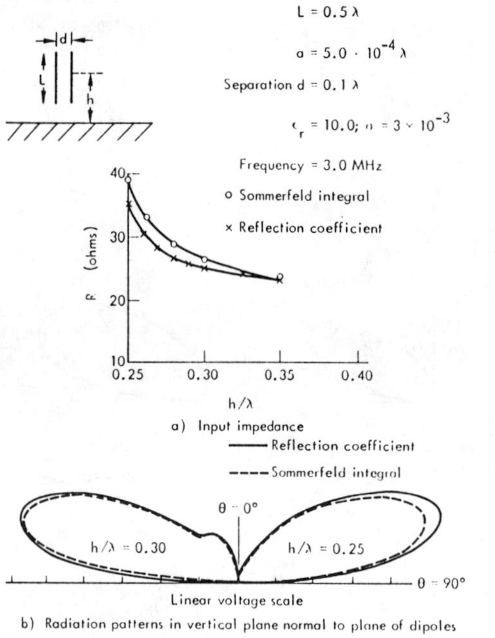

FIG. 5. Input impedance and radiation patterns for a parasitic array of two half-wave dipoles.

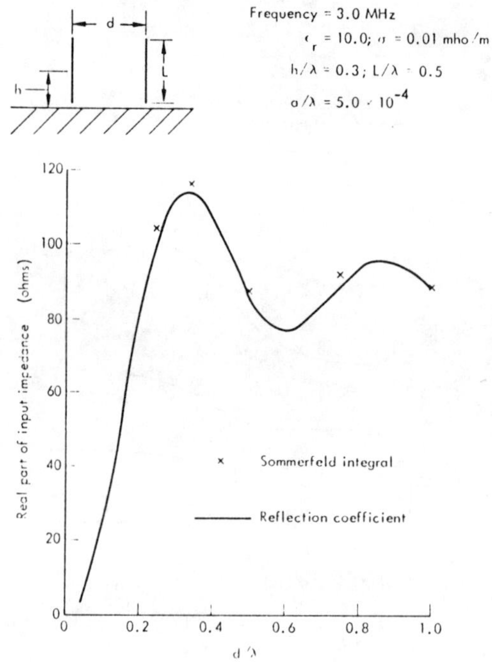

FIG. 6. Input resistance for a parasitic array of two half-wave dipoles.

approach compared with the data presented by Wait (1969).

The $L/\lambda = 1.0$ patterns for the two sets of data are normalized with the remaining patterns plotted relative to these. Although Wait's results are derived from an assumed current distribution and evidently assuming the feedpoint admittance variation with L/λ is the same as for free space, the pattern shapes and their amplitudes relative to $L/\lambda = 1.0$ are very similar.

We conclude our presentation of numerical results by showing some data obtained for an array of two vertical half-wave dipoles. The input resistance of the excited dipole is shown in Fig. 5 as a function of the array center height over a ground with $\varepsilon_r = 10$, $\sigma = 3 \times 10^{-3}$ mhos/m, a frequency of 3 MHz, and a dipole separation distance of 0.1λ. Also shown are the radiation patterns in the vertical plane normal to the antenna plane for heights of 0.25 and 0.3λ. The resistance data is again generally within 10% for the Sommerfeld and reflection coefficient methods, and as remarked above, the radiation patterns are of similar shape, but different amplitude, due to the input impedance variation.

A plot of the excited dipole input resistance as a function of the dipole separation for a fixed height of 0.3λ, $\varepsilon_r = 10$, $\sigma = 0.01$ mhos/m, and a frequency of 3 MHz is shown in Fig. 6. Again, the Sommerfeld and reflection coefficient results compare well. It should be noted that the increased conductivity in the present case is responsible for the closer agreement compared with the $\sigma = 3 \times 10^{-3}$ mhos/m case.

IV. Conclusion

An integral equation method for the treatment of wire antennas located in free space near a half-space of arbitrary electrical parameters has been presented. Results obtained for the impedance and radiation patterns of various vertical antenna configurations have demonstrated the potential usefulness of an approximate method of accounting for the ground based upon plane-wave reflection coefficients. The approximate method has been found to agree reasonably well (generally within 10%) with a rigorous approach based upon the Sommerfeld formulation, at a fraction (on the order of 1%) of the computer cost. Extension of the reflection coefficient analysis for horizontally oriented antennas will be discussed in a subsequent presentation.

BAÑOS, A. 1966. Dipole radiation in the presence of a conducting half-space (Pergamon Press, N.Y.).

CHANG, D. C. and WAIT, J. R. 1970*a*. IEEE Trans. Antennas Propag. **18**, No. 2.

——— 1970*b*. J. Phys. **48**, 737.

FENLON, F. H. 1969. Proc. IEEE, **57**, 291.

GEE, S., MILLER, E. K., POGGIO, A. J., SELDEN, E. S., and BURKE, G. J. 1971. Computer techniques for electromagnetic scattering and radiation analysis. Paper presented at the Electromagnetic Compatibility Meeting, Philadelphia, Penn.

HARRINGTON, R. F. 1968. Field computation by moment methods (MacMillan and Sons, Ltd., N.Y.).

HARRINGTON, R. F. *et al.* 1967. Computations for linear wire antennas and scatterers, Vol. II (Syracuse University, Syracuse, New York). AD639745.

JASIK, H. 1961. Antenna engineering handbook (McGraw-Hill Book Co., N.Y.).

KRAUSS, J. D. 1950. Antennas (McGraw-Hill Book Co., N.Y.).

MA, M. T. and WALTERS, L. C. 1969. ESSA Tech. Rept. IER 104, ITS 74.

MEI, K. K. 1965. IEEE Trans. Antennas Propag. **13**, 374.

MILLER, E. K. 1970. J. Comput. Phys. **5**, No. 2.

MILLER, E. K., BURKE, G. J., and SELDEN, E, S. 1971. IEEE Trans. Antennas Propag. **19**, No. 4.

NEUREUTHER, A. E., FULLER, B. D., HAKKE, G. D., HOHMANN, G., *et al.* 1968. A comparison of numerical methods for thin wire antennas, 1968 Fall URSI Meeting, Dept. Electrical Engineering and Computer Sciences, Univ. of California, Berkeley, California.

POGGIO, A. J. and MILLER, E. K. 1970. *In* Computer methods in electromagnetic scattering and antenna analysis, *edited by* R. Mittra (Pergamon Press, N.Y.).

SOMMERFELD, A. 1909. Ann. Phys. **28**, 663.

——— 1964. Partial differential equations in physics (Academic Press, N.Y.).

WAIT, J. R. 1969. *In* Antenna theory, Part II, *edited by* R. E. Collin and F. J. Zucker (McGraw-Hill Book Co., N.Y.).

YEH, Y. S. and MEI, K. K. 1967. IEEE Trans. Antennas Propag. **15**, 634.

Mutual impedance between vertical dipoles over a flat earth

J. H. Richmond and E. H. Newman

The Ohio State University ElectroScience Laboratory, Department of Electrical Engineering, Columbus, Ohio 43212

(Received January 23, 1979; revised July 11, 1979.)

The fields of a vertical dipole and the mutual impedance between vertical dipoles over a flat earth are expressed in terms of a Sommerfeld integral.

1. INTRODUCTION

The use of vertical wire antennas, or arrays of vertical wires, over the earth is common. This problem can be analyzed by using the method of moments [*Harrington,* 1968], where the effects of the flat earth are accounted for by modifying each element in the impedance matrix. The computation of the modification to a single element in the matrix involves the numerical evaluation of a Sommerfeld integral [*Sommerfeld,* 1964]. With sinusoidal bases and point matching, it has been shown [*Miller et al.*, 1972] that a single evaluation of the Sommerfeld integral suffices for each impedance element. Miller et al. report a computation time of 20 s on a CDC-6600 computer for impedance of a vertical half-wave dipole over the flat earth. In view of this computational expense, they present data based on a 5-segment antenna model, although these data differ by 8% from those with a 21-segment model. They present the theory based on enforcement of the integral equation on the axis of the wire.

In this paper we present the general formulation and show that a single evaluation of the Sommerfeld integral suffices for almost any choice of basis and testing functions. Instead of matching on the axis, we present the theory appropriate for enforcement of the integral equation on the surface of the wire. More accurate data are presented for the impedance of a center-fed vertical wire versus height above the earth, and it is found that our calculations are significantly faster than those quoted above.

2. FIELD OF VERTICAL DIPOLE

For an electric line source located on the z axis in free space the vector potential has only a z component given by

$$A = \int_{z_1'}^{z_2'} \frac{I(z')e^{-jkR}}{4\pi R}\, dz' \tag{1}$$

where z_1' and z_2' denote the end points of the line source and $I(z')$ represents the current distribution with the time dependence $e^{j\omega t}$ suppressed.

By expanding the spherical wave $\exp(-jkR)/R$ in a spectrum of cylindrical waves, we obtain

$$A = \int_0^\infty \beta J_0(\beta\rho) \frac{F(\gamma)\, e^{\gamma z}}{4\pi\gamma}\, d\beta \tag{2}$$

$$\gamma^2 = \beta^2 - k^2 \tag{3}$$

$$F(\gamma) = \int_{z_1'}^{z_2'} I(z') e^{-\gamma z'}\, dz' \tag{4}$$

where z is less than z_1'. The electric field intensity is given by

$$E_z = \frac{1}{j\omega\varepsilon}\left(k^2 A + \frac{\partial^2 A}{\partial z^2}\right) \tag{5}$$

$$E_z = C \int_0^\infty \beta^3 J_0(\beta\rho)(F/\gamma) e^{\gamma z}\, d\beta \tag{6}$$

where $C = -j\eta/(4\pi k)$ and η is the impedance of free space.

Now let the xy plane be the air-earth interface. With the field in (6) incident on the interface the reflected field is given by

$$E_z^r = C \int_0^\infty \beta^3 R J_0(\beta\rho)(F/\gamma) e^{-\gamma z}\, d\beta \tag{7}$$

where the reflection coefficient R is a function of γ and the parameters of the earth; (ρ, ϕ, z) denote cylindrical coordinates with origin on the interface and line source on the z axis.

Now consider a tubular vertical dipole with radius 'a,' current distribution $I(z')$, and surface current density uniformly distributed around the tube. The reflected field for this source is found from (7) by using the addition theorem for the Bessel function and integrating around the surface of the tube:

$$E_z^r = C \int_0^\infty \beta^3 R J_0(\beta a) J_0(\beta\rho)(F/\gamma) e^{-\gamma z} \, d\beta \qquad (8)$$

where ρ is measured from the axis of the tubular source.

3. MUTUAL IMPEDANCE OF VERTICAL DIPOLES

The mutual impedancc between tubular vertical dipoles is given by

$$Z_{12} = -\frac{1}{I_1 I_2} \iint J_2 E_1 \, ds \qquad (9)$$

I_1 and I_2 denote the terminal currents, and the integration extends over the cylindrical surface of dipole 2. Dipole 2 has radius a_2, current distribution $I_2(z)$, and uniform surface current density J_2 around the tube. The z component of the field of dipole 1 is given by

$$E_1 = E_1^0 + E_1^r \qquad (10)$$

where the first term denotes the free-space field and the second term is the field reflected from the interface. From (9) and (10),

$$Z_{12} = Z_{12}^0 + Z_{12}^r \qquad (11)$$

where the first term is the mutual impedance in free space and the second term is the change in mutual impedance arising from interface reflection.

Using (8) for the reflected field, invoking the addition theorem, and integrating around the surface of dipole 2, we find

$$Z_{12}^r = -\frac{C}{I_1 I_2} \int_0^\infty \beta^3 (R/\gamma) \cdot J_0(\beta a_1) J_0(\beta a_2) J_0(\beta d) FG \, d\beta \qquad (12)$$

$$G(\gamma) = \int_{z_1}^{z_2} I_2(z) e^{-\gamma z} \, dz \qquad (13)$$

The horizontal distance between the dipole axes is denoted by d, and z_1 and z_2 are the end points of dipole 2.

In moment method applications the current distributions $I_m(z')$ and $I_n(z)$ represent the basis functions, and (12) determines Z_{mn}^r. For any simple basis functions it is apparent from (4) and (13) that the functions F and G can be obtained in simple closed form. Thus we see from (12) that a single evaluation of the Sommerfeld integral is sufficient to calculate each of the mutual impedance terms Z_{mn}^r. In the sinusoidal Galerkin formulation the basis and testing functions are

$$I_m(z') = \frac{\sin k(h_m - |z' - z_m'|)}{\sin kh_m} \qquad (14)$$

$$I_n(z) = \frac{\sin k(h_n - |z - z_n|)}{\sin kh_n} \qquad (15)$$

where h_m and h_n denote the segment lengths. Function m extends over the two segments which intersect at z_m', and function n extends over the two segments which intersect at z_n.

4. NUMERICAL RESULTS

Figure 1 illustrates the impedance of a half-wave center-fed vertical wire. Here h denotes the distance from the air-earth interface to the center of the wire. These data were calculated with the sinusoidal Galerkin formulation. The wire was divided into 30 segments, so there were 29 equations and 29 unknowns. In Figure 1 the trends are similar to

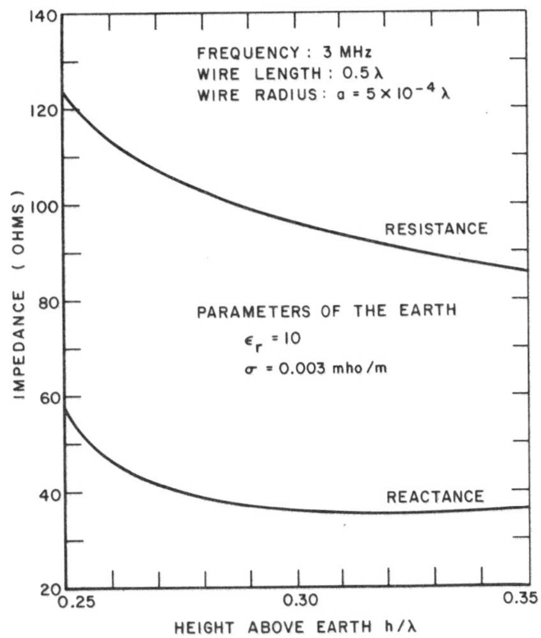

Fig. 1. Impedance of center-fed vertical wire versus height above the earth.

those in Figure 2 of *Miller et al.* [1972], but the data agree more closely with their 21-segment results rather than the 5-segment results they present.

If the 20-s computation time quoted by Miller et al. applies to their 5-segment model, our corresponding 6-segment calculations are 100 times faster. If it applies to their 21-segment model, our 22-segment calculations are 12 times faster. We employed a simple Newton-Cotes integration as opposed to their adaptive integration for the Sommerfeld integral. In addition, our free-space impedance terms Z^0_{mn} are in simple closed form, whereas theirs may require numerical integration.

Surface matching is slower than axis matching, and the two techniques give essentially the same results for wires with small radius as in Figure 1. For larger radii such as the often quoted results for $a = 0.007\ \lambda$, however, surface matching is required for accurate results and, indeed, for convergence.

Acknowledgment. The work reported in this paper was supported in part by contract DAAG29-79-C-0082 between the Department of the Army, U.S. Army Research Office, and The Ohio State University Research Foundation.

REFERENCES

Harrington, R. F. (1968), *Field Computation by Moment Methods,* 229 pp., Macmillan, New York.

Miller, E. K., A. J. Poggio, G. J. Burke, and E. S. Selden (1972), Analysis of wire antennas in the presence of a conducting half-space, I, The vertical antenna in free space, *Can. J. Phys., 50,* 879–888.

Sommerfeld, A. (1964), *Partial Differential Equations in Physics,* 335 pp., Academic, New York.

Part 5:
Apertures, Slots And Arrays

A problem important to EMP or high-power microwave environments is that of field penetration through a slit or aperture in a metallic body. Butler and Umashankar (1976) and Harrington and Mautz (1977) treat a geometry that involves a plane sheet.

Waveguide slot arrays are used widely in aviation and other fields; moment methods have nearly obviated the laborious measurement procedure formerly needed for slot array design. Following the pioneering work of Vu Khac and Carson (1973), Lyon and Sangster (1981) analyze transverse and longitudinal slots. The utility of the shunt representation for longitudinal slots is examined by Stern and Elliott (1985).

Turning to arrays, a design based on moment method, including both self- and mutual-impedance determination, is given by Sangster and McCormick (1989). Pozar (1989) analyzes an infinite array of probe-fed patches, using attachment modes. Another important type of array is the frequency selective surface (FSS). Most analyses use the spectral domain, but two spatial moment method approaches are included. Chen (1971) uses circular guide mode bases for a screen with an array of circular holes. More complex is the application of current modes to the Jerusalem cross FSS by Parker, Hamdy and Langley (1983)

BIBLIOGRAPHY

Apertures

R.F. Wallenberg and R.F. Harrington, "Radiation from Apertures in Conducting Cylinders of Arbitrary Cross Section," *Trans. IEEE,* Vol. AP-17, January 1969, pp. 56–62.

T.B.A. Senior and G.A. Desjardins, "Electromagnetic Field Penetration into a Spherical Cavity," *Trans. IEEE,* Vol. EMC-16, November 1974, pp. 205–208.

C.M. Butler and K.R. Umashankar, "Electromagnetic Excitation of a Wire through an Aperture Perforated Conducting Screen," *Trans. IEEE,* Vol. AP-24, July 1976, pp. 456–462.

R.F. Harrington and J.R. Mautz, "A Generalized Network Formulation for Aperture Problems," *Trans. IEEE,* Vol. AP-24, November 1976, pp. 870–873.

D.T. Auckland and R.F. Harrington, "Electromagnetic Transmission through a Filled Slit in a Conducting Plane of Finite Thickness, TE Case," *Trans. IEEE,* Vol. MTT-26, July 1978, pp. 499–505.

R.F. Harrington and D.T. Auckland, "Electromagnetic Transmission through Narrow Slots in Thick Conducting Screens," *Trans. IEEE,* Vol. AP-28, September 1980, pp. 616–622.

J.R. Mautz and R.F. Harrington, "Boundary Formulations for Aperture Coupling Problems," *AEU,* Vol. 34, September 1980, pp. 377–384.

D.T. Auckland and R.F. Harrington, "A Nonmodal Formulation for Electromagnetic Transmission through a Filled Slot of Arbitrary Cross Section in a Thick Conducting Screen," *Trans. IEEE,* Vol. MTT-28, June 1980, pp. 548–555.

C.M. Butler, "Investigation of a Scatterer Coupled to an Aperture in a Conducting Plane," *Proc. IEE,* Vol. 127H, June 1980, pp. 161–169.

K.Y. Kabalan, R.F. Harrington, J.R. Mautz and H.A. Auda, "Characteristic Modes for a Slot in a Conducting Plane, TM Case," *Trans. IEEE,* Vol. AP-35, March 1987, pp. 331–335.

H.K. Schuman and D.E. Warren, "Aperture Coupling in Bodies of Revolution," *Trans. IEEE,* Vol. AP-26, November 1978, pp. 778–783.

D.B. Seidel, "Aperture Excitation of a Wire in a Rectangular Cavity," *Trans. IEEE,* Vol. MTT-26, November 1978, pp. 908–914.

Y. Rahmat-Samii, "Electromagnetic Pulse Coupling through an Aperture into a Two-parallel-plate Region," *Trans. IEEE,* Vol. EMC-20, August 1978, pp. 436–442.

C.M. Butler, Y. Rahmat-Samii and R. Mittra, "Electromagnetic Penetration through Apertures in Conducting Surfaces," *Trans. IEEE,* Vol. AP-26, January 1978, pp. 82–93.

M.F. Catedra, "A Comparison Between Two Kinds of Equivalent Currents to Analyze Conducting Bodies with Apertures Using Moment Methods," *Trans. IEEE,* Vol. AP-35, July 1987, pp. 782–789.

R.W. Scharstein, "Two Numerical Solutions for the Parallel Plate-Fed Slot Antenna," *Trans. IEEE,* Vol. AP-37, November 1989, pp. 1415–1426.

J.A. Beren, "Diffraction of an H-Polarized Wave by a Circular Cylinder with an Infinite Axial Slot," *Trans. IEEE,* Vol. AP-31, May 1983, pp. 419–425.

J.R. Mautz and R.F. Harrington, "Electromagnetic Penetration into a Conducting Circular Cylinder Through a Narrow Slot, TE Case," *J. Electromag. Waves and Appl.,* Vol. 3, 1989, pp. 307–336.

J.R. Mautz and R.F. Harrington, "Electromagnetic Penetration into a Conducting Circular Cylinder Through a Narrow Slot, TM Case," *J. Electromag. Waves and Appl.,* Vol. 2, 1988, pp. 269–293.

Y. Leviatan, E. Hudis and P.D. Einziger, "A Method of Moments Analysis of Electromagnetic Coupling Through Slots Using a Gaussian Beam Expansion," *Trans. IEEE,* Vol. AP-37, December 1989, pp. 1537–1544.

Waveguide Slots

A.J. Sangster and R.W. Lyon, "Moment-Method Analysis of a T-shaped Slot Radiator in Bifurcated Waveguide," *Proc. IEE-MOA,* Vol. 129, December 1982, pp. 299–306.

A.J. Sangster and A.H.I. McCormick, "Moment Method Applied to Round-ended Slots," *Proc. IEE,* Vol. 134H, June 1987, pp. 310–314.

P.K. Park, G.J. Stern and R.S. Elliott, "An Improved Technique for the Evaluation of Transverse Slot Discontinuities in Rectangular Waveguide," *Trans. IEEE,* Vol. AP-31, January 1983, pp. 148–154.

R. Shavit and R.S. Elliott, "Design of Transverse Slot Arrays Fed by a Boxed Stripline," *Trans. IEEE,* Vol. AP-31, July 1983, pp. 545–552.

L.G. Josefsson, "Analysis of Longitudinal Slots in Rectangular Waveguides," *Trans. IEEE,* Vol. AP-35, December 1987, pp. 1351–1357.

J.R. Parkinson and M.J. Mehler, "Uniqueness of Numerical Solutions to Electric-, Magnetic- and Combined-field Integral Equations for Open-Ended Circular Waveguides," *Proc. IEE,* Vol. 136H, June 1989, pp. 269–275.

Arrays and FSS

B.J. Strait and K. Hirasawa, "Array Design for a Specified Pattern by Matrix Methods," *Trans. IEEE,* Vol. AP-17, March 1969, pp. 237–239.

K. Hirasawa and D.H. Sinnott, "On the Relationship Between Classical and Matrix Design Methods for Arrays of Wire Antennas," *Trans. IEEE,* Vol. AP-20, September 1972, pp. 661–663.

B.J. Strait and K. Hirasawa, "Contrained Optimization of the Gain of an Array of Thin Wire Antennas," *Trans. IEEE,* Vol. AP-20, September 1972, pp. 665–666.

S.M. Sanzgiri and J.A. Cummins, "E-plane Synthesis of Dipole Array Antennas," *Trans. IEEE,* Vol. AP-21, May 1973, pp. 380–382.

K. Siakavara and J.N. Sahalos, "A Simplification of the Synthesis of Parallel Wire Antenna Arrays," *Trans. IEEE,* Vol. AP-37, July 1989, pp. 936–940.

D.H. Sinnott and R.F. Harrington, "Analysis and Design of Circular Antenna Arrays by Matrix Methods," *Trans. IEEE,* Vol. AP-21, September 1973, pp. 610–614.

A.J. Fenn, G.A. Thiele and B.A. Munk, "Moment Method Analysis of Finite Rectangular Waveguide Planar Phased Arrays," *Trans. IEEE,* Vol. AP-30, July 1982, pp. 554–564.

F. Arndt, K.H. Wolff, L. Brunjes, R. Heyen, F. Siefken-Herrlich, W. Bothmer and E. Forgber, "Generalized Moment Method Analysis of Planar Reactively Loaded Rectangular Waveguide Arrays," *Trans. IEEE,* Vol. AP-37, March 1989, pp. 329–338.

A.R. Neureuther and K. Zaki, "Numerical Solutions of Electromagnetic Boundary Value Problems by Means of the Asymptotic Anticipation Method," *Radio Science,* Vol. 3, December 1968, pp. 1158–1167.

C.H. Liang and D.K. Cheng, "Directivity Optimization for Yagi-Uda Arrays of Shaped Dipoles," *Trans. IEEE,* Vol. AP-31, May 1983, pp. 522–525.

A.M.M.A. Allam and E.A. Parker, "Application of Pocklington's Equation to Analysis of Dipole Frequency-Selective Surfaces of Finite Size," *Proc. IEE,* Vol. 134H, December 1987, pp. 521–526.

R. Orta, R. Tascone and R. Zich, "Efficient Choice of Basis Functions in Moment Method Analysis of Frequency-selective Surfaces," *Electronics Letters,* Vol. 23, No. 11, 1987, pp. 586–587.

A. Boag, Y. Leviatan and A. Boag, "Analysis of Two-dimensional Electromagnetic Scattering from Nonplanar Periodic Surfaces Using a Strip Current Model," *Trans. IEEE,* Vol. AP-37, November 1989, pp. 1437–1446.

C. H. Chan, "A Numerically Efficient Technique for the Method of Moments Solution of Electromagnetic Problem Associated with Planar Periodic Structures," *Microw. & Opt. Tech. Letters,* Vol. 1, December 1988, pp. 372–374.

Electromagnetic penetration through an aperture in an infinite, planar screen separating two half spaces of different electromagnetic properties

Chalmers M. Butler and Korada R. Umashankar

Department of Electrical Engineering, University of Mississippi, University, Mississippi 38677

(Received January 19, 1976.)

Integro-differential equations are formulated for the general problem of electromagnetic diffraction by an aperture in a planar conducting screen of infinite extent separating two half spaces of different electromagnetic properties. With the aperture specialized to the case of an infinite slot, equations appropriate for TE and for TM illumination are deduced from those for the general aperture. These slot equations are solved numerically, and results are presented for several cases of interest.

INTRODUCTION

Even though electromagnetic diffraction by an aperture in an infinite conducting screen has been the subject of much research over the years, almost all interest has been limited to the situation in which the screen resides in a homogeneous medium with little attention given to the special case of the screen separating half spaces of different electromagnetic characteristics. It is the purpose of this paper to investigate this special, but interesting, two-media case.

The two-media problem has been investigated for the special case of a narrow slot aperture by *Barakat* [1963] and by *Houlberg* [1967]. They develop equations for the slotted screen problem and present solution techniques applicable when the slot is narrow relative to the wavelengths in the two media; they give data for transmission and reflection coefficients. *Thomas* [1969] derives equations for a circular aperture and provides approximate formulas for transmission and backscatter coefficients with tabulated data valid for circles small relative to the wavelengths in the two media. Bethe theory has been extended to cover the two-media problem and polarizabilities have been calculated for four aperture shapes of practical importance (F. De Meulenaere and J. Van Bladel, personal communication, 1975).

In this paper, integro-differential equations are developed for the problem of an aperture of general shape in a perfectly conducting screen, of vanishing thickness and infinite extent, separating half spaces of different electromagnetic characteristics with illumination incident upon the aperture/screen in both half spaces. The only feasible way to solve the two-media equations is numerically, but, even with the relatively simple rectangular aperture, solving them this way would be very demanding upon computer storage. Therefore, in order to simplify the equations to be solved and yet to retain the essential features of the two-media problem, the aperture is specified to be an infinite slot of uniform width and the illumination is limited to the two cases in which the slot field is either transverse electric (TE) or transverse magnetic (TM) to the slot axis. The general aperture/screen equations for the two-media problem are specialized to the TE and TM slot cases and these reduced equations are solved numerically. The numerical solution procedures employed are valid for any reasonable slot widths, and representative data are presented for several cases of interest.

FORMULATION

In the aperture A and tangential to the aperture/screen plane, there exists some total electric field $\mathbf{E}_t^a$, which is, of course, unknown *a priori.* In terms of $\mathbf{E}_t^a$, the incident fields radiated by the specified sources, the electromagnetic properties of the two half spaces, and the aperture/screen geometry (Figure 1), one can write expressions for the magnetic field in each half space. The enforcement of continuity of transverse magnetic field through the aperture leads directly to the desired integro-differential equation for the problem under investigation.

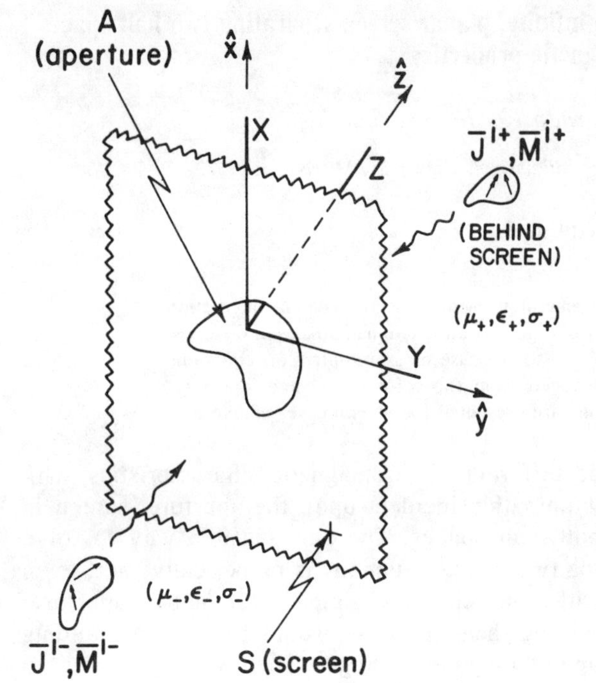

Fig. 1. Aperture in conducting screen separating two media of different electromagnetic properties.

Figure 2 depicts a step-by-step reduction of the left half-space ($z < 0$) problem to a simple equivalent problem in a form which readily suggests how one may develop an expression for the left half-space total magnetic field $\mathbf{H}^-$. This figure is only schematic and directions of vectors shown are not to be interpreted as actual directions. In Figure 2a, one sees the original problem while in Figure 2b the aperture-perforated screen is seen to be replaced by a perfectly conducting (shorted aperture) plane with the original tangential electric field $\mathbf{E}_t^a$ in the aperture restored at $z = 0^-$, $(x,y) \in A$, by an appropriate magnetic surface current $\mathbf{M}_s$ which is specified to have a value $-\mathbf{E}_t^a \times \hat{z}$, i.e., $\mathbf{M}_s = -\mathbf{E}_t^a \times \hat{z}$. Notice that this as-yet-undetermined magnetic current radiates in the presence of the conducting screen and that the left half-space sources $\mathbf{J}^{i-}$ and $\mathbf{M}^{i-}$ of the original problem in Figure 2a illuminate the shorted screen. Next, from image theory, one removes the conducting screen and arrives at the left half-space equivalent problem of Figure 2c.

The total left half-space electromagnetic field $(\mathbf{E}^-,\mathbf{H}^-)$ is the sum of the field radiated by $\mathbf{M}_s$

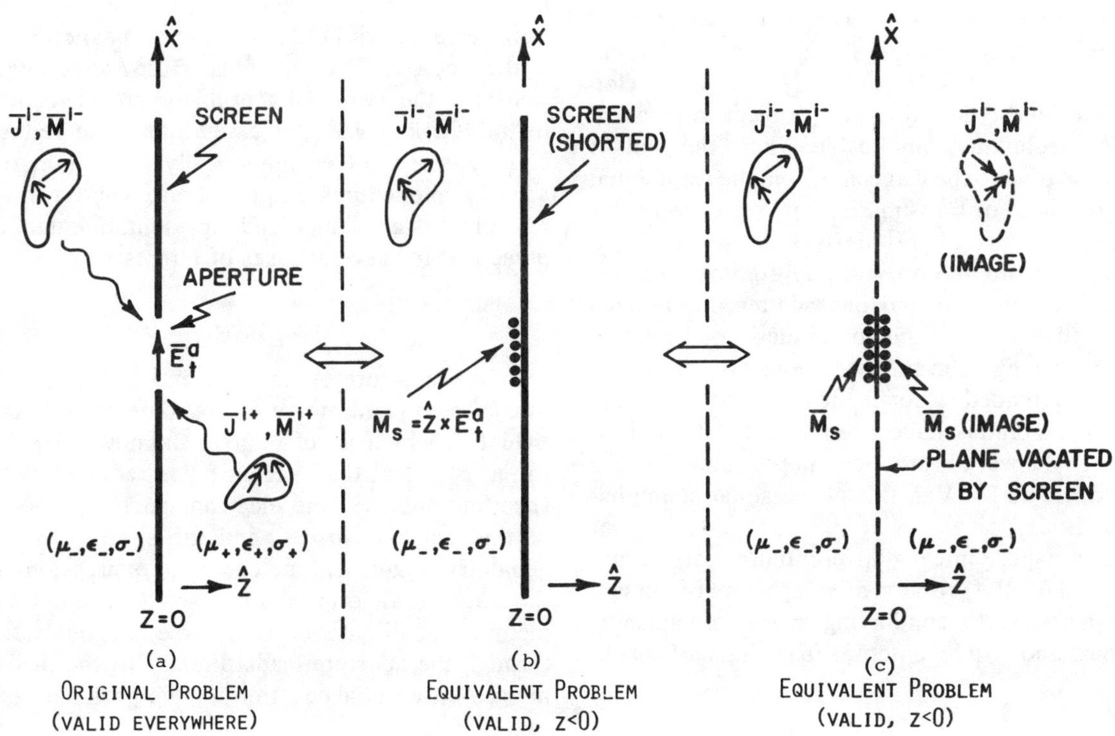

Fig. 2. Left half-space equivalent problems.

plus its image and that radiated by the left half-space sources $(\mathbf{J}^{i-},\mathbf{M}^{i-})$ plus their images, with all radiating in an infinite, homogeneous space characterized by $(\mu_-,\epsilon_-,\sigma_-)$ (Figure 2c). Because the field due to $(\mathbf{J}^{i-},\mathbf{M}^{i-})$ plus their images is that radiated by $(\mathbf{J}^{i-},\mathbf{M}^{i-})$ in the presence of the *shorted* screen, we denote this so-called short-circuit field by $(\mathbf{E}^{sc-},\mathbf{H}^{sc-})$.

One may develop the right half-space equivalences in a way identical to that discussed above for the left half space. $(\mathbf{E}^+,\mathbf{H}^+)$ and $(\mathbf{E}^{sc+},\mathbf{H}^{sc+})$ represent the total and short-circuit fields, respectively, in the right half space.

Aided by Figure 2c and an analogous equivalence for the right half space, one writes $\mathbf{H}^-$ and $\mathbf{H}^+$ as

$$\mathbf{H}^- = \mathbf{H}^{sc-} - j\frac{\omega}{k_-^2}(k_-^2\mathbf{F}^- + \nabla\nabla\cdot\mathbf{F}^-), \quad z<0 \tag{1}$$

and

$$\mathbf{H}^+ = \mathbf{H}^{sc+} + j\frac{\omega}{k_+^2}(k_+^2\mathbf{F}^+ + \nabla\nabla\cdot\mathbf{F}^+), \quad z>0 \tag{2}$$

where ω is the angular frequency of the surpressed harmonic variation in time $e^{j\omega t}$ and where

$$k_\pm^2 = \omega^2\mu_\pm\epsilon_\pm' = \omega^2\mu_\pm\epsilon_\pm - j\omega\mu_\pm\sigma_\pm \tag{3}$$

In (1), $\mathbf{F}^-$ $(\mathbf{F}^+)$ is the electric vector potential due to $\mathbf{M}_s$ $(-\mathbf{M}_s)$ plus its image:

$$\mathbf{F}^\pm(\mathbf{r}) = \frac{\epsilon_\pm'}{4\pi}\iint_A 2\mathbf{M}_s(\mathbf{r}')\frac{\exp(-jk_\pm|\mathbf{r}-\mathbf{r}'|)}{|\mathbf{r}-\mathbf{r}'|}\,ds' \tag{4}$$

where $\mathbf{r}$ locates a point (x,y,z) in space and $\mathbf{r}' \in A$.

Requiring that the magnetic field be continuous along any path through A,

$$\lim_{z\uparrow 0}\mathbf{H}^-\times\hat{z} = \lim_{z\downarrow 0}\mathbf{H}^+\times\hat{z}, \quad \text{in } A$$

and recognizing that $\mathbf{H}^{sc\pm}\times\hat{z} = 2\mathbf{H}^{i\pm}\times\hat{z}$ at $z = 0^\pm$, where $(\mathbf{E}^{i\pm},\mathbf{H}^{i\pm})$ is the incident field in the medium $(\mu_\pm,\epsilon_\pm,\sigma_\pm)$ due to $(\mathbf{J}^{i\pm},\mathbf{M}^{i\pm})$, one arrives at

$$[\mathbf{F}^+ + \mathbf{F}^- + \nabla_t\nabla_t\cdot(\mathbf{F}^+/k_+^2 + \mathbf{F}^-/k_-^2)]\times\hat{z}$$
$$= -j(2/\omega)(\mathbf{H}^{i-} - \mathbf{H}^{i+})\times\hat{z} \quad \text{in } A \tag{5}$$

Equation (5) is an integro-differential equation for the unknown magnetic current $\mathbf{M}_s$ in the aperture. The component of $\mathbf{M}_s$ which is normal to the aperture edge is zero along this edge and the component parallel to the edge is unbounded (according to the edge condition for $\mathbf{E}_t^a$) along the edge.

If the electromagnetic properties of the two half spaces were the same, (5) would reduce immediately to the familiar integro-differential equation for an aperture in a screen. For aperture shapes other than the circle, even this important traditional problem must be solved approximately by numerical techniques [*Mittra et al.*, 1973; *Lin et al.*, 1974]. Such numerical solution procedures demand very high computer storage and run times. Therefore, in order to retain the central features of the two-media problem but to reduce it to one for which numerical solution techniques are practical, we specialize the aperture to be an infinite slot of uniform width. Data which are representative of the behavior of quantities of interest in the general two-media problem are determined from solutions of the special case of the slot.

INFINITE SLOT

For a slot of uniform width w with its axis along the y axis, we consider two cases of excitation in which the problem is two-dimensional: (i) an incident wave having only a y-component of magnetic field and having an electric field transverse to the slot axis, both being independent of y (TE case); and (ii) an incident wave having only a y component (negative) of electric field with the magnetic field transverse to the slot axis, both again independent of y (TM case).

TE case. In the TE case, the only electric field which can exist in the slot is x-directed and independent of y. Thus, $\mathbf{M}_s = M_{s_y}(x)\hat{y}$ and the electric vector potentials have only y components and are independent of y: $\mathbf{F}^\pm = F_y^\pm(x)\hat{y}$ for $z = 0$. In view of the properties above, (5) reduces in the TE case to

$$F_y^+(x) + F_y^-(x) = -j(2/\omega)[H_y^{i-}(x) - H_y^{i+}(x)] \quad \text{in slot} \tag{6}$$

Since $F_y^\pm$ are independent of y,

$$F_y^\pm(x) = (\epsilon_\pm'/2\pi)\int_{x'=-w/2}^{w/2} M_{s_y}(x')$$

$$\cdot \int_{y'=-\infty}^{\infty} \frac{\exp\{-jk_{\pm}[(x-x')^2+y'^2]^{1/2}\}}{[(x-x')^2+y'^2]^{1/2}}\, dy'\, dx' \tag{7}$$

which can be written

$$F_y^{\pm}(x) = \frac{\epsilon'_{\pm}}{2j}\int_{x'=-w/2}^{w/2} M_{s_y}(x')\, H_0^{(2)}(k_{\pm}|x-x'|)\, dx' \tag{8}$$

where $H_0^{(2)}(\xi)$ is the Hankel function of the second kind and zero order. Hence, one now has an integral equation for the unknown M_{s_y} in the TE case:

$$\int_{x'=-w/2}^{w/2} M_{s_y}(x')\,[\epsilon'_+ H_0^{(2)}(k_+|x-x'|) + \epsilon'_- H_0^{(2)}(k_-|x-x'|)]\, dx' = (4/\omega)[H_y^{i-}(x) - H_y^{i+}(x)],\quad x\in(-w/2,w/2) \tag{9}$$

TM case. In the TM case, the electric field in the slot is entirely y-directed and independent of y. Thus, $\mathbf{M}_s = M_{s_x}(x)\hat{x}$ and the electric vector potentials $\mathbf{F}^+$ and $\mathbf{F}^-$ are x-directed and are independent of y: $\mathbf{F}^{\pm} = F_x^{\pm}(x)\hat{x}$ at $z=0$. Hence, in the TM case, (5) reduces to

$$(1/k_+^2)(\partial^2/\partial x^2 + k_+^2)\, F_x^+(x) + (1/k_-^2)(\partial^2/\partial x^2 + k_-^2) \cdot F_x^-(x) = -j(2/\omega)[H_x^{i-}(x) - H_x^{i+}(x)] \quad \text{in slot} \tag{10}$$

where $F_x^{\pm}(x)$ is the right-hand side of (7) with M_{s_y} replaced by M_{s_x}. In view of (8), one sees that (10) reduces to the following integro-differential equation for the TM case:

$$(\partial^2/\partial x^2 + k_+^2)\int_{x'=-w/2}^{w/2} M_{s_x}(x')[(1/\mu_+) \cdot H_0^{(2)}(k_+|x-x'|)]\, dx' + (\partial^2/\partial x^2 + k_-^2)\int_{x'=-w/2}^{w/2} \cdot M_{s_x}(x')[(1/\mu_-) H_0^{(2)}(k_-|x-x'|)]\, dx' = 4\omega\, [H_x^{i-}(x) - H_x^{i+}(x)],\quad x\in(-w/2,w/2) \tag{11}$$

NUMERICAL SOLUTION

Knowledge of the equivalent slot magnetic current enables one to calculate all quantities of interest in either the TE or the TM case. Therefore, in this section, techniques for solving the two equations, (9) and (11), are presented. The integral equation (9) for the TE case introduces no special difficulties and it can be solved numerically quite readily by means of collocation with the magnetic current M_{s_y} approximated by a linear combination of pulses [*Harrington*, 1968]. On the other hand, the integro-differential equation (11) for the TM case warrants careful scrutiny. It cannot be converted to a Hallén-type equation, as is often done in thin-wire analyses, because there are two distinct harmonic differential operators (since $k_+ \neq k_-$) on the left-hand side; the presence of the different k's also precludes the use of the so-called reaction matching technique [*Mittra*, 1973]. There is no reduced kernel as in wire theory available as an approximation of the Hankel functions in (11) so interchange of differentiation and integration, with subsequent application of collocation, would lead to difficulties. However, this equation can be solved numerically by means of a scheme utilizing piecewise linear testing and almost any reasonable subdomain basis set for approximating the magnetic current. This solution procedure is discussed below.

The interval of the slot $(-w/2,w/2)$ is partitioned into $2M$ subintervals each of length $\Delta = w/(2M)$. We define testing functions (triangles) [*Harrington*, 1968],

$$\Lambda_m^l(x) = \begin{cases} \dfrac{\Delta - |x-x_m|}{\Delta}, & x\in(x_{m-1},x_{m+1}) \\ 0, & x\notin(x_{m-1},x_{m+1}) \end{cases} \tag{12}$$

Testing (11) [*Wilton and Butler*, 1976] by Λ_m^l of (12) and integrating by parts twice yields

$$(1/4\omega)\{(1/\Delta\mu_+)[f(k_+x_{m+1}) - 2f(k_+x_m) + f(k_+x_{m-1})] + (1/\Delta\mu_-)[f(k_-x_{m+1}) - 2f(k_-x_m) + f(k_-x_{m-1})] + \int_{x=x_{m-1}}^{x_{m+1}} [(k_+^2/\mu_+)\, f(k_+x) + (k_-^2/\mu_-)\, f(k_-x)] \cdot \Lambda_m^l(x)\, dx\} = \int_{x=x_{m-1}}^{x_{m+1}} [H_x^{i-}(x) - H_x^{i+}(x)] \cdot \Lambda_m^l(x)\, dx,\quad m = 0, \pm1, \pm2,\ldots, \pm(M-1) \tag{13}$$

where

$$f(kx) = \int_{x'=-w/2}^{w/2} M_{s_x}(x')\, H_0^{(2)}(k|x-x'|)\, dx' \quad (14)$$

Even though the numerical solution can be based upon (13), there is an approximation which one may employ to greatly simplify certain computations. In particular, when Δ is sufficiently small, i.e., $\Delta k_- \ll 1$ and $\Delta k_+ \ll 1$ in the present application,

$$\int_{x=x_{m-1}}^{x_{m+1}} u(x)\, \Lambda_m^l(x)\, dx \doteq \Delta u(x_m) \quad (15)$$

Subject to this good approximation, (13) becomes

$$\begin{aligned}(1/4\omega)\big[&(1/\Delta\mu_+)\{f(k_+x_{m+1}) - 2[1-(k_+\Delta)^2/2]\\ &\cdot f(k_+x_m) + f(k_+x_{m-1})\} + (1/\Delta\mu_-)\{f(k_-x_{m+1})\\ &- 2[1-(k_-\Delta)^2/2]\, f(k_-x_m) + f(k_-x_{m-1})\}\big]\\ &\doteq \Delta\,[H_x^{i-}(x_m) - H_x^{i+}(x_m)],\\ &\quad m = 0, \pm 1, \pm 2, \ldots, \pm(M-1) \end{aligned} \quad (16)$$

Even though (16) is the finite-difference approximation of (11), it was obtained by means of piecewise linear testing of (11) subject to the good approximation of (15). As pointed out by *Wilton and Butler* [1976], (16) enjoys all the advantages of a weighted averaging solution procedure plus the simplicity of a collocation scheme. For example, if $\{\Lambda_n^l\}$ of (12) were employed for representation of M_{s_x}, then, subject to (15), (16) would lead to a Galerkin solution.

A number of basis sets would be suitable for approximation of M_{s_x}, but, due to its simplicity and since it yields results as efficiently as any other, we select the set comprising pulses [*Harrington*, 1968] for approximating M_{s_x}:

$$M_{s_x}(x) = \sum_{n=-N}^{N} M_n p_n(x) \quad (17)$$

where $p_n(x) = 1$, $x \in (x_n - \Delta/2, x_n + \Delta/2)$, and $p_n(x) = 0$ outside this interval. The weighting coefficients $\{M_n\}$ are the unknown quantities to be determined. Half pulses are placed at each end of the interval $(-w/2, w/2)$ and are set equal to zero in order to ensure satisfaction of the boundary condition $M_{s_x}(-w/2) = M_{s_x}(w/2) = 0$. With this current approximation introduced into (16), the following system of linear algebraic equations can be constructed:

$$\sum_{n=-N}^{N} M_n Y_{mn} = I_m, \quad m = 0, \pm 1, \pm 2, \ldots, \pm N \quad (18)$$

where

$$I_m = \Delta\,[H_x^{i-}(x_m) - H_x^{i+}(x_m)] \quad (19)$$

and

$$\begin{aligned} Y_{mn} = \frac{1}{4\omega\Delta}\Bigg\{\int_{x'=-\Delta/2}^{\Delta/2} \Bigg[&\frac{1}{\mu_+}\{H_0^{(2)}[k_+|(m-n+1)\Delta\\ &- x'|] - 2[1-(k_+\Delta)^2/2]\, H_0^{(2)}[k_+|(m-n)\Delta\\ &- x'|] + H_0^{(2)}[k_+|(m-n-1)\Delta - x'|]\}\\ &+ (1/\mu_-)\{H_0^{(2)}[k_-|(m-n+1)\Delta - x'|]\\ &- 2[1-(k_-\Delta)^2/2]\, H_0^{(2)}[k_-|(m-n)\Delta - x'|]\\ &+ H_0^{(2)}[k_-|(m-n-1)\Delta - x'|]\}\Bigg] dx'\Bigg\} \end{aligned} \quad (20)$$

with

$$x_m = m\Delta, \quad m = 0, \pm 1, \pm 2, \ldots, \pm N \quad (21)$$

$$x_n = n\Delta, \quad n = 0, \pm 1, \pm 2, \ldots, \pm N \quad (22)$$

The admittance elements Y_{mn} of (20) can be calculated easily, and the linear system of (18) can be solved by standard means to obtain M_n, knowledge of which leads to an approximation for M_{s_x} by means of (17).

RESULTS AND CONCLUSIONS

The equivalent slot magnetic current is presented in Figures 3-11 for the TE and TM cases of interest. It is emphasized that the magnetic current in these figures, M_{s_y} of (9) and M_{s_x} of (11), is the surface current on the left half-space side of the shorted screen which, when radiating in the presence of the shorted screen, produces the total field less the short-circuit field in the region $z < 0$, i.e., $(\mathbf{E}^- - \mathbf{E}^{sc-}, \mathbf{H}^- - \mathbf{H}^{sc-})$. This surface magnetic current $\mathbf{M}_s$ is equal in complex magnitude to $\mathbf{E}_t^a$, the total transverse electric field which exists in the aperture; in the TE case, $E_{t_x}^a = M_{s_y}$, and, in the TM case, $E_{t_y}^a = M_{s_x}$. Also, all data presented are normalized with respect to the left half-space intrinsic impedance η_- and the appropriate component of the incident magnetic field evaluated at the center of the slot $\mathbf{H}^i(0)$, where $\mathbf{H}^i$ is defined for convenience to be the magnetic field incident from the left less that incident from the right $(\mathbf{H}^i = \mathbf{H}^{i-} - \mathbf{H}^{i+})$. The media occupying the half spaces are

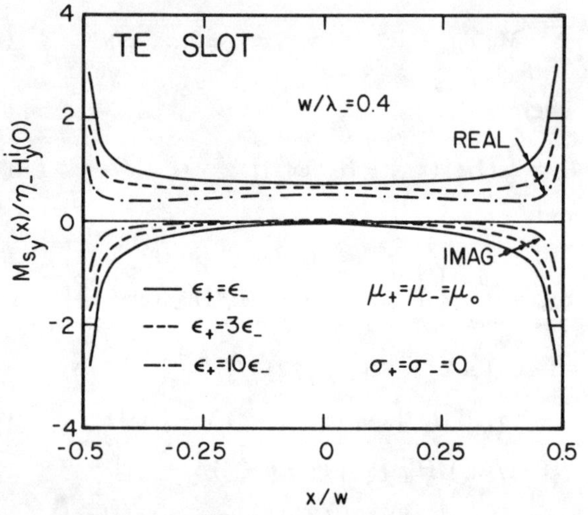

Fig. 3. TE equivalent magnetic current in 0.4-wavelength slot for different right half-space permittivities.

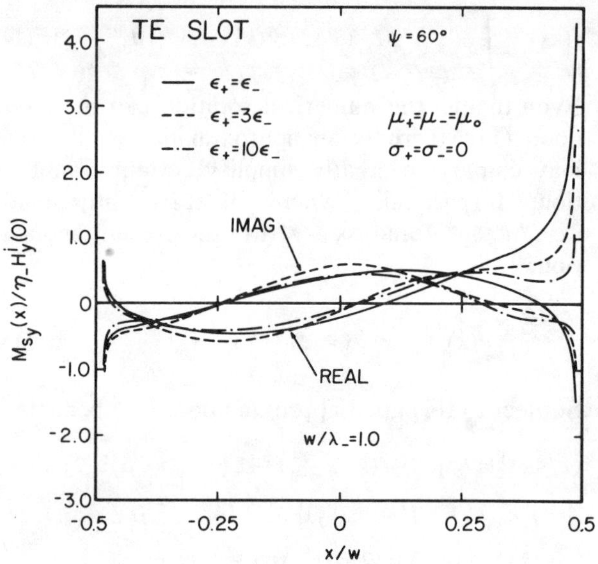

Fig. 5. TE equivalent magnetic current in 1.0-wavelength slot for different right half-space permittivities (60 deg incidence).

lossless ($\sigma_{\pm} = 0$) with $\mu_{\pm} = \mu_0$, and, also, slot widths are measured in numbers of left half-space wavelengths λ_- $(=2\pi/k_-)$.

Figures 3 and 4 display the behavior of the slot magnetic current, or total electric field transverse to the screen, in 0.4- and 1.0-wavelength slots, respectively, for the TE case with H_y^i constant over the slot. One observes two major effects associated with the change in ϵ_+, the right half-space permittivity: first, the magnitude of the slot current reduces with increasing ϵ, and, second, the edge singularity manifests itself closer and closer to the screen edge ($x = \pm w/2$) as ϵ_+ increases. In Figure 5 M_{s_y} is shown for the TE case when the excitation is a non-normally incident plane wave in the left half space, traveling along a path at $\Psi = 60°$ above the z axis of Figure 1.

Turning attention to results in the TM cases, we mention that Figure 6 shows data for M_{s_x} in a slotted screen separating half spaces of the same electromagnetic properties with constant H_x^i; these data are provided here for reference in subsequent dis-

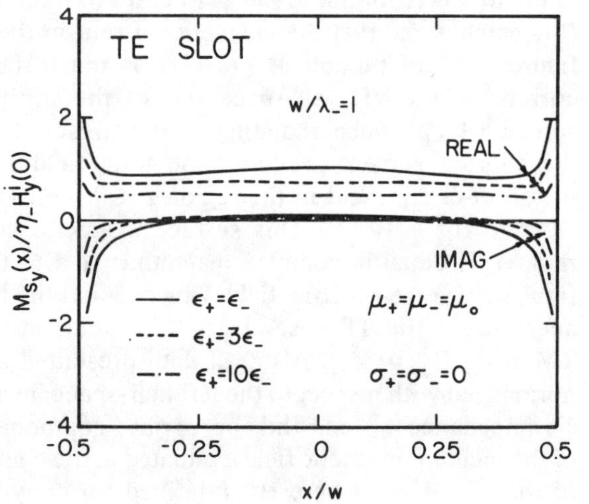

Fig. 4. TE equivalent magnetic current in 1.0-wavelength slot for different right half-space permittivities.

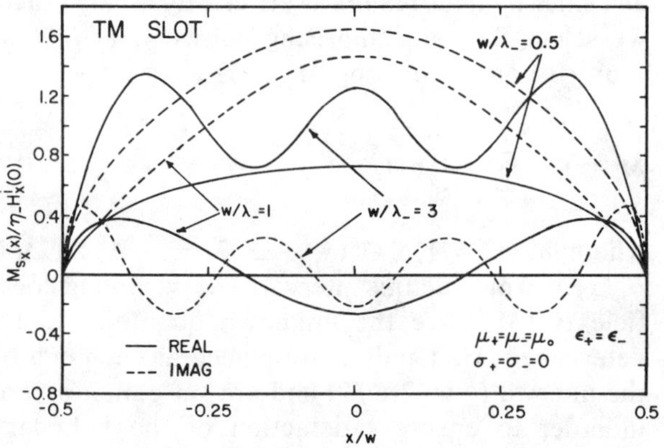

Fig. 6. TM equivalent magnetic current in 0.5-, 1.0-, and 3.0-wavelength slots in screen separating half-spaces of the same media.

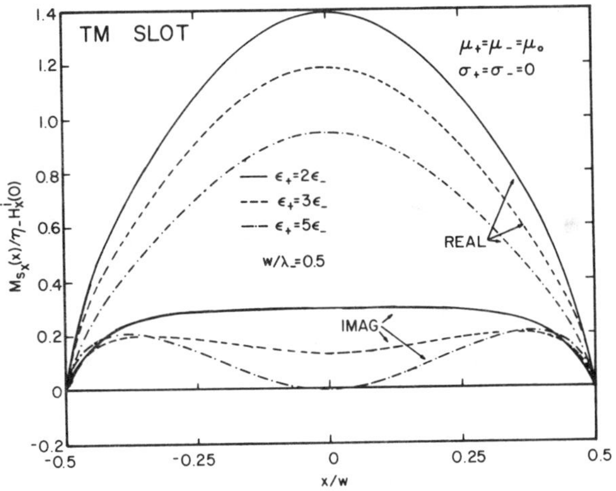

Fig. 7. TM equivalent magnetic current in 0.5-wavelength slot for different right half-space permittivities.

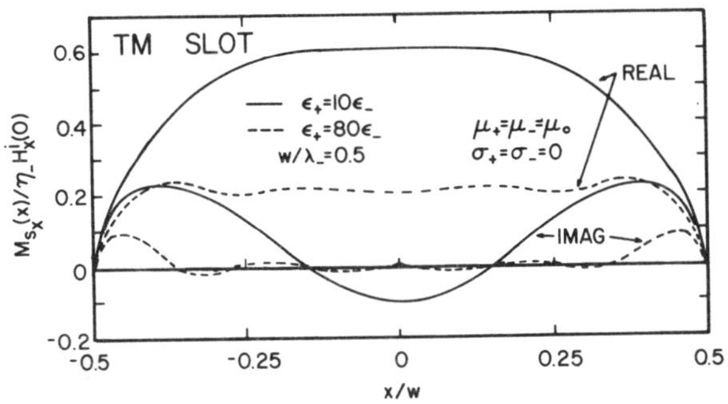

Fig. 8. TM equivalent magnetic current in 0.5-wavelength slot in screen separating half spaces of high dielectric contrast.

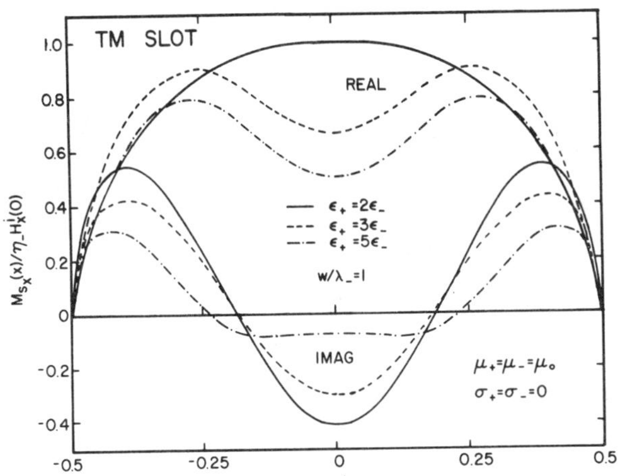

Fig. 9. TM equivalent magnetic current in 1.0-wavelength slot for different right half-space permittivities.

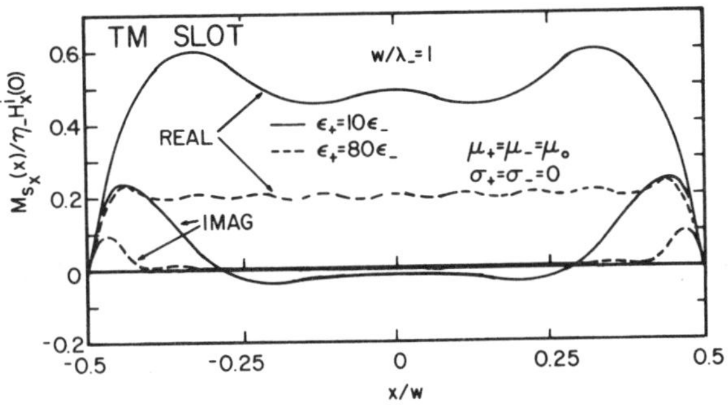

Fig. 10. TM equivalent magnetic current in 1.0-wavelength slot in screen separating half spaces of high dielectric contrast.

cussion of the two-media problem. From Figures 7-10, one observes that not only do changes in magnitude of M_{s_x} occur with changes in ϵ_+, but also significant modifications are introduced in its distribution or shape, especially with high dielectric contrast between the two media (Figures 8 and 10). As one expects from the fact that the kernel of (11) comprises two Hankel functions, one with argument $k_- |x - x'|$ and the other with argument $k_+ |x - x'|$, he recognizes that the distribution of M_{s_x} becomes similar to that of a wider slot (larger w/λ_-) with increases in ϵ_+. Such similarities are apparent from comparisons of curves in Figures 7-10 with those in Figure 6. Figure 11 illustrates the behavior of M_{s_x} due to non-normally ($\Psi = 60°$)

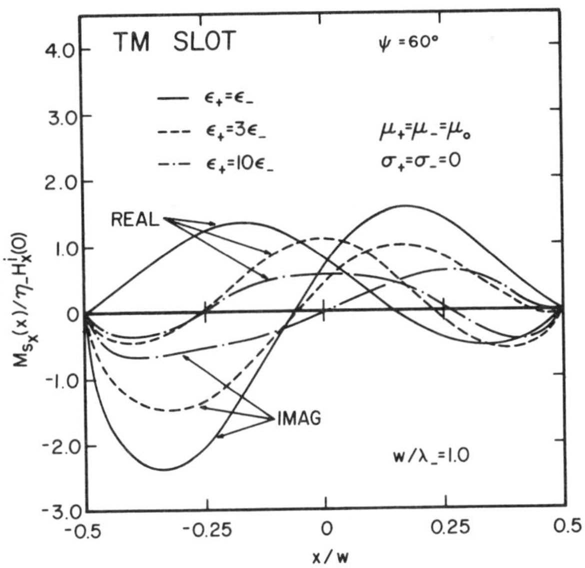

Fig. 11. TM equivalent magnetic current in 1.0-wavelength slot for different right half-space permittivities (60 deg incidence).

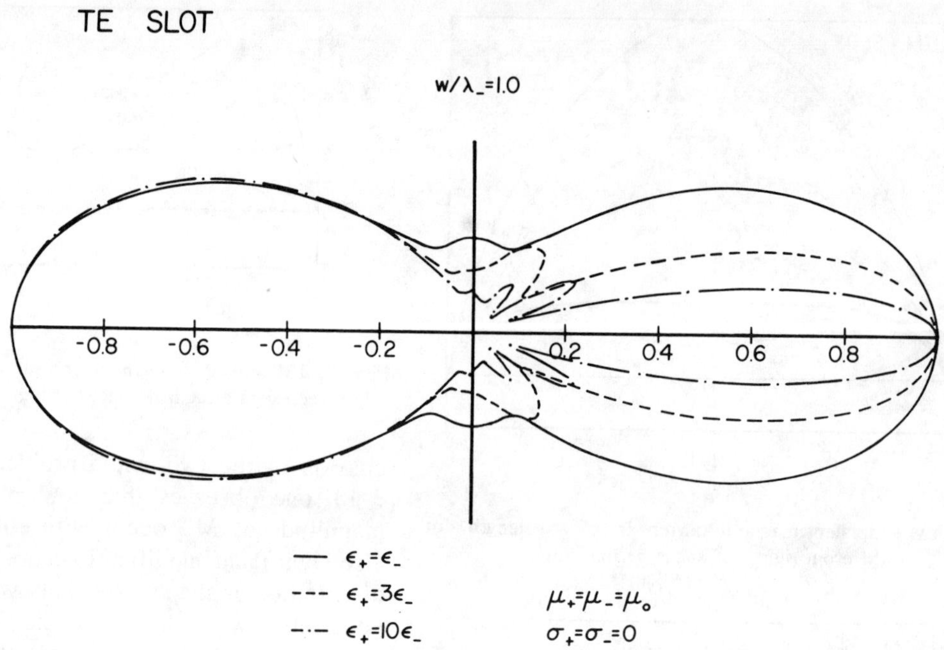

Fig. 12. Far magnetic field due to presence of TE-excited slot in screen for different right half-space permittivities.

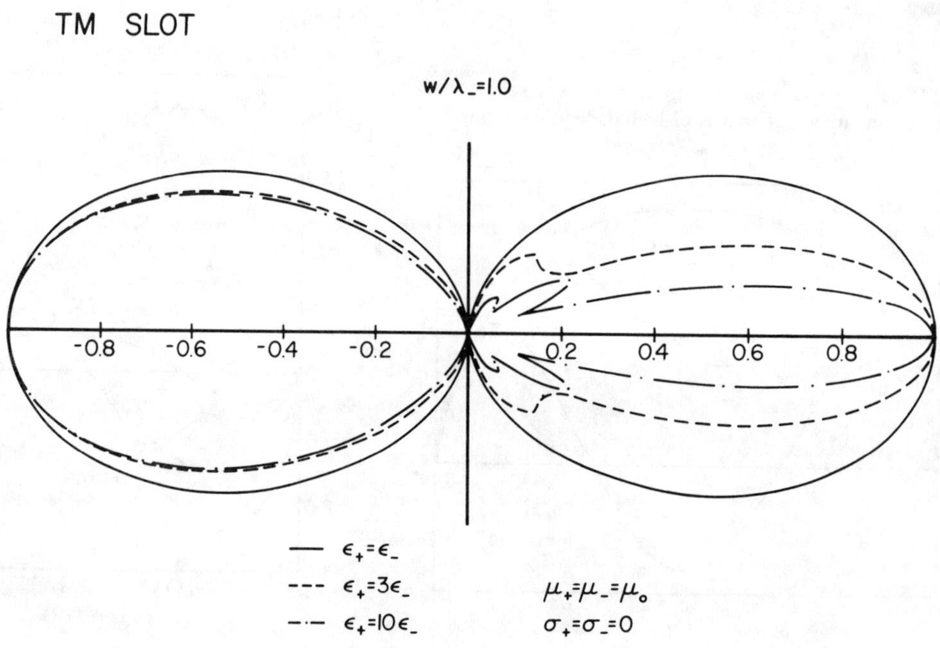

Fig. 13. Far electric field due to presence of TM-excited slot in screen for different right half-space permittivities.

incident TM excitation ($H_x^{i+} = 0$) with different values of ϵ_+.

For a large aperture in a screen separating half spaces of the same media, it is well known that the field in the aperture at locations remote from the aperture/screen edge(s) is essentially equal to the incident field evaluated at such points. Of course, *large* in the present context means that the aperture dimensions are large relative to the single wavelength ($\lambda_+ = \lambda_-$) in the same-media problem. In the two-media problem, however, with different wavelengths in the two half spaces, the size of an aperture must be judged on the basis of both λ_+ and λ_-. In either the same- or the two-media aperture/screen problem, though, the deviation of the total field from the incident field in the aperture is dependent upon the distance in terms of wavelengths that the constituent contributions to the scattered field from locations on the screen must travel to the point in question. In the absence of the conducting screen entirely and with a normally incident plane wave in the left half space, the ratio of the total electric field to the incident electric field at the interface of two dielectric half spaces is readily determined to be $2/[1 + (\epsilon_+/\epsilon_-)^{1/2}]$. In the slot with the screen present at this interface, the ratio of the total electric field $\mathbf{E}_t^a$ ($=\mathbf{M}_s \times \hat{z}$) to the incident electric field $\mathbf{E}^{i-}$ ($=\eta_- \mathbf{H}^{i-} \times \hat{z}$) is $E_{t_x}^a / E_x^{i-} = M_{s_y}/(\eta_- H_y^{i-})$ for the TE case and is $E_{t_y}^a / E_y^{i-} = M_{s_x}/(\eta_- H_x^{i-})$ for the TM case. From an inspection of the data displayed in the figures, one finds the field ratios in the neighborhood of $x = 0$ to be nearly equal, particularly so when the dielectric contrast is high ($\epsilon_+/\epsilon_- = 10, 80$). These observations lead one to conclude that, with sufficiently high contrast, the electric field near the center of a moderately wide slot excited by a normally incident plane wave is approximately equal to the value one would calculate there for the two-media problem with the screen removed.

The final two Figures (12 and 13) show the far fields on the two sides of the screen. To a minor degree, the right half-space patterns differ from those of the left half space because of the different slot distributions for different dielectric contrasts, but the major differences are due to the different electrical widths (w/λ_- or w/λ_+) of the slot in the two media.

In addition to those presented here, data are available for lossy media and it is pointed out that the analysis of this paper can be extended directly to enable one to solve the problem of multiple, parallel slots in a screen separating different media.

REFERENCES

Barakat, R. (1963), Diffraction of plane waves by a slit between two different media, *J. Opt. Soc. Amer.*, *53*, 1231–1243.

Harrington, R. F. (1968), *Field Computation by Moment Methods*, Macmillan, New York.

Houlberg, K. (1967), Diffraction by a narrow slit in the interface between two media, *Can. J. Phys.*, *45*, 57–81.

Lin, J. L., Curtis, W. L., and M. C. Vincent (1974), On the field distribution of an aperture, *IEEE Trans. Antennas Propagat.*, *AP-22*, 608–610.

Mittra, R. (Ed.) (1973), *Computer Techniques for Electromagnetics*, Pergamon, New York.

Mittra, R., Y. Rahmat-Samii, D. V. Jamnejad, and W. A. Davis (1973), A new look at the thin-plate scattering problem, *Radio Sci.*, *8*, 869–875.

Thomas, D. P. (1969), Electromagnetic diffraction by a circular aperture in a plane screen between different media, *Can. J. Phys.*, *47*, 921–930.

Wilton, D. R., and C. M. Butler (1976), Efficient numerical techniques for solving Pocklington's equation and their relationships to other methods, *IEEE Trans. Antennas Propagat.*, *AP-24*, 83–86.

Electromagnetic Transmission through an Aperture in a Conducting Plane

by Roger F. Harrington * and Joseph R. Mautz *

The problem of electromagnetic wave transmission through an aperture in a conducting plane is treated by obtaining an operator equation for the equivalent magnetic current, and then reducing it to a matrix equation via the method of moments. Integral formulas for the matrix elements are given for apertures of arbitrary shape. These integral formulas have been evaluated for rectangular apertures using expansion and testing functions which are the product of triangles and pulses. Sample computations are given for rectangular apertures, and a computer program is available.

Elektromagnetische Transmission durch eine Öffnung in einer leitenden Ebene

Für die elektromagnetische Transmission durch eine Öffnung in einer leitenden Ebene wird eine Operatorengleichung für den äquivalenten magnetischen Strom aufgestellt und mit der Momentenmethode auf eine Matrizengleichung zurückgeführt. Die Matrizenelemente werden für Öffnungen beliebiger Form als Integrale angegeben. Für rechteckige Öffnungen werden diese Integrale durch Entwicklung nach Probefunktionen ausgewertet, welche aus Produkten von Dreiecks- und Impulsverteilungen bestehen. Berechnungsbeispiele werden für rechteckige Öffnungen durchgeführt, dafür gibt es auch ein Rechnerprogramm.

1. Introduction

The problem of electromagnetic transmission through an aperture in a perfectly conducting plane of infinite extent has received considerable attention in the literature [1]—[4]. Most of the papers use Green's theorem to reduce the problem to an integral equation for the unknown tangential electric field in the aperture. We use a more direct, but mathematically equivalent, approach using magnetic currents and the equivalence theorem [5]—[7]. The result is an integral equation for the unknown equivalent magnetic current over the aperture region. This integral equation is reduced to a matrix equation via the method of moments. An interpretation of the solution in terms of generalized network parameters is given in reference [7]. The general formulas are specialized to the case of a rectangular aperture in a conducting plane excited by an incident plane wave. A computer program has been written for this case [12], and some representative computations are given in this paper.

2. General Formulation

Consider a conducting plane covering the $z=0$ plane except for an aperture, as shown in Fig. 1. The excitation is taken to be an incident wave in the region $z<0$. This incident wave can be either a plane wave or some more complicated field from impressed sources. The region $z>0$ is assumed to be source free. The more genreal case of sources in both regions can be treated as the superposition of two problems, one with sources $z<0$ plus one with sources $z>0$.

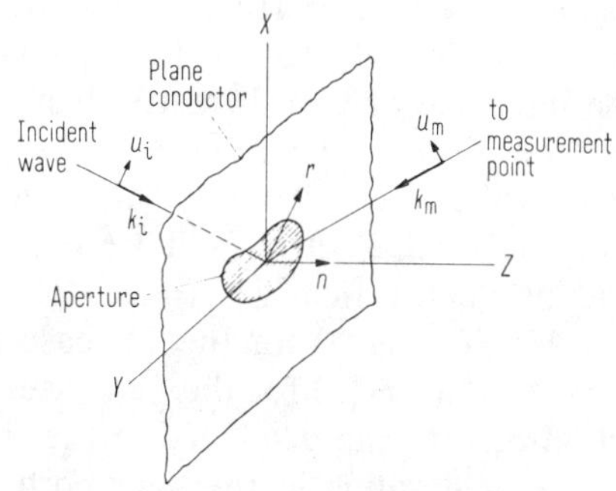

Fig. 1. Aperture in a plane conductor.

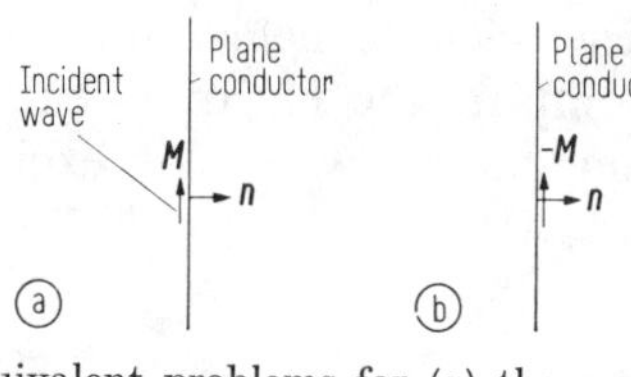

Fig. 2. Equivalent problems for (a) the region $z<0$, and (b) the region $z>0$.

The equivalence principle [5] is applied to obtain separate representations for the field in the regions $z<0$ and $z>0$, as shown in Fig. 2. In the region $z<0$ the field is that produced by the incident wave plus the equivalent magnetic current (**n** denotes a unit z-directed vector)

$$\boldsymbol{M} = \boldsymbol{n} \times \boldsymbol{E} \tag{1}$$

over the aperture region with the aperture covered by an electric conductor. In the region $z>0$ the field is produced by the equivalent magnetic current

---

* Prof. R. F. Harrington, Dr. J. R. Mautz, Department of Electrical and Computer Engineering, 111 Link Hall, Syracuse University, Syracuse, N.Y. 13210, USA.

$-\boldsymbol{M}$ over the aperture region with the aperture covered by an electric conductor. The equivalent currents $+\boldsymbol{M}$ in region $z<0$ and $-\boldsymbol{M}$ in region $z>0$ ensure that the tangential component of electric field is continuous across the aperture. The remaining boundary condition to be applied is the continuity of the tangential component of magnetic field across the aperture.

The tangential component of magnetic field in the region $z<0$ over the aperture will be denoted by $\boldsymbol{H}_t^-$. It is the sum of that due to the impressed sources $\boldsymbol{H}_t^i$ plus that due to the equivalent source $\boldsymbol{H}_t^-(\boldsymbol{M})$, that is

$$\boldsymbol{H}_t^- = \boldsymbol{H}_t^i + \boldsymbol{H}_t^-(\boldsymbol{M}) . \quad (2)$$

Note that $\boldsymbol{H}_t^i$ and $\boldsymbol{H}_t^-(\boldsymbol{M})$ are both computed with a conductor covering the aperture, that is, a complete electric conductor covering the $z=0$ plane. A similar equation holds for the region $z>0$ except that there are no impressed sources and the equivalent source is $-\boldsymbol{M}$. Hence, the tangential component of magnetic field in the region $z>0$ over the aperture, denoted $\boldsymbol{H}_t^+$, is

$$\boldsymbol{H}_t^+ = \boldsymbol{H}_t^+(-\boldsymbol{M}) = -\boldsymbol{H}_t^+(\boldsymbol{M}) . \quad (3)$$

Here $\boldsymbol{H}_t^+(\boldsymbol{M})$ is computed with a conductor covering the aperture, that is, a complete conductor covering the $z=0$ plane. The true solution is obtained when $\boldsymbol{H}_t^-$ of eq. (2) equals $\boldsymbol{H}_t^+$ of eq. (3), or

$$\boldsymbol{H}_t^-(\boldsymbol{M}) + \boldsymbol{H}_t^+(\boldsymbol{M}) = -\boldsymbol{H}_t^i . \quad (4)$$

This is the basic operator equation for determining the equivalent magnetic current $\boldsymbol{M}$, which is related to the tangential component of electric field in the aperture by eq. (1).

We use the method of moments [8] to obtain an approximate solution to eq. (4). Define a set of expansion functions $\{\boldsymbol{M}_n, n=1,2,\ldots,N\}$ and let

$$\boldsymbol{M} = \sum_n V_n \boldsymbol{M}_n \quad (5)$$

where the coefficients V_n are to be determined. Define a symmetric product

$$\langle A, B\rangle = \iint_{\text{apert}} \boldsymbol{A}\cdot\boldsymbol{B}\,\mathrm{d}s \quad (6)$$

and a set of testing functions $\{\boldsymbol{W}_n, n=1,2,\ldots,N\}$. Substitute eq. (5) into eq. (4), and take the symmetric product of the resulting equation with each testing function $\boldsymbol{W}_m$. The result is

$$\sum_n V_n\langle W_m, H_t^-(M_n)\rangle + \sum_n V_n\langle W_m, H_t^+(M_n)\rangle = -\langle W_m, H_t^i\rangle , \quad (7)$$

$m=1,2,\ldots,N$. Solution to this set of linear equations determines the coefficients V_n and the magnetic current $\boldsymbol{M}$ according to eq. (5). Once $\boldsymbol{M}$ is known, the fields and other parameters may be computed in the usual way.

The set of equations (7) can be put into matrix notation as follows. Define an admittance matrix for the region $z<0$ as

$$(Y^-) = (\langle -W_m, H_t^-(M_n)\rangle)_{N\times N} \quad (8)$$

and an admittance matrix for the region $z>0$ as

$$(Y^+) = (\langle -W_m, H_t^+(M_n)\rangle)_{N\times N} . \quad (9)$$

The minus signs are placed in eqs. (8) and (9) on the basis of power considerations. Define a source vector as

$$\vec{I}^{\,i} = (\langle W_m, H_t^i\rangle)_{N\times 1} \quad (10)$$

and a coefficient vector

$$\vec{V} = (V_n)_{N\times 1} . \quad (11)$$

Now the matrix equation equivalent to equations (7) is

$$(Y^- + Y^+)\,\vec{V} = \vec{I}^{\,i} . \quad (12)$$

The two half-space regions $z<0$ and $z>0$ are identical, and hence their admittance matrices are the same. Therefore, we let

$$(Y^- + Y^+) = 2(Y^{\text{hs}}) \quad (13)$$

where (Y^{hs}) denotes the admittance matrix for a half-space region, say $z>0$, bounded by an electric conductor over the plane $z=0$. Now eq. (12) can be written as

$$2(Y^{\text{hs}})\,\vec{V} = \vec{I}^{\,i} . \quad (14)$$

The resultant voltage vector

$$\vec{V} = \tfrac{1}{2}(Y^{\text{hs}})^{-1}\,\vec{I}^{\,i} \quad (15)$$

gives the vector of coefficients (11), and the equivalent magnetic current $\boldsymbol{M}$ according to eq. (5).

Now linear functionals of $\boldsymbol{M}$, such as components of fields or of currents on the conductor, can be computed in the usual way. For this it is convenient to introduce an adjoint field and compute the symmetric product of $\boldsymbol{M}$ with this adjoint field. We illustrate the procedure by the following special case.

Suppose we wish to compute a component H_m of magnetic field at a point $\boldsymbol{r}_m$ in the region $z>0$ of Fig. 1. It is known that this component can be obtained by placing a magnetic dipole $K\boldsymbol{l}_m$ at $\boldsymbol{r}_m$ and applying the reciprocity theorem to its field (the adjoint field) and the original field [5]. This application of the reciprocity theorem yields

$$H_m K l_m = -\iint_{\text{apert}} \boldsymbol{M}\cdot\boldsymbol{H}^{\text{m}}\,\mathrm{d}s . \quad (16)$$

Here $\boldsymbol{H}^{\text{m}}$ is the magnetic field from $K\boldsymbol{l}_m$ in the presence of a complete electric conductor covering the $z=0$ plane, and H_m is the component in the direction of $K\boldsymbol{l}_m$ of $\boldsymbol{H}$ at $\boldsymbol{r}_m$ due to $-\boldsymbol{M}$ in the presence of a complete electric conductor covering the $z=0$ plane. To evaluate eq. (16), substitute for $\boldsymbol{M}$ from eq. (5) and obtain

$$H_m K l_m = \sum_n V_n\langle -M_n, H^{\text{m}}\rangle . \quad (17)$$

This can be written in matrix form as

$$H_m K l_m = \tilde{I}^{\text{m}}\,\vec{V} \quad (18)$$

where $\tilde{I}^{\text{m}}$ denotes the transpose of a "measurement" vector,

$$\vec{I}^{\,\text{m}} = (\langle -M_n, H^{\text{m}}\rangle)_{N\times 1} . \quad (19)$$

Note that the elements of $\vec{I}^{\,\text{m}}$ are similar in form to those of $\vec{I}^{\,i}$, given by eq. (10), except that $-\boldsymbol{M}_n$

replaces $\boldsymbol{W}_m$. Now substitute eq. (15) into eq. (18) to obtain

$$H_m K l_m = \tfrac{1}{2} \tilde{I}^{\mathrm{m}} (Y^{\mathrm{hs}})^{-1} \vec{I}^{\mathrm{i}} . \tag{20}$$

If the magnetic dipole is of unit moment, then eq. (20) gives H_m at $\boldsymbol{r}_m$ directly.

Another parameter of interest is the power transmitted through the aperture, which is a quadratic functional of $\boldsymbol{M}$. The complex power P_{t} transmitted through the aperture can be calculated from

$$P_{\mathrm{t}} = \iint_{\text{apert}} \boldsymbol{E} \times \boldsymbol{H}^* \cdot \boldsymbol{n} \, \mathrm{d}s \tag{21}$$

where the asterisk denotes complex conjugate. Substituting from eq. (1), we have

$$P_{\mathrm{t}} = \iint_{\text{apert}} \boldsymbol{M} \cdot \boldsymbol{H}^* \, \mathrm{d}s . \tag{22}$$

This involves only the tangential component of $\boldsymbol{H}$ over the aperture, which in the region $z > 0$ we denoted by $\boldsymbol{H}_{\mathrm{t}}^{+}(-\boldsymbol{M})$. For $\boldsymbol{M}$, use the linear combination (5) and obtain

$$\boldsymbol{H}_{\mathrm{t}}^{+}(-\boldsymbol{M}) = -\sum_n V_n \boldsymbol{H}_{\mathrm{t}}^{+}(\boldsymbol{M}_n) . \tag{23}$$

Substitute eq. (23) for $\boldsymbol{H}$ and eq. (5) for $\boldsymbol{M}$ into eq. (22), and obtain

$$P_{\mathrm{t}} = -\sum_m \sum_n V_m V_n^* \iint_{\text{apert}} \boldsymbol{M}_m \cdot \boldsymbol{H}_{\mathrm{t}}^{+*}(\boldsymbol{M}_n) \, \mathrm{d}s . \tag{24}$$

If the $\boldsymbol{M}_m$ are real, the conjugate operations can be taken outside the integrals. Moreover, if $\boldsymbol{M}_m = \boldsymbol{W}_m$ (Galerkin's method), then the negatives of the integrals in eq. (24) are $Y_{mn}^{+*} = Y_{mn}^{\mathrm{hs}\,*}$ as defined by eq. (9), and eq. (24) reduces to

$$P_{\mathrm{t}} = \sum_m \sum_n V_m V_n^* Y_{mn}^{\mathrm{hs}\,*} . \tag{25}$$

In matrix form, this can be written as

$$P_{\mathrm{t}} = \tilde{V} (Y^{\mathrm{hs}})^* \vec{V}^* . \tag{26}$$

Note that this is the usual form for the power transmitted into an N-port network having an admittance matrix (Y^{hs}).

3. Evaluation of the Matrix Parameters

Again consider the aperture in a conducting plane as shown in Fig. 1. When the aperture is covered by a conductor, the $z = 0$ plane is a complete conducting plane, and image theory applies. The magnetic current expansion functions are on the surface of the $z = 0$ plane. Their images are equal to them and are also on the $z = 0$ plane [5]. The result is that (Y^{hs}) is the admittance matrix obtained using expansion functions $2\boldsymbol{M}_n$ radiating into free space everywhere. This problem is dual to that for the impedance matrix of a plane conductor, a problem considered recently in the literature [9].

The original excitation of the aperture is by impressed sources $\boldsymbol{J}^{\mathrm{i}}$, $\boldsymbol{M}^{\mathrm{i}}$ in the region $z < 0$. The impressed field $\boldsymbol{H}_{\mathrm{t}}^{\mathrm{i}}$ used in the operator equation (4) is the tangential magnetic field due to $\boldsymbol{J}^{\mathrm{i}}$, $\boldsymbol{M}^{\mathrm{i}}$ with the aperture covered by a conductor (Fig. 2a). In this case the $z = 0$ plane is a complete conductor, and image theory again applies. The result is that the tangential component of $\boldsymbol{H}$ over the $z = 0$ plane when it is covered by a conductor is just twice what it is for the same sources in free space. Hence,

$$\boldsymbol{H}_{\mathrm{t}}^{\mathrm{i}} = 2\boldsymbol{H}_{\mathrm{t}}^{\mathrm{i0}} \tag{27}$$

where $\boldsymbol{H}_{\mathrm{t}}^{\mathrm{i0}}$ is the tangential component of the magnetic field over the aperture due to the sources $\boldsymbol{J}^{\mathrm{i}}$, $\boldsymbol{M}^{\mathrm{i}}$ in free space. The components of the excitation vector $\vec{I}^{\mathrm{i}}$ defined by eq. (10) are now

$$I_m^{\mathrm{i}} = 2 \iint_{\text{apert}} \boldsymbol{W}_m \cdot \boldsymbol{H}_{\mathrm{t}}^{\mathrm{i0}} \, \mathrm{d}s \tag{28}$$

where $\boldsymbol{W}_m$ is the mth testing function.

A case of special interest is that of plane wave excitation. A unit plane wave is given by

$$\boldsymbol{H}^{\mathrm{i0}} = \boldsymbol{u}_{\mathrm{i}} \exp(-\mathrm{j}\, \boldsymbol{k}_i \cdot \boldsymbol{r}) \tag{29}$$

where $\boldsymbol{u}_i$ is a unit vector specifying the direction of $\boldsymbol{H}^{\mathrm{i0}}$, $\boldsymbol{k}_i$ is the propagation vector of magnitude $2\pi/\lambda$ and pointing in the direction of propagation, and $\boldsymbol{r}$ is the radius vector to an arbitrary field point. These vectors are shown in Fig. 1. The components (28) of the plane-wave excitation vector are then

$$P_m^{\mathrm{i}} = 2 \iint_{\text{apert}} \boldsymbol{W}_m \cdot \boldsymbol{u}_i \exp(-\mathrm{j}\, \boldsymbol{k}_i \cdot \boldsymbol{r}) \, \mathrm{d}s . \tag{30}$$

The symbol $\vec{P}^{\mathrm{i}}$ has been used for this particular vector to distinguish it from the more general excitation vector (28).

Similar simplifications apply to the adjoint (measurement) problem. For the evaluation of a component of magnetic field at a point $\boldsymbol{r}_m$, a magnetic dipole $K\boldsymbol{l}_m$ is placed at the measurement point $\boldsymbol{r}_m$. This radiates in the presence of a complete conductor over the $z = 0$ plane, and hence, analogous to eq. (27), we have

$$\boldsymbol{H}_{\mathrm{t}}^{\mathrm{m}} = 2\boldsymbol{H}_{\mathrm{t}}^{\mathrm{m0}} . \tag{31}$$

Here $\boldsymbol{H}_{\mathrm{t}}^{\mathrm{m}}$ denotes the tangential component of $\boldsymbol{H}$ over the aperture from $K\boldsymbol{l}_m$ when the $z = 0$ plane is covered by a conductor, and $\boldsymbol{H}_{\mathrm{t}}^{\mathrm{m0}}$ denotes that from $K\boldsymbol{l}_m$ when it radiates into free space. The components of the measurement vector $\vec{I}^{\mathrm{m}}$ defined by eq. (19) are now

$$I_n^{\mathrm{m}} = -2 \iint_{\text{apert}} \boldsymbol{M}_n \cdot \boldsymbol{H}_{\mathrm{t}}^{\mathrm{m0}} \, \mathrm{d}s \tag{32}$$

where $\boldsymbol{M}_n$ is the nth expansion function.

A case of special interest is that of far-field measurement. This is obtained by a procedure dual to that used for radiation and scattering from conducting wires [10]. To obtain a component of $\boldsymbol{H}$ on the radiation sphere, we take a source $K\boldsymbol{l}_m$ perpendicular to $\boldsymbol{r}_m$ and let $r_m \to \infty$. At the same time we adjust $K\boldsymbol{l}_m$ so that it produces a unit plane wave in the vicinity of the origin. The required dipole moment is given by

$$\frac{1}{K l_m} = \frac{-\mathrm{j}\omega\varepsilon}{4\pi r_m} \exp(-\mathrm{j} k r_m) \tag{33}$$

and the plane-wave field it produces in the vicinity of the origin is

$$\boldsymbol{H}^{\mathrm{m0}} = \boldsymbol{u}_m \exp(-\mathrm{j}\, \boldsymbol{k}_m \cdot \boldsymbol{r}) . \tag{34}$$

Here $\boldsymbol{u}_m$ is a unit vector in the direction of $\boldsymbol{H}^{\mathrm{m0}}$,

$\boldsymbol{k}_m$ is the propagation vector, and $\boldsymbol{r}$ is the radius vector to an arbitrary field point. Again these vectors are shown in Fig. 1. The components (32) of the far-field measurement vector are then

$$P_n^{\mathrm{m}} = -2 \iint_{\text{apert}} \boldsymbol{M}_n \cdot \boldsymbol{u}_m \exp(-\mathrm{j}\, \boldsymbol{k}_m \cdot \boldsymbol{r})\, \mathrm{d}s\,. \tag{35}$$

The symbol $\vec{P}^{\mathrm{m}}$ is used for this particular measurement vector to distinguish it from the more general measurement vector (32). The far-zone magnetic field is now given by eq. (20) with $K\boldsymbol{l}_m$ given by eq. (33), $\vec{I}^{\mathrm{m}} = \vec{P}^{\mathrm{m}}$, and $\vec{I}^{\mathrm{i}} = \vec{P}^{\mathrm{i}}$. Hence

$$H_m = \frac{-\mathrm{j}\,\omega\,\varepsilon}{8\pi r_m} \exp(-\mathrm{j}\,k\,r_m)\, \tilde{P}^{\mathrm{m}} (Y^{\mathrm{hs}})^{-1} \vec{P}^{\mathrm{i}}\,. \tag{36}$$

The usual two radiation components H_θ and H_φ are obtained by orienting $K\boldsymbol{l}_m$ in the θ- and φ-directions, respectively.

A parameter sometimes used to express the transmission characteristics of an aperture is the transmission cross section τ. It is defined as that area for which the incident wave contains sufficient power to produce the radiation field H_m by omnidirectional radiation over half space. For unit incident magnetic field, this is

$$\tau = 2\pi r_m^2 |H_m|^2\,. \tag{37}$$

Substituting from eq. (36), we obtain

$$\tau = \frac{\omega^2\,\varepsilon^2}{32\,\pi} |\tilde{P}^{\mathrm{m}} (Y^{\mathrm{hs}})^{-1} \vec{P}^{\mathrm{i}}|^2\,. \tag{38}$$

Note that τ depends upon the polarization and direction of the incident wave (via $\vec{P}^{\mathrm{i}}$), and upon the polarization measured and direction to the measurement point (via $\tilde{P}^{\mathrm{m}}$).

Another parameter used to express the transmission characteristics of an aperture is the transmission coefficient T, defined as

$$T = P_{\mathrm{trans}}/P_{\mathrm{inc}} \tag{39}$$

where P_{trans} is the time-average power transmitted by the aperture, and P_{inc} is the free space power incident on the aperture. The incident power is

$$P_{\mathrm{inc}} = \eta\, S \cos\theta_{\mathrm{inc}} \tag{40}$$

where $\eta = \sqrt{\mu/\varepsilon}$ is the intrinsic impedance of free space, S is the aperture area, and θ_{inc} is the angle between $\boldsymbol{k}_i$ and $\boldsymbol{n}$. The transmitted power is

$$P_{\mathrm{trans}} = \mathrm{Re}(P_{\mathrm{t}}) \tag{41}$$

where $\mathrm{Re}(P_{\mathrm{t}})$ denotes the real part of P_{t}, given by eq. (26). Hence

$$T = \frac{1}{\eta\, S \cos\theta_{\mathrm{inc}}} \mathrm{Re}\{\tilde{V} (Y^{\mathrm{hs}})^* \vec{V}^*\}\,. \tag{42}$$

Note that T depends on both the direction of incidence and on the polarization of the incident wave.

Finally, because of symmetry about the $z=0$ plane, the difference field $\boldsymbol{H} - \boldsymbol{H}^{\mathrm{i}}$ which exists in the region $z<0$ is simply related to the transmitted field which exists in the region $z>0$. The difference field in the region $z<0$ is produced by an equivalent current $\boldsymbol{M}$ on a plane conductor over the $z=0$ plane. By image theory, it is also the field produced in the region $z<0$ by the source $2\boldsymbol{M}$ in free space. As noted earlier, the transmitted field in the region $z>0$ is produced by the source $-2\boldsymbol{M}$ in free space. Hence, the difference field in the region $z<0$ and the negative of the transmitted field in the region $z>0$ are both produced by the same magnetic current $2\boldsymbol{M}$ radiating in free space.

4. Rectangular Apertures

Consider now a rectangular aperture in a plane conductor, as shown in Fig. 3. The perfectly conducting plate covers the entire $z=0$ plane except for the aperture, which is rectangular in shape with side lengths $L_x\Delta x$ and $L_y\Delta y$ in the x- and y-directions, respectively. The excitation of the aperture is a uniform plane wave incident from the region $z<0$. The field to be computed is the far zone magnetic field in the region $z>0$, at the angles θ, φ.

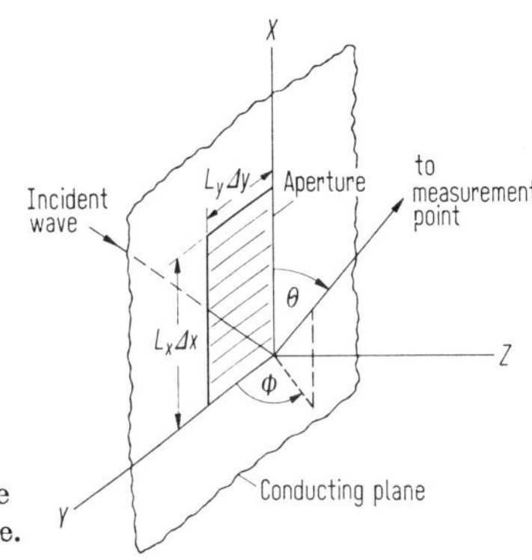

Fig. 3.
Rectangular aperture in a conducting plane.

a) Admittance matrix

According to eq. (9), the elements of the admittance matrix $(Y^- + Y^+) = (2\,Y^{\mathrm{hs}})$ are given by

$$Y_{ij} = -4\langle W_i, H(M_j)\rangle \tag{43}$$

where $\boldsymbol{H}(\boldsymbol{M}_j)$ is the magnetic field produced by $\boldsymbol{M}_j$ radiating in free space. The magnetic field $\boldsymbol{H}(\boldsymbol{M}_j)$ can be expressed in terms of an electric vector potential $\boldsymbol{F}$ and magnetic scalar potential φ as [11]

$$\boldsymbol{H}(\boldsymbol{M}_j) = -\mathrm{j}\,\omega\, \boldsymbol{F}_j - \boldsymbol{\nabla}\varphi_j \tag{44}$$

where

$$\boldsymbol{F}_j = \frac{\varepsilon}{4\pi} \iint_{\text{apert}} \boldsymbol{M}_j \frac{\exp(-\mathrm{j}\,k|\boldsymbol{r}-\boldsymbol{r}'|)}{|\boldsymbol{r}-\boldsymbol{r}'|}\, \mathrm{d}s\,, \tag{45}$$

$$\varphi_j = \frac{1}{4\pi\mu} \iint_{\text{apert}} \varrho_j \frac{\exp(-\mathrm{j}\,k|\boldsymbol{r}-\boldsymbol{r}'|)}{|\boldsymbol{r}-\boldsymbol{r}'|}\, \mathrm{d}s\,, \tag{46}$$

$$\varrho_j = \frac{\boldsymbol{\nabla}\cdot\boldsymbol{M}_j}{-\mathrm{j}\,\omega} \tag{47}$$

where $\boldsymbol{r}$ and $\boldsymbol{r}'$ are respectively the vectors to the field and source points in the aperture, ω is the angular frequency, ε is the capacitivity of free space, μ is the permeability of free space, and $k = \omega\sqrt{\mu\,\varepsilon}$ is the propagation constant in free space. Substituting eqs. (6) and (44) into eq. (43), we obtain

$$Y_{ij} = 4 \iint_{\text{apert}} \boldsymbol{W}_i \cdot (\mathrm{j}\,\omega\, \boldsymbol{F}_j + \boldsymbol{\nabla}\varphi_j)\, \mathrm{d}s\,. \tag{48}$$

Because of the identity

$$0=\iint_{\text{apert}}\nabla\cdot(\varphi_j \boldsymbol{W}_i)\,\mathrm{d}s = \iint_{\text{apert}}\boldsymbol{W}_i\cdot\nabla\varphi_j\,\mathrm{d}s+\iint_{\text{apert}}\varphi_j\nabla\cdot\boldsymbol{W}_i\,\mathrm{d}s \tag{49}$$

eq. (48) becomes

$$Y_{ij}=4\,\mathrm{j}\,\omega\iint_{\text{apert}}(\boldsymbol{F}_j\cdot\boldsymbol{W}_i+\varphi_j\varrho_i)\,\mathrm{d}s \tag{50}$$

where

$$\varrho_i=\frac{\nabla\cdot\boldsymbol{W}_i}{-\mathrm{j}\,\omega}. \tag{51}$$

We choose the set of testing functions $\boldsymbol{W}_i$ equal to the set of expansion functions $\boldsymbol{M}_i$. The rectangular aperture $0\leqq x\leqq L_x\Delta x$, $0\leqq y\leqq L_y\Delta y$ where L_x and L_y are integers is divided into rectangular subareas of length Δx in x and Δy in y. The set $\boldsymbol{M}_i$ of expansion functions is split into a set $\boldsymbol{M}_i^x$ of x directed magnetic currents and a set $\boldsymbol{M}_i^y$ of y directed magnetic currents defined by

$$\boldsymbol{M}^x_{p+(q-1)(L_x-1)}=\hat{\boldsymbol{x}}\,T^x_p(x)\,P^y_q(y)\begin{cases}p=1,2,\ldots,L_x-1\\ q=1,2,\ldots,L_y,\end{cases} \tag{52}$$

$$\boldsymbol{M}^y_{p+(q-1)L_x}=\hat{\boldsymbol{y}}\,T^y_q(y)\,P^x_p(x)\begin{cases}p=1,2,\ldots,L_x\\ q=1,2,\ldots,L_y-1\end{cases} \tag{53}$$

where $T^x_p(x)$ are triangle functions defined by

$$T^x_p(x)=\begin{cases}\dfrac{x-(p-1)\,\Delta x}{\Delta x} & \text{for}\quad (p-1)\,\Delta x\leqq x\leqq p\,\Delta x\\[2mm] \dfrac{(p+1)\,\Delta x-x}{\Delta x} & \text{for}\quad p\,\Delta x\leqq x\leqq (p+1)\,\Delta x\\[2mm] 0 & \text{for}\quad |x-p\,\Delta x|\geqq\Delta x\end{cases} \tag{54}$$

and $T^y_q(y)$ are similarly defined. $P^x_p(x)$ are pulse functions defined by

$$P^x_p(x)=\begin{cases}1 & \text{for}\quad (p-1)\,\Delta x\leqq x<p\,\Delta x\\ 0 & \text{for}\quad \text{all other } x\end{cases} \tag{55}$$

and $P^y_q(y)$ are similarly defined. The magnetic charge sheets associated with $\boldsymbol{M}^x_j$ and $\boldsymbol{M}^y_j$ are obtained from eq. (47), the equation of continuity. Introduction of the two types of expansion functions $\boldsymbol{M}^x_j$ and $\boldsymbol{M}^y_j$ and the two types of testing functions $\boldsymbol{M}^x_i$ and $\boldsymbol{M}^y_i$ into eq. (50) gives rise to four Y submatrices defined by

$$Y^{uv}_{ij}=4\,\mathrm{j}\,\omega\iint_{\text{apert}}(\boldsymbol{F}^v_j\cdot\boldsymbol{M}^u_i+\varphi^v_j\varrho^u_i)\,\mathrm{d}s \tag{56}$$

where u is either x or y and v is either x or y. In eq. (56), $\boldsymbol{F}^v_j$ and φ^v_j are the electric vector and magnetic scalar potentials due to $\boldsymbol{M}^v_j$.

The integrations over the "field" magnetic current and charge $\boldsymbol{M}^u_i$ and ϱ^u_i explicit in eq. (56) are approximated by sampling the integrands at the center of each pulse function. The exponential functions in eqs. (45) and (46) are approximated by the following four-term approximation

$$\mathrm{e}^{-\mathrm{j}kr}\approx\mathrm{e}^{-\mathrm{j}kr_0}\times\left[1-\mathrm{j}\,k(r-r_0)-\frac{k^2}{2}(r-r_0)^2+\frac{\mathrm{j}\,k^3}{6}(r-r_0)^3\right] \tag{57}$$

where

$$r=\sqrt{x^2+y^2}\,,\qquad r_0=\sqrt{(s\,\Delta x)^2+(t\,\Delta y)^2} \tag{58, 59}$$

where s is the difference between the subscripts p of the "source" and "field" pulses. Similarly, t is defined such that $t\Delta y$ is the difference between the y coordinates of the centers of the "source" and "field" pulses $P^y_q(y)$. The resultant integrals can then be evaluated in terms of known integrals. The details can be found in [12].

b) *Plane wave excitation and measurement vectors*

The plane wave excitation vector $\vec{P}^{\mathrm{i}}$ of eq. (30) and the plane wave measurement vector $\vec{P}^{\mathrm{m}}$ of eq. (35) are of the same form except for a minus sign. We therefore need to evaluate only one of them, say the measurement vector $\vec{P}^{\mathrm{m}}$. We specialize it to four principal plane patterns as

$$(P^{\mathrm{m}x}_{p+(q-1)(L_x-1)})_{\theta y}=2\,\Delta x\,\Delta y\sin\theta\left(\frac{\sin\dfrac{k\,\Delta x\cos\theta}{2}}{\dfrac{k\,\Delta x\cos\theta}{2}}\right)^2\exp(\mathrm{j}\,p\,k\,\Delta x\cos\theta)\,,\ \begin{cases}p=1,2,\ldots,L_x-1\\ q=1,2,\ldots,L_y,\end{cases} \tag{60}$$

$$(P^{\mathrm{m}y}_i)_{\theta y}=0\,,\quad i=1,2,\ldots,L_x(L_y-1)\,, \tag{61}$$

$$(P^{\mathrm{m}x}_i)_{yy}=0\,,\quad i=1,2,\ldots,(L_x-1)\,L_y\,, \tag{62}$$

$$(P^{\mathrm{m}y}_{p+(q-1)L_x})_{yy}=-\,2\,\Delta x\,\Delta y\left(\frac{\sin\dfrac{k\,\Delta x\cos\theta}{2}}{\dfrac{k\,\Delta x\cos\theta}{2}}\right)\exp[\mathrm{j}\,k(p-\tfrac{1}{2})\,\Delta x\cos\theta]\,,\ \begin{cases}p=1,2,\ldots,L_x\\ q=1,2,\ldots,L_y-1\,,\end{cases} \tag{63}$$

$$(P^{\mathrm{m}x}_i)_{\varphi x}=0\,,\quad i=1,2,\ldots,(L_x-1)\,L_y\,, \tag{64}$$

$$(P^{\mathrm{m}y}_{p+(q-1)L_x})_{\varphi x}=2\,\Delta x\,\Delta y\sin\varphi\left(\frac{\sin\dfrac{k\,\Delta y\cos\varphi}{2}}{\dfrac{k\,\Delta y\cos\varphi}{2}}\right)^2\exp(\mathrm{j}\,k\,q\,\Delta y\cos\varphi)\,,\ \begin{cases}p=1,2,\ldots,L_x\\ q=1,2,\ldots,L_y-1\,,\end{cases} \tag{65}$$

$$(P^{mx}_{p+(q-1)(L_x-1)})_{xx} = -2\,\Delta x\,\Delta y \left(\frac{\sin \dfrac{k\,\Delta y \cos\varphi}{2}}{\dfrac{k\,\Delta y\cos\varphi}{2}} \right) \exp[\mathrm{j}\,k(q-\tfrac{1}{2})\,\Delta y \cos\varphi]\,, \begin{cases} p = 1, 2, \ldots, L_x - 1 \\ q = 1, 2, \ldots, L_y\,, \end{cases} \quad (66)$$

$$(P^{my}_i)_{xx} = 0\,, \quad i = 1, 2, \ldots, L_x(L_y - 1)\,. \quad (67)$$

The superscripts x and y and the subscripts $p+(q-1)(L_x-1)$ and $p+(q-1)L_x$ on the left-hand sides of eqs. (60) to (67) are the same as those in eqs. (52) and (53). The subscript θy denotes a θ-polarized measurement in the $y=0$ plane, yy denotes a y-polarized measurement in the $y=0$ plane, φx denotes a φ-polarized measurement in the $x=0$ plane and xx denotes an x-polarized measurement in the $x=0$ plane. The coordinate system and angles θ and φ appearing on the right-hand sides of eqs. (60) to (67) are defined in Fig. 3.

5. Representative Computations

A versatile computer program has been developed using the formulas obtained from Section **4**. This program is described and listed in [12]. Some representative computations obtained with this program are given in this section.

The first example is for a narrow slot, of width $\lambda/20$ (λ denotes wavelength) and of variable length L. The far-zone quantity plotted was the transmission cross section, defined by eq. (37). We use the notation

$$\begin{aligned} \tau_{\theta y} &= 2\pi r^2 |H_\theta|^2 \quad \text{in the } y = 0 \text{ plane}, \\ \tau_{xx} &= 2\pi r^2 |H_x|^2 \quad \text{in the } x = 0 \text{ plane}. \end{aligned} \quad (68)$$

For the case being considered, the orthogonal components of $\boldsymbol{H}$ in these two planes were zero. Fig. 4

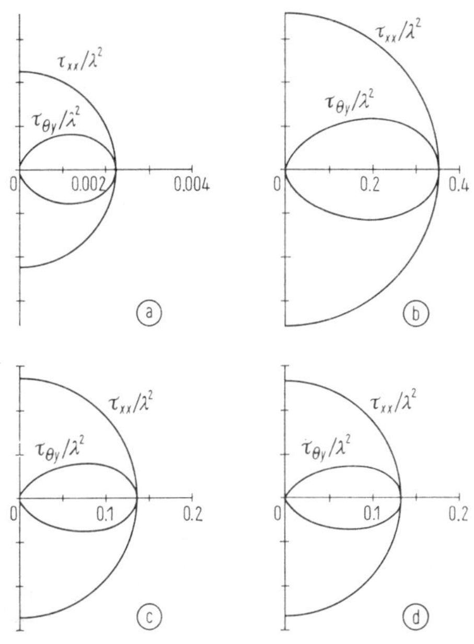

Fig. 4. Transmission cross section for slots of length L in the x-direction and width $\lambda/20$ in the y-direction; (a) $L=\lambda/4$, (b) $L=\lambda/2$, (c) $L=3\lambda/4$, (d) $L=\lambda$. Excitation is by a plane wave normally incident on the conducting plane with magnetic field in the x-direction.

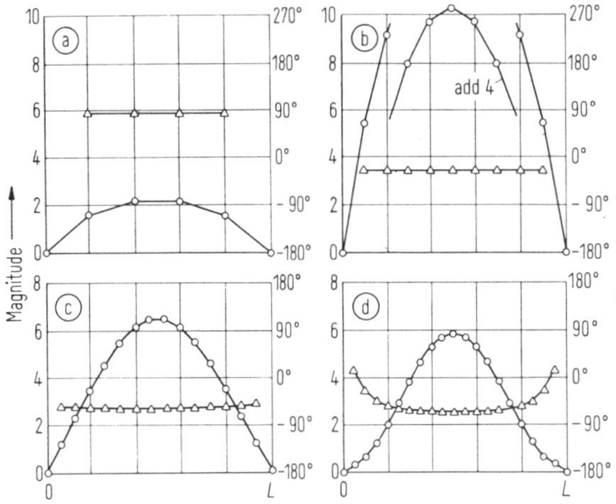

Fig. 5. Magnitude and phase of M/E^i, where M is the x-directed magnetic current and E^i is the incident electric field, for the same slots as for Fig. 4; (a) $L=\lambda/4$, (b) $L=\lambda/2$, (c) $L=3\lambda/4$, (d) $L=\lambda$. Circles denote magnitude, triangles denote phase.

shows plots of $\tau_{\theta y}$ and τ_{xx} for x-directed slots of width $\lambda/20$ and length (a) $L=\lambda/4$, (b) $L=\lambda/2$, (c) $L=3\lambda/4$, and (d) $L=\lambda$. In all cases the excitation was due to a plane wave normally incident on the conducting plane with the magnetic field in the x-direction. Note the large transmission cross section for $L=\lambda/2$, case (b), due to the slot being near resonance. The plots of τ are of the same form as scattering cross section from the complementary conducting strips, as known from Babinet's principle.

Fig. 5 shows plots of the equivalent magnetic current in the aperture region for the same slots. Since $\boldsymbol{M}=\boldsymbol{n}\times\boldsymbol{E}$, they are also plots of the tangential component of $\boldsymbol{E}$ in the slots. Again note the large value of M for the case $L=\lambda/2$, which is near resonance. Note also that, for short slots ($L\leq 3\lambda/4$), the M is almost equiphasal and closely approximated by a half sine wave. The number of data points shown equals the number of expansion functions used. In each case the slot was one subsection wide.

Next, computations were made to test the rate of convergence of the solution as the number of subsections was increased. A slot of width $\lambda/10$ and length 2λ was chosen for the study. Again the excitation is a plane wave normally incident on the conducting plane with the magnetic field in the x-direction. Fig. 6 shows plots of $\tau_{\theta y}$ and τ_{xx} for the cases (a) 39, (b) 19, (c) 9, and (d) 4 triangular expansion functions respectively. Note that the patterns (a) and (b) are essentially the same, and pattern (c) is only slightly different. They differ appreciably from (d), which results from only four expansion functions. The difference in the solutions as the number of expansion functions is decreased is better illustrated by plots of M, as shown in Fig. 7.

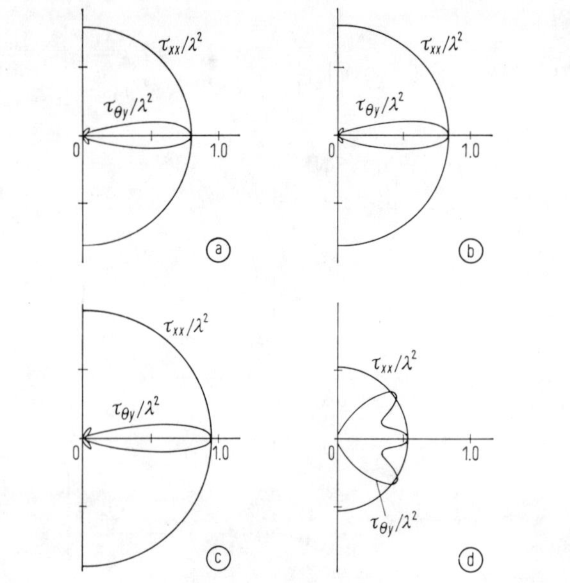

Fig. 6. Transmission cross section when the number of expansion functions is (a) 39, (b) 19, (c) 9, and (d) 4. Computations are for a slot of length 2λ in the x-direction and width $\lambda/10$ in the y-direction. Excitation is by a plane wave normally incident on the conducting plane with magnetic field in the x-direction.

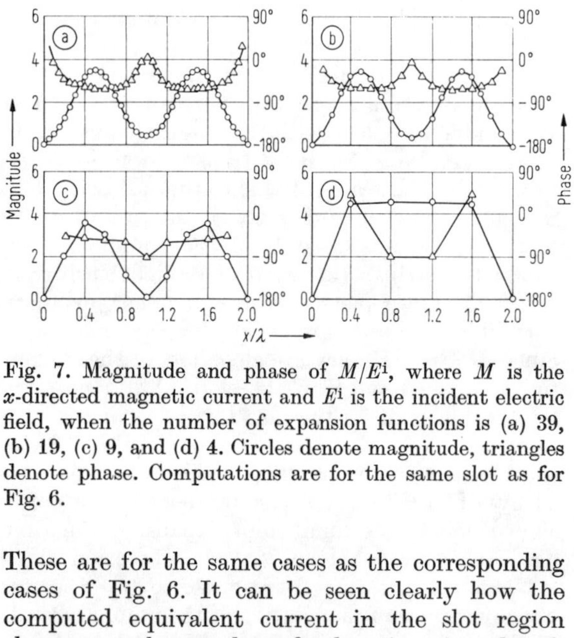

Fig. 7. Magnitude and phase of M/E^i, where M is the x-directed magnetic current and E^i is the incident electric field, when the number of expansion functions is (a) 39, (b) 19, (c) 9, and (d) 4. Circles denote magnitude, triangles denote phase. Computations are for the same slot as for Fig. 6.

These are for the same cases as the corresponding cases of Fig. 6. It can be seen clearly how the computed equivalent current in the slot region changes as the number of subsections is reduced. As a rule of thumb, for near-field quantities (such as M) one should use subareas of length $\lambda/10$ or less and for far-field quantities (such as τ) length $\lambda/5$ or less.

Other computations were made for wider apertures and by waves not normally incident on the conducting plane. Examples of these can be found in [12].

6. Discussion

The problem of electromagnetic transmission through apertures has been formulated in terms of an equivalent magnetic current over the aperture region. The operator equation for the equivalent magnetic current has been reduced to a matrix equation via the method of moments. Integral formulas for the matrix elements have been given for the case of apertures of arbitrary shape. These integral formulas have been evaluated for the special case of rectangular apertures, using expansion and testing functions which are triangles in the direction of current, and pulses in the transverse direction. This solution has been implemented by a computer program, which is available with instructions and sample input-output data [12].

The computer program is written explicitly for rectangular apertures, but the formulas are valid for any aperture composed of rectangular subsections. Other apertures, such as L-shaped, T-shaped, square O-shaped, etc., could be treated by appropriately changing the computer program. Apertures of arbitrary shape could be treated by approximating them by rectangular subsections. As with all moment solutions, the size of the apertures which can be treated depends upon the size of the matrix which can be computed and inverted. The examples indicate that the rectangular subsections should have side lengths not greater than 0.2 wavelengths for reasonable accuracy.

(Received August 27th, 1976.)

References

[1] Bethe, H. A., Theory of diffraction by small holes. Phys. Rev. **66** [1944], 163–182.

[2] Bouwkamp, C. J., Diffraction theory. Repts. Progr. in Phys. **17** [1954], 35–100.

[3] Huang, C., Kodis, R. D., and Levine, H., Diffraction by apertures. J. Appl. Phys. **26** [1955], 151–165.

[4] Bekefi, G., Diffraction of electromagnetic waves by an aperture in a large screen. J. Appl. Phys. **24** [1953], 1123–1130.

[5] Harrington, R. F., Time-harmonic electromagnetic fields. McGraw-Hill Book Co., New York 1961, pp. 106–120.

[6] Butler, C. M., Electromagnetic excitation of a wire through an aperture-perforated conducting screen. Transact. Inst. Elect. Electron. Engrs. AP-**24** [1976], 456–462.

[7] Harrington, R. F. and Mautz, J. R., A generalized network formulation for aperture problems. Transact. Inst. Elect. Electron. Engrs. AP-**24** [1976], 870–873. See also Rept. TR-75-13, ECE Dept., Syracuse University, 1975.

[8] Harrington, R. F., Field computation by moment methods. Macmillan Co., New York 1968.

[9] Wang, N. N., Richmond, J. H., and Gilreath, M. C., Sinusoidal reaction formulation for radiation and scattering from conducting surfaces. Transact. Inst. Elect. Electron. Engrs. AP-**23** [1975], 376–382.

[10] Harrington, R. F., Matrix methods for field problems. Proc. Inst. Elect. Electron. Engrs. **55** [1967], 136–149.

[11] Papas, C. H., Theory of electromagnetic wave propagation. McGraw-Hill Book Co., New York 1965, p. 23.

[12] Mautz, J. R. and Harrington, R. F., Electromagnetic transmission through a rectangular aperture in a perfectly conducting plane. Report TR-76-1, ECE Dept., Syracuse University, 1976.

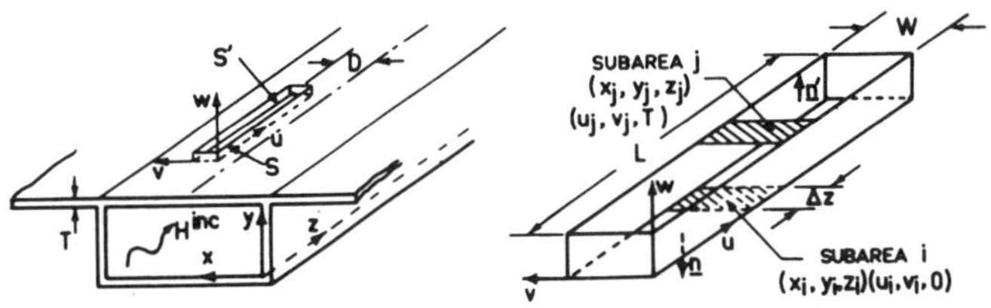

Fig. 1. Slot antenna in broad face of rectangular waveguide.

Impedance Properties of a Longitudinal Slot Antenna in the Broad Face of a Rectangular Waveguide

T. VU KHAC AND C. T. CARSON

Abstract—**The impedance properties of a waveguide longitudinal slot antenna of arbitrary length and offset in guide walls of finite thickness are investigated. These properties are obtained by treating the slot discontinuity as an electromagnetic boundary value problem and the resulting pair of coupled integral equations are solved by numerical techniques. Results are given for slots having lengths from 0.35λ to 1.7λ and are in good agreement with experimental results.**

Introduction

The radiation characteristics of a longitudinal slot antenna in the broad face of a rectangular waveguide are determined by the length of the slot and its displacement from the center of the waveguide. Although slots used in waveguide arrays are normally approximately one-half-wavelength long [1], [2], it is important that their performance at other lengths can be predicted. Longitudinal slots with length varying from 0.4λ to 1.5λ have been investigated theoretically and experimentally by Oliner [3], Das and Sanyal [4]. In their analysis, the integral equation for the tangential components of the electric field at the slot is first derived by imposing the continuity condition for the magnetic field at the slot. It is subsequently cast into variational form for impedance or admittance. Trigonometric functions are then used as trial functions in the variational formulas giving values in reasonable agreement with experimental results.

This communication presents a method for computing the impedance of a longitudinal slot of arbitrary length, with arbitrary off center displacement and takes into full consideration the thickness of the guide wall. The slot in the thick guide wall is described by a pair of coupled integral equations. They are reduced to a system of linear equations by applying the method of weighted residuals with pulses as the basis [5]–[7]. This choice of subsectional basis functions significantly simplifies the integrals which have to be evaluated and keeps to a minimum the additional effort required whenever there is a change in the slot boundary. An approximate aperture field will be obtained upon solving this system of linear equations. Although not carried out in this communication, this approximate solution may be used as the trial function for the variational expressions discussed above. Results for slots with length varying from 0.35λ to 1.7λ for some typical off-center displacements are presented and are in good agreement with experimental results.

Problem Formulation

For an incident travelling wave as shown in Fig. 1, the continuity conditions for the tangential components of the magnetic field at the surfaces S and S' will give rise to the following pair of coupled integral equations:

$$n \times H^{\mathrm{inc}}(r) = j\omega\epsilon n \times \int_S \{\bar{G}_h^{(1)}(r \mid r_0) + \bar{G}_h^{(3)}(r \mid r_0)\} \cdot \{n \times E(r_0)\}\, dS_0 + j\omega\epsilon n \times \int_{S'} \bar{G}_h^{(3)}(r \mid r_0') \cdot \{n' \times E(r_0')\}\, dS_0' \quad (1)$$

for r on S, and

$$0 = j\omega\epsilon n' \times \int_{S'} \{\bar{G}_h^{(2)}(r' \mid r_0') + \bar{G}_h^{(3)}(r' \mid r_0')\} \cdot \{n' \times E(r_0')\}\, dS_0' + j\omega\epsilon n' \times \int_S \bar{G}_h^{(3)}(r' \mid r_0) \cdot \{n \times E(r_0)\}\, dS_0 \quad (2)$$

for r' on S'.

r and r' are the position vectors on the surface S and S' where S is the inner surface of the slot and S' the outer surface. $\bar{G}_h^{(1)}, \bar{G}_h^{(3)}, \bar{G}_h^{(2)}$ are the appropriate magnetic Green's dyadics for the waveguide, cavity and half-space regions, respectively. If a finite number of pulse functions and suitable weighting functions are selected, this pair of coupled integral equations may be represented by the matrix form

$$\begin{bmatrix} [T_{ij}^1] & [C_{ij}^1] \\ [C_{ij}^2] & [T_{ij}^2] \end{bmatrix} \begin{bmatrix} [V_j^1] \\ [V_j^2] \end{bmatrix} = \begin{bmatrix} [U_i^1] \\ [0] \end{bmatrix} \quad (3)$$

where V_j^1 is the integral of E across the slot width at subarea j, U_i^1 is the average value of H_z^{inc} across the slot at subarea i. Superscripts 1,2 refer to field quantities at surface S and S', respectively. Note that implicit in (3) is the assumption that only one component of the electric field at the slot is taken into consideration. This is a reasonable approximation since L/W, the ratio of the slot length to its width, is greater than 7. The weighting functions are such that the average H_z across the slot are matched at S and S'. As H_z^{inc} is nonlinear across the slot width, this is considered to be preferable to point matching at the half-way point. With this choice of basis functions and weighting functions, the elements of the matrices in (3) may be written explicitly as

$$T_{ij}^1 = P_{ij}^1 + S_{ij}^1, \qquad T_{ij}^2 = F_{ij}^2 + S_{ij}^2$$

where P_{ij}^1 is the average H_z across the slot width at subarea i due to a rectangular sheet of magnetic current of unit strength located at subarea j in the absence of the slot and is given by

$$P_{ij}^1 = \frac{1}{W^2} \sum_{n=0}^{\infty} \sum_{m=0}^{\infty} \frac{\epsilon_{0mn}}{ab} \phi_i(n) V(m,n) \cdot \left\{ \exp\left[-\Gamma_{mn} \left| z_i - \left(z_j + \frac{\Delta z}{2} \right) \right| \right] - \exp\left[-\Gamma_{mn} \left| z_i - \left(z_j - \frac{\Delta z}{2} \right) \right| \right] \right\}$$

$$V(m,n) = -\frac{j\omega\epsilon}{\Gamma_{mn}^2} + \frac{1}{j\omega\mu}, \qquad \Gamma_{mn}^2 = \left(\frac{n\pi}{a}\right)^2 + \left(\frac{m\pi}{b}\right)^2 - k^2$$

$$\phi_i(n) = \left(\frac{a}{n\pi}\right)^2 \left[\sin\frac{n\pi(x_i + W/2)}{a} - \sin\frac{n\pi(x_i - W/2)}{a} \right]^2, \qquad \phi_i(0) = W^2$$

for $i \neq j$; $\epsilon_{0mn} = 1$ if m or $n = 0$, $\epsilon_{0mn} = 2$ otherwise. The case when both m and $n = 0$ is excluded. Note that $x_i = x_j$ since the slot is

Manuscript received December 18, 1972; revised February 13, 1973.
The authors are with the Department of Electrical Engineering, Monash University, Clayton, Vic. 3168, Australia.

longitudinal. When $i = j$, care should be exercised in evaluating P_{ii}^1 because this requires the calculation of the magnetic field inside a sheet of magnetic current. It can be shown that

$$P_{ii}^1 = -\frac{2}{W^2}\sum_{n=0}^{\infty}\sum_{m=0}^{\infty}\frac{\epsilon_{0mn}}{ab}\phi_i(n)V(m,n)\exp\left[\frac{-\Gamma_{mn}\Delta z}{2}\right]$$

$$-2\frac{j\omega\epsilon}{ab}\times\frac{1}{W^2}\left\{W^2\frac{b^2}{\pi^2}U_0(\chi_0)-\frac{W^2}{2k^2}\right.$$

$$\left.+\sum_{n=1}^{\infty}\phi_i(n)\left[\frac{a^2}{n^2\pi^2-k^2a^2}+\frac{2b^2}{\pi^2}U_n(\chi_n)\right]\right\}$$

$$U_n(\chi_n) = \frac{1}{2\chi_n^2}-\frac{\pi\cos(\pi\chi_n)}{2\chi_n\sin(\pi\chi_n)}, \qquad \text{for } n = 0,1$$

$$= -\frac{1}{2\chi_n^2}+\frac{\pi\cosh(\pi\chi_n)}{2\chi_n\sinh(\pi\chi_n)}, \qquad \text{for } n \geq 2$$

$$\chi_n = b\left|\frac{n^2}{a^2}-\frac{k^2}{\pi^2}\right|^{1/2}$$

F_{ij}^2 has similar physical interpretation but now relates to subareas i,j at surface S', i.e.,

$$F_{ij}^2 = \frac{2}{W^2}\int_{-W/2}^{W/2}\left\{\left[-j\omega\epsilon+\frac{1}{j\omega\mu}\frac{\partial^2}{\partial^2 z_i}\right]\right.$$

$$\left.\cdot\int_{-W/2}^{W/2}\int_{z_j-\Delta_z/2}^{z_j+\Delta z/2}\frac{\exp(-jkr_1)}{4\pi r_1}\,dx'\,dz'\right\}dx_i$$

$$F_{ii}^2 = -\frac{1}{W^2}\times\frac{\Delta z}{2j\omega\mu\pi}\iint_{-W/2}^{W/2}\exp(-jkr_2)\left[\frac{1}{r_2^3}+\frac{jk}{r_2^2}\right]dx'\,dx$$

$$-\frac{2j\omega\epsilon}{W^2}\iint_{-W/2}^{W/2}\int_{-\Delta_z/2}^{\Delta z/2}\frac{\exp(-jkr_3)}{4\pi r_3}\,dz'\,dx'\,dx$$

$$r_1 = \{(z_i - z')^2 + (x_i - x')^2\}^{1/2}$$

$$r_2 = \left\{(x - x')^2 + \left(\frac{\Delta z}{2}\right)^2\right\}^{1/2}$$

$$r_3 = \{(x - x')^2 + z'^2\}^{1/2}.$$

The elements of the matrices $[S_{ij}^1]$,$[S_{ij}^2]$ and the coupling matrices $[C_{ij}^1]$, $[C_{ij}^2]$ can be shown to be

$$S_{ij}^1 = S_{ij}^2 = \frac{2}{WL}\sum_{n=1}^{\infty}\psi_n(i,j)\left[1+\frac{2\exp(-\Gamma_{0n}'2T)}{1-\exp(-\Gamma_{0n}'2T)}\right]$$

$$C_{ij}^1 = C_{ij}^2 = \frac{4}{WL}\sum_{n=1}^{\infty}\psi_n(i,j)$$

$$\cdot\left[\frac{\exp(-\Gamma_{0n}'T)}{1-\exp(-\Gamma_{0n}'T)}-\frac{\exp(-\Gamma_{0n}'2T)}{1-\exp(-\Gamma_{0n}'2T)}\right]$$

$$\psi_n(i,j) = \frac{1}{\Gamma_{0n}'}\left[-j\omega\epsilon-\frac{1}{j\omega\mu}\left(\frac{n\pi}{L}\right)^2\right]\frac{L}{n\pi}$$

$$\cdot\sin\frac{n\pi u_i}{L}\left[\cos\frac{n\pi(u_j-\Delta z/2)}{L}-\cos\frac{n\pi(u_j+\Delta z/2)}{L}\right]$$

$$\Gamma_{0n}'^2 = \left(\frac{n\pi}{L}\right)^2 - k^2.$$

The solution of the system of linear equations (3) will generate the approximate x component of the electric field at S and S'.

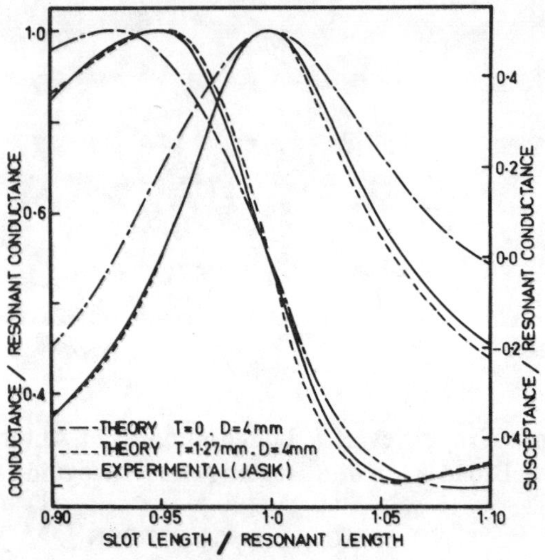

Fig. 2. Admittance of longitudinal slot as function of slot length. Slot width = 1.5875 mm. F = 9.375 GHz.

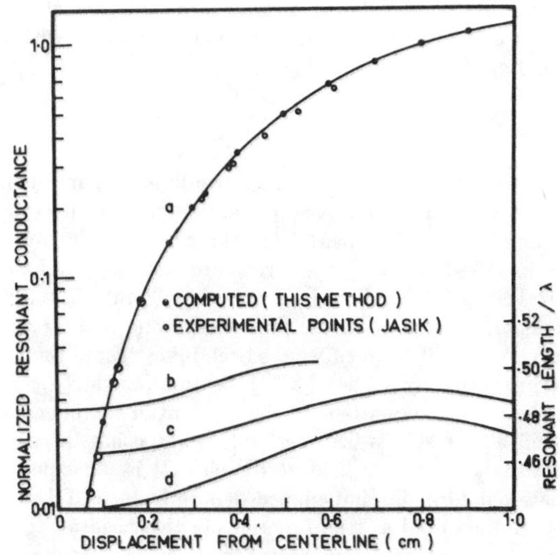

Fig. 3. Resonant conductance and Resonant length of shunt slot as function of slot displacement. (a) Computed from Stevenson's formula. (b) Resonant length experimental. (c) Computed with this method (T = 1.27 mm). (d) Computed with this method (T = 0).

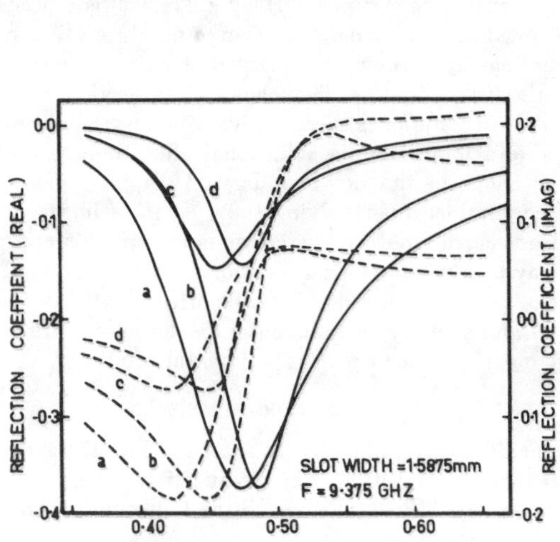

Fig. 4. Reflection coefficient of dominant mode as function of slot length. (a) D = 10 mm, T = 0 mm. (b) D = 10 mm, T = 1.27 mm. (c) D = 4 mm, T = 0 mm. (d) D = 4 mm, T = 1.27 mm.

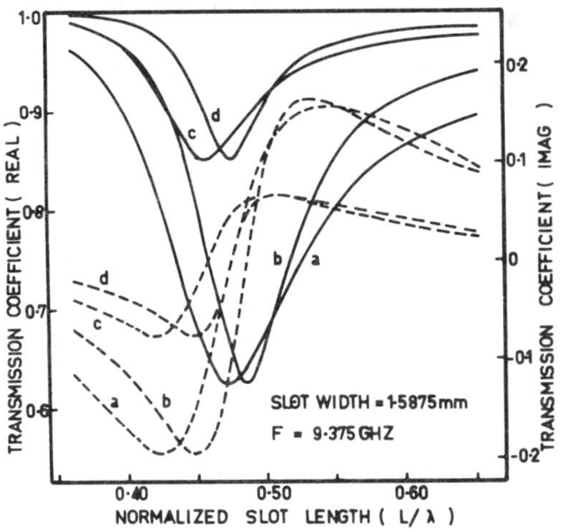

Fig. 5. Transmission coefficient of dominent mode as function of slot length. (a) $D = 10$ mm, $T = 0$ mm. (b) $D = 10$ mm, $T = 1.27$ mm. (c) $D = 4$ mm, $T = 0$ mm. (d) $D = 4$ mm, $T = 1.27$ mm.

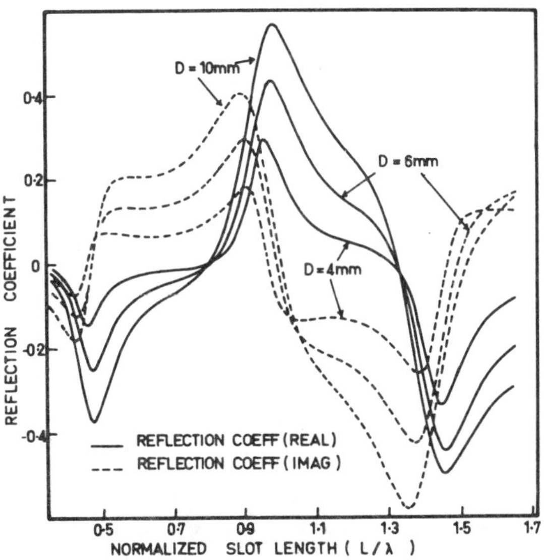

Fig. 6. Reflection coefficient of dominant mode as function of slot length. Wall thickness $T = 0$ mm. $F = 9.375$ GHz. $W = 1.5875$ mm.

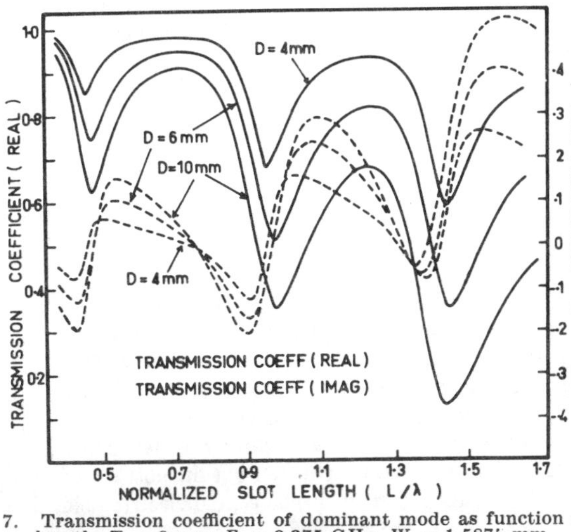

Fig. 7. Transmission coefficient of dominant mode as function of slot length. $T = 0$ mm. $F = 9.375$ GHz. $W = 1.5875$ mm.

The transmission coefficient T_{10} and the reflection coefficient R_{10} may then be evaluated from the formulas [8]

$$T_{10} = 1 - \tfrac{1}{2} \int_S \boldsymbol{H}_{10}^- \cdot (\boldsymbol{n} \times \boldsymbol{E})\, dS$$

$$R_{10} = -\tfrac{1}{2} \int_S \boldsymbol{H}_{10}^+ \cdot (\boldsymbol{n} \times \boldsymbol{E})\, ds.$$

Numerical Results

The variation of normalized conductance and susceptance with slot length for a typical displaced longitudinal slot is shown in Fig. 2. The agreements between computed and experimental curves are significantly improved when the thickness of the guide wall is taken into consideration. The influence of the wall thickness can be seen from the resulting sharper resonance curve and the increase in resonant length of the slot. The magnitude of shunt conductance at resonance is not affected, however. This is demonstrated more clearly in Fig. 3 where the variation of resonant conductance and resonant length with slot offset is shown. Note the qualitative agreements between Fig. 3 and Pounder's calculations [9] for an S-band waveguide. The discrepancies between the experimental curve and the computed curve are about 4 percent and could be due to the rounding of the ends of the experimental slots. The reflection coefficients and the transmission coefficients for some typical displacements are shown in Figs. 4 and 5. For $L/\lambda < 0.55$, it can be seen that the relation $T_{10} = 1 + R_{10}$ is approximately correct and thus justifies the representation of this discontinuity by a shunt impedance. When L is nearly equal to λ, a series impedance would be the correct representation for the slot. These results and others for zero wall thickness are shown in Figs. 6 and 7. Pulse functions with small dimension $= 0.0174\lambda$ are used throughout. It is interesting to note that when the wall thickness T is equal to zero, the matrix that has to be inverted is a special form of the Toeplitz matrix and algorithms for its rapid inversion exist [10]. Although this property has not been exploited here, it suggests that extremely long slots and leaky wave antennas may be fruitfully analyzed with the method described in this communication.

References

[1] R. C. Hansen, *Microwave Scanning Antennas.* New York: Academic Press, 1966.
[2] R. E. Collin and F. J. Zucker. *Antenna Theory.* New York: McGraw-Hill, 1969, ch. 14.
[3] A. A. Oliner, "The impedance properties of narrow radiating slots in the broad face of rectangular waveguide," *IRE Trans. Antennas Propagat.*, vol. AP-5, pp. 4–20, Jan. 1957.
[4] B. N. Das and G. S. Sanyal, "Network parameters of a waveguide broad wall slot radiator," *Proc. Inst. Elec. Eng.*, vol. 117, pp. 41–44, Jan. 1970.
[5] B. A. Finlayson and L. E. Scriven, "The method of weighted residuals—a review," *Appl. Mech. Rev.*, vol. 19, pp. 735–748, Sept. 1966.
[6] R. F. Harrington, *Field Computation by Moment Methods.* New York: Macmillan, 1968.
[7] T. Vu Khac and C. T. Carson, "Coupling by slots in rectangular waveguides with arbitrary wall thickness," *Electron. Lett.*, vol. 8, pp. 456–458, Sept. 1972.
[8] R. E. Collin, *Field Theory of Guided Waves.* New York: McGraw-Hill, 1960, p. 289.
[9] W. H. Watson, *Waveguide Transmission and Antenna Systems.* Oxford, England: Oxford Univ. Press, 1947, p. 200.
[10] D. H. Preis, "The Toeplitz matrix: its occurrence in antenna problems and a rapid inversion algorithm," *IEEE Trans. Antennas Propagat.*, vol. AP-20, pp. 204–206, March 1972.
[11] H. Jasik, *Antenna Engineering Handbook.* New York: McGraw-Hill, 1961, ch. 9.
[12] A. F. Stevenson, "Theory of slots in rectangular waveguides," *J. Appl. Phys.*, vol. 19, pp. 24–38, Jan. 1948.

Efficient moment method analysis of radiating slots in a thick-walled rectangular waveguide

R.W. Lyon, B.Sc., Ph.D., Mem. I.E.E.E., and A.J. Sangster, M.Sc., Ph.D., C.Eng., M.I.E.E.

Indexing terms: *Waveguides, Wave and wave scattering, Electromagnetics*

Abstract: A narrow slot in the broad wall of a rectangular waveguide is analysed using a moment method which takes account of both the waveguide wall thickness, and the influence of side walls. In essence, the method of moments is used to solve a pair of coupled-integral equations, derived from the electromagnetic boundary conditions using Green's function techniques, to find the tangential electric fields on the upper and lower surfaces of the slot. These fields are represented by a sinusoidal Fourier set of basis functions, and it is shown that only a few terms are required to cause the series to converge. The necessity of inverting large, full matrices, which can be a disadvantage of this method, is thus avoided. Theoretical results are presented for the scattering parameters of slots and these are compared with experiment. A computed curve showing the functional form of the tangential electric field in the slot is also included and is compared with some earlier calculated and trial slot-field distributions.

1 Introduction

Although, slotted-waveguide antennas have been in use for many years, it is only recently that comprehensive solutions of the slotted-waveguide boundary value problem have become available. One of the earliest papers dealing with both resonant and nonresonant slots [1] used equivalent circuits for shunt and series slots (displaced longitudinal, transverse and rotated central slots) which were constructed by employing a standing-wave formulation of the waveguide modes [2]. The equivalent circuit immittances were obtained from a variational method using sinusoidal trial functions, with the slot thickness being accounted for using a network approach. More recently, variational methods, [3–5] have been described which treat offset inclined slots and also slots in corrugated waveguides. In Reference 5, the trial function which is employed differs significantly from earlier assumptions.

An inherent weakness of the variational solutions described above is that the effects of finite wall thickness, sidewall proximity and higher-order mode coupling cannot easily be incorporated into the analyses. A moment-method analysis, which does not suffer from these limitations, has been described by Vu Khac and Carson [6–9]. Although only longitudinal slots have been examined by these authors, their method is undoubtedly capable of extension to other aperture alignments. The results obtained agree well with experiment, but, unfortunately, the use of pulse-basis functions results in the inversion of large, full matrices whose entries are infinite series.

In this paper, an alternative moment-method analysis of slots in thick-walled waveguides is presented. The method incorporates, quite naturally, sidewall-proximity effects and higher-order mode coupling within the slot. Although only transverse and longitudinal slots have been examined in detail, the analysis can readily be extended to more general cases such as offset inclined slots. Sinusoidal basis functions are used to form a series expansion of the electric fields in the slots, and it is found that accurate results can be obtained using only a few terms. Consequently the computations involve the manipulation of relatively small matrices. A graph of the computed slot-field distribution is also presented in order to determine the accuracy of various trial functions used in previous variational analyses.

Paper 1392H, first received 14th January and in revised form 16th April 1981

The authors are with the Department of Electrical & Electronic Engineering, Heriot-Watt University, 31–35 Grassmarket, Edinburgh EH1 2HT, Scotland

2 Moment-method formulation

A moment-method solution to the slotted-waveguide problem must ensure that the electric and magnetic field distributions:

(*a*) obey Maxwell's equations

(*b*) satisfy the radiation condition

(*c*) satisfy continuity requirements across the upper and lower surfaces of the aperture.

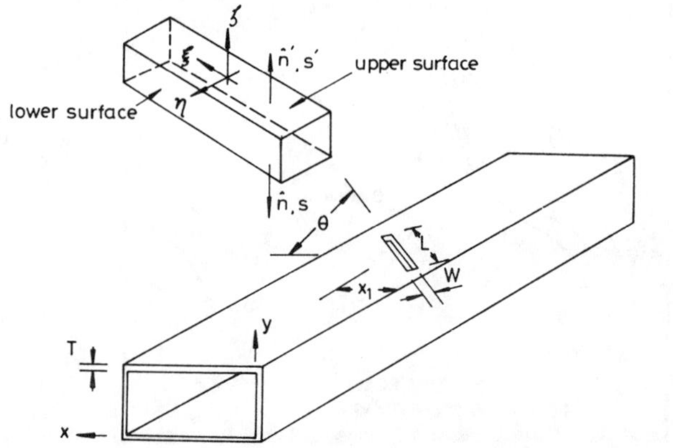

Fig. 1 *Co-ordinate system of slotted waveguide with expanded view of slot*

A diagram of the co-ordinate system used is illustrated, with an expanded view of the slot, in Fig. 1. For clarity, the waveguide is designated region *a*, the slot region *b* and the halfspace region *c*. The fields in regions *a*, *b* and *c* are $(\boldsymbol{E}_a + \boldsymbol{E}_i, \boldsymbol{H}_a + \boldsymbol{H}_i)$, $(\boldsymbol{E}_b, \boldsymbol{H}_b)$ and $(\boldsymbol{E}_c, \boldsymbol{H}_c)$, respectively, where $\boldsymbol{E}_i$ and $\boldsymbol{H}_i$ are the incident fields in the waveguide when the aperture is short-circuited. In order to satisfy the boundary conditions we require that, over the upper and lower surfaces of the aperture (S' and S),

$$\hat{\boldsymbol{n}} \times (\boldsymbol{E}_a - \boldsymbol{E}_b) = \boldsymbol{0} \quad \text{on } S \tag{1a}$$

and

$$\hat{\boldsymbol{n}}' \times (\boldsymbol{E}_c - \boldsymbol{E}_b) = \boldsymbol{0} \quad \text{on } S' \tag{1b}$$

Using field-equivalence principles it is possible to replace an aperture by a short circuit with magnetic currents $(\hat{\boldsymbol{n}} \times \boldsymbol{E})$ and $-(\hat{\boldsymbol{n}} \times \boldsymbol{E})$ on its upper and lower surfaces, respectively [10]. This equivalence principle converts the slotted-waveguide problem into one comprising three connected regions, whose characteristics are known (waveguide, cavity and halfspace), and which are excited by unknown currents at the position of the slot. These currents constrain the tangential electric field to be continuous across the surface of the slot. The situation is illustrated in Fig. 2.

If a Green's function $\tilde{G}(r|r_0)$, which satisfies the following equation

$$\nabla \times \nabla \times \tilde{G}(r|r_0) - k^2 \tilde{G}(r|r_0) = -\bar{I}\delta(r - r_0) \quad (2)$$

with the boundary condition $\hat{n} \times \nabla \times \tilde{G}(r|r_0) = 0$ on perfectly conducting surfaces, is introduced it is possible to immediately calculate the magnetic field in each region [9], using

$$H(r) = j\omega\epsilon_0 \iint_S \tilde{G}(r|r_0) \cdot J_m(r_0)\, dS_0 \quad (3)$$

where $J_m(r_0)$ is a magnetic surface current.

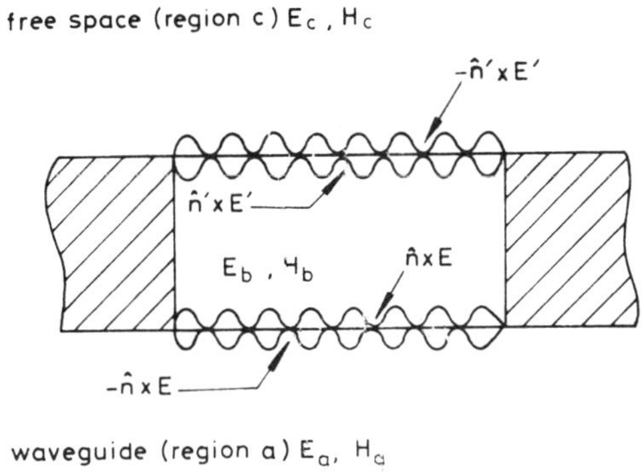

Fig. 2 *Thick slot replaced by three separate regions excited by magnetic currents*

—·— electric wall
~~~~ magnetic current

Imposing the boundary conditions on tangential magnetic fields results in a pair of simultaneous integral equations for $\hat{n} \times E$ and $\hat{n}' \times E'$:

$$\hat{n} \times H_i(r) = j\omega\epsilon_0 \hat{n} \times \iint_S \{\tilde{G}_a(r|r_0) + \tilde{G}_b(r|r_0)\} \cdot \{\hat{n} \times E(r_0)\} dS_0 + j\omega\epsilon_0 \hat{n} \times \iint_{S'} \tilde{G}_b(r|r_0) \cdot \{\hat{n}' \times E'(r_0)\} dS_0 \qquad (r \text{ on } S) \quad (4a)$$

$$0 = j\omega\epsilon_0 \hat{n}' \times \iint_{S'} \{\tilde{G}_b(r|r_0) + \tilde{G}_c(r|r_0)\} \cdot \{\hat{n}' \times E'(r_0)\} dS_0 + j\omega\epsilon_0 \hat{n}' \times \iint_S \tilde{G}_b(r|r_0) \cdot \{\hat{n} \times E(r_0)\} dS_0 \qquad (r \text{ on } S') \quad (4b)$$

where $\tilde{G}_{a,b,c}$ are the dyadic Green's functions of the regions a, b and c, respectively. These simultaneous integral equations were also used by Vu Khac and Carson as the basis of their analysis [8, 9].

It is evident from the complicated nature of eqns. 4*a* and 4*b* that they cannot be solved directly and a numerical method must be used. The most suitable method is the method of moments [11, 12], which has the advantage that it may be applied directly to the integral equations without any rearrangement being required.

In previous work on apertures in waveguides, trigonometric trial functions have been used with a single function proving to be a reasonable approximation. Also Vu Khac, while preferring pulse-basis functions in his analysis of waveguide slots, shows that around the first resonance the fields within the slot are essentially sinusoidal. Consequently trignometric basis functions have been adopted here in the interests of computational efficiency. So, in the slot co-ordinate system, let

$$\hat{n} \times E = \hat{n} \times \sum_{s=1}^{N} a_s f_s \hat{a}_\eta \qquad \text{on } S \quad (5)$$

and

$$n' \times E' = \hat{n}' \times \sum_{s=1}^{N} b_s f_s \hat{a}_\eta \qquad \text{on } S' \quad (6)$$

where

$$f_s(\xi) = \sin \frac{s\pi(\xi + L/2)}{L} \quad (7)$$

The moment method also requires the definition of a set of testing functions and an inner product. However, much work has been done on the evaluation of integral expressions of the type encountered in aperture problems [1], and, in order to build on this, the inner product chosen is

$$\iint_{S \text{ or } S'} p \cdot q\, ds \quad (8)$$

and the testing functions will be of the same type as the basis functions, namely

$$f_i(\xi)\hat{a}_\eta \qquad i = 1, 2, 3, \ldots, N \quad (9)$$

as in Galerkins method. The procedure converts the integral equations into a matrix of the form

$$\begin{bmatrix} [A] & [C] \\ [D] & [B] \end{bmatrix} \begin{bmatrix} [a] \\ [b] \end{bmatrix} = \begin{bmatrix} [h] \\ [0] \end{bmatrix} \quad (10)$$

where

$$A_{is} = -j\omega\epsilon_0 \iint_S f_i \hat{a}_\eta \cdot \{\hat{n} \times \iint_S [\tilde{G}_a(r|r_0) + \tilde{G}_b(r|r_0)] \cdot f_s \hat{a}_\xi\, ds_0\} ds \quad (11a)$$

$$B_{is} = -j\omega\epsilon_0 \iint_{s'} f_i \hat{a}_\eta \cdot \left\{\hat{n}' \times \iint_{s'} [\tilde{G}_b(r|r_0) + G_c(r|r_0)] \cdot f_s \hat{a}_\xi ds_0\right\} ds \quad (11b)$$

$$C_{is} = j\omega\epsilon_0 \iint_s f_i \hat{a}_\eta \cdot \left\{\hat{n} \times \iint_{s'} \tilde{G}_b(r|r_0) \cdot f_s \hat{a}_\xi ds_0\right\} ds \quad (11c)$$

$$D_{is} = j\omega\epsilon_0 \iint_{s'} f_i \hat{a}_\eta \cdot \left\{\hat{n}' \times \iint_s \tilde{G}_b(r|r_0) \cdot f_s \hat{a}_\xi ds_0\right\} ds \quad (11d)$$

$$h_i = -\iint_s f_i \hat{a}_\eta \cdot \{\hat{n} \times H_i(r)\} ds \quad (11e)$$

When the integrals above have been evaluated and the matrix solved, the scattering parameters of the slot in question can be determined. The Green's functions required (waveguide, free space and rectangular cavity) may be written down without difficulty.

3 Evaluation of integrals in halfspaces

Each of the integrals in eqn. 11 has been evaluated, in some cases after not a little mathematical manipulation. In the following Sections, only the results of these integrations will
~~~~

be quoted. Full descriptions of the mathematical methods used are available in Reference 13. The first integration to be considered is the part of B_{is} due to the halfspace Green's function $\tilde{G}_c$ which has the following form [14]:

$$\tilde{G}_c(r|r_0) = -\left(\bar{I} + \frac{\nabla\nabla}{k^2}\right)\frac{e^{-jk|r-r_0|}}{2\pi|r-r_0|} \tag{12}$$

On substituting, it follows that the integral becomes

$$j\omega\epsilon_0 \iint_{s'} f_i(\xi) \iint_{s'} \left(1 + \frac{1}{k^2}\frac{\partial^2}{\partial\xi^2}\right)\frac{e^{-jk|r-r_0|}}{2\pi|r-r_0|} f_s(\xi_0)\,ds_0\,ds \tag{13}$$

where $f_i(\xi)$ and $f_s(\xi)$ are defined in eqns. 7 and 9.

The halfspace Green's function has a real and imaginary part. The imaginary part is continuous, and may be integrated by series expansion and successive integration, but, the real part has an integrable discontinuity where $r = r_0$ which makes it more difficult to treat. However, by using a suitable change of variables, it becomes possible to reduce the integral to the following one, with an integrable discontinuity, which may be easily removed to allow a numerical solution

$$\frac{W^2}{Z_0\pi}\sum_{n=1}^{\infty}\sum_{p=1}^{n}\sum_{q=1}^{n-p+1}\frac{(-1)^{n+p+q}(kL)^{2n}[(-1)^i + (-1)^s - \delta_{q,n-p+1}\{(-1)^{i+s}+1\}]}{(i\pi)^{2q-1}(s\pi)^{2p-1}(2n+1)(2n-1)(2n-2p-2q+2)!}$$
$$+\frac{j\omega\epsilon_0}{\pi k^2}\int_0^L \left\{\begin{matrix} \Delta_{is}\left[R_{is}\sin\dfrac{s\pi\Lambda}{L} + R_{si}\sin\dfrac{i\pi\Lambda}{L}\right] & i \neq s \\ \left[\left\{k^2 - \left(\dfrac{i\pi}{L}\right)^2\right\}(L-\Lambda)\cos\dfrac{i\pi\Lambda}{L} + R_{ii}\sin\dfrac{i\pi\Lambda}{L}\right] & i = s \end{matrix}\right\} \times$$
$$(W[\ln\{W + (\Lambda^2 + W^2)^{1/2}\} - \ln\Lambda - \psi(\Lambda)] + \frac{1}{k}\{\sin k\Lambda - \sin k(\Lambda^2 + W^2)^{1/2}\})d\Lambda \tag{14}$$

where δ_{is} is the Kronecker delta,

$$\Delta_{is} = \begin{cases} 0 & (i-s)\text{ odd} \\ 1 & (i-s)\text{ even} \end{cases} \tag{15}$$

$$\phi(\Lambda) = \int_0^W \frac{1 - \cos k(\Lambda^2 + \sigma^2)^{1/2}\,d\sigma}{(\Lambda^2 + \sigma^2)^{1/2}} \tag{16}$$

$$R_{is} = \begin{cases} \dfrac{\left(\dfrac{2iL}{\pi}\right)\left\{k^2 - \left(\dfrac{s\pi}{L}\right)^2\right\}}{i^2 - s^2} & i \neq s \\ \dfrac{\left(\dfrac{2L}{\pi}\right)\left\{k^2 + (i)^2\left(\dfrac{\pi}{L}\right)^2\right\}}{2i} & i = s \end{cases} \tag{17}$$

Despite its length, eqn. 14 may be evaluated on a digital computer with relative ease.

4 Calculation of coupling factors

Each of the submatrices A, B, C and D, in eqn. 11, involve terms containing the Green's function of the cavity $\tilde{G}_b$. Physically, these terms can be interpreted as determining the field coupled to the upper and lower surfaces of the aperture due to the planar currents within the slot. In this Section, the integrations involved in evaluating these terms will be performed.

Since each of the currents in the cavity is ξ directed, only the component $G_b^{\xi\xi}$ of the Green's function is required. In Reference 13 this function is evaluated and the required term is

$$G_b^{\xi\xi}(r|r_0) = \sum_{l=0}^{\infty}\sum_{n=0}^{\infty}\frac{\epsilon_{l0}\epsilon_{n0}}{WL}\cos\frac{l\pi\left(\eta + \frac{W}{2}\right)}{W}\cos\frac{l\pi\left(\eta_0 + \frac{W}{2}\right)}{W} \times$$
$$\sin\frac{n\pi\left(\xi + \frac{L}{2}\right)}{L}\sin\frac{n\pi\left(\xi_0 + \frac{L}{2}\right)}{L}\frac{\left(1 - \frac{n\pi}{kL}\right)^2}{k_g\sin k_gT} \times$$
$$\begin{cases}\cos k_g\zeta\cos k_g(\zeta_0 + T) \\ \cos k_g(\zeta + T)\cos k_g\zeta_0\end{cases} \quad \zeta \gtrless \zeta_0 \tag{18}$$

where

$$\epsilon_{j0} = \begin{cases} 1 & j = 0 \\ 2 & j \neq 0 \end{cases} \tag{19}$$

and k_g is the cut-off number of the l, n mode of a waveguide of dimensions $W \times L$. This is the most convenient form for the Green's function since terms in the series prove to be orthogonal to the basis functions.

In the B_{is} entry of the matrix equation the following term appears:

$$-j\omega\epsilon_0\iint_{s'} f_i\hat{a}_\eta \cdot \{\hat{n}' \times \iint_{s'} \tilde{G}_b(r|r_0)\cdot f_s\hat{a}_\xi\,ds_0\}ds \tag{20}$$

Substituting the result obtained in eqn. 18 the following is obtained:

$$-j\omega\epsilon_0\sum_{l=0}^{\infty}\sum_{n=0}^{\infty}\int_{-W/2}^{W/2}\int_{-L/2}^{L/2}\sin\frac{i\pi\left(\xi + \frac{L}{2}\right)}{L}\int_{-W/2}^{W/2}\int_{-L/2}^{L/2}$$
$$\sin\frac{s\pi\left(\xi_0 + \frac{L}{2}\right)}{L}\frac{\epsilon_{l0}\epsilon_{n0}}{WL}\sin\frac{n\pi\left(\xi + \frac{L}{2}\right)}{L}\sin\frac{n\pi\left(\xi_0 + \frac{L}{2}\right)}{L} \times$$
$$\cos\frac{l\pi\left(\eta + \frac{W}{2}\right)}{W}\cos\frac{l\pi\left(\eta_0 + \frac{W}{2}\right)}{W} \times$$
$$\frac{\left\{1 - \left(\frac{n\pi}{kL}\right)^2\right\}}{k_g\sin k_gT}\cos k_gT\,d\xi_0\,d\eta_0\,d\xi\,d\eta \tag{21}$$

and this integration may be performed in a straightforward fashion to give

$$-j\frac{WL}{2}\frac{1}{kz_0T}\frac{k_gT}{\sin k_gT}\cos k_gT\,\delta_{is} \tag{22}$$

where

$$k_g^2 = k^2 - \left(\frac{i\pi}{L}\right)^2 \tag{23}$$

There is a discontinuity in eqn. 22 when T is zero, but this is not important since comparison with infinitely thin-wall theories may be made by allowing the thickness to approach a very small, but finite value.

Similarly, the self coupling to the lower surface of the slot, which appears in A_{is} as the term

$$-j\omega\epsilon_0\iint\limits_s f_i\,\hat{a}_\eta\cdot\left\{\hat{n}\times\iint\limits_s \tilde{G}_b(r|r_0)\cdot f_s\,\hat{a}_\xi\,ds_0\right\}ds \tag{24}$$

can be reduced to

$$j\frac{WL}{2}\frac{1}{kz_0T}\frac{k_gT}{\sin k_gT}\cos(k_gT)\,\delta_{is} \tag{25}$$

The evaluation of the mutual coupling terms C_{is} and D_{is} (eqns. 11c and 11d) is also quite straightforward, and the result is

$$C_{is} = -D_{is} = -j\frac{WL}{2}\frac{1}{kz_0T}\left[\frac{k_gT}{\sin k_gT}\right]\delta_{is} \tag{26}$$

where k_g has the same meaning as before.

5 Calculation of integrations within waveguide

The integral-equation formulation described in Section 2 permits the solution of problems involving slots at any angle or location in the waveguide wall. Examination of eqn. 11a, however, reveals that such a general analysis would involve integrating several components of the waveguide Green's function. Since each of these components involves double-infinite summations this would result in a very large amount of analytical and computational effort, so the analysis has so far been restricted to transverse and longitudinal-slot cases. The transverse case is the most straightforward and will be dealt with first.

The waveguide Green's function appears in the components of the A_{is} submatrix as

$$-j\omega\epsilon_0\iint\limits_s f_i\hat{a}_\eta\cdot\left\{\hat{n}\times\iint\limits_s \tilde{G}_a(r|r_0)\cdot f_s\hat{a}_\xi\,ds_0\right\}ds \tag{27}$$

This integral simplifies to become, in the transverse case,

$$j\omega\epsilon_0\iint\limits_s f_i\iint\limits_S G_a^{xx}f_s\,ds_0\,ds \tag{28}$$

in the waveguide co-ordinate system. It is well known that the magnetic-waveguide Green's function [14, 15] is given by

$$\tilde{G}_a(r|r_0) = \begin{cases}\dfrac{1}{2j\omega\epsilon_0}\displaystyle\sum_{n=1}^{\infty}H_n^+(r)H_n^-(r_0) & z > z_0\\[2ex] \dfrac{1}{2j\omega\epsilon_0}\displaystyle\sum_{n=1}^{\infty}H_n^-(r)H_n^+(r_0) & z < z_0\end{cases} \tag{29}$$

where H_n^+ and H_n^- are the H-field components of the nth waveguide mode (TE or TM) propagating in the $+z$ and $-z$ directions, respectively.

Expanding this and extracting the G_a^{xx} component gives

$$G_a^{xx}(r|r_0) = \frac{1}{2j\omega\epsilon_0}\left\{\sum_{n=1}^{\infty}\sum_{m=1}^{\infty}\left[e^{-\Gamma_{nm}|z-z_0|}B_{nm}^2\left(\frac{m\pi}{b}\right)^2(\omega\epsilon_0)^2\times\sin\frac{n\pi x}{a}\sin\frac{n\pi x_0}{a}\cos\frac{m\pi y}{b}\cos\frac{m\pi y_0}{b}\right]\right.$$
$$\left.-\sum_{n=0}^{\infty}\sum_{m=0}^{\infty}e^{-\Gamma_{nm}|z-z_0|}A_{nm}^2\Gamma_{nm}^2\left(\frac{n\pi}{a}\right)^2\times\sin\frac{n\pi x}{a}\sin\frac{n\pi x_0}{a}\cos\frac{m\pi y}{b}\cos\frac{m\pi y_0}{b}\right\} \tag{30}$$

where A_{nm} and B_{nm} are the normalisation factors, given by

$$A_{nm}^2 = \frac{\epsilon_{on}\epsilon_{om}}{j\omega\mu_0\Gamma_{nm}k_c^2\,ab} \tag{31a}$$

$$B_{nm}^2 = \frac{4}{j\omega\epsilon_0\Gamma_{nm}k_c^2ab} \tag{31b}$$

where k_c is the cut-off wavenumber of the n, m mode.

The integral in eqn. 27 is performed using this expression for G_a^{xx}, and the resulting double summation can be reduced to

$$\frac{2j\omega\epsilon_0}{k^2ab}\sum_{n=1}^{\infty}\left[\sigma(i,n)\,\sigma(s,n)\left(WU_n + \frac{e^{-1-\Gamma_{no}W_{-1}}}{\Gamma_{no}} + 2\Gamma_{no}^2\left\{\left(\frac{b}{\pi}\right)^3\left(\frac{1}{2}\frac{j}{\rho_n^3}+\frac{1}{\rho_n^2}\right)+\sum_{m=1}^{\infty}\frac{e^{-\Gamma_{nm}W}}{\Gamma_{nm}^3}\right\}\right)\right] \tag{32}$$

where

$$\sigma(i,n) = \begin{cases}\begin{cases}\dfrac{\frac{2L}{i\pi}}{\alpha^2-1}\sin\dfrac{\alpha i\pi}{2}\cos\dfrac{n\pi x_1}{a} & \alpha \neq 1\\[2ex] \dfrac{L}{2}(-1)^{i/2}\cos\dfrac{n\pi x_1}{a} & \alpha = 1\end{cases} & i \text{ even}\\[6ex] \begin{cases}\dfrac{\frac{2L}{i\pi}}{1-\alpha^2}\cos\dfrac{\alpha i\pi}{2}\sin\dfrac{n\pi x_1}{a} & \alpha \neq 1\\[2ex] \dfrac{L}{2}(-1)^{(i-2)/2}\sin\dfrac{n\pi x_1}{a} & \alpha = 1\end{cases} & i \text{ odd}\end{cases} \tag{33}$$

$$\alpha = \frac{nL}{ia} \tag{34}$$

$$U_n = \begin{cases}\dfrac{\pi\rho_n\cos\pi\rho_n}{\sin\pi\rho_n} & n = 1\\[2ex] \dfrac{\pi\rho_n\cosh\pi\rho_n}{\sinh\pi\rho_n} & n > 1\end{cases} \tag{35}$$

$$\rho_n = b\sqrt{\left(\frac{k}{\pi}\right)^2-\left(\frac{n}{a}\right)^2} \tag{36}$$

A considerable amount of attention has been paid to summing as many of the terms in the series as possible by analytical means. This has been done because experience has shown that around 2500 modes, in eqn. 29, are required to ensure reasonable convergence in the Green's function.

If the longitudinal-slot analysis is developed in the manner described for the transverse slot, a problem arises with regard to the satisfaction of Maxwell's divergence equation for $\boldsymbol{H}$ which states that

$$\nabla \cdot \boldsymbol{H} = \frac{\rho_m}{\mu_0} \tag{37}$$

When the Green's dyadic is used to find the magnetic field inside a waveguide, the relation

$$\boldsymbol{H}(r) = \tfrac{1}{2} \sum_{n=1}^{\infty} \boldsymbol{H}_n(r) \int_V \boldsymbol{H}_n(r_0) \cdot \boldsymbol{J}_m(r_0) dV_0 \tag{38}$$

is used where the $\boldsymbol{H}_n$ are the waveguide modes. However, the magnetic fields $\boldsymbol{H}_n$ are nondivergent, and so the magnetic fields derived will nto satisfy eqn. 37 within the source.

Several authors [16–20] have considered this problem for a region with electric walls excited by an electric current, and have concluded that an additional term [a $\delta(r - r_0)$ function multiplied by a constant] added to the Green's function was sufficient to cause the fields to satisfy Maxwell's equations. (It is worth emphasising that the Green's function used for the transverse case is complete and that only the G_a^{ZZ} term is changed.)

If a summation of $\boldsymbol{H}$ modes is used to form the Green's function for the longitudinal-slot problem, it is found that a term of order n^{-1} appears in the summation for the magnetic field and causes it to be divergent. For this reason the addition of a correction term is difficult and an alternative technique is used.

This involves the introduction of the vector potential defined by

$$\boldsymbol{E} = -\frac{1}{\epsilon_0} \nabla \times \boldsymbol{A}_m \tag{39}$$

Applying Maxwell's equations it is found that $\boldsymbol{A}_m$ satisfies

$$\nabla^2 \boldsymbol{A}_m + k^2 \boldsymbol{A}_m = -\epsilon_0 \boldsymbol{J}_m \tag{40}$$

where $\boldsymbol{E}$ is given by eqn. 39 and $\boldsymbol{H}$ by

$$\boldsymbol{H} = -j\omega \boldsymbol{A}_m + \frac{\nabla\nabla \cdot \boldsymbol{A}_m}{j\omega\mu_0\epsilon_0} \tag{41}$$

as shown by Collin [15].

The Green's function required to solve eqn. 40 satisfies the equation

$$\nabla^2 \tilde{G}_m(r|r_0) + k^2 \tilde{G}_m(r|r_0) = -\tilde{I}\,\delta(r - r_0) \tag{42}$$

together with

$$\hat{n} \times \nabla \times \tilde{G}_m(r|r_0) = 0 \tag{43}$$

$$\hat{n} \cdot \tilde{G}_m(r|r_0) = 0 \tag{44}$$

on perfectly conducting surfaces.

The integral eqn. 27, used in the preceding Section, is equivalent to

$$\iint_s f_i \hat{a}_\eta \cdot \hat{n} \times \boldsymbol{H}\, ds \tag{45}$$

where $\boldsymbol{H}$ is derived using eqn. 41 and $\boldsymbol{A}_m$ is given by

$$\boldsymbol{A}_m = \epsilon_0 \iint_s \boldsymbol{J}_m(r) \cdot \tilde{G}_m(r|r_0) ds_0 \tag{46}$$

The equivalent current on the surface of a longitudinal slot is, in line with previous Sections,

$$\boldsymbol{J}_m = \hat{a}_y \times \hat{a}_x \sin \frac{s\pi\left(z_0 + \frac{L}{2}\right)}{L} = -\hat{a}_z \sin \frac{s\pi\left(z_0 + \frac{L}{2}\right)}{L} \tag{47}$$

Rahmat-Samii [18], in the course of his work on the electric-dyadic Green's function, solved eqn. 43 when the boundary conditions imposed were those given in eqns. 43 and 44. The term of the Green's function which is required here is

$$G_m^{zz} = \sum_{n=0}^{\infty} \sum_{m=0}^{\infty} \left(\frac{\epsilon_{0n}\epsilon_{0m}}{2ab\Gamma_{nm}} e^{-\Gamma_{nm}|z - z_0|} \cos\frac{n\pi x}{a} \times \cos\frac{n\pi x_0}{a} \sin\frac{m\pi y}{b} \sin\frac{m\pi y_0}{b} \right) \tag{48}$$

This expression for G_m^{zz} is used, in conjunction with eqns. 44–47, to give, after some mathematics, the following expression for that part of A_{is} due to the field on the inner surface of a longitudinal slot:

$$\iint_s f_i \hat{a}_x \cdot (-\hat{a}_y \times \boldsymbol{H}_z)\, ds = -\frac{j\omega\epsilon_0}{2ab} \times \left[\sum_{n=0}^{\infty} \epsilon_{0n} \left\{ \left(\frac{a}{n\pi}\right) \left(\sin\frac{n\pi\left(x_1 + \frac{W}{2}\right)}{a} - \sin\frac{n\pi\left(x_1 - \frac{W}{2}\right)}{a} \right) \right\}^2_{W\,(n=0)} \times \left\{ \left(\frac{\delta_{n0}}{k^2} + \left(1 - \left(\frac{s\pi}{kL}\right)^2\right) \left(\frac{b}{\pi}\right)^2 \gamma(n,s) \right) \right\} L\,\delta_{is} + (si)\left(\frac{\pi}{kL}\right)^2 \times \left\{1 + (-1)^{(i+s)}\right\} \sum_{m=0}^{\infty} \frac{\epsilon_{0m}\left\{1 - (-1)^i e^{-\Gamma_{nm}L}\right\}\left\{1 + \left(\frac{k}{\Gamma_{nm}}\right)^2\right\}}{\Gamma_{nm}^3 \left\{1 + \left(\frac{s\pi}{\Gamma_{nm}L}\right)^2\right\}\left\{1 + \left(\frac{i\pi}{\Gamma_{nm}L}\right)^2\right\}} \right] \tag{49}$$

where $n \neq m = 0$ and

$$\gamma(n,s) = \frac{-\pi^2 \cos b\sqrt{k^2 - \left\{\left(\frac{s\pi}{L}\right)^2 + \left(\frac{n\pi}{a}\right)^2\right\}}}{b\sqrt{k^2 - \left\{\left(\frac{s\pi}{L}\right)^2 + \left(\frac{n\pi}{a}\right)^2\right\}} \sin b\sqrt{k^2 - \left\{\left(\frac{s\pi}{L}\right)^2 + \left(\frac{n\pi}{a}\right)^2\right\}}} \tag{50}$$

6 Contribution of incident field

The contribution of the incident field to the matrix eqn. 11 remains to be calculated before it can be solved for vectors a

and b. The incident fields contribute to a column vector h_i which is given by

$$h_i = -\iint_s f_i \hat{a}_\eta \cdot \{\hat{n} \times H_i(r)\} ds \qquad (51)$$

and which depends on the orientation of the slot. The incident fields are given by

$$E_y^+ = -j\omega\mu_0 \phi \sin\frac{\pi x}{a} e^{-j\beta_g z} \hat{a}_y \qquad (52a)$$

$$H_x^+ = j\beta_g \phi \sin\frac{\pi x}{a} e^{-j\beta_g z} \hat{a}_x \qquad (52b)$$

$$H_z^+ = \frac{\pi}{a} \phi \cos\frac{\pi x}{a} e^{-j\beta_g z} \hat{a}_z \qquad (52c)$$

where

$$\phi = (kZ_0\beta_g ab)^{-1/2} \qquad (53)$$

Substituting eqn. 52 into eqn. 51 gives, for an transverse slot

$$h_i = 2j\phi \sin\left(\frac{\beta_g W}{2}\right) \sigma(i, 1) \qquad (54)$$

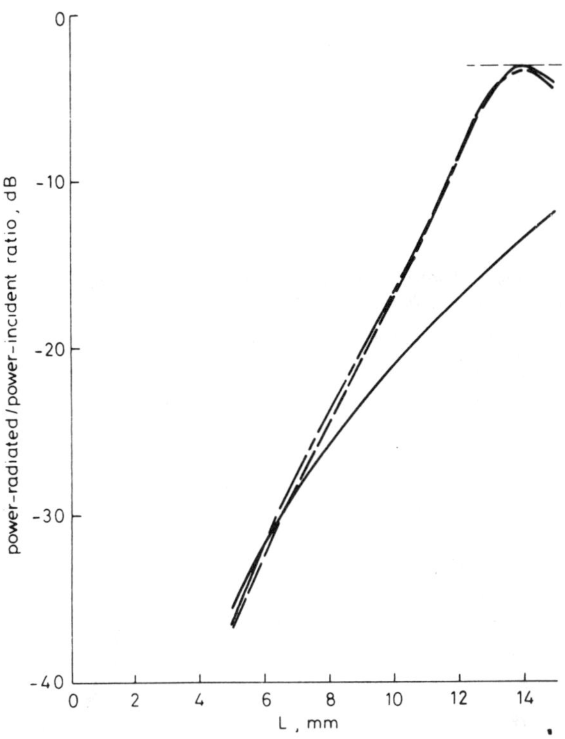

Fig. 3 *Graph of power-radiated/power-incident ratio against slot length for infinitely thin centrally-located transverse slot in WG16*

Predicted by
- - - - Stevenson
——— small-aperture method
— · — Oliner's method
-- -- this paper
where the frequency = 10.0 GHz and W = 1.5 mm

where $\sigma(i, n)$ is given by eqn. 33. For a longitudinal slot

$$h_i = \phi \left[\sin\frac{\pi\left(x_1 + \frac{W}{2}\right)}{a} - \sin\frac{\pi\left(x_1 - \frac{W}{2}\right)}{a}\right] \times \frac{\frac{i\pi}{L}}{\left(\frac{i\pi}{L}\right)^2 - \beta_g^2} \times \begin{cases} 2\cos\beta_g \frac{L}{2} & (i \text{ odd}) \\ 2j\sin\beta_g \frac{L}{2} & (i \text{ even}) \end{cases} \qquad (55)$$

7 Calculation of transmission and reflection coefficients

The transmission and reflection coefficients are calculated by first deducing expressions for the scattered fields, then working out the ratio of the fields transmitted and reflected to the incident waves. We know that the tangential electric field on the inner surface of the slot is

$$\hat{n} \times E(r) = \sum_{s=1}^{N} a_s f_s \hat{a}_\xi \qquad (56)$$

where f_s are the basis functions.

The scattered fields may be obtained by substituting eqn. 56 into eqn. 3 and using the dominant-mode term of the relevant Green's function. The scattering parameters may then be obtained without difficulty. For the transverse slot,

$$S_{11} = \frac{A_{10}\sqrt{2}\pi}{a} \sin\left(\beta_g \frac{W}{2}\right) \sum_{s=1}^{N} a_s \sigma(s, 1) \qquad (57)$$

$$S_{21} = 1 - S_{11} \qquad (58)$$

where A_{nm} and $\sigma(i, n)$ are given by eqns. 31a and 33, respectively, and for the longitudinal slot

$$S_{11} = \frac{A_{10}\pi}{j\sqrt{2}a} \left\{\sin\frac{\pi\left(x_1 + \frac{W}{2}\right)}{a} - \sin\frac{\pi\left(x_1 - \frac{W}{2}\right)}{a}\right\} \sum_{s=1}^{N} \left[a_s \frac{\frac{s\pi}{L}}{\left(\frac{s\pi}{L}\right)^2 - \beta_g^2}\right] \times \begin{cases} 2\cos\beta_g \frac{L}{2} & (s \text{ odd}) \\ 2j\sin\beta_g \frac{L}{2} & (s \text{ even}) \end{cases}$$

$$S_{12} = 1 - (-1)^s S_{11} \qquad (59)$$

The relationship between S_{11} and S_{12} reflects the symmetry of the basis functions in the longitudinal case.

8 Computed results

In this Section computed solutions for the scattering parameters of some typical transverse and longitudinal slots are presented. Initial calculations were directed towards verifying the method which has been evolved here by making comparisons, where possible, with the results of more established theories. These are presented in Fig. 3. In this Figure radiated power has been plotted as a function of slot length for a centrally located transverse slot in a rectangular waveguide of negligible wall thickness. In addition to the moment-method curve, results obtained using Oliner's theory [1] and small-aperture diffraction theory [21] are also shown.

A first-order sinusoidal basis function has been employed in generating the moment-method result, making it more comparable with the Oliner result which is based on the use of a half-sinusoidal trial field in the slot. It can be seen that, for slots which are less than $\lambda/4$ in length, all three methods are in good agreement. Above this slot length the small-aperture results are known to be erroneous because of the onset of resonance. However, the moment method remains in good agreement with Oliner up to and including resonance, where agreement with Stevenson [22] is also indicated. It is concluded, therefore, that for narrow slots of any length in thin-walled rectangular waveguide the moment method is a reliable analytical tool.

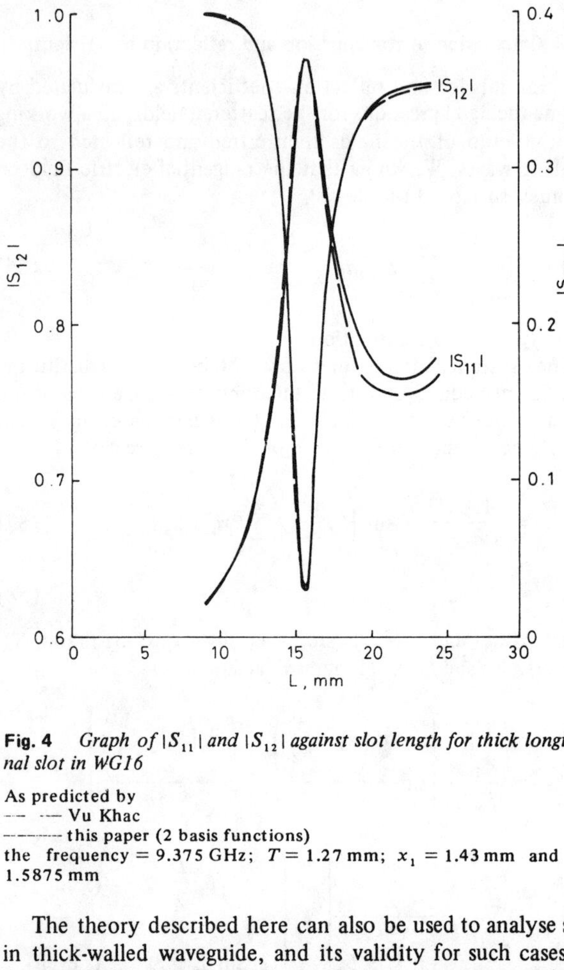

Fig. 4 *Graph of $|S_{11}|$ and $|S_{12}|$ against slot length for thick longitudinal slot in WG16*

As predicted by
— — Vu Khac
——— this paper (2 basis functions)
the frequency = 9.375 GHz; $T = 1.27$ mm; $x_1 = 1.43$ mm and $W = 1.5875$ mm

The theory described here can also be used to analyse slots in thick-walled waveguide, and its validity for such cases has been partially confirmed by making comparisons with the results of Vu Khac [8]. In particular, transmission and reflection coefficient amplitudes have been compared for an offset longitudinal slot located in the broad wall of a thick-walled (1.27 mm) waveguide (Fig. 4). Since the curves have been extended well beyond the first resonance of the slot, two basis functions have been employed in the moment-method computations. Clearly over the entire range of slot lengths our results are in excellent agreement with those of Vu Khac.

For slots in the length range which is of major practical interest, namely from $l = \lambda/8$ to $\lambda/2$, it is found that no more than one basis function is normally required to provide an adequate description of the characteristics of the slot. This is illustrated in Fig. 5, where the radiated power from a longitudinal slot is plotted as a function of slot length. The curves shown represent separately a one-basis-function and a three-basis-function computation. Over their entire length, the amount by which the curves diverge is small and would not be significant in most practical applications, where tolerancing errors in the slot dimensions can produce discrepancies (in the radiated power) of this order.

The nature of the tangential electric-field distribution within a narrow radiating slot has been the subject of some interest in the literature, particularly since such theoretical techniques as the variational and reaction methods require that a good approximation to the aperture field is available in advance. This is not so with the moment method, which can be used to make an accurate estimate of the electric-field distribution within the slot. The results of such a calculation for a longitudinal slot in the broad wall of a rectangular waveguide are shown in Fig. 6 and compared with trial field distributions employed by Oliner [1] and Das [5]. The moment method, using three basis functions, predicts a

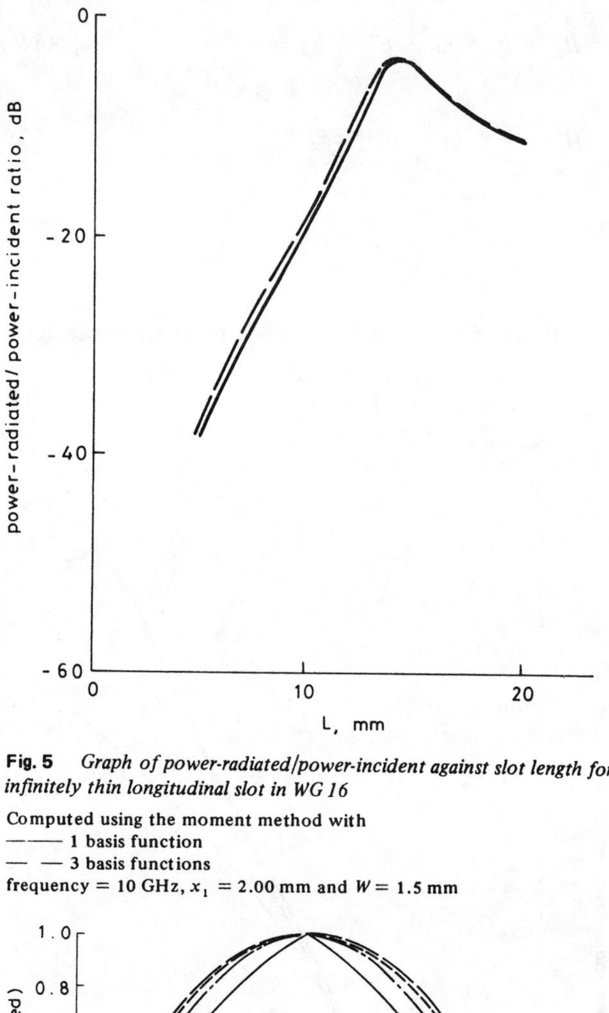

Fig. 5 *Graph of power-radiated/power-incident against slot length for infinitely thin longitudinal slot in WG 16*

Computed using the moment method with
——— 1 basis function
— — 3 basis functions
frequency = 10 GHz, $x_1 = 2.00$ mm and $W = 1.5$ mm

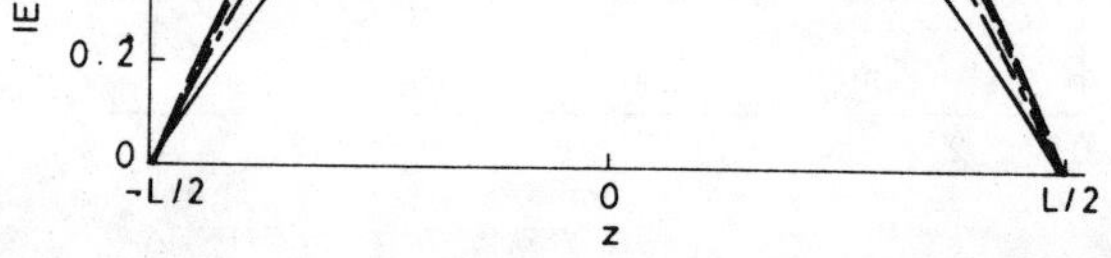

Fig. 6 *Normalised electric-field amplitude in longitudinal slot in WG16*

Calculated by
— — this paper
- - - - Vu Khac
— · — Oliner
——— Das
frequency = 9.375 GHz; $L = 0.361\ \lambda$; $x_1 = 6.43$ mm; $W = 1.5875$ mm and $T = 1.27$ mm

slightly flattened cosine function for the slot field distribution, as does Vu Khac using approximately 30 pulse basis functions (Fig. 6). Whereas Oliner's assumed cosine distribution obviously represents a reasonable approximation to the aperture field shape, the distribution proposed by Das does not appear to be justified, particularly with regard to the discontinuous first derivative which is assumed to exist at the centre of the slot. The distribution of the phase of the slot E-field can also be predicted by the moment method. This is shown in Fig. 7 and compared with the corresponding Vu Khac result. Good agreement is once more indicated. The discrepancies at either end of the slot are due to the larger number of basis functions used by Vu Khac.

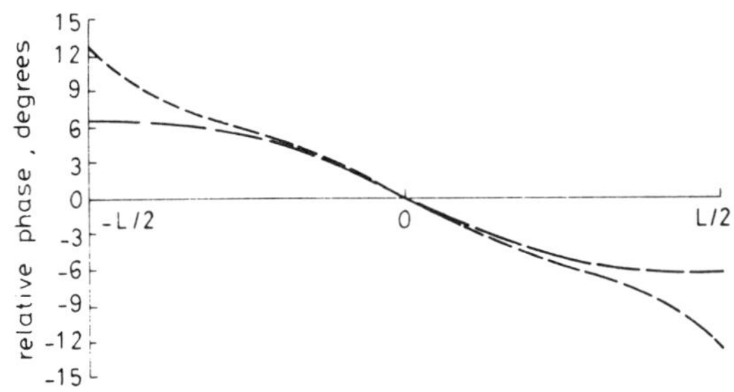

Fig. 7 *Relative phase of electric field in longitudinal slot described in Fig. 6*

—— this paper
- - - - Vu Khac

The characteristics of a typical longitudinal slot radiator in rectangular waveguide have also been determined experimentally using a resonant cavity technique [23]. These are compared with corresponding moment-method predictions in Figs. 8*a* and 8*b*. Although the agreement indicate is generally good, close examination of the curves reveals that the resonant length is, in practice, slightly longer than the predicted value. This discrepancy has been found to be almost wholly attributable to the fact that the experimental slots had rounded ends for mechanical reasons, whereas the theory applies to rectangular-shaped slots.

9 Conclusions

A computationally efficient moment method of analysing narrow radiating slots in thick-walled rectangular waveguide is described. The improved efficiency is achieved by employing 'natural' basis functions for the slot field so that in many practical cases no more than one, or perhaps two, basis functions are required to adequately describe the scattering characteristics of the slot. Calculations have also been performed to take advantage of the inherent ability of the moment method to accommodate waveguide wall-thickness changes. The predicted results have been confirmed experimentally.

For slot configurations in infinitesimally thin-walled waveguide, good agreement with alternative theories has also been demonstrated.

10 References

1 OLINER, A.A.: 'The impedance properties of narrow radiating slots in the broad face of rectangular waveguide', *IRE Trans.*, 1957, **AP-5**, pp. 4–20

2 MARCUWITZ, N., and SCHWINGER, J.: 'On the representation of the fields produced by currents and discontinuities in waveguides', *J. Appl. Phys.*, 1951, **22**, pp. 806–819

3 SANGSTER, A.J., and HAWKINS, D.C.: 'Scattering parameters of an arbitrarily shaped aperture in a waveguide', *Proc. IEE*, 1972, **119**, (10), pp. 1465–1466

4 SANGSTER, A.J., and HAWKINS, D.C.: 'Radiating apertures in a corrugated rectangular waveguide', *Electron. Lett.*, 1973, **9**, pp. 329–331

5 DAS, B.N. and SANYAL, G.S.: 'Network parameters of a waveguide broad wall-slot radiator, *Proc. IEE*, 1970, **117**, (1), pp. 41–44

6 VU KHAC, T.: 'Solution of some waveguide discontinuities by the method of moments', *IEEE Trans.*, 1972, **MTT-20**, pp. 416–417

7 VU KHAC, T., and CARSON, C.T.: 'Coupling by slots in rectangular waveguides with arbitrary wall thickness', *Electron. Lett.*, 1972, **8**, pp. 456–458

8 VU KHAC, T., and CARSON, C.T.: 'Impedance properties of a longitudinal slot antenna in the broad face of rectangular waveguide', *IEEE Trans.*, 1973, **AP-21**, pp. 708–710

9 VU KHAC, T.: 'Studies of slot discontinuities in rectangular waveguides'. Ph.D. thesis, Monash University, 1974

10 BUTLER, C.M., RAHMAT-SAMII, Y., and MITTRA, R.: 'Electromagnetic penetration through apertures in conducting surfaces', *IEEE Trans.*, 1978, **AP-26**, pp. 82–93

11 HARRINGTON, R.F., and MAUTZ, J.R.: 'A generalised network formulation for aperture problems'. Syracuse University Technical Report TR-75-13, Syracuse University, 1975

12 HARRINGTON, R.F., and MAUTZ, J.R.: 'A generalised network formulation for aperture problems', *IEEE Trans.*, 1976, **AP-24**, pp. 870–873

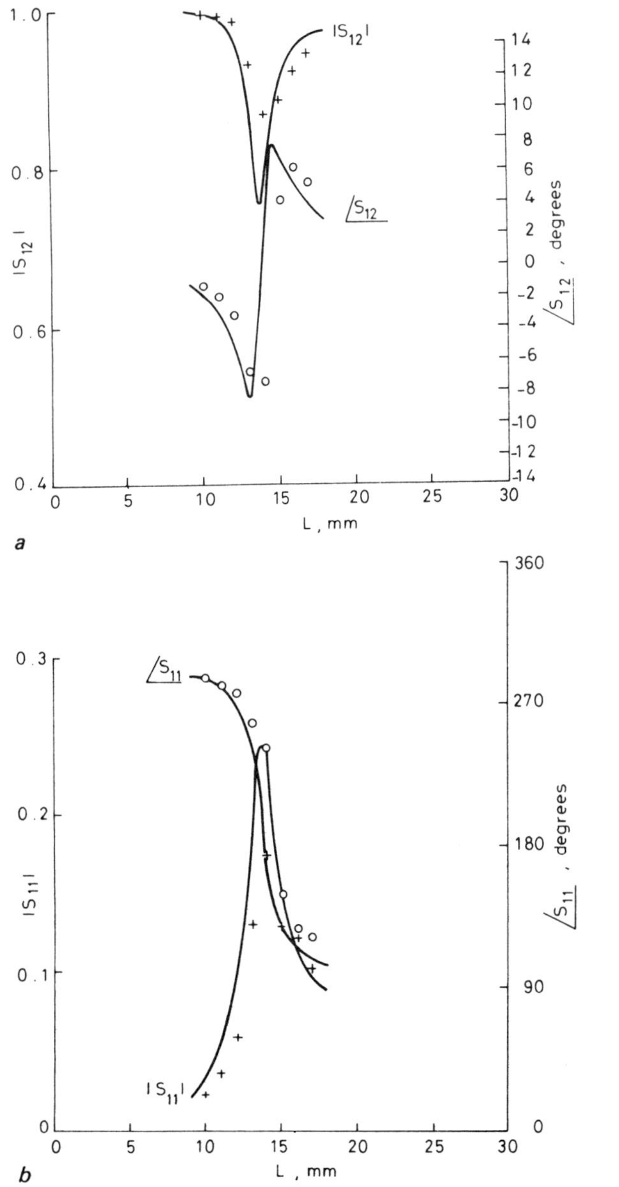

Fig. 8. *Theoretical and experimental graphs of amplitude and phase for a Longitudinal Slot in WG16*

(*a*) S_{12}; (*b*) S_{11};
$T = 1.32$ mm; Frequency = 10.48 GHz; $x_1 = 2.00$ mm and $W = 1.5$ mm
—— theory
+ experiment (amplitude)
o experiment (phase)

13 LYON, R.W.: 'An investigation of apertures in dual mode rectangular waveguide for variable polarisation arrays'. Ph.D. thesis, Heriot-Watt University, 1979

14 COLLIN, R.E. and ZUCKER, F.J.: 'Antenna theory, Pt. 1' (McGraw-Hill, 1969)

15 COLLIN, R.E.: 'Field theory of guided waves' (McGraw-Hill, 1960)

16 VAN BLADEL, J.: 'Some remarks for Green's dyadic for infinite space', *IRE Trans.*, 1961, **AP-9**, pp. 563–566

17 COLLIN, R.E.: 'On the incompleteness of *E* and *H* modes in waveguides', *Can. J. Phys.*, 1973, **51**, pp. 1135–1140

18 RAHMAT-SAMII, Y.: 'On the question of computation of dyadic Green's function at the source region in waveguides and cavities', *IEEE Trans.*, 1975, **MTT-23**, pp. 762–765

19 KUN-MU CHEN: 'A simple physical picture of tensor Green's function in souce region', *Proc. IEEE*, 1977, **65**, pp. 1202–1204

20 YAGHJIAN, A.D.: 'A direct approach to the derivation of electric dyadic Green's functions'. NBS technical note 1000, 1978

21 VAN BLADEL, J.: 'Small-hole coupling of resonant cavities and waveguides', *Proc. IEE*, 1970, **117**, (6), pp. 1098–1104

22 STEVENSON, A.F.: 'Theory of slots in rectangular waveguides', *J. Appl. Phys.*, 1948, **19**, pp. 24–38

23 McGRANE, A.: 'Characterisation of slots in waveguide using a resonant cavity technique'. Plessey Radar, Radar and environmental systems unit report IWR3070, Isle of Wight, 1979

Resonant Length of Longitudinal Slots and Validity of Circuit Representation: Theory and Experiment

GEORGE J. STERN, SENIOR MEMBER, IEEE, AND ROBERT S. ELLIOTT, FELLOW, IEEE

Abstract—**Pertinent theory for the design of longitudinal slot arrays is reviewed and its dependence on the dominant mode scattering off a single slot is pinpointed. The critical need to know resonant length versus slot offset is emphasized and the desirability of determining this information theoretically rather than experimentally is argued. Then method of moments solutions are used to calculate resonant length versus slot offset for given waveguide dimensions and frequency. These theoretical results are compared to new, carefully obtained experimental data. Agreement is found to be so good, it is concluded that one can dispense with the costly gathering of experimental input data when designing longitudinal slot arrays fed by standard rectangular waveguide. A critical look is taken at the validity of representing the longitudinal slot as a shunt element on an equivalent transmission line. This assumption is found to be more and more questionable as the *b* dimension is reduced. For quarter-height guide, an alternate design procedure is suggested as being more accurate.**

INTRODUCTION

IN 1978 a theory was presented [1], [2] that would permit the design of linear and planar arrays of waveguide-fed slots, including the effects of external mutual coupling. That theory was generalized and improved [3] in 1983. Briefly stated, the theory culminates in three design equations which contain the lengths and offsets of the slots as unknowns, with the desired aperture distribution, the desired input match, and the self-admittance of a single slot versus its offset and length as knowns.

Up to the present time, this design procedure has been used by first amassing experimental data on the self-admittance of a single slot, fitting functions of slot length and offset to that data, and then using these fitted functions in the three design equations. This experimental input data-gathering is costly, time-consuming, and inherently less accurate than one would wish when designing high-performance arrays. Thus it would be desirable to replace this *experimentally* obtained input data by *theoretically* obtained input data, if the latter could be generated with sufficient accuracy and at less cost.

That quest has been going on for a considerable number of years. In 1948 Stevenson [4] published a classic paper in which he established the internal Green's functions for a rectangular waveguide and used those functions to analyze scattering off a longitudinal slot excited by an incident TE_{10} mode. He assumed that the waveguide was air-filled, that the slot length was $\lambda_0/2$, that this was the *resonant* length, that the electric field in the slot was totally transverse and possessed an equiphase half-cosinusoid amplitude distribution, that the waveguide walls were vanishingly thin, and that the broad wall containing the slot was imbedded in an infinite ground plane. Under these circumstances Stevenson was able to show that dominant TE_{10} mode scattering off the slot was symmetrical, as a result of which the slot could be represented by a shunt element on an equivalent transmission line. Using a power balance, Stevenson derived a formula for the normalized resonant conductance of a single longitudinal slot as a function of its offset x from the centerline, viz.,

$$g(x)=\frac{G_{\text{res}(x)}}{G_0}=\left[2.09\,\frac{(a/b)}{(\beta_{10}/k_0)}\cos^2\left(\frac{\beta_{10}}{k_0}\frac{\pi}{2}\right)\right]\sin^2\frac{\pi x}{a}. \tag{1}$$

In (1), a and b are the transverse waveguide dimensions, β_{10} is the propagation constant of the TE_{10} mode, and $k_0 = 2\pi/\lambda_0$.

Stevenson then attempted to generalize his result to embrace a nonresonant slot but was unable to find a companion expression for slot susceptance.

In 1951 Stegen [5] carefully performed a series of experiments on longitudinal slots cut in a broad wall of standard X-band waveguide ($a = 0.900$ in, $b = 0.400$ in, $t = 0.050$ in). He used round-ended slots, 0.063 in wide, of various offsets and lengths, and measured relative backscattering B_{10}/A_{10} at a frequency of 9.375 GHz. Stegen wanted to achieve a standard against which theory could be judged, and it is a tribute to his painstaking work that his data served as the standard for 30 years.

Stegen found that he could present the data in a universal form by normalizing slot admittance to resonant conductance, in effect writing

$$\frac{Y(x,\,l)}{G_0}=[h_1(y)+jh_2(y)]g(x). \tag{2}$$

In (2), $Y(x, l)$ is the admittance of the slot, seen as a shunt element on an equivalent lossless transmission line of characteristic conductance G_0; x and $2l$ are the offset and length of the slot; $g(x) = G_{\text{res}}(x)/G_0$ is the normalized conductance of a resonant-length slot whose offset is x; $y = l/l_{\text{res}}$ is actual slot length in ratio to the length of a resonant slot possessing the same offset. Stegen's fitted curves and data points for $h_1(y)$, $h_2(y)$, and $g(x)$ are reproduced as [2, figs. 8.35 and 8.36]. His fitted curve for $g(x)$ is in the form $g(x) = K\sin^2(\pi x/a)$ and is seen to be in good agreement with the experimental data, thus validating the form of (1). However, the value of his constant K, chosen for a best fit, was not quite in agreement

Manuscript received November 20, 1984.

G. S. Stern is with Hughes Missile Systems Group, Canoga Park, CA 91304.

R. S. Elliott is with the Department of Electrical Engineering, University of California, Los Angeles, CA 90024.

with Stevenson's derived constant, contained within the square brackets in equation (1).

Additionally, Stegen found that resonant length, contrary to Stevenson's assumption, was offset-dependent. His fitted curve and data points relating $k_0 l_{res}$ to slot offset are shown as [2, fig. 8.37]. As we shall see, agreement between theory and experiment for $v(x) = k_0 l_{res}(x)$ is critical if one wishes to substitute theoretical input data for experimental data of the type accumulated by Stegen.

Working in parallel with Stegen, Oliner [6] developed the first theory for longitudinal slots which would permit calculation of the susceptance as well as the conductance. He did this by formulating a stationary expression for reactive power and found good agreement with Stegen when he calculated $h_1(y)$, $h_2(y)$, and $g(x)$. However, Oliner had two small obstacles to overcome before he could compare his calculated result for $v(x)$ with Stegen's measured values. First, Oliner's original theory assumed "zero" wall thickness, and Stegen's wall thickness of $t = 0.050$ in was not negligible. Oliner overcame this difficulty by assuming that the rectangular hole which linked the lower and upper slot apertures was a section of waveguide of length t, in which only the dominant TE_{10} mode need be considered as the coupling agent between internal and external fields. With this assumption, Oliner was able to derive a "wall thickness correction," which for Stegen's case amounted to 2 percent. Second, Oliner's theory assumed square-ended slots whereas Stegen's slots were round-ended, having been cut by a vertical milling machine. Oliner made the assumption that round-ended and square-ended slots were equivalent if they had the same area. This led to an additional 2 percent correction. Oliner concluded that his theoretical square-ended slot in a "zero" thickness wall should be 4 percent shorter than Stegen's actual slot. With this adjustment, theoretical and experimental curves for $v(x)$ were in fair-to-good agreement. Unfortunately, in terms of today's design requirements, the agreement was insufficient. More will be said about this point later in the present paper.

The next significant theoretical attack on this problem was undertaken by Khac [7], who took full advantage of the capabilities of the modern computer. Khac assumed a square-ended longitudinal slot in a "zero" thickness wall, assumed that the electric field in the slot aperture was totally transverse, used Stevenson's internal Green's function and the external half-space Green's function to formulate integral expressions (involving the unknown slot aperture electric field) for the internal and external scattered fields, and matched the total longitudinal H-field across the slot aperture. This gave him an integral equation in $E_x(P')$, the unknown aperture field, with the longitudinal component of the incident TE_{10} mode playing the role of driving function. Khac then used the method of moments, employing pulse functions and point matching, to solve for $E_x(P')$ and used that knowledge to find the dominant mode backscattering, and thus $Y(x, l)/G_0$.

Khac's agreement with Stegen on $g(x)$ was excellent but on $h_1(y)$ and $h_2(y)$ only qualitative. Before he could compare $v(x)$ results and improve on his $h_1(y)$ and $h_2(y)$ calculations, he had to overcome the same two obstacles as had confronted Oliner. Khac resorted to coupled integral equations to account for wall thickness and this resulted in good agreement with experiment on $h_1(y)$ and $h_2(y)$. If Khac were then to use the same 2 percent correction for round ends as Oliner, his agreement with Stegen on $v(x)$ could only be characterized as fair. Khac had an additional difficulty in that his use of pulse functions violated the boundary condition that $E_x(P')$ be zero at the two ends of the slot. His ad hoc assumption to add one cell to the length of his theoretical slot brought his $v(x)$ curve within 2 percent of Stegen's, but is difficult to justify on a rigorous basis. As we shall see, a 2 percent error in $v(x)$ is not good enough.

To obtain the theoretical results reported later in this paper, we have used Khac's integral equation and the method of moments, but have chosen basis functions which model $E_x(P')$ more realistically. Also, to overcome some of the difficulties which confronted both Oliner and Khac in comparing their theoretical results to Stegen's experimental data, we have carefully fabricated and tested a family of seven waveguide sections, each containing a longitudinal slot with a different offset. Standard X-band dimensions were used ($a = 0.900$ in, $b = 0.400$ in) but the wall thickness was only $t = 0.005$ in, and the slots were *square*-ended. This brings the theoretical model and the actual test fixtures into closer agreement and permits a more precise comparison. Further, with the modern trend toward thin waveguide walls and slot fabrication via stamping or etching, square-ended slots are just as easy to fabricate as round-ended slots, so this convenient assumption in the theory is no longer academic.

The Design Procedure

The procedure that we use to design linear and planar arrays of longitudinal slots fed by rectanguar waveguide is fully described in [3]. It rests on three design equations:

$$\frac{Y^a_{mn}/G_0}{Y^a_{pq}/G_0}=\frac{f_{mn}V^s_{mn}V_q}{f_{pq}V^s_{pq}V_n} \tag{3}$$

$$Y^a_{mn}/G_0=\frac{2f^2_{mn}}{\dfrac{(2/K)L^2_{mn}}{h_1(y_{mn})+jh_2(y_{mn})}+\mathrm{MC}_{mn}} \tag{4}$$

$$\sum_{m=1}^{M(n)}\frac{Y^a_{mn}}{G_0}=C_n \tag{5}$$

in which

$$f_{mn}=L_{mn}\sin^2\frac{\pi x_{mn}}{a} \tag{6}$$

$$L_{mn}=\frac{[(\pi/2)/y_{mn}v(x_{mn})]\cos[(\beta_{10}/k)y_{mn}v(x_{mn})]}{[(\pi/2)/y_{mn}v(x_{mn})]^2-(\beta_{10}/k)^2} \tag{7}$$

$$\mathrm{MC}_{mn}=j(\beta_{10}/k)(kb)(a/\lambda)^3\sum_r\sum_s{}'\frac{V^s_{rs}}{V^s_{mn}}g_{mnrs}. \tag{8}$$

Many of the symbols which appear in (3)–(8) have been defined in the Introduction. Equation (3) gives the ratio of the active admittances of the mnth slot and the pqth slot in terms of the desired slot voltage ratio V^s_{mn}/V^s_{pq}, the ratio V_q/V_n of the mode voltages in the nth and qth branch lines, and the functions f_{mn} and f_{pq}, these latter depending on the offsets and lengths of the two slots, as can be seen from (6) and (7). Equation (4) contains the mutual coupling term MC_{mn} which, from (8), is seen to involve the summation on all other slots of the functions g_{mnrs}, weighted by the slot voltage ratio. g_{mnrs} is a complex number arising from the calculation of a double integral which depends on the lengths on the mnth and rsth slots and their relative positions. Equation (5) gives the desired admittance level C_n in the nth branch line and assumes a standing wave feed. (A more general form for (5) can be used when a traveling wave feed is employed.)

Briefly stated, the design technique proceeds as follows: One assumes initial offsets and lengths for all the slots and computes every g_{mnrs}. Using the desired aperture distribution V^s_{rs}/V^s_{mn}, one then calculates every MC_{mn}. Next, for the mnth slot, one searches for a couplet (x_{mn}, y_{mn}) that will make the denominator of (4) pure real, that is, one seeks to make

$$\mathcal{I}m\left[\frac{(2/K)L^2_{mn}(x_{mn}, y_{mn})}{h_1(y_{mn})+jh_2(y_{mn})}+\mathrm{MC}_{mn}\right]=0. \tag{9}$$

The realization emerges that there is a *continuum* of couplets that will satisfy (9). Similarly, there is a continuum of couplets (x_{pq}, y_{pq}) that will satisfy (9) for the pqth slot. But, for a particular couplet (x_{mn}, y_{mn}), there is only one (x_{pq}, y_{pq}) that will also satisfy (3). Which particular couplet (x_{mn}, y_{mn}) to choose depends on (5). Thus (3), (4), and (5) jointly determine all the slot offsets and lengths. Of course, one must iterate, since this improved information about offsets and lengths permits improved calculation of MC_{mn}. Iterating can be terminated when all newly found couplets are negligibly different from the previous set or when, for a given set of couplets, the computed slot voltage distribution gives a satisfactory pattern.

The purpose for reviewing the design procedure is so we can focus on the crucial step embodied in satisfying (9), the first term of which can be recognized as an altered form of

$$\frac{2f^2_{mn}}{Y(x_{mn}, l_{mn})/G_0}.$$

We thereby appreciate that (9) is being satisfied by detuning the mnth slot (changing its length away from self-resonance) so as to cancel out the imaginary component of MC_{mn}. But a study of Stegen's $h_1(y_{mn})$ and $h_2(y_{mn})$ curves (and his curves are typical) shows that the useful dynamic range of y_{mn} is about ± 2.5 percent around unity. (This is normally sufficient to achieve a satisfactory design.) However, if our knowledge of resonant length were in error by only 1 percent, that would imply an error in $h_1(y_{mn}) + jh_2(y_{mn})$ far more serious. Indeed, for many slots that error would swamp the desired compensation for mutual coupling. At best, such an error in our knowledge of $v(x)$ would cause the array to perform satisfactorily at other than the design frequency. More likely, it would also cause a degraded sidelobe level and a degraded input admittance over the entire frequency band. It is for this reason that input experimental data on $Y(x, l)/G_0$ must be obtained with great care. For the same reason, if one is planning to substitute a theoretical calculation of $Y(x, l)/G_0$ for experimental data, a high reliability for those calculations must be established. We now address that problem.

An Integral Equation for the Slot Aperture E-Field

The boundary conditions

$$H_x^{\mathrm{inc}}(P)+H_x^{\mathrm{int}}(P)=H_x^{\mathrm{ext}}(P) \tag{10}$$

$$H_z^{\mathrm{inc}}(P)+H_z^{\mathrm{int}}(P)=H_z^{\mathrm{ext}}(P) \tag{11}$$

can be used to develop integral equations for the unknown slot aperture field $\mathbf{E}(P')$ if the wall is vanishingly thin. The slot geometry is shown in Fig. 1; P and P' are points in the slot aperture, $\mathbf{H}^{\mathrm{inc}}$ is the incident TE_{10} mode, $\mathbf{H}^{\mathrm{int}}$ is the internal magnetic field scattered by the slot, and $\mathbf{H}^{\mathrm{ext}}$ is the external field scattered by the slot. Taken together, (10) and (11) represent the statement that the total tangential $\mathbf{H}$-field is being matched across the slot aperture. This is true even when the guide is dielectric-filled.

Stevenson's Green's functions and the external half-space Green's function can be used to formulate $\mathbf{H}^{\mathrm{ext}}_{\mathrm{tang}} - \mathbf{H}^{\mathrm{int}}_{\mathrm{tang}}$ as a pair of integral equations, with the unknown tangential slot aperture electric field components $E_x(P')$, $E_z(P')$ occuring in the integrands. A method of moments type of solution can yield approximations to $E_x(P')$ and $E_z(P')$. We have followed this procedure, dividing the slot aperture into a two dimensional array of rectangular cells, using pulse functions and point matching, and have found for standard X-band waveguide and narrow slots ($w \le 0.063$ in) that the maximum value of $E_z(P')$ in the slot aperture is less than one three thousandth of the maximum value of $E_x(P')$. We therefore decided to ignore the longitudinal component of E-field in the slot, to discard (10), and to cast (11) in the form

$$\begin{aligned} H_z^{\mathrm{inc}}(P)&=jA_{10}\cos\left[\frac{\pi}{a}(x_o+\xi)\right]e^{-j\beta_{10}\zeta}\\ &=H_z^{\mathrm{ext}}(P)-H_z^{\mathrm{int}}(P)=\int_{\mathrm{slot}}E_x(P')G(P, P')\,dS' \end{aligned} \tag{12}$$

where A_{10} is the amplitude of the incident TE_{10} mode and

$$\begin{aligned} G(P, P')=&\left(\frac{\partial^2}{\partial\zeta'^2}+k_0^2\right)\left(\frac{e^{-jk_0R}}{2\pi j\omega\mu_0R}\right)+\frac{2}{j\omega\mu_0ab}\sum_{m=0}^{\infty}\sum_{n=0}^{\infty}\frac{\epsilon^2_{mn}}{\gamma_{mn}}\\ &\cdot\left\{\cos\left[\frac{m\pi}{a}(x_0+\xi)\right]\cos\left[\frac{m\pi}{a}(x_0+\xi')\right]\right.\\ &\left.\cdot\left(\frac{\partial^2}{\partial\zeta'^2}+k^2\right)e^{-\gamma_{mn}|\zeta-\zeta'|}\right\}. \end{aligned} \tag{13}$$

In (13), $\epsilon^2_{00} = 1/4$, $\epsilon^2_{m0} = \epsilon^2_{0n} = 1/2$, $\epsilon^2_{mn} = 1$ otherwise. Also, $k_0^2 = \omega^2\mu_0\epsilon_0$, $k^2 = \omega^2\mu_0\epsilon$ with ϵ the permittivity of the dielectric filling the guide, and

$$\gamma_{mn}=\sqrt{\left(\frac{m\pi}{a}\right)^2+\left(\frac{m\pi}{b}\right)^2-k^2}\qquad R=\sqrt{(\xi-\xi')^2+(\zeta-\zeta')^2}$$

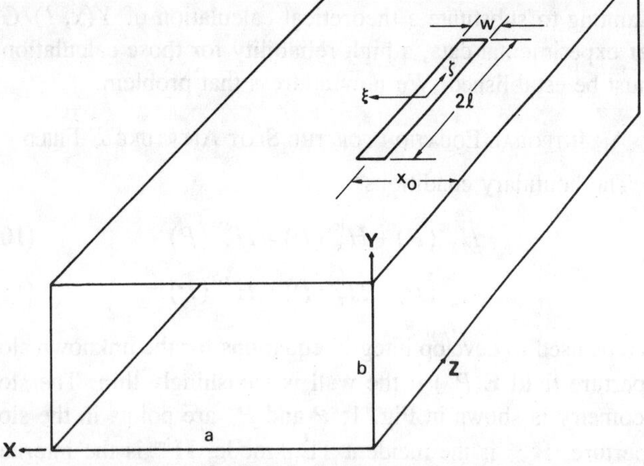

Fig. 1. Geometry of waveguide-fed longitudinal slot.

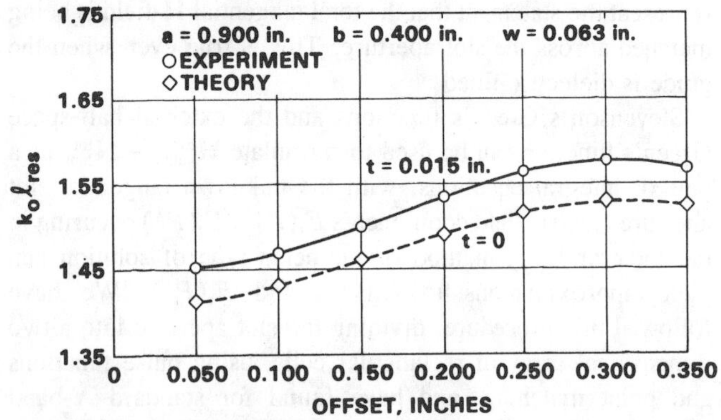

Fig. 2. Khac-type theory versus experiment. Resonant length versus offset for longitudinal slots. Square-ended slots in theory and experiment. No cell added. Wall thickness not taken into account.

where the local coordinates defined in Fig. 1 are used to represent $P(\xi, \zeta)$ and $P'(\xi', \zeta')$.

By dividing the slot aperture into N rectangular cells of size w by $(2l/N)$, by assuming $E_x(P')$ to be constant in a cell, and by satisfying (12) at the central point of each cell, we were able to reproduce Khac's results. The approximate solution obtained for $E_x(P')$ was used in the dominant mode backscattering term of Stevenson's Green's function to determine B_{10}, the amplitude of the backscatter. B_{10}/A_{10} was then used to find $Y(x, l)/G_0$, with resonance occurring when B_{10}/A_{10} was pure real negative. We found, as had Khac, that the computed results were typically in excellent agreement with experiment for $h_1(y)$, $h_2(y)$, and $g(x)$, but satisfactory agreement was not achieved for $v(x)$. For example, Fig. 2 shows the comparison between theory and experiment for longitudinal slots 0.063 in wide, cut in a 0.015 in broad wall of otherwise standard X-band waveguide (a = 0.900 in, b = 0.400 in). The actual slots were square-ended, so the principal difference between theory and experiment was that the theory assumed a "zero" wall thickness. We see that the two curves track each other quite well but that the theoretical curve lies 3 percent below the experimental curve. Even if we were to attribute 1 percent to wall thickness,[1] the remaining 2 percent gap is not acceptable when it comes to using this data in the design of arrays.

[1] Such an attribution would be crude because it turns out that Oliner's wall thickness correction is sensitive to slot length.

Attention next turned to improving on Khac's model. First, it was appreciated that, with a "zero" wall thickness, $E_x(P')$ should become infinite at $\xi = \pm w/2$ [8]. If we continue to make Khac's assumption that $E_x(\xi', \zeta')$ is separable, a better choice than pulse functions is to assume that

$$E_x(P') = \frac{w/2}{\sqrt{(w/2)^2-(\xi')^2}}\, F(\zeta'). \tag{14}$$

This causes (12), after an integration by parts, to adopt the form

$$\begin{aligned} jA_{10} \cos\left[\frac{\pi}{a}(x_0+\xi)\right] e^{-j\beta_{10}\zeta} &= \frac{k_0^2}{2\pi j\omega\mu_0}\int_{-l}^{l}\int_{-w/2}^{w/2} F(\zeta')\frac{w/2}{\sqrt{(w/2)^2-(\xi')^2}} \cdot \frac{e^{-jk_0R}}{R}\, d\xi'\, d\zeta' \\ &- \frac{1}{2\pi j\omega\mu_0}\int_{-l}^{l}\int_{-w/2}^{w/2}\frac{dF(\zeta')}{d\zeta'}\frac{w/2}{\sqrt{(w/2)^2-(\xi')^2}} \cdot (1+jk_0R)(\zeta-\zeta')\frac{e^{-jk_0R}}{R^3}\, d\xi'\, d\zeta' \\ &+ \frac{\pi w}{j\omega\mu_0 ab}\sum_{m=0}^{\infty}\sum_{n=0}^{\infty}\frac{\epsilon_{mn}^2}{\gamma_{mn}} J_0\left(\frac{m\pi w}{2a}\right) \cdot \cos\frac{m\pi x_0}{a}\cos\left[\frac{m\pi}{a}(x_0+\xi)\right] \\ &\cdot \int_{-l}^{l} F(\zeta')\cdot\left(\frac{\partial}{\partial\zeta'^2}+k^2\right)e^{-\gamma_{mn}|\zeta-\zeta'|}\, d\zeta'. \end{aligned} \tag{15}$$

In (15), J_0 is the Bessel function of zero order of the first kind. This equation is the departure point for various models which we employed. The most successful models all satisfied the boundary condition $E_x(P') = 0$ at $\zeta' \pm l$, that is, $F(l) = F(-l) = 0$. Good results were obtained with a piecewise sinusoidal formulation for $F(\zeta')$, viz.,

$$F(\zeta') = \sum_{n=1}^{N} K_n F_n(\zeta') \tag{16}$$

in which

$$\begin{aligned} F_n(\zeta') &= \frac{\sin\,[k_0(\zeta'-\zeta_{n-1})]}{\sin\,[k_0(\zeta_n-\zeta_{n-1})]}, \qquad \zeta_{n-1}\le\zeta'\le\zeta_n \\ &= \frac{\sin\,[k_0(\zeta_{n+1}-\zeta')]}{\sin\,[k_0(\zeta_{n+1}-\zeta_n)]}, \qquad \zeta_n\le\zeta'\le\zeta_{n+1} \end{aligned}$$

with

$$\zeta_n = n\frac{2l}{N+1}, \qquad n=0,\ 1,\ \cdots,\ N+1.$$

TABLE I
RESONANT LENGTH OF SQUARE-ENDED LONGITUDINAL SLOTS VERSUS OFFSET

OFFSET, INCHES	RESONANT LENGTH, INCHES		
	PIECEWISE SINUSOIDAL	GLOBAL GALERKIN	PERCENT DEVIATION
.050	.5935	.5925	.17
.100	.6022	.6017	.09
.150	.6159	.6157	.03
.200	.6328	.6326	.04
.250	.6494	.6486	.13
.300	.6597	.6587	.15
.350	.6595	.6583	.18

Theoretical calculations using (15). a = 0.900 in, b = 0.400 in, w = 0.0625 in, t = 0 in, ν = 9 GHz.

Good results were also obtained using a global Galerkin formulation, viz,

$$F(\zeta')=\sum_{m=1}^{N} K_m \sin\left[\frac{m\pi}{2l}(\zeta'+l)\right]. \tag{17}$$

As with Khac's model, use of (16) or (17) gave satisfactory agreement with experiment for $h_1(y)$, $h_2(y)$, and $g(x)$. But the crucial question remained "how good is the agreement for $v(x)$?" In anticipation of the desire to test a set of fixtures fabricated to address this question, computations were undertaken, using either (16) or (17) in conjunction with (15), for the case a = 0.900 in, b = 0.400 in, w = 0.0625 in, t = 0 in, and ν = 9 GHz. In using (16), we chose $\zeta_{n+1} - \zeta_n = 0.02$ in; in using (17), we chose M = 20.

Table I indicates the deviations between the two calculations for various offsets; the maximum deviation is 0.18 percent; the average deviation is only 0.11 percent. Computations using the piecewise sinusoidal model were less costly.

Test Fixtures

We desired to test the theoretical results for resonant length displayed in Table I, by constructing a family of fixtures, each containing one slot (but with different offsets), and with the slot imbedded in as thin a wall as possible. The technique which proved most satisfactory involved the machining of aluminum mandrels of rectangular cross section, with a = 0.900 in and b = 0.400 in. Each mandrel contained a raised rectangular boss whose position and dimensions (0.0625 in by $2l$) corresponded to the size and placement of the desired slot. Copper was electroplated onto the aluminum mandrel to a depth of 0.005″ after which a central area 1 in by 2 in, and including the boss, was masked. Electroplating was then

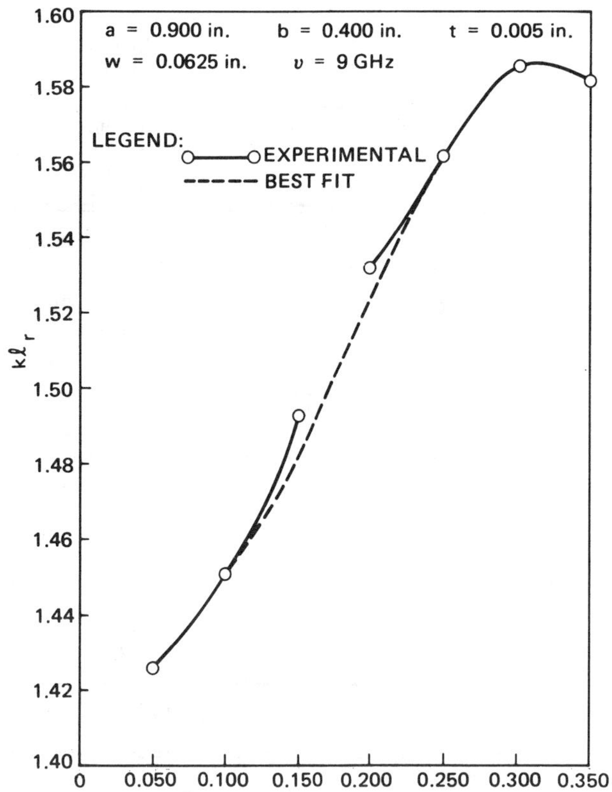

Fig. 3. Experimental values of resonant length versus slot offset.

continued until all unmasked walls were 0.05 in. The result, after the aluminum mandrel was etched away, was a piece of standard X-band waveguide except for a rectangular region around a square-ended slot, that region being only 0.005 in thick.

Seven such fixtures were constructed, each 7.6590 in long ($4\lambda_g$ at 9 GHz), and were fitted with flanges at both ends. The slot offsets relative to the waveguide centerline were to be 0.050 in, 0.100 in, $\cdots$, 0.350 in and the desired lengths $2l$ were read from the piecewise sinusoidal data of Table I.

Two of the fixtures proved defective (those for 0.150 in and 0.200 in offset). Both had rippled walls in the thin wall region and the 0.150 in fixture additionally had a bump in one broad wall near a flange. Added to this visual evidence was the experimental observation that each had a measured length about one percent higher than one would predict using a best-fit curve, as can be seen from Fig. 3.

Thus our sample was reduced from seven fixtures to five. The actual slot offsets and lengths for the remaining five fixtures are entered in the first and second columns of Table II.

The actual lengths of all five fixtures were within 0.001 in of specification and each slot was equidistant from the two ends of its fixture within 0.001 in. Four equispaced measurements of the **a** dimension of each fixture were made and the maximum deviation from 0.900 in was 0.0005 in.

After a fixture was flush-mounted into a 6.5 in by 8 in ground plane, dominant mode forward and backward scatter were measured on an automatic network analyzer over the frequency range 8 to 10 GHz. Column 3 of Table II lists the resonant frequencies for the five slots.

TABLE II
RESONANT LENGTH VERSUS OFFSET FOR SQUARE-ENDED LONGITUDINAL SLOTS; COMPARISON OF THEORY AND EXPERIMENT

ACTUAL SLOT OFFSET, INCHES	ACTUAL SLOT LENGTH, INCHES	MEASURED RESONANT FREQUENCY, MHz	SLOT LENGTH ADJUSTED TO 9 GHz	THEORETICAL COMPUTATION OF RESONANT LENGTH	PERCENT DEVIATION
.0507	.5960	8987	.5951	.5946	-.08
.1000	.6031	9031	.6053	.6035	-.30
.2020	.6486	9034	.6516	.6499	-.26
.3005	.6599	9028	.6625	.6586	-.59
.3498	.6597	8999	.6596	.6597	+0.02

$a = 0.900$ in, $b = 0.400$ in, $w = 0.0625$ in, $t = 0.005$ in, $\nu = 9$ GHz (nominal).

TABLE III
RESONANT LENGTH OF SQUARE-ENDED LONGITUDINAL SLOTS VERSUS OFFSET; THEORETICAL CALCULATIONS

	FULL-HEIGHT GUIDE, b = .400"			QUARTER-HEIGHT GUIDE, b = .100"		
OFFSET, INCHES	$2\ell_{res}$, INCHES	B_{10}/A_{10}	C_{10}/A_{10}	$2\ell_{res}$, INCHES	B_{10}/A_{10}	C_{10}/A_{10}
.050	.5935	-.0229	.0230 $\angle$-179.97°	.6575	-.0832	.0834 $\angle$-176.70°
.100	.6022	-.0832	.0833 $\angle$-177.75°	.7127	-.2546	.2565 $\angle$-174.09°
.150	.6159	-.1615	.1618 $\angle$-177.30°	.8303	-.3960	.4270 $\angle$-164.49°
.200	.6328	-.2400	.2407 $\angle$-176.60°	.9782	-.3877	.6064 $\angle$-142.64°
.250	.6494	-.3077	.3092 $\angle$-175.70°	1.0335	-.2590	.7582 $\angle$-130.68°
.300	.6597	-.3602	.3630 $\angle$-174.40°	1.0291	-.2031	.8295 $\angle$-127.87°
.350	.6595	-.3988	.4018 $\angle$-174.52°	1.0184	-.2072	.8818 $\angle$-128.36°

$w = 0.0625$ in, $t = 0$ in, $\nu = 9$ GHz.

Next, we used computed data for $l_{res}(x)$ at 9 GHz and 10 GHz, plus linear interpolation, to deduce what the actual length of each slot should have been in order for the slot to be resonant at 9 GHz. These values are entered in the fourth column of Table II. Finally, we performed an Oliner-type wall thickness correction [6] on the piecewise sinusoidal curve of Fig. 3, thereby obtaining the entries shown in the fifth column of Table II. The percent deviation between the values shown in columns 4 and 5 is listed in column 6 and can fairly be interpreted as the discrepancy between theory and experiment. We see that the maximum deviation is 0.59 percent; the average deviation is only 0.24 percent. Despite the excessive care we took in measuring back scatter for these five fixtures, we do not believe that our experimental error had an average value as low as one quarter of one percent. Therefore we reach the important conclusion that our theoretical calculations of resonant length versus offset are as trustworthy as the most careful experimental measurements we are able to make. When one pauses to appreciate that the theoretical approach works equally well for a dielectric-filled guide, whereas the experimental approach becomes more difficult, this conclusion becomes even more compelling.

THE SLOT VIEWED AS A SHUNT ELEMENT

It has been common practice to assume that a longitudinal slot is equivalent to a shunt element on a transmission line. This assumption is based on the premise that the dominant mode forward and backward scattering off the slot are equal and in phase, which in turn requires that the electric field in the slot aperture be symmetrical. Our numerical solutions based on (15) can test this assumption. We list in Table III the forward scatter and backward scatter for various slot offsets

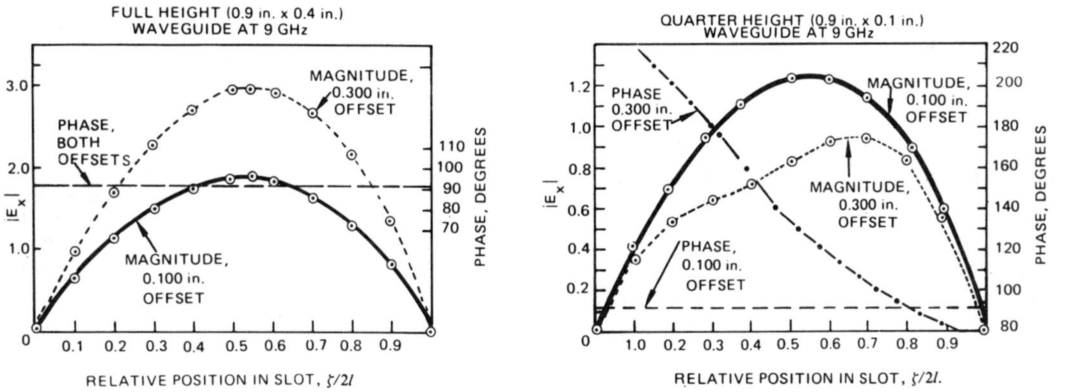

Fig. 4. Electric field in slot aperture versus slot offset for resonant-length slots. Full-height and quarter-height waveguide.

TABLE IV
FORWARD AND BACKWARD SCATTERED TE_{10} WAVES VERSUS SLOT OFFSET AND LENGTH FOR FULL-HEIGHT AND QUARTER-HEIGHT WAVEGUIDE

OFFSET	FULL HEIGHT (0.9" x 0.4") WAVEGUIDE @ 9 GHz				
0.050	LENGTH	0.560	0.580	0.600	0.620
	B/A	-0.015-j.011	-0.021-j.006	-0.023+j.003	-0.019+j.009
	C/A	-0.015-j.011	-0.021-j.007	-0.023+j.002	-0.019+j.008
0.200	LENGTH	0.600	0.620	0.640	0.660
	B/A	-0.2-j.086	-0.234-j.036	-0.24+j.02	-0.22+j.067
	C/A	-0.2-j.097	-0.234-j.05	-0.24+j.005	-0.22+j.05
0.350	LENGTH	0.620	0.640	0.660	0.680
	B/A	-0.35-j.13	-0.39-j.066	-0.4+j.002	-0.39+j.064
	C/A	-0.35-j.15	-0.39-j.098	-0.4-j.035	-0.39+j.021

OFFSET	QUARTER HEIGHT (0.9" x 0.1") WAVEGUIDE @ 9 GHz				
0.050	LENGTH	0.620	0.640	0.660	0.680
	B/A	-0.06-j.036	-0.078-j.020	-0.083+j.003	-0.077+j.023
	C/A	-0.06-j.039	-0.078-j.024	-0.083-j.002	-0.078+j.017
0.200	LENGTH	0.940	0.960	0.980	1.000
	B/A	-0.39-j.044	-0.39-j.021	-0.39+j.002	-0.38+j.026
	C/A	-0.46-j.35	-0.47-j.36	-0.48-j.37	-0.50-j.38
0.350	LENGTH	0.980	1.000	1.020	1.040
	B/A	-0.24-j.037	-0.23-j.017	-0.2+j.0015	-0.18+j.019
	C/A	0.50-j.66	-0.52-j.68	-0.55-j.69	-0.57-j.71

when slot length causes the back scatter to be out of phase with the incident TE_{10} mode. This is done for full-height and quarter-height X-band guide (a = 0.900 in, w = 0.0625 in, t = 0, ν = 9 GHz). We see that B_{10} and C_{10} become more disparate as the slot offset increases and/or as the b dimension decreases.

Concomitant with this is increasing asymmetry in the slot aperture field, as Fig. 4 attests. We see that for full-height guide, the asymmetry is slight enough to justify the equivalent shunt element assumption. However, for quarter-height guide, this is no longer the case. At the modest offset of 0.100 in, the phase of the aperture field is quite flat, but the amplitude is visibly skewed. At the large offset of 0.300 in, both amplitude and phase are strongly asymmetric.

These asymmetries have severe consequences not only at resonance, but also for slots of nonresonant length. Table IV shows B_{10} and C_{10} versus slot offset and length for full-height and quarter-height quide. A study of the entries in this table reveals that, for slot offsets of 0.200 in or more in quarter-height guide, C_{10}/A_{10} does not even go through a resonance, throwing into question the concept of resonant length. Even at 0.050 in offset, C_{10}/A_{10} goes through a resonance at a slot length 1 percent greater than does B_{10}/A_{10}.

The field asymmetries noted in Fig. 4 also have serious implications when one contemplates using an Oliner-type correction for wall thickness, since his theory relies on the assumption that the aperture field is essentially an equiphase half-cosinusoid.

One can conclude from all this that the shunt element assumption is less appropriate at larger offsets and can be more seriously questioned as the **b** dimension is reduced. For quarter-height guide the assumption has doubtful validity.

When it is inappropriate to use the shunt element assumption the design procedure described in [3] can be replaced by one which does away with equivalent circuits and deals directly with scattered waves.

Conclusion

It has been demonstrated that a method of moments type solution can give results for the resonant length of a waveguide-fed longitudinal slot versus its offset that are comparable in accuracy to a carefully performed experiment. This creates the opportunity to dispense with the more costly experimental gathering of input data in the design of slot arrays.

It has also been shown that the standard model in which a longitudinal slot is represented by a shunt element on an equivalent transmission line is less justified at larger slot offsets and for smaller **b** dimensions.

References

[1] R. S. Elliott and L. A. Kurtz, "The design of small slot arrays," *IEEE Trans. Antennas Propagat.*, vol. AP-26, pp. 214–219, Mar. 1978.

[2] R. S. Elliott, *Antenna Theory and Design.* Englewood Cliffs, NJ: Prentice-Hall, 1981, pp. 397–423.

[3] ——, "An improved design procedure for small slot arrays," *IEEE Trans. Antennas Propagat.*, vol. AP-31, pp. 48–53, Jan. 1983.

[4] A. F. Stevenson, "Theory of slots in rectangular waveguides," *J. Appl. Phys.*, vol. 19, pp. 24–38, Jan. 1948.

[5] R. J. Stegen, "Longitudinal shunt slot characteristics," Hughes Tech. Memo. 261, Hughes Aircraft Co., Culver City, CA, Nov. 1951. (Design curves reprinted in [2], pp. 408–410.)

[6] A. A. Oliner, "The impedance properties of narrow radiating slots in the broad face of rectangular waveguide," *IRE Trans. Antennas Propagat.*, vol. AP-5, pp. 1–20, Jan. 1957.

[7] T. V. Khac, "A study of some slot discontinuities in rectangular waveguides," Ph.D. dissertation, Monash Univ., Australia; Nov. 1974.

[8] R. Mittra and S. W. Lee, *Analytical Techniques in the Theory of Guided Waves,* New York: MacMillan, 1971, pp. 4–11.

George J. Stern (S'58–M'61–M'69–SM'84) was born in Los Angeles, CA, in 1931. He received the B.S. and M.S. degrees from the University of Southern California, Los Angeles, in 1959 and 1960, respectively.

From 1960 to 1965 he worked on monopulse avionics antenna problems at Hughes Aircraft Company, Los Angeles, CA, and from 1965 to 1973 was responsible for slot array design at Rantec Corporation, Calabasas, CA. Since 1973, he has been with the Hughes Missile Systems Group. He is currently involved in the analysis and design of small electronically scanned antennas and characterization of waveguide slot radiators.

Mr. Stern is a member of Tau Beta Pi and Eta Kappa Nu.

Robert S. Elliott (S'46–A'52–SM'54–F'61) received the A.B. degree in English literature and the B.S. degree in electrical engineering from Columbia University, New York, NY, in 1942 and 1943, respectively, the M.S. and Ph.D. degrees in electrical engineering from the University of Illinois, in 1947 and 1952, respectively, and the M.A. degree in economics from the University of California, Santa Barbara, in 1971.

His experience includes periods at the Applied Physics Laboratory of John Hopkins and at the Hughes Research Laboratories, where he headed the antenna research activities. He has also been on the faculty of the University of Illinois and was a founder of Rantec Corporation, serving as its first Vice President and Technical Director. He has been a Professor of Electrical Engineering at the University of California, Los Angeles, since 1957. He is currently serving a second two-year stint as a Distinguished Lecturer for the IEEE and was Chairman of the Coordinating Committee for the 1981 IEEE Symposium, held in Los Angeles. He currently serves as a consultant to Hughes, Canoga Park.

Dr. Elliott, is a member of Sigma Xi, Tau Beta Pi, and the New York Academy of Sciences. He has received numerous teaching prizes and is the recipient of two Best Paper awards by the IEEE.

Theoretical design/synthesis of slotted waveguide arrays

A.J. Sangster
A.H.I. McCormick

Indexing terms: Antennas (arrays), Antennas (slot)

Abstract: A computer aided design/synthesis procedure for array antennas of the slotted waveguide type is described. It differs from earlier contributions to this topic in that it uses theoretically evaluated self-admittances, rather than measured self-admittances, for the individual broadwall slots in the array. By adopting the moment method to generate the slot self-admittances, it is shown that accuracies which are comparable with those obtained from a measurement based procedure can be achieved.

1 Introduction

Antennas formed from one or more slotted waveguide arrays are used widely in radar and microwave communication systems because of their flat, compact structure. Furthermore they can be constructed to a high degree of accuracy on modern numerically controlled milling machines, and this permits very low side-lobe designs to be realised.

Design methods for slotted rectangular waveguide array antennas have been evolving for some forty years now. The early developments were largely aimed at understanding and formulating methods for calculating the scattering characteristics of individual apertures in waveguide. This work has now progressed to the point where, through the moment method [1–4], calculated results for broadwall slot characteristics, such as self-admittance, differ by less than 1% from the best measured results [4], for some commonly used slot examples. In parallel with these aperture investigations, effort has also been directed towards evolving procedures for forming arrays of such apertures [5], and in particular at methods of accommodating the mutual effects [6], which become important in low side-lobe level designs. The outcome of this work has been the formulation of a relatively sophisticated synthesis procedure for slotted-waveguide array antennas [5].

However, the synthesis technique referred to above, which requires that the self-admittances of individual slots in the array be known to a high degree of accuracy, has usually been implemented using accurately measured values for the self-admittances. This, of course, implies that it is possible to synthesise effectively only those arrays for which measured results already exist for the constituent slot radiators. It is the purpose of this paper to show that this is becoming an increasingly unnecessary restriction, and that modern high speed computers permit the incorporation, into a synthesis program (e.g. described by Elliott and Kurtz [5]) a moment method computation of the self-admittance of each aperture, as and when demanded by the synthesis program. The technique has been verified, firstly by repeating a broadwall slotted-waveguide array design described by Elliott [7] and secondly by employing it to design a 19 element shunt-slot array on which accurate near-field measurements have been performed.

2 Synthesis of slotted waveguide array antenna

The synthesis procedure described in this Section is based to a large extent on the method described by Elliott and Kurtz [5], but it incorporates a number of essential modifications to accommodate the inclusion of the moment method, which has been introduced to compute the self-admittances of individual apertures. Without these modifications the resultant CAD package would be excessively demanding on computer storage space and processor time.

The array is assumed to be uniform, although this need not be the case, and the required excitation levels of the array elements, to meet the prescribed far-field pattern, are determined using a standard technique. Such techniques exhibit varying degrees of sophistication, depending on the pattern constraint and array geometry. They are well documented in the literature [8–10]. However, for the purposes of demonstrating the principle of slotted-waveguide array synthesis using the moment method, a relatively simple far-field pattern specification is assumed, which can be represented by a Dolph-Chebyshev [11] function. The Schelkunoff unit circle technique is used to determine the number of slots required to meet the specification. The array factor is then deduced from the roots, hence the active scattering parameters for each slot can be evaluated. The active parameter for any given slot includes both the self-effect and the mutual effect of its neighbours.

In traditional slotted-waveguide array synthesis programmes, the self-effect, usually in the form of self-admittance values, is extracted from measured data. Here, theoretical values, obtained using the moment method, are calculated as and when demanded by the synthesis program. Consequently, the program contains three essential components, namely the formulation of the active admittances of the slots, the calculation of the mutual admittances and finally the calculation of the self-admittances. The methods used to perform these calculations are summarised below.

Paper 6377H (E11) first received 16th May, and in revised form 26th August 1988

A.J. Sangster is, and A.H.I. McCormick was formerly, with the Electrical and Electronic Engineering Department, Heriot-Watt University, Grassmarket, Edinburgh, EH1 2HT, United Kingdom

A.H.I. McCormick is now with Plessey Radar Ltd., Newport Road, Cowes, Isle of Wight, United Kingdom

3 Derivation of algorithms

The active admittances of the elements in the array may be determined routinely once the array factor is known. This is achieved by realising that the weighting of the

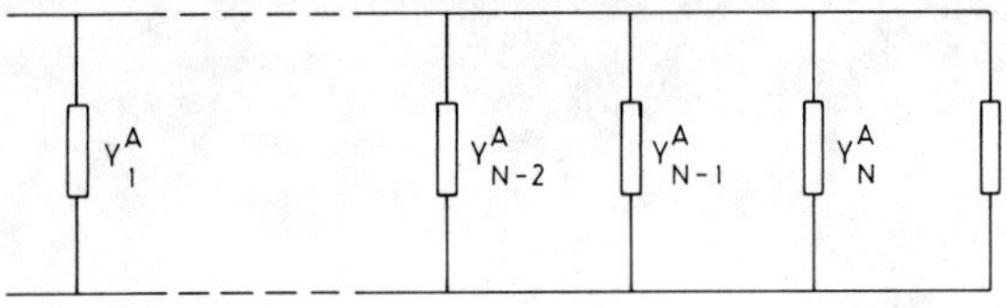

Fig. 1 *Equivalent circuit for travelling wave array*

coefficients of the array factor is the same as the weighting of the slot voltages across the array. The equivalent circuit for a travelling wave array is illustrated in Fig. 1, where Y_L is the load admittance, and $Y^A_{n(n=1,\ldots,N)}$ are the respective slot admittances. Their values are determined by applying scattering theory to ascertain the element reflection coefficient ρ_n, hence by using

$$Y^A_n = \frac{2\rho_n}{1-\rho_n} \tag{1}$$

Equations relating the active admittances Y^A_n to the self-admittances Y_n, and the mutual admittances Y_{mn} between the m and n–th slots, are to be found in References 5–7. To implement these equations, in computational form, preliminary knowledge of slot spacings and off-sets is required in the case of a broadwall slot array. However, the program can be initiated using fairly approximate starting values for these dimensions, and these are deduced from the array factor followed by applications of Stevenson's conductance formula [12]. The active admittance relations can then be implemented to yield first pass values for the self-admittances Y_n.

Once a set of self-admittances for the slot array have been determined, as outlined above, new slot lengths (l_n) and slot off-sets (s_n) are deduced in a two stage process. This involves both the use of the variational method [13], which generates rapidly calculable closed-form solutions, followed by the use of the moment method which does not. This two stage approach is dictated by the need to include the moment method into a corrected Gauss-Newton root-finding routine, which searches for the slot length and slot off-set appropriate to any given value of self-admittance. In order that this process should not be excessively demanding in computing time in a multi-slot array design, it is imperative that 'good' trial values for slot length and slot off-set are available.

For the first pass through the program, and for the first two or three slots, sufficiently 'good' trial values for l_n and s_n can be computed relatively quickly, using the closed-form variational equations. These have been corrected for wall thickness using Reference 14. The procedure adopted in the program is to eliminate s_n from the conductance part of the self-admittance equation, then to solve for l_n using a bisectional root-finding routine. The slot offsets are then deduced by substitution.

The variational values for l_n and s_n are now, as indicated above, used as trial values in a more accurate root-finding routine which incorporates the moment method. The particular method adopted is based on the use of trigonometric basis functions, since with these, computation speeds can be maximised [3]. The program has been implemented using a single cosine function as the basis function. While this involves some sacrifice in accuracy, the time savings which accrue are very significant. Computational speeds can also be enhanced by taking advantage of the symmetry of the moment method equations. This is demonstrated below.

The moment method solution for aperture setting in a waveguide can be summarised as a matrix equation

$$\begin{bmatrix} [A] & [C] \\ [D] & [B] \end{bmatrix} \begin{bmatrix} [a] \\ [b] \end{bmatrix} = \begin{bmatrix} [h] \\ [o] \end{bmatrix} \tag{2}$$

the elements of which can be expressed in the following integral forms:

$$A_{i,s} = -j\omega\varepsilon_0 \iint_S f_i \hat{\underline{a}}_n \cdot \left\{ \hat{n}x \iint_S [\boldsymbol{G}_a(\underline{r}/\underline{r}_0) + \boldsymbol{G}_b(\underline{r}/\underline{r}_0)] \cdot f_s \hat{\underline{a}}_\xi \, dS_0 \right\} dS \tag{3}$$

$$B_{i,s} = -j\omega\varepsilon_0 \iint_{S'} f'_i \hat{\underline{a}}_n \cdot \left\{ \hat{n}'x \iint_{S'} [\boldsymbol{G}_b(\underline{r}/\underline{r}_0) + \boldsymbol{G}_c(\underline{r}/\underline{r}_0)] \cdot f'_s \hat{\underline{a}}_\xi \, dS_0 \right\} dS \tag{4}$$

$$C_{i,s} = j\omega\varepsilon_0 \iint_S f'_i \hat{\underline{a}}_n \cdot \left\{ \hat{n}x \iint_{S'} \boldsymbol{G}_b(\underline{r}/\underline{r}_0) \cdot f_s \hat{\underline{a}}_\xi \, dS_0 \right\} dS \tag{5}$$

$$D_{i,s} = j\omega\varepsilon_0 \iint_{S'} f_i \hat{\underline{a}}_n \cdot \left\{ \hat{n}x \iint_S \boldsymbol{G}_b(\underline{r}/\underline{r}_0) \cdot f'_s \hat{\underline{a}}_\xi \, dS_0 \right\} dS \tag{6}$$

$$h_i = -\iint_S f_i \hat{\underline{a}}_n \cdot \{\hat{n}x\underline{H}_i(\underline{r})\} \, dS \tag{7}$$

$\boldsymbol{G}_{a,b,c}$ are the Green's functions of the waveguide, cavity and halfspace, respectively, while f_i and f_s are the basis functions. H_i is the magnetic field component of the TE_{10} incident mode in the waveguide.

Equations 2–7 display both a considerable degree of repetitiveness, and a high degree symmetry. This can be used to advantage in developing a computer aided design package. Basic solutions have been developed and reported in Reference 3, and these have been used as the starting point for algorithm formulation.

At any given slot in the array, the solution of eqn. 2 involves multiple calculations over the very large number

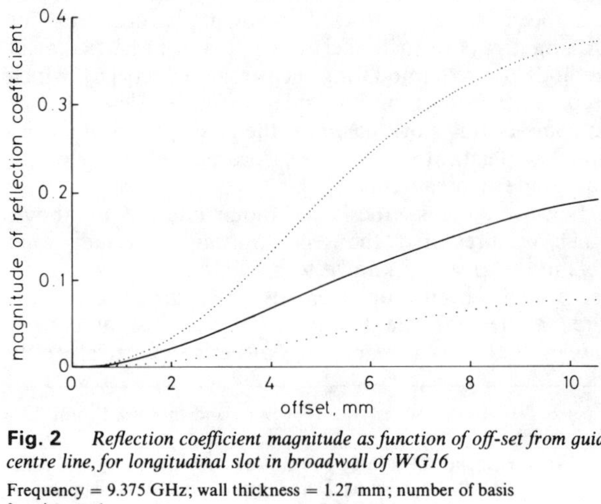

Fig. 2 *Reflection coefficient magnitude as function of off-set from guide centre line, for longitudinal slot in broadwall of WG16*

Frequency = 9.375 GHz; wall thickness = 1.27 mm; number of basis functions = 1
Slot length:
.... 16 mm
——— 14 mm
. . . . . 12 mm

of modes, (typically 2500), associated with the Green's function G_a. Fortunately, however, many modal characteristics are invariant from slot to slot in an array

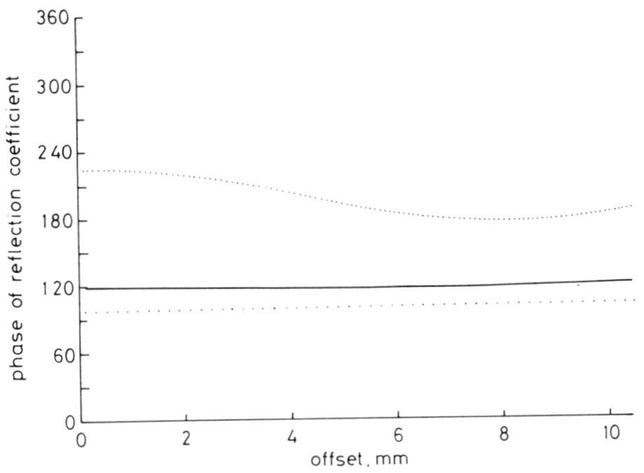

Fig. 3 *Reflection coefficient phase as function of off-set from guide centre line, for longitudinal slot in broadwall of WG16*

Frequency = 9.375 GHz; wall thickness = 1.27 mm; number of basis functions = 1
Slot length:
.... 16 mm
——— 14 mm
· · · · · 12 mm

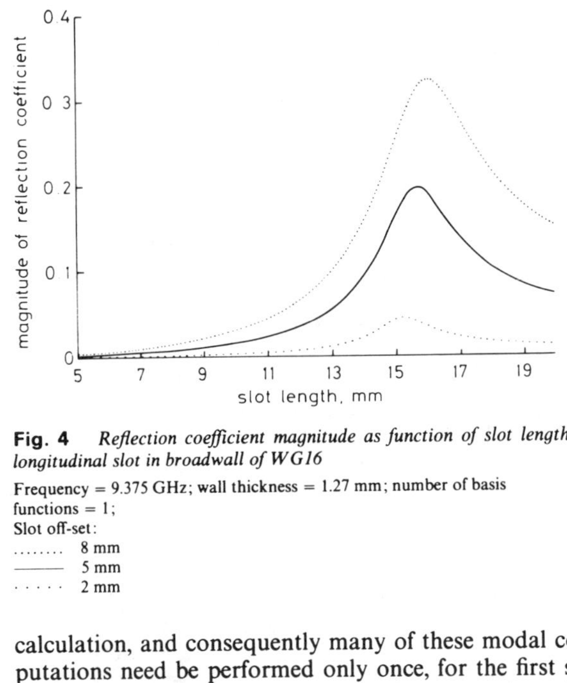

Fig. 4 *Reflection coefficient magnitude as function of slot length, for longitudinal slot in broadwall of WG16*

Frequency = 9.375 GHz; wall thickness = 1.27 mm; number of basis functions = 1;
Slot off-set:
.... 8 mm
——— 5 mm
· · · · · 2 mm

calculation, and consequently many of these modal computations need be performed only once, for the first slot, values being stored for re-use in subsequent slot calculations. Computational speed in evaluating the integrals in eqns. 3–6 is also critically dependent on evolving efficient convergence testing procedures.

Finally, these moment method equations are incorporated into a corrected Gauss-Newton root-finding routine which generates improved values for l_n and s_n. However, rather than search for slot self-admittance in this routine, it has been found that searching for reflection coefficient ρ is more effective. This is because with ρ it is easier to separate the effect of slot length from that of slot off-set, as a result of the differing ways in which these dimensions influence the magnitude and the phase of ρ, (Figs. 2–5). More particularly, the program makes use of the observation that the magnitude of ρ is almost linearly dependent on off-set, while its argument is almost linearly dependent on slot length for small changes in these dimensions near resonance.

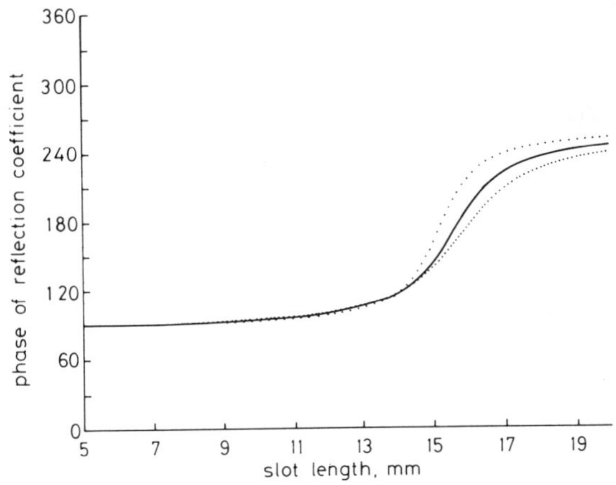

Fig. 5 *Reflection coefficient phase as function of slot length, for longitudinal slot in broadwall of WG16*

Frequency = 9.375 GHz; wall thickness = 1.27 mm; number of basis functions = 1;
Slot off-set:
.... 8 mm
——— 5 mm
· · · · · 2 mm

4 Program implementation

A flow diagram showing the basic steps in the synthesis program is shown in Fig. 6. Most of the steps shown are reasonably self-explanatory, but it is perhaps pertinent to point out some aspects of the flow chart which have been successful in promoting computational efficiency.

On the first pass through the program, and for the first slot in the array, improved values for l_1 and s_1 are deduced using the variational calculation. These are then used as trial estimates in the moment method search routine, to generate the desired first pass values for l_1 and s_1. In subsequent passes trial values for the moment method routine are obtained from the previous pass. At slot 2, on the first pass, l_1 and s_1 are used as trial values in a variational search for s_2, which is used with l_1 as suitable trial values for the moment method search for first pass values for l_2 and s_2. For slot n ($n \geqq 3$) the previous value of slot length and an estimated slot-off-set, deduced by interpolation from the gradient of $|\rho|$, are used as trial values in the Gauss-Newton search for l_n and s_n. The computation is repeated, as indicated on the flow-chart, until the values of l_n and s_n change from their values on the previous pass, by an increment which is set at just less than realistic fabrication tolerances. It has been found that 3 to 4 passes is usually enough to achieve this condition.

The final step in the program involves calculating the small radiation phase and transmission phase deviations which occur at each slot, as a consequence of its finite distributed length. These phase deviations are incorporated into the design by making minor compensating adjustments to the slot separation distances (see Fig. 7).

5 Theoretical and experimental results

The array synthesis package described in the preceding Sections has been tested in two ways. Firstly, it has been used to design an array previously described by Elliott

[7] and for which directly comparable results are available. Secondly, it has been used to design a 19 element shunt slot array which, once constructed, was

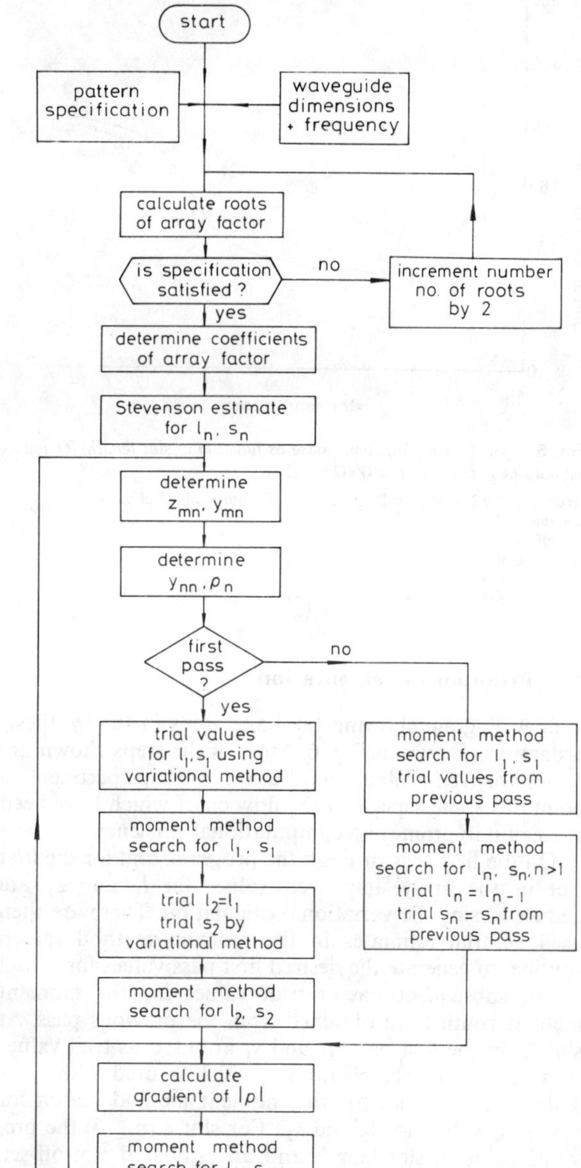

Fig. 6 *Array synthesis program flow diagram*

comprehensively evaluated on an accurate near-field measurement system.

The Elliott array is a Dolph-Chebyshev design with a −30 dB sidelobe specification with a mainbeam at 45° from forward end-fire. The frequency of operation was specified as 9.375 GHz. The synthesis program predicts that a 21 element array is required, dimensioned as indicated in Table 1.

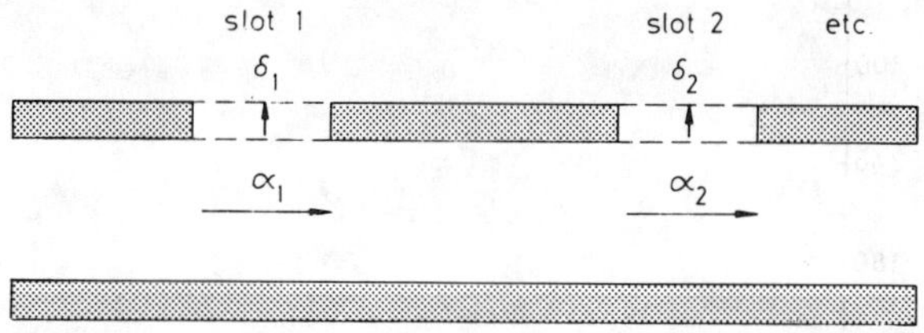

Fig. 7 *Schematic diagram of waveguide slots showing α and δ phase shifts*

α = phase shift passed the slot
δ = phase shift through the slot

Table 1 : Elliott 21 element array

No.	Length, mm,	Offset, mm	No.	Length, mm	Offset, mm
1	15.374	1.689	12	15.167	1.794
2	15.214	1.035	13	15.141	1.606
3	15.273	1.474	14	15.123	1.420
4	15.302	1.833	15	15.094	1.236
5	15.306	2.131	16	15.066	1.044
6	15.313	2.315	17	15.040	0.866
7	15.308	2.375	18	14.996	0.715
8	15.296	2.353	19	14.984	0.551
9	15.270	2.273	20	15.013	0.374
10	15.243	2.144	21	14.930	0.562
11	15.212	1.978			

The slot length and slot off-set magnitudes shown in this Table, are substantially in agreement with the corresponding Elliott values, when due allowance is made for the fact that Elliott's design is based on the use of measured self-admittance values for round-ended slots and also contains a constraint, which ensures that the resultant slot dimensions are consistent with data file values. In Fig. 8, comparison is also made between theoretical and experimental results for this array. The Figure shows the specified pattern (solid curve), a pattern deduced by applying the method of Reference 7 to the slot distribution of Table 1 (dotted curve), and a measured pattern reconstructed from Elliott [7] (chain dotted curve). In connection with this Figure, it is particularly interesting to note the discrepancy which occurs between the specified pattern (solid curve) and the pattern associated with the synthesised design. The discrepancy appears predominantly in the side lobe levels which are generally lower than their 'ideal' value, except for those most remote from the main beam. This error can largely be attributed to the method used to evaluate mutual coupling, which relies on analogy with dipole theory. This analogy necessitates finding a relationship between slot voltages and mode voltages, and currently this involves the use of the approximate formula of Stevenson [12]. A more direct method of calculating the mutual impedances in slot arrays is clearly indicated, if this weakness in the synthesis procedure is to be eliminated. However, when the synthesised results are compared with Elliott's measurements (based on 144 data points; chain linked line, Fig. 8) the degree of agreement indicated is not untypical of synthesised slot array designs.

A nineteen element shunt slot array with staggered slots has also been designed using the synthesis procedure described above. The specification for this array required the main beam to be 9° from broadside, with −24 dB sidelobes and in X-band waveguide (WG16) at a centre frequency of 9.375 GHz. The resultant design which uses square-ended 1.6 mm wide slots is presented in Table 2.

coupling a staggered array suffers. This in turn means that the Stevenson approximation which has been identified as the primary cause of these theoretical discrepancies is much less significant for this array. A measured pattern for this array is compared with the 'ideal' pattern

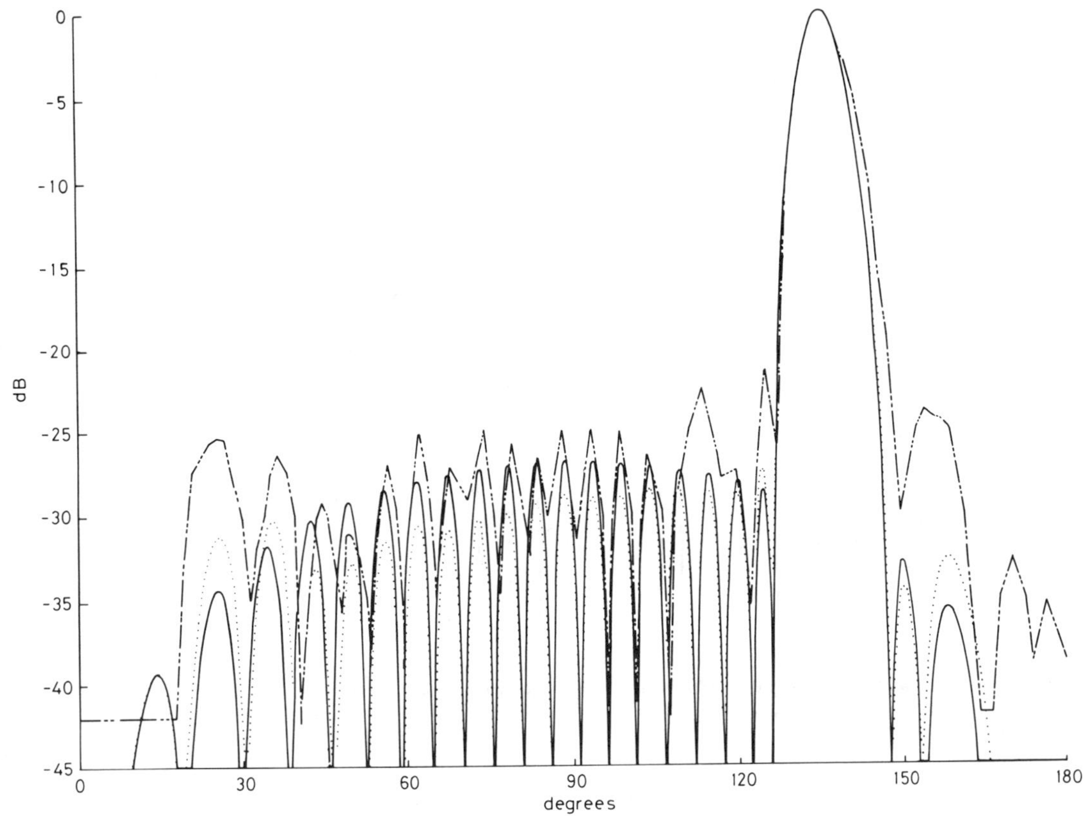

Fig. 8 *Comparison of theoretical and measured far-field patterns for 21 element linear slotted-waveguide array*

Frequency = 9.375 GHz; wall thickness = 1.27 mm; waveguide = WG16; slot width = 1.6 mm
——— ideal pattern
.... predicted pattern
· · — · · measured pattern

in Fig. 10, and shows that while the main beam position and width are in good agreement with theory, the far-out side-lobes in particular are considerably in error. The reasons for this error are predominantly due to machining inaccuracies, observed using an electron microscope, of the square-ended slots. These inaccuracies are highlighted by near-field amplitude and phase patterns for the array (Figs. 11 and 12), which show that slots 10, 15, 18 and 19 are particularly poorly formed, and when they are computer corrected to eliminate the gross near-field pattern deviations, indicated in Figs. 11 and 12, the far-field pattern of the array is modified as shown in Fig. 13. The adjusted pattern is represented by the solid line curve, and shows that the far-out side lobes which were particularly in error, are significantly improved by this simple re-alignment.

Table 2: Shunt slot array

No.	Length, mm	Offset, mm	Separation, mm
1	15.171	+2.034	18.36
2	15.122	−1.485	18.34
3	15.109	+1.701	18.33
4	15.121	−2.046	18.31
5	15.157	+2.280	18.30
6	15.167	−2.467	18.29
7	15.189	+2.564	18.29
8	15.191	−2.595	18.29
9	15.185	+2.549	18.30
10	15.178	−2.447	18.31
11	15.159	+2.301	18.32
12	15.151	−2.122	18.34
13	15.150	+1.918	18.35
14	15.146	−1.702	18.36
15	15.133	+1.466	18.37
16	15.136	−1.226	18.38
17	15.158	+1.006	18.36
18	15.013	−0.826	18.38
19	15.206	+1.086	

The predicted far field pattern for this array (dotted curve) has been compared with the 'ideal' or specified pattern (solid line) in Fig. 9. The differences between the two patterns are less than those observed for the Elliott array, and this is due to the lower level of the mutual

6 Conclusions

A computer aided design package for the synthesis of rectangular waveguide broadwall slot arrays has been described. The package differs from preceding contributions in this area in that the self-admittances of the slots are computed, rather than taken from a data bank of measured values. The self-admittance computations rely on the use of the moment method, which is known to be

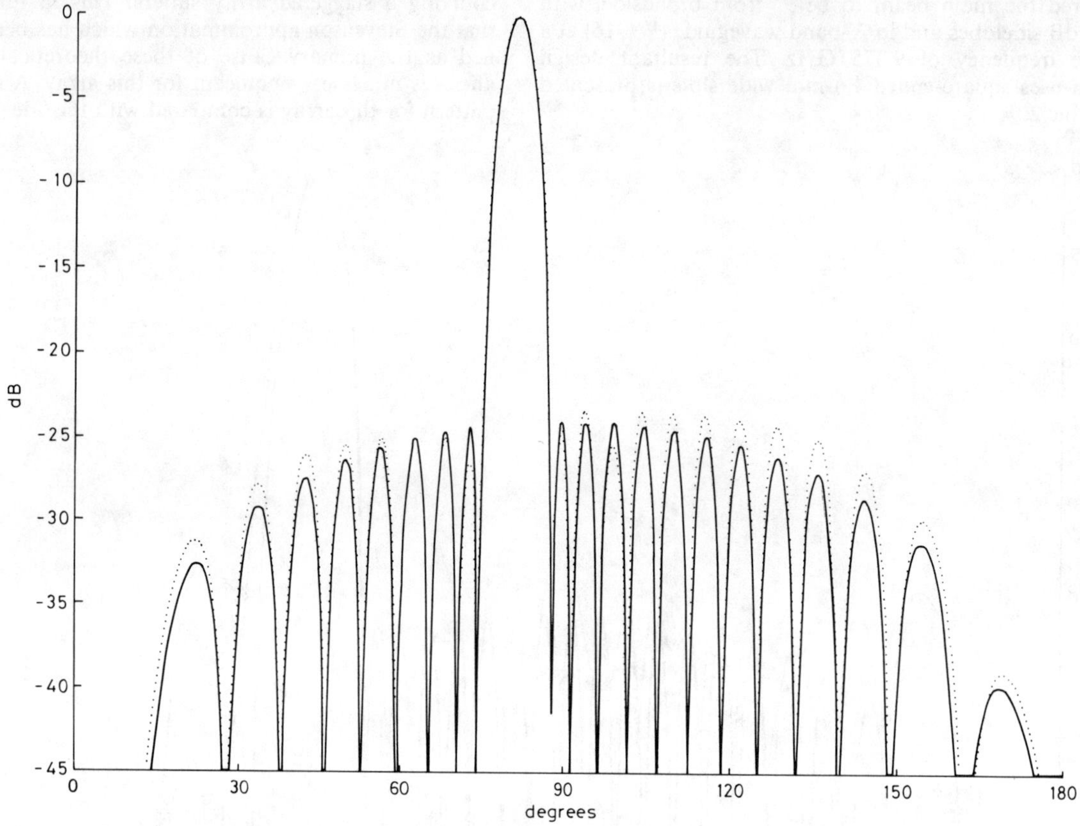

Fig. 9 *Comparison of theoretical far-field patterns for 19 element linear slotted waveguide array*

Frequency = 9.375 GHz; wall thickness = 1.27 mm; waveguide = WG16
——— ideal pattern
······ predicted pattern

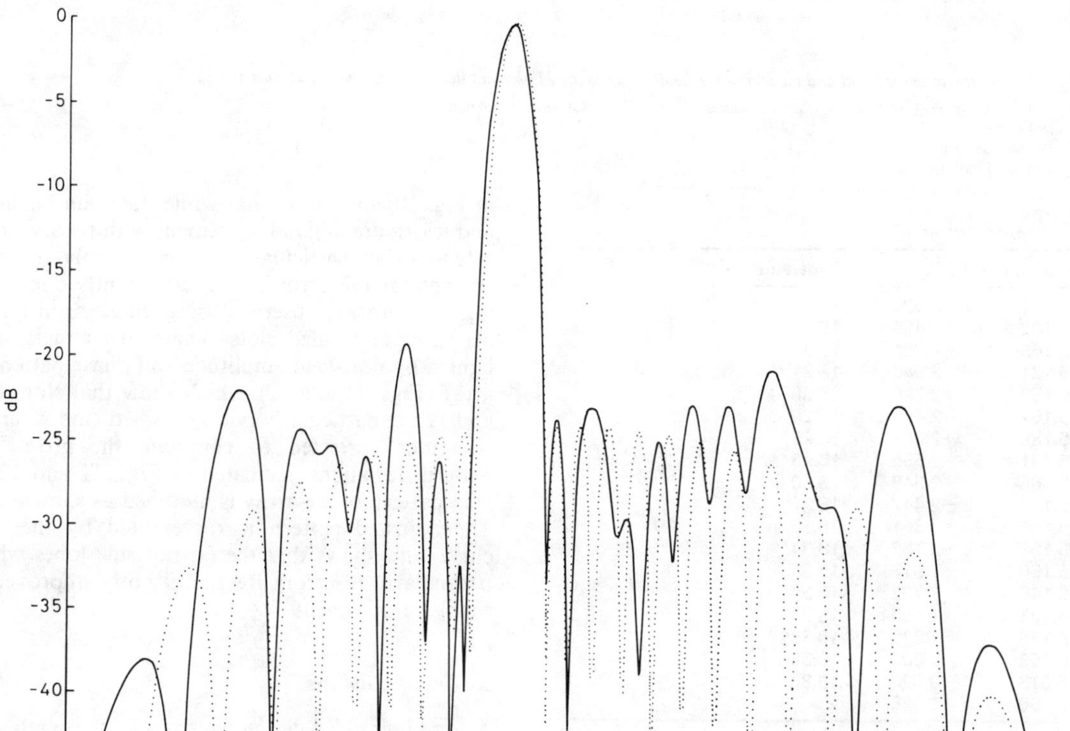

Fig. 10 *Comparison of theoretical and measured far-field patterns for 19 element linear slotted-waveguide array*

Frequency = 9.375 GHz; wall thickness = 1.27 mm; waveguide = WG16
——— measured pattern
······ ideal pattern

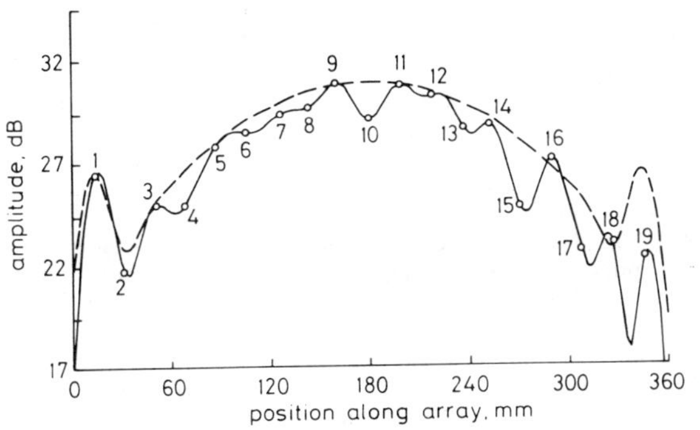

Fig. 11 *Magnitude comparison of theoretical and measured near-field patterns for 19 element slotted waveguide array*

- - - - theory
——— experiment

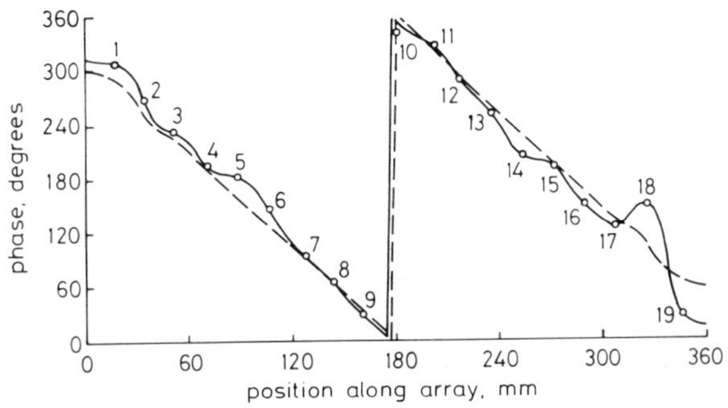

Fig. 12 *Phase comparison of theoretical and measured near-field pattern for 19 element slotted-waveguide array*

- - - - theory
——— experiment

potentially as accurate as the best measurements in evaluating slot self-scattering effects. To accommodate the moment method, which does not, unfortunately, yield rapidly calculable closed-form expressions for the scattering parameters of a slot, considerable care has been exercised in developing time saving computational procedures.

The resultant CAD package has been applied to the design of two linear slotted waveguide arrays, and in one case the design obtained has been shown to be compatible with a more conventional design, which relied on the use of measured slot-admittances. Near-field measurements on the second of the two arrays have been shown to be in reasonable agreement with theory, once certain constructional limitations are taken into consideration.

Some continuing limitations in the CAD package have also been identified, but these could readily be eliminated once computational speeds permit. For example, the use of a more elaborate basis function in the moment method would improve the basic accuracy of the program, as would a more direct method of calculating the mutual impedance between the slots. This would eliminate the disadvantage of requiring to relate slot voltages to mode voltages, which the 'indirect' equivalent dipole technique entails.

7 Acknowledgment

This work was sponsored by the UK Science and Engineering Research Council and the authors wish to record their gratitude for this support. Thanks are also due to Mr. J. Tanner of Ferranti plc, Edinburgh, for the care and effort which he put into the experimental work.

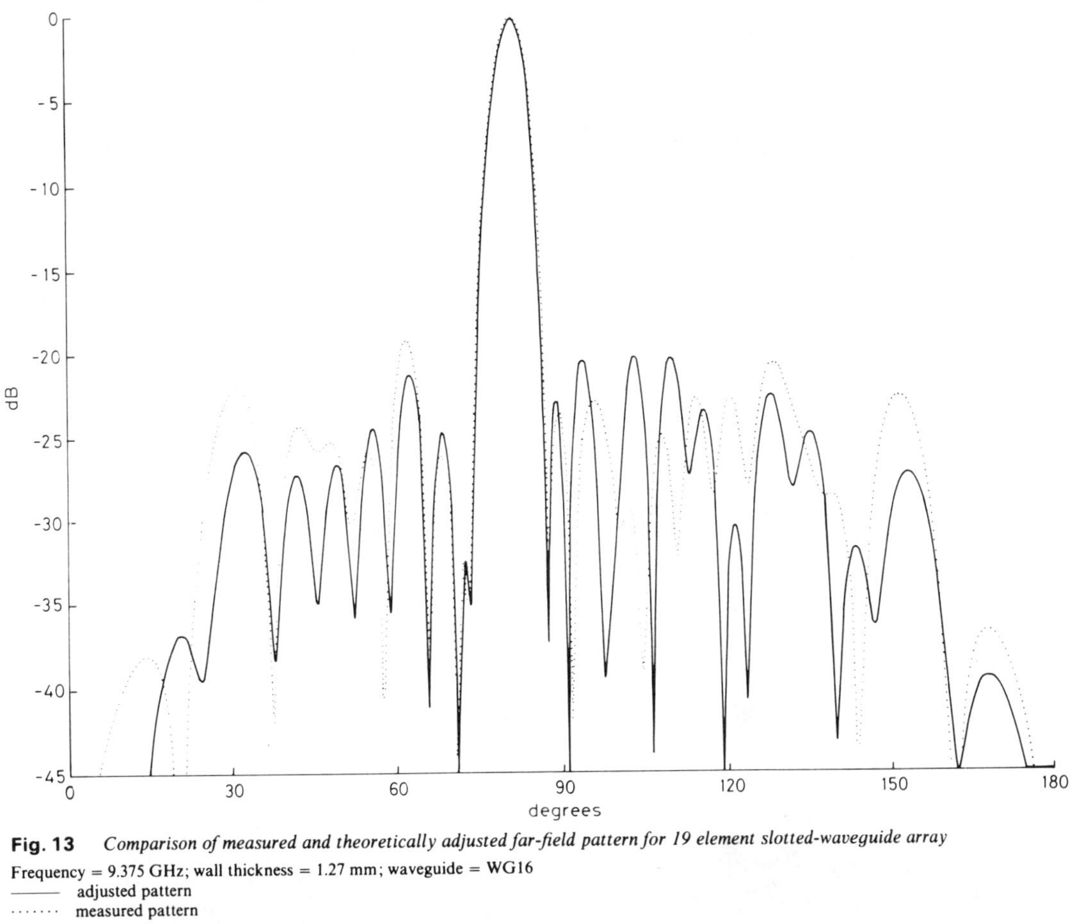

Fig. 13 *Comparison of measured and theoretically adjusted far-field pattern for 19 element slotted-waveguide array*

Frequency = 9.375 GHz; wall thickness = 1.27 mm; waveguide = WG16
——— adjusted pattern
······· measured pattern

8 References

1 HARRINGTON, R.F.: 'Field computation by moment methods' (MacMillan, London, 1968)

2 VU KHAC, T., and CARSON, C.T.: 'Impedance properties of longitudinal slot antenna in the broadface of rectangular waveguide', *IEEE Trans.*, 1973, **AP-21,** pp. 708–710

3 SANGSTER, A.J., and LYON, R.W.: 'An efficient moment method analysis of radiating slots in thick-walled waveguide', *IEE Proc. H, Microwaves, Opt. & Antennas*, 1981, **128,** (4), pp. 197–205

4 STERN, G.T., and ELLIOTT, R.S.: 'Resonant length of longitudinal slots and validity of circuit representation: theory and experiment', *IEEE Trans.*, 1985, **AP-33,** pp. 1264–1271

5 ELLIOTT, R.S., and KURTZ, L.A.: 'The design of small slot arrays', *ibid.*, 1978, **AP-26,** pp. 214–219

6 BAKER, H.C., and LAGRONE, A.H.: 'Digital implementation of the mutual impedance between thin dipoles', *IRE Trans.*, 1962, **AP-10,** pp. 172–178

7 ELLIOTT, R.S.: 'On the design of travelling wave-fed longitudinal shunt slot arrays', *IEEE Trans.*, 1979, **AP-27,** pp. 717–720

8 MAUTZ, J.R., and HARRINGTON, R.F.: 'Computational methods for antenna pattern synthesis', *ibid.*, 1975, **AP-23,** pp. 507–512

9 NARASIMHAN, M.S., VARADARANGAN, K., and CHRISTOPHER, S.: 'A new technique of synthesis of near of far-field patterns of arrays', *ibid.*, 1986, **AP-34,** pp. 773–778

10 NG, T.S.: 'Generalised array pattern synthesis using the projection matrix', *IEE Proc. H, Microwaves, Antennas & Propag.*, 1985, **132,** (1), pp. 44–46

11 DOLPH, C.L.: 'A current distribution for broadside arrays which optimises the relationship between beam width and side-lobe level', *Proc. IRE.*, 1946, **34,** pp. 335–348

12 STEVENSON, A.F.: 'Theory of slots in rectangular waveguide', *J. App. Phys.*, 1948, **19,** pp. 24–38

13 OLINER, A.A.: 'The impedance properties of narrow radiating slots in the broadface of rectangular waveguide', *IRE Trans.*, 1957, **AP-5,** pp. 4–20

14 LEVY, R.: 'Analysis and synthesis of waveguide multi-aperture directional couplers', *IEEE Trans.*, 1968, **MTT-16,** pp. 995–1006

Analysis of infinite arrays of probe-fed rectangular microstrip patches using a rigorous feed model

J.T. Aberle
D.M. Pozar

Indexing terms: Antennas (microstrip), Antennas (arrays)

Abstract: An analysis of infinite arrays of rectangular microstrip patches with each element fed by one or two coaxial probes is presented. The probe self-impedance and the rapidly varying patch current near the feed connection point are rigorously included in the analysis by using an attachment mode derived from the corresponding cavity model solution. Two–probe-fed patches are analysed for both reduced cross-polarisation and circular-polarisation applications, and the effect of probe-to-probe coupling in the unit cell is determined. The theory generates results in excellent agreement with waveguide simulator measurements over a wide range of substrate dielectric constants and thicknesses for both one- and two-probe-fed patches, overcoming the drawbacks of previous solutions that used an idealised probe feed model. Similar solutions can be developed for other types of patch antennas and arrays.

1 Introduction

Quite a variety of techniques exist for feeding microstrip patch antennas and arrays, but one of the earliest developed and still widely used methods is to probe-feed the patch via the centre conductor of a coaxial transmission line. Although this technique is relatively simple to implement practically, it is actually quite difficult to treat theoretically because of the rapidly varying patch current near the point where the probe feed is connected to the patch.

A number of full-wave solutions for the input impedance and radiation properties of probe-fed printed patch antennas and arrays have been proposed [1–9]. The approach of [1–3] is to use an 'idealised' feed model consisting of a uniform electric current filament. No attempt is made to model the rapidly varying patch current near the feed point, nor is the boundary condition that the tangential electric field must vanish on the feed structure enforced. Thus, the self-impedance of the probe is ignored. Because the patch behaves like a high-Q resonator, this idealised probe feed model can yield good agreement with measured impedance data for patches printed on electrically thin substrates. In addition, it is useful for predicting general trends in the scan performance of printed arrays even when the substrate is electrically thick, since these properties are determined primarily by the substrate parameters and grid spacing rather than the particular radiating element or feed structure [2, 3, 10]. But the model fails to accurately predict input impedance when the substrate is thicker than about $0.02\lambda_0$.

The approach used in References 5–7 combines the Green's function/moment method approach of References 1–3 with a two-step technique for modelling the probe feeds [4]. This approach overcomes the deficiencies of the idealised probe feed model, but does not seem to be readily applicable to geometries other than infinite arrays of rectangular microstrip patches. The solution of Reference 8 also overcomes the deficiencies of the idealised feed model, but is limited in application to isolated circular microstrip patches. The approach of Reference 9, for isolated elements, uses a simple attachment mode but does not account for the self-reactance of the feed.

The improved feed model, developed here for infinite arrays of rectangular microstrip patches, can be used for isolated elements and finite arrays as well. In addition, similar approaches can be developed for other canonically-shaped patches. As in References 1–3, the periodic grounded slab Green's function is used, limiting the analysis to one unit cell of the array. The source is assumed to be a delta-gap voltage generator at the base of the probe. The current on both the patch and feed is expanded in a set of vector basis functions with unknown coefficients. One of the basis functions is a special attachment mode that exists on both the patch and feed. It ensures current continuity from the feed to the patch and models the rapidly varying patch currents that exist near the feed point. The patch current of this mode is derived from the corresponding cavity model solution. The electric field integral equation (EFIE) representing the boundary condition that the total tangential electric field must vanish on both the patch and feed is discretised using the method of moments, and the unknown coefficients are evaluated by solving the resulting matrix equation. It is then possible to calculate the input impedance and radiated field of the array.

2 Theory

2.1 Theory of one-probe-fed rectangular patch infinite arrays

The geometry of an infinite periodic planar array of identical one-probe-fed rectangular patches probe-fed via the centre conductor of a coaxial transmission line is shown in Fig. 1. If the $m = 0$, $n = 0$ patch is assumed to be centred at the origin, then the co-ordinates of the centre of the m, nth patch are at $x = ma$ and $y = nb$. The feed point for the m, nth patch is located at $x = ma + x_p$ and $y = nb + y_p$ (in Fig. 1 $y_p = 0$). The driving source is assumed to be a one volt delta-gap generator located at $x = ma + x_p$, $y = nb + y_p$ and $z = 0$.

To formulate a moment method solution for the currents on the feed and patch, the spectral domain

Paper 6550H (E11), first received 13th July and in revised form 28th October 1988

The authors are with the Department of Electrical and Computer Engineering, University of Massachusetts, Amherst, MA 01003, USA grounded slab dyadic Green's function for the electric field due to an infinite periodic phased array of infinitesimal current elements is needed. For notational convenience, it becomes useful to define a dyadic representation of the Green's function as

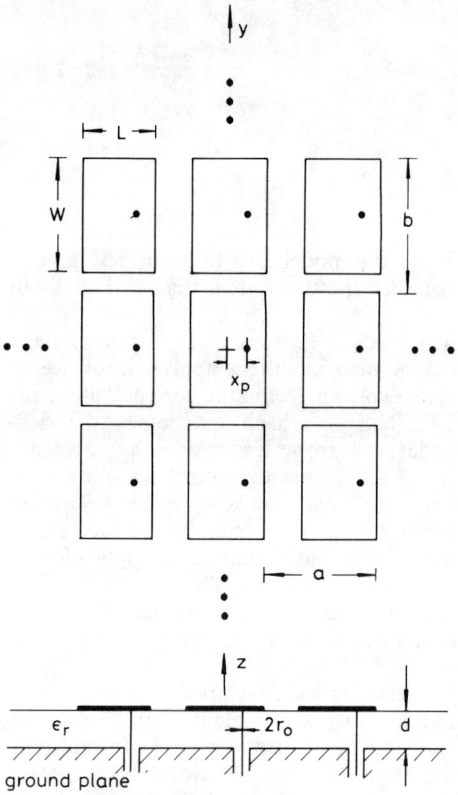

Fig. 1 *Geometry of infinite array of one-probe-fed rectangular microstrip patches (ground plane at z = 0)*

$$\bar{\bar{E}} = \hat{x}E_{xx}\hat{x} + \hat{x}E_{xy}\hat{y} + \hat{x}E_{xz}\hat{z} + \hat{y}E_{yx}\hat{x} + \hat{y}E_{yy}\hat{y} + \hat{y}E_{yz}\hat{z} + \hat{z}E_{zx}\hat{x} + \hat{z}E_{zy}\hat{y} + \hat{z}E_{zz}\hat{z} \tag{1}$$

where

$$E_{pq}(x, y, z \mid x_0, y_0, z_0) = \frac{1}{ab}\sum_{m=-\infty}^{\infty}\sum_{n=-\infty}^{\infty} \times G_{pq}(-k_x, -k_y, z \mid z_0)e^{-jk_x(x-x_0)}e^{-jk_y(y-y_0)} \tag{2}$$

$E_{pq}(x, y, z \mid x_0, y_0, z_0)$ is the $\hat{p}$-directed electric field at the point (x, y, z) due to an infinite periodic phased array of infinitesimal $\hat{q}$-directed current elements, where (x_0, y_0, z_0) is the location of the $m = 0$, $n = 0$ current element, and

$$k_x = \frac{2\pi m}{a} + k_0 u \tag{3}$$

$$k_y = \frac{2\pi n}{b} + k_0 v \tag{4}$$

$$u = \sin\theta\cos\phi \tag{5}$$

$$v = \sin\theta\sin\phi \tag{6}$$

The elements are phased for scanning to (θ, ϕ). The details of the Green's function components are given in Appendix 7.

The EFIE representing the boundary condition that the total tangential electric field must vanish on the feed and patch conductors can be written as

$$\bar{E}^{inc}_{tan} = -\bar{E}^{scat}_{tan} \text{ on patch and feed} \tag{7}$$

where

$$\bar{E}^{scat}(x, y, z) = \iiint \bar{\bar{E}}(x, y, z \mid x_0, y_0, z_0) \cdot \bar{J}(x_0, y_0, z_0)\, dx_0\, dy_0\, dz_0 \tag{8}$$

and $\bar{J}$ is the vector electric volume current density on the feed and patch. Since the driving source is assumed to be a delta-gap generator, the incident electric field is taken to be

$$\bar{E}^{inc}(x, y, z) = \begin{cases} \hat{z}\delta(z), & x = x_p, y = y_p \\ 0, & \text{otherwise} \end{cases} \tag{9}$$

The current density on the feed and patch is expanded in a set of N vector basis functions:

$$\bar{J}(x, y, z) = \sum_{j=1}^{N} I_j \bar{J}_j(x, y, z) \tag{10}$$

Each expansion mode may have two pieces, one of which exists on the feed, the other on the patch, i.e.

$$\bar{J}_j(x, y, z) = \bar{J}^F_j(x, y, z) + \bar{J}^P_j(x, y, z), \quad j = 1, 2, \ldots, N \tag{11}$$

where $\bar{J}^F_j$ is the part of the jth expansion mode existing on the feed, and $\bar{J}^P_j$ is the part of the jth expansion mode existing on the patch.

Since the feed radius is small, a filamentary current approximation is made. That is, the field arising from $\bar{J}^F_j$ is assumed to result from a volume current density given by

$$\hat{z}f_j(z)\delta(x - x_p)\delta(y - y_p) \tag{12}$$

where the function $f_j(z)$ is the current of the jth mode on the feed. However, $\bar{J}^F_j$ itself is assumed to be of the form:

$$\bar{J}^F_j(x, y, z) = \hat{z}\frac{f_j(z)}{2\pi r_0}\delta(\rho' - r_0) \tag{13}$$

where

$$\rho' = \sqrt{(x - x_p)^2 + (y - y_p)^2} \tag{14}$$

and r_0 is the probe radius. Thus, the effect of the probe radius is included in the analysis. Abusing somewhat the notation, one can write

$$\bar{J}^P_j(x, y, z) = \bar{J}^P_j(x, y)\delta(z - d) \tag{15}$$

where now $\bar{J}^P_j(x, y)$ represents a surface current density on the patch. Use of Galerkin's method in the discretisation of eqn. 7 yields a matrix equation for the unknown coefficients:

$$[Z][I] = [V] \tag{16}$$

where the voltage vector elements are given by

$$V_i = f_i(0) \tag{17}$$

and the impedance matrix elements are given by

$$z_{ij} = z^{FF}_{ij} + z^{FP}_{ij} + z^{PF}_{ij} + z^{PP}_{ij} \tag{18}$$

with

$$z^{\alpha\beta}_{ij} \equiv -\iiint \bar{E}^{\beta}_j(x, y, z) \cdot \bar{J}^{\alpha}_i(x, y, z)\, dxdydz \tag{19}$$

where

$$\bar{E}_j^F(x, y, z) = \int_0^d \bar{\bar{E}}(x, y, z \mid x_p, y_p, z_0) \cdot \hat{z} f_j(z_0)\, dz_0 \quad (20)$$

$$\bar{E}_j^P(x, y, z) = \int_{-W/2}^{W/2} \int_{-L/2}^{L/2} \bar{\bar{E}}(x, y, z \mid x_0, y_0, d) \cdot \bar{J}_j^P(x_0, y_0)\, dx_0\, dy_0 \quad (21)$$

i.e. $\bar{E}_j^F$ is the field due to $\bar{J}_j^F$, and $\bar{E}_j^P$ is the field due to $\bar{J}_j^P$. Hence

$$z_{ij}^{FF} = -\frac{1}{ab} \sum_{m=-\infty}^{\infty} \sum_{n=-\infty}^{\infty} Q_{ij}(k_x, k_y) J_0(\beta r_0) \quad (22)$$

$$z_{ij}^{FP} = -\frac{1}{ab} \sum_{m=-\infty}^{\infty} \sum_{n=-\infty}^{\infty} \{R_{xi}(k_x, k_y) F_{xj}(-k_x, -k_y) + R_{yi}(k_x, k_y) F_{yi}(-k_x, -k_y)\} \times J_0(\beta r_0) e^{-jk_x x_p} e^{-jk_y y_p} \quad (23)$$

$$z_{ij}^{PF} = \frac{1}{ab} \sum_{m=-\infty}^{\infty} \sum_{n=-\infty}^{\infty} \{R_{xj}(k_x, k_y) F_{xi}(k_x, k_y) + R_{yj}(k_x, k_y) F_{yi}(k_x, k_y)\} e^{jk_x x_p} e^{jk_y y_p} \quad (24)$$

$$z_{ij}^{PP} = -\frac{1}{ab} \sum_{m=-\infty}^{\infty} \sum_{n=-\infty}^{\infty} \{F_{xi}(k_x, k_y) \times G_{xx}(k_x, k_y, d \mid d) F_{xj}(-k_x, -k_y) + F_{xi}(k_x, k_y) G_{xy}(k_x, k_y, d \mid d) F_{yj}(-k_x, -k_y) + F_{yi}(k_x, k_y) G_{yx}(k_x, k_y, d \mid d) F_{xj}(-k_x, -k_y) + F_{yi}(k_x, k_y) G_{yy}(k_x, k_y, d \mid d) F_{yj}(-k_x, -k_y)\} \quad (25)$$

where

$$Q_{ij}(k_x, k_y) = \int_0^d f_i(z) \int_0^d G_{zz}(k_x, k_y, z \mid z_0) f_j(z_0)\, dz_0\, dz \quad (26)$$

$$R_{xi}(k_x, k_y) = \int_0^d G_{xz}(k_x, k_y, d \mid z_0) f_i(z_0)\, dz_0 \quad (27)$$

$$R_{yi}(k_x, k_y) = \int_0^d G_{yz}(k_x, k_y, d \mid z_0) f_i(z_0)\, dz_0 \quad (28)$$

$$F_{xi}(k_x, k_y) = \int_{-W/2}^{W/2} \int_{-L/2}^{L/2} \hat{x} \bar{J}_i^P(x_0, y_0) \times e^{-jk_x x_0} e^{-jk_y y_0}\, dx_0\, dy_0 \quad (29)$$

$$F_{yi}(k_x, k_y) = \int_{-W/2}^{W/2} \int_{-L/2}^{L/2} \hat{y} \bar{J}_i^P(x_0, y_0) \times e^{-jk_x x_0} e^{-jk_y y_0}\, dx_0\, dy_0 \quad (30)$$

To continue with this development, it is now necessary to choose a set of modes in which to expand the unknown currents on the feed and patch. The first mode is an attachment mode which exists on both the feed and patch. This mode is used to account for the connection of a probe feed to a microstrip patch and to ensure continuity of current from the feed to the patch. The part of the attachment mode on the feed is taken to be

$$f_1(z) = 1 \quad 0 < z < d \quad (31)$$

The part of the attachment mode existing on the patch is taken to be the surface current that would exist on the top conductor of a lossy cavity formed by electric walls on top and bottom and a magnetic wall around its perimeter (the usual cavity model of a rectangular patch), and fed by a uniform current probe. The expectation is that the use of such a mode will accelerate the convergence of the moment method solution by modelling the rapid variation in patch current near the feed connection point. In addition, we expect the cavity model currents to be quite close to the actual current on the patch, so convergence is further enhanced. The patch part of the attachment mode is given by

$$\bar{J}_1^P(x, y) = \frac{-1}{2W} \sum_{n=0}^{\infty} \varepsilon_n \cos\frac{n\pi}{2} \left\{ \hat{x} \tilde{f}_s(Z, L, x_p, x) \times \cos\frac{n\pi}{W}(y + W/2) + \hat{y}\left(\frac{n\pi}{W}\right) \tilde{f}_c(Z, L, x_p, x) \times \sin\frac{n\pi}{W}(y + W/2) \right\} \quad (32)$$

where

$$\tilde{f}_s(Z, L, x_p, x) = \begin{cases} \dfrac{\sin Z(x + x_p) + \sin Z(L - (x_p - x))}{\sin ZL} & x < x_p \\ \dfrac{\sin 2Zx_p}{\sin ZL} & x = x_p \\ \dfrac{\sin Z(x + x_p) - \sin Z(L - (x - x_p))}{\sin ZL} & x > x_p \end{cases} \quad (33)$$

$$\tilde{f}_c(Z, L, x_p, x) = \frac{\cos Z(x + x_p) + \cos Z(L - |x - x_p|)}{Z \sin ZL} \quad (34)$$

$$Z = \sqrt{k^2 - \left(\frac{n\pi}{W}\right)^2} \quad (35)$$

$$k = \omega\sqrt{\varepsilon_r \varepsilon_0 \mu_0} \quad (36)$$

$$\varepsilon_n = \begin{cases} 1, & n = 0 \\ 2, & n > 0 \end{cases} \quad (37)$$

Fig. 2 shows a plot of the $\hat{x}$-directed part of the attachment mode on a square patch fed at $x_p = 0.25L$, $y_p = 0$. The current is plotted against x for several values of y.

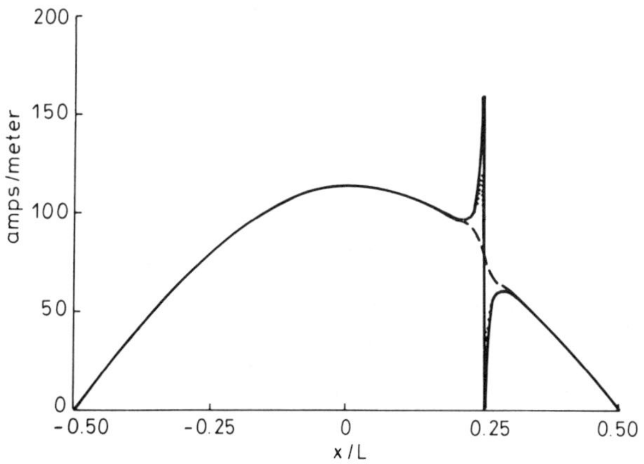

Fig. 2 *Plot of $\hat{x}$-directed component of attachment mode on square patch*

$\varepsilon_r = 1$; $L = W = 0.498\lambda_0$; $x_p = 0.25L$; $y_p = 0$
——— $y = 0.002W$
······ $y = 0.004W$
- - - - $y = 0.02W$

Notice that this current is virtually indistinguishable from the TM_{10} dominant mode current except in the immediate vicinity of the feed point. The sharp spike in the current is due to the injection of current from the probe at the feed point. Two peaks are observed because the injected current is in different directions on each side of the feed point. It can be shown that the charge build up at the probe tip associated with the discontinuous (at $z = d$) $\hat{z}$ – directed feed current of eqn. 31 is exactly cancelled by a charge build up associated with the $\hat{x}$- and $\hat{y}$-directed patch current of eqn. 32. The Fourier transform of eqn 32 may be evaluated in closed form as

$$\bar{F}_1(k_x, k_y) = \frac{-1}{2W} \sum_{n=0}^{\infty} \varepsilon_n \cos \frac{n\pi}{W} (Y_p + W/2) \times \left\{ \hat{x} F_c(n, W, k_y) \bar{F}_s(Z, L, x_p, k_x) + \hat{y}\left(\frac{n\pi}{W}\right) F_s(n, W, k_y) \tilde{F}_c(Z, L, x_p, k_x) \right] \quad (38)$$

where

$$F_s(m, L, k_x) = \frac{\frac{m\pi}{L}}{k_x^2 - \left(\frac{m\pi}{L}\right)^2} \times \{(-1)^m e^{-j(k_x L/2)} - e^{j(k_x L/2)}\} \quad (39)$$

$$F_c(m, L, k_x) = \frac{jk_x}{k_x^2 - \left(\frac{m\pi}{L}\right)^2} \times \{(-1)^m e^{-j(k_x L/2)} - e^{j(k_x L/2)}\} \quad (40)$$

$$\tilde{F}_s(Z, L, x_p, k_x) = \frac{-2j}{(k_x^2 - Z^2) \sin ZL} \times [Z\{e^{jZx_p} \sin (k_x - Z)L/2 + e^{-jZx_p} \times \sin (k_x + Z)L/2\} - k_x e^{-jk_x x_p} \sin ZL] \quad (41)$$

$$\tilde{F}_c(Z, L, x_p, k_2) = \frac{2}{(k_x^2 - Z^2) Z \sin ZL} \times [k_x\{e^{jZx_p} \sin (k_x - Z)L/2 + e^{-jZx_p} \times \sin (k_x + Z)L/2\} - Ze^{-jk_x x_p} \sin ZL] \quad (42)$$

The remaining expansion modes are a complete, orthogonal set that exists on the patch. These are given by

$$\bar{J}_i^P(x, y) = \hat{x}\left(\frac{m_i \pi}{L}\right) \sin \frac{m_i \pi}{L} (x + L/2) \times \cos \frac{n_i \pi}{W} (y + W/2) + \hat{y}\left(\frac{n_i \pi}{W}\right) \times \cos \frac{m_i \pi}{L} (x + L/2) \sin \frac{n_i \pi}{W} (y + W/2) \quad i \leq 2 \quad (43)$$

These modes were also used in Reference 2 and correspond to the eigenmodes of the magnetic wall cavity model of the rectangular patch previously mentioned.

The Fourier transforms of these modes are given by

$$\bar{F}_i(k_x, k_y) = \hat{x}\left(\frac{m_i \pi}{L}\right) F_s(m_i, L, k_x) F_c(n_i, W, k_y) + \hat{y}\left(\frac{n_i \pi}{W}\right) F_c(m_i, L, k_z) F_s(n_i, W, k_y) \quad (44)$$

It is now instructive to examine the structure of the impedance matrix given the expansion modes described above. Eqn. 19 implies that $z_{ij}^{\alpha\beta}$ vanishes if either mode i does not have a piece on structure α or mode j does not have a piece on structure β. Hence,

$$[Z] = \begin{bmatrix} z_{11}^{FF} + z_{11}^{FP} + z_{11}^{PF} + z_{11}^{PP} & z_{12}^{FP} + z_{12}^{PP} & z_{13}^{FP} + z_{13}^{PP} & \cdots & z_{1N}^{FP} + z_{1N}^{PP} \\ z_{21}^{PF} + z_{21}^{PP} & z_{22}^{PP} & z_{23}^{PP} & \cdots & z_{2N}^{PP} \\ z_{31}^{PF} + z_{31}^{PP} & z_{32}^{PP} & z_{33}^{PP} & \cdots & z_{3N}^{PP} \\ \vdots & \vdots & \vdots & & \vdots \\ z_{N1}^{PF} + z_{N1}^{PP} & z_{N2}^{PP} & z_{N3}^{PP} & \cdots & z_{NN}^{PP} \end{bmatrix} \quad (45)$$

The voltage vector is given by

$$[V] = \begin{bmatrix} 1 \\ 0 \\ \vdots \\ 0 \end{bmatrix} \quad (46)$$

A computer program can then be written to evaluate the impedance matrix elements and to compute the current coefficients from eqn. 16. The impedance looking into the patch is then calculated as

$$Z_{in}(\theta, \phi) = 1/I_1 \quad (47)$$

2.2 *Theory of two-probe-fed rectangular patch infinite arrays*

The two-probe-fed rectangular patch in the infinite array environment is treated as a two-port network described by a port admittance matrix $[Y^P]$. The elements of this matrix are defined as

$$y_{11}^p = \frac{I_1^p}{V_1^p} \mid V_2^p = 0 \quad (48)$$

$$y_{12}^p = \frac{I_1^p}{V_2^p} \mid V_1^p = 0 \quad (49)$$

$$y_{21}^p = \frac{I_2^p}{V_1^p} \mid V_2^p = 0 \quad (50)$$

$$y_{22}^p = \frac{I_2^p}{V_2^p} \mid V_1^p = 0 \quad (51)$$

where V_k^p and I_k^p are the terminal voltage and current, respectively, of the kth feed. The superscripts p denote that the quantities involved are port quantities, as opposed to moment method quantities.

The EFIE representing the boundary condition that the total tangential electric field must vanish on the feed and patch conductors when either feed is excited can be written as

$$\bar{E}_k^{inc} = -\bar{E}_k^{scat} \text{ on patch and feeds } k = 1, 2 \quad (52)$$

where

$$\bar{E}_k^{scat}(x, y, z) = \iiint \bar{E}(x, y, z \mid x_0, y_0, z_0) \cdot \bar{J}_k(x_0, y_0, z_0)\, dx_0\, dy_0\, dz_0 \quad k = 1, 2 \quad (53)$$

$\bar{E}_k^{scat}$ is the electric field scattered by the feeds and patch and $\bar{J}_k$ is the vector electric volume current density of the feeds and patch when the kth feed is excited. The driving sources are again assumed to be delta-gap generators with incident fields given by

$$\bar{E}_k^{inc}(x, y, z) = \begin{cases} \hat{z}\delta(z), \; x = x_{pk}, \; y = y_{pk} & k = 1, 2 \\ 0 & \text{otherwise} \end{cases} \quad (54)$$

where the probe positions are now taken to be (x_{p1}, y_{p1}) and (x_{p2}, y_{p2}).

The current density on the feeds and patch is expanded in a set of N vector basis functions:

$$\bar{J}_k(x, y, z) = \sum_{j=1}^{N} I_{kj} \bar{J}_j(x, y, z) \quad k = 1, 2 \quad (55)$$

As for the one-probe-fed case, each expansion mode may consist of a piece which exists on the feeds and another which exists on the patch. Again, a filamentary current approximation is used for the feeds, with $\bar{J}_j^F$ given by

$$\bar{J}_j^F(x, y, z) = \hat{z}\left\{\frac{f_{1j}(z)}{2\pi r_0}\,\delta(\rho'_1 - r_0) + \frac{f_{2j}(z)}{2\pi r_0}\,\delta(\rho'_2 - r_0)\right\} \quad (56)$$

where

$$\rho'_1 = \sqrt{(x - x_{p1})^2 + (y - y_{p1})^2} \quad (57)$$

$$\rho'_2 = \sqrt{(x - x_{p2})^2 + (y - y_{p2})^2} \quad (58)$$

The function $f_{kj}(z)$ is the current of the jth mode on the kth feed.

As for the one-probe-fed case, Galerkin's method is used to discretise the EFIE (52) yielding

$$[Z][I_k] = [V_k] \quad k = 1, 2 \quad (59)$$

where the voltage vector elements are given by

$$V_{ki} = f_{ki}(0) \quad (60)$$

and the impedance matrix elements are given by eqns. 18–25 generalised to the two-probe-fed case.

It is now necessary to choose a set of modes in which to expand the unknown currents on the feed and patch. The first mode is an attachment mode which exists on both feed 1 and the patch with

$$f_{k1}(z) = \begin{cases} 1 & k = 1 \\ 0 & k = 2 \end{cases} \quad (61)$$

The second mode is an attachment mode which exists on both feed 2 and the patch with

$$f_{k2}(z) = \begin{cases} 0 & k = 1 \\ 1 & k = 2 \end{cases} \quad (62)$$

The part of each attachment mode on the patch can be obtained by generalising eqn. 32. The additional expansion modes on the patch are still given by eqn. 43 but now for $i \geq 3$. The structure of the impedance matrix generated by this choice of expansion modes is

$$[Z] = \begin{bmatrix} z_{11}^{FF} + z_{11}^{FP} + z_{11}^{PF} + z_{11}^{PP} & z_{12}^{FF} + z_{12}^{FP} + z_{12}^{PF} + z_{12}^{PP} & z_{13}^{FP} + z_{13}^{PP} & \cdots & z_{1N}^{FP} + z_{1N}^{PP} \\ z_{21}^{FF} + z_{21}^{FP} + z_{21}^{PF} + z_{21}^{PP} & z_{22}^{FF} + z_{22}^{FP} + z_{22}^{PF} + z_{22}^{PP} & z_{23}^{FP} + z_{23}^{PP} & \cdots & z_{2N}^{FP} + z_{2N}^{PP} \\ z_{31}^{PF} + z_{31}^{PP} & z_{32}^{PF} + z_{32}^{PP} & z_{33}^{PP} & \cdots & z_{3N}^{PP} \\ \vdots & \vdots & \vdots & & \vdots \\ z_{N1}^{PF} + z_{N1}^{PP} & z_{N2}^{PF} + z_{N2}^{PP} & z_{N3}^{PP} & \cdots & z_{NN}^{PP} \end{bmatrix} \quad (63)$$

The voltage vectors are given by

$$[V_1] = \begin{bmatrix} 1 \\ 0 \\ \vdots \\ 0 \end{bmatrix} \quad (64)$$

$$[V_2] = \begin{bmatrix} 0 \\ 1 \\ \vdots \\ 0 \end{bmatrix} \quad (65)$$

A computer program can then be written to evaluate the impedance matrix elements and to compute the current coefficients from eqn. 59. Notice that I_{11} and I_{12} are the port currents I_1^p and I_2^p when port 1 is excited by a one volt generator and port 2 is short-circuited. Further, I_{21} and I_{22} are the port currents I_1^p and I_2^p when port 2 is excited by a one volt generator and port 1 is short-circuited. Hence, the two port admittance matrix of the patch may be evaluated as

$$[Y^p] = \begin{bmatrix} I_{11} & I_{21} \\ I_{12} & I_{22} \end{bmatrix} \quad (66)$$

The two-port impedance matrix $[Z^p]$ can then be obtained by inverting $[Y^p]$.

3 Results

3.1 Convergence considerations

There are three separate (but interrelated) convergence issues which need to be addressed in the present solution. First, the patch part of the attachment mode, given by eqn. 32, is expressed as an infinite summation. Although a convergence acceleration scheme is implemented, about ten terms are still required for accurate numerical calculation. Second, the solution to the problem of an infinite array is formulated in terms of impedance matrix elements, given by eqn. 22 through eqn. 25, which are expressed as infinite summations. Each term in the infinite summation corresponds to a Floquet mode of the infinite periodic structure. To calculate the impedance matrix elements with good accuracy, 301 Floquet modes ($-150 < m, n < 150$) were used to evaluate each series. This is significantly more than the number of terms used in the series expressions of Reference 2. The reason for this is that the attachment mode has a wide spectral bandwidth since it models the rapidly varying patch current near the feed point as well as the feed current. Third, an appropriate choice of expansion modes of the form of eqn. 43 must be made. Through numerical convergence checks and comparison of numerical results with measured data, it was found that the $(m, n) = (1, 0)$, (0, 1), (0, 2) (3, 0), (5, 0) and (7, 0) expansion modes yield good results for linearly polarised patches. For circularly polarised patches, the (2, 0), (0, 3), (0, 5) and (0, 7) expansion modes are also included. Hence, all of the improved probe feed model calculations presented here use between seven and twelve expansion modes (including an attachment mode for each probe feed). Therefore, at worst, a 12 × 12 impedance matrix must be inverted to find the unknown coefficients.

3.2 Experimental results

By imaging a small number of elements in the walls of a waveguide, one can evaluate the performance of an element in an infinite array [11]. Such waveguide simulator experiments were used to confirm the theory of one- and two-probe-fed rectangular patch infinite arrays presented here. A number of experiments were done on Duroid 5870, 5880 and 6010.2 substrates of various thickness.

Figs. 3 and 4 show the measured reflection coefficients of one-probe-fed rectangular patches in waveguide simulators compared with calculations using the theory presented here and the idealised probe feed model of Reference 2. In Fig. 3, the simulator is standard *S*-band waveguide operating in the TE_{10} mode. Hence, scanning in the *H*-plane is simulated, and at 3.5 GHz the scan angle is 36.45°. In Fig. 4, the simulator is standard *C*-band waveguide operating in the TE_{10} mode; at 4.0 GHz the simulated scan angle is 52.06° in the *H*-plane. In both cases the new theory agrees quite well with experimental data and is clearly an improvement over the solution of Reference 2. Duroid 6010.2, which is quoted as having a dielectric constant of $\varepsilon_r = 10.2 \pm 0.25$, was the substrate used in the experiment of Fig. 4. In the calculation, a dielectric constant of $\varepsilon_r = 10.6$ was found to give the best agreement with measured data. Although this value is outside the quoted value, the substrate is anisotropic, and thus will appear to have a somewhat higher dielectric constant [12]. Other experiments were done on Duroid 5870 and 5880, where the isotropy is better and dielectric constant tolerance is tighter. In these cases, no adjustment of the dielectric constant was required to achieve excellent agreement between the new theory and experiment.

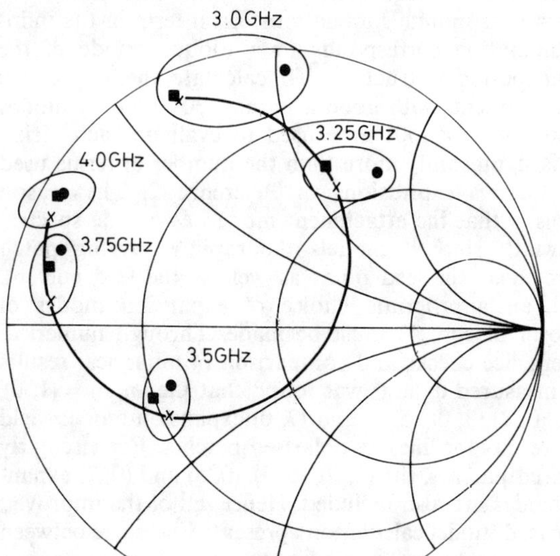

Fig. 3 *Measured and calculated reflection coefficient of one-probe-fed patch in S-band waveguide simulator*

$\varepsilon_r = 2.2$; $d = 0.159$ cm; $L = 2.86$ cm; $W = 2.54$ cm; $x_p = 0.72$ cm; $y_p = 0$; $r_0 = 0.045$ cm; $a = 3.40$ cm; $b = 3.61$ cm
● theory (old)
■ theory (new)
× experiment

Figs. 5 and 6 show measured *S*-parameters of a two-probe-fed rectangular patch in a *C*-band waveguide simulator. In this experiment, a scan angle of 39.12° in the *H*-plane is simulated at 5.0 GHz. The two-probe-fed patch is treated as a two-port device and its scattering parameters are measured. In general, this array element is a nonreciprocal device because of the asymmetry introduced by beam scanning. However, when the array is scanned in the *H*-plane, the balanced two-probe-fed patch retains its symmetry and is a reciprocal two-port. As can be seen, there is excellent agreement between theory and experiment for both S_{11} and S_{21}. In the idealised feed model, an *ad hoc* correction for the probe

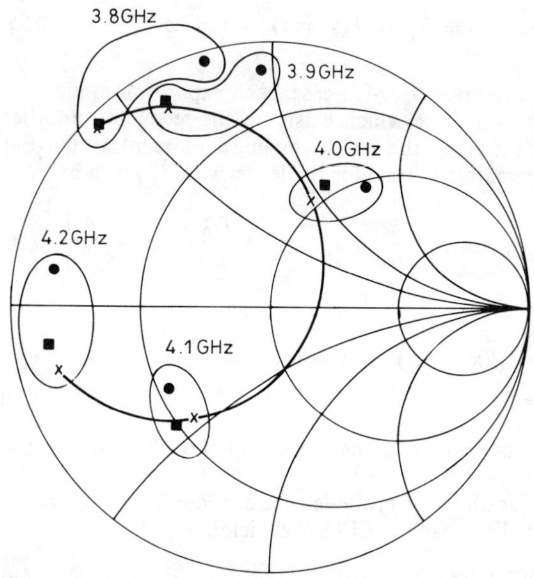

Fig. 4 *Measured and calculated reflection coefficient of one-probe-fed patch in C-band waveguide simulator*

$\varepsilon_r = 10.6$; $d = 0.127$ cm; $L = 1.1$ cm; $W = 1.1$ cm; $x_p = 0.165$ cm; $y_p = 0$; $r_0 = 0.045$ cm; $a = 2.215$ cm; $b = 2.378$ cm
● theory (old)
■ theory (new)
× experiment

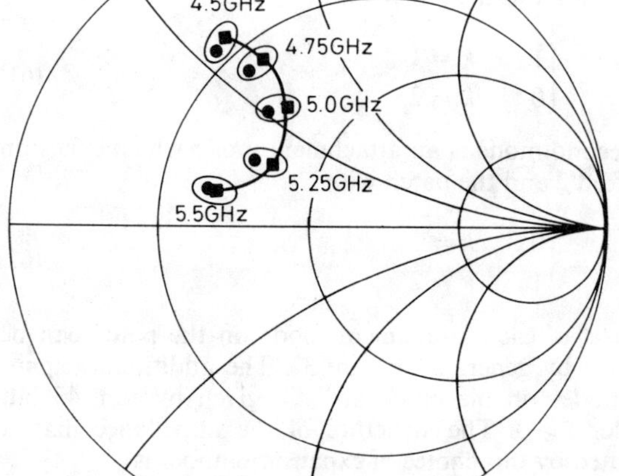

Fig. 5 *Measured and calculated S_{11} of two-probe-fed patch in C-band waveguide simulator*

$\varepsilon_r = 2.2$; $d = 0.159$ cm; $L = 1.8$ cm; $W = 1.8$ cm; $x_{p1} = -x_{p2} = 0.5$ cm; $y_{p1} = y_{p2} = 0$; $r_0 = 0.045$ cm; $a = 2.215$ cm; $b = 2.378$ cm
● theory
■ experiment

mutual impedance was needed, even for thin substrates, in order to obtain good agreement with measured data. No such correction is needed here as the probe-to-probe coupling is included implicitly in the analysis.

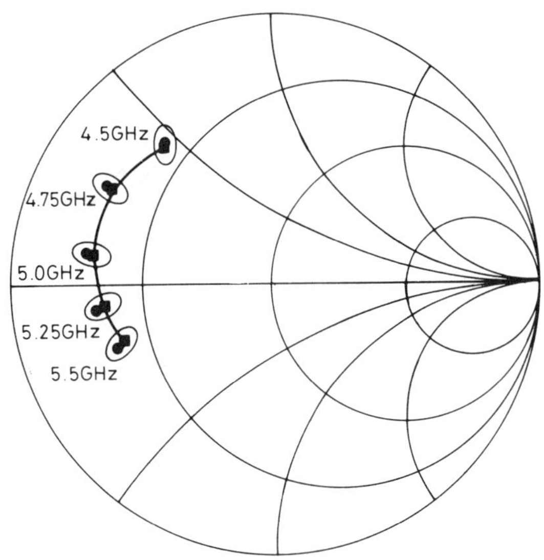

Fig. 6 *Measured and calculated S_{21} of two-probe-fed patch in C-band waveguide simulator (same parameters as Fig. 5)*

● theory
■ experiment

3.3 Theoretical results

The scan performance of the one-probe-fed patch infinite array element is evaluated by calculating the reflection coefficient against scan angle, as carried out in References 2 and 3 using

$$R(\theta, \phi) = \frac{Z_{in}(\theta, \phi) - Z_{in}(0, 0)}{Z_{in}(\theta, \phi) + Z_{in}^*(0, 0)} \tag{67}$$

where Z_{in} (0, 0) is the input impedance of the patch at broadside. Usually, the patch is used at or near its resonance, i.e. its input impedance at broadside is purely resistive. Since the patch is conjugate-matched to its broadside impedance, the reflection coefficient is zero at broadside and increases as the beam is scanned due to the patch impedance variation.

Fig. 7 shows the reflection coefficient magnitude in three planes of scan for a rectangular patch array printed on a substrate with $\varepsilon_r = 2.55$ and $d = 0.06\lambda_0$. (The '*D*-plane' is an intercardinal plane with $\phi = 45°$.) Note that this is the same case as in Fig. 4 of Reference 2, except that the feed point has been moved away from the patch edge. (The reason for doing this is because the convergence acceleration scheme used in calculating the attachment mode transform is not as effective for edge fed patches, hence many more terms in the series would be needed, slowing down the calculation.) Since the element spacing along the *x*-axis is $0.51\lambda_0$, scanning in the *E*-plane to $\theta = 73.90°$ brings a grating lobe into visible space. The condition for surface wave propagation is satisfied for $\theta = 68.89°$ in the *E*-plane and $\theta = 76.43°$ in the *H*-plane. As can be seen the reflection coefficient is large (but not unity) near the *E*-plane scan angle at which a surface wave can be excited. The idealised probe feed model [2] always predicted total reflection at these angles in both the *E*- and *H*-planes. In the improved probe feed model, the probes themselves can also couple to a TM surface wave at the angle determined by the substrate parameters and interelement spacing. (However, they cannot couple to a TE surface wave.)

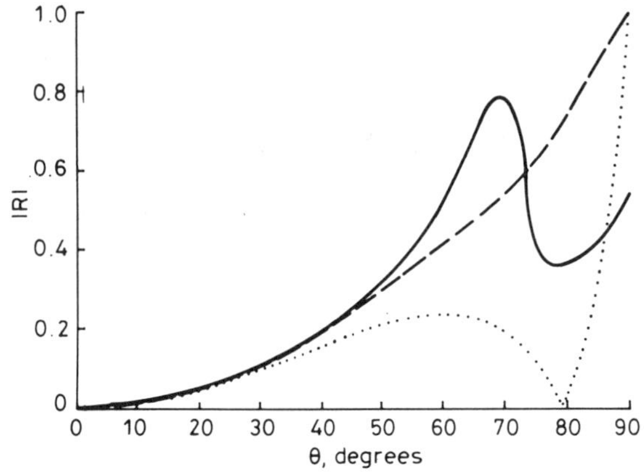

Fig. 7 *Calculated reflection coefficient magnitude of infinite array of one-probe-fed patches*

$\varepsilon_r = 2.55$; $d = 0.06\lambda_0$; $L = 0.28\lambda_0$; $W = 0.3\lambda_0$; $x_p = 0.07\lambda_0$; $y_p = 0$; $r_0 = 0.0004\lambda_0$; $a = 0.51\lambda_0$; $b = 0.5\lambda_0$

——— *E*-plane
······ *D*-plane
- - - - *H*-plane

Fig. 8 shows the level of cross-polarisation against scan angle for the array of Fig. 7. The definition of cross-polarisation used here is Ludwig's third definition [13]. In this Figure and in the other polarisation results to follow, the calculation was not done for θ greater than 80°. The reason for this is that the radiated field of the array is taken to be the $m = n = 0$ Floquet mode. Since this is no longer true as θ approaches 90°, the polarisation results are meaningless near endfire.

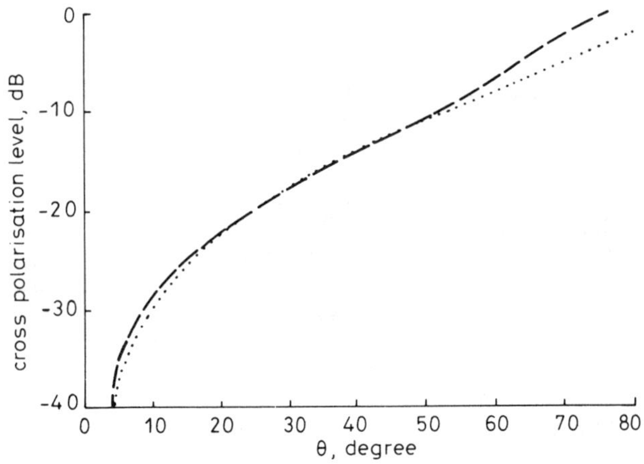

Fig. 8 *Calculated cross polarisation levels of array of Fig. 7*

······ *D*-plane
- - - - *H*-plane

It should be pointed out that both the broadside-matched reflection coefficient magnitudes and cross-polarisation levels calculated using the improved probe feed model are almost identical to those calculated using the idealised probe feed model for the case of a one-probe-fed patch array.

To evaluate the scan performance of the two-probe-fed patch in the infinite array, it is necessary to choose a network with which to feed the patch. Both resistive and reactive power dividers could be used in such an application. However, the use of a reactive power divider is probably a poor choice for elements used in arrays intended to cover a large scan volume, because of the potential for scan blindness caused by parasitic resonances in the feed network [6]. Hence, the feed network shown in Fig. 9 is used here. This feed network was also used in Reference 3, where it was found that the normalised active element gain for a patch fed with this network is given by

$$G_e(\theta, \phi) = \{1 - |R(\theta, \phi)|^2 - P_d(\theta, \phi)\} \cos \theta \qquad (68)$$

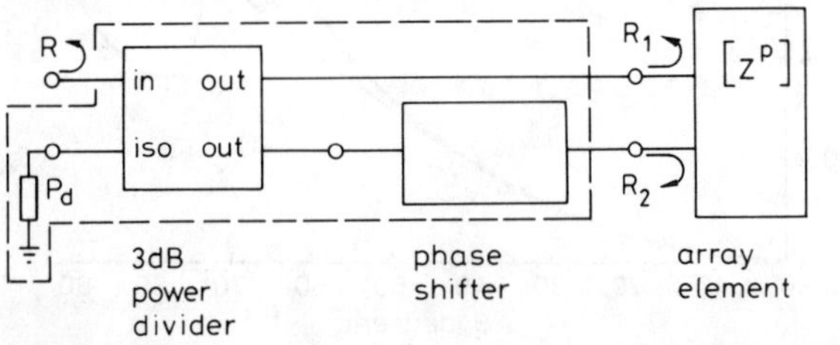

Fig. 9 *Feed network configuration for two-probe-fed patch array element*

where R is the reflection coefficient looking into the feed network given by

$$R(\theta, \phi) = \tfrac{1}{2}\{R_1(\theta, \phi) + R_2(\theta, \phi)e^{-j2\Omega}\} \qquad (69)$$

and P_d is the relative power dissipated in the termination of the isolated port given by

$$P_d(\theta, \phi) = \tfrac{1}{4}|R_1(\theta, \phi) - R_2(\theta, \phi)e^{-j2\Omega}|^2 \qquad (70)$$

and Ω is the phase difference between the probes. Since the active element gain includes power both reflected out of and dissipated in the feed network due to mismatch at each feed, it is the most convenient measure of scan performance for two-probe-fed arrays.

Two feed configurations are of particular interest in the study of arrays comprising radiating elements with two feed points:

(i) balanced feed for linear polarisation with reduced cross polarisation

(ii) orthogonal feed for circular polarisation.

For the balanced feed, the probes are at $(x_{p1}, y_{p1}) = (x_p, 0)$ and $(x_{p2}, y_{p2}) = (-x_p, 0)$, and are excited in antiphase. For the orthogonal feed, the probes are at $(x_{p1}, y_{p1}) = (x_p, 0)$ and $(x_{p2}, y_{p2}) = (0, x_p)$, and are fed in quadrature.

Fig. 10 shows the active element gain in three planes of scan for a balanced-fed patch array printed on a substrate with $\varepsilon_r = 2.55$ and $d = 0.02\lambda_0$. Fig. 11 shows the cross polarisation level for this array. Notice from Fig. 11 that the balanced feed reduces, but does not completely eliminates, cross polarisation in the H-plane.

Fig. 12 shows the active element gain in three planes of scan for a circularly polarised patch array. Except for the orthogonal feed, all of the parameters are the same as for the array of Fig. 10. Fig. 13 shows the axial ratio in dB against scan angle for the array of Fig. 12. Note that even at broadside, the transmitted field is not perfectly circularly polarised (0 dB axial ratio). The same result was obtained in Reference 7. However, calculations made using the idealised probe feed model predict axial ratios at broadside that are very close to 0 dB. The reason for this discrepancy is that the old theory dramatically underestimates the amount of cross polarised current excited by each of the probes in this configuration. Since the cross polarised currents look different for different feeds, the two orthogonal polarisations are unbalanced. Much higher quality circular polarisation can be

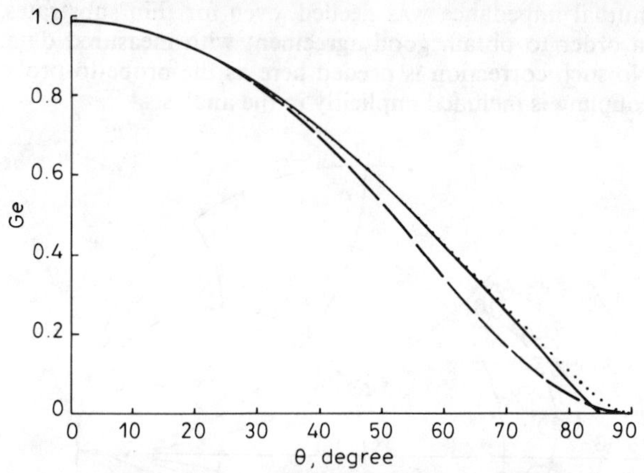

Fig. 10 *Calculated normalised active element gain pattern of infinite array of two-probe-fed linearly polarised patches*

$\varepsilon_r = 2.55$; $d = 0.02\lambda_0$; $L = W = 0.3022\lambda_0$; $x_{p1} = -x_{p2} = 0.0755\lambda_0$; $y_{p1} = y_{p2} = 0$; $a = b = 0.5\lambda_0$

——— *E*-plane
······ *D*-plane
- - - - *H*-plane

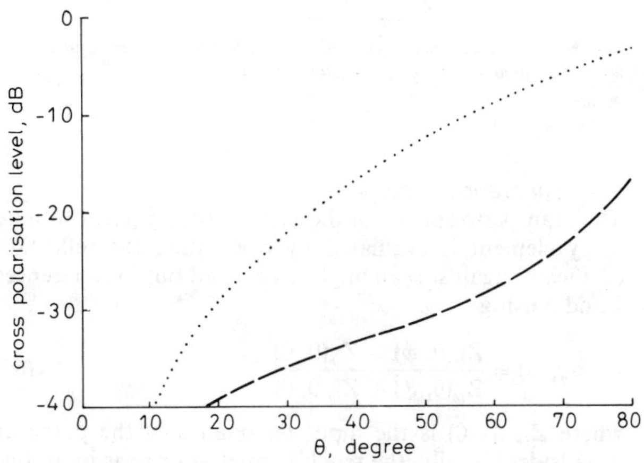

Fig. 11 *Calculated cross-polarisation levels of array of Fig. 10*

······ D-plane
- - - - H-plane

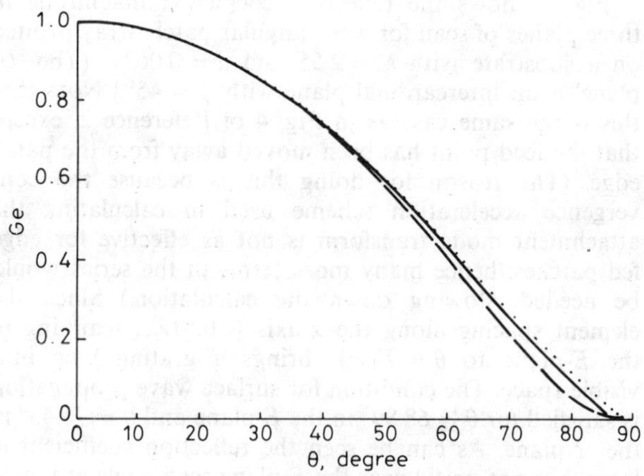

Fig. 12 *Calculated normalised active element gain pattern of infinite array of two-probe-fed circularly polarised patches*

$\varepsilon_r = 2.55$; $d = 0.02\lambda_0$; $L = W = 0.3022\lambda_0$; $x_{p1} = y_{p2} = 0.0755\lambda_0$; $y_{p1} = x_{p2} = 0$; $a = b = 0.5\lambda_0$

——— $\phi = 0°$
······ $\phi = 45°$
- - - - $\phi = 90°$

achieved by feeding the patch with two balanced pairs of probes, as carried out in Reference 7.

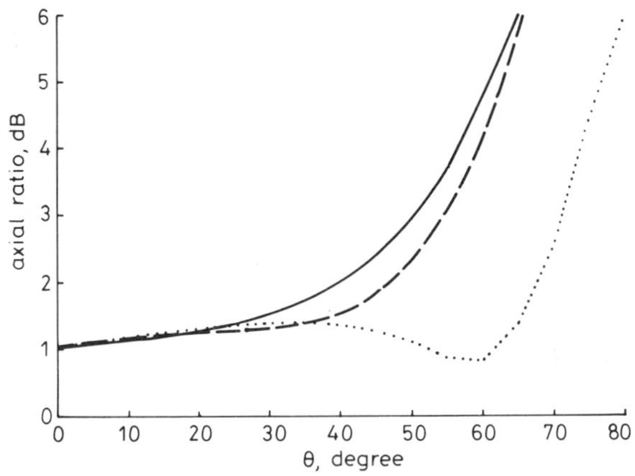

Fig. 13 *Calculated axial ratio for array of Fig. 12*

—— $\phi = 0°$
······ $\phi = 45°$
– – – – $\phi = 90°$

4 Conclusions

This paper has presented a rigorous analysis of infinite arrays of rectangular microstrip patches fed by one or two probes, with results which agreed quite well with measured data over a wide range of substrate thicknesses and dielectric constants. The solution presented here is clearly an improvement over the idealised feed model used in References 1–3 and also possesses some advantages over other solutions [4–9].

In this work, only one expansion mode was used on the feed. However, it may be necessary to use additional expansion modes on the feed for the case of a very thick substrate. This is a relatively straightforward procedure.

There is one drawback to this improved probe feed model in that the resulting numerical code requires a relatively large amount of computer time. For example, in Fig. 2, each frequency point for the idealised feed model takes about 2 minutes of CPU time on a VAX 11/750. For the improved feed model, each frequency point takes about 35 minutes of CPU time. One reason for this is that the series expression for the attachment mode requires about ten terms for accurate numerical calculation, even with convergence acceleration. However, the main reason is the large number of terms required to calculate the impedance matrix elements. Some work has been done in applying a method of convergence acceleration based on the ideas of Reference 14, but the results to date have not been promising.

In the future, the improved probe feed model will be applied to isolated elements and finite arrays of rectangular patches. In addition, similar solutions are being developed for other canonically-shaped (e.g. circular) patches.

5 Acknowledgment

This work was supported by Presidential Young Investigator Grant ECS 8352325 (USA) from the National Science Foundation and by the Hughes Aircraft Corporation.

6 References

1 POZAR, D.M.: 'Input impedance and mutual coupling of rectangular microstrip antennas', *IEEE Trans.*, 1982, **AP-30**, pp. 1191–1196
2 POZAR, D.M., and SCHAUBERT, D.H.: 'Analysis of an infinite array of rectangular Microstrip patches with idealised probe feeds', *ibid.*, 1984, **AP-32**, pp. 1101–1107
3 ABERLE, J.T., and POZAR, D.M.: 'Analysis of infinite arrays of one- and two-probe-fed circular patches', submitted for publication to *IEEE Trans.*, **AP**
4 LIU, C.C., HESSEL, A., and SHMOYS, J.: 'Performance of probe-fed microstrip-patch element phased arrays'. 1985 Phased Arrays Symposium, RADC-TR-85-170, September 1985
5 SCHUSS, J.J., HANFLING, J.D., CONTI, R.J., and HERRICK, B.R.: 'Numerical design of millimeter wave patch radiator arrays', *ibid.*, RADC-TR-85-170, September 1985.
6. SCHUSS, J.J., and HANFLING J.D.: 'Non-reciprocity and scan blindness in phased arrays using balanced-fed radiators', *IEEE Trans.*, 1987, **AP-35**, pp. 134–138
7 SCHUSS, J.J., and BAUER, R.L.: 'Axial ratio of balanced and unbalanced fed circularly polarized patch radiator arrays'. IEEE AP Symposium Digest, June 1987, **1**, pp. 286–289
8 CHEW, W.C., and KONG, J.A.: 'Analysis of a circular microstrip disk antenna with a thick dielectric substrate', *IEEE Trans.*, 1981, **AP-29**, pp. 68–76
9 MOSIG, J.R., and GARDIOL, F.E.: 'General integral equation formulation for microstrip antennas and scatterers', *IEE Proc., H, Microwaves, Antenna of Propag.*, 1985, **132**, pp. 424–434
10 POZAR, D.M.: 'General relations for a phased array of printed antennas derived from infinite current sheets', *IEEE Trans.*, 1985, **AP-33**, pp. 498–504.
11 WHEELER, H.A.: 'A survey of the simulator technique for designing a radiating element in a phased-array antenna', *in* OLINER, A.A., and KNITTEL, G.H. (Eds.): *'Proc of the 1970 symposium, phased array antennas'* (Artech House, Inc., 1972)
12 POZAR, D.M.: 'Radiation and scattering from a microstrip patch on a uniaxial substrate', *IEEE Trans.*, 1987, **AP-35**, pp. 613–621
13 RUDGE, A.W., MILNE, K., OLVER, A.D., and KNIGHT, P. (Eds.): '*The handbook of antenna design*. Vol. 1' (Peter Peregrinus Ltd., 1982)
14 POZAR, D.M.: 'Improved computational efficiency for the moment method solution of printed dipoles and patches', *J Electromag. Soc.*, 1984, **3**, pp. 299–309

7 Appendix

The grounded dielectric slab Green's function for the $\hat{p}$-directed electric field at the point (x, y, z) due to an infinite, periodic phased array of infinitesimal $\hat{q}$-directed current elements, where (x_0, y_0, z_0) is the location of the $m = 0, n = 0$ element, is given by eqn. 2.

Note that it is not necessary to know the Green's function components in complete generality. In particular, the presented analysis requires

$$G_{pq}(k_x, k_y, d \mid d) \quad \text{for} \quad p, q = x, y \tag{71}$$

$$G_{pz}(k_{x_1}, k_{y_1}, d/z_0 \leqslant d)$$

$$= -G_{zp}(k_{x_1}, k_{y_1}, z \leqslant d/d) \quad \text{for} \quad p = x_1, y \tag{72}$$

$$G_{zz}(k_x, k_y, z \leqslant d \mid z_0 \leqslant d) \tag{73}$$

$$G_{xx}(k_x, k_y, d \mid d) = \frac{-jZ_0}{k_0} \frac{(\varepsilon_r k_0^2 - k_x^2)k_2 \cos k_1 d + jk_1(k_0^2 - k_x^2) \sin k_1 d}{T_e T_m} \sin k_1 d \tag{74}$$

$$G_{xy}(k_x, k_y, d \mid d) = G_{yx}(k_x, k_y, d \mid d) = \frac{jZ_0}{k_0} \frac{k_x k_y \sin k_1 d[k_2 \cos k_1 d + jk_1 \sin k_1 d]}{T_e T_m} \tag{75}$$

$$G_{yy}(k_x, k_y, d \mid d) = \frac{-jZ_0}{k_0} \frac{(\varepsilon_r k_0^2 - k_y^2)k_2 \cos k_1 d + jk_1(k_0^2 - k_y^2) \sin k_1 d}{T_e T_m} \sin k_1 d \tag{76}$$

$$G_{xz}(k_x, k_y, d \mid z_0 \leqslant d) = \frac{-Z_0}{k_0} \frac{k_x k_z \cos k_1 z_0}{T_m} \tag{77}$$

$$G_{yz}(k_x, k_y, d \mid z_0 \leqslant d) = \frac{-Z_0}{k_0} \frac{k_y k_z \cos k_1 z_0}{T_m} \tag{78}$$

$$G_{zx}(k_x, k_y, z \leqslant d \mid d) = \frac{Z_0}{k_0} \frac{k_x k_z \cos k_1 z}{T_m} \tag{79}$$

$$G_{zy}(k_x, k_y, z \leqslant d \mid d) = \frac{Z_0}{k_0} \frac{k_y k_z \cos k_1 z}{T_m} \tag{80}$$

$$G_{zz}(k_x, k_y, z \leqslant d \mid z_0 \leqslant d) = \begin{cases} -\delta(z - z_0) \\ + \dfrac{\beta^2}{k_1 T_m} \{\varepsilon_r k_2 \sin k_1(d - z_0) - jk_1 \cos k_1(d - z_0)\} \cos k_1 z & z \leqslant z_0 \\ \dfrac{\beta^2}{k_1 T_m} \{\varepsilon_r k_2 \sin k_1(d - z) - jk_1 \cos k_1(d - z)\} \cos k_1 z_0, & z \leqslant z_0 \end{cases} \tag{81}$$

It should be pointed out that G_{xx}, G_{xy}, G_{yz} and G_{yy} given here are the same as the corresponding quantities in References 2 and 3, but the remaining functions are different because they have not been integrated over a uniform current probe, as carried out in Reference 2. The quantities T_{m1}, T_e, k_1 and k_2 are the same as in Reference 2.

Diffraction of Electromagnetic Waves by a Conducting Screen Perforated Periodically with Circular Holes

CHAO-CHUN CHEN, MEMBER, IEEE

Abstract—The reflection and transmission of a plane wave incident on a thin perfectly conducting screen perforated periodically with circular holes are considered. The spacings between holes may be comparable to or less than the wavelength. The metal screen is assumed to be either backed by a flat dielectric sheet, or sandwiched between two or more dielectric sheets that are used to match the susceptance of the screen. The techniques of transmission line analysis and the method of moments are employed to obtain a solution for this general class of boundary-value problems. Numerical results for several cases of interest in both *X* and *Ku* bands were obtained and shown to be in good agreement with experimental measurements.

INTRODUCTION

A KNOWLEDGE of the reflection and transmission of a plane wave through either a two-dimensional periodic array of metallic disks or the complementary structure consisting of a conduction screen perforated periodically with apertures is important in many applications such as the design of artificial dielectrics, microwave filters, wave polarizers, and antenna reflectors or ground planes.

The behavior of a plane wave propagating through a planar rectangular array of circular disks with radius $ka<1$ has been investigated by Collin and Eggimann [1], [2]. In their work, the diffracted field was evaluated approximately from the induced electric and magnetic dipole moments. Higher order multipole moments and their interactions between disks were considered, but the analysis was restricted to the case in which the geometry of an array of disks and its diffracted field were coincident with the images of a disk-loaded TE_{10} rectangular waveguide. McMillan [3] adapted the above approach of dipole moments for rectangular lattices of conducting disks, to the analysis of lattices of holes in a conducting screen, and also extended the work to allow for arbitrary incidence and polarization.

The problem of mutual coupling between radiating elements in a two-dimensional periodic structure has been investigated extensively for phased arrays because of the practical design problems that arise from element impedance variations with scanning. The aperture reflection coefficient for a waveguide element in an infinite array environment has been determined accurately by matching the tangential field components at the apertures and solving the resulting integral equation for the aperture field by the Ritz–Galerkin method (Borgiotti [4], Amitay and Galindo [5], [6]). Essentially this same method was also employed by the author to solve the problem of a plane wave propagating through a two-dimensional periodic array of slots in a thin conducting screen [7]. The work presented here is an extension of this previous work to an array of circular holes. A metal–dielectric and a dielectric–metal–dielectric laminate wall are assumed. The reflection and transmission coefficients for an incident plane wave are derived by the techniques of transmission-line analysis. The use of equivalent circuit representation provides an insight into the physical phenomena of this structure.

Manuscript received August 19, 1970; revised November 9, 1970.
The author is with the Aerospace Group, Hughes Aircraft Company, Culver City, Calif. 90230.

INTEGRAL FORMULATION

Fig. 1 illustrates a two-dimensional array of circular apertures with radius a arranged periodically along any two arbitrary coordinates S_1 and S_2. The transverse electric field components on both sides of the screen can be expanded in a set of orthonormal mode functions [5] Φ_{pqr} (with two spatial harmonic numbers p and q), which have propagation constants γ_{pq} along the z axis. The third subscript $r=1$ or 2, is used to indicate TE or TM waves, respectively.

A plane wave with unit electric field intensity, incident at an angle θ with respect to the z axis in the ϕ plane can be written in terms of modal voltage $\bar{\Phi}_{00r}$ and modal admittance of $\xi_{00r}{}^f$ as

$$\bar{E}^i = \sum_{r=1}^{2} A_{00r}\bar{\Phi}_{00r} \tag{1}$$

$$\bar{H}^i = \sum_{r=1}^{2} A_{00r}\xi_{00r}{}^f(\hat{z} \times \bar{\Phi}_{00r}) \tag{2}$$

with $e^{-j(\gamma_{00}z-\omega t)}$ variation assumed.

Next let R_{pqr} and B_{pqr} be the reflected and transmitted coefficients for the Φ_{pqr} space harmonic modes. $\xi_{pqr}{}^f$ and $Y_{pqr}{}^+$ are the characteristic modal admittance in free space and the modal admittance at $z=0^+$, which accounts for the dielectric sheet in the positive z region.

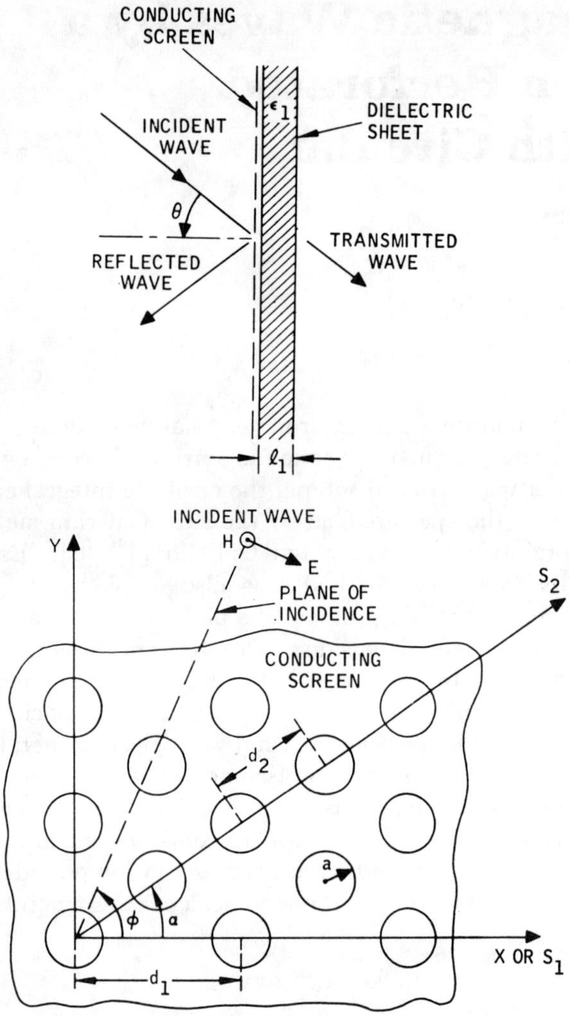

Fig. 1. Geometry of two-dimensional array of circular holes in thin conducting screen.

Then the transverse field components in the $z \leq 0$ region are

$$\overline{E}_1 = \sum_{r=1}^{2} A_{00r}\overline{\Phi}_{00r} + \sum_p \sum_q \sum_{r=1}^{2} R_{pqr}\overline{\Phi}_{pqr} \tag{3}$$

$$\overline{H}_1 = \sum_{r=1}^{2} A_{00r}\xi_{00r}{}^f(\hat{z} \times \overline{\Phi}_{00r}) - \sum_p \sum_q \sum_{r=1}^{2} R_{pqr}\xi_{pqr}{}^f(\hat{z} \times \overline{\Phi}_{pqr}) \tag{4}$$

and in the $z \geq 0$ region

$$\overline{E}_2 = \sum_p \sum_q \sum_{r=1}^{2} B_{pqr}\overline{\Phi}_{pqr} \tag{5}$$

$$\overline{H}_2 = \sum_p \sum_q \sum_{r=1}^{2} B_{pqr}Y^+{}_{pqr}(\hat{z} \times \overline{\Phi}_{pqr}),$$

$$p, q = 0, \pm 1, \pm 2, \cdots, \pm \infty \quad \text{and} \quad r = 1, 2. \tag{6}$$

Matching the tangential field components E_t and H_t at $z=0$, one obtains

$$2\sum_{r=1}^{2} A_{00r}\xi_{00r}{}^f\overline{\Phi}_{00r} = \sum_p \sum_q \sum_{r=1}^{2} (\xi_{pqr}{}^f + Y_{pqr}{}^+)\overline{\Phi}_{pqr} \iint_{\text{aperture}} \overline{E}_t \cdot \overline{\Phi}_{pqr}{}^* \, da \tag{7}$$

and

$$\delta_{pq}A_{00r} + R_{pqr} = B_{pqr} = \iint_{\text{aperture}} \overline{E}_t \cdot \overline{\Phi}_{pqr}{}^* \, da \tag{8}$$

where $\delta_{pq}=1$ for both p and $q=0$, and 0 otherwise. In the absence of the dielectric sheet, the wave admittance $Y_{pqr}{}^+$ is equal to $\xi_{pqr}{}^f$.

To solve the integral equation (7), the method of moments [8] was used. In applying this technique one needs to select a set of linearly independent modes that satisfy the boundary conditions at the individual holes to represent the electric field $\overline{E}_t$. For the circular holes, one may use the circular waveguide mode functions $\overline{\psi}_{mnr}$ and let

$$\overline{E}_t = \sum_m \sum_n \sum_{r=1}^{2} F_{mnr}\overline{\psi}_{mnr} \tag{9}$$

where

$$\overline{\psi}_{mn1} = g\left[\frac{ma}{x_{mn}'\rho} J_m\left(\frac{x_{mn}'}{a}\rho\right) \sin m\phi\hat{\rho} + J_m'\left(\frac{x_{mn}'}{a}\rho\right) \cos m\phi\hat{\phi}\right] \tag{10}$$

$$\overline{\psi}_{mn2} = h\left[J_m'\left(\frac{x_{mn}}{a}\rho\right) \cos m\phi\hat{\rho} + \frac{ma}{x_{mn}\rho} J_m\left(\frac{x_{mn}\rho}{a}\right) \sin m\phi\hat{\phi}\right] \tag{11}$$

and

$$g = \sqrt{\frac{2}{\pi}} \cdot \frac{1}{J_m(x_{mn}')\sqrt{(x_{mn}')^2 - m^2}} \cdot \frac{x_{mn}'}{a}$$

$$h = \sqrt{\frac{2}{\pi}} \cdot \frac{1}{aJ_{m-1}(x_{mn})}.$$

By substituting (9) into (7), multiplying both sides of the equations by the test function $\overline{\psi}_{mnr}$ and integrating over the aperture, (7) can be transformed into a set of linear algebraic equations with the mode coefficients F_{mnr} as unknowns. The aperture field $\overline{E}_t$ is then determined by solving this set of simultaneous equations. Once the aperture field is known, the reflection and transmission coefficients R_{pqr} and B_{pqr} can be obtained from (8).

The equivalent input admittance for a plane wave with unit electric field intensity ($A_{00r}=1$) incident at

$z=0$ is

$$Y_r = G_r + jB_r = \left(\frac{1 - R_{00r}}{1 + R_{00r}}\right)\xi_{00r}{}^f,$$

$$r = \begin{cases} 1 \text{ for TE wave} \\ 2 \text{ for TM wave.} \end{cases} \quad (12)$$

In the absence of the dielectric sheet, the perfectly conducting screen absorbs no energy. If neither the cross-polarized mode is excited nor the higher space harmonic modes are propagated, the equivalent shunt admittance at $z=0$ must be a pure susceptance in parallel with the characteristic admittance of the incident wave. That means, if the approximate solution is valid, the equivalent circuit conductance G_r for a plane wave incident in a plane that is symmetric with the geometry of the array, should be equal or close to $\xi_{00r}{}^f$.

Sandwich Structure

In practical applications it may be preferable to have the conducting screen laminated between two dielectric layers, as shown in Fig. 2, since this structure provides more design parameters for bandwidth and impedance matching. Since the free-space-to-dielectric interface does not excite higher order modes, all spatial harmonics existing are generated by the discontinuity at the conducting screen. The modal voltage coefficients R_{pqr} and B_{pqr} at $z=0$ can be obtained with use of the previous procedure. The integral equation for the unknown aperture field at $z=0$ is

$$2\sum_{r=1}^{2} A_{00r}' \frac{\xi_{00r}{}^f \xi_{00r}{}^{d2} \overline{\Phi}_{00r}}{[\xi_{00r}{}^{d2}\cos(\gamma_{00}l_2) + j\xi_{00r}{}^f \sin(\gamma_{00}l_2)]}$$

$$= \sum_p \sum_q \sum_{r=1}^{2} (Y_{pqr}{}^- + Y_{pqr}{}^+)\overline{\Phi}_{pqr} \cdot \iint_{\text{aperture}} \overline{E}_t \cdot \overline{\Phi}_{pqr}{}^* \, da \quad (13)$$

where $\xi_{00r}{}^{d2}$ is the characteristic admittance of the $\overline{\Phi}_{00r}$ wave in the left dielectric region, and $Y_{pqr}{}^-$ is the modal admittance of the $\overline{\Phi}_{pqr}$ wave looking into the negative z direction at $z=0$. Both magnitude and phase of A_{00r}' are referred to the free-space-to-dielectric interface at $z=-l_2$. The reflection coefficient and the input admittance of a plane wave incident at $z=-l_2$ can be obtained by transmission-line analysis. The results are

$$R_{pqr}' = \frac{\xi_{pqr}{}^{d2}\iint_{\text{aperture}} \overline{E}_t \cdot \overline{\Phi}_{pqr}{}^* da - \delta_{pq} A_{00r}'[\xi_{00r}{}^{d2}\cos(\gamma_{00}l_2) - j\xi_{00r}{}^f \sin(\gamma_{00}l_2)]}{[\xi_{pqr}{}^{d2}\cos(\gamma_{pq}l_2) + j\xi_{pqr}{}^f \sin(\gamma_{pq}l_2)]} \quad (14)$$

$$Y_r' = \left(\frac{1 - R_{00r}'}{1 + R_{00r}'}\right)\xi_{00r}{}^f \quad (15)$$

and the transmission coefficient referred to the right dielectric-to-free-space interface, $z=l_1$, is

$$B_{pqr}' = \frac{\xi_{pqr}{}^{d1}}{[\xi_{pqr}{}^{d1}\cos(\gamma_{pq}l_1) + j\xi_{pqr}{}^f \sin(\gamma_{pq}l_1)]} \cdot \iint_{\text{aperture}} \overline{E}_t \cdot \overline{\Phi}_{pqr}{}^* \, da \quad (16)$$

where $\xi_{pqr}{}^{d1}$ is the characteristic wave admittance of the $\overline{\Phi}_{pqr}$ wave in the right dielectric region. The insertion phase through the metal–dielectric laminated wall is

$$\text{insertion phase} = \underline{/B_{00r}'} - \frac{2\pi}{\lambda}(l_1 + l_2). \quad (17)$$

Numerical Results

The reflection coefficient and the input admittance were evaluated for several cases of interest at both the X and Ku band frequencies. The triangular lattice of holes with element spacings $d_1 = d_2 = 1.73$ cm and $\alpha = 60$ deg as shown in Fig. 1 was assumed. The radius of the circular holes and the thickness of the dielectric sheets are individually shown on the appropriate figures. The data presented are based on a computation which included the lowest ten waveguide modes in (9), and those free-space modes $\overline{\Phi}_{pqr}$ having a propagation constant $|\gamma_{pq}| \leq 3k$ in free space. This choice of the number of modes in the computation of the reflection coefficient and the equivalent circuit parameters appears to be sufficient. This is indicated by the fact that the normalized conductance obtained in the absence of the dielectric sheets is very close to unity for the case of single-mode propagation. This fact is consistent in the sense that the perfectly conducting screen contributes a pure shunt susceptance that is in parallel with the characteristic wave admittance of incident plane wave. The problem of the convergence of the solutions obtained by the method of moments and questions about energy conversation have been discussed in previous papers [7], [11].

Fig. 3 illustrates the equivalent circuit parameters of a conducting screen perforated periodically with circular holes when the screen is illuminated by a normally incident plane wave with E field parallel to the y axis. The shunt susceptance of a conducting screen with small circular holes changes rapidly with frequency. Consequently, it has a high Q factor and possesses a narrow passband. Below the resonant frequency, the shunt susceptance of the screen is inductive; as the fre-

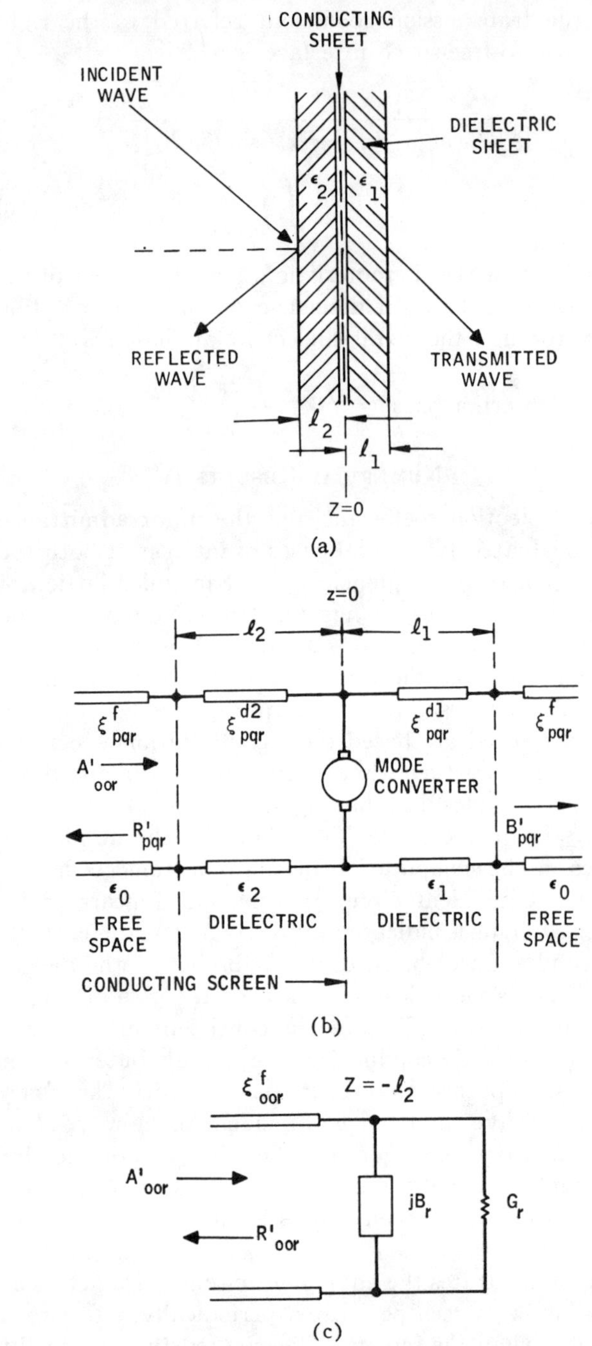

Fig. 2. Geometry of sandwich wall and its equivalent circuit. (a) Sandwich wall construction. (b) Transmission-line representation. (c) Equivalent input admittance at $z=l_2$.

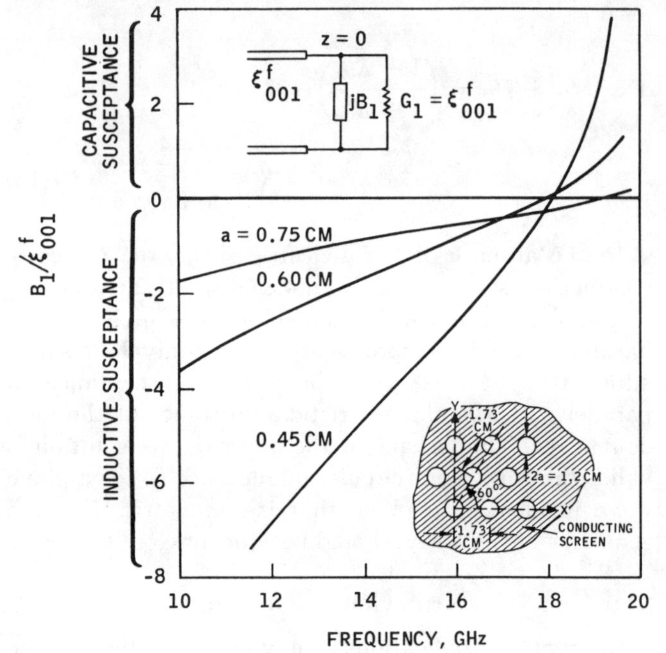

Fig. 3. Shunt susceptance of thin conducting screen for plane wave with E field parallel to y axis at normal incidence.

quency is increased, it first becomes resonant and then capacitive.

Fig. 4 illustrates the effects of dielectric loading on the conducting screen. The reflection coefficient for a plane wave with E field parallel to the y axis at normal incidence is plotted as a function of frequency. Without dielectric loading, the screen is resonant at 17.9 GHz. With the insertion of a 0.04-cm thick dielectric sheet with $\epsilon_r=4$ on one side of the screen, the resonant frequency is lowered to 15.3 GHz. If the screen is sandwiched between two dielectric sheets, the resonant frequency is further lowered to 13.5 GHz. From Fig. 2(b) one can see that insertion of a dielectric sheet in front (or back) of a conducting screen is equivalent to the insertion of a section of transmission line that possesses a different characteristic admittance in the original free-space line. Thus a thin dielectric sheet placed in close contact with the conducting screen can be designed to contribute a capacitive susceptance to match the inductive screen. As a result, a dielectric–metal–dielectric sandwich wall construction can be made to provide a good match over a wide range of incidence angles and frequency.

Measurements of the insertion loss at normal incidence were made on 13- by 16-in flat panels for the three cases of dielectric loading shown in Fig. 4. The tested sample was patched on a 9- by 12-in opening in a wall of microwave absorbing material with the edges of the panel covered by an absorber. Two waveguide horns, one sweep generator, and one X–Y recorder were used in the measurement. The transmitting horn and the receiving horn were placed 8 ft apart with the test panel erected in between. The transmitted power level versus frequency exhibited oscillator ripples; these may be attributed to diffraction from the edges of the panel. Deviation from the calculated results may also be caused by the nonplanar wave used to illuminate the flat panel. Comparison of the measured means and the calculated results are shown in Fig. 5.

Variation of the reflection coefficient as a function of incident angle is shown in Fig. 6. The sandwiched screen has an inductive susceptance for a plane wave at normal incidence. As the beam scans away from broadside, the screen becomes resonant at $\theta=10$ deg and then capacitive in the region of large scan angles. It is noted that a sudden change of phase is observed at the location

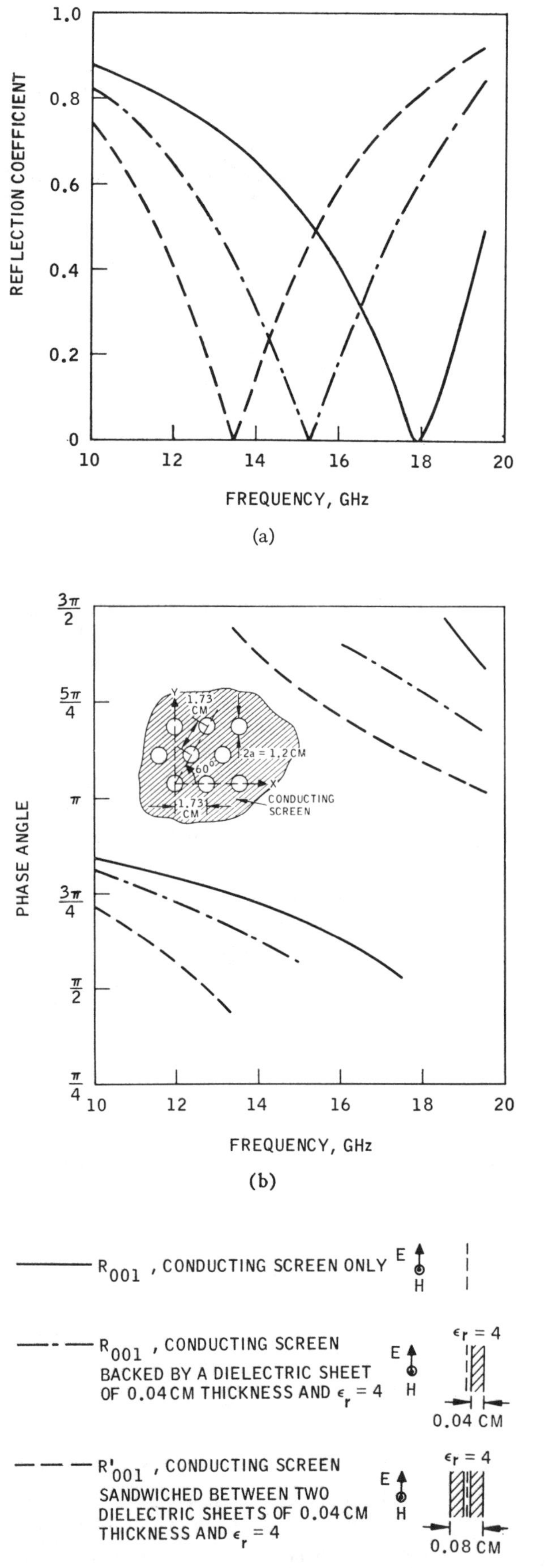

Fig. 4. Reflection coefficients for plane wave with E field parallel to y axis at normal incidence. (a) Magnitude of R_{001} or R_{001}'. (b) Phase of R_{001} or R_{001}'.

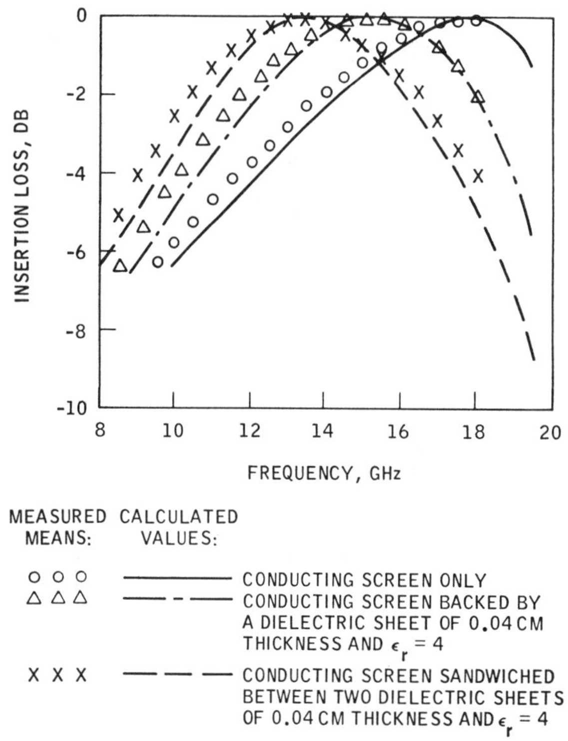

Fig. 5. Comparison of measured and calculated insertion losses for plane wave with E field parallel to y axis at normal incidence for three different cases of dielectric loading as given in Fig. 4.

where resonance occurs. The magnitude of reflection for a parallel-polarized wave incident at a large angle from broadside is generally less than that of a perpendicularly polarized wave, which is to be expected.

Waveguide Discontinuity Problems

The previous analysis can be applied directly to solve a wide class of wave discontinuity problems which arise when a thin metal diaphragm or irises are placed in a rectangular waveguide, provided that the scattering obstacles in the waveguide have certain symmetry [2], [9], and that the multiple images of these obstacles with respect to the guide walls form an infinite planar array of the type discussed. For example, the equivalent circuit parameters of a disk-loaded rectangular waveguide were evaluated and plotted in Fig. 7 for different disk radii a. The waveguide is loaded with two circular conducting disks at $z=0$, in the transverse cross section of the guide, and a dominant TE_{10} mode is incident from $z=-\infty$. This corresponds to an infinite planar array of disks illuminated by two plane waves incident at the opposite angles, $\theta=\sin^{-1}(\lambda/\lambda_c)$, where λ_c is the cut-off wavelength of the incident mode propagating in the rectangular waveguide and λ is the free space wavelength. The shunt susceptance of the disk-loaded waveguide is obtained as follows. First, the transmission coefficient of a parallel-polarized wave incident on the complementary conducting screen of an infinite array of disks is evaluated. The reflection coefficient of the TE_{10} mode for the original problem is obtained by use

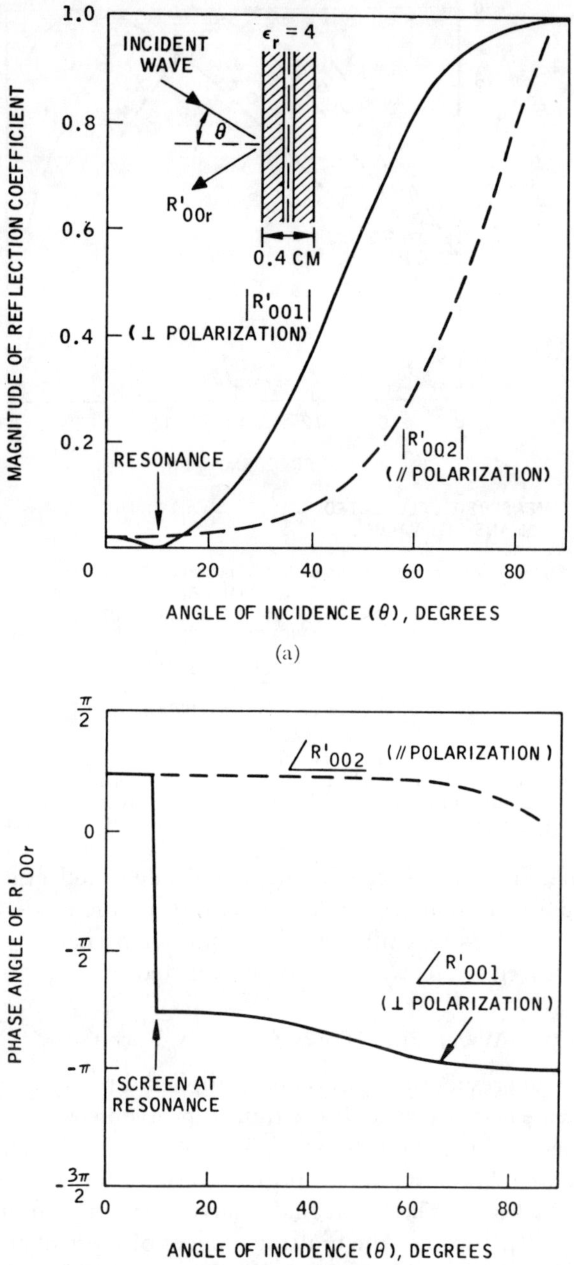

Fig. 6. Reflection coefficients for plane wave incident at oblique angles in x–z plane; $f=8.5$ GHz. (Array dimensions as shown in Fig. 3.) (a) Magnitude of R_{00r}'. (b) Phase of R_{00r}'.

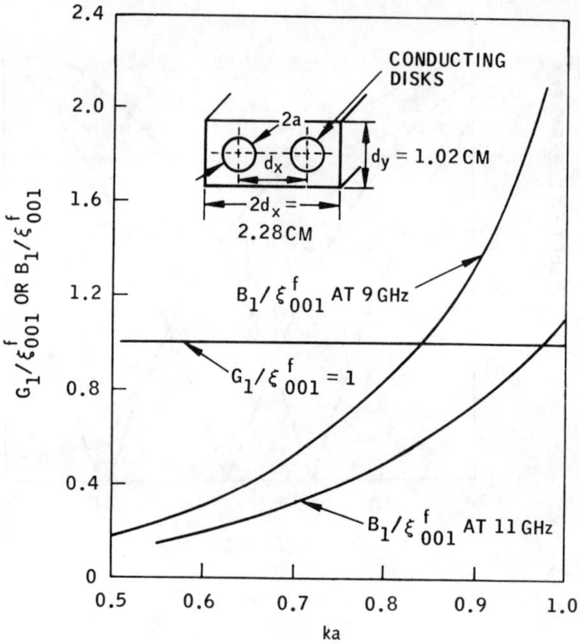

Fig. 7. Input admittance of disk-loaded rectangular waveguide at $z=0$ for different disk radii a, $f=9$ and 11 GHz.

of Babinet's principle. Results are in agreement with the values obtained by Eggimann and Collin [2] in their multipole moment calculations; however, the normalized conductance they obtained does not equal unity because of the difficulty in determining the induced current on the disks.

Discussion

The techniques of transmission-line analysis and the method of moments can be applied to obtain numerical solutions for a wide class of wave discontinuity problems, such as the calculation of the refraction index of either isotropic or anisotropic artificial dielectric media, and the analysis of corrugated surface waveguides. Two sets of eigenfunctions are used in the formulation of the problem, one set which describes the field as a whole, the other which describes the field in the individual apertures. The effects of higher order modes, cross-polarized modes, and interactions between apertures are accounted for in the computations. The numerical results presented here are limited to cases in which only the dominant mode propagates; however, the solution is general and it can be applied to situations with any number of propagating modes.

Acknowledgment

The author wishes to thank D. J. Lewis and Dr. W. H. Kummer for their helpful suggestions and for reviewing this manuscript.

References

[1] R. E. Collin and W. H. Eggimann, "Dynamic interaction fields in a two-dimensional lattice," *IRE Trans. Microwave Theory Tech.*, vol. MTT-9, Mar. 1961, pp. 110–115.

[2] W. H. Eggimann and R. E. Collin, "Electromagnetic diffraction by a planar array of circular disks," *IRE Trans. Microwave Theory Tech.*, vol. MTT-10, Nov. 1962, pp. 528–535.

[3] E. B. McMillan, "Millimeter/microwave radome building block—a dielectric sandwiched metallic circular hole lattice." Resources Development Institute, Final Rep. 93101.

[4] G. V. Borgiotti, "Modal analysis of periodic planar phased arrays of apertures," *Proc. IEEE.* vol. 56, Nov. 1968, pp. 1881–1892.

[5] N. Amitay and V. Galindo, "The analysis of circular waveguide phased arrays," *Bell Syst. Tech. J.*, vol. 47, Nov. 1968, pp. 1903–1932.

[6] N. Amitay and V. Galindo, "Characteristics of dielectric loaded and covered circular waveguide phased arrays," *IEEE Trans.*

Antennas Propagat., vol. AP-17, Nov. 1969, pp. 722–729.
[7] C. C. Chen, "Transmission through a conducting screen perforated periodically with apertures," *IEEE Trans. Microwave Theory Tech.*, vol. MTT-18, Sept. 1970, pp. 627–632.
[8] R. F. Harrington, "Matrix methods for field problems," *Proc. IEEE*, vol. 55, Feb. 1967, pp. 136–149.
[9] P. W. Hannan and M. A. Balfour, "Simulation of a phased-array antenna in waveguide," *IEEE Trans. Antennas Propagat.*, vol. AP-13, May 1965, pp. 342–353.
[10] R. B. Kieburtz and A. Ishimaru, "Scattering by a periodically apertured conducting screen," *IRE Trans. Antennas Propagat.*, vol. AP-9, Nov. 1961, pp. 506–514.
[11] N. Amitay and V. Galindo, "On energy conservation and the method of moments in scattering problems," *IEEE Trans. Antennas Propagat.*, vol. AP-17, Nov. 1969, pp. 747–751.
[12] R. E. Collin, *Field Theory of Guided Waves*. New York: McGraw-Hill, 1960.
[13] J. Brown and W. Jackson, "The relative permittivity of tetragonal arrays of perfectly conducting thin disks," *Proc. Inst. Elec. Eng.* (London), vol. 102, pt. B, Jan. 1955, pp. 37–42.

Modes of resonance of the Jerusalem cross in frequency-selective surfaces

E.A. Parker, M.A., Ph.D., C.Eng., F.R.A.S., M.I.E.E., S.M.A. Hamdy, M.Sc., Ph.D., and R.J. Langley, B.Sc. Ph.D., C.Eng., M.I.E.E.

Indexing terms: Electromagnetics, Antennas, Frequency-selective surfaces

Abstract: Two frequency-selective surfaces consisting of arrays of Jerusalem crosses have been analysed using a modal technique. Calculated plane-wave-transmission coefficients are compared with experimental values in the frequency range 1–40 GHz, for incidence up to 50° in the principal planes. At each reflection resonance the calculated element current distribution is illustrated and discussed. At the main resonance the current is confined largely to the E-plane arm of the cross, whereas, at higher frequencies, asymmetrical resonances of the entire structure are observed. 22 basis functions are required to represent the element current over the entire frequency range, but computer processing time can be reduced by identifying the significant functions in restricted wavebands.

1 Introduction

The Jerusalem cross has a complicated geometry (Fig. 1), but it belongs to a class of array elements for which scalar equivalent circuits are available as design tools for modelling their plane-wave-transmission characteristics. Experimental studies of Jerusalem-cross arrays as frequency selective surfaces were first described by Arnaud and Ruscio [1], and by Arnaud and Pelow [2]. A simple LC equivalent circuit was subsequently proposed by Anderson [3] to predict the location of the main reflection resonance (f_1 in this paper). Since then the circuits have been improved [4, 5] to model more closely a greater range of the transmission curve for plane-wave incidence.

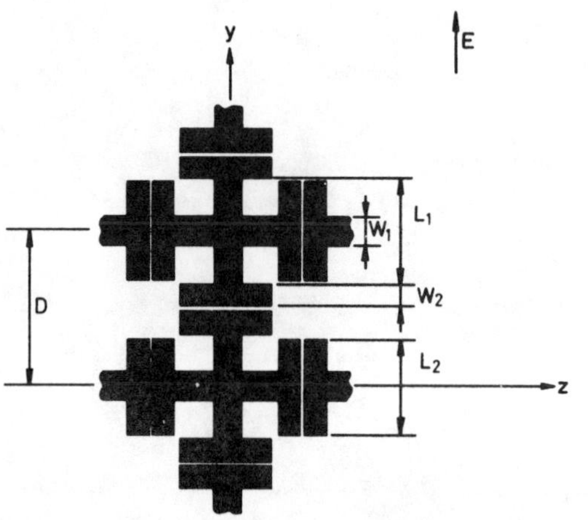

Fig. 1 *Geometry of the Jerusalem cross*

E-field lies in the (y, z) plane
Array 1: $L_1 = 5.69$, $L_2 = 4.96$, $W_1 = 0.90$, $W_2 = 0.30$, $D = 6.50$
Array 2: $L_1 = 4.69$, $L_2 = 4.58$, $W_1 = 0.80$, $W_2 = 0.40$, $D = 5.82$
Dimensions are in mm

An important property of the Jerusalem cross is the closeness with which the transmission and reflection bands can be spaced, if required. Band centre frequency ratios of about 1.5 can be readily attained. As far as we are aware, the only other single layer element described in the published literature and capable of providing such close, or closer, spacings is the concentric ring [6].

As the form of the transmission curve for the Jerusalem cross depends critically on (among other factors) the spacing between the capacitive end caps of the element (Fig. 1), problems can arise when the elements are distributed on curved surfaces such as a hyperbolic Cassegrain subreflector. Jerusalem-cross arrays tend, therefore, to be used as plane reflectors. Recently, Cahill and Parker [7] have compared the linear crosspolar performance of concentric ring and Jerusalem-cross arrays as plane reflectors in a 45° incidence diplexer. Both arrays gave levels of the order of − 33 dB in reflection, but the Jerusalem cross gave somewhat higher levels in part of its transmission band. Prediction of the crosspolar levels to be expected requires a vector analysis, such as that described elsewhere [8, 9] for dichroic plates constructed from stacks of waveguides. Much more computer power is then needed compared with the requirements of the equivalent-circuit approach, but the results provide a better understanding of the fairly complicated transmission curves shown by the Jerusalem cross.

Some results of a modal analysis of a constituent part of the Jerusalem cross, the E-plane arm represented by an array of LC elements, have already been presented [10]. (It should

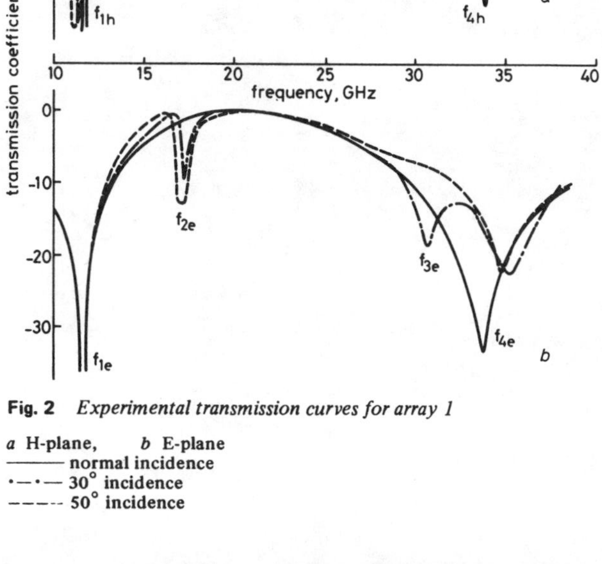

Fig. 2 *Experimental transmission curves for array 1*

a H-plane, *b* E-plane
——— normal incidence
·—·— 30° incidence
----- 50° incidence

Paper 2421 H, first received 13th July 1982 and in revised form 14th January 1983

The authors are with the Electronics Laboratories, The University, Canterbury, Kent CT2 7NT, England be noted that a printer's error has introduced an ambiguity into the key for Fig. 2*a* of Reference 10, where the second pair of transmission curves are for 50° incidence and not 0°.) Eleven current functions were needed to represent the structure. For the full Jerusalem cross the computer time required can become prohibitive. This paper describes the results (summarised briefly elsewhere [11]) of a plane-wave modal analysis made by two of the authors (Hamdy and Parker), compares them with some results of an experimental study made by Langley, and discusses the choice of current functions in the light of the need to minimise the computer requirements.

2 Analysis

The detailed structure of the experimental transmission curves, which this work sets out to account for, is shown for three angles of incidence in the principal planes in Fig. 2. The position of the main reflection resonance is insensitive to angle of incidence and in this particular case occurs at about 11 GHz. At higher frequencies there are several other features which influence the width of the transmission band available near 18 GHz. They are related, in Section 4 below, to specific modes of resonance predicted by the analysis.

In the calculation, infinite arrays of conducting elements were set in the (x, y) plane (Fig. 1) and supported on dielectric substrates 0.021 mm thick, with $\epsilon_r = 2.3$. They were illuminated with linearly polarised plane waves, with the incident E-field in the (y, z) plane. The Jerusalem-cross array was considered to be a superposition of six arrays of conducting rectangular strips. The mutual coupling between them was taken into account through the integration over the unit cell which is required in the modal solution [12]. The elements of these arrays were as follows:

(1) Two horizontal and vertical rectangular bars of length L_1 and width W_1, centred at the origin of the (x, y) plane (Fig. 1). The sinusoidal functions used to expand the induced current density on the conductors were of the form

$$h^{(b)}(x) = \sqrt{\frac{2}{W_1 L_1}} \left[a_n \cos\left(\frac{n\pi}{L_1}\right) x + b_m \sin\left(\frac{m\pi}{L_1}\right) x \right] \hat{x} \qquad (1)$$

for the horizontal bars, where the superscript (b) stands for bar and a_n, b_m $(n, m = 0, 1, 2 \ldots)$ are the complex amplitudes of the current functions. For the vertical bars, x was exchanged for y in Eqn. 1.

(2) Four arrays of rectangular caps of length L_2 and width W_2, equally shifted in the x and y directions from the origin by a distance $|r| = (L_1 + W_2)/2$. Two of these arrays were y-directed caps and the other two x-directed caps. The corresponding sinusoidal functions, used to expand the induced currents on them, were of the form

$$h\binom{x}{y} = h^{(c)}\binom{x}{y} \exp\,[jk_t . r] \qquad (2)$$

where

$$h^{(c)}(x) = \sqrt{\frac{2}{W_2 L_2}} \left[c_n \cos\left(\frac{(2n-1)\pi}{L_2}\right) x + d_m \sin\left(\frac{2m\pi}{L_2}\right) x \right] \hat{x} \qquad (3)$$

for the horizontal caps, and again y substituted for x for the vertical caps. The superscript (c) stands for caps and c_n, d_m $(n, m = 1, 2, \ldots)$ are complex amplitudes. k_t is the transverse component of the wave number k.

A computer program, employing the method of moments [13], was written to calculate these complex amplitudes (and hence the current distribution on the conducting strips); together with the transmission through the arrays in the range 1–40 GHz. The relative importance of the basis functions in Eqns. 1–3 was examined by including them in the expansion of the currents in order of increasing numerical value of their indices, m and n. The functions with lower indices were tested first, with subsequent computer runs including higher orders. Each time the functions with comparatively low complex amplitudes were rejected from the series. The number of Floquet modes [13] required to express the fields adjacent to the conductors was also investigated, using a similar procedure. The result was a set of 22 different current functions, all of which were necessary to describe fully the induced currents on the Jerusalem crosses in the two principal planes at angles of incidence up to 50°. Nevertheless, including these functions all at once in the computer model, over the frequency range of interest, required excessive amounts of computer time.

3 Minimum current function models

The relative importance of the individual basis functions in the series expansion of the induced currents was found to be dependent on the plane of incidence. The two principal planes were therefore treated separately. In each case, the theoretical transmission curves were calculated using effective values for L_1 and L_2, the lengths of the centre section and end caps of the Jerusalem cross. These values were obtained by optimising the fit to the experimental curves for two different arrays, with dimensions given in Fig. 1.

3.1 H-plane model

In the H-plane, only 11 basis functions were sufficient to describe the induced currents at all angles of incidence up to 50° and at all frequencies in the range 1–40 GHz. The currents induced on the horizontal bar were negligibly small compared with the currents induced on the other segments of the Jerusalem cross. This indicates that in the H-plane the two vertical caps are electrically decoupled from the centre section of the cross.

Varying the length of the end caps (L_2) and the vertical centre section (L_1) showed that the model is sensitive to the length of the conductors. A change of one per cent in L_1 altered the value of f_1, the lowest reflection resonant frequency, by five per cent. The optimum fit of the transmission curves for both arrays occurred when L_1 was increased by 1.2 W_2; although a closer fit for array 1 alone was obtained with an increase of only 1.0 W_2. For comparison, with the E-plane arm alone the effective length of the vertical bar was found in Reference 10 to be $L_1 + W_2$ in the H-plane and slightly longer in the E-plane. It is reasonable to expect the effective length to be greater than L_1, to account for fringing of the current as it flows into the end caps. The effective lengths of the latter appear to be close to the physical lengths in the H-plane.

3.2 E-plane model

The situation in the E-plane was rather different from that in the H-plane. We found it difficult to proceed in the same way owing to the fact that the horizontal bar is now contributing to the complicated transmission resonance pattern (Fig. 2*b*) in the E-plane. Four main distinct resonances (f_{1e}, f_{2e}, f_{3e} and f_{4e}) were measured in the transmission plot.

Each of them requires a different set of basis functions to describe the induced current at each resonant frequency.

For this reason, the sets of harmonic functions shown in Table 1 were chosen at three different bands of frequencies, set around the various resonances. The two upper resonant frequencies are included in a common high-frequency band. In all cases, 121 Floquet modes were used to describe the tangential fields close to the arrays. In addition, we found the effective lengths of the vertical and horizontal bars were $L_1 + 1.6\ W_2$ for optimum fit to the observed transmission curves up to 50° incidence.

4 Modes of resonance of the Jerusalem cross

For comparison with the predictions of the minimum current function model the experimental results from Fig. 2 are reproduced for clarity in Figs. 3 and 4. The results are those for array 1 at three angles of incidence: $\theta = 0°$, 30° and 50°. Fig. 4 also shows the frequency bands where different basis functions were used in the E-plane. At each resonant frequency, the complex current functions (eqns. 1–3) have been added vectorially for each segment of the Jerusalem cross. The resultant current distributions are sketched in Fig. 5. They apply to array 1 at 30° incidence; but the results for array 2 are similar.

4.1 *Fundamental resonance at f_1*

The resonance at f_1 is the fundamental mode and corresponds to that observed for arrays of elements consisting of only the E-plane arm of the cross [10]. It is observed in both the E- and H-planes and is insensitive to angle of incidence. It is due to a resonance of the structure formed by the vertical centre section, together with the two horizontal caps. The currents induced in the horizontal section and vertical caps are very small, with amplitudes down by a factor of 10. The current functions used for the vertical centre section add to give an almost uniform current distribution, as can be seen in Figs. 5 and 6*a*. The current is not zero at the ends of this section at f_1. In both planes of incidence, two asymmetrical functions were required for the horizontal caps (band 1 in Table 1). It is possible that these are the two most significant terms in a series of asymmetrical functions which would sum to give a distribution with a phase discontinuity at the centre of each cap, as is sketched in Fig. 6*a*. The representation

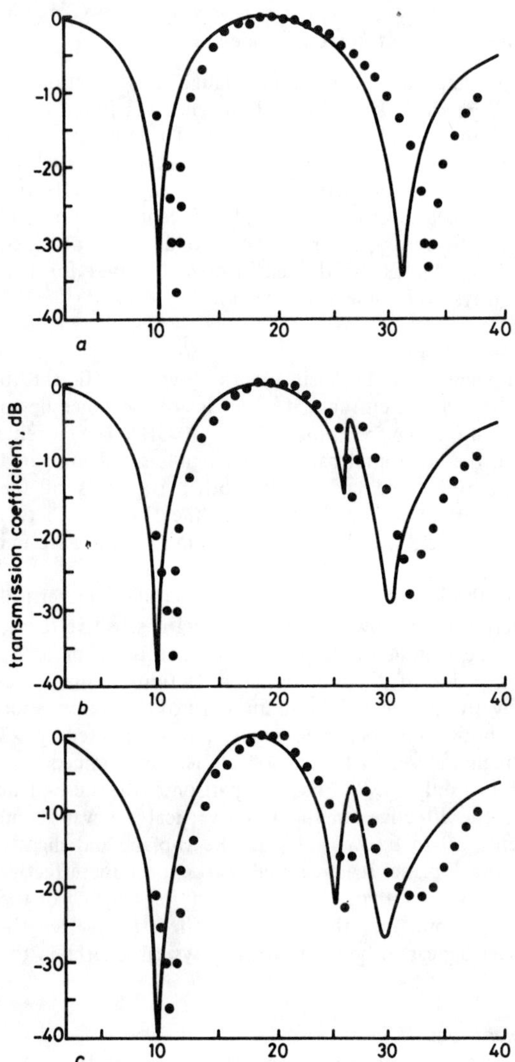

Fig. 3 *Calculated and measured transmission coefficients in the H-plane*

Angles of incidence
a $\theta = 0°$ *b* $\theta = 30°$ *c* $\theta = 50°$
——— calculated results
. . . . experimental results

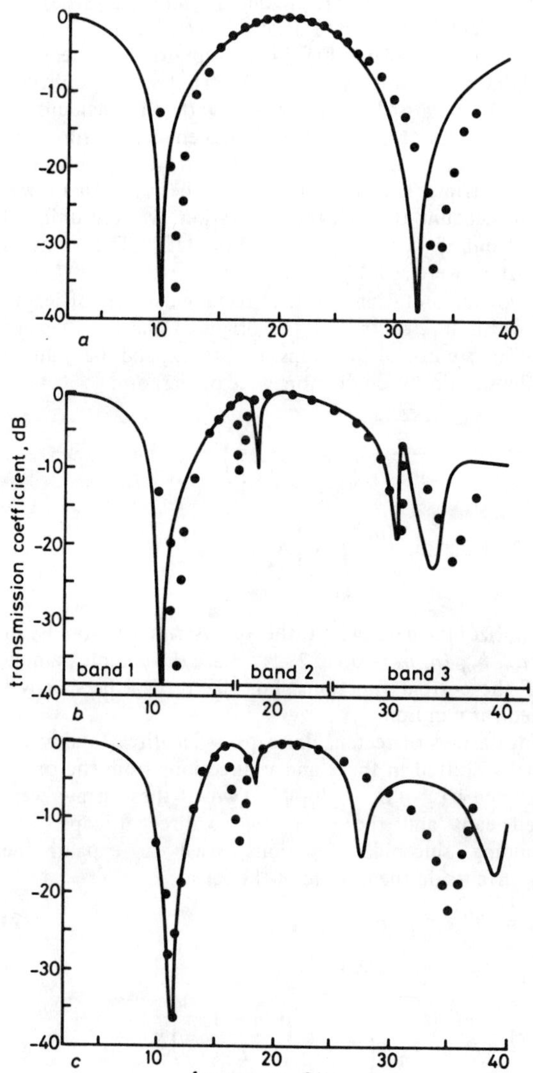

Fig. 4 *E-plane transmission coefficients*

a $\theta = 0°$ *b* $\theta = 30°$ *c* $\theta = 50°$
——— calculated results
. . . . experimental results
Location of frequency bands 1–3 in Table 1 shown in *b*

Table 1: Sets of functions used to describe the induced current on the Jerusalem cross in the E-plane at three bands of frequencies

Frequency range	Number of current functions	Current functions			
f_{1e} band 1: 1–17 GHz	11	sin 2* sin 4*	const cos** cos 2** cos 3** sin** sin 2** sin 3**		
f_{2e} band 2: 17–25 GHz	16	sin 2* sin 4*	sin** sin 2** sin 3** 3.5 sin 4**	cos* sin 2*	sin** sin 2** sin 3** sin 4**
f_{3e}, f_{4e} band 3: 25–40 GHz	13	sin 2* sin 4*	cos** cos 2** cos 3** sin** sin 3**	cos* sin 2*	sin**

* the term π/L_2 $\binom{x}{y}$ is omitted for simplicity
** the term π/L_1 $\binom{x}{y}$ is omitted

of such distributions by orthonormal functions, of the type shown in eqn. 3, is discussed in our analysis of crossed dipoles in Reference 14. The arrows in Fig. 5 show that current flows from the lower cap into the vertical bar and then into the top cap, to charge the capacitative section formed by the adjacent caps of adjacent elements. This is exactly the situation found for the E-plane arm alone [10]. Columns of elements are coupled by this current flow, extending from one end of the array to the other. The H-plane arm of the complete cross is apparently decoupled from the array at f_1.

Since this work was completed, Ohta *et al.* [15] have presented a brief summary of an analysis of what appears to be the main resonance of their Jerusalem-cross arrays. They obtain a reasonable agreement between the measured and computed transmission curves in the vincity of f_1.

4.2 Resonance at f_2 – existing only in E-plane

This narrow feature arises through an asymmetrical resonance of the *x*- and *y*-directed arms of the Jerusalem cross, i.e. a similar behaviour to that observed with crossed dipoles [14, 16]. The upper-half of the vertical dipole (Fig. 5*a*) couples with the right-hand half of the horizontal dipole, with a corresponding current flow between the other two halves. Here, however, the resonance lies a long way from the other reflection resonances, whereas, with crossed dipoles, the symmetrical and asymmetrical resonances are usually close in frequency. It was, therefore, a while before we recognised the explanation for the resonance at f_2 and also a while before we were able to match the predicted and observed frequencies, and so obtain convergence of the series. The predicted frequency was very sensitive to the use of

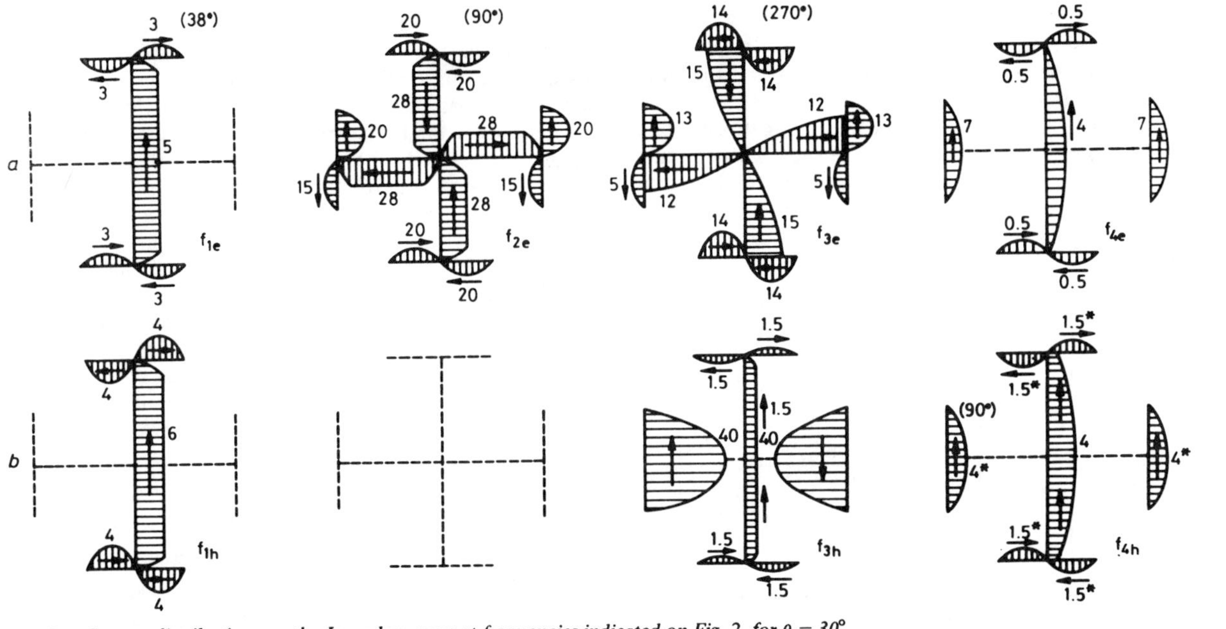

Fig. 5 *Current distributions on the Jerusalem cross at frequencies indicated on Fig. 2, for θ = 30°*
a E-plane
b H-plane
Arrows show direction of current flow
Angles indicate phase relative to rest of element
Relative amplitudes of resultant currents are indicated
*indicates current not passing through resonance
Note that the current in the horizontal centre section is very small for H-plane incidence

insufficient asymmetrical current functions in the arms of the Jerusalem cross. With only one such function in each arm the resonance occurred at about 25 GHz, compared with the observed value of about 17 GHz for both arrays.

Four asymmetrical functions were required in each arm (Table 1, band 2), including two $(\sin \pi/L_1 \binom{x}{y}, \sin 3\pi/L_1 \binom{x}{y})$ which have nonzero values at the junctions with the end caps. However, their sum gives zero at these locations. Fig. 6*b* shows the addition of the components in the horizontal bar. The predicted frequency is about 18.5 GHz for array 1 and 17.5 GHz for array 2, but only if the effective length of the end caps is taken to be 20% greater than the physical value. Otherwise the frequencies rise to 19.5 GHz and 20.0 GHz. It is likely, therefore, that the precise prediction of f_2 requires the inclusion of even more asymmetrical functions. Nevertheless, these results, summarised pictorially in Fig. 5, illustrate the complexity of this resonant mode.

Once again, as for f_1, two asymmetrical functions were needed in the horizontal caps, but in the vertical caps one symmetrical and one asymmetrical function were better; giving the enhanced current in the upper portion of these caps, as seen in Fig. 5. There is also a $\pi/2$ phase difference between the top and bottom caps. Unlike the situation at f_1 and at higher frequencies, the current is zero at the junction of the caps with the centre sections. The diagram for f_{2e} in Fig. 5 is therefore not readily interpreted in terms of current flow between caps and bars.

4.3 High-frequency resonances f_3 and f_4

At normal incidence only one resonance is observed at the upper end of our frequency range and has been referred to as a dipole-like resonance. At oblique incidence it splits into two, f_3 and f_4, as seen in Figs. 2–4. Langley and Drinkwater [5] compared this resonance with that observed in an array of crossed dipoles. However, there are distinct differences between the resonances observed for the Jerusalem cross and the fundamental resonance of crossed-dipole arrays. One such difference is that the crossed-dipole resonance is not split into two distinct sharp resonances in the H-plane at oblique incidence. Another is that, for the dipoles in the E-plane, the main resonant frequency is insensitive to angle of incidence, whereas a second (asymmetrical) resonance appears, usually above it at oblique incidence [14, 16]. The modal analysis helps to clarify this point.

Fig. 5 shows that in the H-plane f_3 is due to a strong symmetrical resonance of the two vertical caps, with currents out of phase by π radians. The currents in the E-plane arm of the cross are also resonant, but the amplitudes are lower by a factor of at least 10. The horizontal arm carries almost no current.

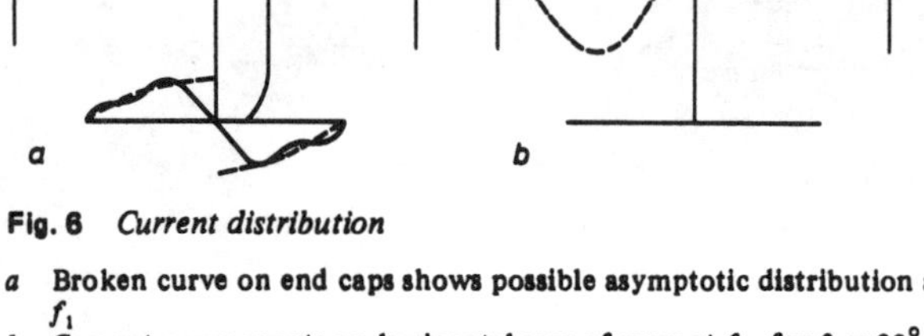

a   b

Fig. 6 *Current distribution*

a **Broken curve on end caps shows possible asymptotic distribution at f_1**
b **Current components on horizontal arm of cross at f_2, for $\theta = 30°$ Continuous curve shows resultant**

Although f_3 is thus similar in the H-plane to a halfwave dipole resonance of two of the caps, in the E-plane f_{3e} corresponds to a mode as complicated as f_2. Here, however, the current is not zero at the ends of the centre sections, but again the top cap is in phase quadrature with the rest of the structure. All four caps resonate in conjunction with the centre sections at f_{3e} to give a second asymmetrical resonance. At present we do not understand why this mode should be excited at a higher frequency than that found at f_{2e}. It should be noted that f_{3e} was not present in the experimental results available for array 1 at $\theta = 50°$ (Fig. 4*c*), although it is predicted to be present for both arrays at $\theta = 30°$ and $50°$. For array 2, the experimental value of f_{3e} was 30.5 GHz at $\theta = 50°$, compared with a computed value of 29.0 GHz.

The high-frequency resonance f_{4h} may perhaps be interpreted as a symmetrical dipole resonance of the centre bar section. This resonance is fairly sharp at normal incidence and, as the angle θ increases, it becomes wider (Fig. 2). The resonant frequency is moderately insensitive to changes in the angle of incidence. The difference between the side cap resonances at f_{3h} and f_{4h} is that the current amplitudes are much smaller and do not pass through maxima at f_{4h}. Also, in Fig. 5, the left-side cap is in phase quadrature with the rest of the structure.

In the E-plane, the position of the resonance at f_4 is again only slightly angle dependent experimentally, but *increases* with increasing angle. It involves the vertical caps and bar, as does f_{3h}.

5 Conclusions

The modal analysis has been able to account for the main features of the transmission curves in terms of the current distributions on the array element. At the fundamental resonance f_1, the current is largely confined to the E-plane arm of the cross along. At higher frequencies, the current configurations are more complex. An asymmetrical resonance occurs at f_{2e}, and , in cases where its presence narrows the upper transmission band, it should be possible to suppress it by uncoupling the E-plane arm from the H-plane arm of the cross. This could be done by printing them on opposite sides of the dielectric. Although the compromise fit of the transmission curves for both arrays was not as close, at the upper resonances, as might be achieved by a further investigation of the convergence of the current expansion, and by a study of a wider range of Jerusalem-cross arrays, it is not clear that the additional computer time involved can be justified. Modifying the array to include an inductive grid [1, 2] would obviously lengthen the analysis. A reduction in computer processing time is required to generate the complete copolar and crosspolar radiation patterns [8, 9] of a nearby feed transmitting through a Jerusalem-cross array, owing to the wide range of angles of incidence involved. Nevertheless, this work demonstrates that a modal analysis of quite complicated element geometries is feasible.

6 Acknowledgments

This work was financed by a research grant from the UK Science and Engineering Research Council.

7 References

1 ARNAUD, J.A., and RUSCIO, J.T.: 'Resonant-grid quasioptical diplexer', *Electron Lett.*, 1973, **9**, (25), pp. 589–590
2 ARNAUD, J.A., and PELOW, F.A.: 'Resonant-grid quasi-optical diplexers', *Bell Syst. Tech. J.*, 1975, **54**, (2), pp. 263–283

3 ANDERSON, I.: 'On the theory of self-resonant grids', *Bell Syst. Tech. J.*, 1975, **54**, pp. 1725–1731
4 LEONARD, T.W., and COFER, J.W.: 'A new equivalent circuit representation for the Jerusalem cross'. *IEE Conf. Publ. 169*, 1978, pp. 65–69
5 LANGLEY, R.J., and DRINKWATER, A.J.: 'Improved empirical model for the Jerusalem cross', *IEE Proc. H., Microwaves, Opt. & Antennas*, 1982, **129**, (1) pp. 1–6
6 PARKER, E.A., HAMDY, S.M.A., and LANGLEY, R.J.: 'Arrays of concentric rings as frequency selective surfaces', *Electron Lett.*, 1981, **17**, (23), pp. 880–881
7 CAHILL, R., and PARKER, E.A.: 'Concentric ring and Jerusalem cross arrays as frequency selective surfaces for a 45° incidence diplexer', *ibid.*, 1982, **18**, (8), pp. 313–314
8 LANGSFORD, P.A., and PARKER, E.A.: 'Feed radiation patterns modified by a dichroic plate', *ibid.*, 1978, **14**, (23), pp. 736–737
9 LANGSFORD, P.A., and PARKER, E.A.: 'Cross polarisation of electromagnetic waves transmitted through waveguide dichroic plates', *IEE Proc. H., Microwaves, Opt. & Antennas*, 1981, **128**, (1), pp. 1–5
10 HAMDY, S.M.A., and PARKER, E.A.: 'Comparison of modal analysis and equivalent circuit representation of E-plane arm of the Jerusalem cross', *Electron Lett.*, 1982, **18**, (2), pp. 94–95
11 HAMDY, S.M.A., CAHILL, R., LANGLEY, R.J., and PARKER, E.A.: 'The Jerusalem cross and related structures in frequency selective surfaces'. Eighth antenna symposium, Queen Mary College, London, March 1982
12 MONTGOMERY, J.P.: 'Scattering by an infinite periodic array of thin conductors on a dielectric sheet', *IEEE Trans.*, 1975, **AP-23**, (1), pp. 70–75
13 AMITAY, N., GALINDO, V., and WU, C.P.: 'Theory and analysis of phased array antennas' (Wiley, New York, 1972)
14 HAMDY, S.M.A., and PARKER, E.A.: 'Influence of lattice geometry on transmission of electromagnetic waves through arrays of crossed dipoles', *IEE Proc. H, Microwaves, Opt. & Antennas*, 1982, **129**, (1), pp. 7–10
15 OHTA, H.H., LANG, K.C., TSAO, C.H., and MITTRA, R.: 'Frequency selective surface for satellite communications antenna applications'. AP-S International Symposium Digest, 1982, pp. 475–478
16 PELTON, E.L., and MUNK, B.A.: 'Scattering from periodic arrays of crossed dipoles', *IEEE Trans.*, 1979, **AP-27**, (3), pp. 323–330

Part 6:
Hybrid Methods

Applicability of moment methods is limited by the number of equations that can realistically be set up and solved, which, in turn, limits the size of the body in wavelengths. These restrictions have been overcome by combining moment method with other techniques such as GTD, physical optics, harmonic series, *et cetera,* so that the former gives precision where the body's fine structure is critical, whereas the other technique allows larger bodies to be handled. These hybrid methods have proved to be extremely powerful. For a GTD overview, see Hansen (1981) as listed in the bibliography.

Thiele and Newhouse (1975) use diffracted (GTD) rays to augment mutual-impedance terms, where a diffracted ray appears at one basis function due to excitation by another basis function. An additional way of utilizing GTD is to match the moment method field to a GTD field at a few carefully selected points as shown by Burnside, Yu and Marhefka (1975); each match point adds one equation. Curved surface impedance matrix modification via GTD is shown by Henderson and Thiele (1981). In many important geometries, the diffraction coefficients are not known. Chuang and Burnside (1980) derive a numerical coefficient in a hybrid scheme. A review paper is by Newman (1988). In addition to references on hybrid methods, the bibliography lists some miscellaneous papers on transients and near-fields.

BIBLIOGRAPHY

MM and GTD

R.C. Hansen, *Geometric Theory of Diffraction,* IEEE Press/Wiley, 1981.

T.K. Wu and L.L. Tsai, "Scattering by a Dielectric Wedge: A Numercial Solution," *Trans. IEEE,* Vol. AP-25, July 1977, pp. 570–571.

G.A. Thiele and G.K. Chan, "Application of the Hybrid Technique to Time Domain Problems," *Trans. IEEE,* Vol. AP-26, January 1978, pp. 151–155.

E.P. Ekelman and G.A. Thiele, "A Hybrid Technique for Combining the Moment Method Treatment of Wire Antennas with GTD for Curved Surfaces," Trans. IEEE, Vol. AP-28, November 1980, pp. 831–839.

J.N. Sahalos and G.A. Thiele, "On the Application of the GTD-MM Techniques and its Limitations," *Trans. IEEE,* Vol. AP-29, September 1981, pp. 780–786.

D.M. Pozar, "Analysis of a Monopole Mounted Near an Edge or a Vertex," *Trans. IEEE,* Vol. AP-30, May 1982, pp. 401–408.

P. Kildal, "Dipole-Disk Antenna with Beam-forming Ring to Improve Radiation Characteristics," *Trans. IEEE,* Vol. AP-30, July 1982, pp. 529–534.

P.E. Zimourtopoulos, J.N. Sahalos, and G.A. Thiele, "Input Impedance of a Monopole Antenna on a Finite Polygonal Ground Plane," *Proc. IEE-MOA,* Vol. 130, June 1983, pp. 276–280.

S.A. Davidson and G.A. Thiele, "A Hybrid Method of Moments—GTD Technique for Computing Electromagnetic Coupling Between Two Monopole Antennas on a Large Cylindrical Surface," *Trans. IEEE,* Vol. EMC-26, May 1984, pp. 90–97.

R. Tiberio, G. Manara and G. Pelosi, "A Hybrid Technique for Analyzing Wire Antennas in the Presence of a Plane Interface," *Trans. IEEE,* Vol. AP-33, August 1985, pp. 881–885.

W.D. Burnside, B. Dewitt and B.Z. Hollman, "Analysis of Time Domain Scattering by a Flat Plate," *Trans. IEEE,* Vol. AP-33, August 1985, pp. 917–922.

K. Siakavara and J.N. Sahalos, "A Hybrid MM—GTD Technique for the Optimization of the Power Gain of Arrays of Wire Antennas Near an Elliptic Cylinder," *Proc. IEEE,* Vol. 73, September 1985, pp. 1426–1428.

L. Srikanth, P.H. Pathak and C.W. Chuang, "Hybrid UTD—MM Analysis of the Scattering by a Perfectly Conducting Semicircular Cylinder," *Trans. IEEE,* Vol. AP-34, October 1986, pp. 1250–1257.

M. Marin and M.F. Catedra, "A Study of a Monopole Arbitrarily Located on a Disk Using Hybrid MM/GTD Techniques," *Trans. IEEE,* Vol. AP-35, March 1987, pp. 287–292.

MM and Physical Optics—Fock Theory

T.K. Wu and L.L. Tsai, "Scattering by a Dielectric Wedge: A Numerical Solution," *Trans. IEEE,* Vol. AP-25, July 1977, pp. 570–571.

T.J. Kim and G.A. Thiele, "A Hybrid Diffraction Technique—General Theory and Applications," *Trans. IEEE,* Vol. AP-30, September 1982, pp. 888–897.

L.W. Henderson and G.A. Thiele, "A Hybrid MM—Geometrical Optics Technique for the Treatment of Wire Antennas Mounted on a Curved Surface," *Trans. IEEE,* Vol. AP-30, November 1982, pp. 1257–1261.

P.K. Murthy, K.C. Hill and G.A. Thiele, "A Hybrid-iterative Method for Scattering Problems," *Trans. IEEE,* Vol. AP-34, October 1986, pp. 1173–1180.

L.N. Medgyesi-Mitschang and D.S. Wang, "Hybrid Solutions for Scattering from Perfectly Conducting Bodies of Revolution," *Trans. IEEE,* Vol. AP-31, July 1983, pp. 570–583.

L.N. Medgyesi-Mitschang and D.S.Y. Wang, "Hybrid Solutions for Scattering from Large Bodies of Revolution with Material Discontinuities and Coatings," *Trans. IEEE,* Vol. AP-32, July 1984, pp. 717–723.

L.N. Medgyesi-Mitschang and D.S. Wang, "Hybrid Solutions at Internal Resonances," *Trans. IEEE,* Vol. AP-33, June 1985, pp. 671–674.

L.N. Medgyesi-Mitschang and D.S. Wang, "Hybrid Solutions for Large-impedance Coated Bodies of Revolution," *Trans. IEEE,* Vol. AP-34, November 1986, pp. 1319–1329.

L.N. Medgyesi-Mitschang and D.S. Wang, "Hybrid Methods for Analysis of Complex Scatterers," *Proc. IEEE,* Vol. 77, May 1989, pp. 770–779.

MM and Green's Function

F.M. Tesche and A.R. Neureuther, "The Analysis of Monopole Antennas Located on a Spherical Vehicle: Part 1, Theory," *Trans. IEEE,* Vol. EMC-18, February 1976, pp. 2–8.

F.M. Tesche, A.R. Neureuther and R.E. Stovall, "The Analysis of Monopole Antennas Located on a Spherical Vehicle: Part 2, Numerical and Experimental Results," Trans. IEEE, Vol. EMC-18, February 1976, pp. 8–15.

C.M. Butler and T.L. Keshavamurthy, "Analysis of a Wire Antenna in the Presence of a Sphere," *Trans. IEEE,* Vol. EMC-22, May 1980, pp. 113–118, Correction: November 1980, p. 329.

D.M. Pozar, "Analysis of a Monopole Mounted Near or at the Edge of a Half-plane," *Trans. IEEE,* Vol. AP-29, May 1981, pp. 488–495.

E.H. Newman, "TM Scattering by a Dielectric Cylinder in the Presence of a Half-plane," *Trans. IEEE,* Vol. AP-33, July 1985, pp. 773–782.

E.H. Newman, "TM and TE Scattering by a Dielectric/Ferrite Cylinder in the Presence of a Half-plane," *Trans. IEEE,* Vol. AP-34, June 1986, pp. 804–813.

E.H. Newman and J.L. Blanchard, "TM Scattering by an Impedance Sheet Extension of a Parabolic Cylinder," *Trans. IEEE,* Vol. AP-36, April 1988, pp. 527–534.

Transients

E.P. Sayre and R.F. Harrington, "Time Domain Radiation and Scattering by Thin Wires," *Appl. Sci. Res.,* Vol. 26, September 1972, pp. 413–444.

E.K. Miller, A.J. Poggio and G.J. Burke, "An Integro-Differential Equation Technique for the Time Domain Analysis of Thin Wire Structures, Part I: Numerical Method," *J. Comp. Phys.,* Vol. 12, May 1973, pp. 24–48.

E.K. Miller, A.J. Poggio and G.J. Burke, "An Integro-Differential Equation Technique for the Time Domain Analysis of Thin Wire Structures, Part II: Numerical Results:," *J. Comp. Phys.,* Vol. 12, June 1973, pp. 210–233.

E.K. Miller and M.L. Van Blaricum, "The Short-Pulse Response of a Straight Wire," *Trans. IEEE,* Vol. AP-21, May 1973, pp. 396–398.

F.M. Tesche, "Calculation of the Exterior Natural Frequencies of a Wire by the Use of Hallen's Integral Equation," *Electronics Letters,* Vol. 9, November 29, 1973, pp. 558–559.

J.A. Landt and E.K. Miller, "Short Pulse Response of a Circular Loop," *Trans. IEEE,* Vol. AP-22, January 1974, pp. 114–116.

H. Schuman, "Time-domain Scattering from Nonlinearly Loaded Wire," *Trans. IEEE,* Vol. AP-22, July 1974, pp. 611–613.

T.K. Sarkar and D.D. Weiner, "Scattering Analysis of Nonlinearly Loaded Antennas," *Trans. IEEE,* Vol. AP-24, March 1976, pp. 125–131.

T.H. Shumpert and D.J. Galloway, "Finite Length Cylindrical Scatterer Near Perfectly Conducting Ground—A Transmission Line Mode Approximation," *Trans. IEEE,* Vol. AP-26, January 1978, pp. 145–151.

J.A. Landt, "Numerical Analysis of Transient Fields near Thin-Wire Antennas and Scatterers," *Radio Science,* Vol. 16, November-December 1981, pp. 1087–1091.

J.A. Landt, "Network Loading of Thin-Wire Antennas and Scatterers in the Time Domain," *Radio Science,* Vol. 16, November-December 1981, pp. 1241–1247.

C.L. Bennett and H. Mieras, "Time Domain Scattering from Open Thin Conducting Surfaces," *Radio Science,* Vol. 16, November-December 1981, pp. 1231–1239.

H. Mieras and C.L. Bennett, "Space-time Integral Equation Approach to Dielectric Targets," *Trans. IEEE,* Vol. AP-30, January 1982, pp. 2–9.

A. Taflove and K. Umashankar, "A Hybrid Moment Method/Finite-Difference Time Domain Approach to Electromagnetic Coupling and Aperture Penetration into Complex Geometries," *Trans. IEEE,* Vol. AP-30, July 1982, pp. 617–627.

Near-Fields

R.J. Mailloux, "Radiation and Near-Field Coupling between Two Collinear Open-Ended Waveguides," *Trans. IEEE,* Vol. AP-17, January 1969, pp. 49–55.

A.T. Adams, B.J. Strait, D.E. Warren, D.C. Kuo and T.E. Baldwin, "Near Fields of Wire Antennas by Matrix Methods," *Trans. IEEE,* Vol. AP-21, September 1973, pp. 602–610.

K.R. Siarkiewicz and A.T. Adams, "Near- and Far-Field Thin Wire Coupling—Theory and Experiment," *Trans. IEEE,* Vol. EMC-19, November 1977, pp. 394–401.

A.T. Adams, T.E. Baldwin, and D.E. Warren, "Near Fields of Thin Wire Antennas—Computation and Experiment," *Trans. IEEE,* Vol. EMC-20, February 1978, pp. 259–266.

K.R. Demarest and R.J. Garbacz, "Anomalous Behavior of Near-Fields Calculated by the Method of Moments," *Trans. IEEE,* Vol. AP-27, September 1979, pp. 609–614.

A.T. Adams, Y. Leviatan and K.S. Nordby, "Electromagnetic Near Fields as a Function of Electrical Size," *Trans. IEEE,* Vol. EMC-25, November 1983, pp. 428–432.

A Hybrid Technique for Combining Moment Methods with the Geometrical Theory of Diffraction

GARY A. THIELE, SENIOR MEMBER, IEEE, AND THOMAS H. NEWHOUSE, STUDENT MEMBER, IEEE

***Abstract*—A technique for combining moment methods with the geometrical theory of diffraction (GTD) is presented, which permits the application of the method of moments to a larger class of problems. The fundamental idea used to develop the hybrid technique is to modify the usual impedance matrix that characterizes, for example, a wire antenna such that a metallic body or discontinuity on that body is properly accounted for. It is shown in general that one can modify the impedance matrix for any basis and/or weighting functions if one can compute the correct modification to the impedance matrix element. The modification is readily accomplished using the geometrical theory of diffraction and/or geometrical optics. Several example problems are considered to illustrate the usefulness of the technique. First, the canonical problem of a monopole near a conducting wedge is investigated. Second, a monopole at the center of a four-sided and an eight-sided flat plate is considered. Impedance results for the latter case are in good agreement with measurements. Third, a monopole at the center of a circular disc is examined and compared with experimental measurements in the literature, and fourth, the problem of a monopole near a conducting step is solved and the dependence of the input impedance upon the step height shown.**

Manuscript received February 1, 1974; revised August 14, 1974. This work was supported in part by Contract N00014-67-A-0232-0018 between the Office of Naval Research and the Ohio State University Research Foundation. A portion of this work was presented at the 1973 URSI/G-AP Symposium, Boulder, Colo., August 21–24.

G. A. Thiele is with the Ohio State University ElectroScience Laboratory, Department of Electrical Engineering, Columbus, Ohio 43212.

T. H. Newhouse was with the Ohio State University ElectroScience Laboratory, Department of Electrical Engineering, Columbus, Ohio. He is now with the Department of Defense, Fort George, Md.

I. INTRODUCTION

THE ELECTROMAGNETIC characterization of wire antennas on or near a three-dimensional metallic surface by the method of moments is significantly limited by computer storage [1], [2]. This is true whether one uses a wire-grid [3] representation or a surface patch [4] representation of the surface. Thus we may refer to the method of moments as a low-frequency technique since its practical use is generally restricted to bodies that are not large in terms of the wavelength. On the other hand, geometrical optics and the geometrical theory of diffraction (GTD) [5], [6] are applicable to bodies that are arbitrarily large in the electrical sense and may thus be referred to as high frequency techniques. Both the method of moments (MM) and GTD are powerful computational methods in their own right. This is due in large measure to the flexibility inherent in each method that permits application to a wide range of radiation and scattering problems.

It is the purpose of this paper to present a new method, which will combine these two computational methods into a hybrid technique. The hybrid technique is developed by considering the general problem of how to extend moment methods to include a class of problems wherein a three-dimensional body, on or near which is located a radiating element, may be arbitrarily large. Thus in developing the hybrid technique in the next section, the approach is to modify the impedance matrix, which in the usual moment method approach characterizes only a portion of the problem (e.g., the radiating element), to properly account for the remainder of the problem to be solved. This is to be contrasted with other work being conducted at this time that considers the extension of GTD via moment methods [7].

II. THE HYBRID TECHNIQUE

In a moment method solution, we expand the current J in a series of basis functions $J_1, J_2, J_3, \cdots$ on the surface of interest and defined in the domain of an operator L. That is,

$$J = \sum_{n=1}^{N} I_n J_n. \tag{1}$$

Using the linearity of the operator, and forming the inner product with a set of weighting functions $W_1, W_2, W_3, \cdots$ in the domain of L, we write

$$\sum_{n=1}^{N} I_n \langle W_m, L(J_n) \rangle = \langle W_m, E^i \rangle. \tag{2}$$

In reality, the preceding expression represents the mth row of equations in a system of N such equations while the quantity $L(J_n)$ represents the electric field from the nth basis function of unit amplitude. In the usual moment method matrix notation, (2) is compactly represented by

$$[Z](I) = (V) \tag{3}$$

where the elements of the generalized impedance matrix are given by

$$Z_{mn} = \langle W_m, L(J_n) \rangle. \tag{4}$$

The symmetric inner product here is defined such that the following three axioms are satisfied:

$$\langle \alpha J_1 + \beta J_2, E \rangle = \alpha \langle J_1, E \rangle + \beta \langle J_2, E \rangle, \quad \alpha \text{ and } \beta \text{ are scalars} \tag{5}$$

$$\langle J, E \rangle = \langle E, J \rangle \tag{6}$$

$$\langle J^*, J \rangle > 0; \quad \text{but} \quad \langle J^*, J \rangle = 0, \quad \text{if and only if } J = 0. \tag{7}$$

Using the first two axioms (i.e., (5) and (6)) it is easily shown that

$$\langle J, aE_1 + bE_2 \rangle = a\langle J, E_1 \rangle + b\langle J, E_2 \rangle \tag{8}$$

where a and b are complex scalars.

Suppose that in (8) the quantity aE_1 represents $L(J_n)$ in (4). That is, $L(J_n)$ is the usual calculation made in a strictly moment method solution. The quantity bE_2 in (8) represents an additional field contribution to Z_{mn} in (4) that is also due to J_n but arrives at observation point (or region) m by a physical process not directly accounted for in the moment method formulation. Thus we may write for the new impedance matrix element

$$Z_{mn}' = \langle W_m, L(J_n) + bL(J_n) \rangle, \qquad a = 1, b = b(m,n) \tag{9}$$

or

$$Z_{mn}' = \langle W_m, L(J_n) \rangle + \langle W_m, bL(J_n) \rangle \tag{10}$$

or

$$Z_{mn}' = Z_{mn} + Z_{mn}^g \tag{11}$$

where the superscript g denotes that Z_{mn}^g is an additional term added to, in general, each impedance matrix element due to a physical process g that also directs energy from the nth basis current function to the mth observation point (or region). Thus we may modify (3) such that

$$[Z'](I') = (V) \tag{12}$$

where $[Z']$ is the generalized impedance matrix properly modified to account for physical processes not accounted for by the moment method formulation, which is assumed to represent only a portion of the problem. Note that modification of (V) is neglected since it has been found to be numerically insignificant in this paper. Solution of (12) can then be written as

$$(I') = [Z']^{-1}(V) \tag{13}$$

where (I') is the current on, for example, an antenna in the presence of other scattering mechanisms that may be accounted for by either geometrical optics techniques or GTD. In the next section we will briefly review some of the current advances in GTD before considering several problems in Section IV that have been treated successfully by the hybrid technique.

III. DIFFRACTION THEORY

Historically, the problem of straight edge diffraction by a perfectly conducting wedge was first solved by Sommerfeld [8]. Later, Pauli [9] introduced the V_B function as a practical formulation in the solution for the fields diffracted by a perfectly conducting wedge. The V_B function is related to the term $D(\beta)$, which Keller [11] has designated as a diffraction coefficient.

More recently, Hutchins and Kouyoumjian [11] have presented a formulation that yields better accuracy than does the Pauli solution in the regions near the incident and reflected shadow boundaries, particularly when $r < \lambda$. Their formulation can be conveniently found in [12, Sect. II] or in [6].

It is the introduction of the distance parameter L by Pathak and Kouyoumjian that permits near-zone to near-zone diffraction to be considered. Experience has thus far shown that accurate results may be obtained for $kL > 1$. Therefore, this is the formulation that will be used in the following section.

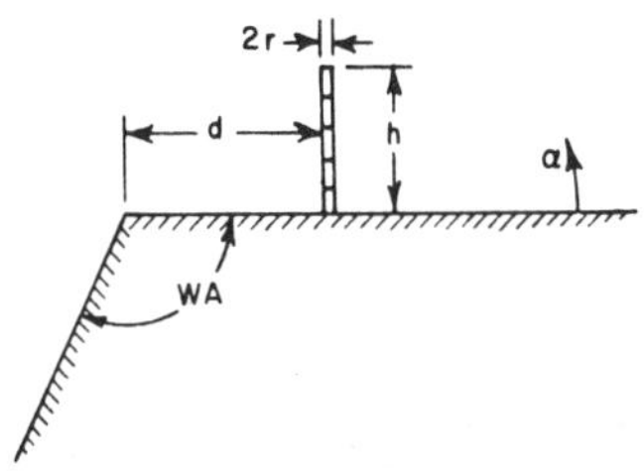

Fig. 1. Monopole on conducting wedge.

IV. EXAMPLES

A. Monopole Near a Wedge

Initially, to combine the method of moments and GTD into a hybrid technique, we have considered the canonical problem of a monopole near a perfectly conducting wedge as shown in Fig. 1. If we describe the monopole on an infinite ground plane strictly by the moment method matrix representation given in (3), then for the monopole near the conducting wedge we utilize (12) where in (11) the term Z^g_{mn} is obtained by considering that energy radiated by the nth basis function on the monopole that is diffracted by the wedge to the mth observation point or region. In our work here we have employed pulse basis functions and point-matching wherein the testing functions are delta functions [2]. However, as indicated in Section II, the choice of basis and testing functions is not so restricted.

To calculate Z^g_{mn} we compute the electric field from the nth pulse basis function incident upon the edge of the wedge at the stationary point. Taking that component of the electric field perpendicular to the edge and to the direction of propagation of the incident field, we then compute the energy diffracted to the observation point at the center of the mth segment on the monopole. The component of this field tangential to segment m is the term Z^g_{mn} of (11) since we are employing delta weighting functions. To compute the diffracted field, we use the formulation of Kouyoumjian and Pathak discussed in the previous section for the case of spherical wave incidence.

Shown in Fig. 2(a) is a calculated curve for the input resistance of a quarter wavelength monopole a distance d from the edge of a perfectly conducting wedge. We note that the resistance oscillates about the value for a quarter wavelength monopole on an infinite ground plane and also that the amount of variation is relatively small being only a few ohms. A similar curve is shown in Fig. 2(b) for the input reactance. Data for both curves was obtained directly from (13) without the need for any *a priori* knowledge of the current distribution or the terminal current value.

In an effort to find a case for which experimental confirmation could readily be obtained, a monopole at the center of a square plate was investigated. In this problem there are four stationary points, one on each edge, that contribute to the Z^g_{mn} term of (11). Diffraction by the four vertices was neglected. A calculated curve for the input resistance of a quarter wavelength monopole on a square plate is shown in Fig. 3. As in the previous case the input resistance oscillates about its value for the infinite ground plane case, but the amount of variation is substantially greater. A similar curve was obtained for the input reactance. While fair agreement between theory and experiment was obtained for the square plate case, it was decided that an octagonal plate would be better for two reasons. First, there would be either stationary points, one on each edge and thus the sensitivity of the input impedance to the plate size would be greater, and second the vertex diffraction would be less since the angle of the vertex would be rather large (i.e., 135° in this case) [13]. A comparison between theory and experiment for a monopole at the center of an octagonal plate showed good agreement for both the input resistance and input reactance.

One objective of the experimental verification for the octagonal plate was to determine how small d/λ can become before inaccurate results are obtained. It was apparent from the measurements that $d > 0.25\lambda$ will yield accurate results. We hypothesize that the hybrid technique will probably permit d to approach $\lambda/8$ but further work is necessary to verify this.

It was thought that perhaps the accuracy of the calculations could be increased if double diffractions were included. That is, energy incident upon an edge that is diffracted to an opposite edge and then to the monopole is also included in the calculation of the Z^g_{mn} term in (11). However, from the calculations it was concluded that, at least for $d > 0.25\lambda$, the effect of double diffractions is negligible.

Thus far the discussion has centered upon the calculation of input impedance. Obviously, if one can accurately compute the input impedance, then quite accurate far-field information can readily be obtained too. For example, in the case of a monopole near a single wedge, as in Fig. 1, there may be as many as three contributors to the far-field. First, there is direct source radiation except in the shadow region. Second, there is the reflected field, which is most conveniently accounted for by using the image in the horizontal surface. Third, there is the diffracted field that contributes in all regions and, of course, is the only source of radiation into the shadow region. A typical far-field pattern is shown in Fig. 4. Note that for $\alpha = 90°$ the field does not go to zero as would be the case if the wedge were not present.

B. Monopole on a Circular Disc

If we wish to investigate a circular ground plane, we obviously cannot extend the method employed for the octagonal plate to an n-sided polygon (n large) since the principle of stationary phase will be violated and the method will fail. The difficulty can be eliminated, however, by replacing the edge of the disc with an equivalent

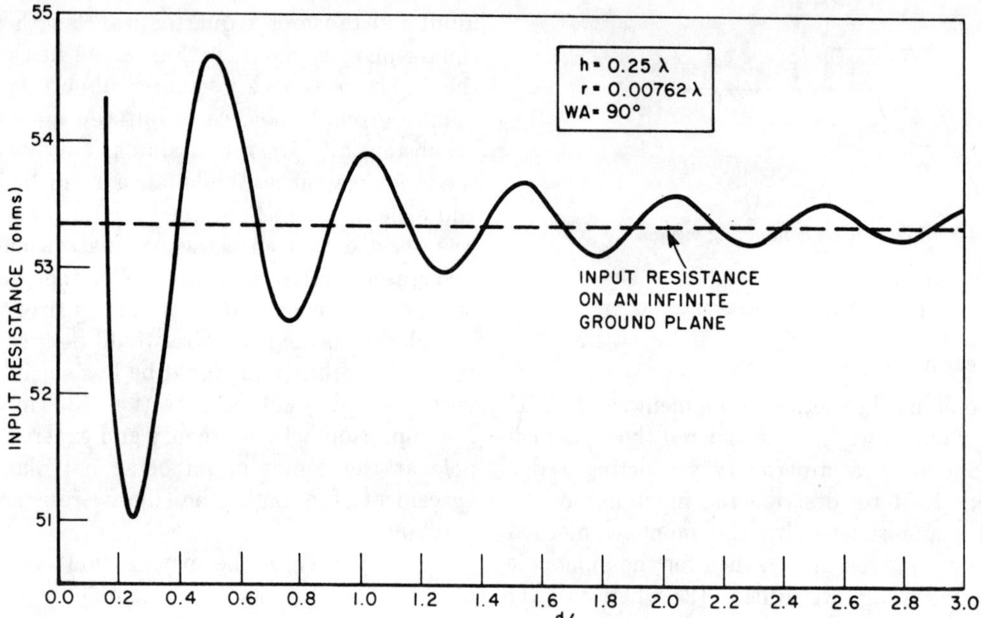

(a)

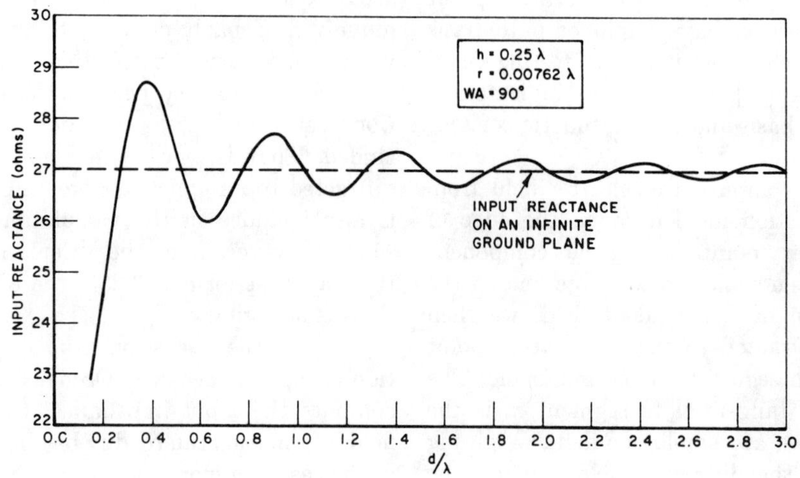

(b)

Fig. 2. (a) Input resistance of monopole on conducting wedge versus distance d as shown in Fig. 1. (b) Input reactance of monopole on conducting wedge versus distance d as shown in Fig. 1.

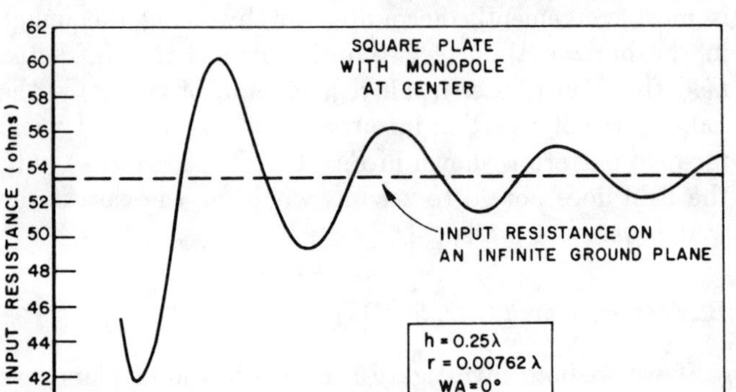

Fig. 3. Input resistance of monopole on square plate versus monopole distance to edges.

magnetic ring current M given by

$$M = -2E_\theta \left(\frac{ss'}{s+s'}\right)^{1/2} \exp\left[j\left(\frac{ss'}{s+s'}\frac{2\pi}{\lambda} - \frac{\pi}{4}\right)\right] \cdot [V_B(\phi - \phi') + V_B(\phi + \phi')]. \quad (14)$$

Equation (14) gives the equivalent magnetic current used to calculate the field at the segment at s due to the current at s' as indicated in Fig. 5. Note that an equivalent magnetic ring current must be calculated for each choice of s and s'.

It is useful for us to break up the equivalent magnetic ring current of Fig. 5 into differential elements dl so that the observation point is in the far-field of each element even though it may be in the near-field of the total ring current [14].

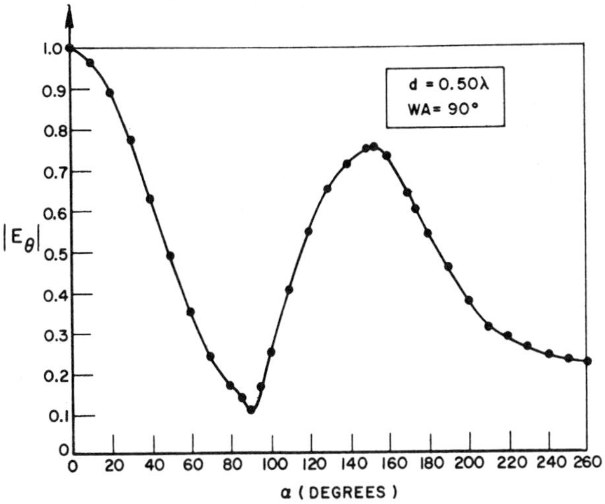

Fig. 4. Normalized far-field pattern of quarter wave monopole near conducting edge.

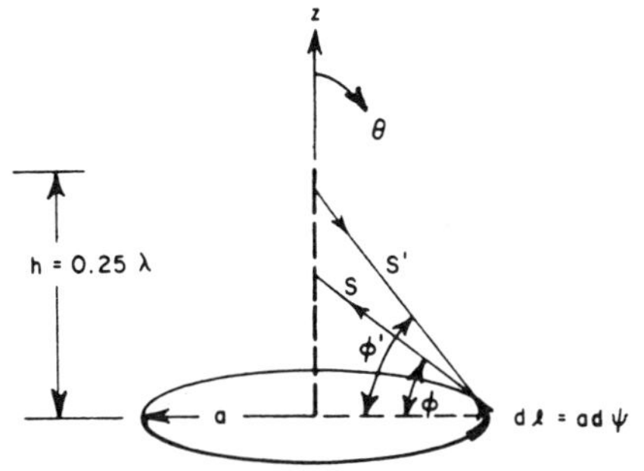

Fig. 5. Segmented monopole encircled by magnetic ring current.

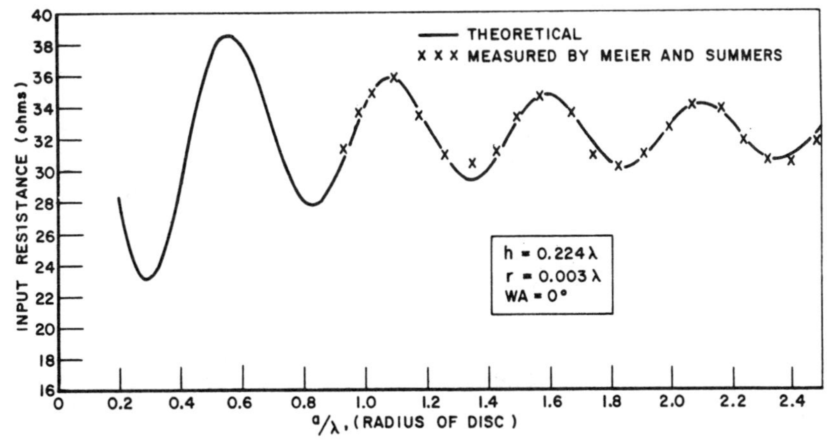

(a)

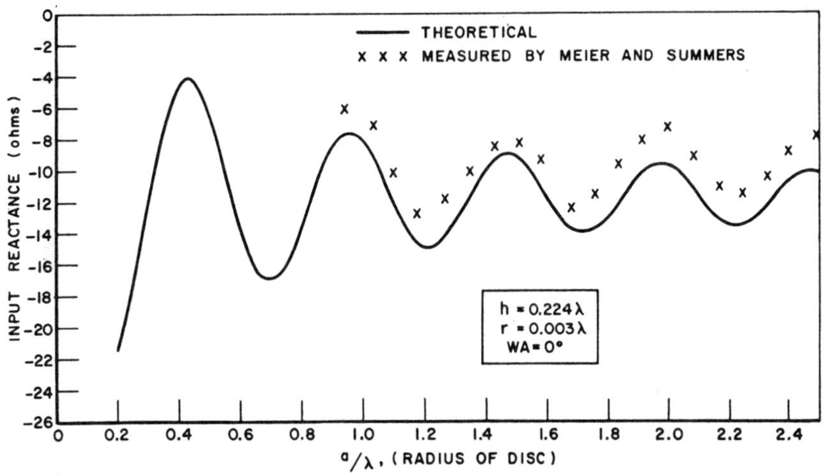

(b)

Fig. 6. (a) Theoretical and experimental input resistance of monopole at center of circular disc. (b) Theoretical and experimental input reactance of monopole at center of circular disc.

The electric field in a plane perpendicular to an element dl is given by [15]

$$dE = \frac{M\,dl}{4\pi}\left[\frac{j\omega}{cr} + \frac{1}{r^2}\right] \exp\,(-jkr) \tag{15}$$

where $r = (a^2 + z^2)^{1/2}$. Letting $dl = a d\psi$, where ψ is the azimuth angle, taking only the z component at the monopole, and integrating over the range $\psi = 0$ to $\psi = 2\pi$ yields

$$E_z = \frac{Ma^2}{2r}\left[\frac{jk}{r} + \frac{1}{r^2}\right] \exp\,(-jkr). \tag{16}$$

The value for E_z is the term Z^g_{mn} that is added to the impedance element obtained for a monopole on an infinite ground plane. This process gives the modified impedance element needed to calculate the modified currents (and hence input impedance) of a monopole on the finite circular ground plane.

Equation (16), without the near-field term, is the one used by Green [16] in his work with a monopole on the base of a large cone. However, in our work we need the near-field term.

Figs. 6(a) and 6(b) show a comparison between calculations made with the equivalent magnetic ring current and measurements made by Meier and Summers [17], [18] for a monopole of length 0.224λ and radius 0.003λ on a circular ground plane of varying radius. It is apparent that the correct variation is accurately predicted for both the real and imaginary parts of the input impedance. For the input resistance the agreement between their measurements and the theory is excellent. For the input reactance the agreement is very good but there is a slight shift in the calculated curve when compared to the measurements. The amount of this shift is sufficiently small that it can be attributed to the usual problems associated with modeling the region in proximity to the driving point. The impedance predicted here is also in agreement with that predicted by Green [16] whose results are based on the measurements by Meier and Summers.

In a partial attempt to resolve the accuracy of our calculations for smaller values of a/λ, measurements were made with a quarter wavelength long monopole of 1/16-in radius on a circular plate. These results are shown in Figs. 7(a) and 7(b). The agreement between theory and experiment for the input resistance and input reactance is fairly good. Most of the difference is attributed to mismatches in the measuring system. In our calculations we use a magnetic frill representation for the coaxial aperture at the base of the monopole [19], which closely models the actual physical geometry and is known to permit prediction of the input impedance as accurately as is meaningful [2].

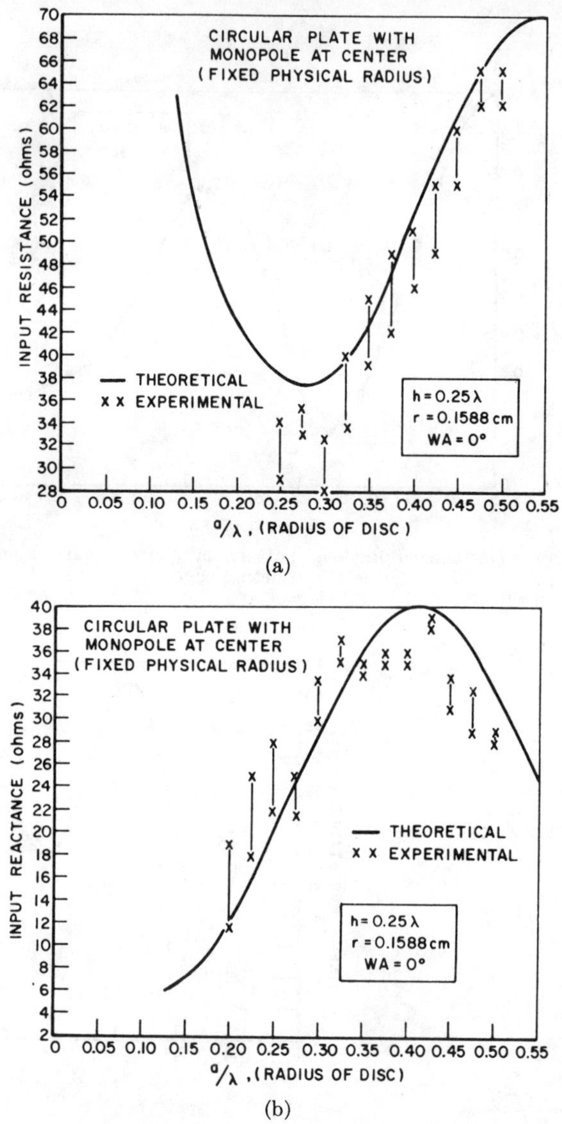

Fig. 7. (a) Theoretical and experimental input resistance of monopole at center of circular disc. (b) Theoretical and experimental input reactance of monopole at center of circular disc.

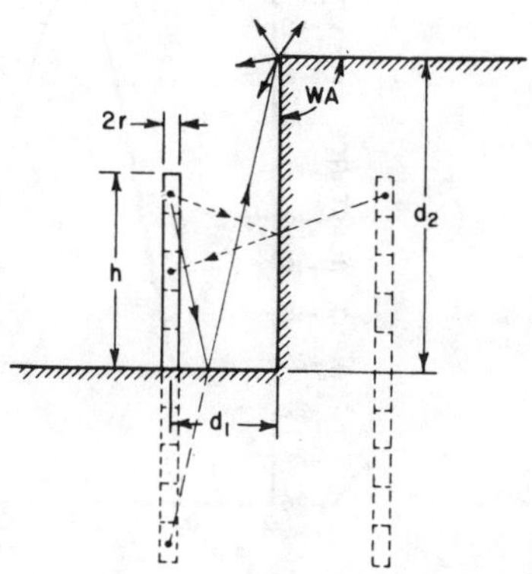

Fig. 8. Monopole near conducting step showing partial use of images.

C. Monopole Near a Conducting Step

Consider the situation shown in Fig. 8 where a monopole of height h is a distance d_1 away from a vertical conducting step. To properly determine the Z^g_{mn} term in (11), it is necessary to determine all the various combinations of reflections that can occur for rays emanating from the monopole and reflecting back to it as well as the diffraction from the top edge of the step. Since the vertical wall

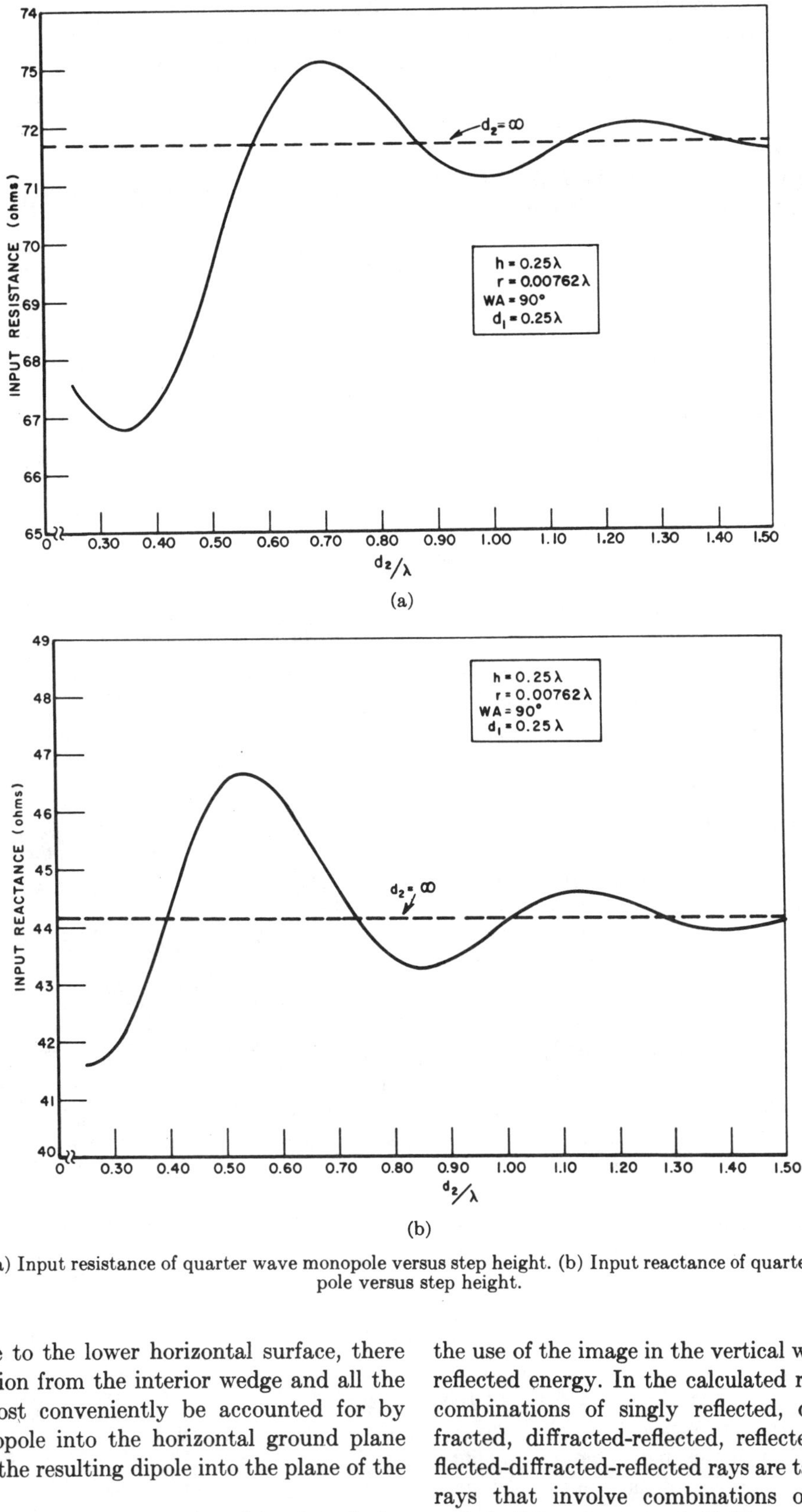

Fig. 9. (a) Input resistance of quarter wave monopole versus step height. (b) Input reactance of quarter wave monopole versus step height.

is at a right angle to the lower horizontal surface, there will be no diffraction from the interior wedge and all the reflections can most conveniently be accounted for by imaging the monopole into the horizontal ground plane and then imaging the resulting dipole into the plane of the vertical wall.

Shown in Fig. 8 are two example situations that depict the utilization of the images. Considering the uppermost segment of the monopole to be the source segment, one set of rays shows the use of the image in the horizontal surface to calculate reflected-diffracted energy reaching the segments of the monopole. The other set of rays shows the use of the image in the vertical wall to calculate singly reflected energy. In the calculated results that follow, all combinations of singly reflected, doubly reflected, diffracted, diffracted-reflected, reflected-diffracted, and reflected-diffracted-reflected rays are taken into account. All rays that involve combinations of double (or higher order) diffractions are negligible.

Fig. 9 shows the calculated input impedance for a quarter wavelength monopole a quarter wavelength away from a vertical wall whose height is $d_2 > 0.25\lambda$. As d_2 increases the impedance oscillates about the value for the case where $d_2 = \infty$. The value of the input impedance for the $d_2 = \infty$ case was verified by an independent computer program employing the piecewise sinusoidal basis [20]. The results of Figs. 9(a) and 9(b) show that as the diffracting edge recedes from the vicinity of the monopole, its effect upon the input impedance rapidly diminishes. Although we have not shown results for the case where the step height is less than the height of the monopole, the same method could be used to investigate such situations.

V. SUMMARY AND CONCLUSIONS

In presenting the hybrid technique we have attempted to show that the idea of modifying the generalized impedance matrix is a useful one. In combining moment methods with GTD we have proceeded from the philosophical viewpoint of extending the method of moments via GTD. In so doing we have shown that modifying the impedance matrix to account for diffraction effects (or geometrical optics effects) enables one to accurately treat a larger class of problems than could be treated by moment methods alone. A potential advantage of such an approach is that in many problems the impedance matrix may be small enough to calculate data at sufficiently many frequencies to obtain time domain solutions via Fourier transformation; whereas, attempting to do so using strictly moment methods may be economically prohibitive or impossible due to the increasing electrical size of the problem with increasing frequency.

An alternative interpretation of the hybrid method is also possible. That is, the procedure employed can be viewed as using GTD to obtain an approximation to the exact dyadic Green's function needed. For example, for the octagonal plate problem, the single and double diffraction terms represent the leading terms in a series expression for the Green's function of the plate. In this context, moment method solutions using other than free space Green's functions are not completely without precedent [21].

While the hybrid method possesses many of the advantages inherent in both MM and GTD, it also has some of the limitations peculiar to each. For example, as in the usual MM problem, one can treat arbitrary configurations of wire antennas (or slot antennas) taking into account lumped loading, finite conductivity, etc., and obtain accurate impedance data and current distributions. Naturally, one still must take the usual precaution of using a sufficient number of basis functions to assure convergence. On the other hand, as in the usual GTD problem, one must take care that the antenna is not too close to a source of diffraction (e.g., $d > 0.2\lambda$). Furthermore, the study described in this paper has mainly addressed itself to the determination of the input impedance of the antenna and hence the near-fields at its location. The determination of the currents and near-fields elsewhere, particularly near a source of diffraction, is a subject for future investigation.

Nevertheless, the examples considered in this paper illustrate well the advantages of combining MM and GTD in a hybrid method since the advantages intrinsic to each technique can be effectively utilized to treat new problems or to treat old ones in a more efficient and accurate manner.

ACKNOWLEDGMENT

The authors wish to express their gratitude to Prof. R. G. Kouyoumjian and R. J. Marhefka at the Ohio State University ElectroScience Laboratory for many stimulating discussions. Constructive reviews of the manuscript by Profs. C. H. Walter and R. J. Garbacz are also appreciated.

REFERENCES

[1] R. F. Harrington, *Field Computation by Moment Methods.* New York: Macmillan, 1968.

[2] G. A. Thiele, "Wire antennas," in ch. 2 of *Computer Techniques for Electromagnetics*, R. Mittra, Ed. London: Pergamon, 1973.

[3] J. H. Richmond, "A wire-grid model for scattering by conducting bodies," *IEEE Trans. Antennas Propagat.*, vol. AP-14, pp. 782–786, Nov. 1966.

[4] J. H. Richmond and N. N. Wang, "Sinusoidal reaction formulation for scattering by conducting bodies of arbitrary shape," presented at the 1973 URSI meeting, Boulder, Colo.

[5] R. G. Kouyoumjian, "Asymptotic high-frequency methods," *Proc. IEEE*, vol. 53, pp. 864–876, Aug. 1965.

[6] ——, Notes for a "Short course on application of GTD and numerical techniques to the analysis of electromagnetic and acoustic radiation and scattering," Ohio State University, Columbis, Sept. 1973.

[7] W. D. Burnside, C. L. Yu, and R. J. Marhefka, "A technique to combine the geometrical theory of diffraction and the moment method," presented at the 1973 URSI meeting, Boulder, Colo.

[8] A. Sommerfeld, *Optics.* New York: Academic, 1954, pp. 245–265.

[9] W. Pauli, "On asymptotic series for functions in the theory of diffraction of light," *Phys. Rev.*, vol. 54, pp. 924–931, Dec. 1, 1968.

[10] J. B. Keller, "Geometrical theory of diffraction," *J. Opt. Soc. Amer.*, vol. 52, pp. 116–130, Feb. 1962.

[11] D. L. Hutchins and R. G. Kouyoumjian, "A new asymptotic solution to the diffraction by a wedge," at *URSI 1967 Spring Meeting*, Ottawa, Ont., Canada, pp. 154–155.

[12] W. D. Burnside, R. J. Marhefka, and C. L. Yu, "Roll-plane analysis of on-aircraft antennas," *IEEE Trans. Antennas Propagat.*, vol. AP-21, pp. 780–786, Nov. 1973.

[13] R. G. Kouyoumjian, private communication.

[14] C. H. Walter, *Traveling Wave Antennas.* New York: Dover, 1970, p. 41.

[15] J. D. Kraus, *Antennas.* New York: McGraw-Hill, 1950, p. 159.

[16] H. E. Green, "Impedance of a monopole on the base of a large cone," *IEEE Trans. Antennas Propagat.*, vol. AP-17, pp. 703–706, Nov. 1969.

[17] A. S. Meier and W. P. Summers, "Measured impedance of vertical antennas over finite ground planes," *Proc. IRE*, vol. 37, pp. 609–616, June 1949.

[18] D. W. Little, D. R. Rhodes, W. P. Summers, and A. S. Meier, "Measured impedance of vertical antennas over finite ground planes," Antenna Laboratory, Department of Electrical Engineering, Ohio State University, Columbus, Rep. 233-3, Oct. 1, 1946.

[19] L. L. Tsai, "A numerical solution for the near and far fields of an annular ring of magnetic current," *IEEE Trans. Antennas Propagat.*, vol. AP-20, pp. 569–576, Sept. 1972.

[20] J. H. Richmond and N. H. Geary, "Mutual impedance between coplanar-skew dipoles," *IEEE Trans. Antennas Propagat.* (Commun.), vol. AP-18, pp. 414–416, May 1970.

[21] F. M. Tesche and A. R. Neureuther, "Radiation patterns for two monopoles on a perfectly conducting sphere," *IEEE Trans. Antennas Propagat.* (Commun.), vol. AP-18, pp. 692–694, Sept. 1970.

A Technique to Combine the Geometrical Theory of Diffraction and the Moment Method

W. D. BURNSIDE, C. L. YU, AND R. J. MARHEFKA

Abstract—**A technique is presented in which the moment method (MM) is combined with the geometrical theory of diffraction (GTD). Since diffraction solutions exist for only relatively few structures, it is very desirable to have a means of obtaining the diffracted field for additional structures. Solutions for many structures can be obtained from this combination of techniques, and thus one is able to handle a wide variety of new problems which could not have been solved previously. The approach is developed and applied to a variety of structures in order to illustrate the approach and its validity.**

INTRODUCTION

For many years, the geometrical theory of diffraction (GTD) has been applied to antenna and scattering problems in which the structure is large in terms of the wavelength. However, diffraction

Manuscript received February 7, 1974; revised February 12, 1975. This work was supported in part by Contract N62269072-C-0354 between Naval Air Development Center and the Ohio State University Research Foundation.

The authors are with the Ohio State University ElectroScience Laboratory, Department of Electrical Engineering, Columbus, Ohio 43212.

solutions exist only for relatively few geometries such that one is limited in theoretically modeling many practical structures. On the other hand, moment method (MM) techniques can be applied to arbitrarily shaped objects provided that the overall size is small in terms of the wavelength. Thus, a combination of the two approaches may lead to a means of handling a whole host of new problems. This paper presents such an approach in terms of simple structures to verify the solutions as well as to solve a new problem which is verified by experimental results.

Morita [1] has shown that one can apply MM to infinite planar structures by first subtracting out the surface current density associated with the incident and reflected fields. Similarly, Tew and Tsai [2] have shown that one can use fewer current samples on large finite structures by first subtracting out the physical optics currents. Based on their results, one should be able to use fewer current samples away from the points of diffraction in that the physical optics current is normally quite reasonable in those regions. However, both of these approaches still require a large number of current samples around the point of diffraction.

Analysis

The following discussion illustrates how one can greatly reduce the number of current samples by including a diffraction current term which is based on GTD solutions. Consider the scattering by a simple wedge with a TE plane wave incident as shown in Fig. 1(a). At the field point (ρ,ψ), the incident and reflected fields are easily computed using geometrical optics and are given, respectively, by

$$H_z^i = e^{jk\rho\cos(\psi-\psi_0)}, \qquad \text{for } |\psi - \psi_0| < 180^\circ$$

$$H_z^r = e^{jk\rho\cos(\psi+\psi_0)}, \qquad \text{for } |\psi + \psi_0| < 180^\circ$$

for fields reflected from the wall along the x axis of Fig. 1(a). The diffracted field is given by [3]

$$H_z^d = V_B(\rho,\psi - \psi_0,n) + V_B(\rho,\psi + \psi_0,n).$$

If the field point is not in the vicinity of a shadow boundary ($\psi \pm \psi_0 = 180^\circ$) and for ρ sufficiently large, the diffracted field can be expressed by [3]

$$H_z^d \simeq [D(\psi - \psi_0, n) + D(\psi + \psi_0, n)]\frac{e^{-jk\rho}}{\sqrt{\rho}}, \qquad \text{for } \rho > \lambda/4 \quad (1)$$

where

$$D(\phi,n) = \frac{e^{-j(\pi/4)}}{\sqrt{2\pi k}}\left[\frac{\frac{1}{n}\sin\frac{\pi}{n}}{\cos\frac{\pi}{n} - \cos\frac{\phi}{n}}\right].$$

Using the above form of the diffracted field, the total H-field can be expressed by

$$H_z^t \simeq H_z^i + H_z^r + C\frac{e^{-jk\rho}}{\sqrt{\rho}} \qquad (2)$$

where

$$C = D(\psi - \psi_0,n) + D(\psi + \psi_0,n)$$

and is independent of the range (ρ). Using the boundary conditions, the total surface current density on a perfectly conducting surface is given by

$$J = (\hat{n} \times \hat{z})H_z^t$$

where $\hat{n}$ is the outward normal to the surface. Using the GTD solution for the total field, the total surface current along the wedge walls is given by

$$J^{\text{GTD}} = (\hat{n} \times \hat{z})[H_z^i + H_z^r + H_z^d]$$

$$\simeq J^i + J^r + (\hat{n} \times \hat{z})C\frac{e^{-jk\rho}}{\sqrt{\rho}}, \qquad \text{for } \rho > \lambda/4. \quad (3)$$

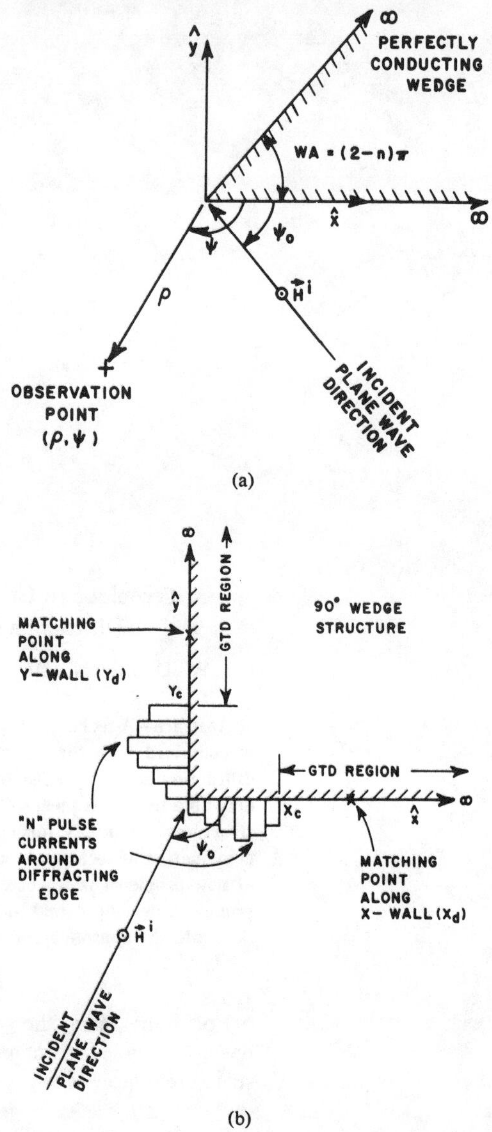

Fig. 1. (a) Basic wedge diffraction geometry. (b) Wedge geometry applied in MM-GTD solution.

Note that C in (3) is, now, assumed to be an unknown constant. Of course, C is known from a canonical solution, but it will be shown that one can solve for C using MM simply by knowing the form of the diffraction current. Once this is verified, the approach can be extended to a whole host of new problems where the diffraction coefficients are not known at the outset. As in the wedge problem, this approach can be extended to new two-dimensional problems with the diffracted field from a structural discontinuity such as plate/cylinder junction represented by the form

$$\left(C\frac{e^{-jk\rho}}{\sqrt{\rho}}\right).$$

Using this form for the surface current, one can solve for the unknown C using the following method.

As given by Harrington [4], the H-field integral equation is given by

$$J = -\left[H_z^i + \hat{z}\cdot\nabla\times\int_S J\,\frac{H_0^{(2)}(k|\rho-\rho'|)}{4j}\,dl'\right] \quad (4)$$

where S is the perfectly conducting surface over which one wishes to find the total surface current density (J). Considering the 90° wedge structure of Fig. 1(b), the total surface current density can be defined by

$$J = \begin{cases} J_y^{\rm GTD}, & y_c \le y \le \infty, \quad x = 0 \\ J_y^{\rm MM}, & 0 \le y < y_c, \quad x = 0 \\ J_x^{\rm MM}, & 0 \le x < x_c, \quad y = 0 \\ J_x^{\rm GTD}, & x_c \le x \le \infty, \quad y = 0 \end{cases}$$

where $J^{\rm GTD}$ is defined by (3) which is only valid away from the diffracting edge. The MM current ($J^{\rm MM}$) around the edge can be defined by simple basis functions such as

$$J^{\rm MM} = \sum_{m=1}^{N} \alpha_m P_m(l - l_m) \quad (5)$$

where $P_m(l - l_m)$ are orthogonal pulse functions with weight α_m. These pulse current samples are illustrated in Fig. 1(b). By point matching the fields at the midpoint of each of the N pulse current segments, one obtains [4]

$$L_n(J) = \sum_{m=1}^{N} l_{mn}\alpha_m + L_n(J_y^{\rm GTD}) + L_n(J_x^{\rm GTD}) = g_n, \quad \text{for } 1 \le n \le N \quad (6)$$

where

$$l_{mn} = \begin{cases} \hat{z}\cdot\nabla\times\displaystyle\int_{dl_m} \frac{H_0^{(2)}(k|\rho_n - \rho_m|)}{4j}\,dl, & \text{for } m \ne n \\ \tfrac{1}{2}, & \text{for } m = n \end{cases}$$

$$g_n = -H_z^i(x_n, y_n)$$

$$L_n(J^{\rm GTD}) = J^{\rm GTD} + \hat{z}\cdot\nabla \times \int_{S_{\rm GTD}} J^{\rm GTD}\,\frac{H_0^{(2)}(k|\rho_n-\rho'|)}{4j}\,dl'.$$

Substituting (3) into (6) one finds that

$$\sum_{m=1}^{N} l_{mn}\alpha_n + L_n\left(J_x^i + J_x^r + D_x\frac{e^{-jkx}}{\sqrt{x}}\right) + L_n\left(J_y^i + J_y^r + D_y\frac{e^{-jky}}{\sqrt{y}}\right) = g_n, \quad \text{for } 1 \le n \le N \quad (7)$$

where the second term exists for $y = 0$, and the third for $x = 0$. Since the incident and reflected currents (J^i and J^r) are known, one can transfer them to the source side. Since the operator (L_n) is linear and D_x and D_y are complex constants, one obtains N equations given by

$$\sum_{m=1}^{N} l_{mn}\alpha_m + D_x L_n\left(\frac{e^{-jkx}}{\sqrt{x}}\right) + D_y L_n\left(\frac{e^{-jky}}{\sqrt{y}}\right) = g_n', \quad \text{for } 1 \le n \le N \quad (8)$$

where

$$g_n' = -H_z^i(x_n, y_n) - L_n(J_x^i + J_x^r) - L_n(J_y^i + J_y^r)$$

$$L_n\left(\frac{e^{-jkx}}{\sqrt{x}}\right) = -\int_{x_c}^{\infty} \frac{jk}{4}\frac{e^{-jkx'}}{\sqrt{x'}} H_1^{(2)}(k|\rho_n - \rho'|)\,dx'$$

$$L_n\left(\frac{e^{-jky}}{\sqrt{y}}\right) = -\int_{\infty}^{y_c} \frac{jk}{4}\frac{e^{-jky'}}{\sqrt{y'}} H_1^{(2)}(k|\rho_n - \rho'|)\,dy'$$

$$L_n(J_x^i + J_x^r) = \begin{cases} -2\displaystyle\int_{x_c}^{\infty}\frac{jk}{4}\,e^{jkx'\cos\psi_0}H_1^{(2)}(k|\rho_n-\rho'|)\,dx', & \text{for } 0 \le \psi_0 \ge 180° \\ 0, & \text{otherwise} \end{cases}$$

$$L_n(J_y^i + J_y^r) = \begin{cases} -2\displaystyle\int_{\infty}^{y_c}\frac{jk}{4}\,e^{jky'\sin\psi_0}H_1^{(2)}(k|\rho_n-\rho'|)\,dy', & \text{for } 90° \le \psi_0 \le 270° \\ 0, & \text{otherwise.} \end{cases}$$

The integrations over the GTD region are infinite in extent; however, in practice, one can simply integrate over a finite section along the x axis from x_c to x_l and along the y axis from y_c to y_l because the Hankel function causes the integrand to become negligible for large ρ'. By introducing two additional matching points in the GTD region at $(x_d,0)$ and $(0,y_d)$, as shown in Fig. 1(b), one obtains two additional equations such as

$$\sum_{m=1}^{N} l_{mn}\alpha_m + \tfrac{1}{2}D_x\frac{e^{jkx_d}}{\sqrt{x_d}} + D_y L_n\left(\frac{e^{-jky}}{\sqrt{y}}\right) = -H_z^i(x_d,0) - \tfrac{1}{2}(J_x^i + J_x^r)_{\text{at } x=x_d} - L_n(J_y^i + J_y^r)$$

for a matching point ($n = N + 1$) along the x-wall at $x = x_d$. A similar equation is obtained for the matching point ($n = N + 2$) along the y-wall at $y = y_d$. This provides the necessary linear equations to solve for the $N + 2$ unknowns [N current pulse functions (α_m) plus two diffraction constants (D_x and D_y)]. This system of simultaneous linear equations can be easily solved using standard numerical techniques.

A comparison of the total surface current density along the x-wall of Fig. 1(b) between the preceding approach and the exact solution is shown for several cases in Fig. 2. In each case, the comparison between the results is quite good except when the plane wave is incident along the x-axis. This discrepancy results, because the x-wall lies on the shadow boundary of the incident field. Note that in most cases only four current samples are used around the edge for these calculations plus two unknown diffraction constants. Thus, only six unknowns are found to successfully predict the total surface current density, in most cases, on an infinite wedge structure.

In order to solve for the diffracted currents along walls which are on or near the shadow boundary of the incident field, one should represent the diffracted current in terms of a series as given by

$$J^d = \sum_{l=0}^{\infty} C_l\,\frac{e^{-jk\rho}}{(\sqrt{\rho})^l}.$$

Based on a brief study of this problem, it appears that one need only consider a few terms. It has been shown, previously, that the second term is sufficient provided that the currents are not being computed near a shadow boundary. If the currents are computed along walls which lie on the shadow boundary of the incident field, then the diffracted current can be simply expressed by

$$J^d = C_0 e^{-jk\rho}.$$

As shown in Fig. 2(c), this current representation compares very favorably with the exact solution when applied along the x-wall with $\psi_0 = 180°$. When the first two terms were applied to find the diffracted current near a shadow boundary, it was found that the solutions were somewhat dependent on the locations of the

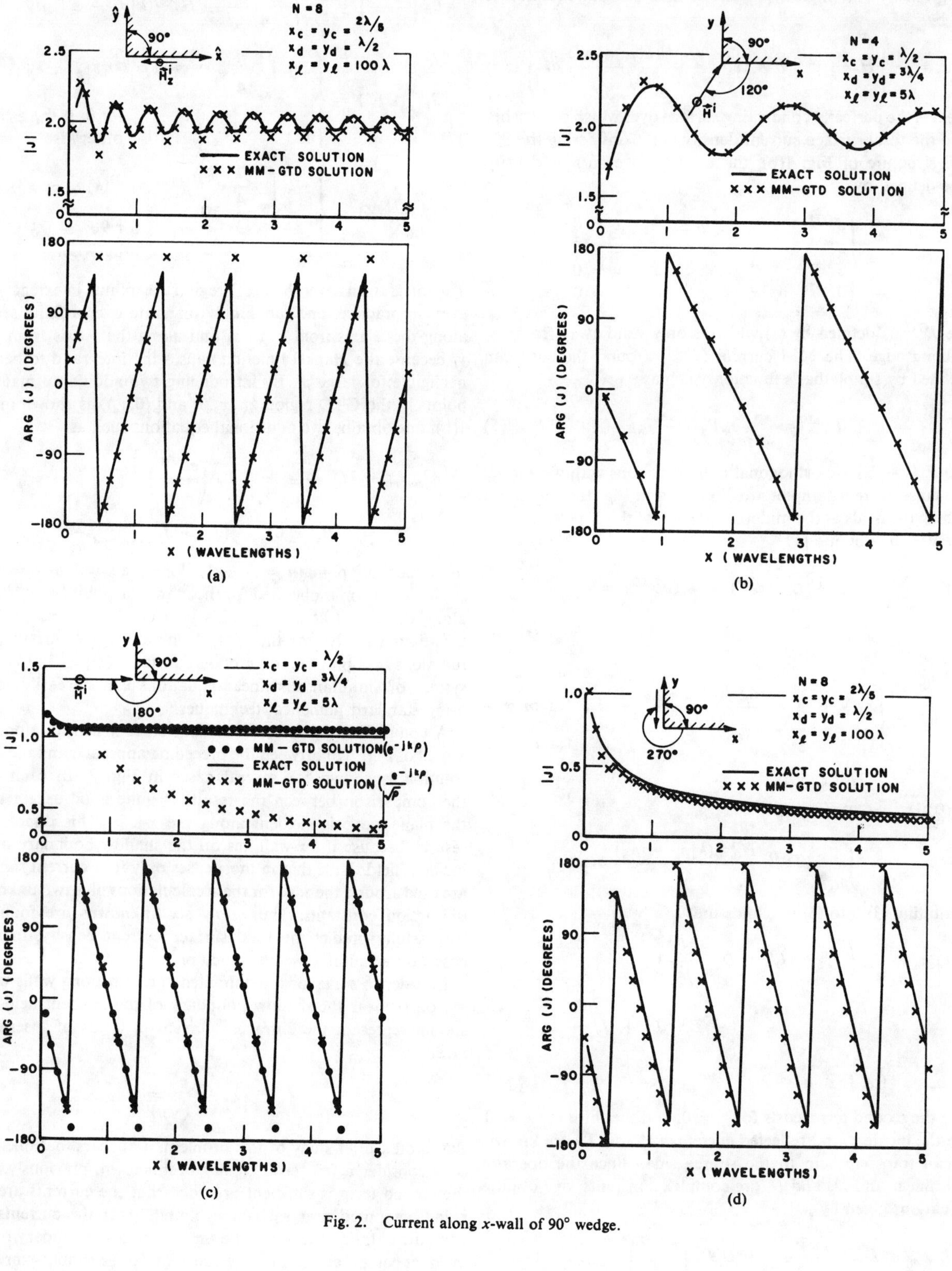

Fig. 2. Current along *x*-wall of 90° wedge.

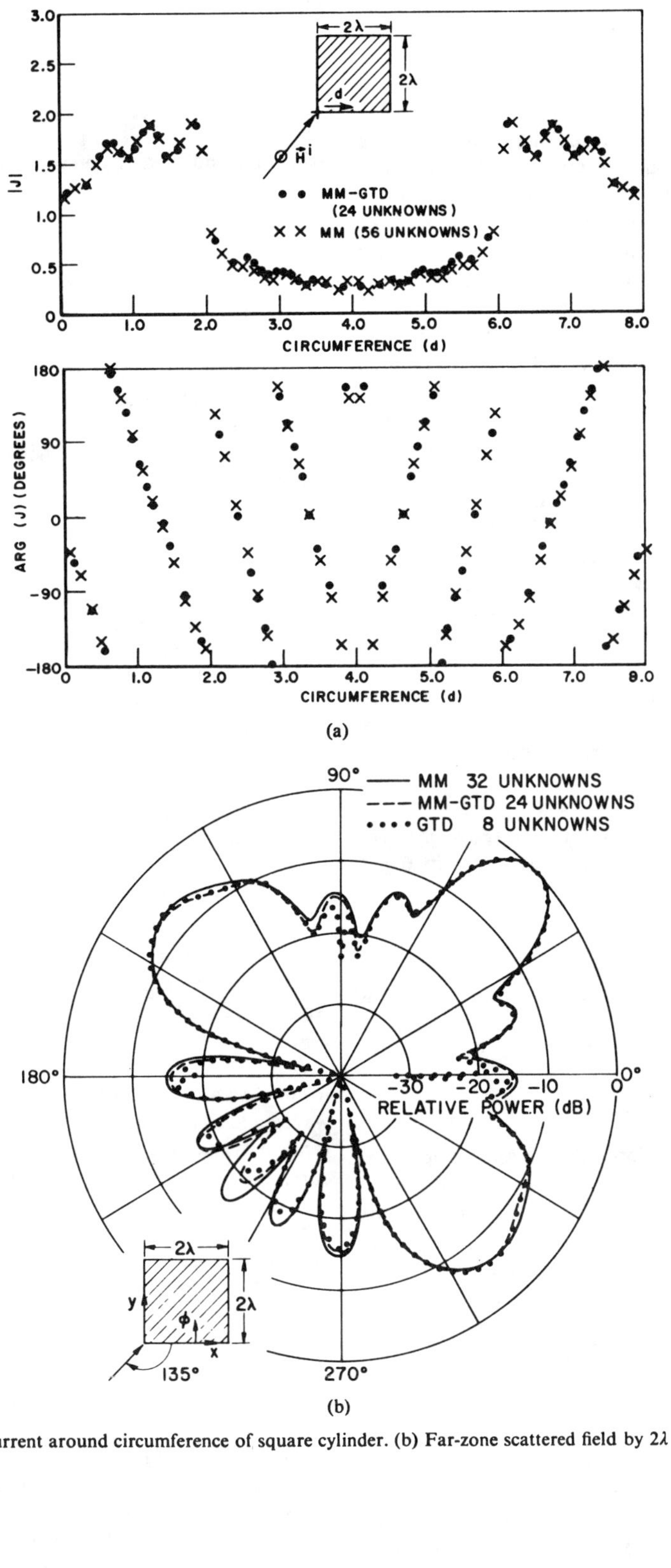

Fig. 3. (a) Current around circumference of square cylinder. (b) Far-zone scattered field by 2λ square cylinder.

sample points. This problem needs to be studied further although this difficulty is only for a small range of incidence angles. However, the first two terms of the above series have been used very successfully by Lee [6] to solve for the currents along walls near the shadow boundary of the incident field for a large wing-shaped structure.

It seems evident from the previous results that one can accurately determine the total currents on an infinite wedge using this approach. In most cases, only four current samples are required around the diffracting edge. The terms $[L_n(J^{\mathrm{GTD}})]$ require the evaluation of infinite integrals; however, in practice one need only integrate a short distance. In fact, if the incident plane wave is not close to grazing along one of the walls, one need only integrate about 5 to 10 wavelengths. On the other hand, if the plane wave is incident along the wall, one must integrate about 100 or more wavelengths. This does not limit the solution in that only a few terms of this type are involved. In reality, one is normally working with a large but finite structure which limits the range of integration.

This technique is next extended to solve for the currents on a square cylinder. Just four current samples are used around each of the diffracting edges. Eight unknown diffraction constants are used to completely describe the GTD currents between the edges using the self-consistent method [7], [8]. The total currents around this structure are shown in Fig. 3(a) using this approach and conventional MM techniques using pulse current samples and employing point matching. The total far-zone scattered field patterns are illustrated in Fig. 3(b) using this approach, conventional GTD, and conventional MM. In every case, the comparisons between the results appear to be very encouraging.

The currents on curved surfaces can be obtained using the same technique. The GTD solution for the currents on large circular cylinders deep in the shadow region as shown in Fig. 4 is given by [5]

$$J_\phi^{\mathrm{GTD}} = \sum_{m=0}^{\infty} D_m(a)A_m(a) \cdot \left[e^{-(\alpha_m+jk)a(\phi-(\pi/2))} + e^{-(\alpha_m+jk)a((3\pi/2)-\phi)}\right] \quad (9)$$

for $90° + \Delta \leq \phi \leq 270° - \Delta$. The angle ($\Delta$) is shown in Fig. 4 and, typically, is defined by

$$\Delta = (ka)^{-1/3}.$$

Note that the mth mode diffraction $[D_m(a)]$ and attachment $[A_m(a)]$ coefficients are known; however, using our new approach they will be determined along with the sample pulse currents around the shadow boundaries ($\phi = \pm 90°$). In this case, normal physical optics currents are applied in the illuminated region, pulse current samples around the shadow boundary, and GTD current in the deep shadow region. Thus, the total current is given using only one mode in the GTD current form by

$$J_\phi^{t} = \begin{cases} 2e^{jka\cos\phi}, & 0 \leq \phi \leq 90° - \delta, \\ & 270° + \delta \leq \phi < 360° \\ \sum_{m=1}^{N} \alpha_m P_m(l - a\phi_m), & 90° - \delta < \phi < 90° + \Delta, \\ & 270° - \Delta < \phi < 270° + \delta \\ C_0\left[e^{-a(\alpha_0+jk)(\phi-(\pi/2))} + e^{-a(\alpha_0+jk)((3\pi/2)-\phi)}\right], & \\ & 90° + \Delta \leq \phi \leq 270° - \Delta. \end{cases} \quad (10)$$

Using this current representation, one obtains a system of simultaneous linear equations by applying (4) as before. In order to analyze this problem the surface is first approximated by equally spaced strips which approximate the circular cylinder. On each of these strips exists a uniform pulse current which is specified by (10). Substituting these pulse current samples into (6), one finds that

$$L_n(J_\phi^t) = \sum_{\text{PO region}} l_{mn}(2e^{jkx_m}) + \sum_{\text{pulse region}} l_{mn}\alpha_n + \sum_{\text{GTD region}} l_{mn}' C_0 = g_n, \qquad \text{for } 1 \leq n \leq N+1$$

where l_{mn} and g_n were defined in (6) and

$$l_{mn}' = \begin{cases} \frac{1}{2}\left[e^{-a(\alpha_0+jk)(\phi_m-(\pi/2))} + e^{-a(\alpha_0+jk)((3\pi/2)-\phi_m)}\right], & \text{for } m = n \\ \hat{z}\cdot\nabla\times\int_{dl_m} J(\phi')\,\frac{H_0^{(2)}(k|\rho_n - \rho_m|)}{4j}\,dl', & \text{for } m \neq n. \end{cases}$$

In this case, just one additional matching point in the GTD region is needed in order to compute the total current in that N pulse currents must be found as well as C_0. Note that α_0 is the well-known attenuation coefficient for a circular cylinder and is given by [5]

$$\alpha_0 = \frac{\bar{q}_0}{a}\,e^{j(\pi/6)}\left(\frac{ka}{2}\right)^{1/3}\left[1 + \left(\frac{2}{ka}\right)^{2/3}\left(\frac{\bar{q}_0}{60} + \frac{1}{10\bar{q}_0^{\,2}}\right)e^{-j(\pi/3)}\right]$$

where $\bar{q}_0 = 1.01879$. Again, using a simple point matching scheme, one can solve for the complete set of unknowns.

The total currents on a circular cylinder using this approach are compared with the exact currents for several cases in Fig. 4. The comparison between the results in each case is quite good.

If necessary, more modes could have been included as originally specified in (9). This would, of course, require more matching points in the GTD region in order to obtain the necessary number of equations to solve for the increased number of unknowns. However, these higher order terms interact such that the solution is somewhat dependent on the location of the matching points. Based on a preliminary study of this problem it appears that perhaps more matching points than unknowns are helpful such that one can apply some averaging scheme. If only the first mode is considered it appears that the best solution is obtained when the matching point is near the pulse region as shown in Fig. 4.

A problem that cannot be handled directly using present GTD techniques alone is illustrated in Fig. 5(a). In this case an axial waveguide is mounted on a two-dimensional structure and the resulting radiation pattern is desired. Using our new approach, current samples are applied around the junction of the flat plates and circular end caps in that these junctions form lines of diffraction. The GTD current forms are applied elsewhere, and the resulting radiation pattern is computed as shown in Fig. 5(b). The solution is compared with experimental results and the agreement is, again, very good.

Conclusions

This study describes a technique which can be used to combine the geometrical theory of diffraction (GTD) and the moment method (MM) solutions. It has been successfully applied to the basic wedge and circular cylinder plane wave scattering problems. Both of these two-dimensional problems have been verified by comparisons made with exact solutions. The square cylinder has been studied in detail with the currents and far-field scattering patterns compared with those obtained using MM and GTD solutions alone. In each case, it appears that one need only use a few current samples around the point or points of diffraction. For a diffracting edge as few as four current samples may be

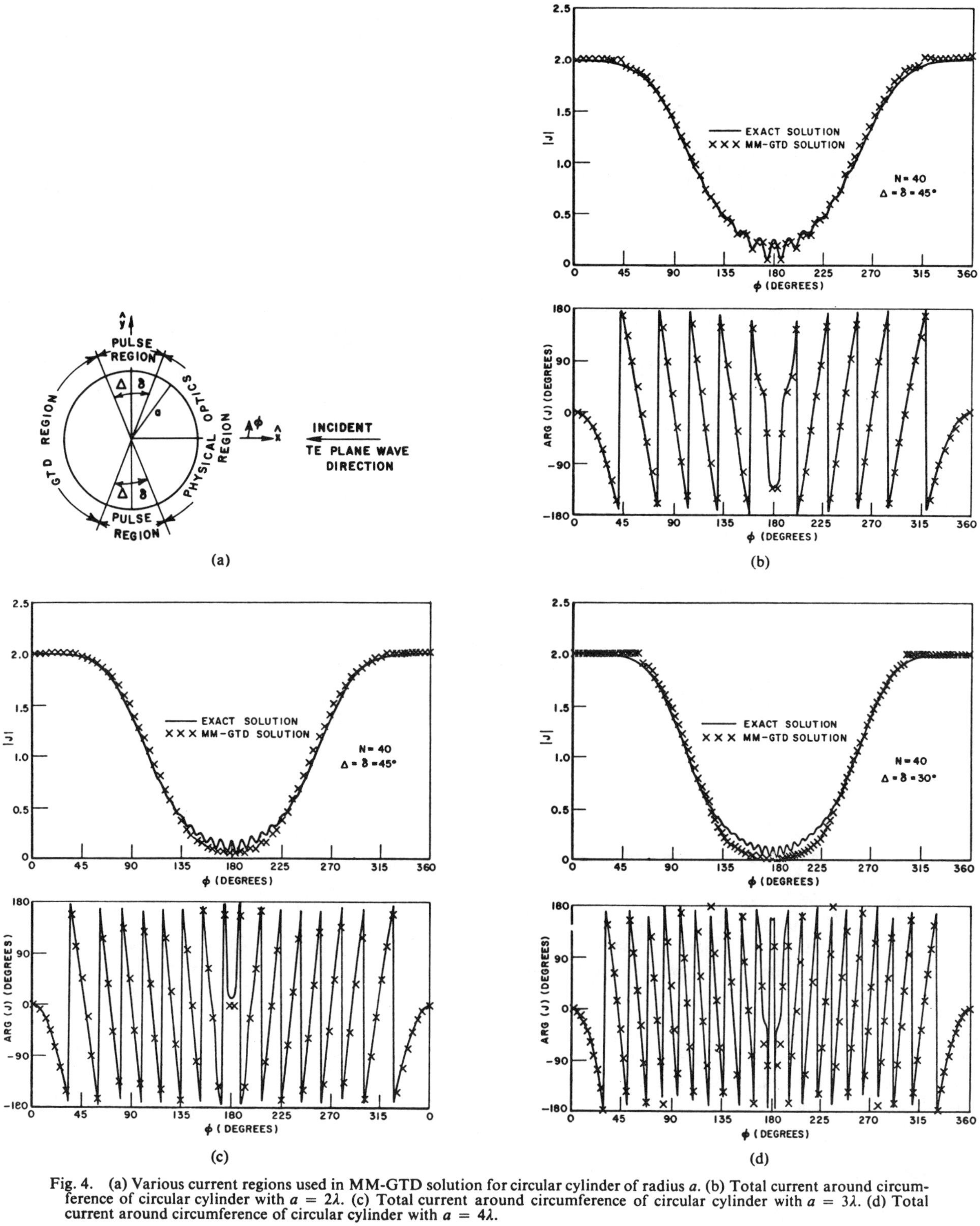

Fig. 4. (a) Various current regions used in MM-GTD solution for circular cylinder of radius a. (b) Total current around circumference of circular cylinder with $a = 2\lambda$. (c) Total current around circumference of circular cylinder with $a = 3\lambda$. (d) Total current around circumference of circular cylinder with $a = 4\lambda$.

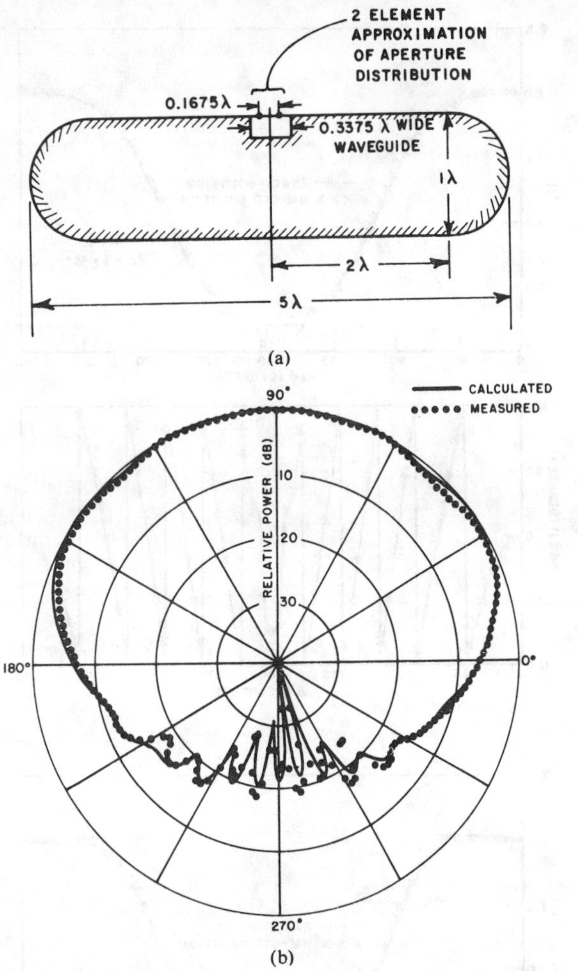

Fig. 5. (a) Axial waveguide mounted in two-dimensional cylinder. (b) Radiation pattern for structure in Fig. 5(a).

used; whereas, for curved surface diffraction it appears that as many as twenty current samples are needed. Consequently, one can use current samples around the various points of diffraction that cannot be handled using present GTD solutions, and use known GTD current forms outside the current sample regions. This allows one to handle a very large or even infinite structure with just a few unknowns.

This approach is, also, verified by a comparison of experimental and theoretical results for the radiation pattern of a waveguide antenna mounted on a complex structure which could not be completely analyzed using present GTD techniques alone. In addition, this solution illustrates that this approach can be successfully applied to near-zone source problems.

This technique can be extremely useful in terms of concave structures. One can use current samples around the concave section and the known GTD curved surface solution in the convex regions. Note that with the currently available attenuation and attachment coefficients, an arbitrary convex surface can be handled using GTD alone. This technique can, also, be applied to surface impedance structures as presented in [6].

Based on the success of the two-dimensional solutions presented here, it appears that this technique can be extended to three-dimensional problems. The diffraction current from a three-dimensional structure, such as a cone tip, has a e^{-jkr}/r range dependence. With this knowledge one should be able to solve for the current samples around the three-dimensional diffraction point as well as the diffraction coefficients which are simply angular dependent.

References

[1] N. Morita, "Diffraction by arbitrary cross-sectional semi-infinite conductor," *IEEE Trans. Antennas Propagat.*, vol. AP-19, pp. 358–364, May 1971.

[2] M. D. Tew and L. L. Tsai, "Use of *a priori* knowledge in moment method solutions," in *1972 GAP Int. Symp.*, Williamsburg, Va., Dec. 11–14, 1972.

[3] R. G. Kouyoumjian and P. H. Pathak, "A uniform geometrical theory of diffraction for an edge in a perfectly conducting surface," *IEEE Proc.*, vol. 62, pp. 1448–1461, Nov. 1974.

[4] R. F. Harrington, Field Computation by Moment Methods. New York: MacMillan, 1968.

[5] P. H. Pathak and R. G. Kouyoumjian, "An analysis of the radiation from apertures in curved surfaces by the geometrical theory of diffraction," *IEEE Proc.*, vol. 62, pp. 1438–1447, Nov. 1974.

[6] S. C. Lee, "Control of electromagnetic scattering by antenna impedance loading," Ohio State Univ. ElectroScience Laboratory, Dep Elec. Eng., Rep. 3424-2, July 1974; prepared under Contract F19628-72-C-0203 for Department of the Air Force.

[7] J. S. Yu and R. C. Rudduck, "Diffraction by conducting walls of finite thickness," Ohio State Univ. ElectroScience Laboratory, Dep. Elec. Eng., Rep. 1767-7, May 1965; prepared under Contract AF 30(602)-3269 for Rome Air Development Center (AD 623077).

[8] ——, "On higher-order diffraction concepts applied to a conducting strip," *IEEE Trans. Antennas Propagat.*, vol. AP-15, pp. 662–668, Sept. 1967.

A hybrid MM-GTD technique for the treatment of wire antennas near a curved surface

L. W. Henderson

ElectroScience Laboratory, The Ohio State University, Columbus, Ohio 43212

G. A. Thiele

Office of Graduate Engineering and Research, University of Dayton, Dayton, Ohio 45469

(Received October 31, 1980; accepted February 25, 1981.)

The method of moments (MM) and the geometrical theory of diffraction (GTD) are formally combined into a hybrid MM-GTD technique. Wire type antennas are treated by the method of moments and their interaction with a large curved surface is accounted for through the GTD. Thus the mutual coupling between two dipoles in the presence of an infinitely long, perfectly conducting elliptic cylinder is analyzed for various orientations of the dipoles with respect to each other and with respect to the cylinder. Results are compared with an independent theory and with experimental data. Excellent agreement is shown, thereby demonstrating the accuracy of the hybrid MM-GTD technique.

1. INTRODUCTION

In recent years several new techniques have been developed for analyzing systems which could not be adequately handled by classical methods. Two of the most powerful of these theories are the moment method (MM) and the geometrical theory of diffraction (GTD).

The hybrid method presented in this paper presents a technique for combining the moment method and the GTD. The work, presented herein, is based on previous papers by *Thiele and Newhouse* [1975] and by *Ekelman and Thiele* [1980]. The approach taken is to cast the small radiator (built of thin wires) in the moment method format of *Richmond* [1973] and then add perturbation terms due to interaction with the nearby large body using GTD. Thus the solution is basically a GTD perturbation of the moment method. The particular form of the GTD used in this work is a new ansatz called the uniform theory of diffraction (UTD) reported by *Pathak et al.* [1980]. The UTD avoids previous difficulties encountered with the GTD when a ray hits a curved edge at grazing incidence.

2. THE HYBRID MM-GTD METHOD

The hybrid method to be presented here is a technique for analyzing thin-wire antennas which are not electrically large in the presence of a body or scatterer which is too large to be handled by the moment method. This is accomplished by computing a second moment method impedance matrix proposed by *Thiele and Newhouse* [1975]. This second matrix $[\Delta Z]$ is a perturbation to the original $[Z]$ and includes the coupling between the test and expansion modes due to rays reflected and diffracted from the nearby structure.

Using GTD ray optics technique, the electric field incident on a wire scatterer due to radiation from another wire scatterer can be written as

$$\bar{E}^{i}_{s} = \bar{E}^{0}_{s} + \bar{E}^{1}_{s} + \bar{E}^{2}_{s} \cdots \bar{E}^{n}_{s} \tag{1}$$

where E^0_s is the incident field due to the direct path (free space case) and $E^1_s \cdots E^n_s$ denote the fields due to the various rays which are reflected or diffracted by the nearby body. Note that for the case where the body hides the source from the observer, $E^0_s = 0$. From the moment method, the component elements Z'_{mn} of the $[Z']$ matrix can be written

$$Z'_{mn} = \int [\bar{E}^0_s(l) + \bar{E}^1_s(l) + \bar{E}^2_s(l) \cdots \bar{E}^n_s(l)] \cdot I(l)\, d\bar{l} \tag{2}$$

which can be written

$$Z'_{mn} = Z_{mn} + \Delta Z_{mn} \tag{3}$$

where

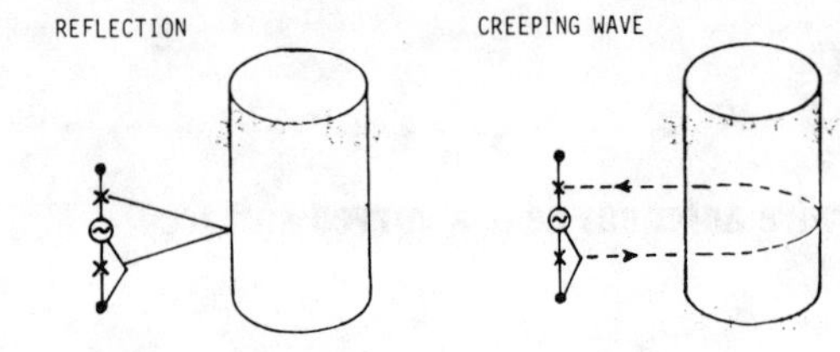

Fig. 1. Scattering mechanisms for dipole interaction with cylinder.

$$Z_{mn} = \int \bar{E}_s^0(l) \cdot I(l)\, d\bar{l} \tag{4}$$

and

$$\Delta Z_{mn} = \int [\bar{E}_s^1(l) + \bar{E}_s^2(l) \cdots \bar{E}_s^n(l)] \cdot I(l)\, d\bar{l} \tag{5}$$

Note that (4) is merely an expression of the usual moment method (only) calculation while (5) requires the use of ray optics. In matrix form,

$$[Z'] = [Z] + [\Delta Z] \tag{6}$$

where $[Z]$ is computed using the direct (free space) path between source and observation point, as in (4), and $[\Delta Z]$ is computed using the ray paths which interact with the large body as in (5). Thus the effects of the large body are included in the moment formulation and the solution obtained will therefore contain these interactions.

It is informative to rewrite (6) in a submatrix form. Thus if there are two antennas in the presence of a scattering object, such as the cylinder used in this paper, and there is no direct (optical) path between them, then

$$\begin{bmatrix} Z'_{11} & Z'_{12} \\ Z'_{21} & Z'_{22} \end{bmatrix} = \begin{bmatrix} Z_{11} & 0 \\ 0 & Z_{22} \end{bmatrix} + \begin{bmatrix} \Delta Z^R_{11} & \Delta Z^C_{12} \\ \Delta Z^C_{21} & \Delta Z^R_{22} \end{bmatrix} \tag{7}$$

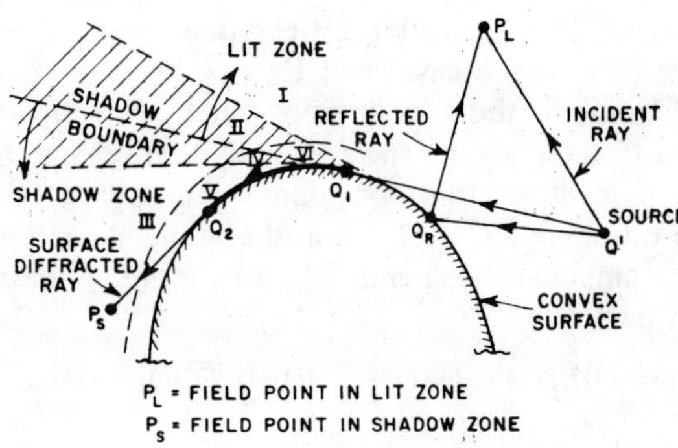

Fig. 2. The rays and regions associated with scattering by a smooth convex cylinder.

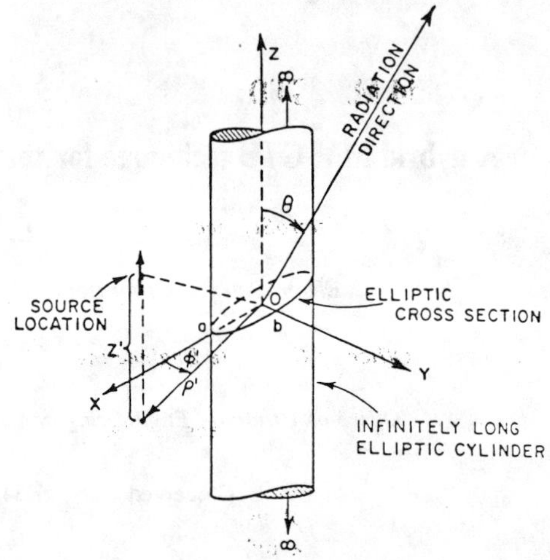

Fig. 3. Scattering configuration involving an infinitely long elliptic cylinder.

where in this equation Z_{11} and Z_{22} are the moment method impedance submatrices of antennas 1 and 2, respectively, when in free space. Viewing Figure 1, ΔZ^R_{11} is the perturbation submatrix for antenna 1 due to reflection. If we imagine a second antenna in Figure 1 positioned, say, on the opposite side of the cylinder, then ΔZ^C_{12} is the perturbation submatrix, due to the creeping wave, which accounts for part of the interaction between antenna

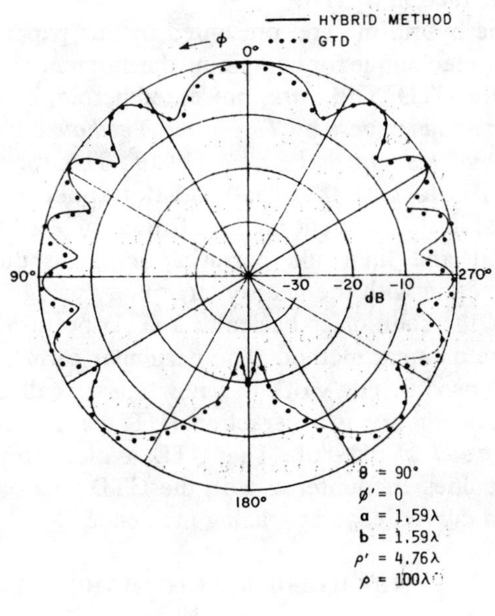

Fig. 4. Comparison of Z_{12} with E_θ radiation of an electric dipole near a circular cylinder.

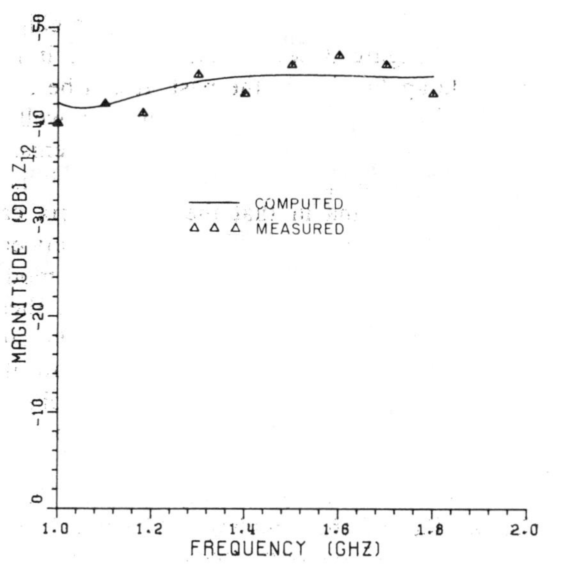

Fig. 5. Magnitude of the transmission coefficient for the monopoles used in the experimental setup.

1 and antenna 2. Similar explanations can be made for ΔZ_{22}^{R} and ΔZ_{21}^{C}.

One problem which arises from (2) is that the GTD requires that the field generated by the sources contain no ray path components. In general this condition cannot be met in the near field. Fortunately, the piecewise sinusoidal (PWS) expansion function used by Richmond can be decomposed to have only transverse field components as shown below.

Ekelman and Thiele [1980] observed that the near fields of a piecewise sinusoidal test dipole can be broken up into contributions from the endpoints and the driving point. These fields have spherical TEM wave fronts and are in the required ray optical form for using them as sources with the GTD (UTD). The exact field of a PWS dipole in this format is given by the sum of the three terms:

$$E_1 = \frac{-j\eta I_0}{4\pi \sin(kd)} \frac{e^{-jkR_1}}{R_1} \tag{8}$$

$$E_2 = \frac{j\eta I_0 2\cos(kd)}{2\pi \sin(kd)} \frac{e^{-jkR_2}}{R_2} \tag{9}$$

$$E_3 = \frac{-j\eta I_0}{4\pi \sin(kd)} \frac{e^{-jkR_3}}{R_3} \tag{10}$$

The above terms are associated with an endpoint, center, and the other endpoint, respectively, of one PWS mode. Note the spherical transverse electi magnetic nature of the three fields which mak(them individually ray optical fields.

3. GEOMETRICAL THEORY OF DIFFRACTION

3.1. *Geometrical optics field.* While the moment method is a technique primarily for low frequency, the geometrical theory of diffraction (GTD) is useful for high frequency, or situations where the scatterer is electrically large. Assuming the scatterer does not hide the source, the dominant field is produced by the direct ray from the source to the observer. The next most important term is the ray reflected from the scatterer, as shown in Figure 1. This field is termed the geometrical optics field.

3.2. *Curved surface diffraction and UTD.* The geometrical optics field by itself is valid only when the source and observation points are not hidden by the structure and the observation point is well away from the shadow boundary. A more general case is when the observation point can be anywhere

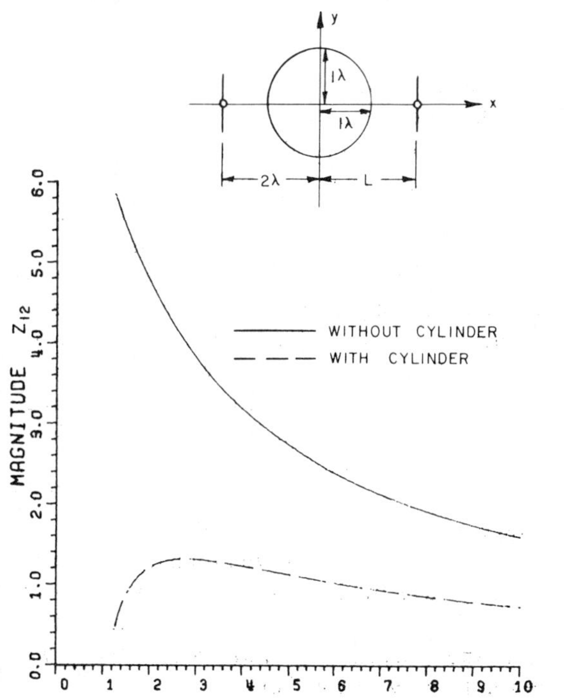

Fig. 6. Comparison between free space and diffracted mutual impedance Z_{12} between two λ/2 dipoles in the deep shadow region.

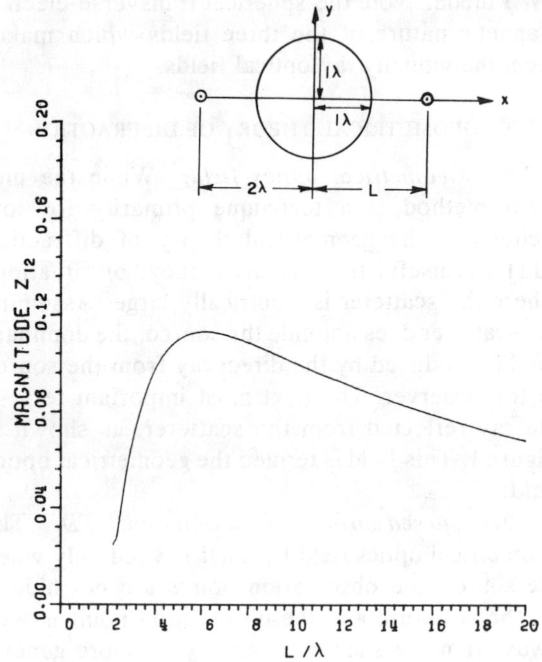

Fig. 7. Mutual impedance Z_{12} between two axial $\lambda/2$ dipoles near an infinite circular cylinder as one dipole moves along the x axis.

in the space around the structure. This problem is much more complex.

Although the theory given in this section is valid for any convex surface, we will limit our discussion to an elliptical cylinder which is infinite in extent. This will serve to demonstrate the essential points of the solution without adding unnecessary complication. The regions of interest around an elliptic cylinder are shown in Figure 2. The uniform UTD ansatz of *Pathak et al.* [1980] was chosen because of its good continuity through the shadow boundary where previous theories failed. This solution is valid when the source and/or observation points are in regions I, II, and III, but not very close to the surface as in regions IV, V, and VI.

In the UTD creeping wave problem, the scattered creeping wave field is computed from the moment method incident field (i.e., (8), (9), and (10)) and the $\bar{\bar{R}}$ and $\bar{\bar{T}}$ dyadics [*Pathak et al.*, 1980] which are too lengthy to give here.

4. DIPOLES NEAR AN ELLIPTIC CYLINDER

The application of the theory presented in sections 2 and 3 is the analysis of dipoles near an infinite, perfectly conducting, elliptic cylinder. The elliptic cylinder of Figure 3 was chosen because it has found wide application in the past. It can be used to model aircraft fuselages and engines as well as antenna masts on ships and land-based systems.

The dipole is of prime interest because it is a typical wire antenna in that the $[Z]$ matrix can be obtained by computing the coupling between the mth test function and the nth expansion mode. In the case of Richmond's thin-wire code, these modes are dipoles with a piecewise sinusoidal current distribution. Thus once one can calculate the near field coupling between dipoles near an object, any thin-wire antenna near the object can be analyzed.

4.1. *Theoretical comparison of results.* Figure 4 shows plots of the coupling between two $\lambda/2$ dipoles using the geometry given in Figure 3. The data are plotted with the mutual coupling computed by the hybrid technique and by an independent GTD code developed at the Ohio State University ElectroScience Laboratory, which computes the far field scattering of electric and magnetic sources in the vicinity of elliptic cylinders and/or finite plates. This code was chosen as a basis for comparison because of its proven accuracy. Since the fields computed by the GTD code are far zone–scattered electric fields, they can be used for comparison only to verify the variation of Z_{12} with respect to

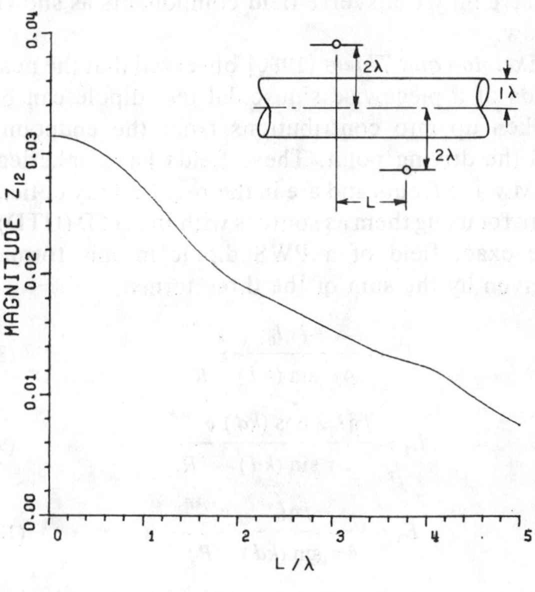

Fig. 8. Mutual impedance Z_{12} between two $\lambda/2$ dipoles as one dipole moves along the z axis.

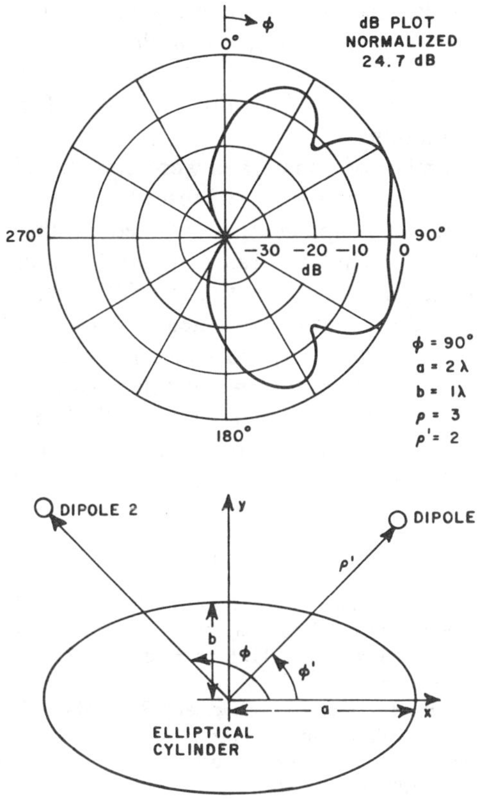

Fig. 9. Mutual impedance Z_{12} between two parallel dipoles in the presence of an elliptic cylinder.

the angle ϕ. The results in Figure 4 show a very close correlation between the two results.

4.2. *Experimental data.* Another method available for comparison was the measurement of actual antenna systems. To avoid the problems associated with balanced antenna feeds, the model was constructed on a ground plane. Monopoles were used instead of dipoles and a very long circular cylinder was set on the ground plane.

The test equipment used was designed to accurately measure the transmission coefficient of a two-port network. The two-antenna system can be viewed as such a network with the input port being the terminals of the transmitting antenna and the output port being the terminals of the receiving antenna. The coupling Z_{12} is related to the transmission coefficient T_{12} by

$$T_{12} = Z_{12}/Z_{22} \tag{11}$$

where Z_{22} is the input impedance of the receiving antenna.

Data were taken over frequencies from 1.0 to 1.8 GHz. In terms of wavelengths this provided a variety of antenna lengths and cylinder sizes. Some measured and computed values are shown in Figure 5. Agreement between the two sets of data is good, with the measured points showing some spread around the computed curve.

4.3. *Other hybrid technique results.* Figures 6, 7, and 8 show the coupling between various orientations of dipoles near infinite cylinders. Note that in Figure 6 the free space and diffracted curves differ only by a constant in the far field. This occurs because the path around the cylinder does not vary in that region. Note also that in Figures 6 and 7 a peak is seen in the coupling. This occurs because in the extreme near field the field is highly attenuated due to the long path around the cylinder. In Figure 8 this peak does not occur because the path around the cylinder is always getting longer. Finally, Figures 9 and 10 show the variation of Z_{12} with the angle ϕ for an elliptic cylinder with an axial ratio

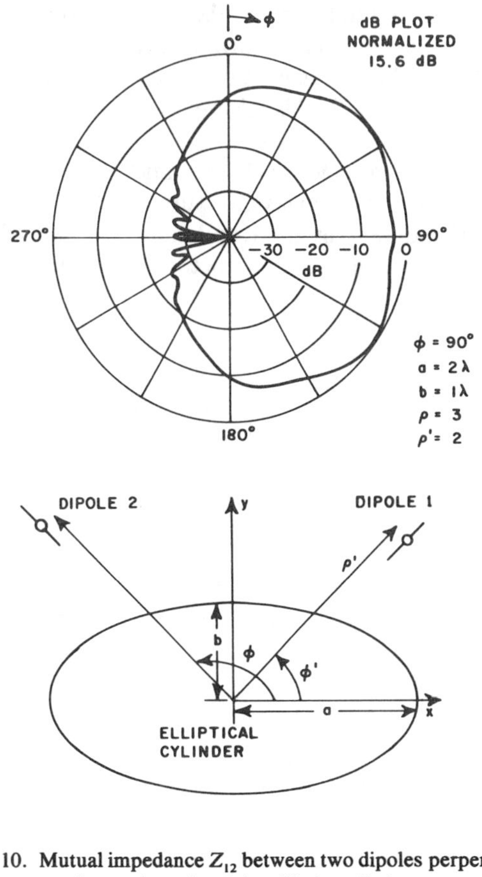

Fig. 10. Mutual impedance Z_{12} between two dipoles perpendicular to the axis of the elliptic cylinder.

of 2 : 1. Figure 9 is for the case when the dipoles are parallel to the axis of the cylinder and Figure 10 is for the case when the half-wave dipoles are perpendicular to the axis of the cylinder. The higher normalization factor of 24.7 dB in Figure 9 indicates weaker mutual coupling relative to Figure 10. The nulls at about 45° and 135° in Figure 9 do not appear in Figure 10 primarily because of the variation in path lengths from one part of one half-wavelength dipole to a part of the other for a given ϕ.

5. SUMMARY AND CONCLUSIONS

The hybrid MM-GTD technique has been extended to handle a more general class of problems than was previously possible. A uniform GTD or UTD for the diffraction of electromagnetic waves by a smooth convex surface was used to provide a means of analyzing antennas arbitrarily oriented near an infinite perfectly conducting elliptic cylinder. The cylinder could be made finite in the analysis by including the effects of curved edge diffraction as was done by *Ekelman and Thiele* [1980]. The moment method part of the analysis retained the piecewise sinusoidal format used by *Richmond* [1973] and thus the method used here is completely compatible with his thin-wire moment method computer code.

In this hybrid MM-GTD technique the free space impedance matrix was computed using the conventional moment method technique. A perturbation matrix was then added to the free space matrix. This delta matrix contained the coupling between elements due to GTD type rays which were reflected or diffracted by the cylinder. Thus the effects of the cylinder were formally accounted for in the final solution.

A number of geometries were considered where two dipoles were in the presence of a cylinder. The mutual coupling Z_{12} computed by the hybrid MM-GTD technique was compared in this paper both with an independent purely GTD analysis and with experimental measurements. The authors have also compared the hybrid technique with an eigenfunction solution for two dipoles parallel to the axis of a circular cylinder. In all these cases, agreement between the hybrid technique results and those of the three other methods was excellent.

REFERENCES

Ekelman, E. P., and G. A. Thiele (1980), A hybrid MM-GTD technique for combining the moment method treatment of wire antennas with the GTD for curved surfaces, *IEEE Trans. Antennas Propag.*, *28*(6). 62–69.

Pathak, P. H., W. D. Burnside, and R. J. Marhefka (1980), A uniform GTD analysis of the diffraction of electromagnetic waves by a smooth convex surface, *IEEE Trans. Antennas Propag.*, *28*(5), 631–642.

Richmond, J. H. (1973), Radiation and scattering by thin-wire structures in the complex frequency domain, *Rep. 2902*, Ohio State Univ. ElectroSci. Lab., Columbus, prepared under grant NGL 36-008-138 for NASA (NASA-CR-2396).

Thiele, G. A., and T. M. Newhouse (1975), A hybrid technique for combining moment methods with the geometrical theory of diffraction, *IEEE Trans. Antennas Propag.*, *23*(1), 62–69.

A Diffraction Coefficient for a Cylindrically Truncated Planar Surface

CHIWEI W. CHUANG, MEMBER, IEEE, AND WALTER D. BURNSIDE, MEMBER, IEEE

***Abstract*—A numerically derived solution of the diffraction coefficient for a source mounted on a perfectly conducting planar surface which is smoothly terminated by a circular cylinder is obtained using the hybrid approach which combines the moment method (MM) with the geometrical theory of diffraction (GTD). This solution is valid not only in the region away from, but also in the region near, the shadow boundary. The accuracy and usefulness of this solution is demonstrated as various structures are treated using it.**

I. INTRODUCTION

As ONE ATTEMPTS to apply ray optical techniques to solve more and more complex structures, it is apparent that more diffraction coefficients are needed in that ray optics require one to simulate a complex structure using simpler configurations which can be analyzed in terms of known diffraction coefficients. Using the wedge and curved surface diffraction coefficients, great strides have been made in the development of simulation models of complex structures. Even those models require one to neglect certain diffraction discontinuities which cannot be handled at present. A procedure was suggested by Burnside *et al.* [1] whereby one could generate new diffraction coefficients using a hybrid approach. That approach is applied in this paper to solve for the diffraction coefficient associated with a perfectly conducting planar surface truncated by a circular cylinder as illustrated in Fig. 1. The usefulness of this new diffraction coefficient will become apparent as various structures are treated using it.

II. DIFFRACTION COEFFICIENT ANALYSIS

Consider the two-dimensional diffraction problem as illustrated in Fig. 1. A magnetic line source is mounted on the planar surface at a distance d from the plate–cylinder junction. The radius of the cylinder is a. Thus a discontinuity in the radius of the curvature occurs at the junction. Senior [2] has derived an analytical solution of the diffraction coefficient for such a structure (curvature discontinuity). However, his solution is valid in the deep lit region only. In lieu of the difficulty in obtaining a good analytical solution of the diffraction coefficient associated with the structure shown in Fig. 1, a numerically derived solution valid not only in the region away from, but also in the region near, the shadow boundary is desirable.

Using the hybrid approach of [1] one begins the solution as any moment method (MM) problem [3] by placing an unknown current on the complete structure. Since the moment method cannot be directly used to solve for the currents on an infinite structure, one must find certain current forms to reduce the number of linear unknowns. A shown in [1], the current on the planar structure can be specified by

Manuscript received May 14, 1979; revised September 25, 1979. This work was supported in part by Contract N00014-78-C-0049 Between the Department of the Navy, Office of Naval Research, and the Ohio State University Research Foundation.

The authors are with the ElectroScience Laboratory, Department of Electrical Engineering, Ohio State University, Columbus, OH 43212.

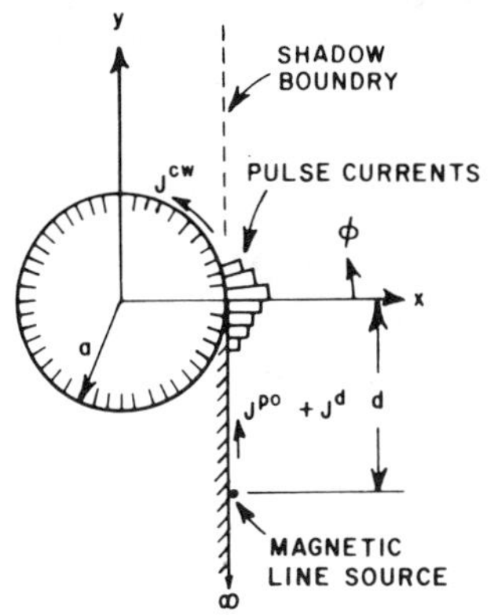

Fig. 1. Geometry associated with planar surface truncated by a circular cylinder.

$$J_p = \begin{cases} \sum_{n=1}^{N_p} \alpha_n p_n(y - y_n), & \text{for the pulse region,} \\ J^{po} + \epsilon_p \dfrac{e^{-jk\rho}}{\sqrt{\rho}}, & \text{for the GTD region,} \end{cases} \tag{1}$$

where p_n are orthogonal pulse functions with weight α_n, J^{po} is the known physical optics current associated with the incident field, and ϵ_p is an unknown complex excitation coefficient for the diffraction current. The current associated with the diffracted field has a range dependence given by $e^{-jk\rho}/\sqrt{\rho}$ where ρ is the distance from the junction to the current sample point. The current on the cylindrical structure can be specified by

$$J_c = \begin{cases} \sum_{n=1}^{N_c} \alpha_n p_n(\phi - \phi_n), & \text{for the pulse region,} \\ \epsilon_c g(\xi) e^{-jkl}, & \text{for the GTD region,} \end{cases} \tag{2}$$

where ϵ_c is an unknown complex excitation coefficient for the creeping current (i.e., $g(\xi)e^{-jkl}$). Note that g is a Fock function, $\xi = (ka/2)^{1/3}\phi$, and l is the length of the arc along the cylindrical surface from the junction to the current sample point. A definition of the function $g(\xi)$ is given in the Appendix. As will be shown later, the use of the creeping current $(g(\xi)e^{-jkl})$ is valid only when $(d/a) > (ka/2)^{-1/3}$, which is the case of interest where the source is not too close to the surface of the cylinder.

Using these linear current forms in the H-field integral equation, one obtains [3]

$$J + \hat{z} \cdot \nabla x \int_{s'} J \frac{H_0^{(2)}(k\,|\bar{\rho} - \bar{\rho}'\,|)}{4j} \, d\bar{l}' = -H_z^{\,i}, \tag{3}$$

where $H_z{}^i$ is the field produced by the source alone and J is the current induced on the perfectly conducting surface s'. Point matching the fields at the sample points, one obtains a system of linear equations which can then be solved using standard numerical techniques. The use of the H-field integral equation (3) assumes that the field vanishes on the shadow side of the planar surface. This assumption is valid when the radius of the cylinder is large. However, a cylinder with a radius of as small as 0.5 λ generally will suffice to yield accurate results.

Even though the unknown currents can be found using the previous procedure, one must integrate these currents on the infinite structure to obtain the field. It has been shown [4] that this situation can be partially alleviated by incorporating the classical half-plane solution into the present problem. Thus (3) becomes

$$\Delta J + \hat{z} \;\; \nabla x \int_{s'} \Delta J \frac{H_0{}^{(2)}(k\,|\bar{\rho}-\bar{\rho}'\,|)}{4j}\, d\bar{l}' = \pm H_z{}^d, \qquad (4)$$

where ΔJ is the total induced current (J) minus the current J^{hp} that would exist on a half-plane (i.e., the cylindrical truncation is removed) and $H_z{}^d$ is the half-plane diffracted field as defined in the geometrical theory of diffraction (GTD) sense [5]. In (4), $+H_z{}^d$ must be taken for matching points on the planar surface and $-H_z{}^d$ for matching points on the circular cylinder. When solving (4) for ΔJ one only needs to include the contribution from a finite section of the perfectly conducting surface which spans an angle of $(a/\lambda)^{-1/3}$ rad on the circular cylinder and a length of 3 λ on the planar surface, because both the Hankel function and ΔJ decrease as one moves away from the plate–cylinder junction.

It is found that the currents on the structure remain stable in terms of adding more unknowns about the diffraction point (the junction) and that it yields accurate results using only three pulse currents, each of 0.1 λ in width, on both sides of the junction. Fig. 2 shows an example of the total current induced on the structure. The ripple in the top curve clearly indicates the diffraction effect of the junction. As the radius of the circular cylinder becomes larger, the diffraction effect and thus the ripple diminish. Hence for $a > 10\ \lambda$ it is possible to approximate the current on the whole planar surface by J^{po} alone and the current on the circular cylinder by $\epsilon_c g(\xi) e^{-jkl}$ extending all the way to the junction without reducing the accuracy of the results.

The far-zone field scattered by the structure can be expressed as the sum of the field scattered by a half-plane plus that contributed by the surface current ΔJ. The former, i.e., the field scattered by a half-plane, is simply given in terms of Fresnel integrals as shown by Sommerfeld [6]. The latter consists of three components. The component contributed by the diffraction current in the GTD region of the planar surface can be integrated analytically, which gives another Fresnel integral. The contribution of the pulse currents about the junction can be numerically integrated very efficiently, as is done in any modest moment method problem. However, the evaluation of the third component which is the contribution of the creeping current on the circular cylinder is more complicated. A brute-force numerical integration in evaluating the third component is far too inefficient if the radius of the cylinder is large. Instead, the method described below is used for performing this task.

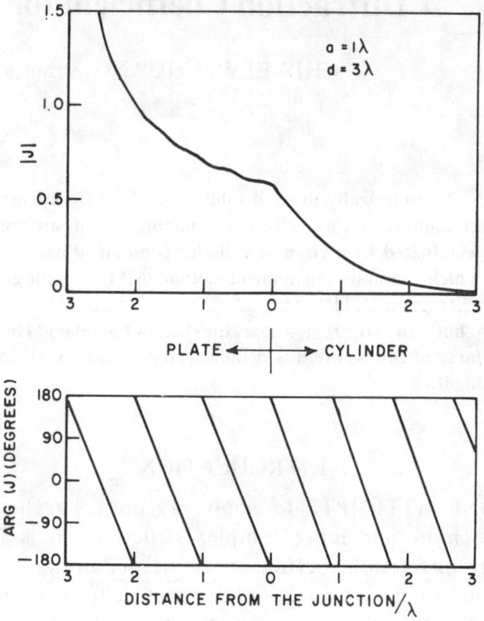

Fig. 2. Total current induced on plate–cylinder geometry.

First, consider the scattered field in the lit region $((-\pi/2) \leqslant \phi \leqslant (\pi/2))$ including the shadow boundary $(\phi = \pi/2)$ contributed by the creeping current. This is given by

$$H_z = \alpha \int_{\phi_0}^{2\pi} g(\xi)\cos(\phi'-\phi)\, e^{jka[\cos(\phi'-\phi)-\phi']}\, d\phi' \qquad (5)$$

where α is a complex constant and ϕ_0 is the starting point of the creeping current. The integration can be split into two parts:

$$\int_{\phi_0}^{2\pi} \{\cdot\}\, d\phi' = \int_{\phi_0}^{\phi_1} \{\cdot\}\, d\phi' + \int_{\phi_1}^{2\pi} \{\cdot\}\, d\phi'. \qquad (6)$$

It is found that if ϕ_1 is chosen equal to $\phi - \pi/2[1-(a/\lambda)^{-1/3}]$ or ϕ_0, whichever is larger, the second integral can be evaluated efficiently using integration by parts. While the first integral must still be evaluated numerically, the amount of calculation is greatly reduced because of the small range $(\phi_0 < \phi < \phi_1)$ it covers.

The approach to evaluating the field in the shadow region $((\pi/2) < \phi < (3\pi/2))$ is quite different. Before proceeding further let us digress for a moment to consider two related scattering and radiation problems. Shown in Fig. 3 is a plane wave incident upon a circular cylinder. A valid asymptotic high-frequency approximation for the current induced on the surface of the cylinder is given by [7] as

$$J_\phi = G(\phi) u \left(\phi + \frac{\pi}{2}\right) + G(-\phi - \pi) u \left(-\phi - \frac{\pi}{2}\right), \qquad (7)$$

where u is a unit step function,

$$G(x) = \begin{cases} -g(\xi_1) e^{-jkax}, & \text{if } x \geqslant 0, \\ -g(\xi_2) e^{-j(\xi_2{}^3/3 + ka \sin x)}, & \text{if } x \leqslant 0, \end{cases}$$

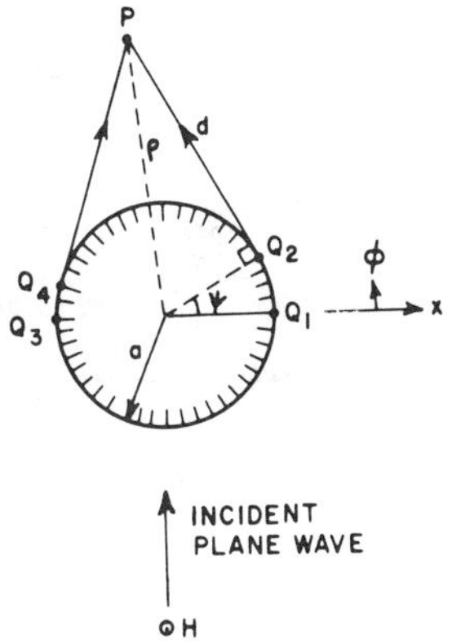

Fig. 3. Scattering of a plane wave by a circular cylinder.

and $\xi_1 = mx$, $\xi_2 = m \sin x$, $m = (ka/2)^{1/3}$. The first term on the right side of (7) is commonly viewed as a creeping wave traveling in the counter clockwise direction and the second term in the clockwise direction. Pathak *et al.* [8], [9] have given a ray interpretation of the field in the shadow region. The field at the point P exterior to the boundary layer due to the ray that creeps around the cylinder from Q_1 to Q_2 and then sheds energy toward the point P is

$$H_z = \sqrt{\frac{2}{k}}\, e^{-j(ka\psi+\pi/4)}\{F(kd\psi^2/2)/(2\psi\sqrt{\pi}) - mq^*(m\psi)\}\frac{e^{-jkd}}{\sqrt{d}}, \tag{8}$$

where the function $F(x)$ involves a Fresnel integral,

$$F(x) = 2j\sqrt{x}\, e^{jx}\int_{\sqrt{x}}^{\infty} e^{-j\tau^2}\, d\tau, \tag{9}$$

and the definition of the function $q(x)$ is given in the Appendix. The boundary layer refers to the region adjacent to the cylinder and is usually characterized by the condition [9]

$$mk(\rho - a) < 1, \tag{10}$$

which is equivalent to

$$d/a < 1/m. \tag{11}$$

Since the problem discussed in this paper excludes the boundary layer, the inequality $d/a > 1/m$ prevails. The surface current that radiates the ray field (8) can be identified with the first term of (7). A similar relation exists between the field due to the ray Q_3Q_4P and the second term of (7).

Another related problem concerns the radiation of a magnetic line source placed at the point Q_1 (Fig. 3). An asymptotic high-frequency approximation for the field at the point P due to the ray Q_1Q_2P is given by [7]

$$H_z = g(m\psi)e^{-jka\psi}\frac{e^{-jkd}}{\sqrt{d}}. \tag{12}$$

A similar expression can be obtained for the field due to the ray $Q_1Q_3Q_4P$.

Now return to the present problem of searching for an efficient method to evaluate the field in the shadow region of the structure shown in Fig. 1. Employing the reciprocity principle in (8), one can write down the total far-zone field in the shadow region as the product of ϵ_c, the excitation coefficient for the creeping current given in (2), and

$$H_{z_1} = -\sqrt{\frac{2}{k}}\, e^{-j(ka\psi+\pi/4)} \cdot \{F(kd\psi^2/2)/(2\psi\sqrt{\pi}) - mq^*(m\psi)\}\frac{e^{-jk\rho}}{\sqrt{\rho}}, \tag{13}$$

where ρ is the distance from the origin to a far-zone field point. This solution is found to agree excellently with that obtained by the accurate but inefficient brute-force numerical integration previously described, except near the shadow boundary. The discrepancy occurs near the shadow boundary because the creeping current given in (2) terminates at $\phi = 0$ and does not exist in the region $\phi < 0$. The termination of the creeping current at $\phi = 0$ results in additional diffracted fields which can be accounted for with a ficticious line source placed at the junction. Thus the total far-zone field in the shadow region can be expressed by

$$H_z = c_1 H_{z_1} + c_2 H_{z_2}, \tag{14}$$

where H_{z_2} is obtained from (12) by replacing d with ρ, and c_1 and c_2 are two unknown complex constants. This solution must satisfy two conditions from which the values of c_1 and c_2 can be determined. First, it must yield a field which is continuous across the shadow boundary, i.e., at $\phi = \pi/2$,

$$c_1 H_{z_1}\left(\frac{\pi}{2}\right) + c_2 H_{z_2}\left(\frac{\pi}{2}\right) = \lim_{\delta\to 0} H_z\left(\frac{\pi}{2} - \delta\right), \tag{15}$$

where the limit on the right side is approached from the lit region toward the shadow boundary. Second, it must blend smoothly with $\epsilon_c H_{z_1}$ in the deep shadow region,

$$c_1 H_{z_1}(\phi) + c_2 H_{z_2}(\phi) \approx \epsilon_c H_{z_1}(\phi), \tag{16}$$

where a good choice of ϕ is $\phi \geqslant (5\pi/6)$. Substituting (15) into (16) one obtains an equation with one unknown (either c_1 or c_2). One can obtain three such equations at $\phi = 5\pi/6$, $11\pi/12$, and π and solve for the unknown approximately in the sense of least squares. The other unknown can then be determined from (15). This approach provides an excellent result, as can be seen in Fig. 4.

With the total far-zone field specified by this hybrid approach, one can consider computing the desired diffraction coefficient. In the total ray optical solution for this problem the diffraction coefficient is the only unknown, and one can easily solve for its complex value through the following equations:

$$H_z(P) = H_z^i(P) + H_z^r(P) + H_z^d(P), \tag{17}$$

$$H_z^d(P) = H_z^i(Q)\cdot D\cdot\frac{e^{-jk|\overline{PQ}|}}{\sqrt{|\overline{PQ}|}}, \tag{18}$$

where H_z is the total field, H_z^i is the incident field, H_z^r is the reflected field, H_z^d is the diffracted field, P is the field point,

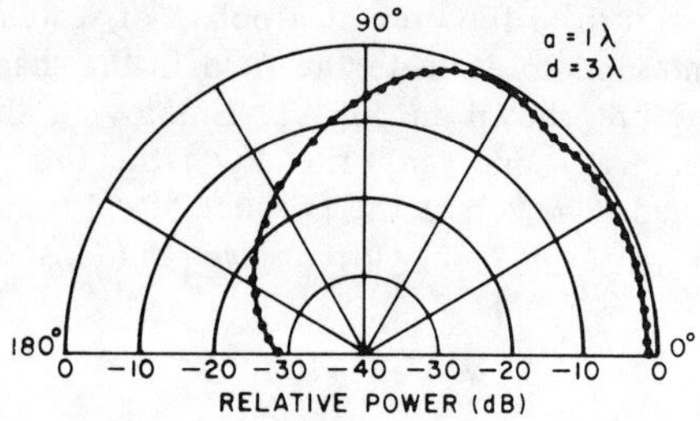

Fig. 4. Radiation pattern of plate-cylinder geometry. Solid curve is obtained using the very efficient method described in text, while dotted points are obtained using an accurate but inefficient brute-force numerical integration.

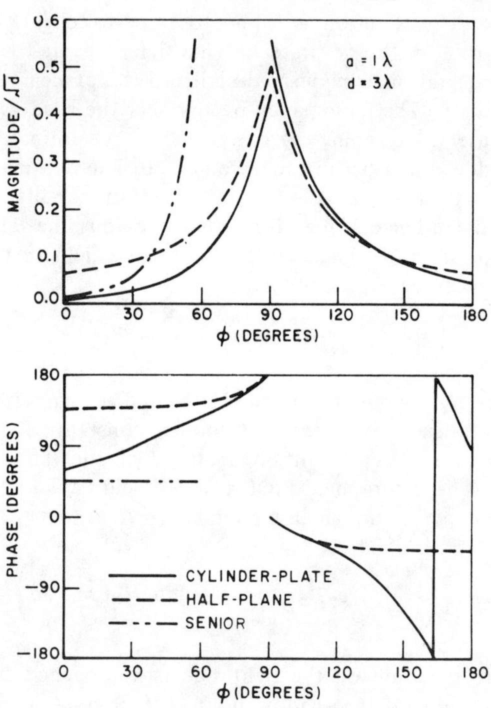

Fig. 5. Example of numerically derived diffraction coefficient compared against half-plane solution and Senior's solution.

Q is the diffraction point, and D is the diffraction coefficient. An example of this numerically derived diffraction coefficient is illustrated in Fig. 5. It is compared with the half-plane solution as well as a solution obtained by Senior [2]. One should note that Senior's solution is not valid in the transition region near the shadow boundary ($\phi = 90^\circ$). Consequently, one can only compare our results in the deep lit region ($\phi \ll 90^\circ$). Also note that the magnitude of the diffraction coefficient for the plate–cylinder geometry is discontinuous; whereas, that for a half-plane is continuous across the shadow boundary.

In order to verify this solution it has been applied to several interesting problems. In the first case it is used to determine the pattern of an axial slot mounted on a finite ground plane as shown in Fig. 6(a). Excellent agreement between measured and calculated results is observed in Fig. 6(b). Next it is used to analyze the E-plane pattern of the horn configuration shown in Fig. 7(a). Measured and calculated results are shown in Fig. 7(b). Again, excellent agreement is obtained in the region $0 \leqslant \phi \leqslant 120^\circ$. The measured pattern for the region ($120^\circ \leqslant \phi \leqslant 180^\circ$) exceeded the range capability such that one could not obtain accurate low-level radiation values.

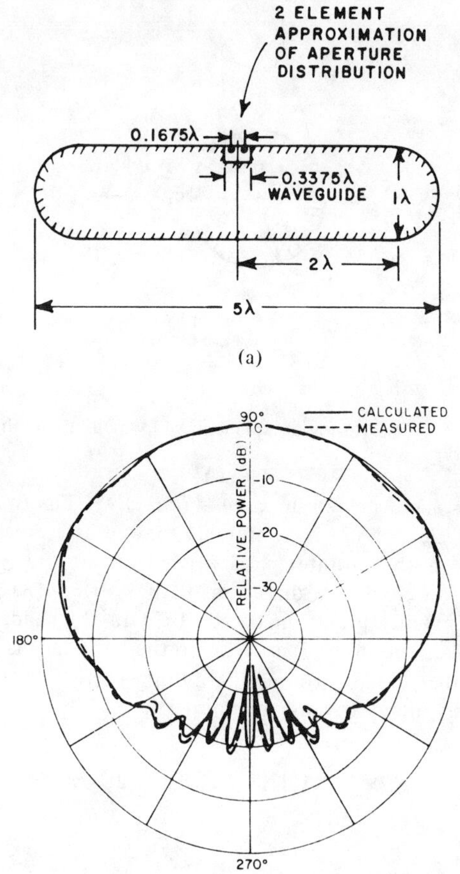

(a)

(b)

Fig. 6. (a) Axial waveguide mounted on a two-dimensional cylinder. (b) Principal plane radiation pattern.

The application of this diffraction coefficient solution to the design of a modified horn is very significant, because a properly designed horn terminated by curved surface sections offers much improved performance compared with conventional horns in terms of pattern, impedance, phase center, and bandwidth. For example, a measured conventional horn pattern is included in Fig. 7(b) for comparison. It is seen that both the sidelobe ripple and the backlobe are greatly reduced in the E-plane pattern of the modified horn. Details of designing such a modified horn will be published in the future.

III. CONCLUSION

Using the hybrid approach of Burnside *et al.* [1], this paper demonstrates that one can successfully solve for new diffraction coefficients. The need for such coefficients is apparent in that the horizons of the ray optical approach need to be expanded.

A two-dimensional problem is treated in this paper. However, since the solution is written in terms of a diffraction coefficient, it can be easily extended to the three-dimensional case as shown by Pathak and Kouyoumjian [10]. To simplify the analysis it was assumed that the curved surface is a circular cylinder. This can also be easily modified to treat the case where the curvature of the curved surface changes slowly, by utilizing the local properties of propagation, scattering, and diffraction of waves at high frequencies as done in the manner described in [11]. The two more serious restrictions imposed on the diffraction coefficient discussed in this paper are that

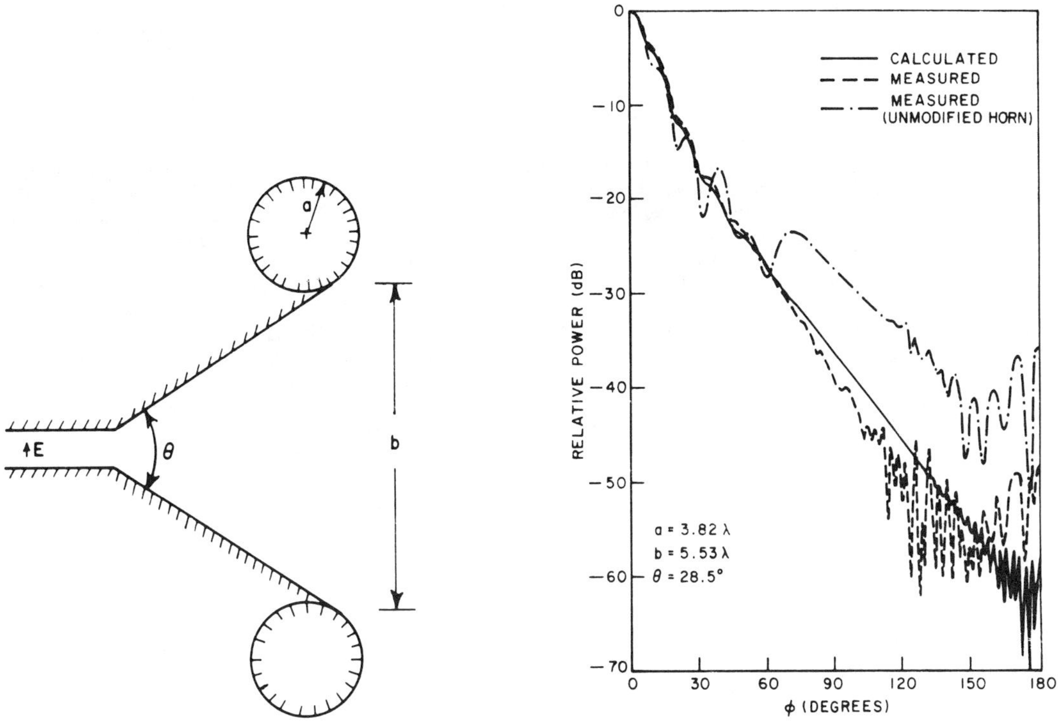

(a) (b)

Fig. 7. (a) Modified horn. (b) E-plane pattern.

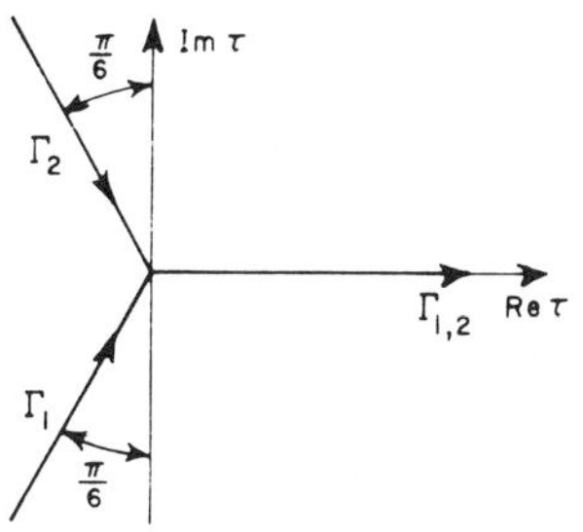

Fig. 8. Contours Γ_1 and Γ_2 in complex τ plane.

$a > 0.5\ \lambda$ and $d/a > (ka/2)^{-1/3}$. Relaxation of these two conditions can be made using the hybrid approach although such modifications are not considered here.

APPENDIX

The definitions of the functions $g(x)$ and $q(x)$ are given in the following:

$$g(x) = \frac{1}{\sqrt{\pi}} \int_{\Gamma_1} \frac{e^{-jx\tau}}{w_2'(\tau)}\, d\tau,$$

$$q^*(x) = \frac{1}{\sqrt{\pi}} \left\{ \frac{1}{2x} + \int_{\Gamma_1} \frac{V'(\tau)}{w_2'(\tau)} e^{-jx\tau}\, d\tau \right\}$$

where

$$V(\tau) = (1/2j)[w_1(\tau) - w_2(\tau)],$$

$$w_{1,2}(\tau) = \frac{1}{\sqrt{\pi}} \int_{\Gamma_{1,2}} e^{\tau t - t^3/3}\, dt,$$

and the contours of integration Γ_1 and Γ_2 are shown in Fig. 8. Both $g(x)$ and $q(x)$ can be easily computed using tabulated data and asymptotic forms as given in [7].

REFERENCES

[1] W. D. Burnside, C. L. Yu, and R. J. Marhefka, "A technique to combine the geometrical theory of diffraction and the moment method," *IEEE Trans. Antennas Propagat.*, vol. AP-23, pp. 551–558, July 1975.

[2] T. B. A. Senior, "The diffraction matrix for a discontinuity in curvature," *IEEE Trans. Antennas Propagat.*, vol. AP-20, pp. 326–333, May 1972.

[3] R. F. Harrington, *Field Computation by Moment Methods*. New York: MacMillan, 1968.

[4] C. L. Yu and W. D. Burnside, "A modified GTD–MM technique to analyze the diffraction by semi-infinite complex structures," presented at the IEEE AP-S Int. Symp., June 1977.

[5] R. G. Kouyoumjian and P. H. Pathak, "A uniform geometrical theory of diffraction for an edge in a perfectly conducting surface," *Proc. IEEE*, vol. 62, pp. 1448–1461, 1974.

[6] A. Sommerfeld, *Optics*. New York: Academic, 1954, pp. 245–265.

[7] N. A. Logan, "General research in diffraction theory," vols. 1 and 2, Missiles and Space Div., Lockheed Aircraft Corp., Rep. LMSD-288087, Dec. 1959.

[8] P. H. Pathak, R. J. Marhefka, and W. D. Burnside, "High frequency scattering by curved surfaces," Ohio State Univ. ElectroSci. Lab., Dep. Elec. Eng., Rep. 3390-5 (prepared under Contract N62269-72-C-0354 for Naval Air Development Center), June 1974.

[9] P. H. Pathak, "An asymptotic result for the scattering of a plane wave by a smooth convex cylinder," Ohio State Univ. ElectroSci. Lab., Dep. Elec. Eng., Rep. 784583-3 (prepared under Contract N62269-76-C-0554 for Naval Air Development Center), Mar. 1978.

[10] P. H. Pathak and R. G. Kouyoumjian, "The dyadic diffraction coefficient for a perfectly conducting wedge," Ohio State Univ. ElectroSci. Lab., Dep. Elec. Eng., Rep. 2183-4 (prepared under Contract AF 19(628)-5929 for Air Force Cambridge Research Laboratory), June 1970.

[11] ——, "An analysis of the radiation from apertures in curved surfaces by the geometrical theory of diffraction," *Proc. IEEE*, vol. 62, pp. 1438–1447, 1974.

Chiwei W. Chuang (M'76) was born in Taiwan, on August 16, 1944. He received the B.S. degree in electrical engineering from National Taiwan University Taipei, Taiwan, in 1967 and the M.S. and Ph.D. degrees in electrical engineering from the Ohio State University, Columbus, in 1970 and 1974, respectively.

Since 1974 he has been working at the Ohio State University ElectroScience Laboratory and has been engaged in research on radiation and scattering problems.

Dr. Chuang is a member of Sigma Xi.

Walter D. Burnside (S'69–M'72) was born in Youngstown, OH, on April 19, 1942. He received the B.S.E., M.S., and Ph.D. degrees in electrical engineering from Ohio State University, Columbus, the first two in 1968 and the last in 1972.

Since 1966 he has been associated with the ElectroScience Laboratory, Department of Electrical Engineering, Ohio State University, where he has been working on applications of wedge diffraction, the geometrical theory of diffraction, broad-band antennas, airborne antenna analysis, hybrid solution, and various high-frequency scattering solutions. From 1966 to 1968 he held a National Science Foundation Traineeship. He is presently a Research Scientist at the laboratory and an Adjunct Associate Professor with the Department of Electrical Engineering.

Dr. Burnside is a member of Tau Beta Pi, Eta Kappa Nu, Phi Eta Sigma, and Sigma Xi. In addition, he is a former Associate Editor of the IEEE TRANSACTIONS ON ANTENNAS AND PROPAGATION.

An Overview of the Hybrid MM/Green's Function Method in Electromagnetics

EDWARD H. NEWMAN, SENIOR MEMBER, IEEE

This paper presents an overview of a hybrid technique for solving electromagnetic radiation and scattering problems by combining the method of moments (MM) with a special Green's function. The method, commonly referred to as an MM/Green's function solution combines the ability of MM solutions to treat geometrically complex bodies with the accuracy and computational efficiency of Green's function solutions. As compared to a standard MM solution, the MM/Green's function solution reduces the number of unknowns, and thus reduces the computer storage requirements. In most, but not all cases, the CPU time for the MM/Green's function solution is considerably less than that for a standard MM solution. The basic formalism of the MM/Green's function solution is presented and contrasted to that of the standard MM solution. The example problem of TM scattering by a semicircular strip in the presence of a circular cylinder is solved by the MM, and by the MM/Green's function technique with a matrix, exact eigenfunction, and high-frequency Green's function.

I. Introduction

This paper will present an overview of a technique which combines the method of moments (MM) [1]-[3] and a Green's function [4] in the solution of electromagnetic radiation and scattering problems. The technique is termed an MM/Green's function solution and combines the flexibility of MM solutions for treating scatterers of complex geometry with the accuracy and computational efficiency of Green's function solutions.

A general problem in electromagnetics is that of finding the fields of known or impressed currents radiating in a vacuum or homogeneous medium in which is present one or more inhomogeneities or scatterers. Traditionally these problems have been solved by what can be called Green's function techniques. That is, one attempts to find either exact or approximate explicit formulas, known as Green's functions, for the fields of an infinitesimal current element radiating in the presence of the scatterers and/or boundaries. The fields of the known impressed current are then found by superposition, that is by integrating through the region of the current the vector dot product of the current distribution and the dyadic Green's function. Green's functions tend to be highly accurate and, as compared to MM solutions, in many cases they are computationally efficient in that they can be evaluated with a minimum of effort. The main limitation of Green's function solutions is that, except for a few simple shapes [4]-[6], the Green's functions are difficult to obtain, with each new geometry requiring a new analysis.

With the widespread availability of high-speed digital computers in the mid-1960s, a numerical technique known as the method of moments (MM) [1]-[3] began to gain popularity. Consider the standard MM solution for a problem involving two scatterers, which we term Scatterer 1 and Scatterer 2. The first step is to use the surface and/or volume equivalence theorems [1], [7] to replace *both* scatterers by free-space and equivalent currents. An exact integral equation for the equivalent currents is then formulated and involves the relatively simple free-space Green's function in its kernel. The unknown equivalent currents are then expanded as a finite sum of N_1 basis or expansion functions on Scatterer 1 and N_2 basis functions on Scatterer 2. N_1 and N_2 weighted averages of the integral equation are enforced on Scatterers 1 and 2, respectively. This transforms the integral equation into an order $N_1 + N_2$ matrix equation, which can be solved for the $N_1 + N_2$ coefficients in the expansion for the equivalent currents. The total fields radiated by the impressed currents in the presence of the two scatterers is the sum of the free-space fields of the impressed currents and the equivalent currents on Scatterers 1 and 2. The main advantages of the MM approach are that it is accurate and extremely versatile as to the geometries which it can treat. For example, while the exact Green's function solution for scattering by a dielectric cylinder appears only to be possible for the circular cross section [6], MM solutions are available for the scattering by dielectric cylinders of essentially arbitrary cross section [8], [9]. However, as compared to Green's function techniques, MM solutions usually require more computational effort.

In an MM/Green's function solution to the same two scatterer problem described above, Scatterer 1, but *not* Scatterer 2, is replaced by free-space and equivalent currents. An exact integral equation for the equivalent currents representing Scatterer 1 is formulated and will involve the rel-

Manuscript received August 31, 1987; revised January 14, 1988. The submission of this paper was encouraged after review of an advance proposal. This work was supported under Contract N00014-78-C-0049 between The Ohio State University Research Foundation and the Joint Service Electronics Program.

The author is with the Department of Electrical Engineering, The Ohio State University, Columbus, OH 43212, USA.

IEEE Log Number 8820076.

atively complicated Scatterer 2 (as compared to the free-space) Green's function in its kernel. The unknown equivalent currents on Scatterer 1 are expanded as a finite sum of N_1 basis functions. N_1 weighted averages of the integral equation are enforced on Scatterer 1. This transforms the integral equation into an order N_1 matrix equation, which can be solved for the N_1 coefficients in the expansion for the current on Scatterer 1. The total fields radiated by the impressed currents in the presence of the two scatterers is the sum of the fields of the impressed currents and the equivalent currents on Scatterer 1, both radiating in the presence of Scatterer 2.

As compared with the Green's function technique, the MM/Green's function method has the advantage that it is applicable to a much wider class of geometries. This is because the MM/Green's function method requires that we know the Green's function for either Scatterer 1 or 2, while the Green's function method requires that we can find the Green's function for the combination of Scatterer 1 in the presence of Scatterer 2. As compared to a standard MM solution, the MM/Green's function method has the advantages that the number of unknowns in the matrix equation is reduced from $N_1 + N_2$ to N_1, thus reducing the required computer storage. The disadvantages of the MM/Green's function technique is that since the kernel of its integral equation involves the relatively complicated Scatterer 2 Green's function, the evaluation of a typical element in the MM/Green's function matrix equation is more difficult and time-consuming than that in a standard MM solution. However, if Scatterer 2 is substantially larger than Scatterer 1, then $N_2 \gg N_1$, and the CPU time for the MM/Green's function solution is almost always considerably less than that of the standard MM solution.

Probably the most common use of MM/Green's function solutions has been for problems involving antennas in the presence of a plane dielectric interface, such as a flat earth. In this case, the antenna is Scatterer 1 and the dielectric half space, representing the earth, is Scatterer 2. Of the many papers in this area we mention the work of the group at Lawrence Livermore Laboratory [10]–[13], much of which has been incorporated into a user-oriented computer code [14], the solution of Chang and Wait for a vertical wire over the earth [15], the use of ray methods by Tiberio *et al.* to represent the energy reflected from the earth [16], and the use by Parhami and Mittra of an approximate but highly accurate representation of the exact half space Green's function [17]. Solutions for obstacles other than wires in the presence of a flat earth have also been presented in [18]–[20]. A closely related problem is that of printed circuit antennas. Here the printed circuit antenna is Scatterer 1 and the dielectric substrate is Scatterer 2. Again, there are a large number of papers, and we mention MM/Green's function solutions for microstrip antennas by Pozar [21], [22], Newman and Forrai [23], and Bailey and Deschamps [24]. In addition, Alexopoulos *et al.* have published several papers on printed circuit dipoles [25], [26]. A second related area is that of periodic arrays in multilayered dielectric slabs of which [27], [28] are representative.

Another popular use of MM/Green's function solutions has been for problems involving antennas or other obstacles in waveguides and cavities. In these cases, Scatterer 1 was the antenna and Scatterer 2 was the waveguide or cavity. Possibly the first MM/Green's function solution was Rao's [29], [30] analysis of a two-element Yagi array in a parallel plate waveguide in 1965. It is interesting that this work was done before Harrington [2] published his description of the MM in 1967. Other work involving wires in waveguides or cavities include Taylor's [31] and Tesche's [32] solutions for a wire in a parallel plate waveguide, and Seidel's [33] solution for a wire in a cavity. MM/Green's function solutions for posts in waveguides have been done by Leviatan *et al.* [34], Auda and Harrington [35], and Jarem [36]. Material bodies in waveguides have been analyzed by Wang [37], Omar and Schunemann [38], and Hsu and Auda [39]. Khac and Carson [40] have analyzed a slot in a waveguide and Thong [41] has analyzed waveguide discontinuities.

Several MM/Green's function solutions have been obtained for cases in which Scatterer 2 is one of the classical shapes which fits into a separable coordinate system, and thus has a well-known Green's function. Solutions for wires in the presence of spheres have been presented by Tesche, Neureuther, and Stovall [42]–[44] and by Butler and Keshavamurthy [45]. MM/Green's function solutions involving circular cylinders have been done by Misra and Chen [46], Steyskal [47], Karunaratne *et al.* [48], and Lamensedorf and Ting [49]. Wire antennas near the edge of a half-plane or a wedge have been analyzed by Pozar and Newman [50], [51]. Newman has analyzed a material cylinder of arbitrary cross section in the presence of a half-plane [52], [53], and Newman and Blanchard [54] have analyzed an impedance sheet in the presence of a parabolic cylinder. Section III of this paper presents an analysis for a semicircular strip in the presence of a circular cylinder.

Most of the above referenced works were natural MM/Green's function solutions in that Scatterer 2 was one of the few shapes for which an exact Green's function was known. Thiele and Newhouse [55] greatly expanded the range of problems which could be solved via the MM/Green's function technique by recognizing that the geometrical theory of diffraction (GTD) [56] could be used to obtain the asymptotic or high-frequency Green's function for a large class of electrically large bodies (Scatterer 2). This important special case of the MM/Green's function technique, where the GTD is used to evaluate the Green's function, is referred to as an MM/GTD solution. For problems involving electrically small and large parts, MM/GTD solutions are ideal since the MM can efficiently treat the electrically small part of the problem (Scatterer 1), while the GTD can efficiently treat the electrically large part (Scatterer 2). The earliest use of the MM with an asymptotic Green's function appears to be that of Green [57] for a monopole on the base of a large cone. Other applications include that by Awadalla and Maclean [58], [59] and by Marin and Catedra [60] to analyze a monopole on a plate. MM/GTD solutions for wires on curved surfaces have been presented by Ekelman and Thiele [61] and by Henderson and Thiele [62], [63]. Thiele and Chan [64] have used the MM/GTD to efficiently generate large amounts of frequency domain data so that the Fourier transform could be used to obtain time domain results. Also, Ko and Mittra [65] have developed an iterative MM/GTD solution which allows for a check as to how well the solution satisfies the boundary conditions.

Harrington and Mautz [66] and Glisson and Butler [67] showed that for a Scatterer 2 geometry which is so complex that it is not feasible to obtain a functional expression for its Green's function, an MM/Green's function solution is still possible. This is done using a numerical Green's function which is evaluated by an MM solution of Scatterer 2. Harrington and Mautz used the MM to find the numerical Green's function for a body of revolution. Glisson and Butler applied the MM/Green's function technique, with numerical Green's function, to the problem of a wire (Scatterer 1) in the presence of a body of revolution (Scatterer 2). The concept of a numerical Green's function is similar to Harker's [68] matrix partitioning method and to the matrix Green's function discussed in the next section.

There are several hybrid techniques, related to the MM/Green's function technique, which will not be treated in this paper. In particular, we mention the so-called GTD/MM technique which uses the MM to numerically evaluate an unknown diffraction coefficient [69]–[71]. Thiele *et al.* have developed a hybrid theory of diffraction (HTD) which uses a high-frequency approximation for the current on a portion of the body and the MM on the remainder [72], [73]. In a related technique, Mitschang and Wang incorporate the high-frequency currents into the MM solution for scattering by a body of revolution [74]. Azarbar and Shafai [75] efficiently analyzed large reflector antennas by employing an MM solution which solved for the difference between the actual and the physical optics currents on the surface of the reflector. Finally, Richmond [76] has developed a physical basis technique for an MM solution of a wide dielectric strip which involves only three basis functions.

Section II of this paper presents the basic formalism for a standard MM solution and an MM/Green's function solution. The two methods are compared and the concept of of a matrix Green's function is introduced. Section III solves the example problem of TM scattering by a perfectly conducting semicircular strip in the presence of a perfectly conducting circular cylinder by a standard MM solution, and by the MM/Green's function method using a matrix, exact eigenfunction, and GTD Green's function. Section IV briefly summarizes the main points of this paper.

II. MM and MM/Green's Function Formulations

A. Problem Geometry

This section will outline the MM and MM/Green's function solution for the scattering by two perfectly conducting scatterers. The basic geometry is shown in Fig. 1(a). Here we have the impressed currents, $(\boldsymbol{J}^i, \boldsymbol{M}^i)$, radiating in the presence of two perfectly conducting obstacles, termed Scatterers 1 and 2, and producing the total fields $(\boldsymbol{E}, \boldsymbol{H})$. In free-space the impressed currents radiate the incident fields $(\boldsymbol{E}^{i0}, \boldsymbol{H}^{i0})$. Scatterer 1 is enclosed by the surface S_1 which has outward unit normal $\hat{\boldsymbol{n}}_1$, and similar definitions apply to Scatterer 2. The ambient medium is free-space with permeability and permittivity (μ_0, ϵ_0). All fields and currents are time harmonic, with the $e^{j\omega t}$ time dependence suppressed. Here $\omega = 2\pi f$ is the radian frequency, λ denotes the free-space wavelength, and $k = 2\pi/\lambda$. For simplicity, we present the MM and MM/Green's function solution for perfectly conducting obstacles, and unless specifically noted, all general remarks apply equally well to problems involving penetrable scatterers.

The MM and MM/Green's function solutions can be notationally complex, involving many quantities, some of which differ only slightly in their meaning. To minimize confusions we introduce a notation based upon subscripts and

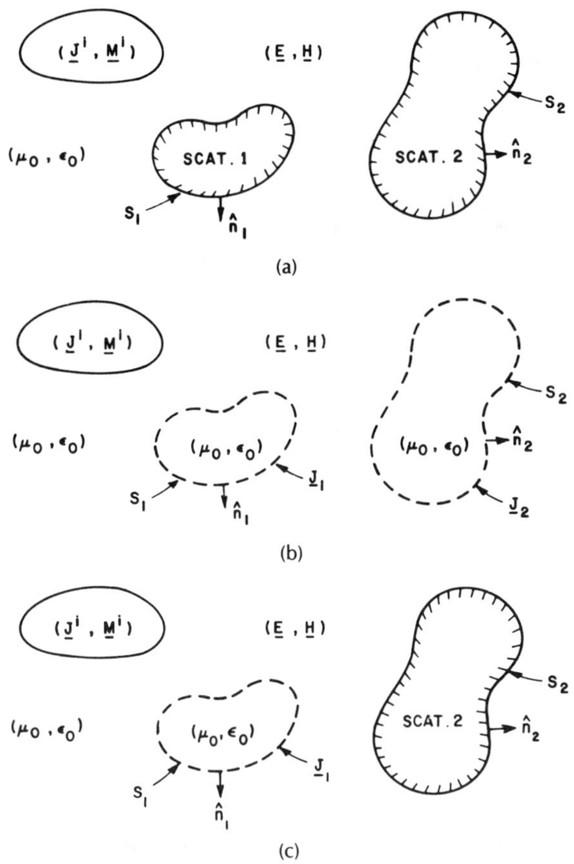

Fig. 1. (a) In the original problem the impressed currents radiate in the presence of two perfectly conducting scatterers. (b) In the standard MM solution, both scatterers are replaced by free-space and equivalent currents. (c) In the MM/Green's function solution, only Scatterer 1 is replaced by free-space and equivalent currents.

superscripts. For example, we use the notation of the superscripts ^{0},^{2}, or ^{S2}, to indicate a quantity computed from the fields of a current in free-space, in the presence of Scatterer 2, or the fields scattered from Scatterer 2, respectively. The subscripts $_1$ or $_2$ will be used to denote a quantity associated with Scatterers 1 or 2, respectively. The integer subscripts $_m$ and $_n$ are used to refer to MM basis function numbers, while the integer subscript $_p$ refers to term p in a cylindrical eigenfunction expansion. Finally, the superscript ^{i} denotes a quantity associated with the impressed currents. This notation is summarized in Table 1.

Table 1 Summary of Superscript and Subscript Notation

Notation	Implies a Quantity Associated with . . .
Superscript 0	the free-space fields of a current
Superscript 2	the fields in the presence of Scatterer 2
Superscript $S2$	the fields scattered from Scatterer 2
Subscript 1	Scatterer 1 or the N_1 modes on Scatterer 1
Subsript 2	Scatterer 2 or the N_2 modes on Scatterer 2
Superscript i	the impressed currents $(\boldsymbol{J}^i, \boldsymbol{M}^i)$
Subscript m or n	MM modes m or n
Subscript p	cylindrical eigenfunction term p

B. MM Solution

As illustrated in Fig. 1(b), the first step in the standard MM solution is to replace all matter, i.e., Scatterers 1 and 2, by free-space and by the equivalent electric surface current density $\boldsymbol{J}$ where

$$\boldsymbol{J} = \begin{cases} \boldsymbol{J}_1 = \hat{\boldsymbol{n}}_1 \times \boldsymbol{H} & \text{on } S_1 \\ \boldsymbol{J}_2 = \hat{\boldsymbol{n}}_2 \times \boldsymbol{H} & \text{on } S_2. \end{cases} \tag{1}$$

Note that $\boldsymbol{J}$ is unknown since $\boldsymbol{H}$ is unknown, however, $\boldsymbol{J}$ will be found in the MM solution. By definition, the scattered fields are the difference between the total fields and the incident fields. In the equivalent problem of Fig. 1(b), the scattered fields are the free-space fields of $\boldsymbol{J}$, denoted ($\boldsymbol{E}^0(\boldsymbol{J})$, $\boldsymbol{H}^0(\boldsymbol{J})$). In this case, the total fields are

$$\boldsymbol{E} = \boldsymbol{E}^{i0} + \boldsymbol{E}^0(\boldsymbol{J}) \tag{2}$$

$$\boldsymbol{H} = \boldsymbol{H}^{i0} + \boldsymbol{H}^0(\boldsymbol{J}). \tag{3}$$

Note that we use the notation of superscript ^{0} to imply the free-space fields of a current distribution.

An electric field integral equation (EFIE) for $\boldsymbol{J}$ is obtained by enforcing the boundary condition that the total electric field tangential to S_1 and S_2 must vanish on the surfaces S_1 and S_2. Then from (2)

$$-\boldsymbol{E}^0(\boldsymbol{J}) = \boldsymbol{E}^{i0} \text{ tangential components on } S_1 \text{ and } S_2. \tag{4}$$

Equation (4) is an integral equation for $\boldsymbol{J}$ since

$$\boldsymbol{E}^0(\boldsymbol{J}) = \iint_{S_1+S_2} \boldsymbol{J} \cdot \overline{\boldsymbol{G}}^0 \, ds' \tag{5}$$

where $\overline{\boldsymbol{G}}^0$ is the dyadic free-space Green's function [4]. Basically $\overline{\boldsymbol{G}}^0$ is a 3×3 matrix which gives the free-space vector electric field of an arbitrarily oriented infinitesimal electric current element. For example, in the rectangular coordinate system, the i, j element of $\overline{\boldsymbol{G}}^0$ is the $\hat{x}$, $\hat{y}$, or $\hat{z}$ polarization (for $j = 1, 2, 3$) of the free-space electric field due to an infinitesimal electric current element with dipole orientation $\hat{x}$, $\hat{y}$, or $\hat{z}$ (for $i = 1, 2, 3$).

Equation (4) must now be solved for $\boldsymbol{J}$. The first step is to expand $\boldsymbol{J}$ in terms of an appropriate set of expansion, basis, or interpolating functions. Thus we write

$$\boldsymbol{J} = \sum_{n=1}^{N} I_n \boldsymbol{J}_n \tag{6}$$

where the $\boldsymbol{J}_n$ are a set of N known linearly independent basis functions defined on S_1 and S_2, and the I_n are a set of N unknown coefficients ($n = 1, 2, \cdots, N$). Denoting $\boldsymbol{E}^0(\boldsymbol{J}_n)$ as the free-space electric field of $\boldsymbol{J}_n$, and substituting (6) into (4) yields

$$-\sum_{n=1}^{N} I_n \boldsymbol{E}^0(\boldsymbol{J}_n) = \boldsymbol{E}^{i0}$$

$$\text{tangential components on } S_1 \text{ and } S_2. \tag{7}$$

Now define a set of N linearly independent vector weighting functions, $\boldsymbol{w}_m$ ($m = 1, 2, \cdots, N$), on and tangential to the surfaces S_1 and S_2. If in (6) the first N_1 expansion functions are on S_1 and the last N_2 are on S_2 ($N = N_1 + N_2$), then the first N_1 weighting functions are on S_1 and the last N_2 are on S_2. Taking the inner product of both sides of (7) with each of the N weighting functions converts (7) into an $N \times N$ system of simultaneous linear algebraic equations which can be written in matrix form as

$$[Z^0][I] = [V^0] \tag{8}$$

where $[Z^0]$ is the $N \times N$ impedance matrix, $[V^0]$ is the length N right-hand-side or voltage vector, and $[I]$ is the length N solution or current vector which contains the $N = N_1 + N_2$ coefficients in the original expansion for the current in (6). Typical elements of $[Z^0]$ and $[V^0]$ are given by

$$Z^0_{mn} = -\iint_m \boldsymbol{E}^0(\boldsymbol{J}_n) \cdot \boldsymbol{w}_m \, ds$$

$$= -\iint_m \left[\iint_n \boldsymbol{J}_n \cdot \overline{\boldsymbol{G}}^0 \, ds' \right] \cdot \boldsymbol{w}_m \, ds \tag{9}$$

$$V^0_m = \iint_m \boldsymbol{E}^{i0} \cdot \boldsymbol{w}_m \, ds. \tag{10}$$

As indicated in (9) and (10) the ds integrals are over the surface of the mth weighting function, while the ds' integrals are over the surface of the nth expansion function.

Assuming N_1 basis functions are employed on Scatterer 1, and N_2 basis functions on Scatterer 2, Fig. 2 shows the matrix equation (8) partitioned into blocks, corresponding to the basis functions on Scatterers 1 and 2. Here $[Z^0_{11}]$ is the

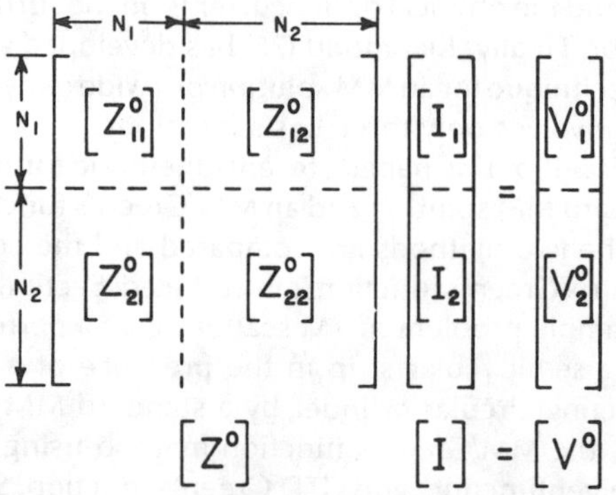

Fig. 2. The standard MM matrix equation is partitioned into blocks related to basis functions on Scatterers 1 and 2.

$N_1 \times N_1$ block of $[Z^0]$ which represents coupling between basis functions on Scatterer 1, $[Z^0_{12}]$ is the $N_1 \times N_2$ block of $[Z^0]$ which represents coupling between basis functions on Scatterer 1 and basis functions on Scatterer 2, $[I_1]$ is the first N_1 elements of $[I]$ containing the coefficients of the expansion functions on Scatterer 1, etc. Note that $[Z^0_{11}]$ and $[Z^0_{22}]$ are the MM impedance matrices for isolated Scatterer 1 or Scatterer 2, respectively, in free-space. Similarly, $[V^0_1]$ and $[V^0_2]$ are the voltage vectors for isolated Scatterer 1 or Scatterer 2, respectively, in free-space. As shown in Table 1, the notation being used is that the subscript $_1$ indicates a quantity associated with the first N_1 basis functions on Scatterer 1, while the subscript $_2$ indicates a quantity associated with the N_2 basis functions on Scatterer 2.

Using standard techniques of matrix algebra, (8) can now be solved for the current vector $[I]$ which when substituted into (6) provides an approximation to the current on the scatterers. In principle, as N is increased, a well-formulated MM solution approaches the exact solution. For problems which are formulated in terms of an unknown surface current density, the number of unknowns, N, which must be retained in the MM solution is proportional to the electrical surface area $(S_1 + S_2)/\lambda^2$, with values from 24 to 100 basis functions per λ^2 being typical.

The CPU time necessary to compute $[Z^0]$ is proportional to N^2. The CPU time to solve (8) tends to be proportional to N^2 for small to modest N, but to N^3 for large N. Thus, if N is not too large, the computer CPU time and storage requirements needed to carry out the MM solution vary roughly as N^2 or as f^4. As the frequency is increased N must be increased, and at some point the computer resources needed to solve the problem become prohibitive. For this reason, MM solutions are often referred to as low-frequency solutions, applicable to bodies which are not too large in terms of a wavelength.

C. *MM/Green's Function Solution*

The fundamental difference between the MM and MM/Green's function solutions is that in the MM/Green's function solution some, but not all, of the matter is replaced by free-space and equivalent currents. In the equivalent problem of Fig. 1(c), Scatterer 1, but not Scatterer 2, is replaced by free-space and equivalent currents. Then, in contrast to (1), the equivalent currents for the MM/Green's function solution are

$$\mathbf{J} = \mathbf{J}_1 = \hat{\mathbf{n}}_1 \times \mathbf{H} \text{ on } S_1 \text{ only.} \tag{11}$$

In Fig. 1(c), the total fields are the superposition of the fields of the impressed currents and $\mathbf{J}_1$, both radiating in the presence of Scatterer 2. Then denoting $(\mathbf{E}^{i2}, \mathbf{H}^{i2})$ and $(\mathbf{E}^2(\mathbf{J}_1), \mathbf{H}^2(\mathbf{J}_1))$ as the fields of the impressed currents and $\mathbf{J}_1$, respectively, radiating in the presence of Scatterer 2, the total fields are

$$\mathbf{E} = \mathbf{E}^{i2} + \mathbf{E}^2(\mathbf{J}_1) \tag{12}$$

$$\mathbf{H} = \mathbf{H}^{i2} + \mathbf{H}^2(\mathbf{J}_1). \tag{13}$$

As indicated in Table 1, the notation of superscript ^{2} implies the fields of a current distribution in the presence of Scatterer 2. Inherent in the MM/Green's function formulation is that we know (to an acceptable accuracy) the Green's function for Scatterer 2. That is, we know the near zone fields of a current element in the vicinity of Scatterer 2.

The EFIE is obtained by enforcing the boundary condition that the total electric field tangential to S_1 must vanish on the surface S_1. Then from (12)

$$-\mathbf{E}^2(\mathbf{J}_1) = \mathbf{E}^{i2} \text{ tangential components on } S_1 \text{ only.} \tag{14}$$

Note that it is not necessary to explicitly satisfy the boundary conditions on Scatterer 2, since it is implicit in (14) that all sources radiate in the presence of Scatterer 2, and by definition these fields satisfy the boundary conditions on Scatterer 2. Equation (14) is an integral equation for $\mathbf{J}_1$ since

$$\mathbf{E}^2(\mathbf{J}_1) = \iint_{S_1} \mathbf{J}_1 \cdot \overline{\mathbf{G}}^2 \, ds' \tag{15}$$

where $\overline{\mathbf{G}}^2$ is the dyadic Green's function for Scatterer 2. $\overline{\mathbf{G}}^2$ is similar to $\overline{\mathbf{G}}^0$ in (5), except the electric current element is radiating in the presence of Scatterer 2.

Equation (14) must now be solved for $\mathbf{J}_1$. We begin by expanding $\mathbf{J}_1$ as

$$\mathbf{J}_1 = \sum_{n=1}^{N_1} I_n \mathbf{J}_n. \tag{16}$$

This expansion is identical to that in (6), except that we include only the first N_1 basis functions on Scatterer 1. Substituting (16) into (14) yields

$$-\sum_{n=1}^{N_1} I_n \mathbf{E}^2(\mathbf{J}_n) = \mathbf{E}^{i2}$$

$$\text{tangential components on } S_1 \text{ only} \tag{17}$$

where $\mathbf{E}^2(\mathbf{J}_n)$ is the electric field of $\mathbf{J}_n$, located on S_1, and radiating in the presence of Scatterer 2. Taking the inner product of (17) with a set of N_1 weighting functions located on S_1 reduces (17) to the matrix equation

$$[Z^2][I_1] = [V^2] \tag{18}$$

where $[Z^2]$ is the $N_1 \times N_1$ impedance matrix, $[I_1]$ is the length N_1 current vector containing the unknown coefficients in the expansion for $\mathbf{J}_1$, and $[V^2]$ is the N_1 element voltage vector. Typical terms of $[Z^2]$ and $[V^2]$ are given by

$$Z^2_{mn} = -\iint_m \mathbf{E}^2(\mathbf{J}_n) \cdot \mathbf{W}_m \, ds$$

$$= -\iint_m \left[\iint_n \mathbf{J}_n \cdot \overline{\mathbf{G}}^2 \, ds' \right] \cdot \mathbf{W}_m \, ds \tag{19}$$

$$V^2_m = \iint_m \mathbf{E}^{i2} \cdot \mathbf{W}_m \, ds. \tag{20}$$

The above expressions for the elements in the MM/Green's function matrix equation are identical to those for the standard MM solution (9), (10), except that we replace the free-space fields with the fields in the presence of Scatterer 2.

Equation (18) can now be solved for $[I_1]$, which when substituted into (16) provides an approximation to the current on Scatterer 1. The total fields are then found using (12) and (13).

The fields of any current distribution in the presence of Scatterer 2 can always be written as the sum of the free-space fields of the current plus a correction term, corresponding to the fields scattered from Scatterer 2. Thus we can write

$$\overline{\mathbf{G}}^2 = \overline{\mathbf{G}}^0 + \overline{\mathbf{G}}^{S2}$$

$$\mathbf{E}^2(\mathbf{J}_n) = \mathbf{E}^0(\mathbf{J}_n) + \mathbf{E}^{S2}(\mathbf{J}_n)$$

$$\mathbf{E}^{i2} = \mathbf{E}^{i0} + \mathbf{E}^{iS2} \tag{21}$$

where the superscript ^{S2} implies the fields scattered from Scatterer 2. Using this notation, (18) can split into its free-space and scattered components as

$$([Z^0_{11}] + [Z^{S2}])[I_1] = [V^0_1] + [V^{S2}_1] \tag{22}$$

where the $[Z^0_{11}]$ and $[V^0_1]$ are the Scatterer 1 matrix elements in the absence of Scatterer 2 as defined by (9), (10), and Fig. 2. $[Z^{S2}]$ and $[V^{S2}]$ are perturbation terms which account for the fields scattered from Scatterer 2, and are given by (19), (20), with the superscript ^{2} replaced by ^{S2}.

In some cases, it is advantageous to split the impedance matrix into its free-space and scattered components, as shown in (22). This is a result of the fact that it is almost always easier and faster to accurately find the elements of $[Z^0_{11}]$ than those of $[Z^2]$ or $[Z^{S2}]$. In many problems the free-space components will be much larger than the scattered components. In this case, it is computationally efficient to first evaluate the elements of $[Z^0_{11}]$ with high accuracy and relatively little CPU time, and then expend as much CPU time as is reasonable to evaluate the elements of $[Z^{S2}]$. If, for example, a typical element of $[Z^{S2}]$ is about 10 percent as large as the corresponding element in $[Z^0_{11}]$, then a 10-percent error in evaluating that element in $[Z^{S2}]$ will result in only about a 1-percent error in the corresponding element in $[Z^2]$. Thus one can save considerable CPU time by requiring only modest accuracy in the computation of those elements of $[Z^2]$. The split shown in (22) is especially advantageous when treating the singularity in the self-impedance (i.e., $m = n$) terms, or other terms where the expansion and weighting functions physically overlap. Since this singularity is entirely in $[Z^0_{11}]$, the split in (22) allows one to analyze the singularity with the relatively simple free-space Green's function, and not the more complex Scatterer 2 Green's function. An exception would be if Scatterers 1 and 2 touch at an edge, or corner of Scatterer 2, in which case $\overline{G}^{S2}$ would be singular at the attachment line or point, respectively. Treatment of this singularity in $\overline{G}^{S2}$ is highly problem dependent, and is beyond the scope of this tutorial paper. The split shown in (22) is also natural for the MM/GTD method where the ray picture of the Scatterer 2 Green's function expresses the fields as the sum of an incident ray plus several scattered or diffracted rays. A further practical advantage of the split in (22) is that computer programs for the evaluation of $[Z^0_{11}]$ and $[V^0_1]$ may already be available. We caution that there are instances where certain elements of $[Z^0_{11}]$ are almost the negative of the corresponding elements of $[Z^{S2}]$, making the split in (22) a numerically unstable method of computing $[Z^2]$ [52], [77].

D. A Comparison Between the MM and MM/Green's Function Solutions

First consider the basic accuracy of the MM and MM/Green's function solutions. The basic integral equations on which the MM and MM/Green's function solutions are based are both exact. If both methods use the same N_1 expansion and weighting functions on Scatterer 1, and the exact Green's function is used in the MM/Green's function solution, then in the limit as N_1 and $N_2 \to \infty$, both solutions should in principle yield the exact result. For the case of finite N_1, but in the limit as $N_2 \to \infty$, both solutions should produce the same approximate result. In particular, the N_1 elements of $[I_1]$ in (18) should be identical to the first N_1 elements of $[I]$ in (8). However, in practice N_1 and N_2 will both be finite, and the two solutions will yield somewhat different results. The most accurate method is the one with the most accurate model of Scatterer 2. The MM solution employs an MM model of Scatterer 2, while the MM/Green's function method employs the Scatterer 2 Green's function. If the exact Scatterer 2 Green's function is used (and we assume that it can be evaluated exactly), then the MM/Green's function solution will be the most accurate. However, if an approximate Green's function is used, then the MM/Green's function solution may not be as accurate as the MM solution. In this case, the relative accuracy will depend upon the accuracy of the approximate Scatterer 2 Green's function and on the number of basis functions N_2 used in the MM model of Scatterer 2.

Next consider the CPU time and storage for the MM and MM/Green's function solutions. The MM solution for the current on Scatterers 1 and 2 involves an order $N = N_1 + N_2$ matrix equation, while the MM/Green's function solution involves an order $N = N_1$ matrix equation. Since N^2 storage locations are required for the impedance matrix ($(N^2 + N)/2$ for a symmetric matrix), the MM/Green's function solution requires less storage, especially if $N_2 \gg N_1$. By the same argument, the matrix solution time is less for the MM/Green's function solution. The CPU time to compute a single element in an MM/Green's function impedance matrix generally requires more CPU time than a corresponding element in a MM impedance matrix. This is because the elements in the MM/Green's function impedance matrix involves the relatively complicated Scatterer 2 Green's function. In both cases, the total CPU time to compute or fill the impedance matrix is proportional to N^2. However, if $N_2 \gg N_1$, then the CPU time to compute the N_1^2 elements in the MM/Green's function impedance matrix will usually be less than the CPU time to compute the $(N_1 + N_2)^2$ elements in the MM impedance matrix.

E. The Matrix Green's Function

As mentioned above, the MM and MM/Green's function solutions produce essentially the same result for $[I_1]$. It follows that there must be a very close relationship between the MM matrix equation (8) and the MM/Green's function matrix equation (18) or (22). We will now obtain this relationship by casting (8) into an order N_1 matrix equation for $[I_1]$, and then comparing the result to (22).

As illustrated in Fig. 2, the MM matrix equation (8) is equivalent to the coupled matrix equations

$$[Z^0_{11}][I_1] + [Z^0_{12}][I_2] = [V^0_1] \tag{23}$$

$$[Z^0_{21}][I_1] + [Z^0_{22}][I_2] = [V^0_2]. \tag{24}$$

Left multiplying (24) by $[Z^0_{12}][Z^0_{22}]^{-1}$ yields

$$[Z^0_{12}][Z^0_{22}]^{-1}[Z^0_{21}][I_1] + [Z^0_{12}][I_2] = [Z^0_{12}][Z^0_{22}]^{-1}[V^0_2]. \tag{25}$$

If we now subtract (25) from (23), the $[I_2]$ terms cancel leaving the order N_1 matrix equation for $[I_1]$

$$([Z^0_{11}] - [Z^0_{12}][Z^0_{22}]^{-1}[Z^0_{21}])[I_1] = [V^0_1] - [Z^0_{12}][Z^0_{22}]^{-1}[V^0_2]. \tag{26}$$

After (26) has been solved for $[I_1]$, then $[I_2]$ can be obtained from (24) as

$$[I_2] = [Z^0_{22}]^{-1}[V^0_2] - [Z^0_{22}]^{-1}[Z^0_{21}][I_1]. \tag{27}$$

Comparison with (22) shows that (26) is in the form of an MM/Green's function solution, where the perturbation terms, caused by Scatterer 2, are given by

$$[Z^{S2}] = -[Z^0_{12}][Z^0_{22}]^{-1}[Z^0_{21}] \tag{28}$$

$$[V^{S2}] = -[Z^0_{12}][Z^0_{22}]^{-1}[V^0_2]. \tag{29}$$

$[Z^{S2}]$ and $[V^{S2}]$ in (28), (29) are not identical to those in (22), since different models are being used for Scatterer 2. In principle, they will become identical if the exact Green's function is used to compute $[Z^{S2}]$ and $[V^{S2}]$ in (22), and if $N_2 \to \infty$ basis functions are used to model Scatterer 2 in the evaluation of (26). For finite N_2, the corresponding elements in these equations will be approximately the same. In either case, (26) can be interpreted as an MM/Green's function solution for the current on Scatterer 1. However, the Green's function is in the form of a matrix rather than a functional expression. Thus, we can loosely refer to $[Z^0_{12}][Z^0_{22}]^{-1}[Z^0_{21}]$ as a matrix Green's function for Scatterer 2, since it is a complete description of the scattering of electromagnetic waves from Scatterer 2 due to the source currents modeling Scatterer 1. The advantage of the matrix Green's function is that it can be obtained for any Scatterer 2 geometry for which a standard MM solution for Scatterer 2 in free-space is available. Employing (26), with the matrix Green's function, is similar to Glisson and Butler's use of a numerical Green's function to analyze a wire antenna in the presence of a body of revolution (BOR) [67] and to Harker's matrix partitioning method [68].

The following physical interpretation can be made for the matrix terms in (26):

- $[Z_{21}^0][I_1]$ is the electric field incident upon Scatterer 2 caused by the current on Scatterer 1 radiating in free-space.
- $[Z_{22}^0]^{-1}[Z_{21}^0][I_1]$ is the current induced on Scatterer 2 by the above incident field.
- $-[Z^{S2}] = [Z_{12}^0][Z_{22}^0]^{-1}[Z_{21}^0][I_1]$ is the electric field incident upon Scatterer 1 caused by the above current on Scatterer 2 radiating in free-space, i.e., this term accounts for the coupling between the currents on Scatterers 1 and 2.
- $[Z_{22}^0]^{-1}[V_2^0]$ is the current induced on Scatterer 2 by the original incident field, $(\boldsymbol{E}^{i0}, \boldsymbol{H}^{i0})$.
- $[V^{S2}] = [Z_{12}^0][Z_{22}^0]^{-1}[V_2^0]$ is the electric field incident upon Scatterer 1 caused by the above current on Scatterer 2, i.e., this term accounts for the perturbation in the electric field incident upon Scatterer 1 caused by the presence of Scatterer 2.

Although the MM and MM/Green's function solution with matrix Green's function are algebraically equivalent, there are times when the MM/matrix Green's function solution can reduce the CPU time. A practical example would be a parameter study to find the length and location of a dipole on an airplane in order to obtain a prescribed input impedance and/or radiation pattern. To solve this design problem we analyze dipoles of various lengths and locations on the airplane, until a length and location is found which meets the pattern and impedance specifications. In the MM model of the dipole and the aircraft, the number of unknowns needed to model the airplane would far exceed that needed for the dipole. If we let the dipole be Scatterer 1 and the airplane Scatterer 2, this problem is an excellent candidate for an MM/Green's function solution since $N_2 \gg N_1$. Using thin wire and surface patch modeling techniques, a standard MM solution for the dipole in the presence of the airplane is available [78]. However, this $(N_1 + N_2)$ basis function MM solution would be time-consuming, especially since it would have to be repeated for many dipole lengths and locations. The MM/Green's function solution, with matrix Green's function, can dramatically reduce the CPU time for the second and subsequent runs in the parameter study by the following four-step procedure:

1) Begin by computing $[Z_{22}^0]$, the MM impedance matrix for the airplane in free-space. Then compute $[Z_{22}^0]^{-1}$ (and store it in the locations originally used for $[Z_{22}^0]$). For large N_2, this can be a very time-consuming process. However, it is only done once, and the remainder of the computation is much faster.

2) Next a dipole location and length is chosen and the matrices $[Z_{11}^0]$, $[Z_{12}^0]$, and $[Z_{21}^0]$ are computed. The CPU time to compute these three matrices will be much less than that to compute $[Z_{22}^0]$, since $N_2 \gg N_1$. The vectors $[V_1^0]$ and $[V_2^0]$ are also computed.

3) Equation (26) is solved for $[I_1]$, which then is inserted into (27) to find $[I_2]$. These operations are relatively fast, as compared to computing and inverting $[Z_{22}^0]$, since they involve only matrix multiplications and the solution of an order N_1 matrix equation.

4) With the current on the dipole and airplane known, the computation of the dipole input impedance and far-zone radiation patterns is straightforward and fast. If the pattern and input impedance meet specifications, then stop; if not, go back to step 2.

The advantage of the above procedure is that the most time-consuming part, i.e., the computation and inversion of $[Z_{22}^0]$, is done only once, regardless of the number of dipole lengths and locations which are analyzed. Steps 2–4 may be fast enough to be done interactively.

As an aside, if the dipole was a monopole which contacted the aircraft, then the MM model for the monopole on the aircraft would contain an attachment mode at the wire to surface junction point [78], [79]. This attachment mode is used to enforce continuity of current at the wire/surface junction and involves currents on the wire and on the aircraft surface. In the MM/matrix Green's function solution, the attachment mode should be considered as part of the wire or Scatterer 1.

III. MM and MM/Green's Function Examples

This section presents MM and MM/Green's function solutions to the two-dimensional problem of TM scattering by a perfectly conducting semicircular strip in the presence of a perfectly conducting circular cylinder. The strip will be Scatterer 1, and the cylinder will be Scatterer 2. As illustrated in Fig. 3, the cylinder has radius a, and the strip is at radius b and extends over the angular range $-\phi_s < \phi \leq \phi_s$.

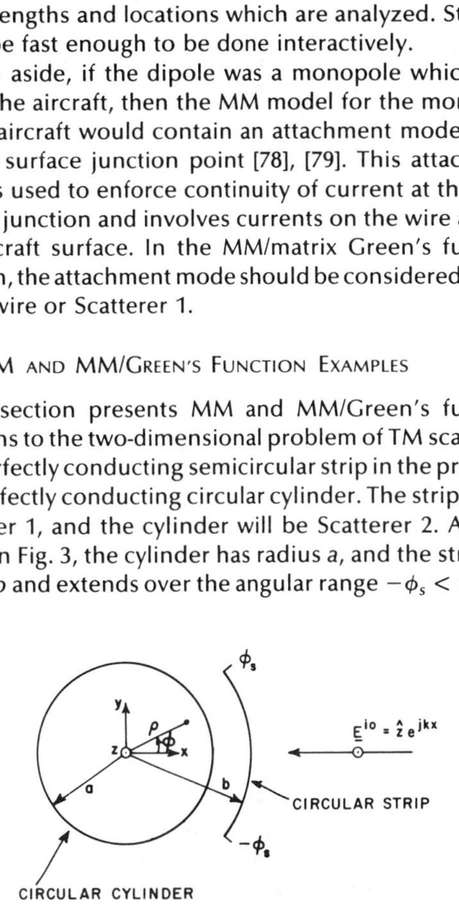

Fig. 3. Geometry for a TM plane wave incident upon a perfectly conducting semicircular strip in the presence of a perfectly conducting circular cylinder.

The incident field is the plane wave

$$E^{i0} = \hat{z}e^{jkx} \tag{30}$$

incident from the $+x$ axis. For this problem, all electric fields and currents will be $\hat{z}$ polarized. Thus for simplicity, in this section we will drop the vector notation, and the $\hat{z}$ component is understood. The solution to this problem is presented using:

1) a straight MM solution,
2) an MM/Green's function solution with a matrix Green's function,
3) an MM/Green's function solution with an exact Green's function,
4) an MM/Green's function solution with a GTD (geometrical optics) Green's function.

All computer CPU times are for the VAX 8550, which is about five times faster than a VAX 11/780. Also, by CPU time it is meant total CPU time, i.e., matrix fill and solution time. Although the matrices for this simple example have the Toeplitz property, we did not use this property in our solution so that the CPU times would be more typical of most problems which do not have Toeplitz matrices.

A. MM and MM/Matrix Green's Function Solution

The MM solution for TM scattering by a perfectly conducting cylinder of arbitrary cross section is well known [1], and the particular solution employed here is based upon that of Wang [80]. As illustrated in Fig. 4, to implement the MM solution, we make a piecewise straight approximation to the circular cylinder and strip. To obtain reasonable

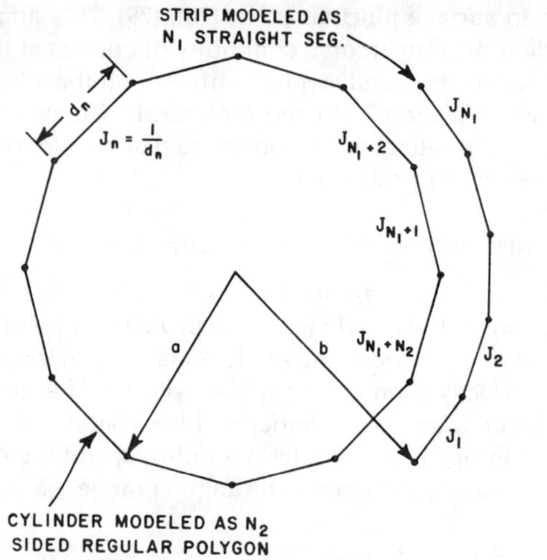

Fig. 4. For the MM solution, a piecewise linear approximation is made to the strip and cylinder contour in Fig. 3.

results, the width of the straight segments should not exceed $\lambda/4$, with $\lambda/10$ being typical. If we denote the strip as Scatterer 1 and the cylinder as Scatterer 2, the strip is modeled as N_1 straight segments of equal length and the circular cylinder is modeled by an N_2-sided regular polygon. Referring to (6), the expansion functions are the piecewise constant or pulse functions

$$J_n = 1/d_n, \qquad n = 1, 2, \cdots, N = N_1 + N_2 \qquad (31)$$

which are numbered so that the first N_1 are on the strip (Scatterer 1) and the last N_2 are on the cylinder (Scatterer 2). The width of basis function n is denoted d_n. The weighting functions are chosen identical to the expression functions, i.e., $w_m = J_m$. This is referred to as a Galerkin solution, and results in a symmetric impedance matrix. The MM matrix equation, given by (8)–(10) and illustrated in Fig. 2, can then be evaluated using computer subroutines supplied by Wang [80].

Consider the specific example of a circular cylinder of radius $a = \lambda$, and a strip with radius $b = 1.01\lambda$ and angular sector $2\phi_s = 10°$. The radius b is chosen slightly larger than a to emphasize the coupling between the strip and the cylinder. For this geometry, Fig. 5 shows the magnitude and phase of the current density at the center of the strip versus N_2 = the number of basis functions used to model the cyl-

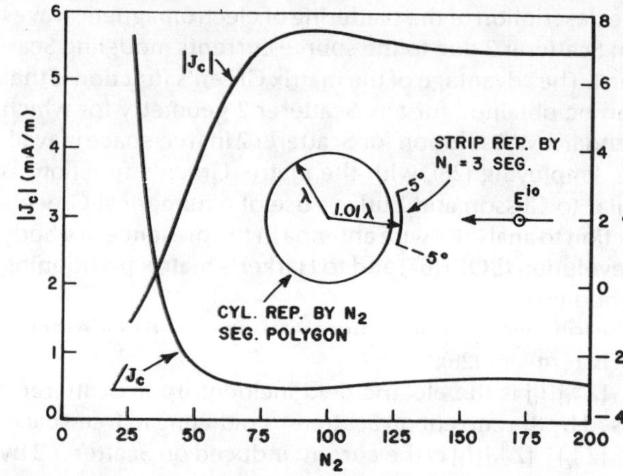

Fig. 5. The MM or MM/matrix Green's function solution for the center current of a circular strip in the presence of a circular cylinder versus N_2 = the number of basis functions on the cylinder.

inder. Numerical experimentation showed that $N_1 = 3$ basis functions on the strip results in a reasonably well converged result, and this will be used in all computations to follow. As is typical of MM solutions, as N_2 increases, the strip center current density converges, in this case to about 5.45 at an angle of −2.8° mA/m. About $N_2 = 100$ basis functions are needed on the cylinder to obtain a reasonably converged result, which corresponds to a segment size of about 0.063λ. The computer CPU times for the MM solutions in Fig. 5 are shown by the solid line in Fig. 6. The CPU time

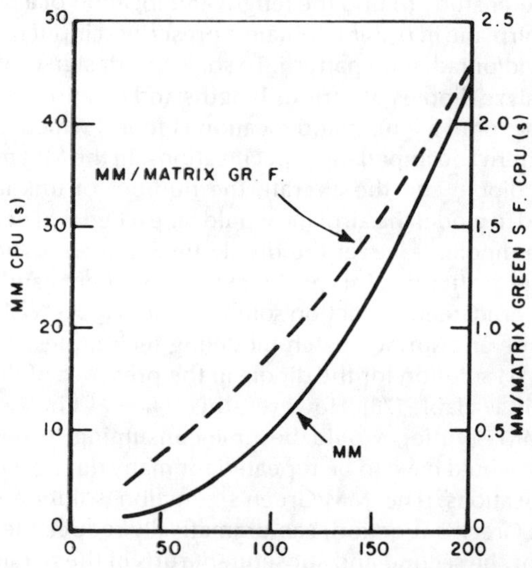

Fig. 6. The CPU time on a VAX 8550 for the MM and MM/matrix Green's function solution to the geometry of Fig. 5.

is roughly proportional to $(N_2)^2$, and about 10 s were required for the $N_2 = 100$ basis function solution.

The MM/matrix Green's function solution, defined by (26), is algebraically equivalent to the straight MM solution. Thus the data in Fig. 5 also applies to the MM/matrix Green's function solution. The corresponding CPU times for the MM/matrix Green's function solution are shown by the dashed line in Fig. 6, and are considerably less than that for the straight MM solution. It is important to point out that these times do not include the time to compute and invert $[Z_{22}^0]$. If a series of problems is being run using the MM/matrix Green's function formulation, with the cylinder fixed but with the strip geometry being varied, these CPU times would be for the second and subsequent runs. Also, if a large number of runs is being made, then these times represent the average CPU time for each run (including the first run where $[Z_{22}^0]$ is computed and inverted).

B. MM/Exact Green's Function Solution

This section describes the MM/Green's function solution to the problem of Fig. 3 using the exact eigenfunction cylinder Green's function. We begin by presenting the exact eigenfunction representation of the cylinder Green's function. Consider a unit amplitude line current located at the source point with cylindrical coordinates (ρ', ϕ') and radiating in the presence of a perfectly conducting circular cylinder of radius a which is located concentric with the z axis. The total field of this line source at the field point (ρ, ϕ), i.e., the cylinder Green's function, is given by [7]

$$G^2(\rho, \phi; \rho', \phi') = C \sum_{p=0}^{\infty} \epsilon_p F_p(\rho, \rho') \cos p(\phi - \phi') \tag{32}$$

in which $C = -k^2/4\omega\epsilon_0$, $\epsilon_p = 1$ if $p = 0$ and $\epsilon_p = 2$ if $p \neq 0$, and

$$\begin{aligned} F_p(\rho, \rho') &= H_p^{(2)}(k\rho_>)[B_p(k\rho_<) + c_p H_p^{(2)}(k\rho_<)] \\ &= F_{-p}(\rho, \rho') \end{aligned} \tag{33}$$

$$c_p = -\frac{B_p(ka)}{H_p^{(2)}(ka)} = c_{-p} \tag{34}$$

$$\begin{aligned} \rho_< &= \text{the minimum of } (\rho, \rho') \\ \rho_> &= \text{the maximum of } (\rho, \rho'). \end{aligned} \tag{35}$$

Here B_p denotes the Bessel function of the first kind and $H_p^{(2)}$ denotes the Hankel function of the second kind, both of order p.

If the same cylinder is illuminated by a unit amplitude plane wave incident from the $+x$ axis (30), then the total field at (ρ, ϕ) is

$$E^{i2} = \sum_{p=-\infty}^{\infty} D_p(\rho) e^{-jp\phi} \tag{36}$$

in which

$$D_p(\rho) = (-j)^{-p}[B_p(k\rho) + c_p H_p^{(2)}(k\rho)]. \tag{37}$$

While in the MM solution it was convenient to make a piecewise linear approximation to the curved cylinder and strip, when using the above eigenfunction expansion of the fields, it is advantageous to retain the true circular nature of the strip. Thus in Fig. 7 the semicircular strip is shown segmented into N_1 smaller semicircular strips. Each of the N_1 smaller strips corresponds to a piecewise constant basis function. The current density on expansion function n is given by

$$J_n = \frac{1}{b(\phi_{n2} - \phi_{n1})} \qquad \phi_{n1} \leq \phi \leq \phi_{n2}. \tag{38}$$

Again we choose the weighting functions identical to the expansion functions, i.e., $W_m = J_m$.

We now show in some detail the evaluation of the elements in the MM/Green's function impedance matrix. Referring to (19), a typical element in our case is given by

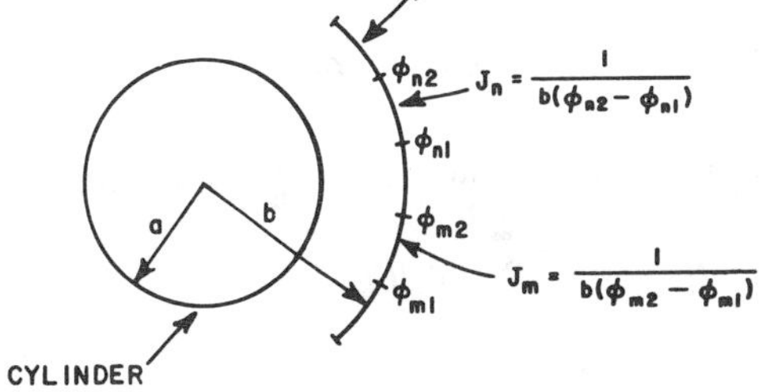

Fig. 7. For the MM/exact Green's function solution the semicircular strip is segmented into N_1 smaller semicircular strips corresponding to the MM basis functions.

$$Z_{mn}^2 = -\int_{\phi_{m1}}^{\phi_{m2}} E^2(J_n)\, J_m b\, d\phi = Z_{nm}^2$$

$$m, n = 1, 2, \cdots, N_1. \tag{39}$$

Here $E^2(J_n)$ is the electric field of expansion function n, located on the semicircular strip, and radiating in the presence of the circular cylinder. $E^2(J_n)$ is computed by integrating over expansion function n the product of J_n and the cylinder Green's function

$$E^2(J_n) = \int_{\phi_{n1}}^{\phi_{n2}} J_n(\phi')\, G^2(b, \phi; b, \phi') b\, d\phi' \tag{40}$$

in which we have set $\rho = \rho' = b$ since in our case the source and field point are on the strip. Recognizing that $J_n(\phi')$ is constant for $\phi_{n1} \leq \phi \leq \phi_{n2}$, and inserting G^2 from (32) into (40) yields

$$E^2(J_n) = CJ_n b \sum_{p=0}^{\infty} \epsilon_p F_p(b, b) \int_{\phi_{n1}}^{\phi_{n2}} \cos p(\phi - \phi')\, d\phi'. \tag{41}$$

Equation (41) can now be easily integrated to yield

$$E^2(J_n) = CJ_n b \sum_{p=0}^{\infty} \epsilon_p F_p(b, b) \cdot \begin{cases} \dfrac{1}{p}[\sin p(\phi - \phi_{n1}) - \sin p(\phi - \phi_{n2})], & p \neq 0 \\ (\phi_{n2} - \phi_{n1}), & p = 0. \end{cases} \tag{42}$$

Z_{mn} is now obtained by inserting $E^2(J_n)$ from (42) into (39). Again the integrations are easily done yielding

$$Z_{mn}^2 = -CJ_m J_n b^2 \sum_{p=0}^{\infty} \epsilon_p F_p(b, b) \cdot \begin{cases} \dfrac{1}{p^2}[\cos p(\phi_{m1} - \phi_{n1}) - \cos p(\phi_{m2} - \phi_{n1}) \\ \quad - \cos p(\phi_{m1} - \phi_{n2}) + \cos p(\phi_{m2} - \phi_{n2})], \\ \qquad p \neq 0 \\ (\phi_{n1} - \phi_{n2})(\phi_{m1} - \phi_{m2}), \qquad p = 0. \end{cases} \tag{43}$$

Similarly, from (20), the elements in the voltage vector are

$$V_m^2 = \int_{\phi_{m1}}^{\phi_{m2}} E^{i2} J_m b \, d\phi, \qquad m = 1, 2, \cdots, N_1. \tag{44}$$

Inserting E^{i2} from (36) into (44) and integrating yields

$$V_m^2 = J_m b \sum_{p=-\infty}^{\infty} D_p(b) \cdot \begin{cases} \frac{j}{p}(e^{-jp\phi_{m2}} - e^{-jp\phi_{m1}}), & p \neq 0 \\ (\phi_{m2} - \phi_{m1}), & p = 0. \end{cases} \tag{45}$$

The ability to carry out the integrations in closed form for the elements in the impedance matrix and the right-hand-side vector is typical for problems where the surface of Scatterer 1 is a constant coordinate surface for the functions being used to expand the Scatterer 2 Green's function [21], [23], [54], [81].

Although (43) for the Z_{mn}^2 is in a relatively simple form, a numerical problem arises in that the self-impedance terms (i.e., $m = n$) are slowly convergent. Referring to (22), it is the $[Z_{11}^0]$ or free-space part of $[Z^2]$ which is slowly convergent when basis functions m and n are electrically close. For this reason, it is numerically efficient to evaluate $[Z_{11}^0]$ separately as in a conventional MM solution for the strip in free-space [1], [80]. The problem then remains to evaluate $[Z^{S2}]$ using the eigenfunction expansion of (43). F_p in (33) and D_p in (37) both contain a factor of the form $[B_p + c_p H_p^{(2)}]$. The term proportional to B_p corresponds to the free-space field, while the term proportional to $c_p H_p^{(2)}$ corresponds to the scattered field. Thus referring to (22), the B_p term contributes solely to $[Z_{11}^0]$ and $[V_1^0]$, while the $c_p H_p^{(2)}$ term contributes solely to $[Z^{S2}]$ and $[V^{S2}]$. It can then be seen that (43) and (45) can be used to compute $[Z^{S2}]$ and $[V^{S2}]$, respectively, by omitting the B_p term.

For the problem illustrated in Fig. 5, the $N_1 = 3$ basis function MM/exact Green's function solution yielded a current density at the center of the strip of 5.9 at an angle of $-2.7°$ mA/m, which is very close to the MM solution for $N_2 \geq 100$. This MM/Green's function solution required only three unknowns and 0.11 s of CPU time. As compared to the $N_1 = 3$ and $N_2 = 100$ basis function MM solution, this is about a factor of 100 reduction in CPU time and 1000 reduction in matrix storage requirements. As the cylinder size increases, the relative advantage of the MM/Green's function solution will also increase. Although the reduction in storage is a general characteristic of MM and MM/Green's function solutions, the reduction of CPU time is not. The fact that we could carry out the integrations in (39) and (44) in closed form greatly reduced the CPU time for the MM/Green's function solution. For most MM/Green's function solutions these integrations must be done numerically and can be very time-consuming. In some cases, the CPU time for the MM/Green's function solution can actually exceed that of the straight MM solution.

C. MM/GTD Green's Function Solution

Using the GTD to obtain the Green's function greatly expands the range of problems which can be treated via MM/Green's function techniques. This section describes the solution to the problem of Fig. 3 using the MM/Green's function technique with a GTD Green's function.

The GTD Green's function is described with the aid of Fig. 8, which shows a unit amplitude line source located at (ρ', ϕ') and radiating in the presence of a perfectly conducting cylinder of radius a. In describing the field of the line source

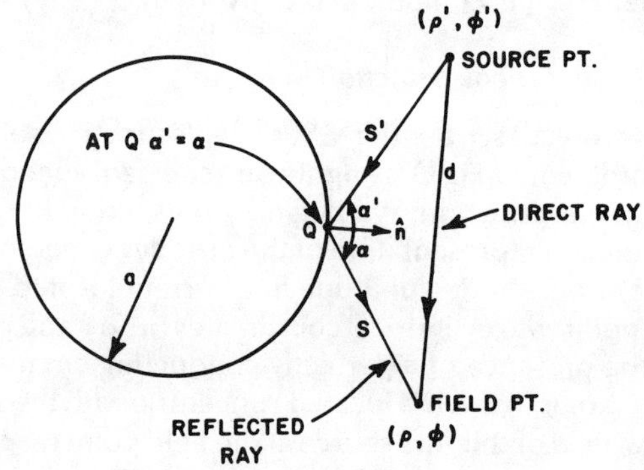

Fig. 8. The direct and reflected rays used in the GTD Green's function for a line source in the presence of a circular cylinder.

it is assumed that the field point (ρ, ϕ) is in the lit region, i.e., a line or ray from the source point to the field point does not pass through the cylinder. It is also assumed that $ka \gg 1$ and that the source and field point are not too near the surface of the cylinder. In the lit region the total field is dominated by two rays, shown in Fig. 8 as the direct ray and the reflected ray. The field of the direct ray is simply the free-space field of the line source. The reflected ray propagates along a straight-line path of length s' from the source point to the point Q on the cylinder, and then along a straight-line path of length s from Q to the field point. The point Q is defined by the requirement that the angle of incidence α' is equal to the angle of reflection, α. Then, using the notation of (21), the total field of the line source is [82]

$$G^2 = G^0 + G^{S2} = CH_0^{(2)}(kd) - CH_0^{(2)}(ks')\sqrt{\frac{\rho^r}{\rho^r + s}}\, e^{-jks} \tag{46}$$

where d is the distance from the source point to the field point, $C = -k^2/4\omega\epsilon_0$, and

$$\frac{1}{\rho^r} = \frac{1}{s'} + \frac{2}{a \cos \alpha}. \tag{47}$$

Similarly, the field of the plane wave of (30) in the presence of the cylinder can be written as

$$E^{i2} = E^{i0} + E^{iS2} = e^{jkx} - e^{jkx_Q}\sqrt{\frac{\rho^r}{\rho^r + s}}\, e^{-jks} \tag{48}$$

where ρ^r and x_Q, the x coordinate of point Q, are evaluated for a source point at ($\rho' = \infty$, $\phi' = 0$).

Referring to (22), the first term in (46) and (48) contribute to $[Z_{11}^0]$ and $[V_1^0]$, respectively, i.e., they result in the MM matrix elements for the strip in free-space. As described above, they are evaluated using computer subroutines developed by Wang [80]. The second term in (46) and (48) contribute to $[Z^{S2}]$ and $[V^{S2}]$, respectively. These elements are evaluated by inserting G^{S2} from (46) into (39) and (40), inserting E^{iS2} from (48) into (44), and performing the integrations numerically. For this simple example, requiring only one ray, these integrations are very fast. However, for a more complicated Scatterer 2 geometry, more rays would be required, and the CPU time needed for the numerical integrations would increase.

The insert in Fig. 9 shows a semicircular strip located $\lambda/2$ above a circular cylinder. With $N_1 = 3$ basis functions

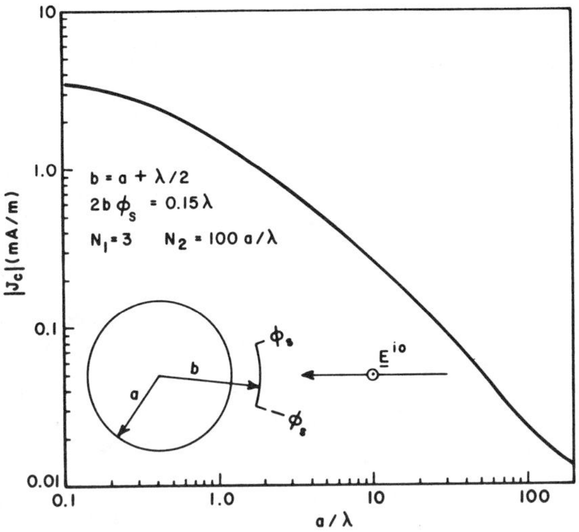

Fig. 9. The magnitude of the center current for a semicircular strip located $\lambda/2$ above a circular cylinder, computed by the MM, the MM/exact Green's function, or the MM/GTD Green's function methods.

on the strip, Fig. 9 shows the magnitude of the current density at the center of the strip versus a/λ. The angular sector of the strip is adjusted so that its width remains constant at $2b\phi_s = 0.15\lambda$. This data was generated with the MM, the MM/exact Green's function and the MM/GTD Green's function techniques, and essentially the same results for the strip current were obtained by each method. For the MM solution, $N_2 = 100a/\lambda$ basis functions were used on the cylinder. The strip current in Fig. 9 goes to zero as $a/\lambda \to \infty$ since the direct ray cancels the reflected ray on the strip, causing $[V_1^0] \to -[V^{S2}]$ and thus $[V^2] \to 0$. The CPU times for these three methods are shown in Fig. 10. Note that the left scale applies for the MM solution, and is 100 times as large as the right scale which is for the MM/Green's function solutions. For $a/\lambda > 1$, the CPU time for the MM solution is two orders of magnitude or more greater than that for the MM/Green's function solutions. The CPU time for the MM/exact Green's function solution increases as a/λ increases, since more terms must be retained in the eigenfunction summations. By comparison, for $a/\lambda > 1$, the CPU time for the MM/GTD Green's function solution is essentially constant at 0.25 s.

As compared with the MM/exact Green's function solution, the advantage of the MM/GTD solution is that it is applicable to a far wider class of Scatterer 2 (in this example, cylinder) geometries. As compared to the MM, the advantage of the MM/GTD solution is that it is much faster and requires far less computer storage as the electrical size of Scatterer 2 increases.

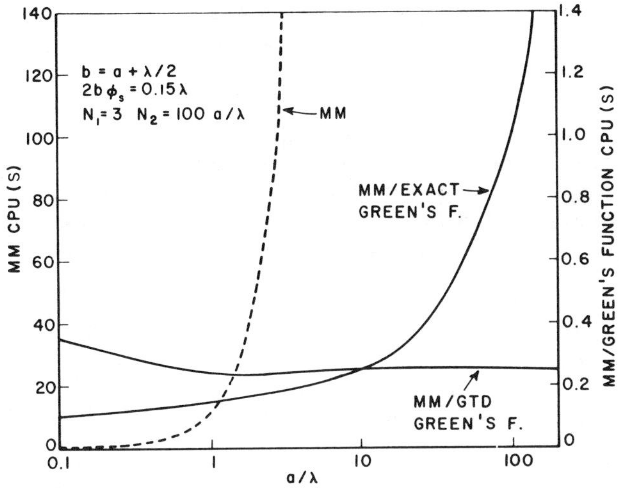

Fig. 10. The CPU times on a VAX 8550 for the data in Fig. 9 computed by the MM (dashed curve) and by the MM/Green's function methods (solid curves).

IV. Summary

This paper describes a technique for combining the MM and Green's function solutions in electromagnetics. The fundamental difference between the standard MM and MM/Green's function solutions is in the application of the equivalence theorems used in obtaining their respective integral equations. In a standard MM solution all matter is replaced by free-space and by equivalent currents. These currents are then found as the solution of an exact integral equation whose kernel contains the free-space Green's function. By comparison, in an MM/Green's function solution some, but not all, matter is replaced by free-space and by equivalent currents. Again the currents are found as the solution of an exact integral equation, but now the kernel contains the Green's function for that portion of the matter which was not replaced by free-space and equivalent currents. Both integral equations are solved by transforming them into a system of simultaneous linear equations, i.e., a matrix equation. The advantage of the MM/Green's function technique is that the number of unknowns, and thus the order of this matrix equation is reduced. This results in a reduction in computer storage and, in most cases, CPU time.

The number of problems which can be solved by MM/Green's function techniques can be tremendously increased by not requiring the Green's functions to be exact. In particular, using the GTD one can obtain the Green's function for a geometry as complicated as an aircraft. This is referred to as an MM/GTD solution [55].

It is shown that, by manipulating the MM matrix equation, a standard MM solution can be put into the format of an MM/Green's function solution. This is referred to as an MM/matrix Green's function solution. Such a solution is algebraically equivalent to the standard MM solution, and can be obtained for any geometry for which a standard MM solution is possible. The advantage of the MM/matrix Green's function solution is that it is faster than a standard MM solution if a series of problems is being analyzed in which only a small portion of the problem geometry changes from one run to the next.

In order to illustrate the MM/Green's function method, and to compare it to a standard MM solution, the problem of plane wave scattering by a perfectly conducting semicircular strip in the presence of a perfectly conducting circular cylinder is solved using the MM and the MM/Green's function technique with a matrix, an exact eigenfunction, and a GTD Green's function. Basically, all methods yield the same numerical results. However, the MM/Green's function solutions are substantially faster, and require far less storage (except for the matrix Green's function), especially as the cylinder size is increased.

References

[1] R. F. Harrington, *Field Computation by Moment Methods.* New York, NY: Macmillan, 1968.

[2] —, "Matrix methods for field problems," *Proc. IEEE*, vol. 55, pp. 136–149, Feb. 1967.

[3] J. H. Richmond, "Digital computer solutions of the rigorous equations for scattering problems," *Proc. IEEE*, vol. 53, pp. 796–804, Aug. 1965.

[4] C.-T. Tai, *Dyadic Green's Functions in Electromagnetic Theory.* Scranton, PA: Intext Educational Publishers, 1971.

[5] J. J. Bowman, T. B. A. Senior, and P. L. E. Uslenghi, *Electromagnetic and Acoustic Scattering by Simple Shapes.* Amsterdam: North-Holland, 1969.

[6] G. T. Ruck, *Radar Cross Section Handbook.* New York, NY: Plenem Press, 1970.

[7] R. F. Harrington, *Time-Harmonic Electromagnetic Fields.* New York, NY: McGraw-Hill, 1961.

[8] J. H. Richmond, "Scattering by a dielectric cylinder of arbitrary cross section shape," *IEEE Trans. Antennas Propagat.*, vol. AP-13, pp. 334–341, May 1965.

[9] —, "TE wave scattering by a dielectric cylinder of arbitrary cross-section shape," *IEEE Trans. Antennas Propagat.*, vol. AP-14, pp. 460–464, July 1966.

[10] E. K. Miller, A. J. Poggio, G. J. Burke, and E. S. Seldon, "Analysis of wire antennas in the presence of a conducting half-space, part I. The vertical antenna in free space," *Can. J. Phys.*, vol. 50, pp. 879–888, 1972.

[11] —, "Analysis of wire antennas in the presence of a conducting half-space, part II. The horizontal antenna in free space," *Can. J. Phys.*, vol. 50, pp. 2614–2627, 1972.

[12] G. J. Burke *et al.*, "Computer-modeling of antennas near the ground," *Electromagnetics*, vol. 1, pp. 29–49, 1981.

[13] G. J. Burke, W. A. Johnson, and E. K. Miller, "Modelling of simple antennas near to and penetrating an interface," *Proc. IEEE*, vol. 71, no. 1, Jan. 1983.

[14] G. J. Burke and A. J. Poggio, "Numerical electromagnetic code (NEC)—method of moments," Naval Oceans Systems Center, Tech Doc. 116, AFWL-TR-320, July 1977.

[15] D. C. Chang and J. R. Wait, "Theory of a vertical tubular antenna located above a conducting half-space," *IEEE Trans. Antennas Propagat.*, vol. AP-18, pp. 182–188, Mar. 1970.

[16] R. Tiberio and G. Manara, "A hybrid technique for analyzing wire antennas in the presence of a plane interface," *IEEE Trans. Antennas Propagat.*, vol. AP-33, pp. 881–885, Aug. 1985.

[17] P. Parhami and R. Mittra, "Wire antennas over a lossy half-space," *IEEE Trans. Antennas Propagat.*, vol. AP-28, pp. 397–403, May 1980.

[18] L. Chommeloux, C. Pichot, and J. Bolomey, "Electromagnetic modeling for microwave imaging of cylindrical buried inhomogeneities," *IEEE Trans. Microwave Theory Tech.*, vol. MTT-34, pp. 1064–1074, Oct. 1986.

[19] C. M. Butler, X. Xu, and A. W. Glisson, "Current on a conducting cylinder located near the planar interface between two semi-infinite half-spaces," *IEEE Trans. Antennas Propagat.*, vol. AP-33, June 1985.

[20] X. Xu and C. M. Butler, "Scattering of TM excitation by coupled and partially buried cylinders at the interface between two media," *IEEE Trans. Antennas Propagat.*, vol. AP-35, pp. 529–538, May 1987.

[21] D. M. Pozar, "Input impedance and mutual coupling of rectangular microstrip antennas," *IEEE Trans. Antennas Propagat.*, vol. AP-30, pp. 1191–1196, Nov. 1982.

[22] —, "A reciprocity method of analysis for printed slot and slot-coupled microstrip antennas," *IEEE Trans. Antennas Propagat.*, vol. AP-34, pp. 1439–1446, 1986.

[23] E. H. Newman and D. Forrai, "Scattering from a microstrip patch," *IEEE Trans. Antennas Propagat.*, vol. AP-35, pp. 245–251, Mar. 1987.

[24] M. C. Bailey and M. D. Deschamps, "Integral equation formulation of microstrip antennas," *IEEE Trans. Antennas Propagat.*, vol. AP-30, pp. 651–656, July 1982.

[25] I. E. Rana and N. G. Alexopoulos, "Current distribution and input impedance of printed dipoles," *IEEE Trans. Antennas Propagat.*, vol. AP-29, pp. 99–105, Jan. 1981.

[26] D. R. Jackson and N. G. Alexopoulos, "Analysis of planar strip geometries in a substrate-superstrate configuration," *IEEE Trans. Antennas Propagat.*, vol. AP-34, pp. 1430–1438, Dec. 1986.

[27] B. A. Munk and R. J. Luebbers, "Reflection properties of two-layer dipole arrays," *IEEE Trans. Antennas and Propagat.*, vol. AP-22, pp. 766–773, Nov. 1974.

[28] B. A. Munk, R. J. Luebbers, and R. D. Fulton, "Transmission through a two-layer array of loaded slots," *IEEE Trans. Antennas Propagat.*, vol. AP-22, pp. 804–809, Nov. 1974.

[29] B. R. Rao, "A two-element Yagi-type array in a parallel-plate waveguide—theoretical and experimental studies," *IEEE Trans. Antennas Propagat.*, vol. AP-13, pp. 675–682, Sept. 1965.

[30] —, "Current distribution and impedance of an antenna in a parallel-plate region," *Proc. Inst. Elec. Eng.* (London), vol. 112, p. 259, Feb. 1965.

[31] C. D. Taylor, "Thin wire receiving antenna in a parallel plate waveguide," *IEEE Trans. Antennas Propagat.*, vol AP-15, pp. 572–574, July 1967.

[32] F. M. Tesche, "On the behavior of thin-wire antennas and scatterers arbitrarily located within a parallel plate region," *IEEE Trans. Antennas Propagat.*, vol. AP-20, pp. 482–486, July 1972.

[33] D. B. Seidel, "Aperture excitation of a wire in a rectangular cavity," *IEEE Trans. Microwave Theory Tech.*, vol. MTT-26, pp. 908–914, Nov. 1978.

[34] Y. Leviatan, P. G. Li, A. T. Adams, and J. Perini, "Single-post inductive obstacle in rectangular waveguide," *IEEE Trans. Microwave Theory Tech.*, vol. MTT-31, pp. 806–811, Oct. 1983.

[35] H. Auda and R. F. Harrington, "Inductive posts and diaphragms of arbitrary shape and number in a rectangular waveguide," *IEEE Trans. Microwave Theory Tech.*, vol. MTT-32, pp. 606–613, June 1984.

[36] J. M. Jarem, "A multifilament method-of-moments solution for the input impedance of a probe-excited semi-infinite waveguide," *IEEE Trans. Microwave Theory Tech.*, vol. MTT-35, pp. 14–19, Jan. 1987.

[37] J. J. H. Wang, "Analysis of a three-dimensional arbitrarily shaped dielectric or biological body inside a rectangular waveguide," *IEEE Trans. Microwave Theory Tech.*, vol. MTT-26, pp. 457–462, July 1978.

[38] A. S. Omar and K. Schunemann, "Scattering by material and conducting bodies inside waveguides, part 1: theoretical formulations," *IEEE Trans. Microwave Theory Tech.*, vol. MTT-34, pp. 266–272, Feb. 1986.

[39] C. G. Hsu and H. A. Auda, "Multiple dielectric posts in a rectangular waveguide," *IEEE Trans. Microwave Theory Tech.*, vol. MTT-34, pp. 883–891, Aug. 1986.

[40] T. V. Khac and C. T Carson, "Impedance properties of a longitudinal slot antenna in the broad face of a rectangular waveguide," *IEEE Trans. Antennas Propagat.*, vol. AP-21, pp. 708–710, Sept. 1973.

[41] V. K. Thong, "Solutions for some waveguide discontinuities by the method of moments," *IEEE Trans. Microwave Theory Tech.*, vol. MTT-20, pp. 416–418, June 1972.

[42] F. M. Tesche and A. R. Neureuther, "Radiation patterns for two monopoles on a perfectly conducting sphere," *IEEE Trans. Antennas Propagat.*, vol. AP-18, pp. 692–694, Sept. 1970.

[43] —, "The analysis of monopole antennas located on a spherical vehicle: Part I—Theory," *IEEE Trans. Electromag. Compat.*, vol. EMC-18, pp. 2–8, Feb. 1976.

[44] F. M. Tesche, A. R. Neureuther, and R. E. Stovall, "The analysis of monopole antennas located on a spherical vehicle: Part II—Numerical and experimental results," *IEEE Trans. Electromag. Compat.*, vol. EMC-18, pp. 8–15, Feb. 1976.

[45] C. M. Butler and T. L. Keshavamurthy, "Analysis of a wire

antenna in the presence of a sphere," *IEEE Trans. Electromag. Compat.*, vol. EMC-22, pp. 113–118, May 1980.
[46] D. Misra and K. Chen, "Responses of electric-field probes near a cylindrical model of the human body," *IEEE Trans. Microwave Theory Tech.*, vol. MTT-33, pp. 447–452, June 1985.
[47] H. Steyskal, "Analysis of circular waveguides arrays on cylinders," *IEEE Trans. Antennas Propagat.*, vol. AP-25, pp. 610–616, Sept. 1977.
[48] M. D. G. Karunaratne, K. A. Michalski, and C. M. Butler, "TM scattering from a conducting strip loaded by a dielectric cylinder," *Proc. Inst. Elec. Eng.* (London), Apr. 1985.
[49] D. Lamensdorf and C. Ting, "An experimental and theoretical study of the monopole embedded in a cylinder of anisotropic dielectric," *IEEE Trans. Antennas Propagat.*, vol. AP-16, pp. 342–349, May 1968.
[50] D. M. Pozar and E. H. Newman, "Analysis of a monopole mounted near or at the edge of a half-plane," *IEEE Trans. Antennas Propagat.*, vol. AP-29, pp. 488–495, May 1981.
[51] —, "Analysis of a monopole mounted near an edge or a vertex," *IEEE Trans. Antennas Propagat.*, vol. AP-30, pp. 401–408, May 1982.
[52] E. H. Newman, "TM scattering by a dielectric cylinder in the presence of half-plane," *IEEE Trans. Antennas Propagat.*, vol. AP-33, pp. 773–782, July 1985.
[53] —, "TM and TE scattering by a dielectric/ferrite cylinder in the presence of a half-plane," *IEEE Trans. Antennas Propagat.*, vol. AP-34, pp. 806–813, June 1986.
[54] E. H. Newman and J. L. Blanchard, "TM scattering by an impedance sheet extension of a parabolic cylinder," *IEEE Trans. Antennas Propagat.*, accepted for publication.
[55] G. A. Thiele and T. H. Newhouse, "A hybride technique for combining moment methods with the geometrical theory of diffraction," *IEEE Trans. Antennas Propagat.*, vol. AP-23, pp. 62–69, Jan. 1975.
[56] R. G. Kouyoumjian, "Asymptotic high-frequency methods," *Proc. IEEE*, vol. 53, pp. 864–876, Aug. 1965.
[57] H. E. Green, "Impedance of a monopole on the base of a large cone," *IEEE Trans. Antennas Propagat.*, vol. AP-17, pp. 703–706, Nov. 1969.
[58] K. H. Awadalla and T. S. M. Maclean, "Input impedance of a monopole antenna at the center of a finite ground plane," *IEEE Trans. Antennas Propagat.*, vol. AP-26, pp. 244–248, Mar. 1978.
[59] —, "Monopole antenna at center of a circular ground plane: Input impedance and radiation pattern," *IEEE Trans. Antennas Propagat.*, vol. AP-27, pp. 151–153, Mar. 1979.
[60] M. Marin and M. F. Catedra, "A study of a monopole arbitrarily located on a disk using hybrid MM/GTD techniques," *IEEE Trans. Antennas Propagat.*, vol. AP-35, pp. 287–292, Mar. 1987.
[61] E. P. Ekelman and G. A. Thiele, "A hybrid technique for combining the moment method treatment of wire antennas with the GTD for curved surfaces," *IEEE Trans. Antennas Propagat.*, vol. AP-28, pp. 831–839, Nov. 1980.
[62] L. W. Henderson and G. A. Thiele, "A hybrid MM-geometrical optics technique for the treatment of wire antennas mounted on a curved surface," *IEEE Trans. Antennas Propagat.*, vol. AP-30, pp. 1257–1261, Nov. 1982.
[63] —, "A hybrid MM-GTD technique for the treatment wire antennas near a curved surface," *Radio Sci.*, vol. 16, pp. 1125–1130, Nov./Dec. 1981.
[64] G. A. Thiele and G. K. Chan, "Application of the hybrid technique to time domain problems," *IEEE Trans. Antennas Propagat.*, vol. AP-26, pp. 151–155, Jan. 1978.
[65] W. L. Ko and R. Mittra, "A new approach based upon a combination of integral equation and asymptoptic techniques for solving electromagnetic scattering problems," *IEEE Trans. Antennas Propagat.*, vol. AP-25, pp. 187–197, Mar. 1977.
[66] R. F. Harrington and J. R. Mautz, "Green's functions for surfaces of revolution," *Radio Sci.*, vol. 7, pp. 603–611, May 1972.
[67] A. W. Glisson and C. M. Butler, "Analysis of wire antenna in the presence of a body of revolution," *IEEE Trans. Antennas Propagat.*, vol. AP-28, pp. 604–609, Sept. 1980.
[68] K. J. Harker, "Partitioning method in method of moments scattering theory," presented at the National Radio Science Meeting, Boston, MA, June 25–29, 1984.
[69] W. D. Burnside, C. L. Yu, and R. J. Marhefka, "A technique to combine the geometrical theory of diffraction and the moment method," *IEEE Trans. Antennas Propagat.*, vol. AP-23, pp. 551–558, July 1975.
[70] C. W. Chuang and W. D. Burnside, "A diffraction coefficient for a truncated planar surface," *IEEE Trans. Antennas Propagat.*, vol. AP-28, pp. 177–182, Mar. 1980.
[71] S. Srikanth, P. H. Pathak, and C. W. Chuang, "Hybrid UTD-MM analysis of the scattering by a perfectly conducting semicircular cylinder," *IEEE Trans. Antennas Propagat.*, vol. AP-34, pp. 1250–1257, Oct. 1986.
[72] T. J. Kim and G. A. Thiele, "A hybrid diffraction technique—general theory and applications," *IEEE Trans. Antennas Propagat.*, vol. AP-30, pp. 888–897, Sept. 1982.
[73] P. K. Murthy, K. C. Hill, and G. A. Thiele, "A hybrid-iterative method for scattering problems," *IEEE Trans. Antennas Propagat.*, vol. AP-34, pp. 1173–1180, Oct. 1986.
[74] L. N. Medgyesi-Mitschang and D.-S. Wang, "Hybrid solution for scattering from perfectly conducting bodies of revolution," *IEEE Trans. Antennas Propagat.*, vol. AP-31, pp. 570–583, July 1983.
[75] B. Azarbar and L. Shafai, "Application of moment method to large cylindrical reflector antennas," *IEEE Trans. Antennas Propagat.*, vol. AP-26, pp. 500–502, May 1978.
[76] J. H. Richmond, "Scattering by thin dielectric strips," *IEEE Trans. Antennas Propagat.*, vol. AP-33, pp. 64–68, Jan. 1985.
[77] E. H. Newman and P. Tulyathan, "Analysis of microstrip antennas using moment methods," *IEEE Trans. Antennas Propagat.*, vol. AP-29, pp. 47–53, Jan. 1981.
[78] E. H. Newman and R. L. Dilsavor, "A user's manual for the electromagnetic surface patch code: ESP version III," Ohio St. Univ. Research Foundation Rep. 716148, prepared under Grant NSG 1613 with the National Aeronautics and Space Administration, May 1987.
[79] E. H. Newman and D. M. Pozar, "Electromagnetic modeling of composite wire and surface geometries," *IEEE Trans. Antennas Propagat.*, vol. AP-26, pp. 784–789, Nov. 1978.
[80] N. N. Wang, "Reaction formulation for radiation and scattering from plates, corner reflectors, and dielectric-coated cylinders," Ph.D. dissertation, The Ohio St. Univ., Dept. of Elec. Engr., Columbus, OH, Apr. 1974.
[81] J. H. Richmond and E. H. Newman, "Mutual impedance between vertical dipoles over a flat earth," *Radio Sci.*, vol. 14, pp. 957–959, Nov.-Dec. 1979.
[82] P. H. Pathak, W. D. Burnside, and R. J. Marhefka, "A uniform GTD analysis of the diffraction of electromagnetic waves by a smooth convex surface," *IEEE Trans. Antennas Propagat.*, vol. AP-28, pp. 631–642, Sept. 1980.

Edward H. Newman (Senior Member, IEEE) was born in Cleveland, OH, on July 9, 1946. He received the B.S.E.E., M.S., and Ph.D. degrees in electrical engineering from The Ohio State University in 1969, 1970, and 1974, respectively.

Since 1974 he has been a member of the Ohio State University, Department of Electrical Engineering, ElectroScience Laboratory. His primary research interest is in the development of method of moments, techniques for the analysis of general antenna or scattering problems, and he is the primary author of the "Electromagnetic Surface Patch Code" (ESP). Other research interests include electrically small antennas, arrays, printed circuit antennas, antennas in inhomogeneous media, and scattering from material coated edges. He has published over 35 journal articles in these areas.

Dr. Newman is a member of Commission B of URSI, and is a past Chairman of the Columbus sections of the IEEE Antennas and Propagation and Microwave Theory and Techniques Societies.

AUTHOR INDEX

The Editor

Robert C. Hansen received his BSEE degree from the Missouri School of Mines and his PhD degree from the University of Illinois in 1955. Since 1971 he has been a consulting engineer for antennas and system-related problems.

From 1967 to 1971 he was with K.M. Siegel at KMS Industries. Between 1960, when part of STL became Aerospace Corporation, and 1967 he worked on USAF satellite programs at Aerospace Corporation, leaving a post as Group Director. He was Senior Staff Engineer at STL (now TRW) during 1960, after leaving the Microwave Laboratory at Hughes, where he headed a section. Six years of part-time work at the University of Illinois Antenna Laboratory included machine language coding for ILLIAC 1. His work has included ferrite loops, direction-finding and homing systems, surface wave and slot antennas, near-fields, electronic scanning and phased arrays, and superdirective and superconducting antennas. Dr. Hansen has undertaken many professional activities including: Chairman of U.S. Commission B of URSI (1967–69); Chairman of the 1958 WESCON Technical Program Committee; Chairman of IEEE APS (1963 and 1980), Chairman of Standards Subcommittee 2.5, which revised Methods of Testing Antennas; Editor of APS Newsletter (1961–63); member APS AdCom (1959–74, and 1980–85); member IEEE Publications Board (1971 and 1974); and Director IEEE 1975. He is a Fellow of IEEE and IEE; a registered Professional Engineer in California and the U.K., and a member of the American Physical Society, Tau Beta Pi, Sigma Xi, Eta Kappa Nu, and Phi Kappa Phi. He was awarded an honorary Doctor of Engineering degree by the University of Missouri-Rolla in 1975. The University of Illinois College of Engineering awarded a Distinguished Alumnus Service Medal in 1986.

Dr. Hansen has written over 100 papers on electromagnetics, has been an Associate Editor of *Microwave Journal* since 1960, was Associate Editor of *Radio Science* (1967–69), Associate Editor of *Microwave Engineer's Handbook* (Artech House, 1971), Editor of *Microwave Scanning Antennas,* Vol. I (1964), Vols, II and III (1966), *Significant Phased Array Papers* (Artech House, 1973), and *Geometric Theory of Diffraction* (1981).